1800

Other Contributors
Pierre de Fermat (1601-1665)
Michel Rolle (1652-1719)
Brook Taylor (1685-1731)
Colin Maclaurin (1698-1746)

Thomas Simpson (1710-1761)
Pierre-Simon de Laplace (1749-1827)
George Green (1793-1841)
George Gabriel Stokes (1819-1903)

Lagrange

Gauss

Cauchy

Riemann

Lebesgue

...ge (1736-1813)

C. Gauss (1777-1855)

A. Cauchy (1789-1857)

K. Weierstrass (1815-1897)

G. Riemann (1826-1866)

J. Gibbs (1839-1903)

S. Kovalevsky (1850-1891)

H. Lebesgue (1875-1941)

Agnesi

Weierstrass

Kovalevsky

Gibbs

6   1799   1821   1854   1873   1902

Gauss proves
Fundamental
Theorem of
Algebra

Riemann integral

Lebesgue integral

begins
ique
ique

Precise notion of
limit (Cauchy)

e is transcendental
(Hermite)

FORMULAS FROM GEOMETRY

Triangle

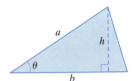

$$\text{Area} = \frac{1}{2}bh$$

$$\text{Area} = \frac{1}{2}ab\sin\theta$$

Parallelogram

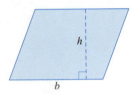

$$\text{Area} = bh$$

Trapezoid

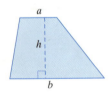

$$\text{Area} = \frac{a+b}{2}h$$

Circle

$$\text{Circumference} = 2\pi r$$

$$\text{Area} = \pi r^2$$

Sector of Circle

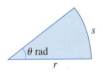

$$\text{Arc length } s = r\theta$$

$$\text{Area} = \frac{1}{2}r^2\theta$$

Polar Rectangle

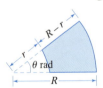

$$\text{Area} = \frac{R+r}{2}(R-r)\theta$$

Right Circular Cylinder

$$\text{Lateral area} = 2\pi rh$$

$$\text{Volume} = \pi r^2 h$$

Ball

$$\text{Area} = 4\pi r^2$$

$$\text{Volume} = \frac{4}{3}\pi r^3$$

Right Circular Cone

$$\text{Lateral area} = \pi rs$$

$$\text{Volume} = \frac{1}{3}\pi r^2 h$$

Frustum of Right Circular Cone

$$\text{Lateral area} = \pi s(r+R)$$

$$\text{Volume} = \frac{1}{3}\pi(r^2 + rR + R^2)h$$

General Cone

$$\text{Volume} = \frac{1}{3}(\text{area } B)h$$

Wedge

$$\text{Area } A = (\text{area } B)\sec\theta$$

Calculus

EIGHTH EDITION

Calculus
Instructor's Edition

Dale Varberg
Hamline University

Edwin J. Purcell
University of Arizona

Steven E. Rigdon
Southern Illinois University Edwardsville

Prentice Hall, Upper Saddle River, New Jersey 07458

Acquisitions Editor: George Lobell
Editor-in-Chief: Jerome Grant
Marketing Manager: Melody Marcus
Marketing Assistant: Vince Jansen
Production Editors: Barbara Mack, Richard DeLorenzo
Senior Managing Editor: Linda Mihatov Behrens
Executive Managing Editor: Kathleen Schiaparelli
Assistant Vice-President of Production and Manufacturing: David W. Riccardi
Manufacturing Buyer: Alan Fischer
Manufacturing Manager: Trudy Pisciotti
Editorial Assistant/Supplements Editor: Gale Epps
Associate Editor, Mathematics/Statistics Media: Audra J. Walsh
Art Director/Cover Designer: Maureen Eide
Associate Creative Director: Amy Rosen
Director of Creative Services: Paul Belfanti
Assistant to Art Director: John Christiana
Art Manager: Gus Vibal
Art Editor: Grace Hazeldine
Cover image: Ben Mangor/SuperStock, Inc. Galleria Umberto I, Naples, Italy
Text Design: Judith A. Matz-Coniglio
Photo Researcher: Mary Teresa Giancoli
Photo Editor: Beth Boyd

Printed in the United States of America.
10 9 8 7 6 5 4 3 2 1

ISBN 0-13-085149-3 (Student Edition ISBN 0-13-081137-8)

Prentice-Hall International (UK) Limited, *London*
Prentice-Hall of Australia Pty. Limited, *Sydney*
Prentice-Hall Canada Inc., *Toronto*
Prentice-Hall Hispanoamericana, S.A., *Mexico*
Prentice-Hall of India Private Limited, *New Delhi*
Prentice-Hall of Japan, Inc., *Tokyo*
Prentice-Hall (*Singapore*) Pte Ltd
Editora Prentice-Hall do Brasil, Ltda., *Rio de Janeiro*

To Earl and Clara Rigdon,
The Greatest Parents in the World

Contents

4 Applications of the Derivative   161

5 The Integral   209

6 Applications of the Integral   273

7 Transcendental Functions   319

Preface

The eighth edition of *Calculus* is a modest revision in content, more substantial in the line by line details. Users of previous editions have reported great success and we have no desire to overhaul a workable text.

Classifiers of calculus books would call this a traditional book. Most theorems are proved or left as exercises to prove; when the proof is too deep for a beginning calculus course, we say so, and in many cases we give an argument to at least make the result plausible. Theorems are stated with all conditions clearly spelled out. Through additional problems and projects, we make more use of available technology, such as graphing calculators and computer algebra systems, but the focus is still on understanding the concepts of calculus. While many revisionists see the emphasis on clear, rigorous presentation as being a distraction to the understanding of the concepts of calculus, we see the two as complementary.

A Brief Text   The eighth edition continues to be the briefest of all the successful, mainstream calculus texts. We have tried to prevent the text ballooning upward with new topics and alternative approaches. In about 800 pages we cover the major topics in calculus, including a preliminary chapter, material on differential equations, and the appendix.

In the last few decades, students have developed some bad habits. They prefer not to read the textbook. They desire to find the appropriate worked-out example so it can be matched to their homework problem. Our goal with this text continues to keep calculus as a course focused on some few basic ideas centered around words, formulas, and graphs. Solving problem sets, while crucial to developing math skills, should not overshadow the goal of understanding calculus. To encourage students to read the textbook with understanding, we begin every problem set with four fill-in-the-blank items. These test mastery of the basic vocabulary, understanding of the theorems, and ability to apply the concepts in the simplest settings. Students should respond to these items before proceeding to the later problems. We encourage this by giving immediate feedback; the correct answers are given at the end of the problem set. These items also make good quiz questions to see whether students have done the required reading.

Sections are roughly of equal length, allowing an instructor to cover about one section per day. This means that some topics are stretched into two sections. Problem sets gradually lead the student from routine exercises to challenging applied problems.

Number sense distinguishes the mature mathematics student from the neophyte. All calculus students make numerical mistakes in solving problems, but the ones with the number sense recognize an absurd answer and rework the problem. To encourage and develop this important ability, we have emphasized a process we call estimation. We suggest how to make mental estimates, how to arrive at ballpark numerical answers to questions. We do this ourselves in the text in many places, and we propose that students do this, especially in problems marked with the symbol ≈.

Use of Technology Many problems in the eighth edition are flagged with one of these symbols:

> $\boxed{\text{C}}$ indicates that an ordinary calculator will be helpful in solving the problem.
> $\boxed{\text{GC}}$ indicates that a graphing calculator is required.
> $\boxed{\text{CAS}}$ indicates that a computer algebra system is required.

Each chapter now has two Technology Projects (except Chapter 11 which has three). Each project is divided into three parts: Preparation, Using Technology, and Reflection. The Preparation section should be done before attempting to use the technology (a graphing calculator or a CAS). These exercises are important because they force the student to think about the project before diving into the technology. The Reflection section usually asks students questions that are a little deeper; some exercises in the Reflection section require additional use of technology and some don't.

This text now has a website that is free to all students who purchase this text. The address is **www.prenhall.com/varberg**

Most text examples that contain a geometric picture are to be found animated with "what-if scenarios." Questions accompany each animation. Each section of each chapter contains links to other interesting websites that explicitly cover the topic at hand. Some of these websites contain other faculties' views of how to teach a topic in a clever or insightful way. Each section has True/False quizzes that require students to read. Also, these quizzes can be forwarded directly from students to faculty. Other useful teaching materials will be added over the course of this edition.

There is now available with this text five interchangeable technology manuals: one each for Maple, Mathematica, MATLAB, the Texas Instruments Graphing Calculators, and the Hewlett-Packard Graphing Calculators. The technology projects and problems are identical in each manual. What differs is the specific keystroke or syntax instruction that is given chapter by chapter. Thus, any school could have one instructor teaching calculus with a TI calculator while another instructor uses Maple, while another uses no technology. Each manual can be wrapped with the text for a small additional fee. An added value of using these manuals is that an instructor doesn't have to instruct the student in how to use technology; the manual will do this. The instructor need only teach mathematics. Some Maple and Mathematica code for the Technology Projects is given in the *Student Solutions Manual* and at the website www.prenhall.com/varberg.

Changes in New Edition The basic structure of the book remains unchanged. Here are the most significant changes in the eighth edition.

- The First Fundamental Theorem of Calculus is now stated as $\dfrac{d}{dx}\displaystyle\int_a^x f(t)\,dt = f(x)$, and the Second Fundamental Theorem of Calculus states that $\displaystyle\int_a^b f(x)\,dx = F(b) - F(a)$ if F is an antiderivative of f.

- The order of Chapters 10 and 11 has been reversed. Now the chapter on infinite series comes before the chapter on numerical methods and approximations. With the new ordering, Taylor and Maclaurin polynomials, the first section of Chapter 11, come immediately after Taylor and Maclaurin series, the last section in Chapter 10. Also, Newton's method is now seen as generating a *sequence* that we hope converges to a solution of the equation $f(x) = 0$.

- Linear first-order differential equations are covered in Chapter 7. We use the integrating factor to find solutions.

- Section 11.5, Approximations for Differential Equations, is new. It includes slope fields, Euler's method, and the improved Euler method.

- One additional Technology Project has been added to each chapter, giving two per chapter, except for Chapter 11 where we added two projects, bringing the total to three.

- There are now over 6500 problems, many of which ask students conceptual questions to test their understanding of the concepts of calculus.

Supplements for the Instructor

- *Instructor's Resource Manual.* Contains worked out solutions to all exercises in the text, as well as a printed test bank. (ISBN 0-13-085140-X)
- *PH Custom Test for Windows.* Fully editable test generator with algorithmic capabilities, which provides an instructor's grade book and allows on-line testing. (ISBN 0-13-085090-X)

Supplements for the Student

- *Student Solutions Manual.* Contains worked out solutions for all odd-numbered exercises in the text. (ISBN 0-13-085151-5)
- *How to Study Calculus Booklet.* Contains strategies, suggestions, and hints for learning and achieving success in calculus. (ISBN 0-13-435116-9)

The following platform-specific manuals offer additional technology problem sets and projects, as well as keystroke instructions.

- *A Maple Approach to Calculus* (ISBN 0-13-010583-X)
- *A Mathematica Approach to Calculus* (ISBN 0-13-010586-4)
- *Calculus with MATLAB* (ISBN 0-13-520354-6)
- *A TI Graphing Calculator Approach to Calculus* (ISBN 0-13-020020-4)
- *Calculus with the Hewlett Packard Calculator* (ISBN 0-13-520339-2)

Acknowledgments   I would like to thank the staff at Prentice Hall, including George Lobell, Barbara Mack, Gale Epps, and Richard DeLorenzo for their encouragement and their careful attention to the many details associated with preparing a calculus book. Nancy Toscano, of M. and N. Toscano, and Joe Will also deserve special thanks for their careful reading of the manuscript and its problem sets. I wish to thank Kevin Bodden for his tireless work in preparing the solutions for the *Instructor's Resource Manual* and the *Student Solutions Manual*, and to Chris Rigdon for inputting the exam questions for the *Instructor's Resource Manual*. I also wish to thank the faculty at Southern Illinois University Edwardsville, especially Paul Phillips, Chung-wu Ho, Rahim Karimpour, and George Pelekanos for helpful comments.

I wish to thank the following faculty who offered useful comments in preparing this edition:

Aleksander Levin, Catholic University of America

Abraham A. Unger, North Dakota State University

Sherif El-Helaly, Catholic University of America

Norberto Kerzman, University of North Carolina, Chapel Hill

Li Guo, Rutgers, New Brunswick

Jennifer Johnson, University of Utah, Salt Lake City

Donald Johnson, North Central College (retired)

Frederick Weening, Edinboro University of Pennsylvania

Rechard H. Reese, Edinboro University of Pennsylvania

Manfred Stoll, University of South Carolina-Columbia

David Protos, California State University Northridge

Finally, I would like to thank my wife Pat, and children Chris, Mary, and Emily for tolerating the many nights and weekends that I spent at the office.

S.E.R.
srigdon@siue.edu
Edwardsville, Illinois

1

Preliminaries

1.1 The Real Number System

Calculus is based on the real number system and its properties. But what are the real numbers and what are their properties? To answer, we start with some simpler number systems.

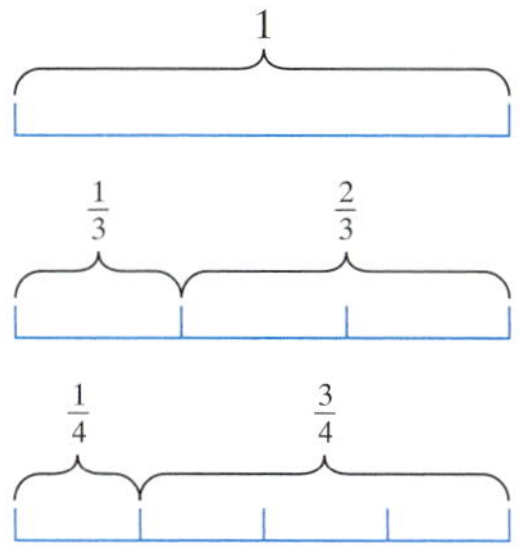

The Integers and the Rational Numbers

The simplest numbers of all are the **natural numbers**,

$$1, 2, 3, 4, 5, 6, \ldots$$

With them we can *count:* our books, our friends, and our money. If we include their negatives and zero, we obtain the **integers**

$$\ldots, -3, -2, -1, 0, 1, 2, 3, \ldots$$

When we *measure* length, weight, or voltage, the integers are inadequate. They are spaced too far apart to give sufficient precision. We are led to consider quotients (ratios) of integers (Figure 1), numbers such as

$$\frac{3}{4}, \frac{-7}{8}, \frac{21}{5}, \frac{19}{-2}, \frac{16}{2}, \text{ and } \frac{-17}{1}$$

Note that we included $\frac{16}{2}$ and $\frac{-17}{1}$, though we would normally write them as 8 and -17 since they are equal to the latter by the ordinary meaning of division. We did not include $\frac{5}{0}$ or $\frac{-9}{0}$ since it is impossible to make sense out of these symbols (see Problem 36). In fact, let us agree once and for all to banish division by zero from this book (Figure 2). Numbers that can be written in the form m/n, where m and n are integers with $n \neq 0$, are called **rational numbers**.

![Figure 1 showing a bar divided into 1, then thirds (1/3, 2/3), then quarters (1/4, 3/4)]

Figure 1

Figure 2

1

Figure 3

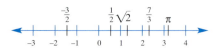

Figure 4

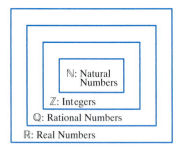

Figure 5

Do the rational numbers serve to measure all lengths? No. This surprising fact was discovered by the ancient Greeks in about the fifth century B.C. They showed that while $\sqrt{2}$ measures the hypotenuse of a right triangle with legs of length 1 (Figure 3), $\sqrt{2}$ cannot be written as a quotient of two integers (see Problem 45). Thus, $\sqrt{2}$ is an **irrational** (not rational) number. So are $\sqrt{3}$, $\sqrt{5}$, $\sqrt[3]{7}$, π, and a host of other numbers.

The Real Numbers Consider all numbers (rational and irrational) that can measure lengths, together with their negatives and zero. We call these numbers the **real numbers**.

The real numbers may be viewed as labels for points along a horizontal line. There they measure the distance to the right or left (the **directed distance**) from a fixed point called the **origin** and labeled 0 (Figure 4). Though we cannot possibly show all the labels, each point does have a unique real number label. This number is called the **coordinate** of the point. And the resulting coordinate line is referred to as the **real line**.

There are standard symbols to identify the classes of numbers so far discussed. From now on, $\mathbb{N}$ will denote the set of natural numbers (positive integers), $\mathbb{Z}$ (from the German *Zahlen*) will denote the set of integers, $\mathbb{Q}$ (quotients of integers) the set of rational numbers, and $\mathbb{R}$ the set of real numbers. As suggested by Figure 5,

$$\mathbb{N} \subseteq \mathbb{Z} \subseteq \mathbb{Q} \subseteq \mathbb{R}$$

where the symbol $\subseteq$ indicates set inclusion. The symbol $\subseteq$ is read "is a subset of." The statement $A \subseteq B$ means that every element of A is also an element of B.

You may remember that the real number system can be enlarged still more—to the **complex numbers**. These are numbers of the form $a + bi$, where a and b are real numbers and $i = \sqrt{-1}$. Complex numbers will rarely be used in this book. In fact, if we say or suggest *number* without any qualifying adjective, you can assume that we mean real number. The real numbers are the principal characters in calculus.

The Four Arithmetic Operations Given two real numbers x and y, we may add them and multiply them to obtain two new real numbers $x + y$ and $x \cdot y$ (also written simply as xy). Addition and multiplication have the following familiar properties. We call them the *field properties*.

The Field Properties

1. **Commutative laws.** $x + y = y + x$ and $xy = yx$.
2. **Associative laws.** $x + (y + z) = (x + y) + z$ and $x(yz) = (xy)z$.
3. **Distributive law.** $x(y + z) = xy + xz$.
4. **Identity elements.** There are two distinct numbers 0 and 1 satisfying $x + 0 = x$ and $x \cdot 1 = x$ for every real number x.
5. **Inverses.** Each number x has an *additive inverse* (also called the *opposite*), $-x$, satisfying $x + (-x) = 0$. Also, each number x except 0 has a *multiplicative inverse* (also called the *reciprocal*), x^{-1}, satisfying $x \cdot x^{-1} = 1$.

Subtraction and divison are defined by

$$x - y = x + (-y)$$

and

$$\frac{x}{y} = x \div y = x \cdot y^{-1}$$

as long as $y \neq 0$. Division by 0 is undefined.

From these basic facts, many others follow. In fact, most of algebra ultimately rests on the five field properties, the definition of subtraction and division, and a bit of logic.

A Bit of Logic   Important results in mathematics are called **theorems**; you will find many theorems in this book. The most important ones occur with the label *Theorem* and are usually given names (e.g., the Pythagorean Theorem). Others occur in the problem sets and are introduced with the words *show that* or *prove that*. In contrast to axioms or definitions, which are taken for granted, theorems require proof.

Many theorems can be stated in the form "If P then Q" or they can be restated in this form. We often abbreviate the statement "If P then Q" by $P \Rightarrow Q$, which is also read "P implies Q." We call P the *hypothesis* and Q the *conclusion* of the theorem. A proof consists of showing that Q must be true whenever P is true.

Beginning students (and some mature ones) may confuse $P \Rightarrow Q$ with its **converse**, $Q \Rightarrow P$. These two statements are not equivalent. "If John is a Missourian then John is an American" is a true statement, but its converse "If John is an American, then John is a Missourian" may not be true.

The **negation** of the statement P is written $\sim P$. For example, if P is the statement "It is raining," then $\sim P$ is the statement "It is not raining." The statement $\sim Q \Rightarrow \sim P$ is called the **contrapositive** of the statement $P \Rightarrow Q$ and it is equivalent to $P \Rightarrow Q$. By "equivalent" we mean that $P \Rightarrow Q$ and $\sim Q \Rightarrow \sim P$ are either both true or both false. For our example about John, the contrapositive of "If John is a Missourian, then John is an American" is "If John is not an American, then John is not a Missourian."

Because a statement and its contrapositive are equivalent, we can prove a theorem of the form "If P then Q" by proving its contrapositive "If $\sim Q$ then $\sim P$." Thus, to prove $P \Rightarrow Q$, we can assume $\sim Q$ and try to deduce $\sim P$. Here is a simple example.

EXAMPLE 1   If n^2 is even, then n is even.

Proof   The contrapositive of this sentence is "If n is not even, then n^2 is not even," which is equivalent to "If n is odd, then n^2 is odd." We will prove the contrapositive. If n is odd, then there exists an integer k such that $n = 2k + 1$. Then,

$$n^2 = (2k + 1)^2 = 4k^2 + 4k + 1 = 2(2k^2 + 2k) + 1$$

Therefore, n^2 is equal to one more than twice an integer. Hence n^2 is odd.   ■

Proof by Contradiction

Proof by contradiction also goes by the name *reductio ad absurdum.* Here is what the great mathematician G. H. Hardy had to say about it.

"Reductio ad absurdum, which Euclid loved so much, is one of a mathematician's finest weapons. It is a far finer gambit than any chess gambit; a chess player may offer the sacrifice of a pawn or even a piece, but a mathematician offers the game."

The *Law of the Excluded Middle* says: Either R or $\sim R$, but not both. Any proof that begins by assuming the conclusion of a theorem is false and proceeds to show this assumption leads to a contradiction is called a **proof by contradiction**.

Occasionally, we will need another type of proof called **mathematical induction**. It would take us too far afield to describe this now, but we have given a complete discussion in Appendix A.1.

Sometimes both the statements $P \Rightarrow Q$ (if P then Q) and $Q \Rightarrow P$ (if Q then P) are true. In this case we write $P \Leftrightarrow Q$, which is read "P if and only if Q."

EXAMPLE 2   In Example 1 we showed that "If n^2 is even, then n is even," but the converse "If n is even, then n^2 is even" is also true. Thus, we would say "n is even if and only if n^2 is even."   ■

Order on the Real Line

To say that $x < y$ means that x is to the left of y on the real line.

Order   The nonzero real numbers separate nicely into two disjoint sets—the positive real numbers and the negative real numbers. This fact allows us to introduce the order relation $<$ (read "is less than") by

$$\boxed{x < y \Leftrightarrow y - x \text{ is positive}}$$

We agree that $x < y$ and $y > x$ shall mean the same thing. Thus, $3 < 4, 4 > 3$, $-3 < -2$, and $-2 > -3$. Note the geometric interpretation of $<$ shown in the box in the margin.

than 1) can be written as the product of a unique set of primes. For example, $45 = 3 \cdot 3 \cdot 5$. Write each of the following as a product of primes. *Note:* The product is trivial if the number is prime—that is, it has only one factor.

(a) 243   (b) 127
(c) 5100   (d) 346

44. Use the Fundamental Theorem of Arithmetic (Problem 43) to show that the square of any natural number (other than 1) can be written as the product of a unique set of primes, with each prime occurring an *even* number of times. For example, $(45)^2 = 3 \cdot 3 \cdot 3 \cdot 3 \cdot 5 \cdot 5$.

45. Show that $\sqrt{2}$ is irrational. *Hint:* Try a proof by contradiction. Suppose that $\sqrt{2} = p/q$, where p and q are natural numbers (necessarily different from 1). Then $2 = p^2/q^2$, and so $2q^2 = p^2$. Now use Problem 44 to get a contradiction.

46. Show that $\sqrt{3}$ is irrational (see Problem 45).

47. Show that the sum of two rational numbers is rational.

48. Show that the product of a rational number (other than 0) and an irrational number is irrational. *Hint:* Try proof by contradiction.

49. Which of the following are rational and which are irrational?

(a) $-\sqrt{9}$ (b) 0.375
(c) $1 - \sqrt{2}$ (d) $(1 + \sqrt{3})^2$
(e) $(3\sqrt{2})(5\sqrt{2})$ (f) $5\sqrt{2}$

50. Is the sum of two irrational numbers necessarily irrational? Explain.

51. Show that if the natural number m is not a perfect square then $\sqrt{m}$ is irrational.

52. Show that $\sqrt{6} + \sqrt{3}$ is irrational.

53. Show that $\sqrt{2} - \sqrt{3} + \sqrt{6}$ is irrational.

54. Show that $\log_{10}5$ is irrational.

55. Write the converse and the contrapositive to the following statements.

(a) If I do all the homework assignments, then I will get an A in this course.
(b) If x is a real number, then x is an integer.
(c) If $\triangle ABC$ is an equilateral triangle, then $\triangle ABC$ is an isosceles triangle.

Answers to Concepts Review: **1.** rational **2.** $\sqrt{2}$; π **3.** real **4.** theorems

---

1.2
Decimals, Calculators, Estimation

Any rational number can be written as a decimal, since by definition it can always be expressed as the quotient of two integers; if we divide the denominator into the numerator, we obtain a decimal (Figure 1). For example,

$$\frac{1}{2} = 0.5 \qquad \frac{3}{8} = 0.375$$

$$\frac{13}{11} = 1.181818 \ldots \qquad \frac{3}{7} = 0.428571428571428571 \ldots$$

Irrational numbers, too, can be expressed as decimals. For instance,

$$\sqrt{2} = 1.4142135623 \ldots, \qquad \sqrt{3} = 1.7320508075 \ldots$$

$$\pi = 3.1415926535 \ldots$$

Repeating and Nonrepeating Decimals The decimal representation of a rational number either terminates (as in $\frac{3}{8} = 0.375$) or else repeats in regular cycles forever (as in $\frac{13}{11} = 1.181818 \ldots$). A little experimenting with the long division algorithm will show you why. (Note that there can be only a finite number of different remainders.) A terminating decimal can be regarded as a repeating decimal with repeating zeros. For instance,

$$\frac{3}{8} = 0.375 = 0.3750000 \ldots$$

Thus, every rational number can be written as a repeating decimal. In other words, if x is a rational number, then x can be written as a repeating decimal. It is a remarkable fact that the converse is also true; if x can be written as a repeating decimal, then x is a rational number. This is obvious in the case of a terminating decimal (for instance, $3.137 = 3137/1000$), and it is easy to show for the case of a nonterminating repeating decimal.

```
  0.375          1.181
8 3.000      11 13.000
  2 4            11
  60             2 0
  56             1 1
  40              90
  40              88
                  20
                  11
                   9
 3/8 = 0.375

 13/11 = 1.181818 ...
```

Figure 1

EXAMPLE 1   (Repeating decimals are rational.) Show that

$$x = 0.136136136\ldots \quad \text{and} \quad y = 0.27171717\ldots$$

represent rational numbers.

Solution   We subtract x from $1000x$ and then solve for x.

$$\begin{aligned}
1000x &= 136.136136\ldots \\
x &= 0.136136\ldots \\
\hline
999x &= 136 \\
x &= \frac{136}{999}
\end{aligned}$$

Similarly,

$$\begin{aligned}
100y &= 27.17171717\ldots \\
y &= 0.27171717\ldots \\
\hline
99y &= 26.9 \\
y &= \frac{26.9}{99} = \frac{269}{990}
\end{aligned}$$

The decimal representations of irrational numbers do not repeat in cycles. Conversely, a nonrepeating decimal must represent an irrational number. Thus, for example,

$$0.101001000100001\ldots$$

must represent an irrational number (note the pattern of more and more 0s between the 1s). The diagram in Figure 2 summarizes what we have said.

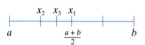

The Real Numbers

| Rational Numbers (the repeating decimals) | Irrational Numbers (the nonrepeating decimals) |

Figure 2

Figure 3

Figure 4

Denseness   Between any two different real numbers a and b, no matter how close together, there is another real number. In particular, the number $x_1 = (a + b)/2$ is a real number that is midway between a and b (Figure 3). Since there is another real number, x_2, between a and x_1, and another real number, x_3, between x_1 and x_2, and since this argument can be repeated ad infinitum, we conclude that there are infinitely many real numbers between a and b. Thus, there is no such thing as "the real number just larger than 3."

Actually, we can say more. Between any two distinct real numbers, there are both a rational number and an irrational number. (In Exercise 29 you are asked to show that there is a rational number between any two real numbers.) Hence, by the preceding argument, there are infinitely many of each.

One way that mathematicians describe the situation we have been discussing is to say that both the rational numbers and the irrational numbers are **dense** along the real line. Every number has both rational and irrational neighbors arbitrarily close to it. The two types of numbers are inseparably intertwined and relentlessly crowded together.

One consequence of the density property is that any irrational number can be approximated as closely as we please by a rational number—in fact, by a rational number with a terminating decimal representation. Take $\sqrt{2}$ as an example. The sequence of rational numbers $1, 1.4, 1.41, 1.414, 1.4142, 1.41421, 1.414213, \ldots$ marches steadily and inexorably toward $\sqrt{2}$ (Figure 4). By going far enough along in this sequence we can get as near to $\sqrt{2}$ as we wish.

Calculators and Computers   There was a time when all budding scientists and engineers walked around campus with mechanical devices called slide rules attached to their belts. By the 1970s, students carried calculators that could perform the basic arithmetic operations and find square roots. Calculators became more powerful over the years, and by the early 1980s an inexpensive calculator could evaluate exponential, logarithmic, and trigonometric functions. Graphing calculators became available in the early 1990s. Today there are calculators that can manipulate algebraic expressions; for example, these calculators can expand $(x - 3y)^{12}$, they can solve $x^3 - 2x^2 + x = 0$, and they can approximate a solution to $x^2 - \cos\sqrt{x} = 0$.

Many problems in this book are marked with a special sysmbol.

$\boxed{\text{C}}$ means USE A CALCULATOR.

$\boxed{\text{GC}}$ means USE A GRAPHING CALCULATOR.

$\boxed{\text{CAS}}$ means USE A COMPUTER ALGEBRA SYSTEM.

$\boxed{\approx}$ means MAKE AN ESTIMATE OF THE ANSWER BEFORE WORKING THE PROBLEM; THEN CHECK YOUR ANSWER AGAINST THIS ESTIMATE.

$\boxed{\text{EXPL}}$ means THE PROBLEM ASKS YOU TO EXPLORE AND GO BEYOND THE EXPLANATIONS GIVEN IN THE BOOK.

There are plenty of uses for a calculator in this book, especially in problems marked with a $\boxed{\text{C}}$.

There are now a number of powerful software packages that can do calculations such as $(\pi - \sqrt{2})^{100}$, symbolic manipulations such as expanding $(2x - 3y)^{22}$, and graphics such as plotting the graph of $y = x \sin x$. These programs can help you in the process of learning and understanding calculus, but you should not rely on them to do calculus for you. Computer software has the advantage over graphing calculators of being more powerful and capable of displaying the results on a high-resolution screen. Graphing calculators have the advantage that they cost less and will fit in your pocket.

Calculators and computers usually work only with rational numbers in the form of decimals of some prescribed length, for example, ten digits. Some software programs are capable of storing some irrational numbers in a symbolic format that, in effect, retains the exact value. For example, both Mathematica and Maple can store $\sqrt{2}$ in such a way that subsequent manipulations use this exact value. Mathematica, for example, will simplify the input `4/Sqrt[2]` and return the value `2 Sqrt[2]`.

Our advice regarding calculators and computers is this: Do calculations that can easily be done by hand without a calculator, especially if this allows an exact answer. For example, we generally prefer the exact answer $\sqrt{3}/2$ for the sine of 60° to the calculator value 0.8660254. However, in any complicated calculation we encourage use of a calculator.

Estimation Given a complicated arithmetic problem, a careless student might quickly press a few keys on a calculator and report the answer, not realizing that a missed parenthesis or a slip of the finger has given an incorrect result. A careful student with a feeling for numbers will press the same keys, immediately recognize that the answer is wrong if it is far too big or far too small, and recalculate it correctly. It is important to know how to make a mental estimate.

EXAMPLE 2 Calculate $(\sqrt{430} + 72 + \sqrt[3]{7.5})/2.75$.

Solution A wise student approximated this as $(20 + 72 + 2)/3$ and said that the answer should be in the neighborhood of 30. Thus, when her calculator gave 93.448 for an answer, she was suspicious (she had actually calculated $\sqrt{430} + 72 + \sqrt[3]{7.5}/2.75$). On recalculating, she got the correct answer: 34.434. ∎

If a man tells you that the volume of his body is 20,000 cubic inches, be suspicious. You might estimate his volume this way. He is about 70 inches tall and his belt length is 30 inches, giving a radius at the waist of about 5 inches. If we approximate his volume by that of a cylinder, we find the volume to be $\pi r^2 h \approx 3(5^2)70 \approx 5000$ cubic inches. He is not as big as he says he is.

Here we have used $\approx$ to mean "approximately equal." Use this symbol in your scratch work when you are making an approximation to an answer. In more formal work you should never use this symbol without knowing how large the error could be. Here is an example more related to calculus.

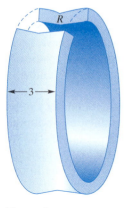

Figure 5

EXAMPLE 3 Suppose that the shaded region R shown in Figure 5 is revolved about the x-axis. Estimate the volume of the resulting solid ring S.

Solution The region R is about 3 units long and 0.9 units high. We estimate its area as $3(0.9) \approx 3$ square units. Imagine the solid ring S to be slit open and laid out flat, forming a box about $2\pi r \approx 2(3)(6) = 36$ units long. The volume of a box is its cross-sectional area times its length. Thus, we estimate the volume of the box to be $3(36) = 108$ cubic units. If you calculate it to be 1000 cubic units, you need to check your work. ∎

The process of *estimation* is just organized common sense combined with reasonable number approximations. We urge you to use it all the time, especially on word problems. Before you attempt to get a precise answer, make an estimate. If your answer is close to your estimate, there is no guarantee that your answer is correct. On the other hand, if your answer and your estimate are far apart, you should check your work. There is probably an error in your answer or in your approximation. Remember that $\pi \approx 3$, $\sqrt{2} \approx 1.4$, $2^{10} \approx 1000$, 1 foot $\approx$ 10 inches, 1 mile $\approx$ 5000 feet, and so on.

Because we think estimation is such an important skill, we will emphasize it in our exposition of calculus. Now and then we will label a problem with the symbol $\boxed{\approx}$. This symbol will suggest that it is a good idea to make a rough estimate of the answer either before you begin serious work or after you have what you think is the correct answer.

Concepts Review

1. Every real number can be represented by an unending decimal. The decimal representation of $\frac{1}{3}$ is _____; the decimal representation of $\frac{1}{5}$ is _____; the decimal representation of π starts out as _____.

2. The decimal representation of a number will repeat in cycles if and only if the number is _____.

3. Between any two real numbers, we can always find both _____ numbers and _____ numbers (and both in infinite supply).

4. Some rational numbers can be represented by terminating decimals (e.g., $\frac{1}{2} = 0.5$, $\frac{9}{8} = 1.125$). Every _____ number can be approximated arbitrarily closely by a terminating decimal.

Problem Set 1.2

In Problems 1–6, change each rational number to a decimal by performing a long division.

1. $\frac{1}{12}$ **2.** $\frac{2}{7}$

3. $\frac{3}{21}$ **4.** $\frac{5}{17}$

5. $\frac{11}{3}$ **6.** $\frac{11}{13}$

In Problems 7–12, change each repeating decimal to a ratio of two integers (see Example 1).

7. $0.123123123\ldots$ **8.** $0.217171717\ldots$

9. $2.56565656\ldots$ **10.** $3.929292\ldots$

11. $0.199999\ldots$ **12.** $0.399999\ldots$

13. Since $0.199999\ldots = 0.200000\ldots$ and $0.399999\ldots = 0.400000\ldots$ (see Problems 11 and 12), we see that certain rational numbers have two different decimal expansions. Which rational numbers have this property?

14. Show that any rational number p/q, for which the prime factorization of q consists entirely of 2s and 5s, has a terminating decimal expansion.

15. Find a positive rational number and a positive irrational number both smaller than 0.00001.

16. What is the smallest positive integer? The smallest positive rational number? The smallest positive irrational number?

17. Find a rational number between 3.14159 and π. Note that $\pi = 3.141592\ldots$).

18. Is there a number between $0.9999\ldots$ (repeating 9s) and 1? How do you resolve this with the statement that between any two different real numbers there is another real number?

19. Is $0.1234567891011121314\ldots$ rational or irrational? (You should see a pattern in the given sequence of digits.)

20. Find two irrational numbers who sum is rational.

$\boxed{\approx}$ *In Problems 21–28, find the best decimal approximation that your calculator allows. Begin by making a mental estimate.*

21. $\left(\sqrt{3} + 1\right)^3$ **22.** $\left(\sqrt{2} - \sqrt{3}\right)^4$

23. $\sqrt[3]{1.123} - \sqrt[3]{1.09}$ **24.** $(3.1415)^{-1/2}$

25. $\dfrac{\sqrt{130} - \sqrt{5}}{3^{1.2} - 3}$

26. $\dfrac{(0.00121)(5.23 \times 10^{-3})}{6.16 \times 10^{-4}}$

27. $\sqrt{8.9\pi^2 + 1} - 3\pi$ **28.** $\sqrt[4]{(6\pi^2 - 2)\pi}$

29. Show that between any two different real numbers there is a rational number. (*Hint:* If $a < b$, then $b - a > 0$, so there is a natural number n such that $1/n < b - a$. Consider the set $\{k \mid k/n > b\}$ and use the fact that a set of integers that is bounded from below contains a least element.) Show that between any two different real numbers there are infinitely many rational numbers.

$\boxed{\approx}$ **30.** Estimate the number of cubic inches in your head.

$\boxed{\approx}$ **31.** Estimate the length of the equator in feet. Assume the radius of the earth to be 4000 miles.

$\boxed{\approx}$ **32.** About how many times has your heart beat by your twentieth birthday?

$\boxed{\approx}$ **33.** The General Sherman tree in California is about 270 feet tall and averages about 16 feet in diameter. Estimate the number of board feet (1 board foot equals 1 inch by 12 inches by 12 inches) of lumber that could be made from this tree, assuming no waste and ignoring the branches.

$\boxed{\approx}$ **34.** Assume that the General Sherman tree (Problem 33) produces an annual growth ring of thickness 0.004 foot. Estimate the resulting increase in the volume of its trunk each year.

$\boxed{\text{C}}$ **35.** Note that

$$2x^3 - 7x^2 + 11x - 2 = \left[(2x - 7)x + 11\right]x - 2$$

To calculate the expression on the right at $x = 3$, press the following keys on an algebraic logic calculator.

$$2 \boxed{\times} 3 \boxed{-} 7 \boxed{=} \boxed{\times} 3 \boxed{+} 11 \boxed{=} \boxed{\times} 3 \boxed{-} 2 \boxed{=}$$

Use this idea to calculate the given expression in each case.

(a) $x = 2\pi$ (b) $x = 2.15$
(c) $x = 2.71828$ (d) $x = 1.1$

C **36.** Use the idea discussed in Problem 35 to evaluate $x^4 - 3x^3 + 4x^2 + 6x - 10$ at each value.

(a) $x = 1$ (b) $x = \pi$
(c) $x = 14.2$ (d) $x = 1.2157$

37. A number b is called an **upper bound** for a set S of numbers if $x \leq b$ for all x in S. For example 5, 6.5, and 13 are upper bounds for the set $S = \{1, 2, 3, 4, 5\}$. The number 5 is the **least upper bound** for S (the smallest of the upper bounds). Similarly, 1.6, 2, and 2.5 are upper bounds for the infinite set $T = \{1.4, 1.49, 1.499, 1.4999, \ldots\}$, whereas 1.5 is its least upper bound. Find the least upper bound of each of the following sets.

(a) $S = \{-10, -8, -6, -4, -2\}$
(b) $S = \{-2, -2.1, -2.11, -2.111, -2.1111, \ldots\}$

(c) $S = \{2.4, 2.44, 2.444, 2.4444, \ldots\}$
(d) $S = \{1 - \frac{1}{2}, 1 - \frac{1}{3}, 1 - \frac{1}{4}, 1 - \frac{1}{5}, \ldots\}$
(e) $S = \{x | x = (-1)^n + 1/n, n \text{ a positive integer}\}$; that is, S is the set of all numbers x that have the form $x = (-1)^n + 1/n$, where n is a positive integer.
(f) $S = \{x : x^2 < 2, x \text{ a rational number}\}$.

EXPL **38. The Axiom of Completeness** for the real numbers says: *Every set of real numbers that has an upper bound has a least upper bound that is a real number.*

(a) Show that the italicized statement is false if the word *real* is replaced by *rational*.
(b) Would the italicized statement be true or false if the word *real* were replaced by *natural*?

Note: The real numbers $\mathbb{R}$ are the only set of numbers simultaneously having the field properties, the order properties, and the completeness property.

Answers to Concepts Review: **1.** 0.333 . . . (3s repeat); 0.200 . . . (0s repeat); 3.14159 **2.** rational **3.** rational; irrational **4.** real

---

1.3
Inequalities

Solving equations (for instance, $3x - 17 = 6$ or $x^2 - x - 6 = 0$) is one of the traditional tasks of mathematics; it will be important in this course and we assume that you remember how to do it. But of almost equal significance in calculus is the notion of solving an inequality (e.g., $3x - 17 < 6$ or $x^2 - x - 6 \geq 0$). To **solve** an inequality is to find the set of all real numbers that make the inequality true. In contrast to an equation, whose solution set normally consists of one number or perhaps a finite set of numbers, the solution set of an inequality is usually an entire interval of numbers or, in some cases, the union of such intervals.

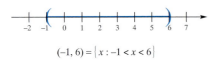

$(-1, 6) = \{x : -1 < x < 6\}$

Figure 1

$[-1, 5] = \{x : -1 \leq x \leq 5\}$

Figure 2

Intervals Several kinds of intervals will arise in our work and we introduce special terminology and notation for them. The double inequality $a < x < b$ describes the **open interval** consisting of all numbers between a and b, not including the end points a and b. We denote it by the symbol (a, b) (Figure 1). In contrast, the inequality $a \leq x \leq b$ describes the corresponding **closed interval,** which does include the end points a and b. It is denoted by $[a, b]$ (Figure 2). The table indicates the wide variety of possibilities and introduces our notation.

> The compound inequality $a < x < b$ means that both $a < x$ and $x < b$. Even though it consists of two inequalities, we will refer to it as "an inequality."

Set Notation	Interval Notation	Graph
$\{x : a < x < b\}$	(a, b)	
$\{x : a \leq x \leq b\}$	$[a, b]$	
$\{x : a \leq x < b\}$	$[a, b)$	
$\{x : a < x \leq b\}$	$(a, b]$	
$\{x : x \leq b\}$	$(-\infty, b]$	
$\{x : x < b\}$	$(-\infty, b)$	
$\{x : x \geq a\}$	$[a, \infty)$	
$\{x : x > a\}$	(a, ∞)	
$\mathbb{R}$	$(-\infty, \infty)$	

Solving Inequalities As with equations, the procedure for solving an inequality consists in transforming the inequality a step at a time until the solution set is obvious. The chief tools are the order properties from Section 1.1. They imply that we may perform certain operations on both sides of an inequality without changing its solution set. In particular:

1. *We may add the same number to both sides of an inequality.*
2. *We may multiply both sides of an inequality by a positive number.*
3. *We may multiply both sides by a negative number, but then we must reverse the direction of the inequality sign.*

EXAMPLE 1 Solve the inequality $2x - 7 < 4x - 2$ and show the graph of its solution set.

Solution

$$2x - 7 < 4x - 2$$
$$2x < 4x + 5 \qquad \text{(adding 7)}$$
$$-2x < 5 \qquad \text{(adding } -4x)$$
$$x > -\tfrac{5}{2} \qquad \text{(multiplying by } -\tfrac{1}{2})$$

The graph appears in Figure 3.

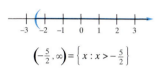

$$\left(-\tfrac{5}{2}, \infty\right) = \left\{x : x > -\tfrac{5}{2}\right\}$$

Figure 3

EXAMPLE 2 Solve $-5 \le 2x + 6 < 4$.

Solution

$$-5 \le 2x + 6 < 4$$
$$-11 \le 2x \quad < -2 \qquad \left(\text{adding } -6\right)$$
$$-\tfrac{11}{2} \le x \quad < -1 \qquad \left(\text{multiplying by } \tfrac{1}{2}\right)$$

Figure 4 shows the corresponding graph.

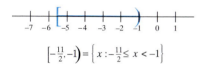

$$\left[-\tfrac{11}{2}, -1\right) = \left\{x : -\tfrac{11}{2} \le x < -1\right\}$$

Figure 4

Before tackling a quadratic inequality, we point out that a linear factor of the form $x - a$ is positive for $x > a$ and negative for $x < a$. It follows that a product $(x - a)(x - b)$ can change from being positive to negative, or vice versa, only at a or b. These points, where a factor is zero, are called **split points.** They are the keys to determining the solution sets of quadratic and other more complicated inequalities.

EXAMPLE 3 Solve the quadratic inequality $x^2 - x < 6$.

Solution As with quadratic equations, we move all nonzero terms to one side and factor.

$$x^2 - x < 6$$
$$x^2 - x - 6 < 0 \qquad \text{(adding } -6)$$
$$(x + 2)(x - 3) < 0 \qquad \text{(factoring)}$$

We see that -2 and 3 are the split points; they divide the real line into the three intervals $(-\infty, -2)$, $(-2, 3)$, and $(3, \infty)$. On each of these intervals, $(x - 3)(x + 2)$ is of one sign; that is, it is either always positive or always negative. To find this sign in each interval, we use the **test points** $-3, 0$, and 5 (any points on the three intervals would do). Our results are shown next.

Test Point	Value of $(x - 3)(x + 2)$	Sign
-3	6	$+$
0	-6	$-$
5	14	$+$

36. Solve $1 + x + x^2 + x^3 + \cdots + x^{99} \leq 0$.

37. The formula

$$\frac{1}{R} = \frac{1}{R_1} + \frac{1}{R_2} + \frac{1}{R_3}$$

gives the total resistance R in an electric circuit due to three resistances, R_1, R_2, and R_3, connected in parallel. If $10 \leq R_1 \leq 20$, $20 \leq R_2 \leq 30$, and $30 \leq R_3 \leq 40$, find the range of values of R.

Answers to Concepts Review: **1.** Interval, intervals
2. $[-1, 5)$, $(-\infty, 2]$ **3.** $b > 0, b < 0$ **4.** 3, $-4, -5$

1.4
Absolute Values, Square Roots, Squares

The concept of absolute value is extremely useful in calculus, and the reader should acquire skill in working with it. The **absolute value** of a real number x, denoted by $|x|$, is defined by

$$\begin{array}{ll} |x| = x & \text{if } x \geq 0 \\ |x| = -x & \text{if } x < 0 \end{array}$$

For example, $|6| = 6$, $|0| = 0$, and $|-5| = -(-5) = 5$.

This two-pronged definition merits careful study. Note that it does not say that $|-x| = x$ (try $x = -5$ to see why). It is true that $|x|$ is always nonnegative; it is also true that $|-x| = |x|$.

One of the best ways to think of the absolute value of a number is as an undirected distance. In particular, $|x|$ is the distance between x and the origin. Similarly, $|x - a|$ is the distance between x and a (Figure 1).

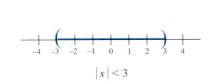

Figure 1

Properties   Absolute values behave nicely under multiplication and division, but not so well under addition and subtraction.

Properties of Absolute Values

1. $|ab| = |a||b|$
2. $\left| \dfrac{a}{b} \right| = \dfrac{|a|}{|b|}$
3. $|a + b| \leq |a| + |b|$ (Triangle Inequality)
4. $|a - b| \geq ||a| - |b||$

Figure 2

Inequalities Involving Absolute Values   If $|x| < 3$, then the distance between x and the origin must be less than 3. In other words, x must be simultaneously less than 3 *and* greater than -3; that is, $-3 < x < 3$. On the other hand, if $|x| > 3$, then the distance between x and the origin must be at least 3. This can happen when $x > 3$ *or* $x < -3$ (Figure 2). These are special cases of the following general statements.

$$|x| < a \Leftrightarrow -a < x < a$$
$$|x| > a \Leftrightarrow x < -a \text{ or } x > a$$

We can use these facts to solve inequalities involving absolute values, since they provide a way of removing absolute value signs.

EXAMPLE 1   Solve the inequality $|x - 4| < 2$ and show the solution set on the real line. Interpret the absolute value as a distance.

Solution   From the equations in (1), with x replaced by $|x - 4|$, we see that

$$|x - 4| < 2 \Leftrightarrow -2 < x - 4 < 2$$

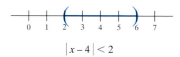

$|x - 4| < 2$

Figure 3

When we add 4 to all three members of this latter inequality, we obtain $2 < x < 6$. The graph is shown in Figure 3.

In terms of distance, the symbol $|x - 4|$ represents the distance between x and 4. The inequality therefore says that the distance between x and 4 must be less than 2. The numbers x with this property are the numbers between 2 and 6; that is, $2 < x < 6$.   ■

The statements in the equations just before Example 1 are valid with $<$ and $>$ replaced by $\le$ and $\ge$, respectively. We need the second statement in this form in our next example.

EXAMPLE 2   Solve the inequality $|3x - 5| \ge 1$ and show its solution set on the real line.

Solution   The given inequality may be written successively as

$$3x - 5 \le -1 \quad \text{or} \quad 3x - 5 \ge 1$$
$$3x \le 4 \quad \text{or} \quad 3x \ge 6$$
$$x \le \tfrac{4}{3} \quad \text{or} \quad x \ge 2$$

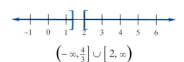

$\left(-\infty, \tfrac{4}{3}\right] \cup [2, \infty)$

Figure 4

The solution set is the union of two intervals; it is the set $\left(-\infty, \tfrac{4}{3}\right] \cup [2, \infty)$ shown in Figure 4.   ■

In Chapter 2 we will need to make the kind of manipulations illustrated by the next two examples. Delta (δ) and epsilon (ε) are the fourth and fifth letters, respectively, of the Greek alphabet and are traditionally used to stand for small positive numbers.

EXAMPLE 3   Let ε (epsilon) be a positive number. Show that

$$|x - 2| < \frac{\varepsilon}{5} \Leftrightarrow |5x - 10| < \varepsilon$$

In terms of distance, this says that the distance between x and 2 is less than $\varepsilon/5$ if and only if the distance between $5x$ and 10 is less than ε.

Solution

$$|x - 2| < \frac{\varepsilon}{5} \Leftrightarrow 5|x - 2| < \varepsilon \quad \text{(multiplying by 5)}$$
$$\Leftrightarrow |5||x - 2| < \varepsilon \quad (|5| = 5)$$
$$\Leftrightarrow |5(x - 2)| < \varepsilon \quad (|a||b| = |ab|)$$
$$\Leftrightarrow |5x - 10| < \varepsilon$$ ■

Finding Delta

Note two facts about our solution to Example 4.

1. The number we find for δ must depend on ε. Our choice is $\delta = \varepsilon/6$.
2. Any positive δ smaller than $\varepsilon/6$ is acceptable. For example $\delta = \varepsilon/7$ or $\delta = \varepsilon/(2\pi)$ are other correct choices.

EXAMPLE 4   Let ε be a positive number. Find a positive number δ (delta) such that

$$|x - 3| < \delta \Rightarrow |6x - 18| < \varepsilon$$

Solution

$$|6x - 18| < \varepsilon \Leftrightarrow |6(x - 3)| < \varepsilon$$
$$\Leftrightarrow 6|x - 3| < \varepsilon \quad (|ab| = |a|\,|b|)$$
$$\Leftrightarrow |x - 3| < \frac{\varepsilon}{6} \quad \left(\text{multiplying by } \frac{1}{6}\right)$$

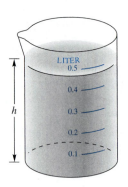

Figure 5

Therefore, we choose $\delta = \varepsilon/6$. Following the implications backward, we see that

$$|x - 3| < \delta \Rightarrow |x - 3| < \frac{\varepsilon}{6} \Rightarrow |6x - 18| < \varepsilon \qquad \blacksquare$$

Here is a practical problem that uses the same type of reasoning.

EXAMPLE 5 A $\frac{1}{2}$-liter (500 cubic centimeter) glass beaker has an inner radius of 4 centimeters. How closely must we measure the height h of water in the glass to be sure we have $\frac{1}{2}$ liter of water with an error of less than 1%, that is, an error of less than 5 cubic centimeters? See Figure 5.

Solution The volume V of water in the glass is given by the formula $V = 16\pi h$. We want $|V - 500| < 5$ or, equivalently, $|16\pi h - 500| < 5$. Now

$$|16\pi h - 500| < 5 \Leftrightarrow \left| 16\pi \left(h - \frac{500}{16\pi} \right) \right| < 5$$

$$\Leftrightarrow \quad 16\pi \left| h - \frac{500}{16\pi} \right| < 5$$

$$\Leftrightarrow \quad \left| h - \frac{500}{16\pi} \right| < \frac{5}{16\pi}$$

$$\Leftrightarrow \quad |h - 9.947| < 0.0947 \approx 0.1$$

Thus, we must measure the height to an accuracy of about 1 millimeter. $\qquad \blacksquare$

Square Roots Every positive number has two square roots. For example, the two square roots of 9 are 3 and -3. We sometimes represent these two numbers as ± 3. For $a \geq 0$, the symbol $\sqrt{a}$, called the **principal square root** of a, denotes the nonnegative square root of a. Thus, $\sqrt{9} = 3$ and $\sqrt{121} = 11$. It is incorrect to write $\sqrt{16} = \pm 4$ because $\sqrt{16}$ means the nonnegative square root of 16, that is, 4. The number 7 has two square roots, which are written as $\pm \sqrt{7}$, but $\sqrt{7}$ represents a single real number.

Here is an important fact worth remembering.

$$\boxed{\sqrt{x^2} = |x|}$$

Most students will recall the **Quadratic Formula**. The solutions to the quadratic equation $ax^2 + bx + c = 0$ are given by

$$\boxed{x = \frac{-b \pm \sqrt{b^2 - 4ac}}{2a}}$$

The number $d = b^2 - 4ac$ is called the **discriminant** of the quadratic equation. This equation has two real solutions if $d > 0$, one real solution if $d = 0$, and no real solutions if $d < 0$.

With the Quadratic Formula, we can easily solve quadratic inequalities even if they do not factor by inspection.

EXAMPLE 6 Solve $x^2 - 2x - 4 \leq 0$.

Solution The two solutions of $x^2 - 2x - 4 = 0$ are

$$x_1 = \frac{-(-2) - \sqrt{4 + 16}}{2} = 1 - \sqrt{5} \approx -1.24$$

and

$$x_2 = \frac{-(-2) + \sqrt{4 + 16}}{2} = 1 + \sqrt{5} \approx 3.24$$

Figure 6

Thus,

$$x^2 - 2x - 4 = (x - x_1)(x - x_2) = (x - 1 + \sqrt{5})(x - 1 - \sqrt{5})$$

The split points $1 - \sqrt{5}$ and $1 + \sqrt{5}$ divide the real line into three intervals (Figure 6). When we test them with the test points $-2, 0$, and 4, we conclude that the solution set for $x^2 - 2x - 4 \leq 0$ is $[1 - \sqrt{5}, 1 + \sqrt{5}]$. ■

We mention that if n is even and $a \geq 0$ the symbol $\sqrt[n]{a}$ denotes the nonnegative nth root of a. When n is odd, there is only one real nth root of a, denoted by the symbol $\sqrt[n]{a}$. Thus, $\sqrt[4]{16} = 2$, $\sqrt[3]{27} = 3$, and $\sqrt[3]{-8} = -2$.

Squares   Turning to squares, we notice that

$$\boxed{|x|^2 = x^2}$$

This follows from the property $|a||b| = |ab|$.

Does the squaring operation preserve inequalities? In general, the answer is no. For instance, $-3 < 2$, but $(-3)^2 > 2^2$. On the other hand, $2 < 3$ and $2^2 < 3^2$. If we are dealing with nonnegative numbers, then $a < b \Leftrightarrow a^2 < b^2$. A useful variant of this is

$$\boxed{|x| < |y| \;\Leftrightarrow\; x^2 < y^2}$$

You are asked to provide a proof of this fact in Problem 31.

EXAMPLE 7   Solve the inequality $|3x + 1| < 2|x - 6|$.

Solution   This inequality is more difficult to solve than our earlier examples, because there are two sets of absolute value signs. We can remove both of them by using the last boxed result.

$$
\begin{aligned}
|3x + 1| < 2|x - 6| &\Leftrightarrow & |3x + 1| &< |2x - 12| \\
&\Leftrightarrow & (3x + 1)^2 &< (2x - 12)^2 \\
&\Leftrightarrow & 9x^2 + 6x + 1 &< 4x^2 - 48x + 144 \\
&\Leftrightarrow & 5x^2 + 54x - 143 &< 0 \\
&\Leftrightarrow & (x + 13)(5x - 11) &< 0
\end{aligned}
$$

The split points for this quadratic inequality are -13 and $\frac{11}{5}$; they divide the real line into the three intervals: $(-\infty, -13)$, $\left(-13, \frac{11}{5}\right)$, and $\left(\frac{11}{5}, \infty\right)$. When we use the test points -14, 0, and 3, we discover that only the points in $\left(-13, \frac{11}{5}\right)$ satisfy the inequality. ■

Concepts Review

1. The inequality $|x - 2| \leq 3$ is equivalent to
_____ $\leq x \leq$ _____.

2. The Triangle Inequality says that _____.

3. Which of the following are always true?
(a) $|-x| = x$ (b) $|x|^2 = x^2$
(c) $|xy| = |x| |y|$ (d) $\sqrt{x^2} = x$

4. To be sure that $|5x - 20| < 0.2$, we would need $|x - 4| <$ _____.

Problem Set 1.4

In Problems 1–12, find the solution sets of the given inequalities (see Examples 1 and 2).

1. $|x + 2| < 1$

2. $|x - 2| \geq 5$

3. $|2x - 1| > 2$

4. $|4x + 5| \leq 10$

5. $\left|\dfrac{x}{4} + 1\right| < 1$

6. $\left|\dfrac{2x}{7} - 5\right| \geq 7$

7. $|2x - 7| > 3$

8. $|5x - 6| > 1$

9. $|4x + 2| \geq 10$

10. $\left|\dfrac{x}{2} + 7\right| \geq 2$

11. $\left|2 + \dfrac{5}{x}\right| > 1$

12. $\left|\dfrac{1}{x} - 3\right| > 6$

In Problems 13–16, solve the given quadratic inequality using the quadratic formula (see Example 6).

13. $x^2 - 3x - 4 \geq 0$

14. $x^2 - 4x + 4 \leq 0$

15. $3x^2 + 17x - 6 > 0$

16. $14x^2 + 11x - 15 \leq 0$

In Problems 17–20, show that the indicated implication is true (see Example 3).

17. $|x - 3| < 0.5 \Rightarrow |5x - 15| < 2.5$

18. $|x + 2| < 0.3 \Rightarrow |4x + 8| < 1.2$

19. $|x - 2| < \dfrac{\varepsilon}{6} \Rightarrow |6x - 12| < \varepsilon$

20. $|x + 4| < \dfrac{\varepsilon}{2} \Rightarrow |2x + 8| < \varepsilon$

In Problems 21–24, find δ (depending on ε) so that the given implication is true (see Example 4).

21. $|x - 5| < \delta \Rightarrow |3x - 15| < \varepsilon$

22. $|x - 2| < \delta \Rightarrow |4x - 8| < \varepsilon$

23. $|x + 6| < \delta \Rightarrow |6x + 36| < \varepsilon$

24. $|x + 5| < \delta \Rightarrow |5x + 25| < \varepsilon$

25. On a lathe, you are to turn out a disk (thin right circular cylinder) of circumference 10 inches. This is done by continually measuring the diameter as you make the disk smaller. How closely must you measure the diameter if you can tolerate an error of at most 0.02 inch in the circumference (see Example 5)?

26. Fahrenheit temperatures and Celsius temperatures are related by the formula $C = \frac{5}{9}(F - 32)$. An experiment requires that a solution be kept at 50°C with an error of at most 3% (or 1.5°). You have only a Fahrenheit thermometer. What error are you allowed on it?

In Problems 27–30, solve the inequalities (see Example 7).

27. $|x - 1| < 2|x - 3|$

28. $|2x - 1| \geq |x + 1|$

29. $2|2x - 3| < |x + 10|$

30. $|3x - 1| < 2|x + 6|$

31. Prove that $|x| < |y| \Leftrightarrow x^2 < y^2$ by giving a reason for each of these steps:

$$|x| < |y| \Rightarrow |x||x| \leq |x||y| \quad \text{and} \quad |x||y| < |y||y|$$
$$\Rightarrow |x|^2 < |y|^2$$
$$\Rightarrow x^2 < y^2$$

Conversely,

$$x^2 < y^2 \Rightarrow |x|^2 < |y|^2$$
$$\Rightarrow |x|^2 - |y|^2 < 0$$
$$\Rightarrow (|x| - |y|)(|x| + |y|) < 0$$
$$\Rightarrow |x| - |y| < 0$$
$$\Rightarrow |x| < |y|$$

32. Use the result of Problem 31 to show that

$$0 < a < b \Rightarrow \sqrt{a} < \sqrt{b}$$

33. Use the properties of the absolute value to show that each of the following is true.

(a) $|a - b| \leq |a| + |b|$ 　　(b) $|a - b| \geq |a| - |b|$
(c) $|a + b + c| \leq |a| + |b| + |c|$

34. Use the Triangle Inequality and the fact that $0 < |a| < |b| \Rightarrow 1/|b| < 1/|a|$ to establish the following chain of inequalities.

$$\left|\frac{1}{x^2 + 3} - \frac{1}{|x| + 2}\right| \leq \frac{1}{x^2 + 3} + \frac{1}{|x| + 2} \leq \frac{1}{3} + \frac{1}{2}$$

35. Show that (see Problem 34)

$$\left|\frac{x - 2}{x^2 + 9}\right| \leq \frac{|x| + 2}{9}$$

36. Show that

$$|x| \leq 2 \Rightarrow \left|\frac{x^2 + 2x + 7}{x^2 + 1}\right| \leq 15$$

37. Show that

$$|x| \leq 1 \Rightarrow \left|x^4 + \tfrac{1}{2}x^3 + \tfrac{1}{4}x^2 + \tfrac{1}{8}x + \tfrac{1}{16}\right| < 2$$

38. Show each of the following:

(a) $x < x^2$ for $x < 0$ or $x > 1$
(b) $x^2 < x$ for $0 < x < 1$

39. Show that $a \neq 0 \Rightarrow a^2 + 1/a^2 \geq 2$. *Hint:* Consider $(a - 1/a)^2$.

40. The number $\frac{1}{2}(a + b)$ is called the average, or **arithmetic mean**, of a and b. Show that the arithmetic mean of two numbers is between the two numbers; that is, prove that

$$a < b \Rightarrow a < \frac{a + b}{2} < b$$

41. The number $\sqrt{ab}$ is called the **geometric mean** of two positive numbers a and b. Prove that

$$0 < a < b \Rightarrow a < \sqrt{ab} < b$$

42. For two positive numbers a and b, prove that

$$\sqrt{ab} \leq \tfrac{1}{2}(a + b)$$

This is the simplest version of a famous inequality called the **geometric mean–arithmetic mean inequality**.

43. Show that, among all rectangles with given perimeter p, the square has the largest area. *Hint:* If a and b denote the lengths of adjacent sides of a rectangle of perimeter p, then the area is ab, and for the square the area is $a^2 = [(a + b)/2]^2$. Now see Problem 42.

44. The radius of a sphere is measured to be about 10 inches. Determine a tolerance δ in this measurement that will ensure an error of less than 0.01 square inch in the calculated value of the surface area of the sphere.

Answers to Concepts Review: **1.** −1; 5 **2.** $|a + b| \leq |a| + |b|$ **3.** b, c **4.** 0.04

1.5
The Rectangular Coordinate System

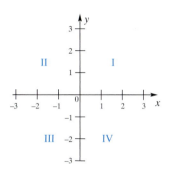

Figure 1

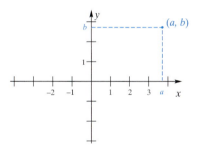

Figure 2

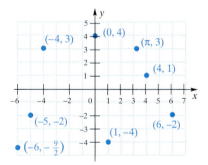

Figure 3

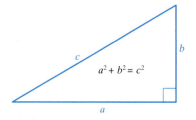

Figure 4

Two Frenchmen deserve credit for the idea of a coordinate system. Pierre de Fermat was a lawyer who made mathematics his hobby. In 1629, he wrote a paper that, in effect, made use of coordinates to describe points and curves. René Descartes was a philosopher who thought mathematics could unlock the secrets of the universe. He published his *La Géométrie* in 1637. It is a famous book and, though it does emphasize the role of algebra in solving geometric problems, one finds only a hint of coordinates there. By virtue of having the idea first and more explicitly, Fermat ought to get the major credit. History can be a fickle friend; coordinates are known as Cartesian coordinates, named after René Descartes.

Cartesian Coordinates   In the plane, produce two copies of the real line, one horizontal and the other vertical, so that they intersect at the zero points of the two lines. The two lines are called **coordinate axes**; their intersection is labeled *O* and is called the **origin**. By convention, the horizontal line is called the **x-axis** and the vertical line is called the **y-axis**. The positive half of the *x*-axis is to the right; the positive half of the *y*-axis is upward. The coordinate axes divide the plane into four regions, called **quadrants**, labeled I, II, III, and IV, as shown in Figure 1.

Each point *P* in the plane can now be assigned a pair of numbers, called its **Cartesian coordinates**. If vertical and horizontal lines through *P* intersect the *x*- and *y*-axes at *a* and *b*, respectively, then *P* has coordinates (a, b) (see Figure 2). We call (a, b) an **ordered pair** of numbers because it makes a difference which number is first. The first number *a* is the **x-coordinate** (or abscissa); the second number *b* is the **y-coordinate** (or ordinate).

Conversely, take any ordered pair (a, b) of real numbers. The vertical line through *a* on the *x*-axis and the horizontal line through *b* on the *y*-axis meet at a point *P*, whose coordinates are (a, b).

Think of it this way: The coordinates of a point are the address of that point. If you have found a house (or a point), you can read its address. Conversely, if you know the address of a house (or a point), you can always locate it. In Figure 3, we show the coordinates of several points.

The Distance Formula   With coordinates in hand, we can introduce a simple formula for the distance between any two points in the plane. It is based on the **Pythagorean Theorem,** which says that, if *a* and *b* measure the two legs of a right triangle and *c* measures its hypotenuse (Figure 4), then

$$a^2 + b^2 = c^2$$

Conversely, this relationship between the three sides of a triangle holds only for a right triangle.

Now consider any two points *P* and *Q*, with coordinates (x_1, y_1) and (x_2, y_2), respectively. Together with *R*, the point with coordinates (x_2, y_1), *P* and *Q* are vertices of a right triangle (Figure 5). The lengths of *PR* and *RQ* are $|x_2 - x_1|$ and $|y_2 - y_1|$, respectively. When we apply the Pythagorean Theorem and take the principal square root of both sides, we obtain the following expression for $d(P, Q)$, the undirected distance between *P* and *Q*.

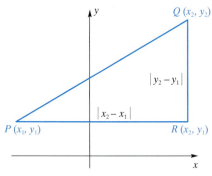

Figure 5

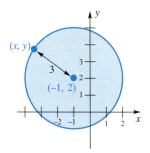

Figure 6

Circle ↔ Equation

To say that

$$(x + 1)^2 + (y - 2)^2 = 9$$

is the equation of the circle of radius 3 with center $(-1, 2)$ means two things:

1. If a point is on this circle, then its coordinates (x, y) satisfy the equation.
2. If x and y are numbers that satisfy the equation, then they are the coordinates of a point on the circle.

$$d(P, Q) = \sqrt{(x_2 - x_1)^2 + (y_2 - y_1)^2}$$

This is called the **Distance Formula**.

EXAMPLE 1   Find the distance between

(a) $P(-2, 3)$ and $Q(4, -1)$ (b) $P(\sqrt{2}, \sqrt{3})$ and $Q(\pi, \pi)$

Solution

(a) $d(P, Q) = \sqrt{(4 - (-2))^2 + (-1 - 3)^2} = \sqrt{36 + 16} = \sqrt{52} \approx 7.21$

(b) $d(P, Q) = \sqrt{(\pi - \sqrt{2})^2 + (\pi - \sqrt{3})^2} \approx \sqrt{4.971} \approx 2.23$ ∎

The formula holds even if the two points lie on the same horizontal line or the same vertical line. Thus, the distance between $P(-2, 2)$ and $Q(6, 2)$ is

$$\sqrt{(-2 - 6)^2 + (2 - 2)^2} = \sqrt{64} = 8$$

The Equation of a Circle   It is a small step from the distance formula to the equation of a circle. A **circle** is the set of points that lie at a fixed distance (the *radius*) from a fixed point (the *center*). Consider, for example, the circle of radius 3 with center at $(-1, 2)$ (Figure 6). Let (x, y) denote any point on this circle. By the Distance Formula,

$$\sqrt{(x + 1)^2 + (y - 2)^2} = 3$$

When we square both sides, we obtain

$$(x + 1)^2 + (y - 2)^2 = 9$$

which we call the equation of this circle.

More generally, the circle of radius r and center (h, k) has the equation

(1)
$$(x - h)^2 + (y - k)^2 = r^2$$

We call this the **standard equation of a circle**.

EXAMPLE 2   Find the standard equation of a circle of radius 5 and center $(1, -5)$. Also find the y-coordinates of the two points on this circle with x-coordinate 2.

Solution   The desired equation is

$$(x - 1)^2 + (y + 5)^2 = 25$$

To accomplish the second task, we substitute $x = 2$ in the equation and solve for y.

$$(2 - 1)^2 + (y + 5)^2 = 25$$
$$(y + 5)^2 = 24$$
$$y + 5 = \pm\sqrt{24}$$
$$y = -5 \pm \sqrt{24} = -5 \pm 2\sqrt{6}$$ ∎

If we expand the two squares in the boxed equation (1) and combine the constants, then the equation takes the form

$$x^2 + ax + y^2 + by = c$$

This suggests asking whether every equation of the latter form is the equation of a circle. The answer is yes, with some obvious exceptions.

EXAMPLE 3   Show that the equation

$$x^2 - 2x + y^2 + 6y = -6$$

represents a circle, and find its center and radius.

Solution   We need to *complete the square,* a process important in many contexts. To complete the square of $x^2 \pm bx$, add $(b/2)^2$. Thus, we add $(-2/2)^2 = 1$ to $x^2 - 2x$ and $(6/2)^2 = 9$ to $y^2 + 6y$, and of course we must add the same numbers to the right side of the equation, to obtain

$$x^2 - 2x + 1 + y^2 + 6y + 9 = -6 + 1 + 9$$
$$(x - 1)^2 + (y + 3)^2 = 4$$

The last equation is in standard form. It is the equation of a circle with center $(1, -3)$ and radius 2. If, as a result of this process, we had come up with a negative number on the right side of the final equation, the equation would not have represented any curve. If we had come up with zero, the equation would have represented the single point $(1, -3)$.   ■

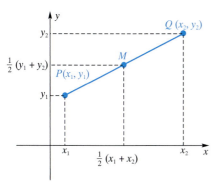

Figure 7

The Midpoint Formula   Consider two points $P(x_1, y_1)$ and $Q(x_2, y_2)$ with $x_1 \le x_2$ and $y_1 \le y_2$, as in Figure 7. The distance between x_1 and x_2 is $x_2 - x_1$. When we add half this distance, $\frac{1}{2}(x_2 - x_1)$, to x_1, we should get the number midway between x_1 and x_2.

$$x_1 + \frac{1}{2}(x_2 - x_1) = x_1 + \frac{1}{2}x_2 - \frac{1}{2}x_1$$
$$= \frac{1}{2}x_1 + \frac{1}{2}x_2$$
$$= \frac{x_1 + x_2}{2}$$

Thus, the point $(x_1 + x_2)/2$ is midway between x_1 and x_2 on the x-axis and, consequently, the midpoint M of the segment PQ has $(x_1 + x_2)/2$ as its x-coordinate. Similarly, we can show that $(y_1 + y_2)/2$ is the y-coordinate of M. Thus, we have the **Midpoint Formula**.

> The midpoint of the line segment joining $P(x_1, y_1)$ and $Q(x_2, y_2)$ is
> $$\left(\frac{x_1 + x_2}{2}, \frac{y_1 + y_2}{2} \right)$$

EXAMPLE 4   Find the equation of the circle having the segment from $(1, 3)$ to $(7, 11)$ as a diameter.

Solution   The center of the circle is at the midpoint of the diameter; thus, the center has coordinates $(1 + 7)/2 = 4$ and $(3 + 11)/2 = 7$. The length of the diameter, obtained from the distance formula, is

$$[(7 - 1)^2 + (11 - 3)^2]^{1/2} = [36 + 64]^{1/2} = 10$$

and so the radius of the circle is 5. The equation of the circle is

$$(x - 4)^2 + (y - 7)^2 = 25$$ ■

Concepts Review

1. The point $(-4, 2)$ lies in quadrant _____, whereas $(2, -4)$ lies in quadrant _____.

2. The distance between the points $(-2, 3)$ and (x, y) is _____.

3. The equation of the circle of radius 5 and center $(-4, 2)$ is _____.

4. The midpoint of the line segment joining $(-2, 3)$ and $(5, 7)$ is _____.

Problem Set 1.5

In Problems 1–6, plot the given points in the coordinate plane and then find the distance between them.

1. $(3, 1), (1, 1)$ **2.** $(-3, 5), (2, -2)$

3. $(4, 5), (5, -8)$ **4.** $(-1, 5), (6, 3)$

5. $(1.345, -1.234), (56.34, 89.56)$

6. $(\sqrt{\pi}, 3.222), (\pi, 8.145)$

7. Show that the triangle whose vertices are $(5, 3), (-2, 4)$, and $(10, 8)$ is isosceles.

8. Show that the triangle whose vertices are $(2, -4), (4, 0)$, and $(8, -2)$ is a right triangle.

9. The points $(3, -1)$ and $(3, 3)$ are two vertices of a square. Give three other pairs of possible vertices.

10. Find the point on the x-axis that is equidistant from $(3, 1)$ and $(6, 4)$.

11. Find the distance between $(-2, 3)$ and the midpoint of the segment joining $(-2, -2)$ and $(4, 3)$.

12. Find the length of the line segment joining the midpoints of the segments AB and CD, where $A = (1, 3)$, $B = (2, 6)$, $C = (4, 7)$, and $D = (3, 4)$.

In Problems 13–18, find the equation of the circle satisfying the given conditions.

13. Center $(1, 1)$, radius 1

14. Center $(-2, 3)$, radius 4

15. Center $(2, -1)$, goes through $(5, 3)$

16. Center $(4, 3)$, goes through $(6, 2)$

17. Diameter AB, where $A = (1, 3)$ and $B = (3, 7)$

18. Center $(3, 4)$ and tangent to x-axis

19. Find the y-coordinates of the two points on the circle of Problem 13 with x-coordinate of $\frac{1}{4}$ (see Example 2).

20. Find the x-coodinates of the two points on the circle of Problem 13 with y-coordinate of 1.

In Problems 21–26, find the center and radius of the circle with the given equation (see Example 3).

21. $x^2 + 2x + 10 + y^2 - 6y - 10 = 0$

22. $x^2 + y^2 - 6y = 16$

23. $x^2 + y^2 - 12x + 35 = 0$

24. $x^2 + y^2 - 10x + 10y = 0$

25. $4x^2 + 16x + 15 + 4y^2 + 6y = 0$

26. $4x^2 + 16x + \frac{105}{16} + 4y^2 + 3y = 0$

27. The points $(2, 3), (6, 3), (6, -1)$, and $(2, -1)$ are corners of a square. Find the equations of the inscribed and circumscribed circles.

28. A belt fits tightly around the two circles, with equations $(x - 1)^2 + (y + 2)^2 = 16$ and $(x + 9)^2 + (y - 10)^2 = 16$. How long is this belt?

29. Cities at A, B, and C are vertices of a right triangle, with the right angle at vertex B. Also, AB and BC are roads of lengths 214 and 179 miles, respectively. An airplane can fly the route AC, which is not a road. It costs \$3.71 per mile to ship a certain product by truck and \$4.82 per mile by plane. Decide whether it is cheaper to ship the product from A to C by truck or plane and find the total cost by the cheaper method.

30. City B is 10 miles downstream from city A and on the opposite side of a river $\frac{1}{2}$ mile wide. Mary will run from city A along the river for 6 miles, then swim diagonally to city B. If she runs at 8 miles per hour and swims at 3 miles per hour, how long will it take her to get from city A to city B? Assume that the current is negligible.

31. Show that the midpoint of the hypotenuse of any right triangle is equidistant from the three vertices.

32. Find the equation of the circle circumscribed about the right triangle whose vertices are $(0, 0), (8, 0)$, and $(0, 6)$.

33. Show that the two circles $x^2 + y^2 - 4x - 2y - 11 = 0$ and $x^2 + y^2 + 20x - 12y + 72 = 0$ do not intersect. *Hint:* Find the distance between their centers.

34. What relationship between a, b, and c must hold if $x^2 + ax + y^2 + by + c = 0$ is the equation of a circle?

35. The ceiling of an attic makes an angle of 30° with the floor. A pipe of radius 2 inches is placed along the edge of the attic in such a way that one side of the pipe touches the ceiling and another side touches the floor (see Figure 8). What is the distance d from the edge of the attic to where the pipe touches the floor?

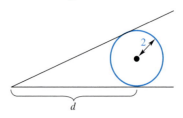

Figure 8

36. A circle of radius R is placed in the first quadrant as shown in Figure 9. What is the radius r of the largest circle that can be placed between the original circle and the origin?

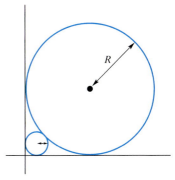

Figure 9

40. Consider a circle C and a point P exterior to the circle. Let line segment PT be tangent to C at T, and let the line through P and the center of C intersect C at M and N. Show that $(PM)(PN) = (PT)^2$.

▱ **41.** A belt fits around the three circles $x^2 + y^2 = 4$, $(x - 8)^2 + y^2 = 4$, and $(x - 6)^2 + (y - 8)^2 = 4$, as shown in Figure 12. Find the length of this belt.

42. Study Problems 28 and 41. Consider a set of nonintersecting circles of radius r with centers at the vertices of a convex n-sided polygon having sides of lengths $d_1, d_2, \ldots, d_n$. How long is the belt that fits around these circles (in the manner of Figure 12)?

▱ C **37.** Find the length of the crossed belt in Figure 10 that fits tightly around the circles $(x - 2)^2 + (y - 2)^2 = 9$ and $(x - 10)^2 + (y - 8)^2 = 9$. *Note:* A little trigonometry is needed to complete this problem.

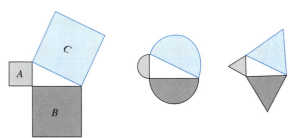

Figure 10

38. Show that the set of points that are twice as far from $(3, 4)$ as from $(1, 1)$ form a circle. Find its center and radius.

39. The Pythagorean Theorem says that the areas A, B, and C of the squares in Figure 11 satisfy $A + B = C$. Show that semicircles and equilateral triangles satisfy the same relation and then guess what a very general theorem says.

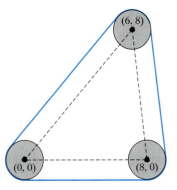

Figure 12

Figure 11

Answers to Concepts Review: **1.** II; IV

2. $\sqrt{(x + 2)^2 + (y - 3)^2}$ **3.** $(x + 4)^2 + (y - 2)^2 = 25$

4. $(1.5, 5)$

1.6
The Straight Line

Of all curves, the straight line is in many ways the simplest. We assume that you have a good intuitive notion of this concept from looking at a taut string or sighting along the edge of a ruler. In any case, let us agree that two points, for example, $A(3, 2)$ and $B(8, 4)$ shown in Figure 1, determine a unique straight line through them. And from now on we use the word *line* as a synonym for *straight line*.

A line is a geometric object. When it is placed in a coordinate plane, it ought to have an equation, just as a circle does. How do we find the equation of a line? To answer, we will need the notion of slope.

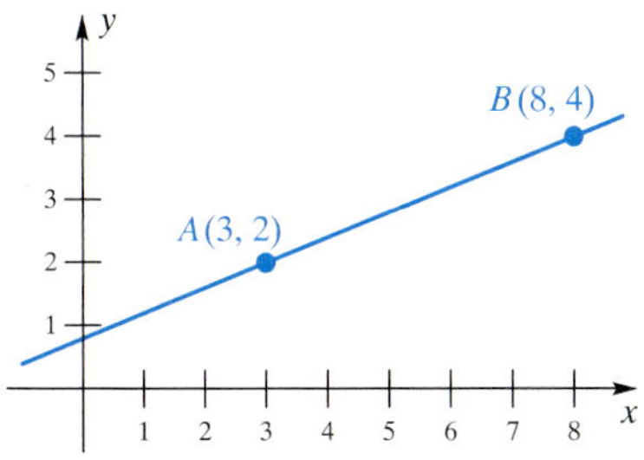

The Slope of a Line   Consider the line in Figure 1. From point A to point B, there is a **rise** (vertical change) of 2 units and a **run** (horizontal change) of 5 units. We say that the line has a slope of $\frac{2}{5}$. In general (Figure 2), for a line through $A(x_1, y_1)$ and $B(x_2, y_2)$, where $x_1 \neq x_2$, we define the **slope** m of that line by

$$m = \frac{\text{rise}}{\text{run}} = \frac{y_2 - y_1}{x_2 - x_1}$$

Figure 1

A (x₁,

Figu

Summary: Equations of Lines

Vertical line: $x = k$

Horizontal line: $y = k$

Point–slope form:

$$y - y_1 = m(x - x_1)$$

Slope–intercept form:

$$y = mx + b$$

General linear equation:

$$Ax + By + C = 0$$

A

A (x₁,

Figu

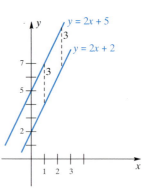

Figure 8

Gr

Th
slo
sho
a p
spo

Ca
A
of

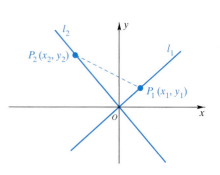

Figure 9

4

2

Figu

These can be rewritten (by taking everything to the left-hand side) as follows:

(1) $4x + y + 6 = 0$

(2) $-5x + y + 3 = 0$

(3) $x + 0y - 5 = 0$

All are of the form

$$\boxed{Ax + By + C = 0, \qquad A \text{ and } B \text{ not both } 0}$$

which we call the **general linear equation**. It takes only a moment's thought to see that the equation of any line can be put in this form. Conversely, the graph of the general linear equation is always a line.

Parallel Lines Two lines that have no points in common are said to be parallel. For example, the lines whose equations are $y = 2x + 3$ and $y = 2x + 5$ are parallel because, for every value of x, the second line is two units above the first (see Figure 8). Similarly, the lines with equations $-2x + 3y + 12 = 0$ and $4x - 6y = 5$ are parallel. To see this, solve each equation for y (i.e., put each in the slope–intercept form). This gives $y = \frac{2}{3}x - 4$ and $y = \frac{2}{3}x - \frac{5}{6}$, respectively. Again, because the slopes are equal, one line will be a fixed number of units above or below the other, so the lines will never intersect. If two lines have the same slope *and* the same intercept, then the two lines are the same, and they are not parallel.

We summarize by stating that two nonvertical lines are parallel if and only if they have the same slope and different intercepts. Two vertical lines are parallel if and only if they are distinct lines.

EXAMPLE 2 Find the equation of the line through $(6, 8)$ that is parallel to the line with equation $3x - 5y = 11$.

Solution When we solve $3x - 5y = 11$ for y, we obtain $y = \frac{3}{5}x - \frac{11}{5}$, from which we read the slope of the line to be $\frac{3}{5}$. The equation of the desired line is

$$y - 8 = \frac{3}{5}(x - 6)$$

or, equivalently, $y = \frac{3}{5}x + \frac{22}{5}$. We know that these lines are distinct because the intercepts are different. ■

Perpendicular Lines Is there a simple slope condition that characterizes perpendicular lines? Yes; *two nonvertical lines are perpendicular if and only if their slopes are negative reciprocals of each other.* To see why this is true, consider two nonvertical intersecting lines l_1 and l_2. Without loss of generality, we may suppose that they intersect at the origin since, if not, we may translate them so that they do without changing their slopes. Let $P_1(x_1, y_1)$ be a point on l_1 and $P_2(x_2, y_2)$ be a point on l_2, as shown in Figure 9. By the Pythagorean Theorem and its converse (Section 1.5), P_1OP_2 is a right angle if and only if

$$[d(P_1, O)]^2 + [d(P_2, O)]^2 = [d(P_1, P_2)]^2$$

that is, if and only if

$$(x_1^2 + y_1^2) + (x_2^2 + y_2^2) = (x_1 - x_2)^2 + (y_1 - y_2)^2$$

After expansion and simplification, this equation becomes $2x_1x_2 + 2y_1y_2 = 0$, or

$$\frac{y_1}{x_1} = -\frac{x_2}{y_2}$$

Now, y_1/x_1 is the slope of l_1, whereas y_2/x_2 is the slope of l_2. Thus, P_1OP_2 is a right angle if and only if the slopes of the two lines are negative reciprocals of each other.

The lines $y = \frac{3}{4}x$ and $y = -\frac{4}{3}x$ are perpendicular. So are $2x - 3y = 5$ and $3x + 2y = -4$, since after solving these equations for y we see that the first line has slope $\frac{2}{3}$ and the second has slope $-\frac{3}{2}$.

EXAMPLE 3   Find the equation of the line through the point of intersection of the lines with equations $3x + 4y = 8$ and $6x - 10y = 7$ that is perpendicular to the first of these two lines (Figure 10).

Solution   To find the point of intersection of the two lines, we multiply the first equation by -2 and add it to the second equation.

$$-6x - 8y = -16$$
$$\underline{6x - 10y = 7}$$
$$-18y = -9$$
$$y = \frac{1}{2}$$

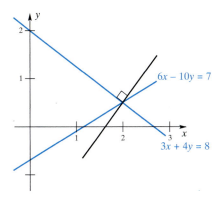

Figure 10

Substituting $y = \frac{1}{2}$ in either of the original equations yields $x = 2$. The point of intersection is $\left(2, \frac{1}{2}\right)$.

When we solve the first equation for y (to put it in slope–intercept form), we get $y = -\frac{3}{4}x + 2$. A line perpendicular to it has slope $\frac{4}{3}$. The equation of the required line is

$$y - \tfrac{1}{2} = \tfrac{4}{3}(x - 2) \qquad \blacksquare$$

Concepts Review

1. The line through (a, b) and (c, d) has slope $m =$ _____ provided $a \neq c$.

2. A horizontal line has slope $m =$ _____, whereas a vertical line does not have slope.

3. The equation of a nonvertical line can always be written in the form _____, whereas the equation of a vertical line can be written in the form _____.

4. Every line has an equation of the form _____, called the general linear equation.

Problem Set 1.6

In Problems 1–8, find the slope of the line containing the given two points.

1. $(1, 1)$ and $(2, 2)$

2. $(3, 5)$ and $(4, 7)$

3. $(2, 3)$ and $(-5, -6)$

4. $(2, -4)$ and $(0, -6)$

5. $(3, 0)$ and $(0, 5)$

6. $(-6, 0)$ and $(0, 6)$

7. $(-1.234, 5.678)$ and $(7.654, -3.456)$

8. $(\pi, \sqrt{3})$ and $(1.642, \sqrt{2})$

In Problems 9–16, find an equation for each line. Then write your answer in the form $Ax + By + C = 0$.

9. Through $(2, 2)$ with slope -1

10. Through $(3, 4)$ with slope -1

11. With y-intercept 3 and slope 2

12. With y-intercept 5 and slope 0

13. Through $(2, 3)$ and $(4, 8)$

14. Through $(4, 1)$ and $(8, 2)$

15. Through $(2, -3)$ and $(2, 5)$

16. Through $(-5, 0)$ and $(-5, 4)$

In Problems 17–20, find the slope and y-intercept of each line.

17. $3y = -2x + 1$

18. $-4y = 5x - 6$

19. $6 - 2y = 10x - 2$

20. $4x + 5y = -20$

21. Write an equation for the line through $(3, -3)$ that is

(a) parallel to the line $y = 2x + 5$;
(b) perpendicular to the line $y = 2x + 5$;
(c) parallel to the line $2x + 3y = 6$;
(d) perpendicular to the line $2x + 3y = 6$;
(e) parallel to the line through $(-1, 2)$ and $(3, -1)$;
(f) parallel to the line $x = 8$;
(g) perpendicular to the line $x = 8$.

22. Find the value of c for which the line $3x + cy = 5$

(a) passes through the point $(3, 1)$;
(b) is parallel to the y-axis;
(c) is parallel to the line $2x + y = -1$;
(d) has equal x- and y-intercepts;
(e) is perpendicular to the line $y - 2 = 3(x + 3)$.

23. Write the equation for the line through $(-2, -1)$ that is perpendicular to the line $y + 3 = -\frac{2}{3}(x - 5)$.

24. Find the value of k such that the line $kx - 3y = 10$

(a) is parallel to the line $y = 2x + 4$;
(b) is perpendicular to the line $y = 2x + 4$;
(c) is perpendicular to the line $2x + 3y = 6$.

25. Does $(3, 9)$ lie above or below the line $y = 3x - 1$?

26. Show that the equation of the line with x-intercept $a \neq 0$ and y-intercept $b \neq 0$ can be written as

$$\frac{x}{a} + \frac{y}{b} = 1$$

In Problems 27–30, find the coordinates of the point of intersection. Then write an equation for the line through that point perpendicular to the line given first (see Example 3).

27. $2x + 3y = 4$
$-3x + y = 5$

28. $4x - 5y = 8$
$2x + y = -10$

29. $3x - 4y = 5$
$2x + 3y = 9$

30. $5x - 2y = 5$
$2x + 3y = 6$

It can be shown that the distance d from the point (x_1, y_1) to the line $Ax + By + C = 0$ is

$$d = \frac{|Ax_1 + By_1 + C|}{\sqrt{A^2 + B^2}}$$

Use this result to find the distance from the given point to the given line.

31. $(-3, 2); 3x + 4y = 6$

32. $(4, -1); 2x - 2y + 4 = 0$

33. $(-2, -1); 5y = 12x + 1$

34. $(3, -1); y = 2x - 5$

In Problems 35 and 36, find the (perpendicular) distance between the given parallel lines. Hint: First find a point on one of the lines.

35. $2x + 4y = 7, 2x + 4y = 5$

36. $7x - 5y = 6, 7x - 5y = -1$

37. A bulldozer costs \$120,000 and each year it depreciates 8% of its original value. Find a formula for the value V of the bulldozer after t years.

38. The graph of the answer to Problem 37 is a straight line. What is its slope, assuming the t-axis to be horizontal? Interpret the slope.

39. Past experience indicates that egg production in Matlin County is growing linearly. In 1980 it was 700,000 cases, and in 1990 it was 820,000 cases. Write a formula for the number N of cases produced n years after 1980 and use it to predict egg production in the year 2005.

40. A piece of equipment purchased today for \$80,000 will depreciate linearly to a scrap value of \$2000 after 20 years. Write a formula for its value V after n years.

41. Suppose that the profit P that a company realizes in selling x items of a certain commodity is given by $P = 450x - 2000$ dollars.

(a) Interpret the value of P when $x = 0$.
(b) Find the slope of the graph of the above equation. This slope is called the **marginal profit**. What is its economic interpretation?

42. The cost C of producing x items of a certain commodity is given by $C = 0.75x + 200$ dollars.

(a) The slope of its graph is called the **marginal cost**. Find it and give an economic interpretation.
(b) Find and interpret the value of C when $x = 0$.

43. Show that for each value of k the equation

$$2x - y + 4 + k(x + 3y - 6) = 0$$

represents a line through the intersection of the two lines $2x - y + 4 = 0$ and $x + 3y - 6 = 0$. *Hint:* It is not necessary to find the point of intersection (x_0, y_0).

44. Find an equation for the line through $(2, 3)$ that has equal x- and y-intercepts. *Hint:* Use Problem 26.

45. Find the equation for the line that bisects the line segment from $(-2, 3)$ to $(1, -2)$ and is at right angles to this line segment.

46. The center of the circumscribed circle of a triangle lies on the perpendicular bisectors of the sides. Use this fact to find the center of the circle that circumscribes the triangle with vertices $(0, 4)$, $(2, 0)$, and $(4, 6)$.

47. Find the radius of the circle that is inscribed in a triangle with sides of lengths 3, 4, and 5 (see Figure 11).

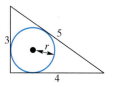

Figure 11

48. Suppose that (a, b) is on the circle $x^2 + y^2 = r^2$. Show that the line $ax + by = r^2$ is tangent to the circle at (a, b).

49. Find the equations of the two tangent lines to the circle $x^2 + y^2 = 36$ that go through $(12, 0)$. *Hint:* See Problem 48.

50. Express the perpendicular distance between the parallel lines $y = mx + b$ and $y = mx + B$ in terms of m, b, and B. *Hint:* The required distance is the same as that between $y = mx$ and $y = mx + B - b$.

51. Show that the line through the midpoints of two sides of a triangle is parallel to the third side. *Hint:* You may assume that the triangle has vertices at $(0, 0)$, $(a, 0)$, and (b, c).

52. Show that the line segments joining the midpoints of adjacent sides of any quadrilateral (four-sided polygon) form a parallelogram.

53. A wheel whose rim has equation $x^2 + (y - 6)^2 = 25$ is rotating rapidly in the counterclockwise direction. A speck of dirt on the rim came loose at the point $(3, 2)$ and flew toward the wall $x = 11$. About how high up on the wall did it hit? *Hint:* The speck of dirt flies off on a tangent so fast that the effects of gravity are negligible by the time it has hit the wall.

Answers to Concepts Review: **1.** $(d - b)/(c - a)$ **2.** 0 **3.** $y = mx + b; x = k$ **4.** $Ax + By + C = 0$

1.7
Graphs of Equations

The use of coordinates for points in the plane allows us to describe a curve (a geometric object) by an equation (an algebraic object). We saw how this was done for circles and lines in the previous sections. Now we want to consider the reverse process: graphing an equation. The **graph of an equation** in x and y consists of those points in the plane whose coordinates (x, y) satisfy the equation, that is, make it a true equality.

The Graphing Procedure   To graph an equation, for example, $y = 2x^3 - x + 19$, by hand, we can follow a simple three-step procedure:

Step 1:   Obtain the coordinates of a few points that satisfy the equation.

Step 2:   Plot these points in the plane.

Step 3:   Connect the points with a smooth curve.

The best way to do Step 1 is to make a table of values. Assign values to one of the variables, such as x, and determine the corresponding values of the other variable, listing the results in tabular form.

A graphing calculator or a computer algebra system will follow much the same procedure, although its procedure is transparent to the user. A user simply defines the function and asks the graphing calculator or computer to plot it.

EXAMPLE 1   Graph the equation $y = x^2 - 3$.

Solution   The three-step procedure is shown in Figure 1.

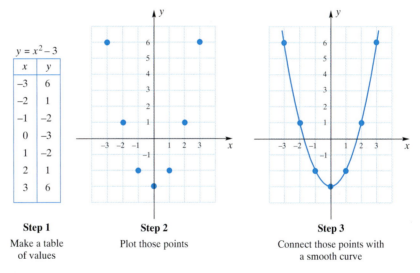

$y = x^2 - 3$

x	y
-3	6
-2	1
-1	-2
0	-3
1	-2
2	1
3	6

Step 1
Make a table
of values

Step 2
Plot those points

Step 3
Connect those points with
a smooth curve

Figure 1

Of course, you need to use common sense and even a little faith. When you have a point that seems out of place, check your calculations. When you connect the points you have plotted with a smooth curve, you are assuming that the curve behaves nicely between consecutive points, which is faith. This is why you should plot enough points so that the outline of the curve seems very clear; the more points you plot, the less faith you will need. Also, you should recognize that you can seldom display the whole curve. In our example, the curve has infinitely long arms, opening wider and wider. But our graph does show the essential features. This is our goal in graphing. Show enough of the graph so that the essential features are visible. Later (Section 4.6) we will use the tools of calculus to refine and improve our understanding of graphs.

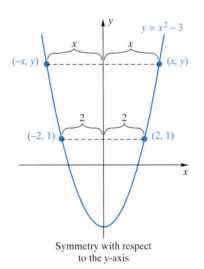

$y = x^2 - 3$

$(-x, y)$ (x, y)

$(-2, 1)$ $(2, 1)$

Symmetry with respect
to the y-axis

Figure 2

Symmetry of a Graph   We can sometimes cut our graphing effort in half by recognizing certain symmetries of the graph as revealed by its equation. Look at the graph of $y = x^2 - 3$, drawn above and again in Figure 2. If the coordinate

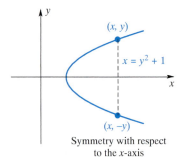

Symmetry with respect
to the x-axis

Figure 3

plane is folded along the y-axis, the two branches of the graph will coincide. For example, $(3, 6)$ will coincide with $(-3, 6)$, $(2, 1)$ will coincide with $(-2, 1)$, and, more generally, (x, y) will coincide with $(-x, y)$. Algebraically, this corresponds to the fact that replacing x by $-x$ in the equation $y = x^2 - 3$ results in an equivalent equation.

Consider an arbitrary graph. It is **symmetric with respect to the y-axis** if, whenever (x, y) is on the graph, $(-x, y)$ is also on the graph (Figure 2). Similarly, it is **symmetric with respect to the x-axis** if, whenever (x, y) is on the graph, $(x, -y)$ is also on the graph (Figure 3). Finally, a graph is **symmetric with respect to the origin** if, whenever (x, y) is on the graph, $(-x, -y)$ is also on the graph (see Example 2).

In terms of equations, we have three simple tests. The graph of an equation is

1. symmetric with respect to the y-axis if replacing x by $-x$ gives an equivalent equation (e.g., $y = x^2$);

2. symmetric with respect to the x-axis if replacing y by $-y$ gives an equivalent equation (e.g., $x = y^2 + 1$);

3. symmetric with respect to the origin if replacing x by $-x$ and y by $-y$ gives an equivalent equation ($y = x^3$ is a good example since $-y = (-x)^3$ is equivalent to $y = x^3$).

EXAMPLE 2   Sketch the graph of $y = x^3$.

Solution   We note, as pointed out above, that the graph will be symmetric with respect to the origin, so we need only get a table of values for nonnegative x's; we can find matching points by symmetry. For example, $(2, 8)$ being on the graph tells us that $(-2, -8)$ is on the graph, $(3, 27)$ being on the graph tells us that $(-3, -27)$ is on the graph, and so on. See Figure 4.   ∎

In graphing $y = x^3$, we used a smaller scale on the y-axis than on the x-axis. This made it possible to show a larger portion of the graph (it also distorted the graph by flattening it). When you graph by hand we suggest that before putting scales on the two axes you should examine your table of values. Choose scales so that all or most of your points can be plotted and still keep your graph of reasonable size. A graphing calculator or a CAS will often choose the scale for the y's once you have chosen the x's to be used. The first choice you make, therefore, is the x values to plot. Most graphing calculators and CASs allow you to override the automatic y-axis scaling. In some cases you may want to use this option.

Intercepts   The points where the graph of an equation crosses the two coordinate axes play a significant role in many problems. Consider, for example,

$$y = x^3 - 2x^2 - 5x + 6 = (x + 2)(x - 1)(x - 3)$$

Notice that $y = 0$ when $x = -2, 1, 3$. The numbers $-2, 1$, and 3 are called **x-intercepts**. Similarly, $x = 0$ when $y = 6$, and so 6 is called the **y-intercept**.

EXAMPLE 3   Find all intercepts of the graph of $y^2 - x + y - 6 = 0$.

Solution   Putting $y = 0$ in the given equation, we get $x = -6$, and so the x-intercept is -6. Putting $x = 0$ in the equation, we find that $y^2 + y - 6 = 0$, or $(y + 3)(y - 2) = 0$; the y-intercepts are -3 and 2. A check on symmetries indicates that the graph has none of the three types discussed earlier. The graph is displayed in Figure 5.   ∎

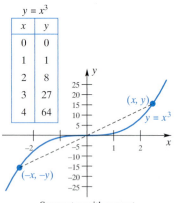

Symmetry with respect
to the origin

Figure 4

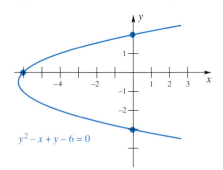

Figure 5

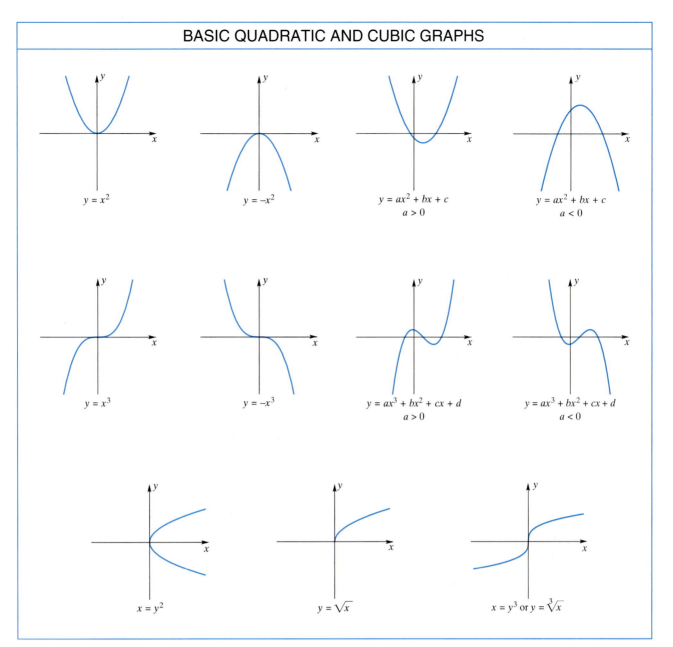

BASIC QUADRATIC AND CUBIC GRAPHS

$y = x^2$

$y = -x^2$

$y = ax^2 + bx + c$
$a > 0$

$y = ax^2 + bx + c$
$a < 0$

$y = x^3$

$y = -x^3$

$y = ax^3 + bx^2 + cx + d$
$a > 0$

$y = ax^3 + bx^2 + cx + d$
$a < 0$

$x = y^2$

$y = \sqrt{x}$

$x = y^3$ or $y = \sqrt[3]{x}$

Figure 6

Since quadratic and cubic equations will often be used as examples in later work, we display their typical graphs in Figure 6.

The graphs of quadratic equations are cup-shaped curves called **parabolas**. If an equation has the form $y = ax^2 + bx + c$ or $x = ay^2 + by + c$ with $a \neq 0$, its graph is a parabola. In the first case, the graph opens up if $a > 0$ and opens down if $a < 0$. In the second case, the graph opens right if $a > 0$ and opens left if $a < 0$. Note that the equation of Example 3 can be put in the form $x = y^2 + y - 6$.

Intersections of Graphs Frequently, we need to know the points of intersection of two graphs. These points are found by solving the two equations for the graphs simultaneously, as illustrated in the next example.

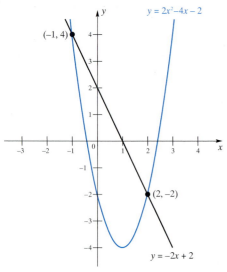

Figure 7

EXAMPLE 4 Find the points of intersection of the line $y = -2x + 2$ and the parabola $y = 2x^2 - 4x - 2$, and sketch both graphs on the same coordinate plane.

Solution We must solve the two equations simultaneously. This is easy to do by substituting the expressions for y from the first equation into the second equation and then solving the resulting equation for x.

$$-2x + 2 = 2x^2 - 4x - 2$$
$$0 = 2x^2 - 2x - 4$$
$$0 = 2(x + 1)(x - 2)$$
$$x = -1, \quad x = 2$$

By substitution, we find the corresponding values of y to be 4 and -2; the intersection points are therefore $(-1, 4)$ and $(2, -2)$.

The two graphs are shown in Figure 7. ∎

Concepts Review

1. If whenever (x, y) is on a graph $(-x, y)$ is also on the graph, then the graph is symmetric with respect to the _____.

2. If $(-4, 2)$ is on a graph that is symmetric with respect to the origin, then _____ is also on the graph.

3. The graph of $y = (x + 2)(x - 1)(x - 4)$ has y-intercept _____ and x-intercepts _____.

4. The graph of $y = ax^2 + bx + c$ is a _____ if $a = 0$ and a _____ if $a \neq 0$.

Problem Set 1.7

In Problems 1–30, plot the graph of each equation. Begin by checking for symmetries and be sure to find all x- and y-intercepts.

1. $y = -x^2 + 1$

2. $x = -y^2 + 1$

3. $x = -4y^2 - 1$

4. $y = 4x^2 - 1$

5. $x^2 + y = 0$

6. $y = x^2 - 2x$

7. $7x^2 + 3y = 0$

8. $y = 3x^2 - 2x + 2$

9. $x^2 + y^2 = 4$

10. $3x^2 + 4y^2 = 12$

11. $y = -x^2 - 2x + 2$

12. $4x^2 + 3y^2 = 12$

13. $x^2 - y^2 = 4$

14. $x^2 + (y - 1)^2 = 9$

15. $4(x - 1)^2 + y^2 = 36$

16. $x^2 - 4x + 3y^2 = -2$

17. $x^2 + 9(y + 2)^2 = 36$

GC **18.** $x^4 + y^4 = 1$

GC **19.** $x^4 + y^4 = 16$

GC **20.** $y = x^3 - x$

GC **21.** $y = \dfrac{1}{x^2 + 1}$

GC **22.** $y = \dfrac{x}{x^2 + 1}$

GC **23.** $2x^2 - 4x + 3y^2 + 12y = -2$

GC **24.** $4(x - 5)^2 + 9(y + 2)^2 = 36$

GC **25.** $y = (x - 1)(x - 2)(x - 3)$

GC **26.** $y = x^2(x - 1)(x - 2)$

GC **27.** $y = x^2(x - 1)^2$

GC **28.** $y = x^4(x - 1)^4(x + 1)^4$

GC **29.** $|x| + |y| = 1$

GC **30.** $|x| + |y| = 4$

GC *In Problems 31–38, plot the graphs of both equations on the same coordinate plane. Find and label the points of intersection of the two graphs (see Example 4). You will need the quadratic formula in Problems 35–38.*

31. $y = -x + 1$
$y = (x + 1)^2$

32. $y = 2x + 3$
$y = -(x - 1)^2$

33. $y = -2x + 3$
$y = -2(x - 4)^2$

34. $y = -2x + 3$
$y = 3x^2 - 3x + 12$

35. $y = x$
$x^2 + y^2 = 4$

36. $y = x - 1$
$2x^2 + 3y^2 = 12$

37. $y - 3x = 1$
$x^2 + 2x + y^2 = 15$

38. $y = 4x + 3$
$x^2 + y^2 = 81$

39. Choose the equation that best represents each graph in Figure 8.

(a) $y = x^2$
(b) $y = ax^3 + bx^2 + cx + d$, with $a > 0$
(c) $y = ax^3 + bx^2 + cx + d$, with $a < 0$
(d) $y = ax^3$, with $a > 0$

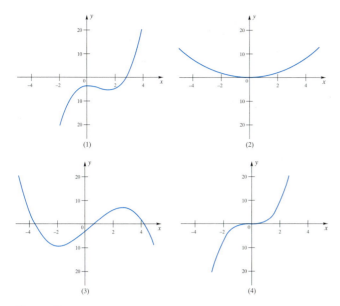

(1)

(2)

(3)

(4)

Figure 8

≈ **40.** Find the distance between the points on the circle $x^2 + y^2 = 13$ with the x-coordinates -2 and 2. How many such distances are there?

≈ **41.** Find the distance between the points on the circle $x^2 + 2x + y^2 - 2y = 20$ with the x-coordinates -2 and 2. How many such distances are there?

Answers to Concepts Review: **1.** y-axis **2.** $(4, -2)$
3. 8; $-2, 1, 4$ **4.** line; parabola

1.8 Chapter Review

Concepts Test

Respond with true or false to each of the following assertions. Be prepared to justify your answer. Normally, this means that you should supply a reason if you answer true and provide a counterexample if you answer false.

1. Any number that can be written as a fraction p/q is rational.

2. The difference of any two rational numbers is rational.

3. The difference of any two irrational numbers is irrational.

4. Between two distinct irrational numbers there is always another irrational number.

5. $0.999\ldots$ (repeating 9s) is less than 1.

6. The operation of exponentiation is commutative; that is, $(a^m)^n = (a^n)^m$.

7. The operation $*$ defined by $m * n = m^n$ is associative.

8. The inequalities $x \le y$, $y \le z$, and $z \le x$ together imply that $x = y = z$.

9. If $|x| < \varepsilon$ for every positive number ε, then $x = 0$.

10. If x and y are real numbers, then $(x - y)(y - x) \le 0$.

11. If $a < b < 0$, then $1/a > 1/b$.

12. It is possible for two closed intervals to have exactly one point in common.

13. If two open intervals have a point in common, then they have infinitely many points in common.

14. If $x < 0$, then $\sqrt{x^2} = -x$.

15. The value of $|x|$ is the number x without its sign.

16. If $|x| < |y|$, then $x < y$.

17. If $|x| < |y|$, then $x^4 < y^4$.

18. If x and y are both negative, then $|x + y| = |x| + |y|$.

19. If $|r| < 1$, then $\dfrac{1}{1 + |r|} \le \dfrac{1}{1 - r} \le \dfrac{1}{1 - |r|}$.

20. If $|r| > 1$, then $\dfrac{1}{1 - |r|} \le \dfrac{1}{1 - r} \le \dfrac{1}{1 + |r|}$.

21. It is always true that $\|x| - |y\| \le |x + y|$.

22. For every positive real number y, there exists a real number x such that $x^2 = y$.

23. For every real number y, there exists a real number x such that $x^3 = y$.

24. It is possible to have an inequality whose solution set consists of exactly one number.

25. The equation $x^2 + y^2 + ax + y = 0$ represents a circle for every real number a.

26. The equation $x^2 + y^2 + ax + by = c$ represents a circle for all real numbers a, b, c.

27. If (a, b) is on a line with slope $\frac{3}{4}$, then $(a + 4, b + 3)$ is also on that line.

28. If (a, b), (c, d) and (e, f) are on the same line, then $\dfrac{a - c}{b - d} = \dfrac{a - e}{b - f} = \dfrac{e - c}{f - d}$ provided all three points are different.

29. If $ab > 0$, then (a, b) lies in either the first or third quadrant.

30. For every $\varepsilon > 0$ there exists a positive number x such that $x < \varepsilon$.

31. If $ab = 0$, then (a, b) lies on either the x-axis or the y-axis.

32. If $\sqrt{(x_2 - x_1)^2 + (y_2 - y_1)^2} = |x_2 - x_1|$, then (x_1, y_1) and (x_2, y_2) lie on the same horizontal line.

33. The distance between $(a + b, a)$ and $(a - b, a)$ is $|2b|$.

34. The equation of any line can be written in point–slope form.

35. The equation of any line can always be written in the general linear form $ax + by + C = 0$.

36. If two nonvertical lines are parallel, they have the same slope.

37. It is possible for two lines to have positive slopes and be perpendicular.

38. If the x- and y-intercepts of a line are rational and nonzero, then the slope of the line is rational.

39. The lines $ax + y = c$ and $ax - y = c$ are perpendicular.

40. $(3x - 2y + 4) + m(2x + 6y - 2) = 0$ is the equation of a line for each real number m.

Sample Test Problems

1. Calculate each value for $n = 1, 2$, and -2.

(a) $\left(n + \dfrac{1}{n}\right)^n$
(b) $(n^2 - n + 1)^2$

(c) $4^{3/n}$
(d) $\sqrt[n]{\left|\dfrac{1}{n}\right|}$

2. Simplify.

(a) $\left(1 + \dfrac{1}{m} + \dfrac{1}{n}\right)\left(1 - \dfrac{1}{m} + \dfrac{1}{n}\right)^{-1}$

(b) $\dfrac{\dfrac{2}{x + 1} - \dfrac{x}{x^2 - x - 2}}{\dfrac{3}{x + 1} - \dfrac{2}{x - 2}}$

(c) $\dfrac{t^3 - 1}{t - 1}$

3. Show that the average of two rational numbers is a rational number.

4. Write the repeating decimal 4.1282828 . . . as a ratio of two integers.

5. Find an irrational number between $\frac{1}{2}$ and $\frac{13}{25}$.

⊏C⊐ **6.** Calculate $\left(\sqrt[3]{8.15 \times 10^4} - 1.32\right)^2/3.24$.

⊏C⊐ **7.** Calculate $\left(\pi - \sqrt{2.0}\right)^{2.5} - \sqrt[3]{2.0}$.

⊏C⊐ **8.** Calculate $\sin^2(2.45) + \cos^2(2.40) - 1.00$.

In Problems 9–18, find the solution set, graph this set on the real line, and express this set in interval notation.

9. $1 - 3x > 0$
10. $6x + 3 > 2x - 5$

11. $3 - 2x \le 4x + 1 \le 2x + 7$

12. $2x^2 + 5x - 3 < 0$
13. $21t^2 - 44t + 12 \le -3$

14. $\dfrac{2x - 1}{x - 2} > 0$

15. $(x + 4)(2x - 1)^2(x - 3) \le 0$

16. $|3x - 4| < 6$
17. $\dfrac{3}{1 - x} \le 2$

18. $|12 - 3x| \ge |x|$

19. Find a value of x for which $|-x| \ne x$.

20. For what values of x does the equation $|-x| = x$ hold?

21. For what values of t does the equation $|t - 5| = 5 - t$ hold?

22. For what values of a and t does the equation $|t - a| = a - t$ hold?

23. Suppose $|x| \le 2$. Use properties of absolute values to show that

$$\left|\frac{2x^2 + 3x + 2}{x^2 + 2}\right| \le 8$$

24. Write a sentence involving the word *distance* to express the following algebraic sentences:

(a) $|x - 5| = 3$
(b) $|x + 1| \le 2$
(c) $|x - a| > b$

25. Sketch the triangle with vertices $A(-2, 6)$, $B(1, 2)$, and $C(5, 5)$, and show that it is a right triangle.

26. Find the distance from $(3, -6)$ to the midpoint of the line segment from $(1, 2)$ to $(7, 8)$.

27. Find the equation of the circle with diameter AB if $A = (2, 0)$ and $B = (10, 4)$.

28. Find the center and radius of the circle with equation $x^2 + y^2 - 8x + 6y = 0$.

29. Find the distance between the centers of the circles with equations

$$x^2 - 2x + y^2 + 2y = 2 \quad \text{and} \quad x^2 + 6x + y^2 - 4y = -7$$

30. Find the equation of the line through the indicated point that is parallel to the indicated line, and sketch both lines.

(a) $(3, 2): 3x + 2y = 6$
(b) $(1, -1): y = \frac{2}{3}x + 1$
(c) $(5, 9): y = 10$
(d) $(-3, 4): x = -2$

31. Write the equation of the line through $(-2, 1)$ that

(a) goes through $(7, 3)$;
(b) is parallel to $3x - 2y = 5$;
(c) is perpendicular to $3x + 4y = 9$;
(d) is perpendicular to $y = 4$;
(e) has y-intercept 3.

32. Show that $(2, -1)$, $(5, 3)$, and $(11, 11)$ are on the same line.

33. Figure 9 can be represented by which equation?

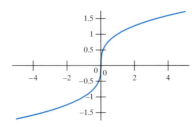

Figure 9

(a) $y = x^3$
(b) $x = y^3$
(c) $y = x^2$
(d) $x = y^2$

34. Figure 10 can be represented by which equation?

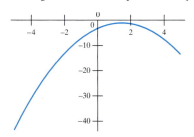

Figure 10

(a) $y = ax^2 + bx + c$. with $a > 0, b > 0$ and $c > 0$
(b) $y = ax^2 + bx + c$, with $a < 0, b > 0$ and $c > 0$
(c) $y = ax^2 + bx + c$, with $a < 0, b > 0$ and $c < 0$
(d) $y = ax^2 + bx + c$, with $a > 0, b > 0$, and $c < 0$

In Problems 35–38, sketch the graph of each equation.

35. $3y - 4x = 6$

36. $x^2 - 2x + y^2 = 3$

GC **37.** $y = \dfrac{2x}{x^2 + 2}$

GC **38.** $x = y^2 - 3$

GC **39.** Find the points of intersection of the graphs of $y = x^2 - 2x + 4$ and $y - x = 4$.

40. Among all lines perpendicular to $4x - y = 2$, find the equation of the one that, together with the positive x- and y-axes, forms a triangle of area 8.

TECHNOLOGY PROJECT 1.1

Graphing

I. PREPARATION

One of the simplest, yet most useful, tools that technology provides us is the ability to display graphs. We can often use this utility to experiment by changing one or two parameters of the expression being graphed and observing the effect on the graph. Much of this project deals with this kind of experimentation. In preparation for this project, learn how to use your software or calculator to draw two or more graphs in the same window.

II. USING TECHNOLOGY

Exercise 1   When using a graphing calculator or computer algebra system (CAS) to graph an equation, you want to choose a graphing window that gives all the important details. The following equations will give you some experience in choosing a good window. In each case, you should experiment with windows of various sizes to make sure that you can see all the details of the graph.

(a) $y = 3x^2 + 10$
(b) $y = x^3 - 9x^2 - 22x + 120$
(c) $y = (x - 1)(x - 2)$
    $(x - 3)(x - 4)$
(d) $y = \dfrac{x^2 + 2x}{x - 1}$

Exercise 2   Consider the lines $y = mx + 4$. In the same graphing window, draw these lines for $m = 0.1$, 0.5, 1.0, 2.0, and 4.0. What do these lines have in common? Explain why.

Exercise 3   Consider the lines $y = -2x + b$. In the same graphing window, draw these lines for $b = 1.0, -1.0, 2.0, -2.0, 3.0, -3.0$, 4.0, and -4.0. What do these lines have in common? Give an algebraic reason for what the graphs display.

III. REFLECTION

Exercise 4   Explain how the value of b in

$$y = x^3 + bx^2 + x + 1$$

affects the shape of the curve and the number of times that the curve crosses the x-axis.

Solving Equations by Zooming

I. PREPARATION

In high school you learned the Quadratic Formula for solving the quadratic equations $ax^2 + bx + c = 0$. However, you may not know any methods for solving more complicated equations. You can often approximate the solutions to such an equation by "zooming in" on the desired root.

Exercise 1 Write an equation of the line L that passes through $(-2, 4)$ and is perpendicular to the line with equation $x + 2y = 17$.

Exercise 2 Solve the following *generalized* version of Exercise 1. Write the equation of the line L that passes through the point (p, q) and is perpendicular to the line with equation $ax + by = c$. Check that your answer "works" by substituting the specific numbers from Exercise 1.

In Chapter 3 you will learn the following fact:

> The slope of the line tangent to the parabola $y = x^2$ at the point (a, a^2) is $2a$.

Thus, the slope of the line tangent to the parabola at the point $(1, 1^2)$ is 2, at $(3, 3^2)$ the slope is 6, and so on. For now, accept this fact; we will derive it, along with many others like it, in Chapter 3.

Exercise 3 Use the following hints to find the equation of the line through the point $P(2, 0)$ that is perpendicular to the tangent line of the parabola $y = x^2$.

Hints: Draw a figure (see Figure 1) that includes the perpendicular from the given point. Use a to denote the x-coordinate of the point where the perpendicular line cuts the parabola. Thus, the point (a, a^2) is on both the parabola and the line perpendicular to the parabola. The slope of the parabola at $x = a$ is $2a$, so the slope of the perpendicular line is the negative reciprocal of $2a$, or $-\frac{1}{2a}$.

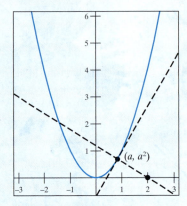

Figure 1

Sketch for Exercise 3.

Now equate the slope of the line connecting (a, a^2) and the given point $(2, 0)$ to the negative reciprocal of the slope of the line tangent to the parabola at (a, a^2). Show that equating these two expressions leads to the *cubic* equation.

$$2a^3 + a - 2 = 0$$

Figure 1 indicates that the solution to this cubic equation is approximately 1, but this is not the exact solution (try it!). Remember, when you find a value of a that satisfies $2a^3 + a - 2 = 0$, you are finding a value of a such that the line connecting (a, a^2) and $(2, 0)$ is perpendicular to the tangent line to the parabola at (a, a^2).

II. USING TECHNOLOGY

Exercise 4 Instead of resorting to trial and error to find the solution of $2a^3 + a - 2 = 0$, we will graph this equation near $a = 1$ and then "zoom in" on the root. You might begin by plotting the graph over the interval $(0, 2)$; the root looks to be somewhere near $a = 0.8$. Next, plot $2a^3 + a - 2 = 0$ over the interval $(0.8, 0.9)$. Keep zooming in until you can get an approximation to the root that is correct to two decimal places.

Exercise 5 One problem-solving maxim we have ignored in the above solution is that of trying to avoid the use of specific numbers. To maintain generality, let (p, q), instead of $(2, 0)$, denote the given point.

Equate the slope of the line connecting (a, a^2) and the given point (p, q) to the negative reciprocal of the slope of the line tangent to the parabola at (a, a^2). Show that equating these two expressions leads to the *cubic* equation

$$2a^3 - (2q - 1)a - p = 0$$

Check to see that this agrees with the specific case worked out in Exercise 3.

Exercise 6 Find the equation of the line through $(p, q) = (4, 0)$ that is perpendicular to the tangent line to $y = x^2$.

Exercise 7 Show that there are three points (a_1, a_1^2), (a_2, a_2^2), and (a_3, a_3^2) on the parabola, with the property that the tangent line at (a_i, a_i^2) is perpendicular to the line connecting (a_i, a_i^2) and $(1, 17/4)$. Notice that the given point is inside the parabola. (For a parabola that opens upward, the "inside of the parabola" or the "interior of the parabola" means those points that are above the parabola.) Determine a_1, a_2, and a_3 by zooming. Then find the ordered pairs (a_1, a_1^2), (a_2, a_2^2), and (a_3, a_3^2). Find the equations of the three lines and graph them along with the parabola in the same graphing window.

III. REFLECTION

In light of the results obtained in the preceding exercises, a natural conjecture is that there are three perpendicular lines for points interior to the parabola and only one for points exterior to the parabola.

Exercise 8 Investigate a special case of this conjecture for the case in which the point is on the y-axis (i.e., $p = 0$ and $q > 0$).

Exercise 9 Investigate the general conjecture by testing points (p, q) that are inside the parabola, but very close to it. Is it true that there are three perpendicular lines for each point interior to the parabola?

2

Functions and Limits

2.1
Functions and Their Graphs

The concept of function is one of the most basic in all mathematics, and it plays an indispensable role in calculus.

Definition

A **function** f is a rule of correspondence that associates with each object x in one set, called the **domain**, a single value $f(x)$ from a second set. The set of all values so obtained is called the **range** of the function. (See Figure 1.)

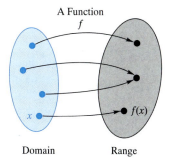

A Function
f

Domain   Range

Figure 1

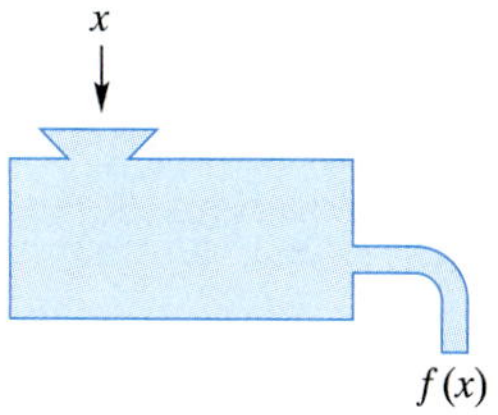

Think of a function as a machine that takes as its input a value x and produces an output $f(x)$. (See Figure 2.) Each input value is matched with a *single* output value, but it can happen that several different input values give the same output value.

The definition puts no restriction on the domain and range sets. The domain might consist of the set of people in your calculus class, the range the set of grades $\{A, B, C, D, F\}$ that will be given, and the rule of correspondence the assignment of grades. Nearly all functions you encounter in this book will be functions of one or more real numbers. For example, the function g might take a real number x and square it, producing the real number x^2. In this case we have a formula that gives the rule of correspondence, that is, $g(x) = x^2$. A schematic diagram for this function is shown in Figure 3.

x
↓

$f(x)$

Figure 2

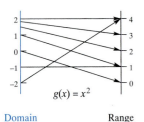

$g(x) = x^2$

Domain   Range

Figure 3

Function Notation   A single letter like f (or g or F) is used to name a function. Then $f(x)$, read "f of x" or "f at x," denotes the value that f assigns to x. Thus, if $f(x) = x^3 - 4$, then

$$f(2) = 2^3 - 4 = 4$$

$$f(-1) = (-1)^3 - 4 = -5$$

$$f(a) = a^3 - 4$$

$$f(a + h) = (a + h)^3 - 4 = a^3 + 3a^2h + 3ah^2 + h^3 - 4$$

Study the following examples carefully. Although some of these examples may look odd now, they will play an important role in the next chapter.

EXAMPLE 1   For $f(x) = x^2 - 2x$, find and simplify

(a) $f(4)$ (b) $f(4 + h)$

(c) $f(4 + h) - f(4)$ (d) $[f(4 + h) - f(4)]/h$

Solution

(a) $f(4) = 4^2 - 2 \cdot 4 = 8$

(b) $f(4 + h) = (4 + h)^2 - 2(4 + h) = 16 + 8h + h^2 - 8 - 2h$

$$= 8 + 6h + h^2$$

(c) $f(4 + h) - f(4) = 8 + 6h + h^2 - 8 = 6h + h^2$

(d) $\dfrac{f(4 + h) - f(4)}{h} = \dfrac{6h + h^2}{h} = \dfrac{h(6 + h)}{h} = 6 + h$ ■

EXAMPLE 2   For $g(x) = 1/x$, find and simplify $[g(a + h) - g(a)]/h$.

Solution

$$\frac{g(a + h) - g(a)}{h} = \frac{\dfrac{1}{a + h} - \dfrac{1}{a}}{h} = \frac{\dfrac{a - (a + h)}{(a + h)a}}{h}$$

$$= \frac{-h}{(a + h)a} \cdot \frac{1}{h} = \frac{-1}{(a + h)a} = \frac{-1}{a^2 + ah}$$ ■

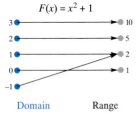

$F(x) = x^2 + 1$

Domain   Range

Figure 4

Domain and Range   To specify a function completely, we must state, in addition to the rule of correspondence, the domain of the function. For example, if F is the function defined by $F(x) = x^2 + 1$ with domain $\{-1, 0, 1, 2, 3\}$ (Figure 4), then the range is $\{1, 2, 5, 10\}$. The rule of correspondence, together with the domain, determines the range.

When no domain is specified for a function, we assume that it is the largest set of real numbers for which the rule for the function makes sense. This is called the **natural domain**. Numbers that you should remember to exclude from the natural domain are those values that would cause division by zero or the square root of a negative number.

EXAMPLE 3   Find the natural domains for

(a) $f(x) = 1/(x - 3)$ (b) $g(t) = \sqrt{9 - t^2}$

(c) $h(w) = 1/\sqrt{9 - w^2}$

Solution

(a) We must exclude 3 from the domain because it would require division by zero. Thus, the natural domain is $\{x \in \mathbb{R} : x \neq 3\}$. This may be read "the set of x's in $\mathbb{R}$ (the real numbers) such that x is not equal to 3."

(b) To avoid the square root of a negative number, we must choose t so that $9 - t^2 \geq 0$. Thus, t must satisfy $|t| \leq 3$. The natural domain is therefore $\{t \in \mathbb{R} : |t| \leq 3\}$, which we can write using interval notation as $[-3, 3]$.

(c) Now we must avoid division by zero *and* square roots of negative numbers, so we must exclude -3 and 3 from the natural domain. The natural domain is therefore the interval $(-3, 3)$. ■

When the rule for a function is given by an equation of the form $y = f(x)$, we call x the **independent variable** and y the **dependent variable**. *Any* value in the domain may be substituted for the independent variable. Once selected, this value of x completely determines the corresponding value of the dependent variable y.

The input for a function need not be a single real number. In many important applications, a function depends on more than one independent variable. For example, the amount A of a monthly car payment depends on the loan's principal P, the rate of interest r, and the required number n of monthly payments. We could write such a function as $A(P, r, n)$. The value of $A(12,000, 0.14, 48)$, that is, the required monthly payment to retire a \$12,000 loan in 48 months at an annual interest rate of 14%, is \$327.92. In this situation, there is no single mathematical formula that gives the output A in terms of the input variables P, r, and n.

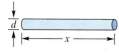

Figure 5

EXAMPLE 4   Let $V(x, d)$ denote the volume of a cylindrical rod of length x and diameter d. (See Figure 5.) Find

(a) a formula for $V(x, d)$ (b) the domain and range of V

(c) $V(4, 0.1)$

Solution

(a) $V(x, d) = x \cdot \pi \left(\dfrac{d}{2}\right)^2 = \dfrac{\pi x d^2}{4}$

(b) Because the length and diameter of the rod must be positive, the domain is $\{(x, d) \mid x \in \mathbb{R}, d \in \mathbb{R}, x > 0, d > 0\}$

(c) $V(4, 0.1) = \dfrac{\pi \cdot 4 \cdot 0.1^2}{4} = 0.01\pi$ ■

Graphing Calculator

Remember, use your graphing calculator to reproduce the figures in this book. Experiment with various graphing windows until you are convinced that you understand all important aspects of the graph.

Chapters 2 through 14 will deal mostly with functions of a single independent variable. Beginning in Chapter 15 we will study properties of functions of two or more independent variables.

Graphs of Functions   When both the domain and range of a function are sets of real numbers, we can picture the function by drawing its graph on a coordinate plane. The **graph of a function** f is simply the graph of the equation $y = f(x)$.

EXAMPLE 5   Sketch the graphs of

(a) $f(x) = x^2 - 2$ (b) $g(x) = 2/(x - 1)$

Solution   The natural domains of f and g are, respectively, all real numbers and all real numbers except 1. Following the procedure described in Section 1.7 (make a table of values, plot the corresponding points, connect these points with a smooth curve), we obtain the two graphs shown in Figures 6 and 7a. ■

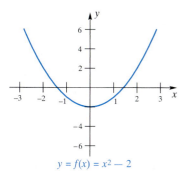

$y = f(x) = x^2 - 2$

Figure 6

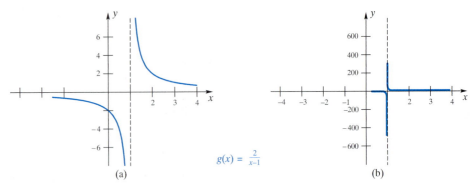

$g(x) = \frac{2}{x-1}$

(a)   (b)

Figure 7

Pay special attention to the graph of g; it points to an oversimplification that we have made and now need to correct. When connecting the plotted points by a smooth curve, do not do so in a mechanical way that ignores special features that may be apparent from the formula for the function. In the case of $g(x) = 2/(x-1)$, something dramatic happens as x nears 1. In fact, the values of $|g(x)|$ increase without bound; for example, $g(0.99) = 2/(0.99 - 1) = -200$ and $g(1.001) = 2000$. We have indicated this by drawing a vertical line, called an **asymptote**, at $x = 1$. As x approaches 1, the graph gets closer and closer to this line, though this line itself is not part of the graph. Rather, it is a guideline. Notice that the graph of g also has a horizontal asymptote, the x-axis.

Functions like $g(x) = 2/(x-1)$ can even cause problems when you graph them on a CAS. For example, Maple, when asked to plot $g(x) = 2/(x-1)$ over the domain $[-4, 4]$ responded with the graph shown in Figure 7b. Maple uses an algorithm much like that described in Chapter 1; it chooses a number of x-values over the stated domain, finds the corresponding y-values, and plots these points with connecting lines. When Maple chose a number near 1, the resulting output was large, leading to the y-axis scaling in the figure. Maple also connected the points right across the break at $x = 1$. Always be cautious and careful when you use a graphing calculator or a CAS to plot functions.

The domains and ranges for the functions f and g are shown in the following table.

Function	Domain	Range
$f(x) = x^2 - 2$	$\mathbb{R}$	$\{y \in \mathbb{R} : y \geq -2\}$
$g(x) = \dfrac{2}{x-1}$	$\{x \in \mathbb{R} : x \neq 1\}$	$\{y \in \mathbb{R} : y \neq 0\}$

Even and Odd Functions   We can often predict the symmetries of the graph of a function by inspecting the formula for the function. If $f(-x) = f(x)$ for all x, then the graph is symmetric with respect to the y-axis. Such a function is called an **even function**, probably because a function that specifies $f(x)$ as a sum of only even powers of x is even. The function $f(x) = x^2 - 2$ (graphed in Figure 6) is even; so are $f(x) = 3x^6 - 2x^4 + 11x^2 - 5$, $f(x) = x^2/(1 + x^4)$, and $f(x) = (x^3 - 2x)/3x$.

If $f(-x) = -f(x)$ for all x, the graph is symmetric with respect to the origin. We call such a function an **odd function**. A function that gives $f(x)$ as a sum of only odd powers of x is odd. Thus, $g(x) = x^3 - 2x$ (graphed in Figure 8) is odd. Note that

$$g(-x) = (-x)^3 - 2(-x) = -x^3 + 2x = -(x^3 - 2x) = -g(x)$$

Consider the function $g(x) = 2/(x-1)$ from Example 2, which we graphed in Figure 7. It is neither even nor odd. To see this, observe that $g(-x) = 2/(-x - 1)$, which is not equal to either $g(x)$ or $-g(x)$. Note that the graph of $y = g(x)$ is neither symmetric with respect to the y-axis nor the origin.

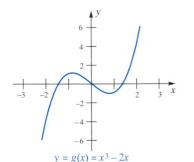

$y = g(x) = x^3 - 2x$

Figure 8

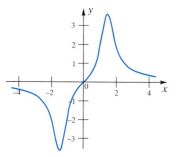

Figure 9

EXAMPLE 6   Is $f(x) = \dfrac{x^3 + 3x}{x^4 - 3x^2 + 4}$ even, odd, or neither?

Solution   Since

$$f(-x) = \frac{(-x)^3 + 3(-x)}{(-x)^4 - 3(-x)^2 + 4} = \frac{-(x^3 + 3x)}{x^4 - 3x^2 + 4} = -f(x)$$

f is an odd function. The graph of $y = f(x)$ (Figure 9) is symmetric with respect to the origin. ∎

Two Special Functions   Among the functions that will often be used as examples are two very special ones: the **absolute value function,** $|\ \ |$, and the **greatest integer function,** $[\![\ \ ]\!]$. They are defined by

$$|x| = \begin{cases} x & \text{if } x \geq 0 \\ -x & \text{if } x < 0 \end{cases}$$

and

$$[\![x]\!] = \text{the greatest integer less than or equal to } x$$

Thus, $|-3.1| = |3.1| = 3.1$, while $[\![-3.1]\!] = -4$ and $[\![3.1]\!] = 3$. We show the graphs of these two functions in Figures 10 and 11. The absolute value function is even, since $|-x| = |x|$. The greatest integer function is neither even nor odd, as you can see from its graph.

We will often appeal to the following special features of these graphs. The graph of $|x|$ has a sharp corner at the origin, while the graph of $[\![x]\!]$ takes a jump at each integer.

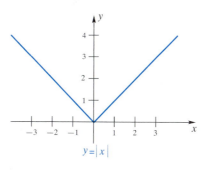

Figure 10

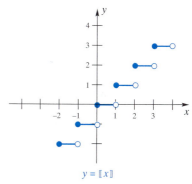

Figure 11

Concepts Review

1. The set of allowable inputs for a function is called the _____ of the function; the set of outputs that are obtained is called the _____ of the function.

2. If $f(x) = 3x^2$, then $f(2u) = $ _____ and $f(x + h) = $ _____.

3. If $f(x)$ gets closer and closer to L as $|x|$ increases indefinitely, then the line $y = L$ is a(an) _____ for the graph of f.

4. If $f(-x) = f(x)$ for all x in the domain of f, then f is called a(an) _____ function; if $f(-x) = -f(x)$ for all x in the domain of f, then f is called a(an) _____ function. In the first case, the graph of f is symmetric with respect to the _____; in the second case, it is symmetric with respect to the _____.

Problem Set 2.1

1. For $f(x) = 1 - x^2$, find each value.

(a) $f(1)$ (b) $f(-2)$ (c) $f(0)$

(d) $f(k)$ (e) $f(-5)$ (f) $f(\frac{1}{4})$

(g) $f(3t)$ (h) $f(2x)$ (i) $f(\frac{1}{t})$

2. For $F(x) = x^3 + 3x$, find each value.

(a) $F(1)$ (b) $F(\sqrt{2})$ (c) $F(\frac{1}{4})$

(d) $F(\pi)$ (e) $F(\frac{1}{t})$ (f) $F(2.718)$

3. For $G(y) = 1/(y - 1)$, find each value.

(a) $G(0)$ (b) $G(0.999)$ (c) $G(1.01)$

(d) $G(y^2)$ (e) $G(-x)$ (f) $G\left(\dfrac{1}{x^2}\right)$

4. For $\Phi(u) = \dfrac{u + u^2}{\sqrt{u}}$, find each value. ($\Phi$ is the uppercase Greek letter phi.)

(a) $\Phi(1)$ (b) $\Phi(-t)$ (c) $\Phi(\frac{1}{2})$

(d) $\Phi(u + 1)$ (e) $\Phi(x^2)$ (f) $\Phi(x^2 + x)$

5. For

$$f(x) = \frac{1}{\sqrt{x - 3}}$$

find each value.

(a) $f(0.25)$ (b) $f(\pi)$ (c) $f(3 + \sqrt{2})$

C **6.** For $f(x) = \sqrt{x^2 + 9}/(x - \sqrt{3})$, find each value.

(a) $f(0.79)$ (b) $f(12.26)$ (c) $f(\sqrt{3})$

7. Which of the following determine a function f with formula $y = f(x)$? For those that do, find $f(x)$. *Hint*: Solve for y in terms of x and note that the definition of function requires a single y for each x.

(a) $x^2 + y^2 = 1$ (b) $xy + y + x = 1, x \neq -1$

(c) $x = \sqrt{2y + 1}$ (d) $x = \dfrac{y}{y + 1}$

8. Which of the graphs in Figure 12 are graphs of functions? (Is there a single y for each x?)

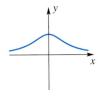
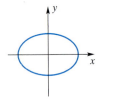
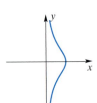
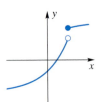

Figure 12

This problem suggests a rule: *For a graph to be the graph of a function, each vertical line must meet the graph in at most one point.*

9. For $f(x) = 2x^2 - 1$, find and simplify $[f(a + h) - f(a)]/h$. (See Examples 1 and 2.)

10. For $F(t) = 4t^3$, find and simplify $[F(a + h) - F(a)]/h$.

11. For $g(u) = 3/(u - 2)$, find and simplify $[g(x + h) - g(x)]/h$.

12. For $G(t) = t/(t + 4)$, find and simplify $[G(a + h) - G(a)]/h$.

13. Find the natural domain for each of the following. (See Example 3.)

(a) $F(z) = \sqrt{2z + 3}$ (b) $g(v) = 1/(4v - 1)$

(c) $\psi(x) = \sqrt{x^2 - 9}$ (d) $H(y) = -\sqrt{625 - y^4}$

14. Find the natural domain in each case.

(a) $f(x) = \dfrac{4 - x^2}{x^2 - x - 6}$ (b) $G(y) = \sqrt{(y + 1)^{-1}}$

(c) $\phi(u) = |2u + 3|$ (d) $F(t) = t^{2/3} - 4$

In Problems 15–30, specify whether the given function is even, odd, or neither, and then sketch its graph. (See Examples 4 and 5.)

15. $f(x) = -4$ **16.** $f(x) = 3x$

17. $F(x) = 2x + 1$ **18.** $F(x) = 3x - \sqrt{2}$

19. $g(x) = 3x^2 + 2x - 1$ **20.** $g(u) = \dfrac{u^3}{8}$

21. $g(x) = \dfrac{x}{x^2 - 1}$ **22.** $\phi(z) = \dfrac{2z + 1}{z - 1}$

23. $f(w) = \sqrt{w - 1}$ **24.** $h(x) = \sqrt{x^2 + 4}$

25. $f(x) = |2x|$ **26.** $F(t) = -|t + 3|$

27. $g(x) = \left[\!\!\left[\dfrac{x}{2}\right]\!\!\right]$ **28.** $G(x) = [\![2x - 1]\!]$

29. $g(t) = \begin{cases} 1 & \text{if } t \leq 0 \\ t + 1 & \text{if } 0 < t < 2 \\ t^2 - 1 & \text{if } t \geq 2 \end{cases}$

30. $h(x) = \begin{cases} -x^2 + 4 & \text{if } x \leq 1 \\ 3x & \text{if } x > 1 \end{cases}$

31. A plant has the capacity to produce from 0 to 100 computers per day. The daily overhead for the plant is $5000, and the direct cost (labor and materials) of producing one computer is $805. Write a formula for $T(x)$, the total cost of producing x computers in one day, and also for the unit cost $u(x)$ (average cost per computer). What are the domains of these functions?

C **32.** It costs the ABC Company $400 + 5\sqrt{x(x - 4)}$ dollars to make x toy stoves that sell for $6 each.

(a) Find a formula for $P(x)$, the total profit in making x stoves.

(b) Evaluate $P(200)$ and $P(1000)$.

(c) How many stoves does ABC have to make to just break even?

C **33.** Find the formula for the amount $E(x)$ by which a number x exceeds its square. Plot a graph of $E(x)$ for $0 \leq x \leq 1$. Use the graph to estimate the positive number less than or equal to 1 that exceeds its square by the maximum amount.

34. Let p denote the perimeter of an equilateral triangle. Find a formula for $A(p)$, the area of such a triangle.

35. The Acme Car Rental Agency charges $24 a day for the rental of a car plus $0.40 per mile.

(a) Write a formula for the total rental expense $E(x)$ for one day, where x is the number of miles driven.

(b) If you rent a car for one day, how many miles can you drive for $120?

36. A right circular cylinder of radius r is inscribed in a sphere of radius $2r$. Find a formula for $V(r)$, the volume of the cylinder, in terms of r.

37. A 1-mile track has parallel sides and equal semicircular ends. Find a formula for the area enclosed by the track, $A(d)$, in terms of the diameter d of the semicircles. What is the natural domain for this function?

38. Let $A(c)$ denote the area of the region bounded from above by the line $y = x + 1$, from the left by the y-axis, from below by the x-axis, and from the right by the line $x = c$. Such a function is called an **accumulation function**. (See Figure 13.) Find

(a) $A(1)$ (b) $A(2)$

(c) $A(0)$ (d) $A(c)$

(e) Sketch the graph of $A(c)$.   (f) What are the domain and range of A?

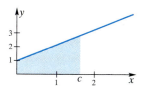

Figure 13

39. Let $B(c)$ denote the area of the region bounded from above by the graph of the curve $y = x(1 - x)$, from below by the x-axis, and from the right by the line $x = c$. The domain of B is the interval $[0, 1]$. (See Figure 14.) Given that $B(1) = \frac{1}{6}$,

(a) Find $B(0)$ (b) Find $B(\frac{1}{2})$

(c) As best you can, sketch a graph of $B(c)$.

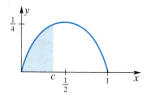

Figure 14

40. Which of the following functions satisfies $f(x + y) = f(x) + f(y)$ for all x and y in $\mathbb{R}$?

(a) $f(t) = 2t$ (b) $f(t) = t^2$

(c) $f(t) = 2t + 1$ (d) $f(t) = -3t$

41. Let $f(x + y) = f(x) + f(y)$ for all x and y in $\mathbb{R}$. Prove that there is a number m such that $f(t) = mt$ for all rational numbers t. *Hint*: First decide what m has to be. Then proceed in steps, starting with $f(0) = 0$, $f(p) = mp$ for p in $\mathbb{N}$, $f(1/p) = m/p$, and so on.

42. A baseball diamond is a square with sides of 90 feet. A player, after hitting a home run, loped around the diamond at 10 feet per second. Let s represent the player's distance from home plate after t seconds.

(a) Express s as a function of t by means of a four-part formula.

(b) Express s as a function of t by means of a three-part formula.

GC *To use technology efficiently, you need to discover its capabilities, its strengths, and its weaknesses. We urge you to practice graphing functions of various types using your own computer package or calculator. Problems 43–48 are designed for this purpose.*

43. Let $f(x) = (x^3 + 3x - 5)/(x^2 + 4)$.

(a) Evaluate $f(1.38)$ and $f(4.12)$.

(b) Construct a table of values for this function corresponding to $x = -4, -3, \ldots, 3, 4$.

44. Follow the instructions in Problem 43 for $f(x) = (\sin^2 x - 3 \tan x)/\cos x$.

45. Draw the graph of $f(x) = x^3 - 5x^2 + x + 8$ on the domain $[-2, 5]$.

(a) Determine the range of f.

(b) Where on this domain is $f(x) \geq 0$?

46. Superimpose the graph of $g(x) = 2x^2 - 8x - 1$ with domain $[-2, 5]$ on the graph of $f(x)$ of Problem 45.

(a) Estimate the x-values where $f(x) = g(x)$.

(b) Where on $[-2, 5]$ is $f(x) \geq g(x)$?

(c) Estimate the largest value of $|f(x) - g(x)|$ on $[-2, 5]$.

47. Graph $f(x) = (3x - 4)/(x^2 + x - 6)$ on the domain $[-6, 6]$.

(a) Determine the x- and y-intercepts.

(b) Determine the range of f for the given domain.

(c) Determine the vertical asymptotes of the graph.

(d) Determine the horizontal asymptote for the graph when the domain is enlarged to the whole real line.

48. Follow the directions in Problem 47 for $g(x) = (3x^2 - 4)/(x^2 + x - 6)$

Answers to Concepts Review: **1.** domain, range **2.** $12u^2$; **3.** $(x + h)^2 = 3x^2 + 6xh + 3h^2$ **3.** asymptote **4.** even; odd; y-axis; origin

---

2.2
Operations
on Functions

Functions are not numbers. But just as two numbers a and b can be added to produce a new number $a + b$, so two functions f and g can be added to produce a new function $f + g$. This is just one of several operations on functions that we will describe in this section.

Sums, Differences, Products, Quotients, Powers   Consider functions f and g with formulas

$$f(x) = \frac{x - 3}{2}, \qquad g(x) = \sqrt{x}$$

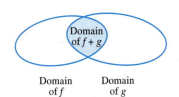

Domain
of f   Domain
              of g

Figure 1

We can make a new function $f + g$ by having it assign to x the value $f(x) + g(x) = (x - 3)/2 + \sqrt{x}$; that is,

$$(f + g)(x) = f(x) + g(x) = \frac{x - 3}{2} + \sqrt{x}$$

Of course, we must be a little careful about domains. Clearly, x must be a number on which both f and g can work. In other words, the domain of $f + g$ is the intersection (common part) of the domains of f and g (Figure 1).

The functions $f - g, f \cdot g$, and f/g are introduced in a completely analogous way. Assuming that f and g have their natural domains, we have the following:

Formula	Domain
$(f + g)(x) = f(x) + g(x) = \dfrac{x - 3}{2} + \sqrt{x}$	$[0, \infty)$
$(f - g)(x) = f(x) - g(x) = \dfrac{x - 3}{2} - \sqrt{x}$	$[0, \infty)$
$(f \cdot g)(x) = f(x) \cdot g(x) = \dfrac{x - 3}{2} \sqrt{x}$	$[0, \infty)$
$\left(\dfrac{f}{g}\right)(x) = \dfrac{f(x)}{g(x)} = \dfrac{x - 3}{2\sqrt{x}}$	$(0, \infty)$

We had to exclude 0 from the domain of f/g to avoid division by 0.

We may also raise a function to a power. By f^n, we mean the function that assigns to x the value $[f(x)]^n$. Thus,

$$f^2(x) = [f(x)]^2 = \left[\frac{x - 3}{2}\right]^2 = \frac{x^2 - 6x + 9}{4}$$

and

$$g^3(x) = [g(x)]^3 = \left(\sqrt{x}\right)^3 = x^{3/2}$$

There is one exception to the above agreement on exponents, namely, when $n = -1$. We reserve the symbol f^{-1} for the inverse function, which will be discussed in Section 7.2. Thus, f^{-1} does not mean $1/f$.

EXAMPLE 1 Let $F(x) = \sqrt[4]{x + 1}$ and $G(x) = \sqrt{9 - x^2}$, with respective natural domains $[-1, \infty)$ and $[-3, 3]$. Find formulas for $F + G, F - G, F \cdot G, F/G$, and F^5 and give their natural domains.

Solution

Formula	Domain
$(F + G)(x) = F(x) + G(x) = \sqrt[4]{x + 1} + \sqrt{9 - x^2}$	$[-1, 3]$
$(F - G)(x) = F(x) - G(x) = \sqrt[4]{x + 1} - \sqrt{9 - x^2}$	$[-1, 3]$
$(F \cdot G)(x) = F(x) \cdot G(x) = \sqrt[4]{x + 1} \sqrt{9 - x^2}$	$[-1, 3]$
$\left(\dfrac{F}{G}\right)(x) = \dfrac{F(x)}{G(x)} = \dfrac{\sqrt[4]{x + 1}}{\sqrt{9 - x^2}}$	$[-1, 3)$
$F^5(x) = [F(x)]^5 = \left(\sqrt[4]{x + 1}\right)^5 = (x + 1)^{5/4}$	$[-1, \infty)$

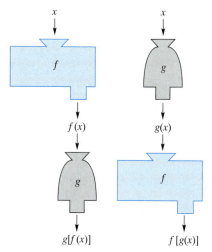

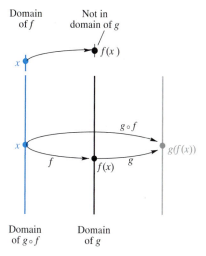

Figure 2

Figure 3

Composition of Functions   Earlier, we asked you to think of a function as a machine. It accepts x as input, works on x, and produces $f(x)$ as output. Two machines may often be put together in tandem to make a more complicated machine; so may two functions f and g (Figure 2). If f works on x to produce $f(x)$ and g then works on $f(x)$ to produce $g(f(x))$, we say that we have *composed* g with f. The resulting function, called the **composition** of g with f, is denoted by $g \circ f$. Thus,

$$(g \circ f)(x) = g(f(x))$$

In our previous examples we had $f(x) = (x - 3)/2$ and $g(x) = \sqrt{x}$. We may compose these functions in two ways:

$$(g \circ f)(x) = g(f(x)) = g\left(\frac{x-3}{2}\right) = \sqrt{\frac{x-3}{2}}$$

$$(f \circ g)(x) = f(g(x)) = f(\sqrt{x}) = \frac{\sqrt{x} - 3}{2}$$

Right away we notice that $g \circ f$ does not equal $f \circ g$. Thus, we say that the composition of functions is not commutative.

We must be careful in describing the domain of a composite function. The domain of $g \circ f$ is equal to the set of those values x that satisfy the following properties:

1. x is in the domain of f.
2. $f(x)$ is in the domain of g.

In other words, x must be a valid input for f, and $f(x)$ must be a valid input for g. In our example, the value $x = 2$ is in the domain of f, but it is not in the domain of $g \circ f$ because this would lead to the square root of a negative number.

$$g(f(2)) = g((2 - 3)/2) = g\left(-\frac{1}{2}\right) = \sqrt{-\frac{1}{2}}$$

The domain for $g \circ f$ is the interval $[3, \infty)$ because $f(x)$ is nonnegative on this interval, and the input to g must be nonnegative. The domain for $f \circ g$ is the interval $[0, \infty)$ (why?), so we see that the domains of $g \circ f$ and $f \circ g$ can be different. Figure 3 shows how the domain of $g \circ f$ excludes those values of x for which $f(x)$ is not in the domain of g.

EXAMPLE 2   Let $f(x) = 6x/(x^2 - 9)$ and $g(x) = \sqrt{3x}$, with their natural domains. First, find $(f \circ g)(12)$; then find $(f \circ g)(x)$ and give its domain.

Solution

$$(f \circ g)(12) = f(g(12)) = f(\sqrt{36}) = f(6) = \frac{6 \times 6}{6^2 - 9} = \frac{4}{3}$$

$$(f \circ g)(x) = f(g(x)) = f(\sqrt{3x}) = \frac{6\sqrt{3x}}{(\sqrt{3x})^2 - 9}$$

The expression $\sqrt{3x}$ appears in both the numerator and denominator. Any negative number for x will lead to the square root of a negative number. Thus, all negative numbers must be excluded from the domain of $f \circ g$. For $x \geq 0$, we have $(\sqrt{3x})^2 = 3x$, allowing us to write

$$(f \circ g)(x) = \frac{6\sqrt{3x}}{3x - 9} = \frac{2\sqrt{3x}}{x - 3}$$

We must also exclude $x = 3$ from the domain of $f \circ g$ because it is not in the domain of f. (It would cause division by 0.) Thus, the domain of $f \circ g$ is $[0, 3) \cup (3, \infty)$. ■

In calculus, we will often need to take a given function and write it as the composition of two simpler functions. Usually, this can be done in a number of ways. For example, $p(x) = \sqrt{x^2 + 4}$ can be written as

$$p(x) = g(f(x)), \qquad \text{where } g(x) = \sqrt{x} \quad \text{and} \quad f(x) = x^2 + 4$$

or as

$$p(x) = g(f(x)), \qquad \text{where } g(x) = \sqrt{x + 4} \quad \text{and} \quad f(x) = x^2$$

(You should check that both of these compositions give $p(x) = \sqrt{x^2 + 4}$ with domain $\mathbb{R}$.) The decomposition $p(x) = g(f(x))$ with $f(x) = x^2 + 4$ and $g(x) = \sqrt{x}$ is regarded as simpler and is usually preferred. We can therefore view $p(x) = \sqrt{x^2 + 4}$ as the square root of a function of x. This way of looking at functions will be important in the next chapter.

EXAMPLE 3   Write the function $p(x) = (x + 2)^5$ as a composite function $g \circ f$.

Solution   The most obvious way to decompose p is to write

$$p(x) = g(f(x)), \qquad \text{where } g(x) = x^5 \quad \text{and} \quad f(x) = x + 2$$

We thus view $p(x) = (x + 2)^5$ as the fifth power of a function of x.   ∎

Translations   Observing how a function is built up from simpler ones can be a big aid in graphing. We may ask this question: How are the graphs of

$$y = f(x) \qquad y = f(x - 3) \qquad y = f(x) + 2 \qquad y = f(x - 3) + 2$$

related to each other? Consider $f(x) = |x|$ as an example. The corresponding four graphs are displayed in Figure 4.

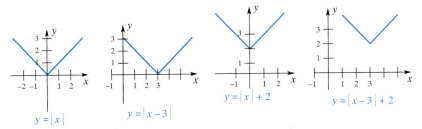

$$y = |x| \qquad\qquad y = |x - 3| \qquad\qquad y = |x| + 2 \qquad\qquad y = |x - 3| + 2$$

Figure 4

Notice that all four graphs have the same shape; the last three are just translations of the first. Replacing x by $x - 3$ translates the graph 3 units to the right; adding 2 translates it upward by 2 units.

What happened with $f(x) = |x|$ is typical. Figure 5 offers an illustration for the function $f(x) = x^3 + x^2$.

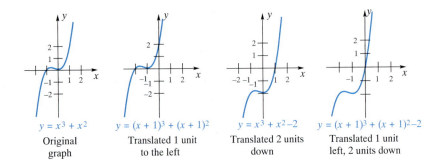

$y = x^3 + x^2$	$y = (x + 1)^3 + (x + 1)^2$	$y = x^3 + x^2 - 2$	$y = (x + 1)^3 + (x + 1)^2 - 2$
Original graph	Translated 1 unit to the left	Translated 2 units down	Translated 1 unit left, 2 units down

Figure 5

Exactly the same principles apply in the general situation. They are illustrated in Figure 6 with both h and k positive. If $h < 0$, the translation is to the left; if $k < 0$, the translation is downward.

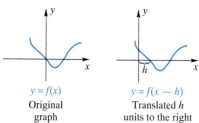

$y = f(x)$
Original
graph

$y = f(x - h)$
Translated h
units to the right

$y = f(x) + k$
Translated k
units up

$y = f(x - h) + k$
Translated h units
to the right
and k units up

Figure 6

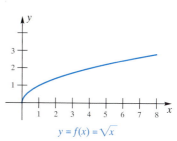

$y = f(x) = \sqrt{x}$

Figure 7

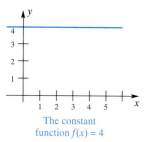

$y = g(x) = \sqrt{x + 3} + 1$

Figure 8

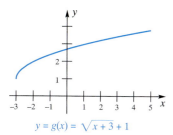

The constant
function $f(x) = 4$

Figure 9

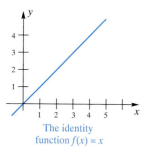

The identity
function $f(x) = x$

Figure 10

EXAMPLE 4 Sketch the graph of $g(x) = \sqrt{x + 3} + 1$ by first graphing $f(x) = \sqrt{x}$ and then making appropriate translations.

Solution By translating the graph of f (Figure 7) 3 units left and 1 unit up, we obtain the graph of g (Figure 8). ∎

Partial Catalog of Functions A function of the form $f(x) = k$, where k is a constant (real number), is called a **constant function**. Its graph is a horizontal line (Figure 9). The function $f(x) = x$ is called the **identity function**. Its graph is a line through the origin of slope 1 (Figure 10). From these simple functions, we can build many important functions.

Any function that can be obtained from the constant functions and the identity function by use of the operations of addition, subtraction, and multiplication is called a **polynomial function**. This amounts to saying that f is a polynomial function if it is of the form.

$$f(x) = a_n x^n + a_{n-1} x^{n-1} + \cdots + a_1 x + a_0$$

where the a's are real numbers and n is a nonnegative integer. If $a_n \neq 0$, n is the **degree** of the polynomial function. In particular, $f(x) = ax + b$ is a first-degree polynomial function, or **linear function**, and $f(x) = ax^2 + bx + c$ is a second-degree polynomial function, or **quadratic function**.

Quotients of polynomial functions are called rational functions. Thus, f is a **rational function** if it is of the form

$$f(x) = \frac{a_n x^n + a_{n-1} x^{n-1} + \cdots + a_1 x + a_0}{b_m x^m + b_{m-1} x^{m-1} + \cdots + b_1 x + b_0}$$

The domain of a rational function consists of those real numbers for which the denominator is nonzero.

The functions so far listed, together with the trigonometric, inverse trigonometric, exponential, and logarithmic functions (to be introduced later), are the basic raw material for calculus.

Concepts Review

1. If $f(x) = x^2 + 1$, then $f^3(x) = $ _____.

2. The value of the composite function $f \circ g$ at x is given by $(f \circ g)(x) = $ _____.

3. Compared to the graph of $y = f(x)$, the graph of $y = f(x + 2)$ is translated _____ units to the _____.

4. A rational function is defined as _____.

Problem Set 2.2

1. For $f(x) = x + 3$ and $g(x) = x^2$, find each value (if possible).

(a) $(f + g)(2)$ (b) $(f \cdot g)(0)$ (c) $(g/f)(3)$

(d) $(f \circ g)(1)$ (e) $(g \circ f)(1)$ (f) $(g \circ f)(-8)$

2. For $f(x) = x^2 + x$ and $g(x) = 2/(x + 3)$, find each value.

(a) $(f - g)(2)$ (b) $(f/g)(1)$ (c) $g^2(3)$

(d) $(f \circ g)(1)$ (e) $(g \circ f)(1)$ (f) $(g \circ g)(3)$

3. For $\Phi(u) = u^3 + 1$ and $\Psi(v) = 1/v$, find each value.

(a) $(\Phi + \Psi)(t)$ (b) $(\Phi \circ \Psi)(r)$

(c) $(\Psi \circ \Phi)(r)$ (d) $\Phi^3(z)$

(e) $(\Phi - \Psi)(5t)$ (f) $((\Phi - \Psi) \circ \Psi)(t)$

4. If $f(x) = \sqrt{x^2 - 1}$ and $g(x) = 2/x$, find formulas for the following and state their domains.

(a) $(f \cdot g)(x)$ (b) $f^4(x) + g^4(x)$

(c) $(f \circ g)(x)$ (d) $(g \circ f)(x)$

5. If $f(s) = \sqrt{s^2 - 4}$ and $g(w) = |1 + w|$, find formulas for $(f \circ g)(x)$ and $(g \circ f)(x)$.

6. If $g(x) = x^2 + 1$, find formulas for $g^3(x)$ and $(g \circ g \circ g)(x)$.

[C] **7.** Calculate $g(3.141)$ if $g(u) = \dfrac{\sqrt{u^3 + 2u}}{2 + u}$.

[C] **8.** Calculate $g(2.03)$ if $g(x) = \dfrac{(\sqrt{x} - \sqrt[3]{x})^4}{1 - x + x^2}$.

[C] **9.** Calculate $[g^2(\pi) - g(\pi)]^{1/3}$ if $g(v) = |11 - 7v|$.

[C] **10.** Calculate $[g^3(\pi) - g(\pi)]^{1/3}$ if $g(x) = 6x - 11$.

11. Find f and g so that $F = g \circ f$. (See Example 3.)

(a) $F(x) = \sqrt{x + 7}$ (b) $F(x) = (x^2 + x)^{15}$

12. Find f and g so that $p = f \circ g$.

(a) $p(x) = \dfrac{2}{(x^2 + x + 1)^3}$ (b) $p(x) = \log(x^3 + 3x)$

13. Write $p(x) = \log \sqrt{x^2 + 1}$ as a composite of three functions in two different ways.

14. Write $p(x) = \log \sqrt{x^2 + 1}$ as a composite of four functions.

15. Sketch the graph of $f(x) = \sqrt{x - 2} - 3$ by first sketching $g(x) = \sqrt{x}$ and then translating. (See Example 4.)

16. Sketch the graph of $g(x) = |x + 3| - 4$ by first sketching $h(x) = |x|$ and then translating.

17. Sketch the graph of $f(x) = (x - 2)^2 - 4$ using translations.

18. Sketch the graph of $g(x) = (x + 1)^3 - 3$ using translations.

19. Sketch the graphs of $f(x) = (x - 3)/2$ and $g(x) = \sqrt{x}$ using the same coordinate axes. Then sketch $f + g$ by adding ordinates.

20. Follow the directions of Problem 19 for $f(x) = x$ and $g(x) = |x|$.

21. Sketch the graph of $F(t) = \dfrac{|t| - t}{t}$.

22. Sketch the graph of $G(t) = t - [\![t]\!]$.

23. State whether each of the following is an odd function, an even function, or neither. Prove your statements.

(a) The sum of two even functions

(b) The sum of two odd functions

(c) The product of two even functions

(d) The product of two odd functions

(e) The product of an even function and an odd function

24. Let F be any function whose domain contains $-x$ whenever it contains x. Prove each of the following.

(a) $F(x) - F(-x)$ is an odd function.

(b) $F(x) + F(-x)$ is an even function.

(c) F can always be expressed as the sum of an odd and an even function.

25. Is every polynomial of even degree an even function? Is every polynomial of odd degree an odd function? Explain.

26. Classify each of the following as a PF (polynomial function), RF (rational function but not a polynomial function), or neither.

(a) $f(x) = 3x^{1/2} + 1$ (b) $f(x) = 3$

(c) $f(x) = 3x^2 + 2x^{-1}$ (d) $f(x) = \pi x^3 - 3\pi$

(e) $f(x) = \dfrac{1}{x + 1}$ (f) $f(x) = \dfrac{x + 1}{\sqrt{x + 3}}$

27. The relationship between the unit price P (in cents) for a certain product and the demand D (in thousands of units) appears to satisfy

$$P = \sqrt{29 - 3D + D^2}$$

On the other hand, the demand has risen over the t years since 1970 according to $D = 2 + \sqrt{t}$.

(a) Express P as a function of t.

(b) Evaluate P when $t = 15$.

28. After being in business for t years, a manufacturer of cars is making $120 + 2t + 3t^2$ units per year. The sales price in dollars per unit has risen according to the formula $6000 + 700t$. Write a formula for the manufacturer's yearly revenue $R(t)$ after t years.

29. Starting at noon, airplane A flies due north at 400 miles per hour. Starting 1 hour later, airplane B flies due east at 300 miles per hour. Neglecting the curvature of the Earth and assuming that they fly at the same altitude, find a formula for $D(t)$, the distance between the two airplanes t hours after noon. *Hint*: There will be two formulas for $D(t)$, one if $0 < t < 1$ and the other if $t > 1$.

[≈][C] **30.** Find the distance between the airplanes of Problem 29 at 2:30 P.M.

31. Let $f(x) = \dfrac{ax + b}{cx - a}$. Show that $f(f(x)) = x$, provided $a^2 + bc \neq 0$ and $x \neq a/c$.

32. Let $f(x) = \dfrac{x - 3}{x + 1}$. Show that $f(f(f(x))) = x$, provided $x \neq \pm 1$.

33. Let $f(x) = \dfrac{x}{x - 1}$. Find and simplify each value.

(a) $f(1/x)$ (b) $f(f(x))$ (c) $f(1/f(x))$

34. Let $f(x) = \dfrac{x}{\sqrt{x} - 1}$. Find and simplify.

(a) $f\left(\dfrac{1}{x}\right)$
        (b) $f(f(x))$

35. Let $f_1(x) = x$, $f_2(x) = 1/x$, $f_3(x) = 1 - x$, $f_4(x) = 1/(1 - x)$, $f_5(x) = (x - 1)/x$, and $f_6(x) = x/(x - 1)$. Note that $f_3(f_4(x)) = f_3(1/(1 - x)) = 1 - 1/(1 - x) = x/(x - 1) = f_6(x)$; that is, $f_3 \circ f_4 = f_6$. In fact, the composition of any two of these functions is another one in the list. Fill in the composition table in Figure 11.

$\circ$	f_1	f_2	f_3	f_4	f_5	f_6
f_1						
f_2						
f_3			f_6			
f_4						
f_5						
f_6						

Figure 11

Then use this table to find each of the following. From Problem 36, you know that the associative law holds.

(a) $f_3 \circ f_3 \circ f_3 \circ f_3 \circ f_3$
    (b) $f_1 \circ f_2 \circ f_3 \circ f_4 \circ f_5 \circ f_6$
(c) F if $F \circ f_6 = f_1$
             (d) G if $G \circ f_3 \circ f_6 = f_1$
(e) H if $f_2 \circ f_5 \circ H = f_5$

36. Prove that the operation of composition of functions is associative; that is, $f_1 \circ (f_2 \circ f_3) = (f_1 \circ f_2) \circ f_3$.

GC Use a computer or a graphing calculator in Problems 37–40.

37. Let $f(x) = x^2 - 3x$. Using the same axes, draw the graphs of $y = f(x)$, $y = f(x - 0.5) - 0.6$, and $y = f(1.5x)$, all on the domain $[-2, 5]$.

38. Let $f(x) = |x^3|$. Using the same axes, draw the graphs of $y = f(x)$, $y = f(3x)$, and $y = f(3(x - 0.8))$, all on the domain $[-3, 3]$.

39. Let $f(x) = 2\sqrt{x} - 2x + 0.25x^2$. Using the same axes, draw the graphs of $y = f(x)$, $y = f(1.5x)$, and $y = f(x - 1) + 0.5$, all on the domain $[0, 5]$.

40. Let $f(x) = 1/(x^2 + 1)$. Using the same axes, draw the graphs of $y = f(x)$, $y = f(2x)$, and $y = f(x - 2) + 0.6$, all on the domain $[-4, 4]$.

CAS **41.** Your computer algebra system (CAS) may allow the use of parameters in defining functions. In each case, draw the graph of $y = f(x)$ for the specified values of the parameter k, using the same axes and $-5 \le x \le 5$.

(a) $f(x) = |kx|^{0.7}$ for $k = 1, 2, 0.5$, and 0.2.

(b) $f(x) = |x - k|^{0.7}$ for $k = 0, 2, -0.5$, and -3.

(c) $f(x) = |x|^k$ for $k = 0.4, 0.7, 1$, and 1.7.

CAS **42.** Using the same axes, draw the graph of $f(x) = |k(x - c)|^n$ for the following choices of parameters.

(a) $c = -1, k = 1.4, n = 0.7$
    (b) $c = 2, k = 1.4, n = 1$

(c) $c = 0, k = 0.9, n = 0.6$

Answers to Concepts Review: **1.** $(x^2 + 1)^3$ **2.** $f(g(x))$
3. 2, left **4.** a quotient of two polynomial functions

2.3
The Trigonometric Functions

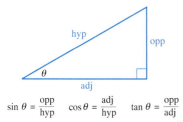

$$\sin\theta = \frac{\text{opp}}{\text{hyp}} \qquad \cos\theta = \frac{\text{adj}}{\text{hyp}} \qquad \tan\theta = \frac{\text{opp}}{\text{adj}}$$

Figure 1

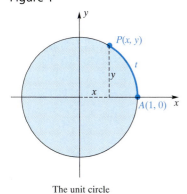

The unit circle

Figure 2

You have probably seen the definitions of the trigonometric functions based on right triangles. Figure 1 summarizes the definitions of the sine, cosine, and tangent functions. You should review Figure 1 carefully, because these concepts are needed for many applications later in this book.

More generally, we define the trigonometric functions based on the unit circle. The unit circle, which we denote C, is the circle with radius 1 and center at the origin; it has equation $x^2 + y^2 = 1$. Let A be the point $(1, 0)$ and let t be a positive number. There is a single point P on the circle C such that the distance, measured in the *counterclockwise* direction around the arc AP, is equal to t. (See Figure 2.) Recall that the circumference of a circle with radius r is $2\pi r$, so the circumference of C is 2π. Thus, if $t = \pi$, then the point P is exactly halfway around the circle from the point A; in this case, P is the point $(-1, 0)$. If $t = 3\pi/2$, then P is the point $(0, -1)$, and if $t = 2\pi$, then P is the point A. If $t > 2\pi$, then it will take more than one complete circuit of the circle C to trace the arc AP.

When $t < 0$, we trace the circle in a *clockwise* direction. There will be a single point P on the circle C such that the arc length measured in the clockwise direction from A is t. Thus, for every real number t, we can associate a unique point $P(x, y)$ on the unit circle. This allows us to make the key definitions of the sine and cosine functions. The functions sine and cosine are written as sin and cos, rather than as a single letter such as f or g. Parentheses around the independent variable are usually omitted unless there is some ambiguity.

Definition   Sine and Cosine Functions

Let t be a real number that determines the point $P(x, y)$ as indicated above. Then

$$\sin t = y \quad \text{and} \quad \cos t = x$$

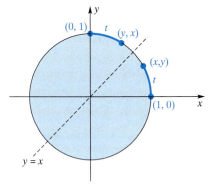

Figure 3

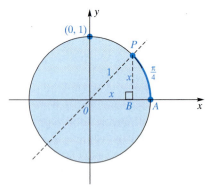

Figure 4

Basic Properties of Sine and Cosine A number of facts follow almost immediately from the definitions given above. First, since t can be any real number, the domain for both the sine and cosine functions is $\mathbb{R}$. Second, x and y are always between -1 and 1. Thus, the range for both the sine and cosine functions is the interval $[-1, 1]$.

Because the unit circle has circumference 2π, the values t and $t + 2\pi$ determine the *same* point $P(x, y)$. Thus,

$$\sin(t + 2\pi) = \sin t \quad \text{and} \quad \cos(t + 2\pi) = \cos t$$

(Notice that parentheses are needed to make it clear that we mean $\sin(t + 2\pi)$, rather than $(\sin t) + 2\pi$. The expression $\sin t + 2\pi$ would be ambiguous.)

The points P_1 and P_2 that correspond to t and $-t$, respectively, are symmetric about the x-axis (Figure 3). Thus, the x-coordinates for P_1 and P_2 are the same, and the y-coordinates differ only in sign. Consequently,

$$\sin(-t) = -\sin t \quad \text{and} \quad \cos(-t) = \cos t$$

In other words, sine is an odd function and cosine is an even function.

The points corresponding to t and $\pi/2 - t$ are symmetric with respect to the line $y = x$ and thus they have their coordinates interchanged (Figure 4). This means that

$$\sin\left(\frac{\pi}{2} - t\right) = \cos t \quad \text{and} \quad \cos\left(\frac{\pi}{2} - t\right) = \sin t$$

Finally, we mention an important identity connecting the sine and cosine function:

$$\sin^2 t + \cos^2 t = 1$$

for every real number t. This identity follows from the fact that the point (x, y) is on the unit circle, hence x and y satisfy $x^2 + y^2 = 1$.

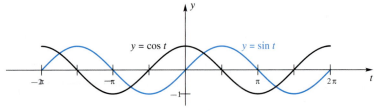

Figure 5

Graphs of Sine and Cosine To graph $y = \sin t$ and $y = \cos t$, we follow our usual procedure of making a table of values, plotting the corresponding points, and connecting these points with a smooth curve. So far, however, we know the values of sine and cosine for only a few values of t. A number of other values can be determined from geometric arguments. For example, if $t = \pi/4$, then t determines the point half of the way counterclockwise around the unit circle between the points $(1, 0)$ and $(0, 1)$. By symmetry, x and y will be on the line $y = x$, so $y = \sin t$ and $x = \cos t$ will be equal. Thus, the two legs of the right triangle OBP are equal, and the hypotenuse is 1 (Figure 5). The Pythagorean Theorem can be applied to give

$$1 = x^2 + x^2 = \cos^2 \frac{\pi}{4} + \cos^2 \frac{\pi}{4}$$

From this we conclude that $\cos(\pi/4) = 1/\sqrt{2} = \sqrt{2}/2$. Similarly, $\sin(\pi/4) = \sqrt{2}/2$. We can determine $\sin t$ and $\cos t$ for a number of other values of t. Some of these are shown in the following table. Using these results, along with a number of results from a calculator (in radian mode), we obtain the graphs shown in Figure 6.

t	$\sin t$	$\cos t$
0	0	1
$\pi/6$	$1/2$	$\sqrt{3}/2$
$\pi/4$	$\sqrt{2}/2$	$\sqrt{2}/2$
$\pi/3$	$\sqrt{3}/2$	$1/2$
$\pi/2$	1	0
$2\pi/3$	$\sqrt{3}/2$	$-1/2$
$3\pi/4$	$\sqrt{2}/2$	$-\sqrt{2}/2$
$5\pi/6$	$1/2$	$-\sqrt{3}/2$
π	0	-1

Figure 6

Four things are noticeable from these graphs:

1. Both $\sin t$ and $\cos t$ range from -1 to 1.

2. Both graphs repeat themselves on adjacent intervals of length 2π.

3. The graph of $y = \sin t$ is symmetric about the origin, and $y = \cos t$ is symmetric about the y-axis. (Thus, the sine function is odd and the cosine function is even.)

4. The graph of $y = \sin t$ is the same as that of $y = \cos t$, but translated $\pi/2$ units to the right.

The next example deals with functions of the form $\sin(at)$ or $\cos(at)$, which occur frequently in applications.

EXAMPLE 1   Sketch the graphs of

(a) $y = \sin(2\pi t)$ (b) $y = \cos(2t)$

Solution

(a) As t goes from 0 to 1, the argument $2\pi t$ goes from 0 to 2π. Thus, the graph of this function will repeat itself on adjacent intervals of length 1. From the entries in the following table, we can sketch a graph of $y = \sin(2\pi t)$.

t	$\sin(2\pi t)$	t	$\sin(2\pi t)$
0	$\sin(2\pi \cdot 0) = 0$	$\dfrac{5}{8}$	$\sin\left(2\pi \cdot \dfrac{5}{8}\right) = -\dfrac{\sqrt{2}}{2}$
$\dfrac{1}{8}$	$\sin\left(2\pi \dfrac{1}{8}\right) = \dfrac{\sqrt{2}}{2}$	$\dfrac{3}{4}$	$\sin\left(2\pi \cdot \dfrac{3}{4}\right) = -1$
$\dfrac{1}{4}$	$\sin\left(2\pi \dfrac{1}{4}\right) = 1$	$\dfrac{7}{8}$	$\sin\left(2\pi \cdot \dfrac{7}{8}\right) = -\dfrac{\sqrt{2}}{2}$
$\dfrac{3}{8}$	$\sin\left(2\pi \dfrac{3}{8}\right) = \dfrac{\sqrt{2}}{2}$	1	$\sin(2\pi \cdot 1) = 0$
$\dfrac{1}{2}$	$\sin\left(2\pi \dfrac{1}{2}\right) = 0$	$\dfrac{9}{8}$	$\sin\left(2\pi \dfrac{9}{8}\right) = \dfrac{\sqrt{2}}{2}$

Figure 7 shows a sketch of the graph of $y = \sin(2\pi t)$.

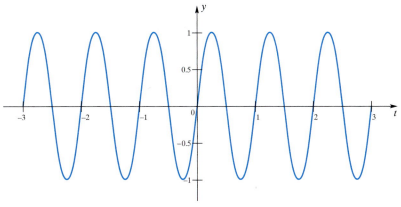

Figure 7

(b) As t goes from 0 to π, the argument $2t$ goes from 0 to 2π. Thus, the graph of $y = \cos(2t)$ will repeat itself on adjacent intervals of length π. Once we construct a table we can sketch a plot of $y = \cos(2t)$. Figure 8 shows a graph of $y = \cos(2t)$.

t	$\cos(2t)$	t	$\cos(2t)$
0	$\cos(2 \cdot 0) = 1$	$\dfrac{5\pi}{8}$	$\cos\left(2 \cdot \dfrac{5\pi}{8}\right) = -\dfrac{\sqrt{2}}{2}$
$\dfrac{\pi}{8}$	$\cos\left(2 \cdot \dfrac{\pi}{8}\right) = \dfrac{\sqrt{2}}{2}$	$\dfrac{3\pi}{4}$	$\cos\left(2 \cdot \dfrac{3\pi}{4}\right) = 0$
$\dfrac{\pi}{4}$	$\cos\left(2 \cdot \dfrac{\pi}{4}\right) = 0$	$\dfrac{7\pi}{8}$	$\cos\left(2 \cdot \dfrac{7\pi}{8}\right) = \dfrac{\sqrt{2}}{2}$
$\dfrac{3\pi}{8}$	$\cos\left(2 \cdot \dfrac{3\pi}{8}\right) = -\dfrac{\sqrt{2}}{2}$	π	$\cos(2 \cdot \pi) = 1$
$\dfrac{\pi}{2}$	$\cos\left(2 \cdot \dfrac{\pi}{2}\right) = -1$	$\dfrac{9\pi}{8}$	$\cos\left(2 \cdot \dfrac{9\pi}{8}\right) = \dfrac{\sqrt{2}}{2}$

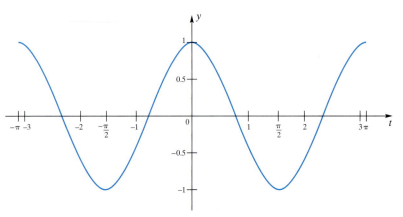

Figure 8

Period and Amplitude of the Trigonometric Functions

Period and Amplitude of the Trigonometric Functions   A function f is **periodic** if there is a number p such that

$$f(x + p) = f(x)$$

for all real numbers x in the domain of f. The smallest such positive number p is called the **period** of f. The sine function is periodic because $\sin(x + 2\pi) = \sin x$ for all x. It is also true that

$$\sin(x + 4\pi) = \sin(x - 2\pi) = \sin(x + 12\pi) = \sin x$$

for all x. Thus, 4π, -2π, and 12π are all numbers p with the property $\sin(x + p) = \sin x$. The period is defined to be the smallest such positive number p. For the sine function, the smallest positive p with the property that $\sin(x + p) = \sin x$ is $p = 2\pi$. We therefore say that the sine function is periodic with period 2π. The cosine function is also periodic with period 2π.

The function $\sin(at)$ has period $2\pi/a$ since

$$\sin\left[a\left(t + \frac{2\pi}{a}\right)\right] = \sin[at + 2\pi] = \sin(at)$$

The period of the function $\cos(at)$ is also $2\pi a$.

EXAMPLE 2   What are the periods of the following functions?

(a) $\sin(2\pi t)$ (b) $\cos(2t)$ (c) $\sin(2\pi t/12)$

Solution

(a) Because the function $\sin(2\pi t)$ is of the form $\sin(at)$ with $a = 2\pi$, its period is
$$p = \frac{2\pi}{2\pi} = 1.$$

(b) The function $\cos(2t)$ is of the form $\cos(at)$ with $a = 2$. Thus, the period of $\cos(2t)$ is $p = \dfrac{2\pi}{2} = \pi.$

(c) The function $\sin(2\pi t/12)$ has period $p = \dfrac{2\pi}{2\pi/12} = 12.$ ■

If the periodic function f attains a minimum and a maximum, we define the **amplitude** A as half the distance between the lowest point and the highest point on the graph.

EXAMPLE 3 Find the amplitude of the following periodic functions.

(a) $\sin(2\pi t/12)$ (b) $3\cos(2t)$

(c) $50 + 21\sin(2\pi t/12 + 3)$

Solution

(a) Since the range of the function $\sin(2\pi t/12)$ is $[-1, 1]$, its amplitude is $A = 1$.

(b) The function $3\cos(2t)$ will take on values from -3 (which occurs when $t = \pm\dfrac{\pi}{2}, \pm\dfrac{3\pi}{2}, \dots$) to 3 (which occurs when $t = 0, \pm\pi, \pm 2\pi, \dots$). The amplitude is therefore $A = 3$.

(c) The function $21\sin(2\pi t/12 + 3)$ takes on values from -21 to 21. Thus, $50 + 21\sin(2\pi t/12 + 3)$ takes on values from $50 - 21 = 29$ to $50 + 21 = 71$. The amplitude is therefore 21. ■

In general,

$$C + A\sin\big(a(t + b)\big) \text{ and } C + A\cos\big(a(t + b)\big) \text{ have period } \frac{2\pi}{a} \text{ and amplitude } A.$$

Trigonometric functions can be used to model a number of physical phenomena, including daily tide levels and yearly temperatures.

EXAMPLE 4 The normal high temperature for St. Louis, Missouri, ranges from 37°F for January 15 to 89°F for July 15. The average daily high temperature follows roughly a sinusoidal curve.

(a) Find values of C, A, a, and b such that
$$T(t) = C + A\sin\big(a(t + b)\big)$$
where t, expressed in months since January 1, is a reasonable model for the average daily high temperature.

(b) Use this model to approximate the average daily high temperature for May 15.

Solution

(a) The required function must have period $t = 12$ since the seasons repeat every 12 months. Thus, $\dfrac{2\pi}{a} = 12$, so we have $a = \dfrac{2\pi}{12}$. The amplitude is half the difference between the lowest and highest points; in this case, $A = \dfrac{1}{2}(89 - 37) = 26$. The value of C is equal to the midpoint of the low and high temperatures, so $C = \dfrac{1}{2}(89 + 37) = 63$. The function $T(t)$ must therefore be of the form

$$T(t) = 63 + 26\sin\left(\frac{2\pi}{12}(t + b)\right)$$

The only constant left to find is b. The lowest normal high temperature is 37, which occurs on January 15, roughly in the middle of January. Thus, our function must satisfy $T(1/2) = 37$, and the function must reach its minimum of 37 when $t = 1/2$. Figure 9 summarizes the information that we have so far. The function $63 + 26\sin(2\pi t/12)$ reaches its minimum when $2\pi t/12 = -\pi/2$, that is, when $t = -3$. We must therefore translate the curve defined by $y = 63 + 26\sin(2\pi t/12)$ to the right by the amount $1/2 - (-3) = 7/2$. In Section 2.2 we showed that replacing x with $x - c$ translates the graph of $y = f(x)$ to the right by c units. Thus, in order to translate the graph of $y = 63 + 26\sin(2\pi t/12)$ to the right by $7/2$ units, we must replace t with $t - 7/2$. Thus,

$$T(t) = 63 + 26\sin\left(\frac{2\pi}{12}\left(t - \frac{7}{2}\right)\right)$$

Figure 10 shows a plot of the normal high temperature T as a function of time t, where t is given in months.

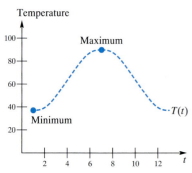

Temperature

Figure 9

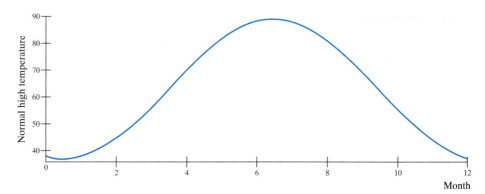

Normal high temperature

Month

Figure 10

(b) To estimate the normal high temperature in May, we substitute $t = 4.5$ (because the middle of May is four and one-half months into the year) and obtain

$$T(4.5) = 63 + 26\sin\left(2\pi(4.5 - 3.5)/12\right) \approx 76.0$$

The normal high temperature for St. Louis on May 15 is actually 75°F. Thus, our model overpredicts by 1°, which is remarkably accurate considering how little information was given. This brings up the important point that all models such as this are simplifications of reality. (That is why they are called *models*.) Although such models are inherently simplifications of reality, many of them are still useful for prediction. ∎

Four Other Trigonometric Functions   We could get by with just the sine and cosine functions, but it is convenient to introduce four additional trigonometric functions: tangent, cotangent, secant, and cosecant.

$$\tan t = \frac{\sin t}{\cos t} \qquad \cot t = \frac{\cos t}{\sin t}$$

$$\sec t = \frac{1}{\cos t} \qquad \csc t = \frac{1}{\sin t}$$

What we know about sine and cosine will automatically give us knowledge about these four new functions.

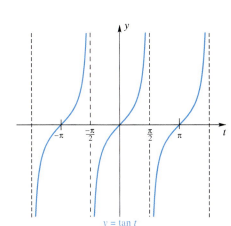

Figure 11

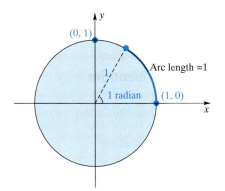

Figure 12

Degrees	Radians
0	0
30	$\pi/6$
45	$\pi/4$
60	$\pi/3$
90	$\pi/2$
120	$2\pi/3$
135	$3\pi/4$
150	$5\pi/6$
180	π
360	2π

Figure 13

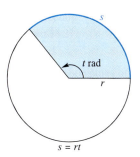

Figure 14

EXAMPLE 5   Show that tangent is an odd function.

Solution

$$\tan(-t) = \frac{\sin(-t)}{\cos(-t)} = \frac{-\sin t}{\cos t} = -\tan t$$ ∎

EXAMPLE 6   Verify that the following are identities.

$$1 + \tan^2 t = \sec^2 t \qquad 1 + \cot^2 t = \csc^2 t$$

Solution

$$1 + \tan^2 t = 1 + \frac{\sin^2 t}{\cos^2 t} = \frac{\cos^2 t + \sin^2 t}{\cos^2 t} = \frac{1}{\cos^2 t} = \sec^2 t$$

$$1 + \cot^2 t = 1 + \frac{\cos^2 t}{\sin^2 t} = \frac{\sin^2 t + \cos^2 t}{\sin^2 t} = \frac{1}{\sin^2 t} = \csc^2 t$$ ∎

When we study the tangent function (Figure 11), we are in for two minor surprises. First, we notice that there are vertical asymptotes at $\pm\pi/2, \pm3\pi/2$, and so on. We should have anticipated this since $\cos t = 0$ at these values of t, which means that $(\sin t)/(\cos t)$ would involve a division by zero. Second, it appears that the tangent is periodic (which we expected), but with period π (which we might not have expected). You will see the analytic reason for this in Problem 33.

Relation to Angle Trigonometry   Angles are commonly measured either in degrees or in radians. One radian is by definition the angle corresponding to an arc of length 1 on the unit circle. See Figure 12. The angle corresponding to a complete revolution measures 360°, but only 2π radians. Equivalently, a straight angle measures 180° or π radians, a fact worth remembering.

$$180° = \pi \text{ radians} \approx 3.1415927 \text{ radians}$$

This leads to the results

$$1 \text{ radian} \approx 57.29578° \qquad 1° \approx 0.0174533 \text{ radian}$$

Figure 13 shows some other common conversions between degrees and radians.

The division of a revolution into 360 parts is quite arbitrary (due to the ancient Babylonians, who liked multiples of 60). The division into 2π parts is more fundamental and lies behind the almost universal use of radian measure in calculus. Notice, in particular, that the length s of the arc cut off on a circle of radius r by a central angle of t radians satisfies (see Figure 14)

$$\frac{s}{2\pi r} = \frac{t}{2\pi}$$

That is, the fraction of the total circumference $2\pi r$ corresponding to an angle t is the same as the fraction of the unit circle corresponding to the same angle t. This implies that $s = rt$.

When $r = 1$, this gives $s = t$. This means that *the length of arc on the unit circle cut off by a central angle of t radians is t.* This is correct even if t is negative, provided that we interpret length to be negative when measured in the clockwise direction.

EXAMPLE 7   Find the distance traveled by a bicycle with wheels of radius 30 centimeters when the wheels turn through 100 revolutions.

Solution   We use the fact that $s = rt$, recognizing that 100 revolutions correspond to $100 \cdot (2\pi)$ radians.

$$s = (30)(100)(2\pi) = 6000\pi$$

$$\approx 18849.6 \text{ centimeters}$$

$$\approx 188.5 \text{ meters}$$ ∎

(a) $y = \sin\left(x + \dfrac{\pi}{2}\right)$ (b) $y = \cos\left(x + \dfrac{\pi}{2}\right)$

(c) $y = -\sin(x + \pi)$ (d) $y = \cos(x - \pi)$

(e) $y = -\sin(\pi - x)$ (f) $y = \cos\left(x - \dfrac{\pi}{2}\right)$

(g) $y = -\cos(\pi - x)$ (h) $y = \sin\left(x - \dfrac{\pi}{2}\right)$

25. Which of the following are odd functions? Even functions? Neither?

(a) $t \sin t$ (b) $\sin^2 t$ (c) $\csc t$

(d) $|\sin t|$ (e) $\sin(\cos t)$ (f) $x + \sin x$

26. Which of the following are odd functions? Even functions? Neither?

(a) $\cot t + \sin t$ (b) $\sin^3 t$ (c) $\sec t$

(d) $\sqrt{\sin^4 t}$ (e) $\cos(\sin t)$ (f) $x^2 + \sin x$

Use the half-angle identities to find the exact values in Problems 27–31.

27. $\cos^2 \dfrac{\pi}{3} =$

28. $\sin^2 \dfrac{\pi}{6} =$

29. $\sin^3 \dfrac{\pi}{6} =$

30. $\cos^2 \dfrac{\pi}{12} =$

31. $\sin^2 \dfrac{\pi}{8} =$

32. Find identities analogous to the addition identities for each expression.

(a) $\sin(x - y)$ (b) $\cos(x - y)$ (c) $\tan(x - y)$

33. Use the addition identity for the tangent to show that $\tan(t + \pi) = \tan t$ for all t in the domain of $\tan t$.

34. Show that $\cos(x - \pi) = -\cos x$ for all x.

35. Suppose that a tire on a truck has an outer radius of 2.5 feet. How many revolutions per minute does the tire make when the truck is traveling 60 miles per hour?

36. How far does a wheel of radius 2 feet roll along level ground in making 150 revolutions? (See Example 3.)

37. A belt passes around two wheels, as shown in Figure 15. How many revolutions per second does the small wheel make when the large wheel makes 21 revolutions per second?

6 in.   8 in.

Figure 15

38. The **angle of inclination** α of a line is the smallest positive angle from the positive x-axis to the line ($\alpha = 0$ for a horizontal line). Show that the slope m of the line is equal to $\tan \alpha$.

39. Find the angle of inclination of the following lines (see Problem 38).

(a) $y = \sqrt{3}\, x - 7$ (b) $\sqrt{3}\, x + 3y = 6$

40. Let ℓ_1 and ℓ_2 be two nonvertical lines with slopes m_1 and m_2, respectively. If θ, the angle from ℓ_1 to ℓ_2, is not a right angle, then

$$\tan \theta = \frac{m_2 - m_1}{1 + m_1 m_2}$$

Show this using the fact that $\theta = \theta_2 - \theta_1$ in Figure 16.

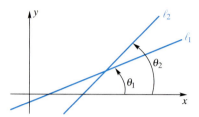

Figure 16

41. Find the angle (in radians) from the first line to the second (see Problem 40).

(a) $y = 2x,\ y = 3x$ (b) $y = \dfrac{x}{2},\ y = -x$

(c) $2x - 6y = 12,\ 2x + y = 0$

42. Derive the formula $A = \frac{1}{2} r^2 t$ for the area of a sector of a circle. Here r is the radius and t is the radian measure of the central angle (see Figure 17).

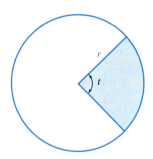

Figure 17

43. Find the area of the sector of a circle of radius 5 centimeters and central angle 2 radians (see Problem 42).

44. A regular polygon of n sides is inscribed in a circle of radius r. Find formulas for the perimeter, P, and area, A, of the polygon in terms of n and r.

45. An isosceles triangle is topped by a semicircle, as shown in Figure 18. Find a formula for the area A of the whole figure in terms of the side length r and angle t (radians). (We say that A is a function of the two independent variables r and t.)

Figure 18

46. From a product identity, we obtain

$$\cos \frac{x}{2} \cos \frac{x}{4} = \frac{1}{2}\left[\cos\left(\frac{3}{4}x\right) + \cos\left(\frac{1}{4}x\right)\right]$$

Find the corresponding sum of cosines for

$$\cos \frac{x}{2} \cos \frac{x}{4} \cos \frac{x}{8} \cos \frac{x}{16}$$

Do you see a generalization?

In Chapter 5 we will have occasion to study functions of an integer variable. In preparation for this material, we explore some aspects of functions of an integer variable in Problems 47–50.

47. Find a formula for the sum of the first n positive integers. *Hint:* We can generate a closed-form representation as a rational function of n for this sum,

$$S_1(n) = 1 + 2 + 3 + 4 + \cdots + n$$

by rearranging the terms in a decreasing order and then adding them to the original sum:

$$S_1(n) = 1 + 2 + 3 + \cdots + n$$

$$S_1(n) = n + (n-1) + (n-2) + \cdots + 1$$

48. Another simple formula may be derived for the integer function

$$S_2(n) = 1^2 + 2^2 + 3^2 + \cdots + n^2$$

by using the formula

$$(x+1)^3 - x^3 = 3x^2 + 3x + 1$$

(a) By setting x equal to the values of $0, 1, 2, 3, \ldots, n$ and adding the results, establish the formula

$$(n+1)^3 = 3S_2(n) + 3S_1(n) + n + 1$$

(b) Using the result derived in Problem 47 for $S_1(n)$, and solving for $S_2(n)$ in part (a), show that

$$S_2(n) = \frac{1}{6}n(n+1)(2n+1)$$

49. Find a formula for $S_3(n)$ using a similar technique.

50. Find a formula for $S_4(n)$ using a similar technique.

51. The normal high temperature for Las Vegas, Nevada, is 55°F for January 15 and 105° for July 15. Assuming that these are the extreme high and low temperatures for the year, use this information to approximate the average high temperature for November 15.

52. Tides are often measured by arbitrary height markings at some location. Suppose that a high tide occurs at noon when the water level is at 12 feet. Six hours later, a low tide with a water level of 5 feet occurs, and by midnight another high tide with a water level of 12 feet occurs. Assuming that the water level is periodic, use this information to find a formula that gives the water

level as a function of time. Then use this function to approximate the water level at 5:30 P.M.

EXPL 53. Circular motion can be modeled by using the parametric representations of the form $x(t) = \sin t$ and $y(t) = \cos t$. (A *parametric representation* means that a variable, t in this case, determines both $x(t)$ and $y(t)$.) This will give the full circle for $0 \le t \le 2\pi$. If we consider a 4-foot-diameter wheel rotating clockwise once every 10 seconds, show that the motion of a point on the rim of the wheel can be represented by $x(t) = 2\sin(\pi t/5)$ and $y(t) = 2\cos(\pi t/5)$.

(a) Find the positions of the point on the rim of the wheel when $t = 2$ seconds, 6 seconds, and 10 seconds. Where was this point when the wheel started to rotate at $t = 0$?

(b) How will the formulas giving the motion of the point change if the wheel is rotating *counterclockwise*.

(c) At what value of t is the point at $(2, 0)$?

EXPL 54. The circular frequency v of oscillation of a point is given by $v = \dfrac{2\pi}{\text{period}}$. What happens when you add two motions that have the same frequency or period? To investigate, we can graph the functions $y(t) = 2\sin(\pi t/5)$ and $y(t) = \sin(\pi t/5) + \cos(\pi t/5)$ and look for similarities. Armed with this information, we can investigate by graphing the following functions over the interval $[-5, 5]$:

(a) $y(t) = 3\sin(\pi t/5) - 5\cos(\pi t/5) + 2\sin((\pi t/5) - 3)$

(b) $y(t) = 3\cos(\pi t/5 - 2) + \cos(\pi t/5) + \cos((\pi t/5) - 3)$

EXPL 55. We now explore the relationship between $A\sin(\omega t) + B\cos(\omega t)$ and $C\sin(\omega t + \phi)$.

(a) By expanding $\sin(\omega t + \phi)$ using the sum of the angles formula, show that $A = C\cos\phi$ and $B = C\sin\phi$.

(b) Consequently, show that $A^2 + B^2 = C^2$ and that ϕ then satisfies the equation $\tan\phi = \dfrac{B}{A}$.

(c) Generalize your result to state a proposition about $A_1\sin(\omega t + \phi_1) + A_2\sin(\omega t + \phi_2) + A_3\sin(\omega t + \phi_3)$.

(d) Write an essay, in your own words, that expresses the importance of the identity between $A\sin(\omega t) + B\cos(\omega t)$ and $C\sin(\omega t + \phi)$. Be sure to note that $|C| \ge \max(|A|, |B|)$ and that the identity only holds when you are forming a linear combination (adding and/or subtracting multiples of single powers of) sine and cosine of the same frequency.

Trigonometric functions that have high frequencies pose special problems for graphing. We now explore how to plot such functions.

GC 56. Graph the function $f(x) = \sin 50x$ using the window given by a y range of $-1.5 \le y \le 1.5$ and the x range given by

(a) $[-15, 15]$ (b) $[-10, 10]$ (c) $[-8, 8]$

(d) $[-1, 1]$ (e) $[-0.25, 0.25]$

Indicate briefly which x-window shows the true behavior of the function, and discuss reasons why the other x-windows give results that are different.

57. Graph the function $f(x) = \cos x + \dfrac{1}{50}\sin 50x$ using the windows given by the following ranges of x and y.

(a) $-5 \leq x \leq 5, -1 \leq y \leq 1$

(b) $-1 \leq x \leq 1, 0.5 \leq y \leq 1.5$

(c) $-0.1 \leq x \leq 0.1, 0.9 \leq y \leq 1.1$

Indicate briefly which (x, y)-window shows the true behavior of the function, and discuss reasons why the other (x, y)-windows give results that are different. In this case is it true that only one window gives the important behavior, or do we need more than one window to graphically communicate the behavior of this function?

58. Let $f(x) = \dfrac{3x + 2}{x^2 + 1}$ and $g(x) = \dfrac{1}{100}\cos(100x)$.

(a) Use functional composition to form $h(x) = (f \circ g)(x)$, as well as $j(x) = (g \circ f)(x)$.

(b) Find the appropriate window or windows that give a clear picture of $h(x)$.

(c) Find the appropriate window or windows that give a clear picture of $j(x)$.

Answers to Concepts Review: **1.** $(-\infty, \infty); [-1, 1]$ **2.** $2\pi; 2\pi;$ π **3.** odd, even **4.** $-4/5$

2.4
Introduction to Limits

The topics discussed so far are part of what is called *precalculus*. They provide the foundation for calculus, but they are not calculus. Now we are ready for an important new idea, the notion of *limit*. It is this idea that distinguishes calculus from other branches of mathematics. In fact, we might define calculus as *the study of limits*.

An Intuitive Understanding Consider the function defined by

$$f(x) = \frac{x^3 - 1}{x - 1}$$

Note that it is not defined at $x = 1$ since at this point $f(x)$ has the form $\frac{0}{0}$, which is meaningless. We can, however, still ask what is happening to $f(x)$ as x approaches 1. More precisely, is $f(x)$ approaching some specific number as x approaches 1? To get at the answer, we can do three things. We can calculate some values of $f(x)$ for x near 1, we can show these values in a schematic diagram, and we can sketch the graph of $y = f(x)$. All this has been done, and the results are shown in Figure 1.

x	$y = \dfrac{x^3 - 1}{x - 1}$
1.25	3.813
1.1	3.310
1.01	3.030
1.001	3.003
↓	↓
1.000	?
↑	↑
0.999	2.997
0.99	2.970
0.9	2.710
0.75	2.313

Table
of values

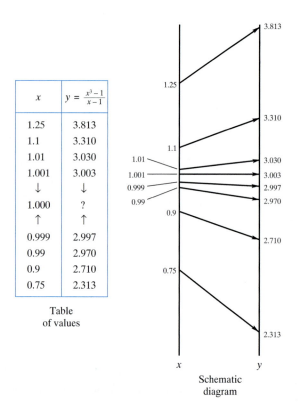

Schematic
diagram

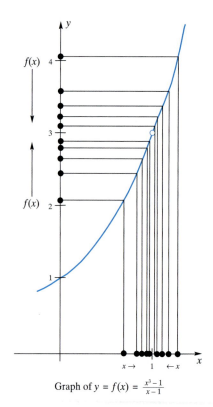

Graph of $y = f(x) = \dfrac{x^3 - 1}{x - 1}$

Figure 1

All the information we have assembled seems to point to the same conclusion: $f(x)$ approaches 3 as x approaches 1. In mathematical symbols, we write

$$\lim_{x \to 1} \frac{x^3 - 1}{x - 1} = 3$$

This is read "the limit as x approaches 1 of $(x^3 - 1)/(x - 1)$ is 3."

Being good algebraists (thus knowing how to factor the difference of cubes), we can provide more and better evidence.

$$\lim_{x \to 1} \frac{x^3 - 1}{x - 1} = \lim_{x \to 1} \frac{(x - 1)(x^2 + x + 1)}{x - 1}$$

$$= \lim_{x \to 1} (x^2 + x + 1) = 1^2 + 1 + 1 = 3$$

Note that $(x - 1)/(x - 1) = 1$ as long as $x \neq 1$. This justifies the second step. The third step should seem reasonable; a rigorous justification will come later.

To be sure that we are on the right track, we need to have a clearly understood meaning for the word *limit*. Here is our first attempt at a definition.

Definition   Intuitive Meaning of Limit

To say that $\lim_{x \to c} f(x) = L$ means that when x is near but different from c then $f(x)$ is near L.

Notice that we do not require anything *at c*. The function *f* need not even be defined at *c*; it was not in the example $f(x) = (x^3 - 1)/(x - 1)$ just considered. The notion of limit is associated with the behavior of a function *near c*, not *at c*.

A cautious reader is sure to object to our use of the word *near*. What does *near* mean? How near is near? For precise answers, you will have to study the next section; however, some further examples will help to clarify the idea.

More Examples   Our first example is almost trivial but is nonetheless important.

EXAMPLE 1   Find $\lim_{x \to 3} (4x - 5)$.

Solution   When x is near $3, 4x - 5$ is near $4 \cdot 3 - 5 = 7$. We write

$$\lim_{x \to 3} (4x - 5) = 7 \qquad \blacksquare$$

EXAMPLE 2   Find $\lim_{x \to 3} \dfrac{x^2 - x - 6}{x - 3}$.

Solution   Note that $(x^2 - x - 6)/(x - 3)$ is not defined at $x = 3$, but this is all right. To get an idea of what is happening as x approaches 3, we could use a calculator to evaluate the given expression, for example, at $3.1, 3.01, 3.001$, and so on. But it is much better to use a little algebra to simplify the problem.

$$\lim_{x \to 3} \frac{x^2 - x - 6}{x - 3} = \lim_{x \to 3} \frac{(x - 3)(x + 2)}{x - 3} = \lim_{x \to 3} (x + 2) = 3 + 2 = 5$$

The cancellation of $x - 3$ in the second step is legitimate because the definition of limit ignores the behavior *at* $x = 3$. Remember, $\dfrac{x - 3}{x - 3} = 1$ as long at x is not equal to 3. $\qquad \blacksquare$

EXAMPLE 3   Find $\lim_{x \to 1} \dfrac{x - 1}{\sqrt{x} - 1}$.

Solution

$$\lim_{x \to 1} \frac{x - 1}{\sqrt{x} - 1} = \lim_{x \to 1} \frac{(\sqrt{x} - 1)(\sqrt{x} + 1)}{\sqrt{x} - 1} = \lim_{x \to 1} (\sqrt{x} + 1) = \sqrt{1} + 1 = 2 \qquad \blacksquare$$

x	$\frac{\sin x}{x}$
1.0	0.84147
0.5	0.95885
0.1	0.99833
0.01	0.99998
↓	↓
0	?
↑	↑
−0.01	0.99998
−0.1	0.99833
−0.5	0.95885
−1.0	0.84147

Figure 2

EXAMPLE 4   Find $\lim\limits_{x \to 0} \dfrac{\sin x}{x}$.

Solution   No algebraic trick will simplify our task; certainly, we cannot cancel the x's. A calculator will help us to get an idea of the limit. Use your own calculator (radian mode) to check the values in the table of Figure 2. Figure 3 shows a plot of $y = \sin x / x$. Our conclusion, though we admit it is a bit shaky, is that

$$\lim_{x \to 0} \frac{\sin x}{x} = 1$$

We will give a rigorous demonstration in Section 2.7.

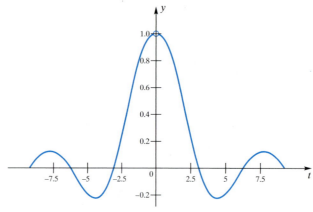

Figure 3

Some Warning Flags   Things are not quite as simple as they may appear. Calculators may mislead us; so may our own intuition. The examples that follow suggest some possible pitfalls.

x	$x^2 - \frac{\cos x}{10000}$
± 1	0.99995
± 0.5	0.24991
± 0.1	0.00990
± 0.01	0.000000005
↓	↓
0	?

Figure 4

EXAMPLE 5   **(Your calculator may fool you.)** Find $\lim\limits_{x \to 0} \left[x^2 - \dfrac{\cos x}{10{,}000} \right]$.

Solution   Following the procedure used in Example 4, we construct the table of values shown in Figure 4. The conclusion it suggests is that the desired limit is 0. But this is wrong. If we recall the graph of $y = \cos x$, we realize that $\cos x$ approaches 1 as x approaches 0. Thus,

$$\lim_{x \to 0} \left[x^2 - \frac{\cos x}{10{,}000} \right] = 0^2 - \frac{1}{10{,}000} = -\frac{1}{10{,}000}$$

EXAMPLE 6   **(No limit at a jump)** Find $\lim\limits_{x \to 2} [\![x]\!]$.

Solution   Recall that $[\![x]\!]$ denotes the greatest integer less than or equal to x (see Section 2.1). The graph of $y = [\![x]\!]$ is shown in Figure 5. For all numbers x less than 2 but near 2, $[\![x]\!] = 1$, but for all numbers x greater than 2 but near 2, $[\![x]\!] = 2$. Is $[\![x]\!]$ near to a single number L when x is near 2? No. No matter what number we propose for L, there will be x's arbitrarily close to 2 on one side or the other, where $[\![x]\!]$ differs from L by at least $\frac{1}{2}$. Our conclusion is that $\lim\limits_{x \to 2} [\![x]\!]$ does not exist. If you check back, you will see that we have not claimed that every limit we can write must exist.

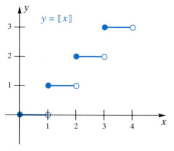

Figure 5

EXAMPLE 7   **(Too many wiggles)** Find $\lim\limits_{x \to 0} \sin(1/x)$.

Solution   This example poses the most subtle limit question asked yet. Since we do not want to make too big a story out of it, we ask you to do two things. First, pick a sequence of x-values approaching 0. Use your calculator to evaluate $\sin(1/x)$ at

x	$\sin\frac{1}{x}$
$2/\pi$	1
$2/(2\pi)$	0
$2/(3\pi)$	−1
$2/(4\pi)$	0
$2/(5\pi)$	1
$2/(6\pi)$	0
$2/(7\pi)$	−1
$2/(8\pi)$	0
$2/(9\pi)$	1
$2/(10\pi)$	0
$2/(11\pi)$	−1
$2/(12\pi)$	0
↓	↓
0	?

Figure 6

these x's. Unless you happen on some very lucky choices, your values will oscillate wildly.

Second, consider trying to graph $y = \sin(1/x)$. No one will ever do this very well, but the table of values in Figure 6 gives a good clue about what is happening. In any neighborhood of the origin, the graph wiggles up and down between −1 and 1 infinitely many times (Figure 7). Clearly, $\sin(1/x)$ is not near a single number L when x is near 0. We conclude that $\lim_{x\to 0} \sin(1/x)$ does not exist. ■

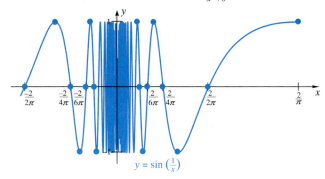

$$y = \sin\left(\frac{1}{x}\right)$$

Figure 7

One-Sided Limits   When a function takes a jump (as does $[\![x]\!]$ at each integer in Example 6), then the limit does not exist at the jump points. For such functions, it is natural to introduce **one-sided limits**. Let the symbol $x \to c^+$ mean that x approaches c from the right, and let $x \to c^-$ mean that x approaches c from the left.

Definition   Right- and Left-Hand Limits

To say that $\lim_{x\to c^+} f(x) = L$ means that when x is near but on the right of c then $f(x)$ is near L. Similarly, to say that $\lim_{x\to c^-} f(x) = L$ means that when x is near but on the left of c then $f(x)$ is near L.

Thus, while $\lim_{x\to 2}[\![x]\!]$ does not exist, it is correct to write (look at the graph in Figure 5)

$$\lim_{x\to 2^-}[\![x]\!] = 1 \quad \text{and} \quad \lim_{x\to 2^+}[\![x]\!] = 2$$

We believe that you will find the following theorem quite reasonable.

Theorem A

$\lim_{x\to c} f(x) = L$ if and only if $\lim_{x\to c^-} f(x) = L$ and $\lim_{x\to c^+} f(x) = L$.

Figure 8 should give additional insight. Two of the limits do not exist, although all but one of the one-sided limits exist.

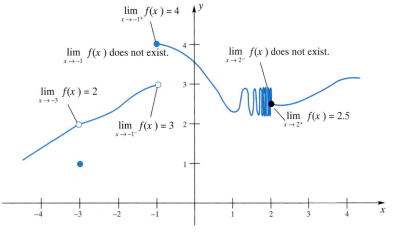

Figure 8

Concepts Review

1. $\lim_{x \to c} f(x) = L$ means that $f(x)$ gets near to _____ when x gets sufficiently near to (but is different from) _____.

2. Let $f(x) = (x^2 - 9)/(x - 3)$ and note that $f(3)$ is undefined. Nevertheless, $\lim_{x \to 3} f(x) =$ _____.

3. $\lim_{x \to c^+} f(x) = L$ means that $f(x)$ gets near to _____ when x approaches c from the _____.

4. If both $\lim_{x \to c^-} f(x) = M$ and $\lim_{x \to c^-} f(x) = M$, then _____.

Problem Set 2.4

In Problems 1–6, find the indicated limit.

1. $\lim_{x \to 3} (x - 5)$

2. $\lim_{t \to -1} (1 - 2t)$

3. $\lim_{x \to -2} (x^2 + 2x - 1)$

4. $\lim_{x \to -2} (x^2 + 2t - 1)$

5. $\lim_{t \to -1} \dfrac{1 - 2t}{\sqrt{3t + 21}}$

6. $\lim_{t \to -1} \dfrac{\sqrt{1 - 2t}}{(3t + 2)^3}$

In Problems 7–18, find the indicated limit. In most cases, it will be wise to do some algebra first (see Examples 2 and 3).

7. $\lim_{x \to 2} \dfrac{x^2 - 4}{x - 2}$

8. $\lim_{t \to -7} \dfrac{t^2 + 4t - 21}{t + 7}$

9. $\lim_{x \to -1} \dfrac{x^3 - 4x^2 + x + 6}{x + 1}$

10. $\lim_{x \to 0} \dfrac{x^4 + 2x^3 - x^2}{x^2}$

11. $\lim_{x \to -t} \dfrac{x^2 - t^2}{x + t}$

12. $\lim_{x \to 3} \dfrac{x^2 - 9}{x - 3}$

13. $\lim_{t \to 2} \dfrac{\sqrt{(t + 4)(t - 2)^4}}{(3t - 6)^2}$

14. $\lim_{t \to 7} \dfrac{\sqrt{(t - 7)^3}}{t - 7}$

15. $\lim_{x \to 3} \dfrac{x^4 - 18x^2 + 81}{(x - 3)^2}$

16. $\lim_{u \to 1} \dfrac{(3u + 4)(2u - 2)^3}{(u - 1)^2}$

17. $\lim_{h \to 0} \dfrac{(2 + h)^2 - 4}{h}$

18. $\lim_{h \to 0} \dfrac{(x + h)^2 - x^2}{h}$

GC *In Problems 19–28, use a calculator to find the indicated limit. Use a graphing calculator to plot the function near the limit point.*

19. $\lim_{x \to 0} \dfrac{\sin x}{2x}$

20. $\lim_{t \to 0} \dfrac{1 - \cos t}{2t}$

21. $\lim_{x \to 0} \dfrac{(x - \sin x)^2}{x^2}$

22. $\lim_{x \to 0} \dfrac{(1 - \cos x)^2}{x^2}$

23. $\lim_{t \to 1} \dfrac{t^2 - 1}{\sin(t - 1)}$

24. $\lim_{x \to 3} \dfrac{x - \sin(x - 3) - 3}{x - 3}$

25. $\lim_{x \to \pi} \dfrac{1 + \sin(x - 3\pi/2)}{x - \pi}$

26. $\lim_{t \to 0} \dfrac{1 - \cot t}{\dfrac{1}{t}}$

27. $\lim_{x \to \frac{\pi}{4}} \dfrac{(x - \pi/4)^2}{(\tan x - 1)^2}$

28. $\lim_{u \to \frac{\pi}{2}} \dfrac{2 - 2\sin u}{3u}$

29. For the function f graphed in Figure 9, find the indicated limit or function value, or state that it does not exist.

(a) $\lim_{x \to -3} f(x)$

(b) $f(-3)$

(c) $f(-1)$

(d) $\lim_{x \to -1} f(x)$

(e) $f(1)$

(f) $\lim_{x \to 1} f(x)$

(g) $\lim_{x \to 1^-} f(x)$

(h) $\lim_{x \to 1^+} f(x)$

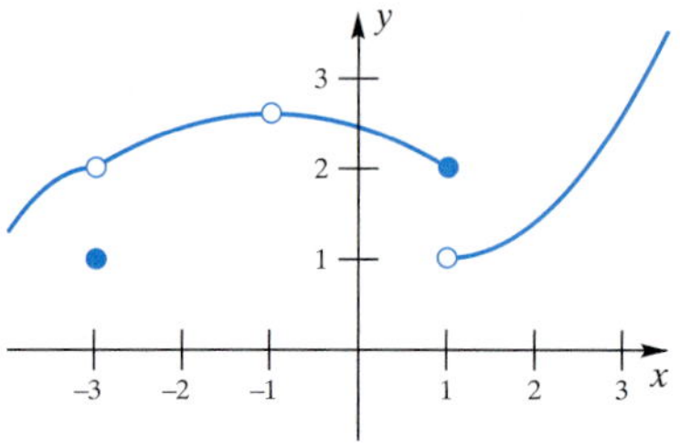

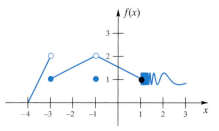

Figure 9

30. Follow the directions of Problem 29 for the function f graphed in Figure 10.

Figure 10

31. Sketch the graph of

$$f(x) = \begin{cases} -x & \text{if } x < 0 \\ x & \text{if } 0 \le x < 1 \\ 1 + x & \text{if } x \ge 1 \end{cases}$$

Then find each of the following or state that it does not exist.

(a) $\lim_{x \to 0} f(x)$

(b) $\lim_{x \to 1} f(x)$

(c) $f(1)$

(d) $\lim_{x \to 1^+} f(x)$

32. Sketch the graph of

$$g(x) = \begin{cases} -x + 1 & \text{if } x < 1 \\ x - 1 & \text{if } 1 < x < 2 \\ 5 - x^2 & \text{if } x \ge 2 \end{cases}$$

Then find each of the following or state that it does not exist.

(a) $\lim_{x \to 1} g(x)$

(b) $g(1)$

(c) $\lim_{x \to 2} g(x)$

(d) $\lim_{x \to 2^+} g(x)$

33. Sketch the graph of $f(x) = x - [\![x]\!]$; then find each of the following or state that it does not exist.

(a) $f(0)$

(b) $\lim_{x \to 0} f(x)$

(c) $\lim_{x \to 0^-} f(x)$

(d) $\lim_{x \to 1/2} f(x)$

Now sup
|(2x + 1) −
soning used a

Thus, |(2x +
This kind
is 7. The defin
$\varepsilon > 0$). Emily
limit is 7. Da
small).
David op
itive real num
instead of 0.0

David can c
$|x − 3| < \varepsilon/2$
in $\varepsilon/2$ of 3. N
has therefore

Some Limit
with what we
of δ in each p
the kind of w
through the p
it, but cover u
proof seems t

EXAMPLE 1

PRELIMINARY
such that

Consider the i

Now we see he

FORMAL PROC
plies that

|(3x

If you read thi
transitive prop

---

34. Follow the directions of Problem 33 for $f(x) = x/|x|$.

35. Find $\lim_{x \to 1} (x^2 − 1)/|x − 1|$ or state that it does not exist.

36. Evaluate $\lim_{x \to 0} (\sqrt{x + 2} − \sqrt{2})/x$. *Hint:* Rationalize the numerator by multiplying the numerator and denominator by $\sqrt{x + 2} + \sqrt{2}$.

37. Let

$$f(x) = \begin{cases} x & \text{if } x \text{ is rational} \\ -x & \text{if } x \text{ is irrational} \end{cases}$$

Find each value, if possible.

(a) $\lim_{x \to 1} f(x)$ (b) $\lim_{x \to 0} f(x)$

38. Sketch, as best you can, the graph of a function f that satisfies all the following conditions.
(a) Its domain is the interval $[0, 4]$.
(b) $f(0) = f(1) = f(2) = f(3) = f(4) = 1$
(c) $\lim_{x \to 1} f(x) = 2$ (d) $\lim_{x \to 2} f(x) = 1$
(e) $\lim_{x \to 3^-} f(x) = 2$ (f) $\lim_{x \to 3^+} f(x) = 1$

39. Let

$$f(x) = \begin{cases} x^2 & \text{if } x \text{ is rational} \\ x^4 & \text{if } x \text{ is irrational} \end{cases}$$

For what values of a does $\lim_{x \to a} f(x)$ exist?

40. The function $f(x) = x^2$ had been carefully graphed, but during the night a mysterious visitor changed the values of f at a million different places. Did this affect the value of $\lim_{x \to a} f(x)$ at any a? Explain.

41. Find each of the following limits or state that it does not exist.

(a) $\lim_{x \to 1} \dfrac{|x − 1|}{x − 1}$ (b) $\lim_{x \to 1^-} \dfrac{|x − 1|}{x − 1}$

(c) $\lim_{x \to 1^-} \dfrac{x^2 − |x − 1| − 1}{|x − 1|}$ (d) $\lim_{x \to 1^-} \left[\dfrac{1}{x − 1} − \dfrac{1}{|x − 1|} \right]$

---

42. Find each of the following limits or state that it does not exist.

(a) $\lim_{x \to 1^+} \sqrt{x − [\![x]\!]}$ (b) $\lim_{x \to 0^+} [\![1/x]\!]$

(c) $\lim_{x \to 0^+} x(-1)^{[\![1/x]\!]}$ (d) $\lim_{x \to 0^+} [\![x]\!](-1)^{[\![1/x]\!]}$

(e) $\lim_{x \to 0^+} x[\![1/x]\!]$ (f) $\lim_{x \to 0^+} x^2[\![1/x]\!]$

CAS *Many software packages have programs for calculating limits, although you should be warned that they are not infallible. To develop confidence in your program, use it to recalculate some of the limits in Problems 1–28. Then for each of the following find the limit or state that it does not exist.*

43. $\lim_{x \to 0} \sqrt{x}$ **44.** $\lim_{x \to 0^+} x^x$

45. $\lim_{x \to 0} \sqrt{|x|}$ **46.** $\lim_{x \to 0} |x|^x$

47. $\lim_{x \to 0} (\sin 2x)/4x$ **48.** $\lim_{x \to 0} (\sin 5x)/3x$

49. $\lim_{x \to 0} \cos(1/x)$ **50.** $\lim_{x \to 0} x \cos(1/x)$

51. $\lim_{x \to 1} \dfrac{x^3 − 1}{\sqrt{2x + 2} − 2}$ **52.** $\lim_{x \to 0} \dfrac{x \sin 2x}{\sin(x^2)}$

53. $\lim_{x \to 2^-} \dfrac{x^2 − x − 2}{|x − 2|}$ **54.** $\lim_{x \to 1^+} \dfrac{2}{1 + 2^{1/(x−1)}}$

CAS **55.** Since calculus software packages find $\lim_{x \to a} f(x)$ by sampling a few values of $f(x)$ for x near a, they can be fooled. Find a function f for which $\lim_{x \to 0} f(x)$ fails to exist but for which your software gives a value for the limit.

Answers to Concepts Review: **1.** $L; c$ **2.** 6 **3.** L; right
4. $\lim_{x \to c} f(x) = M$

Two Different Limits?

A natural question to ask is "Can a function have two different limits at c?" The obvious intuitive answer is no. If a function is getting closer and closer to L as $x \to c$, it cannot also be getting closer and closer to a different number M. You are asked to show this rigorously in Problem 19.

2.5
Rigorous Study
of Limits

You should not believe everything that you are told. It is prudent to be skeptical— not so skeptical that you will not believe anything, but skeptical enough to check a statement before you accept it. Mathematicians tend to be very skeptical people. Tell a mathematician that something is true and you will probably get the response: Prove it. But to be able to prove something requires that we be very clear about the meaning of the words that we are using. This is especially true of the word *limit*, because all of calculus rests on the meaning of that word.

We gave an informal definition of limit in the previous section. Here is a slightly better, but still informal, rewording of that definition. *To say that* $\lim_{x \to c} f(x) = L$

means that the difference between $f(x)$ and L can be made arbitrarily small by restricting x to be sufficiently close to but different from c. Now let us try to pin this down.

Making the Definition Precise First, we follow a long tradition in using the Greek letters ε (epsilon) and δ (delta) to stand for arbitrary positive numbers. Think of ε and δ as small positive numbers.

To say that $f(x)$ differs from L by less than ε is to say that

$$|f(x) − L| < \varepsilon$$

or, equivalently,

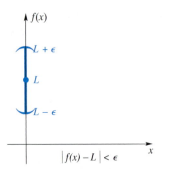

Figure 1

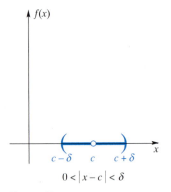

Figure 2

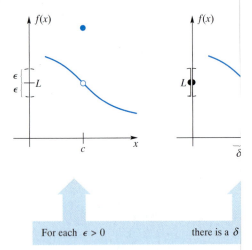

Figure 3

This means
in Figure 1.
   Next, to
some δ, x is i
to say this is

Note that |.
$0 < |x - c|$
scribing is sh
   We are a
nition in calc

Definition

To say that l
there is a
$0 < |x - c|$

The pictures

We mus
to be produc
to Emily tha

chooses (e.g
apply David

jecture that
whenever 0

Thus, the ar
smaller valu
$0 < |x - 3|$
vided that x

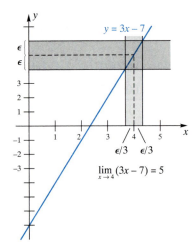

$$\lim_{x \to 4} (3x - 7) = 5$$

Figure 4

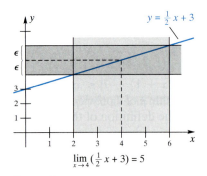

$$\lim_{x \to 4} \left(\tfrac{1}{2} x + 3\right) = 5$$

Figure 5

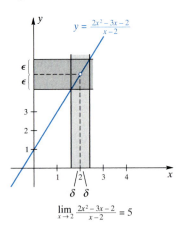

$$\lim_{x \to 2} \frac{2x^2 - 3x - 2}{x - 2} = 5$$

Figure 6

Now, David knows a rule for choosing the value of δ given Emily's challenge. If Emily were to challenge David with $\varepsilon = 0.01$, then David would respond with $\delta = 0.01/3$. If Emily said $\varepsilon = 0.000003$, then David would say $\delta = 0.000001$. If he gave a smaller value for δ, that would be fine, too.

Of course, if you think about the graph of $y = 3x - 7$ (a line with slope 3, as in Figure 4), you know that to force $3x - 7$ to be close to 5 you had better make x even closer (closer by a factor of one-third) to 4. ∎

Look at Figure 5. Then convince yourself that $\delta = 2\varepsilon$ would be an appropriate choice for δ in showing that $\lim_{x \to 4} \left(\tfrac{1}{2} x + 3\right) = 5$.

EXAMPLE 2   Prove that $\lim_{x \to 2} \dfrac{2x^2 - 3x - 2}{x - 2} = 5$.

PRELIMINARY ANALYSIS   We are looking for δ such that

$$0 < |x - 2| < \delta \Rightarrow \left| \frac{2x^2 - 3x - 2}{x - 2} - 5 \right| < \varepsilon$$

Now, for $x \neq 2$,

$$\left| \frac{2x^2 - 3x - 2}{x - 2} - 5 \right| < \varepsilon \Leftrightarrow \left| \frac{(2x + 1)(x - 2)}{x - 2} - 5 \right| < \varepsilon$$
$$\Leftrightarrow |(2x + 1) - 5| < \varepsilon$$
$$\Leftrightarrow |2(x - 2)| < \varepsilon$$
$$\Leftrightarrow |2||x - 2| < \varepsilon$$
$$\Leftrightarrow |x - 2| < \frac{\varepsilon}{2}$$

This indicates that $\delta = \varepsilon/2$ will work (see Figure 6).

FORMAL PROOF   Let $\varepsilon > 0$ be given. Choose $\delta = \varepsilon/2$. Then $0 < |x - 2| < \delta$ implies that

$$\left| \frac{2x^2 - 3x - 2}{x - 2} - 5 \right| = \left| \frac{(2x + 1)(x - 2)}{x - 2} - 5 \right| = |2x + 1 - 5|$$
$$= |2(x - 2)| = 2|x - 2| < 2\delta = \varepsilon$$

The cancellation of the factor $x - 2$ is legitimate because $0 < |x - 2|$ implies that $x \neq 2$, and $\dfrac{x - 2}{x - 2} = 1$ as long as $x \neq 2$. ∎

EXAMPLE 3   Prove that $\lim_{x \to c} (mx + b) = mc + b$.

PRELIMINARY ANALYSIS   We want to find δ such that

$$0 < |x - c| < \delta \Rightarrow |(mx + b) - (mc + b)| < \varepsilon$$

Now

$$|(mx + b) - (mc + b)| = |mx - mc| = |m(x - c)| = |m||x - c|$$

It appears that $\delta = \varepsilon/|m|$ should do as long as $m \neq 0$. (Note that m could be positive or negative, so we need to keep the absolute bars. Recall from Chapter 1 that $|ab| = |a||b|$.)

FORMAL PROOF   Let $\varepsilon > 0$ be given. Choose $\delta = \varepsilon/|m|$. Then $0 < |x - c| < \delta$ implies that

$$|(mx + b) - (mc + b)| = |mx - mc| = |m||x - c| < |m|\delta = \varepsilon$$

And in case $m = 0$, any δ will do just fine since

$$|(0x + b) - (0c + b)| = |0| = 0$$

The latter is less than ε for all x.   ∎

EXAMPLE 4   Prove that if $c > 0$ then $\lim\limits_{x \to c} \sqrt{x} = \sqrt{c}$.

PRELIMINARY ANALYSIS   Refer to Figure 7. We must find δ such that

$$0 < |x - c| < \delta \Rightarrow |\sqrt{x} - \sqrt{c}| < \varepsilon$$

Now

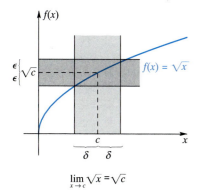

$$|\sqrt{x} - \sqrt{c}| = \left|\frac{(\sqrt{x} - \sqrt{c})(\sqrt{x} + \sqrt{c})}{\sqrt{x} + \sqrt{c}}\right| = \left|\frac{x - c}{\sqrt{x} + \sqrt{c}}\right|$$

$$= \frac{|x - c|}{\sqrt{x} + \sqrt{c}} \leq \frac{|x - c|}{\sqrt{c}}$$

To make the latter less than ε requires that we have $|x - c| < \varepsilon\sqrt{c}$.

FORMAL PROOF   Let $\varepsilon > 0$ be given. Choose $\delta = \varepsilon\sqrt{c}$. Then $0 < |x - c| < \delta$ implies that

$$|\sqrt{x} - \sqrt{c}| = \left|\frac{(\sqrt{x} - \sqrt{c})(\sqrt{x} + \sqrt{c})}{\sqrt{x} + \sqrt{c}}\right| = \left|\frac{x - c}{\sqrt{x} + \sqrt{c}}\right|$$

$$= \frac{|x - c|}{\sqrt{x} + \sqrt{c}} \leq \frac{|x - c|}{\sqrt{c}} < \frac{\delta}{\sqrt{c}} = \varepsilon$$

There is one technical point here. We began with $c > 0$, but it could happen that c sits very close to 0 on the x-axis. We should insist that $\delta \leq c$, for then $|x - c| < \delta$ implies that $x > 0$ so that $\sqrt{x}$ is defined. Thus, for absolute rigor, choose δ to be the smaller of c and $\varepsilon\sqrt{c}$.   ∎

Our demonstration in Example 4 depended on *rationalizing the numerator*, a trick frequently useful in calculus.

EXAMPLE 5   Prove that $\lim\limits_{x \to 3}(x^2 + x - 5) = 7$.

PRELIMINARY ANALYSIS   Our task is to find δ such that

$$0 < |x - 3| < \delta \Rightarrow |(x^2 + x - 5) - 7| < \varepsilon$$

Now

$$|(x^2 + x - 5) - 7| = |x^2 + x - 12| = |x + 4||x - 3|$$

The factor $|x - 3|$ can be made as small as we wish and we know that $|x + 4|$ will be about 7. We therefore seek an upper bound for $|x + 4|$. To do this, we first agree to make $\delta \leq 1$. Then $|x - 3| < \delta$ implies that

$$|x + 4| = |x - 3 + 7|$$

$$\leq |x - 3| + |7| \qquad \text{(Triangle Inequality)}$$

$$< 1 + 7 = 8$$

(Figure 8 offers an alternative demonstration of this fact.) If we also require that $\delta \leq \varepsilon/8$, the product $|x + 4||x - 3|$ will be less than ε.

FORMAL PROOF   Let $\varepsilon > 0$ be given. Choose $\delta = \min\{1, \varepsilon/8\}$; that is, choose δ to be the smaller of 1 and $\varepsilon/8$. Then $0 < |x - 3| < \delta$ implies that

$$|(x^2 + x - 5) - 7| = |x^2 + x - 12| = |x + 4||x - 3| < 8 \cdot \frac{\varepsilon}{8} = \varepsilon$$ ∎

Figure 7

$$\lim\limits_{x \to c} \sqrt{x} = \sqrt{c}$$

$$|x - 3| < 1 \Rightarrow 2 < x < 4$$
$$\Rightarrow 6 < x + 4 < 8$$
$$\Rightarrow |x + 4| < 8$$

Figure 8

EXAMPLE 6 Prove that $\lim_{x \to c} x^2 = c^2$.

PROOF We mimic the proof in Example 5. Let $\varepsilon > 0$ be given. Choose $\delta = \min\{1, \varepsilon/(1 + 2|c|)\}$. Then $0 < |x - c| < \delta$ implies that

$$|x^2 - c^2| = |x + c||x - c| = |x - c + 2c||x - c|$$
$$\leq (|x - c| + 2|c|)|x - c| \qquad \text{(Triangle Inequality)}$$
$$< \frac{(1 + 2|c|) \cdot \varepsilon}{1 + 2|c|} = \varepsilon \qquad \blacksquare$$

Although appearing incredibly insightful, we did not pull δ "out of the air" in Example 6. We simply did not show you the preliminary analysis this time.

EXAMPLE 7 Prove that $\lim_{x \to c} \frac{1}{x} = \frac{1}{c}, c \neq 0$.

PRELIMINARY ANALYSIS Study Figure 9. We must find δ such that

$$0 < |x - c| < \delta \Rightarrow \left| \frac{1}{x} - \frac{1}{c} \right| < \varepsilon$$

Now

$$\left| \frac{1}{x} - \frac{1}{c} \right| = \left| \frac{c - x}{xc} \right| = \frac{1}{|x|} \cdot \frac{1}{|c|} \cdot |x - c|$$

The factor $1/|x|$ is troublesome, especially if x is near 0. We can bound this factor if we can keep x away from 0. To that end, note that

$$|c| = |c - x + x| \leq |c - x| + |x|$$

so

$$|x| \geq |c| - |x - c|$$

Thus, if we choose $\delta \leq |c|/2$, we succeed in making $|x| \geq |c|/2$. Finally, if we also require $\delta \leq \varepsilon c^2/2$, then

$$\frac{1}{|x|} \cdot \frac{1}{|c|} \cdot |x - c| < \frac{1}{|c|/2} \cdot \frac{1}{|c|} \cdot \frac{\varepsilon c^2}{2} = \varepsilon$$

FORMAL PROOF Let $\varepsilon > 0$ be given. Choose $\delta = \min\{|c|/2, \varepsilon c^2/2\}$. Then $0 < |x - c| < \delta$ implies

$$\left| \frac{1}{x} - \frac{1}{c} \right| = \left| \frac{c - x}{xc} \right| = \frac{1}{|x|} \cdot \frac{1}{|c|} \cdot |x - c| < \frac{1}{|c|/2} \cdot \frac{1}{|c|} \cdot \frac{\varepsilon c^2}{2} = \varepsilon \qquad \blacksquare$$

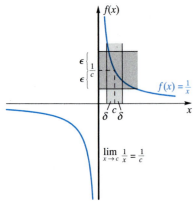

Figure 9

One-Sided Limits It does not take much imagination to give the ε–δ definitions of right- and left-hand limits.

Definition Right-Hand Limit

To say $\lim_{x \to c^+} f(x) = L$ means that for each $\varepsilon > 0$ there is a corresponding $\delta > 0$ such that

$$0 < x - c < \delta \Rightarrow |f(x) - L| < \varepsilon$$

We leave the ε–δ definition for the left-hand limit to the reader. (See Problem 5.)

The ε–δ concept presented in this section is probably the most intricate and elusive topic in a calculus course. It may take you some time to grasp this concept, but it is worth the effort. Calculus is the study of limits, so a clear understanding of the concept of limit is a worthy goal.

The discovery of calculus is usually attributed to Isaac Newton (1642–1727) and Gottfried Wilhelm von Leibniz (1646–1716), who worked independently in the late 1600s. Although Newton and Leibniz, along with their successors, discovered a number of properties of calculus, and calculus was found to have many applications in the physical sciences, it was not until the nineteenth century that a precise definition of a limit was proposed. Augustin Louis Cauchy (1789–1857), a French engineer and mathematician, gave this definition: "If the successive values attributed to the same variable approach indefinitely a fixed value, such that they finally differ from it by as little as one wishes, this latter is called the limit of all the others." Even Cauchy, a master at rigor, was somewhat vague in his definition of a limit. What are "successive values," and what does it mean to "finally differ"? The phrase "finally differ from it by as little as one wishes" contains the seed of the ε–δ definition, because for the first time it indicates that the difference between $f(x)$ and its limit L can be made smaller than any given number, the number we labeled ε. The German mathematician Karl Weierstrass (1815–1897) first put together the definition that is equivalent to our ε–δ definition of a limit.

Concepts Review

1. The inequality $|f(x) - L| < \varepsilon$ is equivalent to _____ $< f(x) <$ _____.

2. The precise meaning of $\lim_{x \to a} f(x) = L$ is this: Given any positive number ε, there is a corresponding positive number δ such that _____ implies _____.

3. To be sure that $|3x - 3| < \varepsilon$, we would require that $|x - 1| <$ _____.

4. $\lim_{x \to a} (mx + b) =$ _____.

Problem Set 2.5

In Problems 1–6, give the appropriate ε–δ definition of each statement.

1. $\lim_{t \to a} f(t) = M$

2. $\lim_{u \to b} g(u) = L$

3. $\lim_{z \to d} h(z) = P$

4. $\lim_{y \to e} \phi(y) = B$

5. $\lim_{x \to c^-} f(x) = L$

6. $\lim_{t \to a^+} g(t) = D$

In Problems 7–18, give an ε–δ proof of each limit fact (see Examples 1–5).

7. $\lim_{x \to 0} (2x - 1) = -1$

8. $\lim_{x \to -21} (3x - 1) = -64$

9. $\lim_{x \to 5} \dfrac{x^2 - 25}{x - 5} = 10$

10. $\lim_{x \to 0} \left(\dfrac{2x^2 - x}{x} \right) = -1$

11. $\lim_{x \to 5} \dfrac{2x^2 - 11x + 5}{x - 5} = 9$

12. $\lim_{x \to 1} \sqrt{2x} = \sqrt{2}$

13. $\lim_{x \to 4} \dfrac{\sqrt{2x - 1}}{\sqrt{x - 3}} = \sqrt{7}$

14. $\lim_{x \to 1} \dfrac{14x^2 - 20x + 6}{x - 1} = 8$

15. $\lim_{x \to 1} \dfrac{10x^3 - 26x^2 + 22x - 6}{(x - 1)^2} = 4$

16. $\lim_{x \to 1} (2x^2 + 1) = 3$

17. $\lim_{x \to -1} (x^2 - 2x - 1) = 2$

18. $\lim_{x \to 0} x^4 = 0$

19. Prove that if $\lim_{x \to c} f(x) = L$ and $\lim_{x \to c} f(x) = M$ then $L = M$.

20. Let F and G be functions such that $0 \le F(x) \le G(x)$ for all x near c, except possibly at c. Prove that if $\lim_{x \to c} G(x) = 0$ then $\lim_{x \to c} F(x) = 0$.

21. Prove that $\lim_{x \to 0} x^4 \sin^2(1/x) = 0$. *Hint*: Use Problems 18 and 20.

22. Prove that $\lim_{x \to 0^+} \sqrt{x} = 0$.

23. By considering left- and right-hand limits, prove that $\lim_{x \to 0} |x| = 0$.

24. Prove that if $|f(x)| < B$ for $|x - a| < 1$ and $\lim_{x \to a} g(x) = 0$ then $\lim_{x \to a} f(x)g(x) = 0$.

25. Suppose that $\lim_{x \to a} f(x) = L$ and that $f(a)$ exists (though it may be different from L). Prove that f is bounded on some interval containing a; that is, show that there is an interval (c, d) with $c < a < d$ and a constant M such that $|f(x)| \le M$ for all x in (c, d).

26. Prove that if $f(x) \le g(x)$ for all x in some deleted interval about a and if $\lim_{x \to a} f(x) = L$ and $\lim_{x \to a} g(x) = M$ then $L \le M$.

limit
(a) l
|
(b) l

(c) l
   i

(d) l
   (

Having given the definitions of these new limits, we must face the question of whether the Main Limit Theorem (Theorem 2.6A) holds for them. The answer is yes, and the proof is similar to the original one. Note how we use this theorem in the following examples.

EXAMPLE 2   Prove that $\lim\limits_{x\to\infty} \dfrac{x}{1+x^2} = 0$.

Solution   Here we use a standard trick: divide numerator and denominator by the highest power of x that appears in the denominator, that is, x^2.

$$\lim_{x\to\infty} \frac{x}{1+x^2} = \lim_{x\to\infty} \frac{\dfrac{x}{x^2}}{\dfrac{1+x^2}{x^2}} = \lim_{x\to\infty} \frac{\dfrac{1}{x}}{\dfrac{1}{x^2}+1}$$

$$= \frac{\lim\limits_{x\to\infty} \dfrac{1}{x}}{\lim\limits_{x\to\infty} \dfrac{1}{x^2} + \lim\limits_{x\to\infty} 1} = \frac{0}{0+1} = 0 \qquad\blacksquare$$

EXAMPLE 3   Find $\lim\limits_{x\to-\infty} \dfrac{2x^3}{1+x^3}$.

Solution   The graph of $f(x) = 2x^3/(1+x^3)$ is shown in Figure 4. To find the limit, divide numerator and denominator by x^3.

$$\lim_{x\to-\infty} \frac{2x^3}{1+x^3} = \lim_{x\to-\infty} \frac{2}{1/x^3 + 1} = \frac{2}{0+1} = 2$$

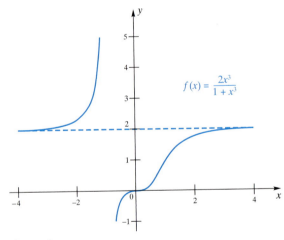

$f(x) = \dfrac{2x^3}{1+x^3}$

Figure 4

Infinite Limits   Consider the graph of $f(x) = 1/(x-2)$, which is shown in Figure 5. It makes no sense to ask for $\lim\limits_{x\to2} 1/(x-2)$, but we think it is reasonable to write

$$\lim_{x\to2^-} \frac{1}{x-2} = -\infty \qquad \lim_{x\to2^+} \frac{1}{x-2} = \infty$$

Here is the definition that relates to this situation.

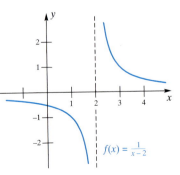

$f(x) = \dfrac{1}{x-2}$

Figure 5

Definition   Infinite Limit

We say that $\lim\limits_{x\to c^+} f(x) = \infty$ if for each positive number M there corresponds a $\delta > 0$ such that

$$0 < x - c < \delta \Rightarrow f(x) > M$$

There are corresponding definitions of

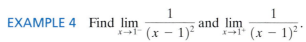

$$\lim_{x \to c^+} f(x) = -\infty \qquad \lim_{x \to c^-} f(x) = \infty \qquad \lim_{x \to c} f(x) = -\infty$$

$$\lim_{x \to \infty} f(x) = -\infty \qquad \lim_{x \to \infty} f(x) = -\infty \qquad \lim_{x \to -\infty} f(x) = -\infty \qquad \lim_{x \to -\infty} f(x) = \infty$$

(See Problems 45 and 46.)

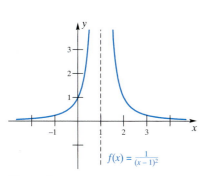

Figure 6

EXAMPLE 4   Find $\displaystyle \lim_{x \to 1^-} \frac{1}{(x-1)^2}$ and $\displaystyle \lim_{x \to 1^+} \frac{1}{(x-1)^2}$.

Solution   The graph of $f(x) = 1/(x-1)^2$ is shown in Figure 6. As $x \to 1^+$, the denominator remains positive but goes to zero, while the numerator is 1 for all x. Thus, the ratio $1/(x-1)^2$ can be made arbitrarily large by restricting x to be near, but above, 1. Similarly, as $x \to 1^-$, the denominator is positive and can be made arbitrarily close to 0. Thus $1/(x-1)^2$ can be made arbitrarily large by restricting x to be near, but to the left of, 1. We therefore conclude that

$$\lim_{x \to 1^+} \frac{1}{(x-1)^2} = \infty \qquad \text{and} \qquad \lim_{x \to 1^-} \frac{1}{(x-1)^2} = \infty$$

Since both limits are ∞, we could also write

$$\lim_{x \to 1} \frac{1}{(x-1)^2} = \infty \qquad \blacksquare$$

EXAMPLE 5   Find $\displaystyle \lim_{x \to 2^+} \frac{x+1}{x^2 - 5x + 6}$.

Solution

$$\lim_{x \to 2^+} \frac{x+1}{x^2 - 5x + 6} = \lim_{x \to 2^+} \frac{x+1}{(x-3)(x-2)}$$

As $x \to 2^+$ we see that $x + 1 \to 3$, $x - 3 \to -1$, and $x - 2 \to 0^+$; thus, the numerator is approaching 3, but the denominator is negative and approaching 0. We conclude that

$$\lim_{x \to 2^+} \frac{x+1}{(x-3)(x-2)} = -\infty \qquad \blacksquare$$

Relation to Asymptotes   Asymptotes were discussed briefly in Section 2.1, but now we can say more about them. The line $x = c$ is a **vertical asymptote** of the graph of $y = f(x)$ if any of the following four statements is true.

1. $\displaystyle \lim_{x \to c^+} f(x) = \infty$ 2. $\displaystyle \lim_{x \to c^+} f(x) = -\infty$

3. $\displaystyle \lim_{x \to c^-} f(x) = \infty$ 4. $\displaystyle \lim_{x \to c^-} f(x) = -\infty$

Thus, in Figure 5, the line $x = 2$ is a vertical asymptote. Likewise, the lines $x = 2$ and $x = 3$, although not shown graphically, are vertical asymptotes in Example 5.

In a similar vein, the line $y = b$ is a **horizontal asymptote** of the graph of $y = f(x)$ if either

$$\lim_{x \to \infty} f(x) = b \quad \text{or} \quad \lim_{x \to -\infty} f(x) = b$$

The line $y = 0$ is a horizontal asymptote in both Figures 5 and 6.

EXAMPLE 6   Find the vertical and horizontal asymptotes of the graph of $y = f(x)$ if

$$f(x) = \frac{2x}{x-1}$$

Solution   We often have a vertical asymptote at a point where the denominator is zero, and in this case we do because

$$\lim_{x \to 1^+} \frac{2x}{x-1} = \infty \quad \text{and} \quad \lim_{x \to 1^-} \frac{2x}{x-1} = -\infty$$

On the other hand,

$$\lim_{x \to \infty} \frac{2x}{x-1} = \lim_{x \to \infty} \frac{2}{1-1/x} = 2 \quad \text{and} \quad \lim_{x \to -\infty} \frac{2x}{x-1} = 2$$

and so $y = 2$ is a horizontal asymptote. The graph of $y = 2x/(x-1)$ is shown in Figure 7.

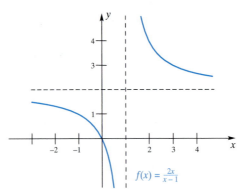

$f(x) = \frac{2x}{x-1}$

Figure 7

Concepts Review

1. To say that $x \to \infty$ means that _____; to say that $\lim_{x \to \infty} f(x) = L$ means that _____. Give your answers in informal language.

2. To say that $\lim_{x \to c^+} f(x) = \infty$ means that _____; to say that $\lim_{x \to c^-} f(x) = -\infty$ means that _____. Give your answers in informal language.

3. If $\lim_{x \to \infty} f(x) = 6$, then the line _____ is a _____ asymptote of the graph of $y = f(x)$.

4. If $\lim_{x \to 6^+} f(x) = \infty$, then the line _____ is a _____ asymptote of the graph of $y = f(x)$.

Problem Set 2.8

In Problems 1–36, find the limits.

1. $\lim_{x \to \infty} \dfrac{x}{x-5}$

2. $\lim_{x \to \infty} \dfrac{x^2}{5-x^3}$

3. $\lim_{t \to -\infty} \dfrac{t^2}{7-t^2}$

4. $\lim_{t \to -\infty} \dfrac{t}{t-5}$

5. $\lim_{x \to \infty} \dfrac{x^2}{(x-5)(3-x)}$

6. $\lim_{x \to \infty} \dfrac{x^2}{x^2 - 8x + 15}$

7. $\lim_{x \to \infty} \dfrac{x^3}{2x^3 - 100x^2}$

8. $\lim_{\theta \to -\infty} \dfrac{\pi \theta^5}{\theta^5 - 5\theta^4}$

9. $\lim_{x \to \infty} \dfrac{3x^3 - x^2}{\pi x^3 - 5x^2}$

10. $\lim_{\theta \to \infty} \dfrac{\sin^2 \theta}{\theta^2 - 5}$

11. $\lim_{x \to \infty} \dfrac{3\sqrt{x^3} + 3x}{\sqrt{2x^3}}$

12. $\lim_{x \to \infty} \sqrt[3]{\dfrac{\pi x^3 + 3x}{\sqrt{2} \, x^3 + 7x}}$

13. $\lim_{x \to \infty} \sqrt[3]{\dfrac{1 + 8x^2}{x^2 + 4}}$

14. $\lim_{x \to \infty} \sqrt{\dfrac{x^2 + x + 3}{(x-1)(x+1)}}$

15. $\lim_{x \to \infty} \dfrac{2x+1}{\sqrt{x^2 + 3}}$. *Hint*: Divide numerator and denominator by x. Note that, for $x > 0$, $\sqrt{x^2 + 3}/x = \sqrt{(x^2 + 3)/x^2}$.

16. $\lim_{x \to \infty} \dfrac{\sqrt{2x + 1}}{x + 4}$

17. $\lim_{x \to \infty} \left(\sqrt{2x^2 + 3} - \sqrt{2x^2 - 5} \right)$. *Hint*: Multiply and divide by $\sqrt{2x^2 + 3} + \sqrt{2x^2 - 5}$.

18. $\lim_{x \to \infty} \left(\sqrt{x^2 + 2x} - x \right)$

19. $\lim_{y \to -\infty} \dfrac{9y^3 + 1}{y^2 - 2y + 2}$. *Hint*: Divide numerator and denominator by y^2.

20. $\lim_{x \to \infty} \dfrac{a_0 x^n + a_1 x^{n-1} + \cdots + a_{n-1} x + a_n}{b_0 x^n + b_1 x^{n-1} + \cdots + b_{n-1} x + b_n}$, where $a_0 \neq 0$, $b_0 \neq 0$, and n is in $\mathbb{N}$.

21. $\lim\limits_{x\to 4^+} \dfrac{x}{x-4}$

22. $\lim\limits_{t\to -3^+} \dfrac{t^2-9}{t+3}$

23. $\lim\limits_{t\to 3^-} \dfrac{t^2}{9-t^2}$

24. $\lim\limits_{x\to \sqrt[3]{5}^+} \dfrac{x^2}{5-x^3}$

25. $\lim\limits_{x\to 5^-} \dfrac{x^2}{(x-5)(3-x)}$

26. $\lim\limits_{\theta\to \pi^+} \dfrac{\theta^2}{\sin\theta}$

27. $\lim\limits_{x\to 3^-} \dfrac{x^3}{x-3}$

28. $\lim\limits_{\theta\to (\pi/2)^+} \dfrac{\pi\theta}{\cos\theta}$

29. $\lim\limits_{x\to 3^-} \dfrac{x^2-x-6}{x-3}$

30. $\lim\limits_{x\to 2^+} \dfrac{x^2+2x-8}{x^2-4}$

31. $\lim\limits_{x\to 0^+} \dfrac{[\![x]\!]}{x}$

32. $\lim\limits_{x\to 0^-} \dfrac{[\![x]\!]}{x}$

33. $\lim\limits_{x\to 0^-} \dfrac{|x|}{x}$

34. $\lim\limits_{x\to 0^+} \dfrac{|x|}{x}$

35. $\lim\limits_{x\to 0^-} \dfrac{1+\cos x}{\sin x}$

36. $\lim\limits_{x\to \infty} \dfrac{\sin x}{x}$

GC *In Problems 37–42, find the horizontal and vertical asymptotes for the graphs of the indicated functions. Then sketch their graphs.*

37. $f(x) = \dfrac{3}{x+1}$

38. $f(x) = \dfrac{3}{(x+1)^2}$

39. $F(x) = \dfrac{2x}{x-3}$

40. $F(x) = \dfrac{3}{9-x^2}$

41. $g(x) = \dfrac{14}{2x^2+7}$

42. $g(x) = \dfrac{2x}{\sqrt{x^2+5}}$

43. The line $y = ax + b$ is called an **oblique asymptote** to the graph of $y = f(x)$ if either $\lim\limits_{x\to \infty}\big[f(x) - (ax+b)\big] = 0$ or $\lim\limits_{x\to -\infty}\big[f(x) - (ax+b)\big] = 0$. Find the oblique asymptote for

$$f(x) = \frac{2x^4 + 3x^3 - 2x - 4}{x^3 - 1}$$

Hint: Begin by dividing the denominator into the numerator.

44. Find the oblique asymptote for

$$f(x) = \frac{3x^3 + 4x^2 - x + 1}{x^2 + 1}$$

45. Using the symbols M and δ, give precise definitions of each expression.
(a) $\lim\limits_{x\to c^+} f(x) = -\infty$

(b) $\lim\limits_{x\to c^-} f(x) = \infty$

46. Using the symbols M and N, give precise definitions of each expression.
(a) $\lim\limits_{x\to \infty} f(x) = \infty$

(b) $\lim\limits_{x\to -\infty} f(x) = \infty$

47. Give a rigorous proof that if $\lim\limits_{x\to \infty} f(x) = A$ and $\lim\limits_{x\to \infty} g(x) = B$, then

$$\lim\limits_{x\to \infty} \big[f(x) + g(x)\big] = A + B$$

48. We have given meaning to $\lim\limits_{x\to A} f(x)$ for $A = a, a^-, a^+, -\infty,$ ∞. Moreover, in each case, this limit may be L (finite), $-\infty, \infty,$ or may fail to exist in any sense. Make a table illustrating each of the 20 possible cases.

49. Find each of the following limits or indicate that it does not exist even in the infinite sense.

(a) $\lim\limits_{x\to \infty} \sin x$

(b) $\lim\limits_{x\to \infty} \sin \dfrac{1}{x}$

(c) $\lim\limits_{x\to \infty} x \sin \dfrac{1}{x}$

(d) $\lim\limits_{x\to \infty} x^{3/2} \sin \dfrac{1}{x}$

(e) $\lim\limits_{x\to \infty} x^{-1/2} \sin x$

(f) $\lim\limits_{x\to \infty} \sin\left(\dfrac{\pi}{6} + \dfrac{1}{x}\right)$

(g) $\lim\limits_{x\to \infty} \sin\left(x + \dfrac{1}{x}\right)$

(h) $\lim\limits_{x\to \infty} \left[\sin\left(x + \dfrac{1}{x}\right) - \sin x\right]$

50. Einstein's Special Theory of Relativity says that the mass $m(v)$ of an object is related to its velocity v by

$$m(v) = \frac{m_0}{\sqrt{1 - v^2/c^2}}$$

Here m_0 is the rest mass and c is the velocity of light. What is $\lim\limits_{v\to c^-} m(v)$?

GC *Use a computer or a graphing calculator to find the limits in Problems 51–58. Begin by plotting the function in an appropriate window.*

51. $\lim\limits_{x\to \infty} \dfrac{3x^2 + x + 1}{2x^2 - 1}$

52. $\lim\limits_{x\to -\infty} \sqrt{\dfrac{2x^2 - 3x}{5x^2 + 1}}$

53. $\lim\limits_{x\to -\infty} \left(\sqrt{2x^2 + 3x} - \sqrt{2x^2 - 5}\right)$

54. $\lim\limits_{x\to \infty} \dfrac{2x + 1}{\sqrt{3x^2 + 1}}$

55. $\lim\limits_{x\to \infty} \left(1 + \dfrac{1}{x}\right)^{10}$

56. $\lim\limits_{x\to \infty} \left(1 + \dfrac{1}{x}\right)^{x}$

57. $\lim\limits_{x\to \infty} \left(1 + \dfrac{1}{x}\right)^{x^2}$

58. $\lim\limits_{x\to \infty} \left(1 + \dfrac{1}{x}\right)^{\sin x}$

CAS *Find the one-sided limits in Problems 59–65. Begin by plotting the function in an appropriate window. Your computer may indicate that some of these limits do not exist, but, if so, you should be able to interpret the answer as either ∞ or $-\infty$.*

59. $\lim\limits_{x\to 3^-} \dfrac{\sin|x-3|}{x-3}$

60. $\lim\limits_{x\to 3^-} \dfrac{\sin|x-3|}{\tan(x-3)}$

61. $\lim\limits_{x\to 3^-} \dfrac{\cos(x-3)}{x-3}$

62. $\lim\limits_{x\to \frac{\pi}{2}^+} \dfrac{\cos x}{x - \pi/2}$

63. $\lim\limits_{x\to 0^+} (1 + \sqrt{x})^{1/\sqrt{x}}$

64. $\lim\limits_{x\to 0^+} (1 + \sqrt{x})^{1/x}$

65. $\lim\limits_{x\to 0^+} (1 + \sqrt{x})^{x}$

Answers to Concepts Review: **1.** x increases without bound; $f(x)$ gets close to L as x increases without bound   **2.** $f(x)$ increases without bound as x approaches c from the right; $f(x)$ decreases without bound as x approaches c from the left   **3.** $y = 6$; horizontal   **4.** $x = 6$; vertical.

2.9
Continuity of Functions

In mathematics and science, we use the word *continuous* to describe a process that goes on without abrupt changes. In fact, our experience leads us to assume this as an essential feature of many natural processes. It is this notion as it pertains to functions that we now want to make precise. In the three graphs shown in Figure 1, only the third graph exhibits continuity at c. In the first two graphs, either $\lim_{x \to c} f(x)$ does not exist, or it exists but does not equal $f(c)$. Only in the third graph does $\lim_{x \to c} f(x) = f(c)$.

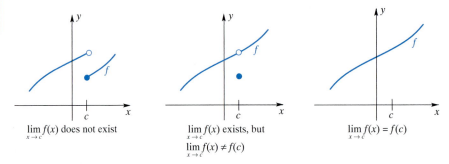

$\lim_{x \to c} f(x)$ does not exist $\lim_{x \to c} f(x)$ exists, but $\lim_{x \to c} f(x) \neq f(c)$ $\lim_{x \to c} f(x) = f(c)$

Figure 1

Here is the formal definition.

Definition   Continuity at a Point

Let f be defined on an open interval containing c. We say that f is **continuous** at c if

$$\lim_{x \to c} f(x) = f(c)$$

We mean by this definition to require three things: (1) $\lim_{x \to c} f(x)$ exists, (2) $f(c)$ exists (i.e., c is in the domain of f), and (3) $\lim_{x \to c} f(x) = f(c)$. If any one of these three fails, then f is **discontinuous** at c. Thus, the functions represented by the first and second graphs of Figure 1 are discontinuous at c. They do appear, however, to be continuous at other points of their domains.

EXAMPLE 1   Let $f(x) = \dfrac{x^2 - 4}{x - 2}$, $x \neq 2$. How should f be defined at $x = 2$ in order to make it continuous there?

Solution

$$\lim_{x \to 2} \frac{x^2 - 4}{x - 2} = \lim_{x \to 2} \frac{(x - 2)(x + 2)}{x - 2} = \lim_{x \to 2} (x + 2) = 4$$

Therefore, we define $f(2) = 4$. The graph of the resulting function is shown in Figure 2. In fact, we see that $f(x) = x + 2$ for all x. ∎

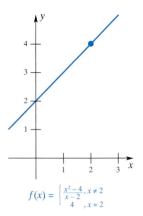

$f(x) = \begin{cases} \frac{x^2-4}{x-2}, & x \neq 2 \\ 4, & x = 2 \end{cases}$

Figure 2

Continuity of Familiar Functions   Most functions that we will meet in this book are either (1) continuous everywhere or (2) continuous everywhere except at a few points. In particular, Theorem 2.6B implies the following result.

Theorem A   Continuity of Polynomial and Rational Functions

A polynomial function is continuous at every real number c. A rational function is continuous at every real number c in its domain, that is, everywhere except where its denominator is zero.

Recall the absolute value function $f(x) = |x|$; its graph is shown in Figure 3. For $x < 0$, $f(x) = -x$, a polynomial; for $x > 0$, $f(x) = x$, another polynomial. Thus, $|x|$ is continuous at all numbers different from 0 by Theorem A. But

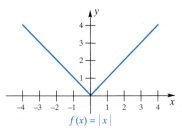

$f(x) = |x|$

Figure 3

$$\lim_{x \to 0} |x| = 0 = |0|$$

(see Problem 23 of Section 2.5). Therefore, $|x|$ is also continuous at 0; it is continuous everywhere.

By the Main Limit Theorem (Theorem 2.6A)

$$\lim_{x \to c} \sqrt[n]{x} = \sqrt[n]{\lim_{x \to c} x} = \sqrt[n]{c}$$

provided $c > 0$ when n is even. This means that $f(x) = \sqrt[n]{x}$ is continuous at each point where it makes sense to talk about continuity. In particular, $f(x) = \sqrt{x}$ is continuous at each real number $c > 0$ (Figure 4). We summarize.

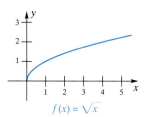

Figure 4

f(x) = √x

Theorem B Continuity of Absolute Value and nth Root Functions

The absolute value function is continuous at every real number c. If n is odd, the nth root function is continuous at every real number c; if n is even, the nth-root function is continuous at every positive real number c.

Continuity under Function Operations Do the standard function operations preserve continuity? Yes, according to the next theorem. In it, f and g are functions, k is a constant, and n is a positive integer.

Theorem C

If f and g are continuous at c, then so are $kf, f + g, f - g, f \cdot g, f/g$ (provided that $g(c) \neq 0$), f^n, and $\sqrt[n]{f}$ (provided that $f(c) > 0$ if n is even).

Proof All these results are easy consequences of the corresponding facts for limits from Theorem 2.6A. For example, that theorem, combined with the fact that f and g are continuous at c, gives

$$\lim_{x \to c} f(x)g(x) = \lim_{x \to c} f(x) \cdot \lim_{x \to c} g(x) = f(c)g(c)$$

This is precisely what it means to say that $f \cdot g$ is continuous at c. ♦

EXAMPLE 2 At what numbers is $F(x) = (3|x| - x^2)/(\sqrt{x} + \sqrt[3]{x})$ continuous?

Solution We need not even consider nonpositive numbers, since F is not defined at such numbers. For any positive number, the functions $\sqrt{x}, \sqrt[3]{x}, |x|$, and x^2 are all continuous (Theorems A and B). It follows from Theorem C that $3|x|, 3|x| - x^2$, $\sqrt{x} + \sqrt[3]{x}$, and finally,

$$\frac{(3|x| - x^2)}{(\sqrt{x} + \sqrt[3]{x})}$$

are continuous at each positive number. ■

The continuity of the trigonometric functions follows from Theorem C and from Theorem 2.7A.

Theorem D

The sine and cosine functions are continuous at every real number c. The functions $\tan x, \cot x, \sec x$, and $\csc x$ are continuous at every real number c in their domains.

Proof Theorem 2.7A says that for every real number c

$$\lim_{x \to c} \sin x = \sin c \quad \text{and} \quad \lim_{x \to c} \cos x = \cos c$$

These are exactly the conditions required for $\sin x$ and $\cos x$ to be continuous at c.

Since $\sin x$ and $\cos x$ are continuous at every real number c, Theorem C implies that the quotient $\dfrac{\sin x}{\cos x} = \tan x$ is continuous as long as the denominator, $\cos x$, is not

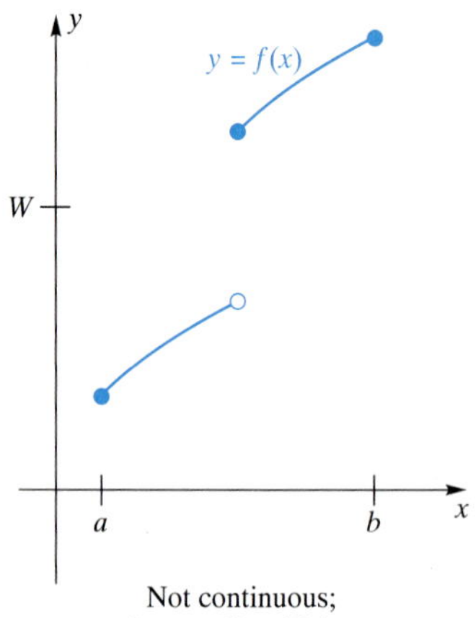

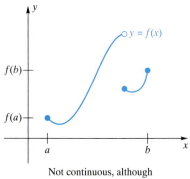

Not continuous;
Intermediate Value
property fails.

Figure 8

Not continuous, although
Intermediate value property holds

Figure 9

in $[a, b]$ such that $f(c) = W$. In other words, f takes on every value between $f(a)$ and $f(b)$. Continuity is needed for this theorem, for otherwise it is possible to find a function f and a number W between $f(a)$ and $f(b)$ such that there is no c in $[a, b]$ that satisfies $f(c) = W$. Figure 8 shows an example of such a function.

It seems clear that continuity is sufficient, although a formal proof of this result turns out to be difficult. We leave the proof to more advanced works.

The converse of this theorem, which is not true in general, says that if f takes on every value between $f(a)$ and $f(b)$ then f is continuous. Figures 7 and 9 show functions that take on all values between $f(a)$ and $f(b)$, but the function in Figure 9 is not continuous on $[a, b]$. Just because a function has the intermediate value property does not mean that it must be continuous.

The Intermediate Value Theorem can be used to tell us something about the solutions of equations, as the next example shows.

EXAMPLE 7   Use the Intermediate Value Theorem to show that the equation $x - \cos x = 0$ has a solution between $x = 0$ and $x = \pi/2$.

Solution   Let $f(x) = x - \cos x$, and let $W = 0$. Then $f(0) = 0 - \cos 0 = -1$ and $f(\pi/2) = \pi/2 - \cos \pi/2 = \pi/2$. Since f is continuous on $[0, \pi/2]$ and since $W = 0$ is between $f(0)$ and $f(\pi/2)$, the Intermediate Value Theorem implies the existence of a c in the interval $(0, \pi/2)$ with the property that $f(c) = 0$. Such a c is a solution to the equation $x - \cos x = 0$. Figure 10 suggests that there is exactly one such c.

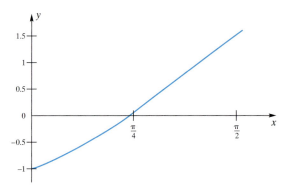

Figure 10

We can go one step further. The midpoint of the interval $[0, \pi/2]$ is the point $x = \pi/4$. When we evaluate $f(\pi/4)$, we get

$$f(\pi/4) = \frac{\pi}{4} - \cos \frac{\pi}{4} = \frac{\pi}{4} - \frac{\sqrt{2}}{2} \approx 0.0782914$$

which is greater than 0. Thus, $f(0) < 0$ and $f(\pi/4) > 0$, so another application of the Intermediate Value Theorem tells us that there exists a c between 0 and $\pi/4$ such that $f(c) = 0$. We have thus narrowed down the interval containing the desired c from $[0, \pi/2]$ to $[0, \pi/4]$. There is nothing stopping us from selecting the midpoint of $[0, \pi/4]$ and evaluating f at that point, thereby narrowing even further the interval containing c. This process could be continued indefinitely until we find that c is in a sufficiently small interval. This method of zeroing in on a solution is called the *bisection method*, and we will study it further in Section 11.3. ■

The Intermediate Value Theorem can also lead to some surprising results.

EXAMPLE 8   Use the Intermediate Value Theorem to show that on a circular wire ring there are always two points opposite from each other with the same temperature.

Solution   Choose coordinates for this problem so that the center of the ring is the origin, and let r be the radius of the ring. (See Figure 11.) Define $T(x, y)$ to be the temperature at the point (x, y). Consider a diameter of the circle that makes an angle θ with the x-axis, and define $f(\theta)$ to be the temperature difference between the points that make angles of θ and $\theta + \pi$; that is,

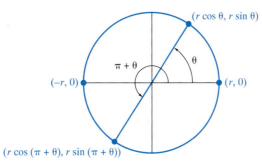

Figure 11

$$f(\theta) = T(r\cos\theta, r\sin\theta) - T(r\cos(\theta + \pi), r\sin(\theta + \pi))$$

With this definition

$$f(0) = T(r, 0) - T(-r, 0)$$

$$f(\pi) = T(-r, 0) - T(r, 0) = -[T(r, 0) - T(-r, 0)] = -f(0)$$

Thus, either $f(0)$ and $f(\pi)$ are both zero, or one is positive and the other is negative. If both are zero, then we have found the required two points. Otherwise, we can apply the Intermediate Value Theorem. Assuming that temperature varies continuously, we conclude that there exists a c between 0 and π such that $f(c) = 0$. Thus, for the two points at the angles c and $c + \pi$, the temperatures are the same. ∎

Concepts Review

1. A function f is continuous at c if _____ = $f(c)$.

2. The function $f(x) = [\![x]\!]$ is discontinuous at _____ .

3. A function f is said to be continuous on a closed interval $[a, b]$ if it is continuous at every point of (a, b) and if _____ and _____ .

4. The Intermediate Value Theorem says that if a function f is continuous on $[a, b]$ and W is a number between $f(a)$ and $f(b)$, then there is a number c between _____ and _____ such that _____ .

Problem Set 2.9

In Problems 1–15, state whether the indicated function is continuous at 3. If it is not continuous, tell why.

1. $f(x) = (x - 3)(x - 4)$

2. $g(x) = x^2 - 9$

3. $h(x) = \dfrac{3}{x - 3}$

4. $g(t) = \sqrt{t - 4}$

5. $h(t) = \dfrac{|t - 3|}{t - 3}$

6. $h(t) = \dfrac{|\sqrt{(t - 3)^4}|}{t - 3}$

7. $f(t) = |t|$

8. $g(t) = |t - 2|$

9. $h(x) = \dfrac{x^2 - 9}{x - 3}$

10. $f(x) = \dfrac{21 - 7x}{x - 3}$

11. $r(t) = \begin{cases} \dfrac{t^3 - 27}{t - 3} & \text{if } t \neq 3 \\ 27 & \text{if } t = 3 \end{cases}$

12. $r(t) = \begin{cases} \dfrac{t^3 - 27}{t - 3} & \text{if } t \neq 3 \\ 23 & \text{if } t = 3 \end{cases}$

13. $f(t) = \begin{cases} t - 3 & \text{if } t \leq 3 \\ 3 - t & \text{if } t > 3 \end{cases}$

14. $f(t) = \begin{cases} t^2 - 9 & \text{if } t \leq 3 \\ (3 - t)^2 & \text{if } t > 3 \end{cases}$

15. $f(t) = \begin{cases} -3x + 4 & \text{if } x \leq 3 \\ -2 & \text{if } x > 3 \end{cases}$

16. From the graph of g (see Figure 12), indicate the values where g is discontinuous. For each of the values state whether g is continuous from the right, left, or neither.

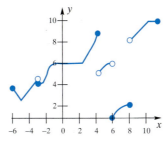

Figure 12

17. From the graph of h given in Figure 13, indicate the intervals on which h is continuous.

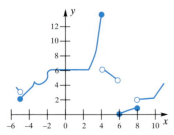

Figure 13

In Problems 18–23, the given function is not defined at a certain point. How should it be defined in order to make it continuous at this point? (See Example 1.)

18. $f(x) = \dfrac{x^2 - 49}{x - 7}$

19. $f(x) = \dfrac{2x^2 - 18}{3 - x}$

20. $g(\theta) = \dfrac{\sin(\theta)}{\theta}$

21. $H(t) = \dfrac{\sqrt{t} - 1}{t - 1}$

22. $\phi(x) = \dfrac{x^4 + 2x^2 - 3}{x + 1}$

23. $F(x) = \sin\dfrac{x^2 - 1}{x + 1}$

In Problems 24–35, at what points, if any, are the functions discontinuous?

24. $f(x) = \dfrac{3x + 7}{(x - 30)(x - \pi)}$

25. $f(x) = \dfrac{33 - x^2}{x\pi + 3x - 3\pi - x^2}$

26. $h(\theta) = |\sin\theta + \cos\theta|$

27. $r(\theta) = \tan\theta$

28. $f(u) = \dfrac{2u + 7}{\sqrt{u + 5}}$

29. $g(u) = \dfrac{u^2 + |u - 1|}{\sqrt[3]{u + 1}}$

30. $F(x) = \dfrac{1}{\sqrt{4 + x^2}}$

31. $G(x) = \dfrac{1}{\sqrt{4 - x^2}}$

32. $f(x) = \begin{cases} x & \text{if } x < 0 \\ x^2 & \text{if } 0 \le x \le 1 \\ 2 - x & \text{if } x > 1 \end{cases}$

33. $g(x) = \begin{cases} x^2 & \text{if } x < 0 \\ -x & \text{if } 0 \le x \le 1 \\ x & \text{if } x > 1 \end{cases}$

34. $f(t) = [\![t]\!]$

35. $g(t) = [\![t + \frac{1}{2}]\!]$

36. Sketch the graph of a function f that satisfies all the following conditions.

(a) Its domain is $[-2, 2]$.

(b) $f(-2) = f(-1) = f(1) = f(2) = 1$.

(c) It is discontinuous at -1 and 1.

(d) It is right continuous at -1 and left continuous at 1.

37. Let

$$f(x) = \begin{cases} x & \text{if } x \text{ is rational} \\ -x & \text{if } x \text{ is irrational} \end{cases}$$

Sketch the graph of this function as best you can and decide where it is continuous.

38. Use the Intermediate Value Theorem to prove that $x^3 + 3x - 2 = 0$ has a real solution between 0 and 1.

39. Use the Intermediate Value Theorem to prove that $(\cos t)t^3 + 6\sin^5 t - 3 = 0$ has a real solution between 0 and 2π.

40. Show that the equation $x^5 + 4x^3 - 7x + 14 = 0$ has at least one real solution. *Hint*: Intermediate Value Theorem.

41. Prove that f is continuous at c if and only if $\lim_{t \to 0} f(t + c) = f(c)$.

42. Prove that if f is continuous at c and $f(c) > 0$ there is an interval $(c - \delta, c + \delta)$ such that $f(x) > 0$ on this interval.

43. Prove that, if f is continuous on $[0, 1]$ and satisfies $0 \le f(x) \le 1$ there, then f has a *fixed point*; that is, there is a number c in $[0, 1]$ such that $f(c) = c$. *Hint*: Apply the Intermediate Value Theorem to $g(x) = x - f(x)$.

44. Find the values of a and b so that the following function is continuous everywhere.

$$f(x) = \begin{cases} x + 1 & \text{if } x < 1 \\ ax + b & \text{if } 1 \le x < 2 \\ 3x & \text{if } x \ge 2 \end{cases}$$

45. A stretched elastic string covers the interval $[0, 1]$. The ends are released and the string contracts so that it covers the interval $[a, b], a \ge 0, b \le 1$. Prove that this results in one point of the string (actually exactly one point) being where it was originally. See Problem 43.

46. Let $f(x) = \dfrac{1}{x - 1}$. Then $f(-2) = -\dfrac{1}{3}$ and $f(2) = 1$. Does the Intermediate Value Theorem imply the existence of a number c between -2 and 2 such that $f(c) = 0$? Explain.

47. Starting at 4 A.M., a hiker slowly climbed to the top of a mountain, arriving at noon. The next day, he returned along the same path, starting at 5 A.M. and getting to the bottom at 11 A.M. Show that at some point along the path his watch showed the same time on both days.

48. Let D be a bounded but otherwise arbitrary region in the first quadrant. Given an angle $\theta, 0 \le \theta \le \pi/2$, D can be circumscribed by a rectangle whose base makes angle θ with the x-axis as shown in Figure 14. Prove that at some angle this rectangle is a square. (This means that *any* bounded region can be circumscribed by a *square*.)

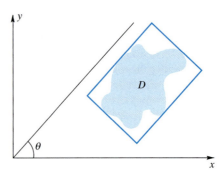

Figure 14

49. Let $f(x + y) = f(x) + f(y)$ for all x and y in $\mathbb{R}$ and suppose that f is continuous at $x = 0$.

(a) Prove that f is continuous everywhere.

(b) Prove that there is a constant m such that $f(t) = mt$ for all t in $\mathbb{R}$ (see Problem 41 of Section 2.1).

In Problems 50–53, we will study linear functions. Such functions have the form y(x) = mx + b, where m and b are constants.

50. Show that the sum of two linear functions is also a linear function.

51. Show that the composition of two linear functions is also a linear function.

52. Show that the product of two linear functions is generally not a linear function.

53. Show that the quotient of two linear functions is generally not a linear function.

54. Prove that if $f(x)$ is a continuous function on an interval then so is the function $|f(x)| = \sqrt{(f(x))^2}$.

55. Show that if $g(x) = |f(x)|$ is continuous it is not necessarily true that $f(x)$ is continuous.

56. Sometimes continuity of a function f is said to be defined by being able to pass the $\lim_{x \to c}$ "through" the function. For example, if f is continuous at c, then $\lim_{x \to c} f(x) = f(\lim_{x \to c} x)$. Prove or disprove this statement.

57. (Famous Problem) Let $f(x) = 0$ if x is irrational and let $f(x) = 1/q$ if x is the rational number p/q in reduced form ($q > 0$).
(a) Sketch (as best you can) the graph of f on $(0, 1)$.
(b) Show that f is continuous at each irrational number in $(0, 1)$, but is discontinuous at each rational number in $(0, 1)$.

58. A thin equilateral triangular block of side length 1 unit has its face in the vertical xy-plane with a vertex V at the origin. Under the influence of gravity, it will rotate about V until a side hits the x-axis floor (Figure 15). Let x denote the initial x-coordinate of the midpoint M of the side opposite V, and let $f(x)$ denote the final x-coordinate of this point. Assume that the block balances when M is directly above V.

(a) Determine the domain and range of f.

(b) Where on this domain is f discontinuous?

(c) Identify any fixed points of f (see Problem 43).

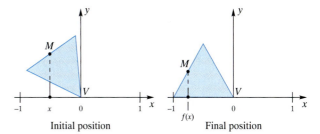

Initial position   Final position

Figure 15

Answers to Concepts Review: **1.** $\lim_{x \to c} f(x)$ **2.** every integer
3. $\lim_{x \to a^-} f(x) = f(a);\ \lim_{x \to b^-} f(x) = f(b)$ **4.** $a; b; f(c) = W$

2.10 Chapter Review

Concepts Test

Respond with true or false to each of the following assertions. Be prepared to justify your answer.

1. The equation $xy + x^2 = 3y$ determines a function with formula of the form $y = f(x)$.

2. The equation $xy^2 + x^2 = 3x$ determines a function with formula of the form $y = f(x)$.

3. The equation $\theta \sin \theta + t - \cos \theta = 0$ determines t as a function of θ.

4. The equation $\Phi + \Psi = |\Phi + \Psi|$ determines Φ as a function of Ψ.

5. The equation $T = \sin(\theta)$ determines θ as a function of T.

6. The natural domain of

$$f(x) = \sqrt{\frac{x}{4 - x}}$$

is the interval $[0, 4)$.

7. The natural domain of

$$f(x) = \sqrt{-(x^2 + 4x + 3)}$$

is the interval $-3 \le x \le -1$.

8. The natural domain of $T(\theta) = \sec(\theta) + \cos(\theta)$ is all θ.

9. The range of $f(x) = x^2 - 6$ is the interval $[-6, \infty)$.

10. The range of the function $f(x) = \tan x - \sec x$ is the interval $(-\infty, -1] \cup [1, \infty)$.

11. The range of the function $f(x) = \csc x - \sec x$ is the interval $(-\infty, -1] \cup [1, \infty)$.

12. The sum of two even functions is an even function.

13. The sum of two odd functions is an odd function.

14. The product of two odd functions is an odd function.

15. The product of an even function with an odd function is an odd function.

16. The composition of an even function with an odd function is an odd function.

17. The composition of two odd functions is an even function.

18. The function $f(x) = (2x^3 + x)/(x^2 + 1)$ is odd.

19. The function

$$f(t) = \frac{(\sin t)^2 + \cos t}{\tan t \csc t}$$

is even.

20. If the range of a function consists of just one number, then its domain also consists of just one number.

21. If the domain of a function contains at least two numbers, then the range also contains at least two numbers.

22. If $g(x) = [x/2]$, then $g(-1.8) = -1$.

23. If $f(x) = x^2$ and $g(x) = x^3$, then $f \circ g = g \circ f$.

24. If $f(x) = x^2$ and $g(x) = x^3$, then $(f \circ g)(x) = f(x) \cdot g(x)$.

25. If f and g have the same domain, then f/g also has that domain.

26. If the graph of $y = f(x)$ has an x-intercept at $x = a$, then the graph of $y = f(x + h)$ has an x-intercept at $x = a - h$.

27. The cotangent is an odd function.

28. The natural domain of the tangent function is the set of all real numbers.

29. If $\cos s = \cos t$, then $s = t$.

30. If $\lim_{x \to c} f(x) = L$, then $f(c) = L$.

31. If $f(c)$ is not defined, then $\lim_{x \to c} f(x)$ does not exist.

32. The coordinates of the hole in the graph of $y = \dfrac{x^2 - 25}{x - 5}$ are $(5, 10)$.

33. If $p(x)$ is a polynomial, then $\lim_{x \to c} p(x) = p(c)$.

34. $\lim_{x \to 0} \dfrac{\sin x}{x}$ does not exist.

35. For every real number c, $\lim_{x \to c} \tan x = \tan c$.

36. The function $f(x) = 2 \sin^2 x - \cos x$ is continuous at every real number.

37. $\lim_{x \to \infty} \dfrac{\sin x}{x} = 1$.

38. If f is a continuous function such that $A \le f(x) \le B$, then $\lim_{x \to \infty} f(x)$ exists and it satisfies $A \le \lim_{x \to \infty} f(x) \le B$.

39. $\lim_{t \to 1^+} \dfrac{2t}{t - 1} = \infty$.

40. If $\lim_{x \to c^-} f(x) = \lim_{x \to c^+} f(x)$, then f is continuous at $x = c$.

41. If $\lim_{x \to c} f(x) = f\left(\lim_{x \to c} x\right)$, then f is continuous at $x = c$.

42. The function $f(x) = [\![x/2]\!]$ is continuous at $x = 2.3$.

43. If $\lim_{x \to 2} f(x) = f(2) > 0$, then $f(x) < 1.001 f(2)$ for all x in some interval containing 2.

44. If $\lim_{x \to c}[f(x) + g(x)]$ exists, then $\lim_{x \to c} f(x)$ and $\lim_{x \to c} g(x)$ both exist.

45. If $0 \le f(x) \le 3x^2 + 2x^4$ for all x, then $\lim_{x \to 0} f(x) = 0$.

46. If $\lim_{x \to a} f(x) = L$ and $\lim_{x \to a} f(x) = M$, then $L = M$.

47. If $f(x) \ne g(x)$ for all x, then $\lim_{x \to c} f(x) \ne \lim_{x \to c} g(x)$.

48. If $f(x) < 10$ for all x and $\lim_{x \to 2} f(x)$ exists, then $\lim_{x \to 2} f(x) < 10$.

49. If $\lim_{x \to a} f(x) = b$, then $\lim_{x \to a} |f(x)| = |b|$.

50. If f is continuous and positive on $[a, b]$, then $1/f$ must assume every value between $1/f(a)$ and $1/f(b)$.

Sample Test Problems

1. For $f(x) = 1/(x + 1) - 1/x$, find each value (if possible).
(a) $f(1)$ 　　　(b) $f(-\frac{1}{2})$ 　　　(c) $f(-1)$
(d) $f(t - 1)$ 　　　(e) $f\left(\dfrac{1}{t}\right)$

2. For $g(x) = (x + 1)/x$, find and simplify.
(a) $g(2)$ 　　　(b) $g(\frac{1}{2})$
(c) $g(\frac{1}{10})$ 　　　(d) $\dfrac{g(2 + h) - g(2)}{h}$

3. Describe the natural domain of each function.
(a) $f(x) = \dfrac{x}{x^2 - 1}$ 　　　(b) $g(x) = \sqrt{4 - x^2}$
(c) $h(x) = \dfrac{\sqrt{1 + x^2}}{|2x + 3|}$

4. Which of the following functions are odd? Even? Neither even nor odd?
(a) $f(x) = \dfrac{3x}{x^2 + 1}$ 　　　(b) $g(x) = |\sin x| + \cos x$
(c) $h(x) = x^3 + \sin x$ 　　　(d) $k(x) = \dfrac{x^2 + 1}{|x| + x^4}$

5. Sketch the graph of each function.
(a) $f(x) = x^2 - 1$ 　　　(b) $g(x) = \dfrac{x}{x^2 + 1}$
(c) $h(x) = \begin{cases} x^2 & \text{if } 0 \le x \le 2 \\ 6 - x & \text{if } x > 2 \end{cases}$

6. Suppose that f is an even function satisfying $f(x) = -1 + \sqrt{x}$ for $x \ge 0$. Sketch the graph of f for $-4 \le x \le 4$.

7. An open box is made by cutting squares of side x inches from the four corners of a sheet of cardboard 24 inches by 32 inches and then turning up the sides. Express the volume $V(x)$ in terms of x. What is the domain for this function?

8. Let $f(x) = x - 1/x$ and $g(x) = x^2 + 1$. Find each value.
(a) $(f + g)(2)$ 　　(b) $(f \cdot g)(2)$ 　　(c) $(f \circ g)(2)$
(d) $(g \circ f)(2)$ 　　(e) $f^3(-1)$ 　　(f) $f^2(2) + g^2(2)$

9. Sketch the graph of each of the following, making use of translations.
(a) $y = \frac{1}{4}x^2$ 　　　(b) $y = \frac{1}{4}(x + 2)^2$
(c) $y = -1 + \frac{1}{4}(x + 2)^2$

10. Let $f(x) = \sqrt{16 - x}$ and $g(x) = x^4$. What is the domain of each of the following?
(a) f 　　　(b) $f \circ g$ 　　　(c) $g \circ f$

11. Write $F(x) = \sqrt{1 + \sin^2 x}$ as the composite of four functions, $f \circ g \circ h \circ k$.

12. Calculate each of the following without using a calculator.
(a) $\sin(570°)$ 　　　(b) $\cos\left(\dfrac{9\pi}{2}\right)$
(c) $\sin^2(5) + \cos^2(5)$ 　　　(d) $\cos\left(\dfrac{-13\pi}{6}\right)$

13. If $\sin t = 0.8$ and $\cos t < 0$, find each value.
(a) $\sin(-t)$ 　　(b) $\cos t$ 　　(c) $\sin 2t$
(d) $\tan t$ 　　(e) $\cos\left(\dfrac{\pi}{2} - t\right)$ 　　(f) $\sin(\pi + t)$

14. Write $\sin 3t$ in terms of $\sin t$. *Hint:* $3t = 2t + t$.

15. A fly sits on the rim of a wheel spinning at the rate of 20 revolutions per minute. If the radius of the wheel is 9 inches, how far does the fly travel in 1 second?

In Problems 16–27, find the indicated limit or state that it does not exist.

16. $\lim\limits_{u \to 1} \dfrac{u^2 - 1}{u + 1}$

17. $\lim\limits_{u \to 1} \dfrac{u^2 - 1}{u - 1}$

18. $\lim\limits_{u \to 1} \dfrac{u + 1}{u^2 - 1}$

19. $\lim\limits_{x \to 2} \dfrac{1 - 2/x}{x^2 - 4}$

20. $\lim\limits_{z \to 2} \dfrac{z^2 - 4}{z^2 + z - 6}$

21. $\lim\limits_{x \to 0} \dfrac{\tan x}{\sin 2x}$

22. $\lim\limits_{y \to 1} \dfrac{y^3 - 1}{y^2 - 1}$

23. $\lim\limits_{x \to 4} \dfrac{x - 4}{\sqrt{x} - 2}$

24. $\lim\limits_{x \to 0} \dfrac{\cos x}{x}$

25. $\lim\limits_{x \to 0^-} \dfrac{|x|}{x}$

26. $\lim\limits_{x \to 1/2^+} [\![4x]\!]$

27. $\lim\limits_{t \to 2^-} ([\![t]\!] - t)$

28. Let $f(x) = \begin{cases} x^3 & \text{if } x < -1 \\ x & \text{if } -1 < x < 1 \\ 1 - x & \text{if } x \geq 1 \end{cases}$

Find each value.

(a) $f(1)$

(b) $\lim\limits_{x \to 1^+} f(x)$

(c) $\lim\limits_{x \to 1^-} f(x)$

(d) $\lim\limits_{x \to -1} f(x)$

29. Refer to f of Problem 28. (a) What are the values of x at which f is discontinuous? (b) How should f be defined at $x = -1$ to make it continuous there?

30. Give the ε–δ definition in each case.

(a) $\lim\limits_{u \to a} g(u) = M$

(b) $\lim\limits_{x \to a^-} f(x) = L$

31. If $\lim\limits_{x \to 3} f(x) = 3$ and $\lim\limits_{x \to 3} g(x) = -2$ and if g is continuous at $x = 3$, find each value.

(a) $\lim\limits_{x \to 3} [2f(x) - 4g(x)]$

(b) $\lim\limits_{x \to 3} g(x) \dfrac{x^2 - 9}{x - 3}$

(c) $g(3)$

(d) $\lim\limits_{x \to 3} g(f(x))$

(e) $\lim\limits_{x \to 3} \sqrt{f^2(x) - 8g(x)}$

(f) $\lim\limits_{x \to 3} \dfrac{|g(x) - g(3)|}{f(x)}$

32. Sketch the graph of a function f that satisfies all the following conditions.

(a) Its domain is $[0, 6]$.

(b) $f(0) = f(2) = f(4) = f(6) = 2$.

(c) f is continuous except at $x = 2$.

(d) $\lim\limits_{x \to 2^-} f(x) = 1$ and $\lim\limits_{x \to 5^+} f(x) = 3$.

33. Let $f(x) = \begin{cases} -1 & \text{if } x \leq 0 \\ ax + b & \text{if } 0 < x < 1 \\ 1 & \text{if } x \geq 1 \end{cases}$

Determine a and b so that f is continuous everywhere.

34. Use the Intermediate Value Theorem to prove that the equation $x^5 - 4x^3 - 3x + 1 = 0$ has at least one solution between $x = 2$ and $x = 3$.

2.11 Additional Problems

1. Use the language of functions to describe the indicated relationship as accurately as possible. State clearly the domain and range of each function that you define. Sketch the graph.

(a) How does the area of a right triangle with a hypotenuse of a fixed length depend on the acute angle?

(b) How does the length of one leg of a right triangle depend on the length of the other leg, if we assume that the hypotenuse is a fixed length?

(c) How does the temperature in degrees Celsius depend on the temperature in degrees Fahrenheit?

(d) Given the angle of a piece of pie from a 10-inch pie, find the area of the piece.

2. (a) Suppose that a continuous function is periodic with period 1 and is linear between 0 and 0.25 and linear between -0.75 and 0. In addition, it has the value 1 at 0 and 2 at 0.25. Sketch the function over the domain -1 to 1, and give a piecewise definition of the function.

(b) Suppose that a continuous function is periodic with period 2 and is quadratic between -0.25 and 0.25 and linear between -1.75 and -0.25. In addition, it has the value 0 at 0 and 0.0625 at ± 0.25. Sketch the function over the domain -2 and 2, and give a piecewise definition of the function.

EXPL **3.** Let α, β, and γ be postive constants. Characterize the changes in the graph of $y = \gamma \cos(\alpha x + \beta)$ if:

(a) α, γ are fixed and β is halved.

(b) γ, β are fixed and α is doubled.

(c) α, β, and γ are doubled.

(d) γ is fixed and α and β are doubled.

EXPL **GC** **4.** We will explore how a parabola depends on the coefficient of the linear term. We consider the specific parabola $y = P_c(x) = 2x^2 + cx - 1$ and investigate how changes in c affect the graph.

(a) Using some sort of graphing device, graph $y = P_c(x)$ for positive values of c on the same set of axes. Try the values $c = 1$, $2, 3, 4$. How does the position of the vertex of the graph change?

(b) Graph $y = P_c(x)$ for $c = -1, -2, -3, -4$. How does the position of the vertex change?

(c) Formulate a rule that summarizes the effect that changing c has on the graph of $P_c(x)$. Your rule should be valid no matter whether c is positive, negative, or zero.

(d) Notice that all the graphs have the same y-intercept. Explain why that must be the case.

EXPL **GC** **5.** Consider the function $g(t) = t^3 + t$ and the function $h(t) = \alpha t$. We will consider the result of composing g with h.

(a) Compare the graphs of $g(\alpha t) = (g \circ h)(t)$ with $\alpha g(t) = (h \circ g)(t)$ for $\alpha = 3, 2, 1/2, 1/4$.

(b) Formulate a rule that summarizes the effect that changing a positive constant α has on the graph of $g(\alpha t)$ and $\alpha g(t)$ compared to the graph of $g(t)$.

(c) Formulate a rule that summarizes the effect that changing a **negative** constant α has on the graph of $g(\alpha t)$ and $\alpha g(t)$ compared to the graph of $g(t)$.

[EXPL] [GC] **6.** Consider the function $\Phi(x) = \dfrac{x^3 + x}{2x - 1}$. We will consider the effect of composing this function with $\Psi(x) = |x|$.

(a) Compare the graph of $(\Phi \circ \Psi)(x) = \dfrac{|x|^3 + |x|}{2|x| - 1}$ with the graph of $\Phi(x)$.

(b) Compare the graph of $(\Psi \circ \Phi)(x) = \left| \dfrac{x^3 + x}{2x - 1} \right|$ with the graph of $\Phi(x)$.

(c) Formulate rules that summarize the effects of composing the absolute value function with that of other functions. That is, give the relationship between the graphs of $\Phi(x), |\Phi(x)|$, and $\Phi(|x|)$.

[GC] **7.** At this stage in the development of calculus, some limits are hard to compute exactly. Nevertheless, it is possile to make an intelligent guess as to the limit by using numerical and graphical methods. First, using your calculator with the indicated values, plot the function near the indicated point. Then, make an educated guess at the value of the following limits:

(a) $\displaystyle \lim_{x \to 0} 12 \left(\dfrac{1 - x^2/2 - \cos x}{x^4} \right)$ evaluated at $x = 0.1, \ 0.01,$ and 0.001.

(b) $\displaystyle \lim_{x \to 0} 12 \left(\dfrac{x - x^3/6 - \sin x}{x^5} \right)$ evaluated at $x = 0.5, 0.3,$ and 0.1.

(c) $\displaystyle \lim_{x \to 0} 12 \left(\dfrac{2x + \frac{8}{3}x^3 - \tan 2x}{256x^5} \right)$ evaluated at $x = 0.5, \ 0.3, \ 0.1,$ and 0.05.

Shifting and Scaling the Graph of a Function

I. Preparation

Make sure that you know how to use the appropriate software to define and plot functions. Also, investigate the capability of your software to animate graphs.

II. Using Technology

The purpose of this project is to recognize the effects of a constants a and b on the graph of $f(x + a), f(x) + a$, $f(bx)$, and $bf(x)$.

Exercise 1

(a) Let $f(x) = \sin x$. Plot the graph of $f(x + a) = \sin(x + a)$ for various values of a from -4 to 4. If your software has the capability to animate graphs, plot and animate $f(x + a)$ for $a = -4$ to $a = 4$ in increments of 0.1.

(b) Repeat part (a) for $f(x) + a = \sin x + a$. Make sure that you understand the difference between $\sin(x + a)$ and $\sin x + a$.

(c) Repeat part (a) for $f(bx) = \sin(bx)$.

(d) Repeat part (a) for $bf(x) = b \sin x$.

Exercise 2 Repeat all four parts of Exercise 1 for the function $g(x) = x \sin x$.

Exercise 3 The graphs of the functions

$$f_1(x) = \sin(2x) + 1$$
$$f_2(x) = \sin(x + 1)$$
$$f_3(x) = 3\sin(2x)$$
$$f_4(x) = 3\sin(x + 1)$$
$$f_5(x) = 3\sin x$$

are shown in Figure 1 in a scrambled order. Match each function with its graph.

Exercise 4 Identify the functions whose graphs are shown in Figure 2.

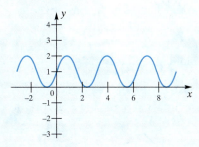
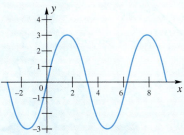
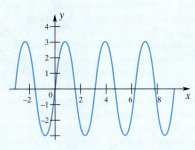
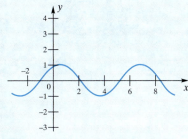
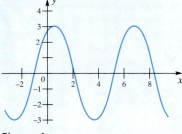

Figure 1

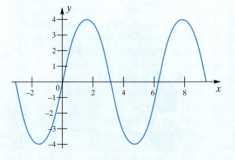
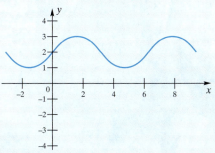
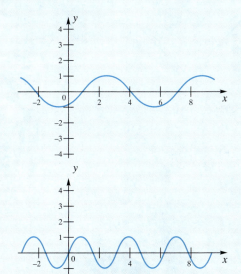

Figure 2

III. Reflection

Write a brief report that explains the effects of the constants a and b on the graphs of the functions $f(x + a)$, $f(x) + a, f(bx)$, and $bf(x)$.

97

Limits

Preparation

Review the definition of a limit in Section 2.5 and the limit theorems in Section 2.6.

Using Technology

Exercise 1 Use factoring to help find the following limits:

(a) $\lim\limits_{x \to 13} \dfrac{x^3 - 9x^2 - 45x - 91}{x - 13}$

(b) $\lim\limits_{x \to 13} \dfrac{x^3 - 9x^2 - 39x - 86}{x - 13}$

(c) $\lim\limits_{x \to 13}$

$\dfrac{x^4 - 26x^3 + 178x^2 - 234x + 1521}{x - 13}$

Exercise 2 Find

$$\lim\limits_{x \to 0} \dfrac{\sqrt{25 + 3x} - \sqrt{25 - 2x}}{x}$$

by each of the following methods:

(a) Make a plot of the fraction near the point $x = 0$. This often "works"

even when, as in this case, the function is not defined at the limit point.

(b) Make a table near $x = 0$.

Exercise 3 Evaluate each of the following limits as $x \to 0$ using both tables and plots.

(a) $\dfrac{x^3 - x^2 - 4x + 4}{x - 1}$

(b) $\dfrac{\sin x}{x}$ (This is an important limit.)

(c) $\dfrac{1 - \cos x}{x}$ (So is this.)

(d) $\dfrac{\sin 5x}{x}$

(e) $(1 + x)^{1/x}$ (Another important limit.)

Exercise 4 As we have seen, sometimes standard tools such as factoring can be used advantageously. Use the trick of multiplying the top and bottom by a sum of square roots ("rationalizing the numerator") to algebraically check the result of Exercise 2 and check that your results were correct.

Reflection

Exercise 5 You have noticed that many limits can be obtained just by evaluating the given function at the given point. State precisely a condition under which this is valid. Which of the following limits can be evaluated in this simple way? Justify your answers.

(a) $\lim\limits_{x \to 5} \dfrac{x + 5}{x^4 + x^2 + 1}$

(b) $\lim\limits_{x \to 5} \dfrac{x^2 - 25}{x - 5}$

(c) $\lim\limits_{x \to 5} \sqrt{x^6 - 5}$

Exercise 6 Give examples of the following:

(a) A rational function (not a polynomial) whose limit at c can be evaluated by substitution

(b) A rational function (not a polynomial) whose limit at c cannot be evaluated by substitution

(c) A rational function whose limit at c is $-\infty$

(d) A nonrational function whose limit at 0 exists

3

The Derivative

3.1
Two Problems with One Theme

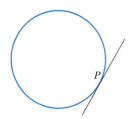

Tangent line at P

Figure 1

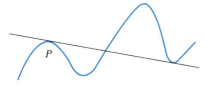

Tangent line at P

Figure 2

Our first problem is very old; it dates back to the great Greek scientist Archimedes (287–212 B.C.). We refer to the problem of the *slope of the tangent line*.

Our second problem is newer. It grew out of attempts by Kepler (1571–1630), Galileo (1564–1642), Newton (1642–1727), and others to describe the speed of a moving body. It is the problem of *instantaneous velocity*.

The two problems, one geometric and the other mechanical, appear to be quite unrelated. In this case, appearances are deceptive. The two problems are identical twins.

The Tangent Line Euclid's notion of a tangent as a line touching a curve at just one point is all right for circles (Figure 1) but completely unsatisfactory for most other curves (Figure 2). The idea of a tangent to a curve at P as the line that best approximates the curve near P is better, but is still too vague for mathematical precision. The concept of limit provides a way of getting the best description.

Let P be a point on a curve and let Q be a nearby *movable point* on that curve. Consider the line through P and Q, called a **secant line**. The **tangent line** at P is the limiting position (if it exists) of the secant line as Q moves toward P along the curve (Figure 3).

Suppose that the curve is the graph of the equation $y = f(x)$. Then P has coordinates $(c, f(c))$, a nearby point Q has coordinates $(c + h, f(c + h))$, and the secant line through P and Q has slope m_{sec} given by (Figure 4):

$$m_{\text{sec}} = \frac{f(c + h) - f(c)}{h}$$

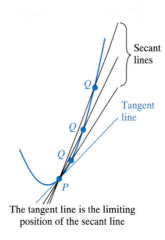

The tangent line is the limiting
position of the secant line

Figure 3

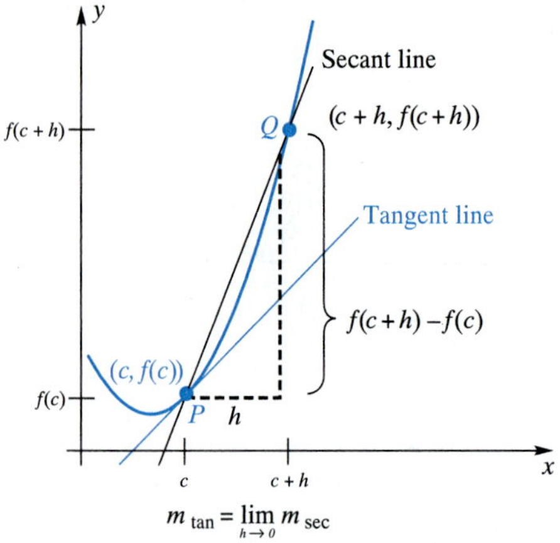

$$m_{\tan} = \lim_{h \to 0} m_{\sec}$$

Figure 4

Using the concept of limit, which we studied in the last chapter, we can now give a formal definition of the tangent line.

Definition   Tangent Line

The **tangent line** to the curve $y = f(x)$ at the point $P(c, f(c))$ is that line through P with slope

$$m_{\tan} = \lim_{h \to 0} m_{\sec} = \lim_{h \to 0} \frac{f(c + h) - f(c)}{h}$$

provided that this limit exists and is not ∞ or $-\infty$.

EXAMPLE 1   Find the slope of the tangent line to the curve $y = f(x) = x^2$ at the point $(2, 4)$.

Solution   The line whose slope we are seeking is shown in Figure 5. Clearly it has a large positive slope.

$$m_{\tan} = \lim_{h \to 0} \frac{f(2 + h) - f(2)}{h}$$

$$= \lim_{h \to 0} \frac{(2 + h)^2 - 2^2}{h}$$

$$= \lim_{h \to 0} \frac{4 + 4h + h^2 - 4}{h}$$

$$= \lim_{h \to 0} \frac{\cancel{h}(4 + h)}{\cancel{h}}$$

$$= 4 \qquad\blacksquare$$

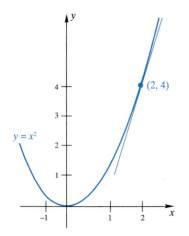

Figure 5

EXAMPLE 2   Find the slopes of the tangent lines to the curve $y = f(x) = -x^2 + 2x + 2$ at the points with x-coordinates $-1, \frac{1}{2}, 2$, and 3.

Solution   Rather than make four separate calculations, it seems wise to calculate the slope at the point with x-coordinate c and then obtain the four desired answers by substitution.

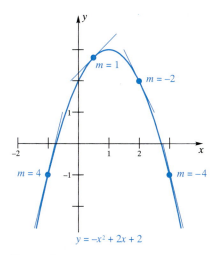

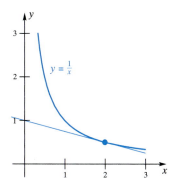

$y = -x^2 + 2x + 2$

Figure 6

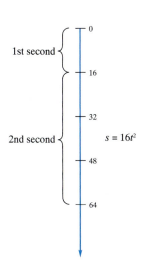

$s = 16t^2$

Figure 8

$$m_{\text{tan}} = \lim_{h \to 0} \frac{f(c + h) - f(c)}{h}$$

$$= \lim_{h \to 0} \frac{-(c + h)^2 + 2(c + h) + 2 - (-c^2 + 2c + 2)}{h}$$

$$= \lim_{h \to 0} \frac{-c^2 - 2ch - h^2 + 2c + 2h + 2 + c^2 - 2c - 2}{h}$$

$$= \lim_{h \to 0} \frac{h(-2c - h + 2)}{h}$$

$$= -2c + 2$$

The four desired slopes (obtained by letting $c = -1, \frac{1}{2}, 2, 3$) are 4, 1, −2, and −4. These answers do appear to be consistent with the graph in Figure 6. ∎

EXAMPLE 3   Find the equation of the tangent line to the curve $y = 1/x$ at $\left(2, \frac{1}{2}\right)$ (see Figure 7).

Solution   Let $f(x) = 1/x$.

$$m_{\text{tan}} = \lim_{h \to 0} \frac{f(2 + h) - f(2)}{h}$$

$$= \lim_{h \to 0} \frac{\dfrac{1}{2 + h} - \dfrac{1}{2}}{h}$$

$$= \lim_{h \to 0} \frac{\dfrac{2}{2(2 + h)} - \dfrac{2 + h}{2(2 + h)}}{h}$$

$$= \lim_{h \to 0} \frac{2 - (2 + h)}{2(2 + h)h}$$

$$= \lim_{h \to 0} \frac{-h}{2(2 + h)h}$$

$$= \lim_{h \to 0} \frac{-1}{2(2 + h)} = -\frac{1}{4}$$

Knowing that the slope of the line is $-\frac{1}{4}$ and that the point $\left(2, \frac{1}{2}\right)$ is on it, we can easily write its equation by using the point–slope form $y - y_0 = m(x - x_0)$. The result is $y - \frac{1}{2} = -\frac{1}{4}(x - 2)$. ∎

Average Velocity and Instantaneous Velocity   If we drive an automobile from one town to another 80 miles away in 2 hours, our average velocity is 40 miles per hour. *Average velocity* is the distance from the first position to the second position divided by the elapsed time.

But during our trip the speedometer reading was often different from 40. At the start, it registered 0; at times it rose as high as 57; at the end it fell back to 0 again. Just what does the speedometer measure? Certainly, it does not indicate average velocity.

Consider the more precise example of an object P falling in a vacuum. Experiment shows that if it starts from rest, P falls $16t^2$ feet in t seconds. Thus, it falls 16 feet in the first second and 64 feet during the first 2 seconds (Figure 8); clearly, it falls faster and faster as time goes on. Figure 9 shows the distance traveled (on the vertical axis) as a function of time (on the horizontal axis).

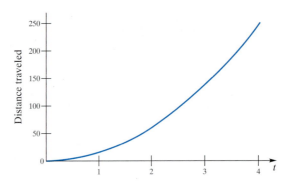

Figure 9

During the second second (i.e., in the time interval from $t = 1$ to $t = 2$), P fell $(64 - 16)$ feet. Its average velocity was

$$v_{avg} = \frac{64 - 16}{2 - 1} = 48 \text{ feet per second}$$

During the time interval from $t = 1$ to $t = 1.5$, it fell $16(1.5)^2 - 16 = 20$ feet. Its average velocity was

$$v_{avg} = \frac{16(1.5)^2 - 16}{1.5 - 1} = \frac{20}{0.5} = 40 \text{ feet per second}$$

Similarly, on the time intervals $t = 1$ to $t = 1.1$ and $t = 1$ to $t = 1.01$, we calculate the respective average velocities to be

$$v_{avg} = \frac{16(1.1)^2 - 16}{1.1 - 1} = \frac{3.36}{0.1} = 33.6 \text{ feet per second}$$

$$v_{avg} = \frac{16(1.01)^2 - 16}{1.01 - 1} = \frac{0.3216}{0.01} = 32.16 \text{ feet per second}$$

What we have done is to calculate the average velocity over shorter and shorter time intervals, each starting at $t = 1$. The shorter the time interval is, the better we should approximate the *instantaneous velocity* at the instant $t = 1$. Looking at the numbers 48, 40, 33.6, and 32.16, we might guess 32 feet per second to be the instantaneous velocity.

But let us be more precise. Suppose that an object P moves along a coordinate line so that its position at time t is given by $s = f(t)$. At time c the object is at $f(c)$; at the nearby time $c + h$, it is at $f(c + h)$ (see Figure 10). Thus the **average velocity** on this interval is

$$v_{avg} = \frac{f(c + h) - f(c)}{h}$$

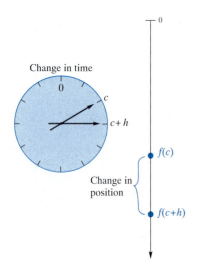

Figure 10

We can now define instantaneous velocity.

Definition   Instantaneous Velocity

If an object moves along a coordinate line with position function $f(t)$, then its **instantaneous velocity** at time c is

$$v = \lim_{h \to 0} v_{avg} = \lim_{h \to 0} \frac{f(c + h) - f(c)}{h}$$

provided that the limit exists and is not ∞ or $-\infty$.

In the case where $f(t) = 16t^2$, the instantaneous velocity at $t = 1$ is

$$v = \lim_{h \to 0} \frac{f(1 + h) - f(1)}{h}$$

$$= \lim_{h \to 0} \frac{16(1 + h)^2 - 16}{h}$$

$$= \lim_{h \to 0} \frac{16 + 32h + 16h^2 - 16}{h}$$

$$= \lim_{h \to 0} (32 + 16h) = 32$$

This confirms our earlier guess.

Now you can see why we called the *slope of the tangent line* and *instantaneous velocity* identical twins. Look at the two definitions in this section. They give the twins different names for the same mathematical concept.

EXAMPLE 4   An object, initially at rest, falls due to gravity. Find its instantaneous velocity at $t = 3.8$ seconds and at $t = 5.4$ seconds.

Solution   We calculate the instantaneous velocity at $t = c$ seconds. Since $f(t) = 16t^2$,

$$v = \lim_{h \to 0} \frac{f(c + h) - f(c)}{h}$$

$$= \lim_{h \to 0} \frac{16(c + h)^2 - 16c^2}{h}$$

$$= \lim_{h \to 0} \frac{16c^2 + 32ch + 16h^2 - 16c^2}{h}$$

$$= \lim_{h \to 0} (32c + 16h) = 32c$$

Thus, the instantaneous velocity at 3.8 seconds is $32(3.8) = 121.6$ feet per second; at 5.4 seconds, it is $32(5.4) = 172.8$ feet per second.   ■

EXAMPLE 5   How long will it take the falling object of Example 4 to reach an instantaneous velocity of 112 feet per second?

Solution   We learned in Example 4 that the instantaneous velocity after c seconds is $32c$. Thus, we must solve the equation $32c = 112$. The solution is $c = \frac{112}{32} = 3.5$ seconds.   ■

EXAMPLE 6   A particle moves along a coordinate line and s, its directed distance in centimeters from the origin at the end of t seconds, is given by $s = f(t) = \sqrt{5t + 1}$. Find the instantaneous velocity of the particle at the end of 3 seconds.

Solution   Figure 11 shows the distance traveled as a function of time. The instantaneous velocity at time $t = 3$ is equal to the slope of the tangent line at $t = 3$.

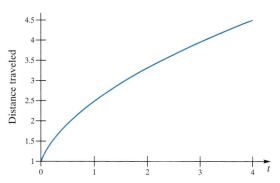

Figure 11

19. A wire of length 8 centimeters is such that the mass between its left end and a point x centimeters to the right is x^3 grams (Figure 12).

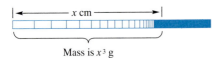

Figure 12

(a) What is the average density of the middle 2-centimeter segment of this wire? *Note*: Average density equals mass/length.
(b) What is the actual density at the point 3 centimeters from the left end?

20. Suppose that the revenue $R(n)$ in dollars from producing n computers is given by $R(n) = 0.4n - 0.001n^2$. Find the instantaneous rates of change of revenue when $n = 10$ and $n = 100$. (The instantaneous rate of change of revenue with respect to the amount of product produced is called the *marginal revenue*.)

21. The rate of change of velocity with respect to time is called **acceleration**. Suppose that the velocity at time t of a particle is given by $v(t) = 2t^2$. Find the instantaneous acceleration when $t = 1$ second.

22. A city is hit by an Asian flu epidemic. Officials estimate that t days after the beginning of the epidemic the number of persons sick with the flu is given by $p(t) = 120t^2 - 2t^3$, when $0 \le t \le 40$. At what rate is the flu spreading at time $t = 10$; $t = 20$; $t = 40$?

23. The graph in Figure 13 shows the amount of water in a city water tank during one day when no water was pumped into the tank. What was the average rate of water usage during the day? How fast was water being used at 8 A.M.?

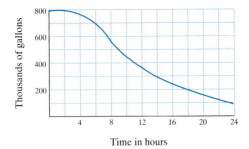

Figure 13

24. Passengers board an elevator at the ground floor (i.e., the zeroth floor) and take it to the seventh floor, which is 84 feet above the ground floor. The elevator's position s as a function of time t (measured in seconds) is shown in Figure 14.

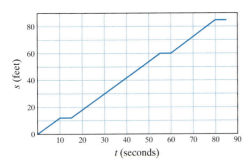

Figure 14

(a) What is the average velocity of the elevator from the time the elevator began moving until the time that it reached the seventh floor?
(b) What was the elevator's approximate velocity at time $t = 20$?
(c) How many stops did the elevator make between the ground floor and the seventh floor (excluding the ground and seventh floors)? On which floors do you think the elevator stopped?

25. Figure 15 shows the normal high temperature for St. Louis, Missouri, as a function of time (measured in days since January 1).

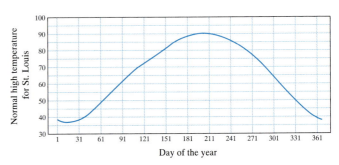

Figure 15

(a) What is the approximate rate of change in the normal high temperture on March 2 (i.e., on day number 61)? What are the units of this rate of change?
(b) What is the approximate rate of change in the normal high temperature on July 10 (i.e., on day number 191)?
(c) In what months is there a moment when the rate of change is equal to 0?
(d) In what months is the absolute value of the rate of change the greatest?

26. Figure 16 shows the population in millions of a developing country for the years 1900 to 1999. What is the approximate rate of change of the population in 1930? In 1990? The percentage growth is often a more appropriate measure of population growth. This is the rate of growth divided by the population size at that time. For this population, what was the approximate percentage growth in 1930? In 1990?

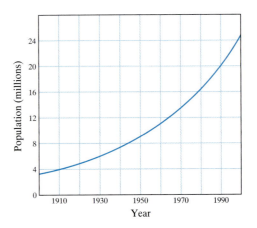

Figure 16

27. Figures 17a and 17b show the position s as a function of time t for two particles that are moving along a line. For each particle, is the velocity increasing or decreasing? Explain.

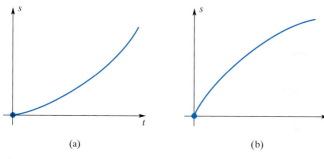

(a)   (b)

Figure 17

28. The rate of change of electric charge with respect to time is called **current**. Suppose that $\frac{1}{3}t^3 + t$ coulombs of charge flow through a wire in t seconds. Find the current in amperes (coulombs per second) after 3 seconds. When will a 20-ampere fuse in the line blow?

29. The radius of a circular oil spill is growing at a constant rate of 2 kilometers per day. At what rate is the area of the spill growing 3 days after it began?

30. The radius of a spherical balloon is increasing at the rate of 0.25 inch per second. If the radius is 0 at time $t = 0$, find the rate of change in the volume at time $t = 3$.

GC *Use a graphing calculator or a CAS to do Problems 31–34.*

31. Draw the graph of $y = f(x) = x^3 - 2x^2 + 1$. Then find the slope of the tangent line at

(a) -1 (b) 0   (c) 1   (d) 3.2

32. Draw the graph of $y = f(x) = \sin x \sin^2 2x$. Then find the slope of the tangent line at

(a) $\pi/3$ (b) 2.8   (c) π (d) 4.2

33. If a point moves along a line so that its distance s (in feet) from 0 is given by $s = t + t\cos^2 t$ at time t seconds, find its instantaneous velocity at $t = 3$.

34. If a point moves along a line so that its distance s (in meters) from 0 is given by $s = (t + 1)^3/(t + 2)$ at time t minutes, find its instantaneous velocity at $t = 1.6$.

Answers to Concepts Review: **1.** tangent line   **2.** secant   **3.** $[f(c + h) - f(c)]/h$ **4.** average velocity

---

3.2
The Derivative

We have seen that *slope of the tangent line* and *instantaneous velocity* are manifestations of the same basic idea. Rate of growth of an organism (biology), marginal profit (economics), density of a wire (physics), and dissolution rates (chemistry) are other versions of the same basic concept. Good mathematical sense suggests that we study this concept independently of these specialized vocabularies and diverse applications. We choose the neutral name *derivative*. Add it to *function* and *limit* as one of the key words in calculus.

Definition   Derivative

The **derivative** of a function f is another function f' (read "f prime") whose value at any number c is

$$f'(c) = \lim_{h \to 0} \frac{f(c + h) - f(c)}{h}$$

provided that this limit exists and is not ∞ or $-\infty$.

If this limit does exist, we say that f is **differentiable** at c. Finding a derivative is called **differentiation**; the part of calculus associated with the derivative is called **differential calculus**.

Finding Derivatives   We illustrate with several examples.

EXAMPLE 1   Let $f(x) = 13x - 6$. Find $f'(4)$.

Solution

$$f'(4) = \lim_{h \to 0} \frac{f(4 + h) - f(4)}{h} = \lim_{h \to 0} \frac{[13(4 + h) - 6] - [13(4) - 6]}{h}$$

$$= \lim_{h \to 0} \frac{13h}{h} = \lim_{h \to 0} 13 = 13 \qquad \blacksquare$$

EXAMPLE 2   If $f(x) = x^3 + 7x$, find $f'(c)$.

Solution

$$
\begin{aligned}
f'(c) &= \lim_{h \to 0} \frac{f(c + h) - f(c)}{h} \\[2mm]
&= \lim_{h \to 0} \frac{\left[(c + h)^3 + 7(c + h)\right] - \left[c^3 + 7c\right]}{h} \\[2mm]
&= \lim_{h \to 0} \frac{3c^2 h + 3ch^2 + h^3 + 7h}{h} \\[2mm]
&= \lim_{h \to 0} (3c^2 + 3ch + h^2 + 7) \\[2mm]
&= 3c^2 + 7
\end{aligned}
$$
∎

EXAMPLE 3   If $f(x) = 1/x$, find $f'(x)$.

Solution   Note a subtle change in the way this example is worded. So far we have used the letter c to denote a fixed number at which a derivative is to be evaluated. Accordingly, we have calculated $f'(c)$. To calculate $f'(x)$, we simply think of x as a fixed, but arbitrary, number and proceed as before.

$$
\begin{aligned}
f'(x) &= \lim_{h \to 0} \frac{f(x + h) - f(x)}{h} = \lim_{h \to 0} \frac{\dfrac{1}{x + h} - \dfrac{1}{x}}{h} \\[2mm]
&= \lim_{h \to 0} \left[\frac{x - (x + h)}{(x + h)x} \cdot \frac{1}{h} \right] = \lim_{h \to 0} \left[\frac{-h}{(x + h)x} \cdot \frac{1}{h} \right] \\[2mm]
&= \lim_{h \to 0} \frac{-1}{(x + h)x} = \frac{-1}{x^2}
\end{aligned}
$$

Thus, f' is the function given by $f'(x) = -1/x^2$. Its domain is all real numbers except $x = 0$. ∎

Since changes in notation can cause confusion, we emphasize the formula for $f'(x)$.

$$
\boxed{\; f'(x) = \lim_{h \to 0} \frac{f(x + h) - f(x)}{h} \;}
$$

This formula says exactly the same thing as the formula in the definition at the beginning of this section.

EXAMPLE 4   Find $F'(x)$ if $F(x) = \sqrt{x}$, $x > 0$.

Solution

$$
\begin{aligned}
F'(x) &= \lim_{h \to 0} \frac{F(x + h) - F(x)}{h} \\[2mm]
&= \lim_{h \to 0} \frac{\sqrt{x + h} - \sqrt{x}}{h}
\end{aligned}
$$

By this time you will have noticed that finding a derivative always involves taking the limit of a quotient where both numerator and denominator are approaching zero. Our task is to simplify this quotient so that we can cancel a factor h from numerator and denominator, thereby allowing us to evaluate the limit by substitution. In the present example, this can be accomplished by rationalizing the numerator.

$$F'(x) = \lim_{h \to 0} \left[\frac{\sqrt{x+h} - \sqrt{x}}{h} \cdot \frac{\sqrt{x+h} + \sqrt{x}}{\sqrt{x+h} + \sqrt{x}} \right]$$

$$= \lim_{h \to 0} \frac{x + h - x}{h(\sqrt{x+h} + \sqrt{x})}$$

$$= \lim_{h \to 0} \frac{h}{h(\sqrt{x+h} + \sqrt{x})}$$

$$= \lim_{h \to 0} \frac{1}{\sqrt{x+h} + \sqrt{x}}$$

$$= \frac{1}{\sqrt{x} + \sqrt{x}} = \frac{1}{2\sqrt{x}}$$

Thus, F', the derivative of F, is given by $F'(x) = 1/(2\sqrt{x})$. Its domain is $(0, \infty)$. ∎

Equivalent Forms for the Derivative There is nothing sacred about use of the letter h in defining $f'(c)$. Notice, for example, that

$$f'(c) = \lim_{h \to 0} \frac{f(c+h) - f(c)}{h}$$

$$= \lim_{p \to 0} \frac{f(c+p) - f(c)}{p}$$

$$= \lim_{s \to 0} \frac{f(c+s) - f(c)}{s}$$

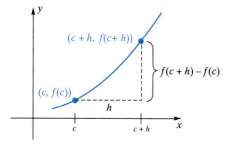

Figure 1

A more radical change, but still just a change of notation, may be understood by comparing Figures 1 and 2. Note how x takes the place of $c + h$, and so $x - c$ replaces h. Thus,

$$\boxed{f'(c) = \lim_{x \to c} \frac{f(x) - f(c)}{x - c}}$$

In a similar vein, we may write

$$f'(x) = \lim_{t \to x} \frac{f(t) - f(x)}{t - x} = \lim_{p \to x} \frac{f(p) - f(x)}{p - x}$$

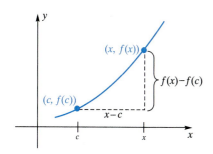

Figure 2

Note that in all cases the number at which f' is evaluated is held fixed during the limit operation.

EXAMPLE 5 Use the last boxed result to find $g'(c)$ if $g(x) = 2/(x+3)$.

Solution

$$g'(c) = \lim_{x \to c} \frac{g(x) - g(c)}{x - c} = \lim_{x \to c} \frac{\dfrac{2}{x+3} - \dfrac{2}{c+3}}{x - c}$$

$$= \lim_{x \to c} \left[\frac{2(c+3) - 2(x+3)}{(x+3)(c+3)} \cdot \frac{1}{x - c} \right]$$

$$= \lim_{x \to c} \left[\frac{-2(x - c)}{(x+3)(c+3)} \cdot \frac{1}{x - c} \right]$$

$$= \lim_{x \to c} \frac{-2}{(x+3)(c+3)} = \frac{-2}{(c+3)^3}$$

Here we manipulated the quotient until we could cancel a factor of $x - c$ from numerator and denominator. Then we could evaluate the limit. ∎

EXAMPLE 6 Each of the following is a derivative, but of what function and at what point?

(a) $\displaystyle \lim_{h \to 0} \frac{(4 + h)^2 - 16}{h}$

(b) $\displaystyle \lim_{x \to 3} \frac{\dfrac{2}{x} - \dfrac{2}{3}}{x - 3}$

Solution

(a) This is the derivative of $f(x) = x^2$ at $x = 4$.

(b) This is the derivative of $f(x) = 2/x$ at $x = 3$. ∎

Differentiability Implies Continuity If a curve has a tangent line at a point, then that curve cannot take a jump or wiggle too badly at the point. The precise formulation of this fact is an important theorem.

> **Theorem A** Differentiability Implies Continuity
>
> If $f'(c)$ exists, then f is continuous at c.

Proof We need to show that $\lim_{x \to c} f(x) = f(c)$. We begin by writing $f(x)$ in a fancy way.

$$f(x) = f(c) + \frac{f(x) - f(c)}{x - c} \cdot (x - c), \qquad x \neq c$$

Therefore,

$$\lim_{x \to c} f(x) = \lim_{x \to c} \left[f(c) + \frac{f(x) - f(c)}{x - c} \cdot (x - c) \right]$$

$$= \lim_{x \to c} f(c) + \lim_{x \to c} \frac{f(x) - f(c)}{x - c} \cdot \lim_{x \to c} (x - c)$$

$$= f(c) + f'(c) \cdot 0$$

$$= f(c) \quad ♦$$

The converse of this theorem is false. If a function f is continuous at c, it does not follow that f has a derivative at c. This is easily seen by considering $f(x) = |x|$ at the origin (Figure 3). This function is certainly continuous at zero. However, it does not have a derivative there, as we now show. Note that

$$\frac{f(0 + h) - f(0)}{h} = \frac{|0 + h| - |0|}{h} = \frac{|h|}{h}$$

Thus,

$$\lim_{h \to 0^+} \frac{f(0 + h) - f(0)}{h} = \lim_{h \to 0^+} \frac{|h|}{h} = \lim_{h \to 0^+} \frac{h}{h} = 1$$

whereas

$$\lim_{h \to 0^-} \frac{f(0 + h) - f(0)}{h} = \lim_{h \to 0^-} \frac{|h|}{h} = \lim_{h \to 0^-} \frac{-h}{h} = -1$$

Since the right- and left-hand limits are different,

$$\lim_{h \to 0} \frac{f(0 + h) - f(0)}{h}$$

does not exist. Therefore, $f'(0)$ does not exist.

A similar argument shows that at any point where the graph of a continuous function has a sharp corner the function is not differentiable. The graph in Figure 4 indicates a number of ways for a function to be nondifferentiable at a point.

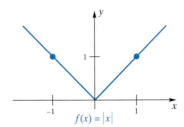

$f(x) = |x|$

Figure 3

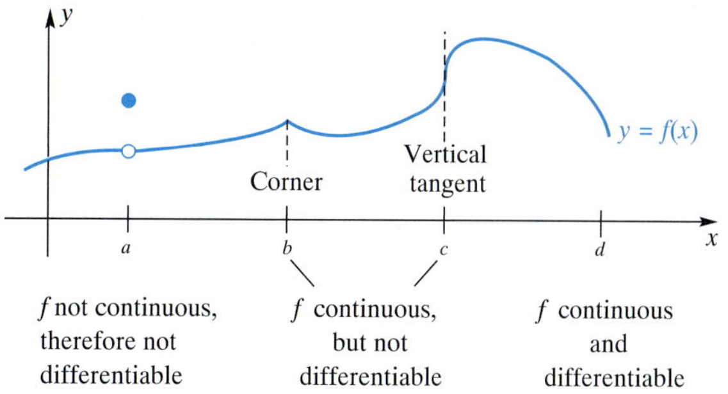

Figure 4

We claim in Figure 4 that the derivative does not exist at the point c where the tangent line is vertical. This is because

$$\lim_{h \to 0} \frac{f(c + h) - f(c)}{h} = \infty$$

This corresponds to the fact that the slope of a vertical line is undefined.

Concepts Review

1. The derivative of f at c is given by $f'(c) = \lim\limits_{h \to 0}$ _____.
Equivalently, $f'(c) = \lim\limits_{t \to c}$ _____.

2. The derivative of f at x is given by $f'(x) = \lim\limits_{h \to 0}$ _____.

3. If f is differentiable at c, then f is _____ at c. The converse is false, as is shown by the example $f(x) =$ _____.

4. $\lim\limits_{h \to 0} \dfrac{2(c + h)^2 - 2c^2}{h}$ is the derivative of $f(x) =$ _____

at $x =$ _____.

Problem Set 3.2

In Problems 1–4, use the definition

$$f'(c) = \lim_{h \to 0} \frac{f(c + h) - f(c)}{h}$$

to find the indicated derivative (see Examples 1 and 2).

1. $f'(1)$ if $f(x) = x^2$

2. $f'(2)$ if $f(t) = (2t)^2$

3. $f'(3)$ if $f(t) = t^2 - t$

4. $f'(4)$ if $f(s) = \dfrac{1}{s - 1}$

In Problems 5–22, use $f'(x) = \lim\limits_{h \to 0} [f(x + h) - f(x)]/h$ to find the derivative at x (see Examples 3 and 4).

5. $s(x) = 2x + 1$

6. $f(x) = \alpha x + \beta$

7. $r(x) = 3x^2 + 4$

8. $f(x) = x^2 + x + 1$

9. $f(x) = ax^2 + bx + c$

10. $f(x) = x^4$

11. $f(x) = x^3 + 2x^2 + 1$

12. $g(x) = x^4 + x^2$

13. $h(x) = \dfrac{2}{x}$

14. $S(x) = \dfrac{1}{x + 1}$

15. $F(x) = \dfrac{6}{x^2 + 1}$

16. $F(x) = \dfrac{x - 1}{x + 1}$

17. $G(x) = \dfrac{2x - 1}{x - 4}$

18. $G(x) = \dfrac{2x}{x^2 - x}$

19. $g(x) = \sqrt{3x}$

20. $g(x) = \dfrac{1}{\sqrt{3x}}$

21. $H(x) = \dfrac{3}{\sqrt{x - 2}}$

22. $H(x) = \sqrt{x^2 + 4}$

In Problems 23–26, use $f'(x) = \lim\limits_{t \to x} [f(t) - f(x)]/[t - x]$ to find $f'(x)$ (see Example 5).

23. $f(x) = x^2 - 3x$

24. $f(x) = x^3 + 5x$

25. $f(x) = \dfrac{x}{x - 5}$

26. $f(x) = \dfrac{x + 3}{x}$

In Problems 27–36, the given limit is a derivative, but of what function and at what point? (See Example 6.)

27. $\lim\limits_{h \to 0} \dfrac{2(5 + h)^3 - 2(5)^3}{h}$

28. $\lim\limits_{h \to 0} \dfrac{(3 + h)^2 + 2(3 + h) - 15}{h}$

29. $\lim\limits_{x \to 2} \dfrac{x^2 - 4}{x - 2}$

30. $\lim\limits_{x \to 3} \dfrac{x^3 + x - 30}{x - 3}$

31. $\lim\limits_{t \to x} \dfrac{t^2 - x^2}{t - x}$

32. $\lim\limits_{p \to x} \dfrac{p^3 - x^3}{p - x}$

33. $\lim\limits_{x \to t} \dfrac{\dfrac{2}{x} - \dfrac{2}{t}}{x - t}$

34. $\lim\limits_{x \to y} \dfrac{\sin x - \sin y}{x - y}$

35. $\lim\limits_{h \to 0} \dfrac{\cos(x + h) - \cos x}{h}$

36. $\lim\limits_{h \to 0} \dfrac{\tan(t + h) - \tan t}{h}$

37. From Figure 5, estimate $f'(0), f'(2), f'(5)$, and $f'(7)$.

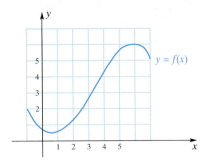

Figure 5

38. From Figure 6, estimate $g'(-1), g'(1), g'(4)$, and $g'(6)$.

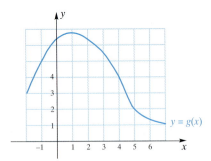

Figure 6

39. Sketch the graph of $y = f'(x)$ on $-1 < x < 7$ for the function f of Problem 37.

40. Sketch the graph of $y = g'(x)$ on $-1 < x < 7$ for the function g of Problem 38.

41. Consider the function $y = f(x)$, whose graph is sketched in Figure 7.

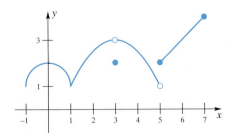

Figure 7

(a) Estimate $f(2), f'(2), f(0.5)$, and $f'(0.5)$.
(b) Estimate the average rate of change in f on the interval $0.5 \le x \le 2.5$.
(c) Where on the interval $-1 < x < 7$ does $\lim_{u \to x} f(u)$ fail to exist?
(d) Where on the interval $-1 < x < 7$ does f fail to be continuous?
(e) Where on the interval $-1 < x < 7$ does f fail to have a derivative?
(f) Where on the interval $-1 < x < 7$ is $f'(x) = 0$?
(g) Where on the interval $-1 < x < 7$ is $f'(x) = 1$?

42. Figure 14 in Section 3.1 shows the position s of an elevator as a function of time t. At what points does the derivative exist? Sketch the derivative of s.

43. Figure 15 in Section 3.1 shows the normal high temperature for St. Louis, Missouri. Sketch the derivative.

44. Figure 8 shows two functions. One is the function f, and the other is its derivative f'. Which one is which?

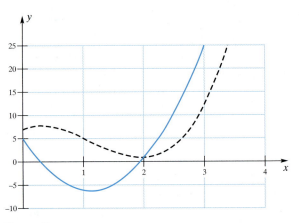

Figure 8

45. Figure 9 shows three functions. One is the function f; another is its derivative f', which we will call g; and the third is the derivative of g. Which one is which?

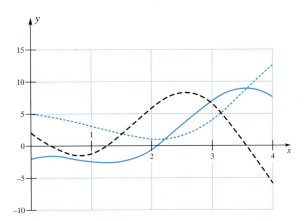

Figure 9

EXPL **46.** Suppose that $f(x + y) = f(x)f(y)$ for all x and y. Show that if $f'(0)$ exists then $f'(a)$ exists and $f'(a) = f(a)f'(0)$.

47. Let $f(x) = \begin{cases} mx + b & \text{if } x < 2 \\ x^2 & \text{if } x \ge 2 \end{cases}$

Determine m and b so that f is differentiable everywhere.

EXPL **48.** The **symmetric derivative** $f_s(x)$ is defined by

$$f_s(x) = \lim_{h \to 0} \frac{f(x + h) - f(x - h)}{2h}$$

Show that if $f'(x)$ exists then $f_s(x)$ exists, but that the converse is false.

49. Let f be differentiable and let $f'(x_0) = m$. Find $f'(-x_0)$ if
(a) f is an odd function
(b) f is an even function.

50. Prove that the derivative of an odd function is an even function and that the derivative of an even function is an odd function.

GC *Use a graphing calculator or a computer to do Problems 51 and 52.*

EXPL **51.** Draw the graphs $f(x) = x^3 - 4x^2 + 3$ and its derivative $f'(x)$ on the interval $[-2, 5]$ using the same axes.

(a) Where on this interval is $f'(x) < 0$?

(b) Where on this interval is $f(x)$ decreasing as x increases?

(c) Make a conjecture. Experiment with other intervals and other functions to support this conjecture.

EXPL **52.** Draw the graphs of $f(x) = \cos x - \sin(x/2)$ and its derivative $f'(x)$ on the interval $[0, 9]$ using the same axes.

(a) Where on this interval is $f'(x) > 0$?

(b) Where on this interval is $f(x)$ increasing as x increases?

(c) Make a conjecture. Experiment with other intervals and other functions to support this conjecture.

Answers to Concepts Review: **1.** $[f(c + h) - f(c)]/h$, $[f(t) - f(c)]/(t - c)$ **2.** $[f(x + h) - f(x)]/h$ **3.** continuous; $|x|$ **4.** $2x^2; c$

3.3 Rules for Finding Derivatives

The process of finding the derivative of a function directly from the definition of the derivative, that is, by setting up the difference quotient

$$\frac{f(x + h) - f(x)}{h}$$

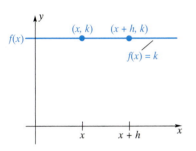

Figure 1

and evaluating its limit, can be time consuming and tedious. We are going to develop tools that will allow us to shortcut this lengthy process—that will, in fact, allow us to find derivatives of the most complicated looking functions.

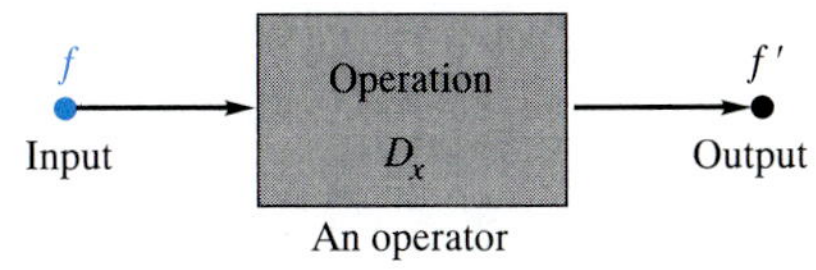

Recall that the derivative of a function f is another function f'. We saw in the previous section that, if $f(x) = x^3 + 7x$ is the formula for f, then $f'(x) = 3x^2 + 7$ is the formula for f'. When we take the derivative of f, we say that we are differentiating f. The derivative *operates* on f to produce f'. We often use the symbol D_x to indicate the operation of differentiating (Figure 1). The D_x symbol says that we are to take the derivative (with respect to the variable x) of what follows. Thus, we write $D_x f(x) = f'(x)$ or (in the case just mentioned) $D_x(x^3 + 7x) = 3x^2 + 7$. This D_x is an example of an **operator**. As Figure 1 suggests, an operator is a function whose input is a function and whose output is another function.

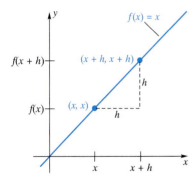

Figure 2

The Constant and Power Rules The graph of the constant function $f(x) = k$ is a horizontal line (Figure 2), which therefore has slope zero everywhere. This is one way to understand our first theorem.

Theorem A Constant Function Rule

If $f(x) = k$, where k is a constant, then for any x, $f'(x) = 0$; that is,

$$D_x(k) = 0$$

Proof

$$f'(x) = \lim_{h \to 0} \frac{f(x + h) - f(x)}{h} = \lim_{h \to 0} \frac{k - k}{h} = \lim_{h \to 0} 0 = 0 \; \blacklozenge$$

The graph of $f(x) = x$ is a line through the origin with slope 1 (Figure 3); so we should expect the derivative of this function to be 1 for all x.

Figure 3

Theorem B Identity Function Rule

If $f(x) = x$, then $f'(x) = 1$; that is,

$$D_x(x) = 1$$

Proof

$$f'(x) = \lim_{h \to 0} \frac{f(x + h) - f(x)}{h} = \lim_{h \to 0} \frac{x + h - x}{h} = \lim_{h \to 0} \frac{h}{h} = 1 \; \blacklozenge$$

Before starting our next theorem, we recall something from algebra: how to raise a binomial to a power.

$$(a + b)^2 = a^2 + 2ab + b^2$$

$$(a + b)^3 = a^3 + 3a^2b + 3ab^2 + b^3$$

$$(a + b)^4 = a^4 + 4a^3b + 6a^2b^2 + 4ab^3 + b^4$$

$$\vdots$$

$$(a + b)^n = a^n + na^{n-1}b + \frac{n(n-1)}{2}a^{n-2}b^2 + \cdots + nab^{n-1} + b^n$$

Theorem C   Power Rule

If $f(x) = x^n$, where n is a positive integer, then $f'(x) = nx^{n-1}$; that is,

$$D_x(x^n) = nx^{n-1}$$

Proof

$$f'(x) = \lim_{h \to 0} \frac{f(x + h) - f(x)}{h} = \lim_{h \to 0} \frac{(x + h)^n - x^n}{h}$$

$$= \lim_{h \to 0} \frac{x^n + nx^{n-1}h + \frac{n(n-1)}{2}x^{n-2}h^2 + \cdots + nxh^{n-1} + h^n - x^n}{h}$$

$$= \lim_{h \to 0} \frac{h\left[nx^{n-1} + \frac{n(n-1)}{2}x^{n-2}h + \cdots + nxh^{n-2} + h^{n-1}\right]}{h}$$

Within the brackets, all terms except the first have h as a factor, and so for every value of x each of these terms has limit zero as h approaches zero. Thus,

$$f'(x) = nx^{n-1} \quad \blacklozenge$$

As illustrations of Theorem C, note that

$$D_x(x^3) = 3x^2 \qquad D_x(x^9) = 9x^8 \qquad D_x(x^{100}) = 100x^{99}$$

D_x Is a Linear Operator   The operator D_x behaves very well when applied to constant multiples of functions or to sums of functions.

Theorem D   Constant Multiple Rule

If k is a constant and f is a differentiable function, then $(kf)'(x) = k \cdot f'(x)$; that is,

$$D_x[k \cdot f(x)] = k \cdot D_x f(x)$$

In words, *a constant multiplier k can be passed across the operator D_x.*

Proof   Let $F(x) = k \cdot f(x)$. Then

$$F'(x) = \lim_{h \to 0} \frac{F(x + h) - F(x)}{h} = \lim_{h \to 0} \frac{k \cdot f(x + h) - k \cdot f(x)}{h}$$

$$= \lim_{h \to 0} k \cdot \frac{f(x + h) - f(x)}{h} = k \cdot \lim_{h \to 0} \frac{f(x + h) - f(x)}{h}$$

$$= k \cdot f'(x)$$

The next-to-last step was the critical one. We could shift k past the limit sign because of the Main Limit Theorem Part 3. $\blacklozenge$

Examples that illustrate this result are

$$D_x(-7x^3) = -7D_x(x^3) = -7 \cdot 3x^2 = -21x^2$$

and

$$D_x\left(\tfrac{4}{3}x^9\right) = \tfrac{4}{3}D_x(x^9) = \tfrac{4}{3} \cdot 9x^8 = 12x^8$$

Theorem E   Sum Rule

If f and g are differentiable functions, then $(f + g)'(x) = f'(x) + g'(x)$; that is,

$$D_x[f(x) + g(x)] = D_x f(x) + D_x g(x)$$

In words, *the derivative of a sum is the sum of the derivatives.*

Proof   Let $F(x) = f(x) + g(x)$. Then

$$F'(x) = \lim_{h \to 0} \frac{[f(x + h) + g(x + h)] - [f(x) + g(x)]}{h}$$

$$= \lim_{h \to 0} \left[\frac{f(x + h) - f(x)}{h} + \frac{g(x + h) - g(x)}{h} \right]$$

$$= \lim_{h \to 0} \frac{f(x + h) - f(x)}{h} + \lim_{h \to 0} \frac{g(x + h) - g(x)}{h}$$

$$= f'(x) + g'(x)$$

Again, the next-to-last step was the critical one. It is justified by the Main Limit Theorem Part 4. ◆

Any operator L with the properties stated in Theorems D and E is called *linear*; that is, L is a **linear operator** if for all functions f and g:

1. $L(kf) = kL(f)$, for every constant k;
2. $L(f + g) = L(f) + L(g)$.

Linear operators will appear again and again in this book; D_x is a particularly important example. A linear operator always satisfies the difference rule $L(f - g) = L(f) - L(g)$ stated next for D_x.

Theorem F   Difference Rule

If f and g are differentiable functions, then $(f - g)'(x) = f'(x) - g'(x)$; that is,

$$D_x[f(x) - g(x)] = D_x f(x) - D_x g(x)$$

The proof of Theorem F is left as an exercise (Problem 54).

EXAMPLE 1   Find the derivatives of $5x^2 + 7x - 6$ and $4x^6 - 3x^5 - 10x^2 + 5x + 16$.

Solution

$$D_x(5x^2 + 7x - 6) = D_x(5x^2 + 7x) - D_x(6) \qquad \text{(Theorem F)}$$

$$= D_x(5x^2) + D_x(7x) - D_x(6) \qquad \text{(Theorem E)}$$

$$= 5D_x(x^2) + 7D_x(x) - D_x(6) \qquad \text{(Theorem D)}$$

$$= 5 \cdot 2x + 7 \cdot 1 + 0 \qquad \text{(Theorems C, B, A)}$$

$$= 10x + 7$$

To find the next derivative, we note that the theorems on sums and differences extend to any finite number of terms. Thus,

---

Linear Operator

The fundamental meaning of the word *linear*, as used in mathematics, is that given in this section. An operator L is linear if it satisfies the two key conditions:

- $L(ku) = kL(u)$
- $L(u + v) = L(u) + L(v)$

Linear operators play a central role in the *linear algebra* course, which many readers of this book will take.

Functions of the form $f(x) = mx + b$ are called *linear functions* because of their connections with lines. This terminology can be confusing because linear functions are not linear in the operator sense. To see this, note that

$$f(kx) = m(kx) + b$$

whereas

$$kf(x) = k(mx + b)$$

Thus, $f(kx) \neq kf(x)$ unless b happens to be zero.

$$D_x(4x^6 - 3x^5 - 10x^2 + 5x + 16)$$

$$= D_x(4x^6) - D_x(3x^5) - D_x(10x^2) + D_x(5x) + D_x(16)$$

$$= 4D_x(x^6) - 3D_x(x^5) - 10D_x(x^2) + 5D_x(x) + D_x(16)$$

$$= 4(6x^5) - 3(5x^4) - 10(2x) + 5(1) + 0$$

$$= 24x^5 - 15x^4 - 20x + 5 \qquad \blacksquare$$

The method of Example 1 allows us to find the derivative of any polynomial. If you know the Power Rule and do what comes naturally, you are almost sure to get the right result. Also, with practice, you will find that you can write the derivative immediately, without having to write any intermediate steps.

Product and Quotient Rules   Now we are in for a surprise. So far, we have seen that the limit of a sum or difference is equal to the sum or difference of the limits. (Theorem 2.6A, Parts 4 and 5), the limit of a product or quotient is the product or quotient of the limits (Theorem 2.6A, Parts 6 and 7), and the derivative of a sum or difference is the sum or difference of the derivatives (Theorems E and F). So what could be more natural than to have the derivative of a product be the product of the derivatives?

This may seem natural, but it is wrong. To see why, let's look at the following example.

EXAMPLE 2   Let $g(x) = x$, $h(x) = 1 + 2x$, and $f(x) = g(x) \cdot h(x) = x(1 + 2x)$. Find $D_x f(x)$, $D_x g(x)$, and $D_x h(x)$, and show that $D_x f(x) \neq [D_x g(x)][D_x h(x)]$.

Solution

$$D_x f(x) = D_x[x(1 + 2x)]$$

$$= D_x(x + 2x^2)$$

$$= 1 + 4x$$

$$D_x g(x) = D_x x = 1$$

$$D_x h(x) = D_x(1 + 2x) = 2$$

Notice that

$$D_x(g(x))D_x(h(x)) = 1 \cdot 2 = 2$$

whereas

$$D_x f(x) = D_x[g(x)h(x)] = 1 + 4x$$

Thus, $D_x f(x) \neq [D_x g(x)][D_x h(x)]$. $\qquad \blacksquare$

That the derivative of a product should be the product of the derivatives seemed so natural that it even fooled Gottfried Wilhelm von Leibniz, one of the discoverers of calculus. In a manuscript of November 11, 1675, he computed the product of two functions and said (without checking) that it was equal to the derivative of the product. Ten days later, he caught the error and gave the correct product rule, which we present as Theorem G.

Theorem G   Product Rule

If f and g are differentiable functions, then

$$(f \cdot g)'(x) = f(x)g'(x) + g(x)f'(x)$$

That is,

$$D_x[f(x)g(x)] = f(x)D_x g(x) + g(x)D_x f(x)$$

This rule should be memorized in words as follows: *The derivative of a product of two functions is the first times the derivative of the second plus the second times the derivative of the first.*

Proof   Let $F(x) = f(x)g(x)$. Then

$$F'(x) = \lim_{h \to 0} \frac{F(x + h) - F(x)}{h}$$

$$= \lim_{h \to 0} \frac{f(x + h)g(x + h) - f(x)g(x)}{h}$$

$$= \lim_{h \to 0} \frac{f(x + h)g(x + h) - f(x + h)g(x) + f(x + h)g(x) - f(x)g(x)}{h}$$

$$= \lim_{h \to 0} \left[f(x + h) \cdot \frac{g(x + h) - g(x)}{h} + g(x) \cdot \frac{f(x + h) - f(x)}{h} \right]$$

$$= \lim_{h \to 0} f(x + h) \cdot \lim_{h \to 0} \frac{g(x + h) - g(x)}{h} + g(x) \cdot \lim_{h \to 0} \frac{f(x + h) - f(x)}{h}$$

$$= f(x)g'(x) + g(x)f'(x)$$

The derivation just given relies first on the trick of adding and subtracting the same thing, that is, $f(x + h)g(x)$. Second, at the very end, we use the fact that

$$\lim_{h \to 0} f(x + h) = f(x)$$

This is just an application of Theorem 3.2A (which says that differentiability at a point implies continuity there) and the definition of continuity at a point. ♦

EXAMPLE 3   Find the derivative of $(3x^2 - 5)(2x^4 - x)$ by use of the Product Rule. Check the answer by doing the problem a different way.

Solution

$$D_x[(3x^2 - 5)(2x^4 - x)] = (3x^2 - 5)D_x(2x^4 - x) + (2x^4 - x)D_x(3x^2 - 5)$$
$$= (3x^2 - 5)(8x^3 - 1) + (2x^4 - x)(6x)$$
$$= 24x^5 - 3x^2 - 40x^3 + 5 + 12x^5 - 6x^2$$
$$= 36x^5 - 40x^3 - 9x^2 + 5$$

To check, we first multiply and then take the derivative.

$$(3x^2 - 5)(2x^4 - x) = 6x^6 - 10x^4 - 3x^3 + 5x$$

Thus,

$$D_x[(3x^2 - 5)(2x^4 - x)] = D_x(6x^6) - D_x(10x^4) - D_x(3x^3) + D_x(5x)$$
$$= 36x^5 - 40x^3 - 9x^2 + 5 \qquad \blacksquare$$

Theorem H   Quotient Rule

Let f and g be differentiable functions with $g(x) \neq 0$. Then

$$\left(\frac{f}{g} \right)'(x) = \frac{g(x)f'(x) - f(x)g'(x)}{g^2(x)}$$

That is,

$$D_x\left(\frac{f(x)}{g(x)} \right) = \frac{g(x)D_x f(x) - f(x)D_x g(x)}{g^2(x)}$$

We strongly urge you to memorize this in words, as follows: *The derivative of a quotient is equal to the denominator times the derivative of the numerator minus the numerator times the derivative of the denominator, all divided by the square of the denominator.*

Memorization

Some people say that memorization is passé, that only logical reasoning is important in mathematics. They are wrong. Some things (including the rules of this section) must become so much a part of our mental apparatus that we can use them without stopping to reflect.

"Civilization advances by extending the number of important operations which we can perform without thinking about them."

Alfred N. Whitehead

Proof Let $F(x) = f(x)/g(x)$. Then

$$F'(x) = \lim_{h \to 0} \frac{F(x + h) - F(x)}{h}$$

$$= \lim_{h \to 0} \frac{\dfrac{f(x + h)}{g(x + h)} - \dfrac{f(x)}{g(x)}}{h}$$

$$= \lim_{h \to 0} \frac{g(x)f(x + h) - f(x)g(x + h)}{h} \cdot \frac{1}{g(x)g(x + h)}$$

$$= \lim_{h \to 0} \left[\frac{g(x)f(x + h) - g(x)f(x) + f(x)g(x) - f(x)g(x + h)}{h} \cdot \frac{1}{g(x)g(x + h)} \right]$$

$$= \lim_{h \to 0} \left\{ \left[g(x)\frac{f(x + h) - f(x)}{h} - f(x)\frac{g(x + h) - g(x)}{h} \right] \frac{1}{g(x)g(x + h)} \right\}$$

$$= \left[g(x)f'(x) - f(x)g'(x) \right] \frac{1}{g(x)g(x)} \; \blacklozenge$$

EXAMPLE 4   Find the derivative of $\dfrac{(3x - 5)}{(x^2 + 7)}$.

Solution

$$D_x\left[\frac{3x - 5}{x^2 + 7} \right] = \frac{(x^2 + 7)D_x(3x - 5) - (3x - 5)D_x(x^2 + 7)}{(x^2 + 7)^2}$$

$$= \frac{(x^2 + 7)(3) - (3x - 5)(2x)}{(x^2 + 7)^2}$$

$$= \frac{-3x^2 + 10x + 21}{(x^2 + 7)^2} \qquad \blacksquare$$

EXAMPLE 5   Find $D_x y$ if $y = \dfrac{2}{x^4 + 1} + \dfrac{3}{x}$.

Solution

$$D_x y = D_x\left(\frac{2}{x^4 + 1} \right) + D_x\left(\frac{3}{x} \right)$$

$$= \frac{(x^4 + 1)D_x(2) - 2D_x(x^4 + 1)}{(x^4 + 1)^2} + \frac{xD_x(3) - 3D_x(x)}{x^2}$$

$$= \frac{(x^4 + 1)(0) - (2)(4x^3)}{(x^4 + 1)^2} + \frac{(x)(0) - (3)(1)}{x^2}$$

$$= \frac{-8x^3}{(x^4 + 1)^2} - \frac{3}{x^2} \qquad \blacksquare$$

EXAMPLE 6   Show that the Power Rule holds for negative integral exponents; that is,

$$\boxed{D_x(x^{-n}) = -nx^{-n-1}}$$

Solution

$$D_x(x^{-n}) = D_x\left(\frac{1}{x^n} \right) = \frac{x^n \cdot 0 - 1 \cdot nx^{n-1}}{x^{2n}} = \frac{-nx^{n-1}}{x^{2n}} = -nx^{-n-1} \qquad \blacksquare$$

We saw as part of Example 5 that $D_x(3/x) = -3/x^2$. Now we have another way to see the same thing.

$$D_x\left(\frac{3}{x}\right) = D_x(3x^{-1}) = 3D_x(x^{-1}) = 3(-1)x^{-2} = -\frac{3}{x^2}$$

Concepts Review

1. The derivative of a product of two functions is the first times _____ plus the _____ times the derivative of the first. In symbols, $D_x[f(x)g(x)] =$ _____.

2. The derivative of a quotient is the _____ times the derivative of the numerator minus the numerator times the derivative of the _____, all divided by the _____. In symbols, $D_x[f(x)/g(x)] =$ _____.

3. The second term (the term involving h) in the expansion of $(x + h)^n$ is _____. It is this fact that leads to the formula $D_x[x^n] =$ _____.

4. L is called a linear operator if $L(kf) =$ _____ and $L(f + g) =$ _____. The derivative operator denoted by _____ is such an operator.

Problem Set 3.3

In Problems 1–44, find $D_x y$ using the rules of this section.

1. $y = 2x^2$

2. $y = 3x^3$

3. $y = \pi x$

4. $y = \pi x^3$

5. $y = 2x^{-2}$

6. $y = -3x^{-4}$

7. $y = \dfrac{\pi}{x}$

8. $y = \dfrac{\alpha}{x^3}$

9. $y = \dfrac{100}{x^5}$

10. $y = \dfrac{3\alpha}{4x^5}$

11. $y = x^2 + 2x$

12. $y = 3x^4 + x^3$

13. $y = x^4 + x^3 + x^2 + x + 1$

14. $y = 3x^4 - 2x^3 - 5x^2 + \pi x + \pi^2$

15. $y = \pi x^7 - 2x^5 - 5x^{-2}$

16. $y = x^{12} + 5x^{-2} - \pi x^{-10}$

17. $y = \dfrac{3}{x^3} + x^{-4}$

18. $y = 2x^{-6} + x^{-1}$

19. $y = \dfrac{2}{x} - \dfrac{1}{x^2}$

20. $y = \dfrac{3}{x^3} - \dfrac{1}{x^4}$

21. $y = \dfrac{1}{2x} + 2x$

22. $y = \dfrac{2}{3x} - \dfrac{2}{3}$

23. $y = x(x^2 + 1)$

24. $y = 3x(x^3 - 1)$

25. $y = (2x + 1)^2$

26. $y = (-3x + 2)^2$

27. $y = (x^2 + 2)(x^3 + 1)$

28. $y = (x^4 - 1)(x^2 + 1)$

29. $y = (x^2 + 17)(x^3 - 3x + 1)$

30. $y = (x^4 + 2x)(x^3 + 2x^2 + 1)$

31. $y = (5x^2 - 7)(3x^2 - 2x + 1)$

32. $y = (3x^2 + 2x)(x^4 - 3x + 1)$

33. $y = \dfrac{1}{3x^2 + 1}$

34. $y = \dfrac{2}{5x^2 - 1}$

35. $y = \dfrac{1}{4x^2 - 3x + 9}$

36. $y = \dfrac{4}{2x^3 - 3x}$

37. $y = \dfrac{x - 1}{x + 1}$

38. $y = \dfrac{2x - 1}{x - 1}$

39. $y = \dfrac{2x^2 - 1}{3x + 5}$

40. $y = \dfrac{5x - 4}{3x^2 + 1}$

41. $y = \dfrac{2x^2 - 3x + 1}{2x + 1}$

42. $y = \dfrac{5x^2 + 2x - 6}{3x - 1}$

43. $y = \dfrac{x^2 - x + 1}{x^2 + 1}$

44. $y = \dfrac{x^2 - 2x + 5}{x^2 + 2x - 3}$

45. If $f(0) = 4, f'(0) = -1, g(0) = -3$, and $g'(0) = 5$, find
(a) $(f \cdot g)'(0)$ (b) $(f + g)'(0)$ (c) $(f/g)'(0)$

46. If $f(3) = 7, f'(3) = 2, g(3) = 6$, and $g'(3) = -10$, find
(a) $(f - g)'(3)$ (b) $(f \cdot g)'(3)$ (c) $(g/f)'(3)$

47. Use the Product Rule to show that $D_x[f(x)]^2 = 2 \cdot f(x) \cdot D_x f(x)$.

EXPL 48. Develop a rule for $D_x[f(x)g(x)h(x)]$.

49. Find the equation of the tangent line to $y = x^2 - 2x + 2$ at the point $(1, 1)$.

50. Find the equation of the tangent line to $y = 1/(x^2 + 4)$ at the point $(1, 1/5)$.

51. Find all points on the graph of $y = x^3 - x^2$ where the tangent line is horizontal.

52. Find all points on the graph of $y = \frac{1}{3}x^3 + x^2 - x$ where the tangent line has slope 1.

53. Find all points on the graph of $y = 100/x^5$ where the tangent line is perpendicular to the line $y = x$.

54. Prove Theorem F in two ways.

55. The height s in feet of a ball above the ground at t seconds is given by $s = -16t^2 + 40t + 100$.
(a) What is its instantaneous velocity at $t = 2$?
(b) When is its instantaneous velocity 0?

56. A ball rolls down a long inclined plane so that its distance s from its starting point after t seconds is $s = 4.5t^2 + 2t$ feet. When will its instantaneous velocity be 30 feet per second?

57. There are two tangent lines to the curve $y = 4x - x^2$ that go through $(2, 5)$. Find the equations of both of them. *Hint:* Let (x_0, y_0) be a point of tangency. Find two conditions that (x_0, y_0) must satisfy. See Figure 4.

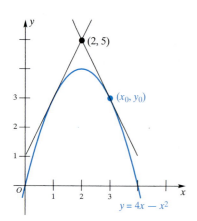

Figure 4

58. A space traveler is moving from left to right along the curve $y = x^2$. When she shuts off the engines, she will continue traveling along the tangent line at the point where she is at that time. At what point should she shut off the engines in order to reach the point $(4, 15)$?

59. A fly is crawling from left to right along the top of the curve $y = 7 - x^2$ (Figure 5). A spider waits at the point $(4, 0)$. Find the distance between the two insects when they first see each other.

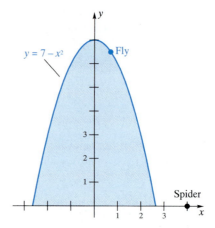

Figure 5

60. Let $P(a, b)$ be a point on the first quadrant portion of the curve $y = 1/x$ and let the tangent line at P intersect the x-axis at A. Show that triangle AOP is isosceles and determine its area.

61. The radius of a spherical watermelon is growing at a constant rate of 2 centimeters per week. The thickness of the rind is always one-tenth of the radius. How fast is the volume of the rind growing at the end of the fifth week? Assume that the radius is initially 0.

CAS **62.** Redo Problems 29–44 on a computer and compare your answers with those you get by hand.

Answers to Concepts Review: **1.** the derivative of the second; second; $f(x)D_x g(x) + g(x)D_x f(x)$ **2.** denominator, denominator, square of the denominator; $[g(x)D_x f(x) - f(x)D_x g(x)]/g^2(x)$ **3.** $nx^{n-1} h; nx^{n-1}$ **4.** $kL(f); L(f) + L(g); D_x$

3.4
Derivatives of Trigonometric Functions

Our modern world runs on wheels. Questions about rotating wheels and velocities of points on them lead inevitably to the study of sines and cosines and their derivatives. Other periodic phenomena that are related to sines and cosines are weather and tides. To prepare for this study, it would be well to review Sections 2.3 and 2.7. Figure 1 reminds us of the definition of the sine and cosine functions. In what follows, t should be thought of as a number measuring the length of an arc on the unit circle or, equivalently, as the number of radians in the corresponding angle. Thus, $f(t) = \sin t$ and $g(t) = \cos t$ are functions for which both domain and range are sets of real numbers. We may consider the problem of finding their derivatives.

The Derivative Formulas We choose to use x rather than t as our basic variable. To find $D_x(\sin x)$, we appeal to the definition of derivative and use the addition identity for $\sin(x + h)$.

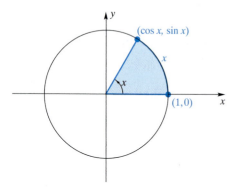

Figure 1

$$D_x(\sin x) = \lim_{h \to 0} \frac{\sin(x + h) - \sin x}{h}$$

$$= \lim_{h \to 0} \frac{\sin x \cos h + \cos x \sin h - \sin x}{h}$$

$$= \lim_{h \to 0} \left(-\sin x \frac{1 - \cos h}{h} + \cos x \frac{\sin h}{h} \right)$$

$$= (-\sin x)\left[\lim_{h \to 0} \frac{1 - \cos h}{h} \right] + (\cos x)\left[\lim_{h \to 0} \frac{\sin h}{h} \right]$$

Notice that the two limits in this last expression are exactly the limits we studied in Section 2.7. In Theorem 2.7B we proved that

$$\lim_{h \to 0} \frac{\sin h}{h} = 1 \quad \text{and} \quad \lim_{h \to 0} \frac{1 - \cos h}{h} = 0$$

Thus,

$$D_x(\sin x) = (-\sin x) \cdot 0 + (\cos x) \cdot 1 = \cos x$$

Similarly,

$$D_x(\cos x) = \lim_{h \to 0} \frac{\cos(x + h) - \cos x}{h}$$

$$= \lim_{h \to 0} \frac{\cos x \cos h - \sin x \sin h - \cos x}{h}$$

$$= \lim_{h \to 0} \left(-\cos x \frac{1 - \cos h}{h} - \sin x \frac{\sin h}{h} \right)$$

$$= (-\cos x) \cdot 0 - (\sin x) \cdot 1$$

$$= -\sin x$$

We summarize these results in an important theorem.

Theorem A

The functions $f(x) = \sin x$ and $g(x) = \cos x$ are both differentiable. In fact,

$$D_x(\sin x) = \cos x \qquad D_x(\cos x) = -\sin x \; \blacklozenge$$

Could You Have Guessed?

The solid curve below is the graph of $y = \sin x$. Note that the slope is 1 at 0, 0 at $\pi/2$, -1 at π, and so on. When we graph the slope function (the derivative), we obtain the dashed curve. Could you have guessed that $D_x \sin x = \cos x$?

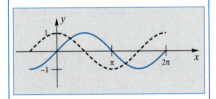

Try plotting these two functions in the same window on your CAS or graphing calculator.

EXAMPLE 1 Find $D_x(3 \sin x - 2 \cos x)$.

Solution

$$D_x(3 \sin x - 2 \cos x) = 3 D_x(\sin x) - 2 D_x(\cos x)$$

$$= 3 \cos x + 2 \sin x \qquad \blacksquare$$

EXAMPLE 2 Find the equation of the tangent line to the graph of $y = 3 \sin 2x$ at the point $(\pi/2, 0)$ (see Figure 2).

Solution We need the derivative of $\sin 2x$; unfortunately, at this point we know how to find only the derivative of $\sin x$. However, $\sin 2x = 2 \sin x \cos x$. Thus,

$$D_x(3 \sin 2x) = D_x(6 \sin x \cos x)$$

$$= 6 D_x(\sin x \cos x)$$

$$= 6 \left[\sin x \, D_x(\cos x) + \cos x \, D_x(\sin x) \right]$$

$$= 6 \left[(\sin x)(-\sin x) + \cos x \cos x \right]$$

$$= 6 \left[\cos^2 x - \sin^2 x \right]$$

$$= 6 \cos 2x$$

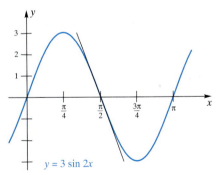

$y = 3 \sin 2x$

Figure 2

At $x = \pi/2$, this derivative has the value -6, which is therefore the slope of the desired tangent line. The equation of this line is

$$y - 0 = -6\left(x - \frac{\pi}{2} \right) \qquad \blacksquare$$

EXAMPLE 3 Consider a Ferris wheel of radius 30 feet, which is rotating counterclockwise with an *angular velocity* of 2 radians per second. How fast is a seat on the rim rising (in the vertical direction) when it is 15 feet above the horizontal line through the center of the wheel?

Solution We may suppose that the wheel is centered at the origin and that the seat P was at $(30, 0)$ at time $t = 0$ (Figure 3). Thus, at time t, P has moved through

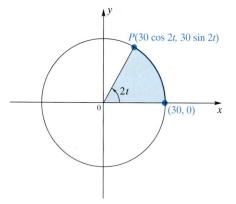

Figure 3

an angle of $2t$ radians and so it has coordinates $(30 \cos 2t, 30 \sin 2t)$. The rate at which P is rising is just the derivative of the vertical coordinate $30 \sin 2t$ measured at an appropriate t-value. By Example 2,

$$D_x(30 \sin 2t) = 60 \cos 2t$$

The appropriate t at which to evaluate this derivative is $t = \pi/12$, since $30 \sin(2 \cdot \pi/12) = 15$. We conclude that at $t = \pi/12$ the seat P is rising at

$$60 \cos\left(2 \cdot \frac{\pi}{12}\right) = 60 \sqrt{3}/2 \approx 51.96 \text{ feet per second} \qquad \blacksquare$$

Once we know the derivatives of the sine and cosine functions, the derivatives of the other trigonometric functions can be found by applying the quotient rule. The results are summarized in Theorem B. For proofs, see Problems 5–8.

Theorem B

$$D_x \tan x = \sec^2 x \qquad\qquad D_x \cot x = -\csc^2 x$$

$$D_x \sec x = \sec x \tan x \qquad\qquad D_x \csc x = -\csc x \cot x$$

Concepts Review

1. By definition, $D_x(\sin x) = \lim\limits_{h \to 0}$ _____.

2. To evaluate the limit in the preceding statement, we first use the addition identity for the sine function and then do a little algebra to obtain

$$D_x(\sin x) = (-\sin x)\left(\lim_{h \to 0} \frac{1 - \cos h}{h}\right) + (\cos x)\left(\lim_{h \to 0} \frac{\sin h}{h}\right)$$

The two displayed limits have the values _____ and _____, respectively.

3. The result of the calculation in the preceding statement is the important derivative formula $D_x(\sin x) =$ _____. The corresponding derivative formula $D_x(\cos x) =$ _____ is obtained in a similar manner.

4. At $x = \pi/3$, $D_x(\sin x)$ has the value _____. Thus, the equation of the tangent line to $y = \sin x$ at $x = \pi/3$ is _____.

Problem Set 3.4

In Problems 1–14, find $D_x y$.

1. $y = 2 \sin x + 3 \cos x$

2. $y = \sin^2 x$

3. $y = \sin^2 x + \cos^2 x$

4. $y = 1 - \cos^2 x$

5. $y = \sec x = 1/\cos x$

6. $y = \csc x = 1/\sin x$

7. $y = \tan x = \dfrac{\sin x}{\cos x}$

8. $y = \cot x = \dfrac{\cos x}{\sin x}$

9. $y = \dfrac{\sin x + \cos x}{\cos x}$

10. $y = \dfrac{\sin x + \cos x}{\tan x}$

11. $y = x^2 \cos x$

12. $y = \dfrac{x \cos x + \sin x}{x^2 + 1}$

13. $y = \tan^2 x$

14. $y = \sec^3 x$

[C] **15.** Find the equation of the tangent line to $y = \cos x$ at $x = 1$.

16. Find the equation of the tangent line to $y = \cot x$ at $x = \dfrac{\pi}{4}$.

17. Consider the Ferris wheel of Example 3. At what rate is the seat on the rim moving horizontally when $t = \pi/4$ seconds (i.e., when the seat reaches the very top of the wheel)?

18. A Ferris wheel of radius 20 feet is rotating counterclockwise at an angular velocity of 1 radian per second. One seat on the rim is at $(20, 0)$ at $t = 0$.

(a) What are its coordinates at $t = \pi/6$?

(b) How fast is it rising (vertically) at $t = \pi/6$?

(c) How fast is it rising (vertically) when it is rising at the fastest rate?

19. Find the equation of the tangent line to $y = \tan x$ at $x = 0$.

20. Find all points on the graph of $y = \tan^2 x$ where the tangent line is horizontal.

21. Find all points on the graph of $y = 9 \sin x \cos x$ where the tangent line is horizontal.

22. Let $f(x) = x - \sin x$. Find all points on the graph of $y = f(x)$ where the tangent line is horizontal. Find all points on the graph of $y = f(x)$ where the tangent line has slope 2.

23. Show that the curves $y = \sqrt{2} \sin x$ and $y = \sqrt{2} \cos x$ intersect at right angles at a certain point with $0 < x < \pi/2$.

24. At time t seconds, the center of a bobbing cork is $2 \sin t$ centimeters above (or below) water level. What is the velocity of the cork at $t = 0, \pi/2, \pi$?

25. Use the definition of derivative to show that $D_x(\sin x^2) = 2x \cos x^2$.

26. Use the definition of derivative to show that $D_x(\sin 5x) = 5 \cos 5x$.

27. Let x_0 be the smallest positive value of x at which the curves $y = \sin x$ and $y = \sin 2x$ intersect. Find x_0 and also the acute angle at which the two curves intersect at x_0 (see Problem 40 of Section 2.3).

28. An isosceles triangle is topped by a semicircle, as shown in Figure 4. Let D be the area of triangle AOB and E be the area of the shaded region. Find a formula for D/E in terms of t and then calculate

$$\lim_{t \to 0^+} \frac{D}{E} \quad \text{and} \quad \lim_{t \to \pi^-} \frac{D}{E}$$

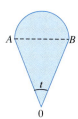

Figure 4

GC *Problems 29 and 30 are computer or graphing calculator exercises.*

29. Let $f(x) = x \sin x$.

(a) Draw the graphs of $f(x)$ and $f'(x)$ on $[\pi, 6\pi]$.

(b) How many solutions does $f(x) = 0$ have on $[\pi, 6\pi]$? How many solutions does $f'(x) = 0$ have on this interval?

(c) What is wrong with the following conjecture? If f and f' are both continuous and differentiable on $[a, b]$, if $f(a) = f(b) = 0$, and if $f(x) = 0$ has exactly n solutions on $[a, b]$, then $f'(x) = 0$ has exactly $n - 1$ solutions on $[a, b]$.

(d) Determine the maximum value of $|f(x) - f'(x)|$ on $[\pi, 6\pi]$.

30. Let $f(x) = \cos^3 x - 1.25 \cos^2 x + 0.225$. Find $f'(x_0)$ at that point x_0 in $[\pi/2, \pi]$ where $f(x_0) = 0$.

Answers to Concepts Review: **1.** $[\sin(x + h) - \sin x]/h$
2. $0; 1$ **3.** $\cos x; -\sin x$ **4.** $\frac{1}{2}; y - \sqrt{3}/2 = \frac{1}{2}(x - \pi/3)$

3.5
The Chain Rule

Imagine trying to find the derivative of

$$F(x) = \left(2x^2 - 4x + 1\right)^{60}$$

We could find the derivative, but we would first have to multiply together the 60 quadratic factors of $2x^2 - 4x + 1$ and then differentiate the resulting polynomial. Or, how about trying to find the derivative of

$$G(x) = \sin 3x$$

We might be able to use some trigonometric identities to reduce it to something that depends on $\sin x$ and $\cos x$ and then use the rules from the previous section.

Fortunately, there is a better way. After learning the *Chain Rule*, we will be able to write the answers

$$F'(x) = \left(2x^2 - 4x + 1\right)^{59}(4x - 4)$$

and

$$G'(x) = 3 \cos 3x$$

The Chain Rule is so important that we will seldom again differentiate any function without using it. In order to state the rule properly, we need to emphasize the significance of x in our D_x notation.

The D_x Notation
The symbol $D_x y$ means the derivative of y with respect to x; it tells how fast y is changing with respect to x. The subscript x indicates that x is being treated as the basic variable. Thus, if $y = s^2 x^3$, we may write

$$D_x y = 3s^2 x^2 \quad \text{and} \quad D_s y = 2sx^3$$

In the first case, s is treated as a constant and x is the basic variable; in the second case, x is constant and s is the basic variable.

More important is the following example. Suppose that $y = u^{60}$ and $u = 2x^2 - 4x + 1$. Then $D_u y = 60u^{59}$ and $D_x u = 4x - 4$. But notice that when we substitute $u = 2x^2 - 4x + 1$ in $y = u^{60}$, we obtain

$$y = \left(2x^2 - 4x + 1\right)^{60}$$

so it makes sense to ask for $D_x y$. What is $D_x y$ and how is it related to $D_u y$ and $D_x u$? More generally, how do we differentiate a composite function?

Differentiating a Composite Function If David can type twice as fast as Mary and Mary can type three times as fast as Jack, then David can type $2 \cdot 3 = 6$ times as fast as Jack. The two rates are multiplied.

Consider the composite function $y = f(g(x))$. Since a derivative indicates a rate of change, we can say that

$$y \text{ changes } D_u y \text{ times as fast as } u$$

$$u \text{ changes } D_x u \text{ times as fast as } x$$

It seems reasonable to conclude that

$$y \text{ changes } D_u y \cdot D_x u \text{ times as fast as } x$$

This is in fact true, and we will suggest a formal proof in the next section. The result is called the **Chain Rule**.

Theorem A Chain Rule

Let $y = f(u)$ and $u = g(x)$. If g is differentiable at x and f is differentiable at $u = g(x)$, then the composite function $f \circ g$, defined by $(f \circ g)(x) = f(g(x))$, is differentiable at x and

$$(f \circ g)'(x) = f'(g(x))g'(x)$$

That is,

$$D_x(f(g(x))) = f'(g(x))g'(x)$$

or

$$D_x y = (D_u y)(D_x u)$$

You can remember the chain rule this way: *The derivative of a composite function is the derivative of the outer function evaluated at the inner function, times the derivative of the inner function.*

Applications of the Chain Rule We begin with the example $(2x^2 - 4x + 1)^{60}$ introduced at the beginning of this section.

EXAMPLE 1 If $y = (2x^2 - 4x + 1)^{60}$, find $D_x y$.

Solution We think of y as the 60th power of a function of x; that is

$$y = u^{60} \quad \text{and} \quad u = 2x^2 - 4x + 1$$

The outer function is u^{60} and the inner function is $2x^2 - 4x + 1$. Thus,

$$D_x y = D_u y \cdot D_x u$$
$$= (60u^{59})(4x - 4)$$
$$= 60(2x^2 - 4x + 1)^{59}(4x - 4) \qquad \blacksquare$$

EXAMPLE 2 If $y = 1/(2x^5 - 7)^3$, find $D_x y$.

Solution Think of it this way.

$$y = \frac{1}{u^3} = u^{-3} \quad \text{and} \quad u = 2x^5 - 7$$

Thus,

$$D_x y = D_u y \cdot D_x u$$
$$= (-3u^{-4})(10x^4)$$
$$= \frac{-3}{u^4} \cdot 10x^4$$
$$= \frac{-30x^4}{(2x^5 - 7)^4} \qquad \blacksquare$$

EXAMPLE 3 If $y = \sin(x^3 - 3x)$, find $D_x y$.

Solution We may write
$$y = \sin u \quad \text{and} \quad u = x^3 - 3x$$
Hence,
$$D_x y = D_u y \cdot D_x u$$
$$= (\cos u) \cdot (3x^2 - 3)$$
$$= [\cos(x^3 - 3x)] \cdot (3x^2 - 3)$$
$$= (3x^2 - 3)\cos(x^3 - 3x) \quad \blacksquare$$

EXAMPLE 4 Find $D_t\left(\dfrac{t^3 - 2t + 1}{t^4 + 3}\right)^{13}$.

Solution Think of this as finding $D_t y$, where
$$y = u^{13} \quad \text{and} \quad u = \frac{t^3 - 2t + 1}{t^4 + 3}$$

Then the Chain Rule followed by the Quotient Rule gives
$$D_t y = D_u y \cdot D_t u$$
$$= 13u^{12} \frac{(t^4 + 3)(3t^2 - 2) - (t^3 - 2t + 1)(4t^3)}{(t^4 + 3)^2}$$
$$= 13\left(\frac{t^3 - 2t + 1}{t^4 + 3}\right)^{12} \cdot \frac{-t^6 + 6t^4 - 4t^3 + 9t^2 - 6}{(t^4 + 3)^2} \quad \blacksquare$$

Soon you will learn to make a mental introduction of the middle variable without actually writing it. Thus, an expert immediately writes
$$D_x(\cos 3x) = (-\sin 3x) \cdot 3 = -3\sin 3x$$
$$D_x(x^3 + \sin x)^6 = 6(x^3 + \sin x)^5(3x^2 + \cos x)$$
$$D_t\left(\frac{t}{\cos 3t}\right)^4 = 4\left(\frac{t}{\cos 3t}\right)^3 \frac{\cos 3t - t(-\sin 3t)3}{\cos^2 3t}$$
$$= \frac{4t^3(\cos 3t + 3t\sin 3t)}{\cos^5 3t}$$

Applying the Chain Rule More than Once

Sometimes when we apply the Chain Rule to a composite function we find that differentiation of the inner function also requires the Chain Rule. In cases like this, we simply have to use the Chain Rule a second time.

EXAMPLE 5 Find $D_x \sin^3(4x)$.

Solution Remember, $\sin^3(4x) = [\sin(4x)]^3$, so we view this as the cube of a function of x. Thus, using our rule "derivative of the outer function evaluated at the inner function times the derivative of the inner function," we have
$$D_x \sin^3(4x) = D_x[\sin(4x)]^3 = 3[\sin(4x)]^{3-1} D_x[\sin(4x)]$$

Now we apply the Chain Rule once again for the derivative of the inner function.

$$D_x \sin^3(4x) = 3[\sin(4x)]^{3-1} D_x \sin(4x)$$

$$= 3[\sin(4x)]^2 \cos(4x) D_x(4x)$$

$$= 3[\sin(4x)]^2 \cos(4x)4$$

$$= 12 \cos(4x) \sin^2(4x)$$ ∎

EXAMPLE 6   Find $D_x \sin[\cos(x^2)]$.

Solution

$$D_x \sin[\cos(x^2)] = \cos[\cos(x^2)] \cdot [-\sin(x^2)] \cdot 2x$$

$$= -2x \sin(x^2) \cos[\cos(x^2)]$$ ∎

EXAMPLE 7   As the sun sets behind a 120-foot building, the building's shadow grows. How fast is the shadow growing (in feet per second) when the sun's rays make an angle of $\pi/4$? (See Figure 1.)

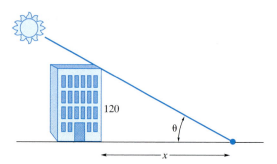

Figure 1

Solution   Let x be the length of the shadow in feet, and let θ be the angle of the sun's ray. Let t denote the time measured in seconds. Then x is a function of θ, and θ is a function of t. We are asked to find $D_t x$. From Figure 1, we see that $x = 120 \cot \theta$, and since the earth rotates once every 24 hours, or 86,400 seconds, we have $D_t \theta = -2\pi/86{,}400$. (The negative sign is used because θ *decreases* as the sun sets.) Using the Chain Rule and the rule for the derivative of the cotangent (Theorem 3.4B), we have

$$D_t x = D_\theta x \cdot D_t \theta = D_\theta(120 \cot \theta) \cdot D_t \theta$$

$$= 120(-\csc^2 \theta)\left(-\frac{2\pi}{86{,}400}\right)$$

$$= \frac{\pi}{360} \csc^2 \theta$$

When $\theta = \pi/4$, we have

$$D_t x = \frac{\pi}{360} \csc^2 \frac{\pi}{4} \approx 0.0175 \, \frac{\text{ft}}{\text{sec}}$$

Notice that as the sun sets, θ is *decreasing* (hence, $D_t \theta$ is negative), while the shadow x is *increasing* (hence, $D_t x$ is positive.) ∎

Concepts Review

1. If $y = f(u)$, where $u = g(t)$, then $D_t y = D_u y \cdot$ _____.
In function notation, $(f \circ g)'(t) =$ _____ _____.

2. If $w = G(v)$, where $v = H(s)$, then $D_s w =$ _____ $D_s v$.
In function notation $(G \circ H)'(s) =$ _____ _____.

3. $D_x \cos[(f(x))^2] = -\sin($_____$) \cdot D_x($_____$)$

4. If $y = (2x + 1)^3 \sin(x^2)$, then $D_x y =$
$(2x + 1)^3 \cdot$ _____ $+ \sin(x^2) \cdot$ _____

Problem Set 3.5

In Problems 1–22, find $D_x y$.

1. $y = (1 + x)^{15}$

2. $y = (7 + x)^5$

3. $y = (3 - 2x)^5$

4. $y = (4 + 2x^2)^7$

5. $y = (x^3 - 2x^2 + 3x + 1)^{11}$

6. $y = (x^5 - 5x^3 + \pi x + 1)^{101}$

7. $y = (x^3 - 2x^2 + 3x + 1)^{111}$

8. $y = (x^2 - x + 1)^{-7}$

9. $y = \dfrac{1}{(x + 3)^5}$

10. $y = \dfrac{1}{(3x^2 + x - 3)^9}$

11. $y = \sin(x^2 + x)$

12. $y = \cos(3x^2 - 2x)$

13. $y = \cos^3 x$

14. $y = \sin^4(3x^2)$

15. $y = \left(\dfrac{x + 1}{x - 1}\right)^3$

16. $y = \left(\dfrac{x - 2}{x - \pi}\right)^{-3}$

17. $y = \cos\left(\dfrac{3x^2}{x + 2}\right)$

18. $y = \cos^3\left(\dfrac{x^2}{1 - x}\right)$

19. $y = (3x - 2)^2(3 - x^2)^2$

20. $y = (2 - 3x^2)^4(x^7 + 3)^3$

21. $y = \dfrac{(x + 1)^2}{3x - 4}$

22. $y = \dfrac{2x - 3}{(x^2 + 4)^2}$

In Problems 23–28, find the indicated derivative.

23. $D_t\left(\dfrac{3t - 2}{t + 5}\right)^3$

24. $D_s\left(\dfrac{s^2 - 9}{s + 4}\right)$

25. $D_t\left(\dfrac{(3t - 2)^3}{t + 5}\right)$

26. $D_\theta(\sin^3 \theta)$

27. $D_x\left(\dfrac{\sin x}{\cos 2x}\right)^3$

28. $D_t[\sin t \tan(t^2 + 1)]$

In Problems 29–32, evaluate the indicated derivative.

29. $f'(3)$ if $f(x) = \left(\dfrac{x^2 + 1}{x + 2}\right)^3$

30. $G'(1)$ if $G(t) = (t^2 + 9)^3(t^2 - 2)^4$

<u>C</u> **31.** $F'(1)$ if $F(t) = \sin(t^2 + 3t + 1)$

32. $g'(\tfrac{1}{2})$ if $g(s) = \cos \pi s \sin^2 \pi s$

In Problems 33–40, apply the Chain Rule more than once (Examples 5 and 6) to find the indicated derivative.

33. $D_x[\sin^4(x^2 + 3x)]$

34. $D_t[\cos^5(4t - 19)]$

35. $D_t[\sin^3(\cos t)]$

36. $D_u\left[\cos^4\left(\dfrac{u + 1}{u - 1}\right)\right]$

37. $D_\theta[\cos^4(\sin \theta^2)]$

38. $D_x[x \sin^2(2x)]$

39. $D_x\{\sin[\cos(\sin 2x)]\}$

40. $D_t\{\cos^2[\cos(\cos t)]\}$

41. Find the equation of the tangent line to $y = (x^2 + 1)^3(x^4 + 1)^2$ at $(1, 32)$.

42. A point P is moving in the plane so that its coordinates after t seconds are $(4 \cos 2t, 7 \sin 2t)$, measured in feet.

(a) Show that P is following an elliptical path. *Hint:* Show that $(x/4)^2 + (y/7)^2 = 1$, which is an equation of an ellipse.

(b) Obtain an expression for L, the distance of P from the origin at time t.

(c) How fast is the distance between P and the origin changing when $t = \pi/8$? You will need the fact that $D_u(\sqrt{u}) = 1/(2\sqrt{u})$ (see Example 4 of Section 3.2).

43. A wheel centered at the origin and of radius 10 centimeters is rotating counterclockwise at a rate of 4 revolutions per second. A point P on the rim is at $(10, 0)$ at $t = 0$.

(a) What are the coordinates of P at time t seconds?

(b) At what rate is P rising (or falling) at time $t = 1$?

44. Consider the wheel–piston device in Figure 2. The wheel has radius 1 foot and rotates counterclockwise at 2 radians per second. The connecting rod is 5 feet long. The point P is at $(1, 0)$ at time $t = 0$.

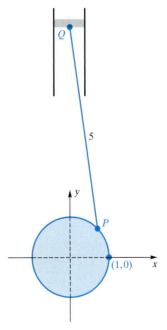

Figure 2

(a) Find the coordinates of P at time t.

(b) Find the y-coordinate of Q at time t (the x-coordinate is always zero).

(c) Find the velocity of Q at time t. You will need the fact that $D_u(\sqrt{u}) = 1/(2\sqrt{u})$.

45. Do Problem 44 assuming that the wheel is rotating at 60 revolutions per minute and t is measured in seconds.

46. Air is being pumped into a spherical balloon in such a way that the radius r is increasing at the constant rate of 0.75 centimeter per second. How fast is the volume of the balloon increasing when the radius is 6 centimeters?

47. Air is being pumped into a spherical balloon at a constant rate of 3 cubic centimeters per second. How fast is the radius of the balloon changing when the volume is 60 cubic centimeters?

48. Show that $D_x|x| = |x|/x$, $x \neq 0$. *Hint*: Write $|x| = \sqrt{x^2}$ and use the Chain Rule with $u = x^2$.

49. Apply the result in Problem 48 to find each derivative.
(a) $D_x|x^2 - 1|$ (b) $D_x|\sin x|$

50. In Chapter 7 we will study a function L satisfying $L'(x) = 1/x$. Find each of the following derivatives.

(a) $D_x L\left(\dfrac{x}{x+1}\right)$ (b) $D_x L(\cos^4 x)$

51. Let $f(0) = 0$ and $f'(0) = 2$. Find the derivative of $f(f(f(f(x))))$ at $x = 0$.

52. Use the Chain Rule to show that the derivative of an odd function is even and the derivative of an even function is odd.

GC **53.** Let $f(x) = \sin(\sin(\sin(\sin x)))$ on $[-3\pi, 3\pi]$.

(a) Draw its graph and use it to guess whether f is even, odd, or neither. Now justify your guess algebraically.

(b) Draw the graph of f' and use it to guess whether f' is even, odd, or neither. Justify your guess (see Problem 52).

(c) Estimate the greatest value of $f(x)$.

(d) Estimate the greatest value of $|f'(x)|$.

GC **54.** Follow the instructions of Problem 53 for $f(t) = \cos(t^3 - 3t)$ on $[-2, 2]$.

Answers to Concepts Review: **1.** $D_t u; f'(g(t))g'(t)$ **2.** $D_v w;$ $G'(H(s))H'(s)$ **3.** $(f(x))^2, (f(x))^2$ **4.** $2x \cos(x^2); 6(2x+1)^2$

3.6
Leibniz Notation

Gottfried Wilhelm von Leibniz was one of the two principal founders of the calculus (the other was Isaac Newton). Leibniz's notation for the derivative is still widely used, especially in applied fields such as physics, chemistry, and economics. Its attraction lies in its form, which often suggests true results and sometimes suggests how to prove them. After we have mastered the Leibniz notation, we will use it to restate the Chain Rule and then actually prove that rule.

Increments If the value of a variable x changes from x_1 to x_2, then $x_2 - x_1$, the change in x, is called an **increment** of x and is commonly denoted by Δx (read "delta x"). Note immediately that Δx does *not* mean Δ times x. If $x_1 = 4.1$ and $x_2 = 5.7$, then

$$\Delta x = x_2 - x_1 = 5.7 - 4.1 = 1.6$$

If $x_1 = c$ and $x_2 = c + h$, then

$$\Delta x = x_2 - x_1 = c + h - c = h$$

Suppose next that $y = f(x)$ determines a function. If x changes from x_1 to x_2, then y changes from $y_1 = f(x_1)$ to $y_2 = f(x_2)$. Thus, corresponding to the increment $\Delta x = x_2 - x_1$ in x, there is an increment in y given by

$$\Delta y = y_2 - y_1 = f(x_2) - f(x_1)$$

EXAMPLE 1 Let $y = f(x) = 2 - x^2$. Find Δy when x changes from 0.4 to 1.3 (see Figure 1).

Solution

$$\Delta y = f(1.3) - f(0.4) = \left[2 - (1.3)^2\right] - \left[2 - (0.4)^2\right] = -1.53 \quad\blacksquare$$

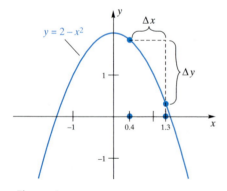

$y = 2 - x^2$

Figure 1

The dy/dx Symbol for the Derivative Suppose now that the independent variable changes from x to $x + \Delta x$. The corresponding change in the dependent variable, y, will be

$$\Delta y = f(x + \Delta x) - f(x)$$

and the ratio

$$\frac{\Delta y}{\Delta x} = \frac{f(x + \Delta x) - f(x)}{\Delta x}$$

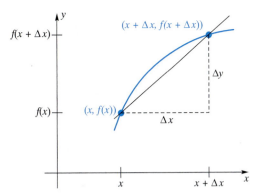

Figure 2

represents the slope of a secant line through $(x, f(x))$, as shown in Figure 2. As $\Delta x \to 0$, the slope of this secant line approaches that of the tangent line, and for this latter slope Leibniz used the symbol dy/dy. Thus,

$$\frac{dy}{dx} = \lim_{\Delta x \to 0} \frac{\Delta y}{\Delta x} = \lim_{\Delta x \to 0} \frac{f(x + \Delta x) - f(x)}{\Delta x} = f'(x)$$

Leibniz called dy/dx a quotient of two infinitesimals. The meaning of the word *infinitesimal* is vague, and we will not use it. However, dy/dx is a standard symbol for the derivative and we will use it frequently from now on. For the present, think of d/dx as an operator symbol with the same meaning as D_x, and read it "the derivative with respect to x." All our past theorems on derivatives continue to apply; only our notation is different.

EXAMPLE 2   Find dy/dx if $y = x^3 - 3x^2 + 7x$.

Solution

$$\frac{dy}{dx} = \frac{d}{dx}\left(x^3 - 3x^2 + 7x\right)$$

$$= \frac{d(x^3)}{dx} - 3\frac{d(x^2)}{dx} + 7\frac{d(x)}{dx}$$

$$= 3x^2 - 3(2x) + 7(1)$$

$$= 3x^2 - 6x + 7 \qquad \blacksquare$$

EXAMPLE 3   Find $\dfrac{d}{dt}\left(\dfrac{3t}{t^2 + 1}\right)$.

Solution   By the Quotient Rule,

$$\frac{d}{dt}\left(\frac{3t}{t^2 + 1}\right) = \frac{(t^2 + 1)(3) - (3t)(2t)}{(t^2 + 1)^2} = \frac{-3t^2 + 3}{(t^2 + 1)^2} \qquad \blacksquare$$

The Chain Rule Again   Suppose that $y = f(u)$ and $u = g(x)$. In Leibniz notation, the Chain Rule takes the particularly elegent form

$$\frac{dy}{dx} = \frac{dy}{du}\frac{du}{dx}$$

It is elegant because it is easy to remember. Just "cancel" the du's on the right side and you have the left side. Do not try to make mathematical sense out of this cancellation, but use it as a memory aid if it helps.

EXAMPLE 4 Find $\dfrac{dy}{dx}$ if $y = (x^3 - 2x)^{12}$.

Solution Let $u = x^3 - 2x$. Then $y = u^{12}$ and

$$\frac{dy}{dx} = \frac{dy}{du}\frac{du}{dx}$$

$$= 12u^{11}(3x^2 - 2)$$

$$= 12(x^3 - 2x)^{11}(3x^2 - 2) \qquad \blacksquare$$

After some practice you will make these substitutions in your head; thus, you will be able to do most problems without explicitly defining the substitution for u.

EXAMPLE 5 Find $\dfrac{d}{dx}\cos(x^2)$.

Solution

$$\frac{d}{dx}\cos(x^2) = -\sin(x^2)\left(\frac{d}{dx}x^2\right) = -\sin(x^2)(2x) = -2x\sin(x^2) \qquad \blacksquare$$

When you must apply the Chain Rule more than once, the Leibniz notation still takes on an elegant form. If $y = f(u), u = g(v)$, and $v = h(x)$, then

$$\frac{dy}{dx} = \frac{dy}{du}\frac{du}{dv}\frac{dv}{dx}$$

EXAMPLE 6 Find $\dfrac{dy}{dx}$ if $y = \cos^3(x^2 + 1)$.

Solution Remember, $\cos^3 x$ means $(\cos x)^3$. Let $y = u^3, u = \cos v$, and $v = x^2 + 1$.

$$\frac{dy}{dx} = \frac{dy}{du}\frac{du}{dv}\frac{dv}{dx}$$

$$= (3u^2)(-\sin v)(2x)$$

$$= (3\cos^2 v)[-\sin(x^2 + 1)](2x)$$

$$= -6x\cos^2(x^2 + 1)\sin(x^2 + 1)$$

After some practice you should be able to reason as follows:

$$\frac{dy}{dx} = [3\cos^2(x^2 + 1)] \cdot \left[\frac{d}{dx}\cos(x^2 + 1)\right]$$

$$= [3\cos^2(x^2 + 1)] \cdot [-\sin(x^2 + 1)]\left[\frac{d}{dx}(x^2 + 1)\right]$$

$$= [3\cos^2(x^2 + 1)] \cdot [-\sin(x^2 + 1)] \cdot 2x$$

$$= -6x\cos^2(x^2 + 1)\sin(x^2 + 1) \qquad \blacksquare$$

A Partial Proof of the Chain Rule In the previous section we stated the Chain Rule as $D_x f(g(x)) = f'(g(x))g'(x)$. Using Leibniz notation, the Chain Rule says that $dy/dx = dy/du \cdot du/dx$. We can now give a sketch of the proof.

Proof We suppose that $y = f(u)$ and $u = g(x)$, that g is differentiable at x, and that f is differentiable at $u = g(x)$. When x is given an increment Δx, there are corresponding increments in u and y given by

$$\Delta u = g(x + \Delta x) - g(x)$$

$$\Delta y = f(g(x + \Delta x)) - f(g(x))$$

$$= f(u + \Delta u) - f(u)$$

Thus,

$$\frac{dy}{dx} = \lim_{\Delta x \to 0} \frac{\Delta y}{\Delta x} = \lim_{\Delta x \to 0} \frac{\Delta y}{\Delta u} \frac{\Delta u}{\Delta x}$$

$$= \lim_{\Delta x \to 0} \frac{\Delta y}{\Delta u} \cdot \lim_{\Delta x \to 0} \frac{\Delta u}{\Delta x}$$

Since g is differentiable at x, it is continuous there (Theorem 3.2A), and so $\Delta x \to 0$ forces $\Delta u \to 0$. Hence,

$$\frac{dy}{dx} = \lim_{\Delta u \to 0} \frac{\Delta y}{\Delta u} \cdot \lim_{\Delta x \to 0} \frac{\Delta u}{\Delta x} = \frac{dy}{du} \cdot \frac{du}{dx}$$

This proof was very slick, but unfortunately it contains a subtle flaw. There are functions $u = g(x)$ that have the property that $\Delta u = 0$ for some points in every neighborhood of x (the constant function $g(x) = k$ is a good example). This means the division by Δu at our first step might not be legal. There is no simple way to get around this difficulty, though the Chain Rule is valid even in this case. We give a complete proof of the Chain Rule in the appendix (Section A.2, Theorem B). ◆

Concepts Review

1. The symbol Δw denotes a(an) _____ (or small change) in the variable w. In terms of this symbol, we may define the derivative of y with respect to x as $\lim_{\Delta x \to 0}$ _____, which led Leibniz to denote this derivative by the symbol _____.

2. If $y = f(x)$, we now have three different symbols for the derivative of y with respect to x. They are _____, _____, and _____.

3. Let $y = f(u)$, where $u = g(x)$. In Leibniz notation, the Chain Rule says that $\frac{dy}{dx} =$ _____.

4. Let $w = f(t)$, where $t = g(s)$ and where $s = h(r)$. In Leibniz notation, the Chain Rule applied twice gives $\frac{dw}{dr} =$ _____.

Problem Set 3.6

In Problems 1–4, find Δy for the given values of x_1 and x_2 (see Example 1).

1. $y = 3x + 2, x_1 = 1, x_2 = 1.5$

2. $y = 3x^2 + 2x + 1, x_1 = 0.0, x_2 = 0.1$

C **3.** $y = \dfrac{3}{x + 1}, x_1 = 2.34, x_2 = 2.31$

C **4.** $y = \cos 2x, x_1 = 0.571, x_2 = 0.573$

In Problems 5–8, first find and simplify

$$\frac{\Delta y}{\Delta x} = \frac{f(x + \Delta x) - f(x)}{\Delta x}$$

Then find dy/dx by taking the limit of your answer as $\Delta x \to 0$.

5. $y = x^2$

6. $y = x^3 - 3x^2$

7. $y = \dfrac{1}{x + 1}$

8. $y = \dfrac{x^3 + x^2}{x^3}$

In Problems 9–20, use the Chain Rule to find dy/dx.

9. $y = u^2$ and $u = \sin x$

10. $y = \cos u$ and $u = \dfrac{1}{x + 1}$

11. $y = \tan(x^2)$

12. $y = \tan^2 x$

13. $y = \left(\dfrac{x^2 + 1}{\cos x}\right)^4$

14. $y = \left[(x^2 + 1)\sin x\right]^3$

15. $y = \cos(x^2)\sin^2 x$

16. $y = \dfrac{(x^3 + 2x)^4}{x^4 + 1}$

17. $y = \sin^4(x^2 + 3)$

18. $y = \sin[(x^2 + 3)^4]$

19. $y = \cos^2\left(\dfrac{x^2 + 2}{x^2 - 2}\right)$

20. $y = \sin^2[\cos^2(x^2)]$

In Problems 21–26, find the indicated derivatives.

21. $\dfrac{d}{dt}(\sin^3 t + \cos^3 t)$

22. $\dfrac{d}{ds}[(s^2 + 3)^3 - (s^2 + 3)^{-3}]$

23. $D_r[\pi(r + 3)^2 - 3\pi r(r + 2)^2]$

24. $D_t[u^3 + 3u]$ if $u = t^2$

25. $f'(2)$ if $f(x) = \left(x + \dfrac{1}{x}\right)^4$

26. $F'(0)$ if $F(t) = \cos(t^2)\sin 3t$

27. Suppose that $f(3) = 2$, $f'(3) = -1$, $g(3) = 3$, and $g'(3) = -4$. Calculate each value.

(a) $(f + g)'(3)$ (b) $(f \cdot g)'(3)$

(c) $(f/g)'(3)$ (d) $(f \circ g)'(3)$

28. If $f(2) = 4, f'(4) = 6$, and $f'(2) = -2$, calculate each value.

(a) $\dfrac{d}{dx}[f(x)]^3$ at $x = 2$ (b) $\dfrac{d}{dx}\left[\dfrac{3}{f(x)}\right]$ at $x = 2$

(c) $(f \circ f)'(2)$

Problems 29 and 30 refer to the graphs in Figures 3 and 4.

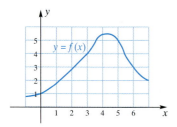

Figure 3

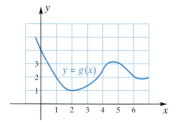

Figure 4

29. Find each value approximately.

(a) $(f + g)'(4)$ (b) $(f \circ g)'(6)$

30. Find each value approximately.

(a) $(f/g)'(2)$ (b) $(g \circ f)'(3)$

31. Each edge of a cube is increasing at the constant rate of 16 centimeters per minute.

(a) Find the rate at which the volume of the cube is increasing at the instant when the edge is 20 centimeters.

(b) Find the rate at which the total surface area of the cube is increasing at the instant when the edge is 15 centimeters.

32. Ships A and B start from the origin at the same time. Ship A travels due east at a rate of 20 miles per hour and ship B travels due north at the rate of 12 miles per hour. How fast are they separating after 3 hours? After 6 hours?

33. Where does the tangent line to the curve $y = x^2 \sin^2(x^2)$ at $x = \sqrt{\dfrac{\pi}{2}}$ intersect the x-axis?

34. The dial of a standard clock has a 10-centimeter radius. One end of an elastic string is attached to the rim at 12 and the other to the tip of the 10-centimeter minute hand. At what rate is the string stretching at 12:15 (assuming that the clock is not slowed down by this stretching)?

35. Suppose that f is differentiable and that there are points x_1 and x_2 such that $f(x_1) = x_2$ and $f(x_2) = x_1$. Let $g(x) = f(f(f(f(x))))$. Show that $g'(x_1) = g'(x_2)$.

36. Let $f(x) = \begin{cases} x^2 \sin \dfrac{1}{x} & \text{if } x \neq 0 \\ 0 & \text{if } x = 0 \end{cases}$

(a) Find $f'(x)$ for $x \neq 0$ by using derivative rules.

(b) Find $f'(0)$ from the definition of the derivative.

(c) Show that $f'(x)$ is discontinuous at $x = 0$.

C **37.** The hour and minute hands of a clock are 6 and 8 inches long, respectively. How fast are the tips of the hands separating at 12:20 (see Figure 5). *Hint*: Law of Cosines.

Figure 5

≈ GC **38.** Find the approximate time between 12:00 and 1:00 when the distance s between the tips of the hands of the clock of Figure 5 is increasing most rapidly, that is, when the derivative ds/dt is largest.

39. Give a second proof of the Quotient Rule. Write

$$D_x\left(\frac{f(x)}{g(x)}\right) = D_x\left(f(x)\frac{1}{g(x)}\right)$$

and use the Product Rule and the Chain Rule.

EXPL **40.** Suppose that f is a differentiable function.

(a) Find $\dfrac{d}{dx} f(f(x))$.   (b) Find $\dfrac{d}{dx} f(f(f(x)))$.

(c) Find $\dfrac{d}{dx} f(f(f(f(x))))$.

(d) Let $f^{[n]}$ denote the function defined as follows: $f^{[1]} = f$ and $f^{[n]} = f \circ f^{[n-1]}$ for $n \geq 2$. Thus, $f^{[2]} = f \circ f, f^{[3]} = f \circ f \circ f$, and so on. Based on your results from parts (a) through (c), make a conjecture regarding $\dfrac{d}{dx} f^{[n]}(x)$. Prove your conjecture.

Answers to Concepts Review: **1.** increment; $\Delta y/\Delta x; dy/dx$

2. $f'(x); D_x y; dy/dx$ **3.** $\dfrac{dy}{du}\dfrac{du}{dx}$ **4.** $\dfrac{dw}{dt}\dfrac{dt}{ds}\dfrac{ds}{dr}$

3.7 Higher-Order Derivatives

The operation of differentiation takes a function f and produces a new function f'. If we now differentiate f', we produce still another function, denoted by f'' (read "f double prime") and called the **second derivative** of f. It, in turn, may be differentiated, thereby producing f''', which is called the **third derivative** of f, and so on. The **fourth derivative** is denoted $f^{(4)}$, the **fifth derivative** is denoted $f^{(5)}$, and so on.

If, for example

$$f(x) = 2x^3 - 4x^2 + 7x - 8$$

then

$$f'(x) = 6x^2 - 8x + 7$$

$$f''(x) = 12x - 8$$

$$f'''(x) = 12$$

$$f^{(4)}(x) = 0$$

Since the derivative of the zero function is zero, the fourth derivative and all *higher-order derivatives* of f will be zero.

We have introduced three notations for the derivative (now also called the *first derivative*) of $y = f(x)$. They are

$$f'(x) \qquad D_x y \qquad \frac{dy}{dx}$$

called, respectively, the *prime notation*, the *D notation*, and the *Leibniz notation*. There is a variation of the prime notation, y', that we will also use occasionally. All these notations have extensions for higher-order derivates, as shown in the accompanying table. Note especially the Leibniz notation, which, though complicated, seemed most appropriate to Leibniz. What, thought he, is more natural than to write

$$\frac{d}{dx}\left(\frac{dy}{dx}\right) \quad \text{as} \quad \frac{d^2 y}{dx^2}$$

Leibniz's notation for the second derivative is read *the second derivative of y with respect to x.*

Notations for Derivatives of $y = f(x)$

Derivative	f' Notation	y' Notation	D Notation	Leibniz Notation
First	$f'(x)$	y'	$D_x y$	$\frac{dy}{dx}$
Second	$f''(x)$	y''	$D_x^2 y$	$\frac{d^2 y}{dx^2}$
Third	$f'''(x)$	y'''	$D_x^3 y$	$\frac{d^3 y}{dx^3}$
Fourth	$f^{(4)}(x)$	$y^{(4)}$	$D_x^4 y$	$\frac{d^4 y}{dx^4}$
Fifth	$f^{(5)}(x)$	$y^{(5)}$	$D_x^5 y$	$\frac{d^5 y}{dx^5}$
Sixth	$f^{(6)}(x)$	$y^{(6)}$	$D_x^6 y$	$\frac{d^6 y}{dx^6}$
$\vdots$	$\vdots$	$\vdots$	$\vdots$	$\vdots$
nth	$f^{(n)}(x)$	$y^{(n)}$	$D_x^n y$	$\frac{d^n y}{dx^n}$

EXAMPLE 1 If $y = \sin 2x$, find d^3y/dx^3, d^4y/dx^4, and $d^{12}y/dx^{12}$.

Solution

$$\frac{dy}{dx} = 2\cos 2x$$

$$\frac{d^2y}{dx^2} = -2^2 \sin 2x$$

$$\frac{d^3y}{dx^3} = -2^3 \cos 2x$$

$$\frac{d^4y}{dx^4} = 2^4 \sin 2x$$

$$\frac{d^5y}{dx^5} = 2^5 \cos 2x$$

$$\vdots$$

$$\frac{d^{12}y}{dx^{12}} = 2^{12} \sin 2x$$
■

Velocity and Acceleration   In Section 3.1, we used the notion of instantaneous velocity to motivate the definition of the derivative. Let's review this notion by means of an example. Also, from now on we will use the single word *velocity* in place of the more cumbersome phrase *instantaneous velocity*.

EXAMPLE 2   An object moves along a coordinate line so that its position s satisfies $s = 2t^2 - 12t + 8$, where s is measured in centimeters and t in seconds with $t \geq 0$. Determine the velocity of the object when $t = 1$ and when $t = 6$. When is the velocity 0? When is it positive?

Solution   If we use the symbol $v(t)$ for the velocity at time t, then

$$v(t) = \frac{ds}{dt} = 4t - 12$$

Thus,

$$v(1) = 4(1) - 12 = -8 \text{ centimeters per second}$$

$$v(6) = 4(6) - 12 = 12 \text{ centimeters per second}$$

The velocity is 0 when $4t - 12 = 0$, that is, when $t = 3$. The velocity is positive when $4t - 12 > 0$, or when $t > 3$. All this is shown schematically in Figure 1.

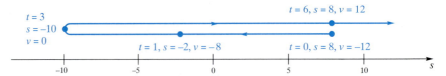

Figure 1

The object is, of course, moving along the s-axis, not on the colored path above it. But the colored path shows what happens to the object. Between $t = 0$ and $t = 3$, the velocity is negative; the object is moving to the left (backing up). By the time $t = 3$, it has "slowed" to a zero velocity. It then starts moving to the right as its velocity becomes positive. Thus, negative velocity corresponds to moving in the direction of decreasing s; positive velocity corresponds to moving in the direction of increasing s. A rigorous discussion of these points will be given in Chapter 4.   ■

There is a technical distinction between the words *velocity* and *speed*. Velocity has a sign associated with it; it may be positive or negative. **Speed** is defined to be the absolute value of the velocity. Thus, in the example above, the speed at $t = 1$ is $|-8| = 8$ centimeters per second. The meter in most cars is a *speed*ometer; it always give nonnegative values.

Now we want to give a physical interpretation of the second derivative d^2s/dt^2. It is, of course, just the first derivative of the velocity. Thus, it measures the rate of change of velocity with respect to time, which has the name **acceleration**. If it is denoted by a, then

$$a = \frac{dv}{dt} = \frac{d^2s}{dt^2}$$

In Example 2, $s = 2t^2 - 12t + 8$. Thus,

$$v = \frac{ds}{dt} = 4t - 12$$

$$a = \frac{d^2s}{dt^2} = 4$$

This means that the velocity is increasing at a constant rate of 4 centimeters per second every second, which we write as 4 centimeters per second per second, or as 4 cm/sec².

EXAMPLE 3   A point moves along a horizontal coordinate line in such a way that its position at time t is specified by

$$s = t^3 - 12t^2 + 36t - 30$$

Here s is measured in feet and t in seconds.

(a) When is the velocity 0?
(b) When is the velocity positive?
(c) When is the point moving to the left (that is, in the negative direction)?
(d) When is the acceleration positive?

Solution

(a) $v = ds/dt = 3t^2 - 24t + 36 = 3(t - 2)(t - 6)$. Thus, $v = 0$ at $t = 2$ and $t = 6$.
(b) $v > 0$ when $(t - 2)(t - 6) > 0$. We learned how to solve quadratic inequalities in Section 1.3. The solution is $\{t: t < 2$ or $t > 6\}$ or, in interval notation, $(-\infty, 2) \cup (6, \infty)$; see Figure 2.
(c) The point is moving to the left when $v < 0$; that is, when $(t - 2)(t - 6) < 0$. This inequality has as its solution the interval $(2, 6)$.
(d) $a = dv/dt = 6t - 24 = 6(t - 4)$. Thus, $a > 0$ when $t > 4$. The motion of the point is shown schematically in Figure 3.

Measuring Time

If $t = 0$ corresponds to the present moment, then $t < 0$ corresponds to the past, and $t > 0$ to the future. In many problems, it will be obvious that we are concerned only with the future. However, since the statement of Example 3 does not specify this, it seems reasonable to allow t to have negative as well as positive values.

Figure 2

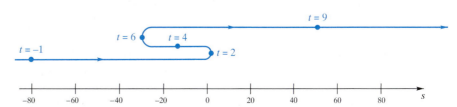

Figure 3

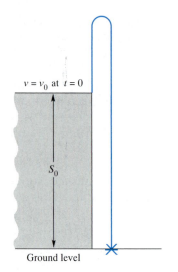

$v = v_0$ at $t = 0$

S_0

Ground level

Figure 4

Falling-Body Problems   If an object is thrown straight upward (or downward) from an initial height of s_0 feet with an initial velocity of v_0 feet per second and if s is its height above the ground in feet after t seconds, then

$$s = -16t^2 + v_0t + s_0$$

This assumes that the experiment takes place near sea level and that air resistance can be neglected. The diagram in Figure 4 portrays the situation we have in mind. Notice that positive velocity means that the object is moving upward.

EXAMPLE 4   From the top of a building 160 feet high, a ball is thrown upward with an initial velocity of 64 feet per second.

(a) When does it reach maximum height?

(b) What is its maximum height?

(c) When does it hit the ground?

(d) With what speed does it hit the ground?

(e) What is its acceleration at $t = 2$?

Solution   Let $t = 0$ correspond to the instant when the ball was thrown. Then $s_0 = 160$ and $v_0 = 64$ (v_0 is positive because the ball was thrown *upward*). Thus,

$$s = -16t^2 + 64t + 160$$

$$v = \frac{ds}{dt} = -32t + 64$$

$$a = \frac{dv}{dt} = -32$$

(a) The ball reached its maximum height at the time its velocity was 0, that is, when $-32t + 64 = 0$ or when $t = 2$ seconds.

(b) At $t = 2$, $s = -16(2)^2 + 64(2) + 160 = 224$ feet.

(c) The ball hit the ground when $s = 0$, that is, when

$$-16t^2 + 64t + 160 = 0$$

Dividing by -16 yields

$$t^2 - 4t - 10 = 0$$

The quadratic formula then gives

$$t = \frac{4 \pm \sqrt{16 + 40}}{2} = \frac{4 \pm 2\sqrt{14}}{2} = 2 \pm \sqrt{14}$$

Only the positive answer makes sense. Thus, the ball hit the ground at $t = 2 + \sqrt{14} \approx 5.74$ seconds.

(d) At $t = 2 + \sqrt{14}$, $v = -32(2 + \sqrt{14}) + 64 \approx -119.73$. Thus, the ball hit the ground with a speed of 119.73 feet per second.

(e) The acceleration is always -32 feet per second per second. This is the acceleration of gravity near sea level.   ■

The Book of Nature

"The great book of Nature lies ever open before our eyes and the true philosophy is written in it.... But we cannot read it unless we have first learned the language and the characters in which it is written.... It is written in mathematical language and the characters are triangles, circles, and other geometrical figures."

Galileo Galilei

Mathematical Modeling   Galileo may have been right in claiming that the book of nature is written in mathematical language. Certainly, the scientific enterprise seems largely an effort to prove him correct. The task of taking a physical phenomenon and representing it in mathematical symbols is called **mathematical modeling**. One of its basic elements is translating word descriptions into mathematical language. Doing this, especially in connection with rates of change, will become increasingly important as we go on. Here are some simple illustrations.

Word Description	*Mathematical Model*
Water is leaking from a cylindrical tank at a rate proportional to the depth of the water.	If V denotes the volume of the water at time t, then $\dfrac{dV}{dt} = -kh$.
A wheel is spinning at a constant rate of 6 revolutions per minute, that is, at $6(2\pi)$ radians per minute.	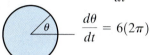 $\dfrac{d\theta}{dt} = 6(2\pi)$
The density (in grams per centimeter) of a wire at a point is twice its distance from the left end.	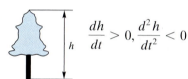 If m denotes the mass of the left x centimeters of the wire, then $\dfrac{dm}{dx} = 2x$.
The height of a tree continues to increase but at a slower and slower rate.	$\dfrac{dh}{dt} > 0, \dfrac{d^2 h}{dt^2} < 0$

The use of mathematical language is not limited to the physical sciences; it is also appropriate in the social sciences, especially in economics.

EXAMPLE 5 A news agency reported in May 1998 that unemployment in eastern Asia was continuing to increase at an increasing rate. On the other hand, the price of food was increasing, but at a slower rate than before. Interpret these statements in mathematical language.

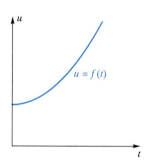

Figure 5

Solution Let $u = f(t)$ denote the number of people unemployed at time t. Although u actually jumps by unit amounts, we will follow standard practice in representing u by a smooth curve as in Figure 5. To say unemployment is increasing is to say that $du/dt > 0$. To say that it is increasing at an increasing rate is to say that the function du/dt is *increasing*; but this means that the derivative of du/dt must be positive. Thus, $d^2u/dt^2 > 0$. In Figure 5, notice that the slope of the tangent line increases as t increases.

Similarly, if $p = g(t)$ represents the price of food (e.g., the typical cost of one day's groceries for one person) at time t, then dp/dt is positive but *decreasing*. Thus, the derivative of dp/dt is negative, so $d^2p/dt^2 < 0$. In Figure 6, notice that the slope of the tangent line decreases as t increases. ∎

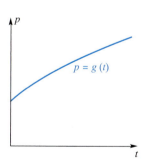

Figure 6

Concepts Review

1. If $y = f(x)$, then the third derivative of y with respect to x can be denoted by any one of the following three symbols: _____ .

2. If $s = f(t)$ denotes the position of a particle on a coordinate line at time t, then its velocity is given by _____ , its speed is given by _____ , and its acceleration is given by _____ .

3. Assume that an object is thrown straight upward so that its height s at time t is given by $s = f(t)$. The object reaches its maximum height when $ds/dt =$ _____ , after which ds/dt _____ .

4. If the amount W of water in a tank at time t is increasing, but at a slower and slower rate, then dW/dt is _____ and d^2W/dt^2 is _____ .

Problem Set 3.7

In Problems 1–8, find d^3y/dx^3.

1. $y = x^3 + 3x^2 + 6x$

2. $y = x^5 + x^4$

3. $y = (3x + 5)^3$

4. $y = (3 - 5x)^5$

5. $y = \sin(7x)$

6. $y = \sin(x^3)$

7. $y = \dfrac{1}{x - 1}$

8. $y = \dfrac{3x}{1 - x}$

In Problems 9–16, find $f''(2)$.

9. $f(x) = x^2 + 1$

10. $f(x) = 5x^3 + 2x^2 + x$

11. $f(t) = \dfrac{2}{t}$

12. $f(u) = \dfrac{2u^2}{5 - u}$

13. $f(\theta) = (\cos \theta\pi)^{-2}$

14. $f(t) = t \sin(\pi/t)$

15. $f(s) = s(1 - s^2)^3$

16. $f(x) = \dfrac{(x + 1)^2}{x - 1}$

17. Let $n! = n(n - 1)(n - 2) \cdots 3 \cdot 2 \cdot 1$. Thus, $4! = 4 \cdot 3 \cdot 2 \cdot 1 = 24$ and $5! = 5 \cdot 4 \cdot 3 \cdot 2 \cdot 1$. We give $n!$ the name **n factorial**. Show that $D_x^n(x^n) = n!$.

18. Using the factorial symbol of Problem 17, find a formula for

$$D_x^n(a_{n-1}x^{n-1} + \cdots + a_1 x + a_0)$$

19. Without doing any calculating, find each derivative.

(a) $D_x^4(3x^3 + 2x - 19)$ (b) $D_x^{12}(100x^{11} - 79x^{10})$

(c) $D_x^{11}(x^2 - 3)^5$

20. Find a formula for $D_x^n(1/x)$.

21. If $f(x) = x^3 + 3x^2 - 45x - 6$, find the value of f'' at each zero of f', that is, at each point c where $f'(c) = 0$.

22. Suppose that $g(t) = at^2 + bt + c$ and $g(1) = 5, g'(1) = 3$, and $g''(1) = -4$. Find $a, b,$ and c.

In Problems 23–28, an object is moving along a horizontal coordinate line according to the formula $s = f(t)$, where s, the directed distance from the origin, is in feet and t is in seconds. In each case, answer the following questions (see Examples 2 and 3).

(a) *What are $v(t)$ and $a(t)$, the velocity and acceleration, at time t?*

(b) *When is the object moving to the right?*

(c) *When is it moving to the left?*

(d) *When is its acceleration negative?*

(e) *Draw a schematic diagram that shows the motion of the object.*

23. $s = 12t - 2t^2$

24. $s = t^3 - 6t^2$

25. $s = t^3 - 9t^2 + 24t$

26. $s = 2t^3 - 6t + 5$

27. $s = t^2 + \dfrac{16}{t}, t > 0$

28. $s = t + \dfrac{4}{t}, t > 0$

29. If $s = \frac{1}{2}t^4 - 5t^3 + 12t^2$, find the velocity of the moving object when its acceleration is zero.

30. If $s = \frac{1}{10}(t^4 - 14t^3 + 60t^2)$, find the velocity of the moving object when its acceleration is zero.

31. Two particles move along a coordinate line. At the end of t seconds their directed distances from the origin, in feet, are given by $s_1 = 4t - 3t^2$ and $s_2 = t^2 - 2t$, respectively.

(a) When do they have the same velocity?

(b) When do they have the same speed?

(c) When do they have the same position?

32. The positions of two particles, P_1 and P_2, on a coordinate line at the end of t seconds are given by $s_1 = 3t^3 - 12t^2 + 18t + 5$ and $s_2 = -t^3 + 9t^2 - 12t$, respectively. When do the two particles have the same velocity?

33. An object thrown directly upward is at a height $s = -16t^2 + 48t + 256$ feet after t seconds (see Example 4).

(a) What is its initial velocity?

(b) When does it reach maximum height?

(c) What is its maximum height?

[C] (d) When does it hit the ground?

[C] (e) With what speed does it hit the ground?

34. An object thrown directly upward from ground level with an initial velocity of 48 feet per second is approximately $s = 48t - 16t^2$ feet high at the end of t seconds.

(a) What is the maximum height attained?

(b) How fast is the object moving, and in which direction, at the end of 1 second?

(c) How long does it take to return to its original position?

[C] **35.** A projectile is fired directly upward from the ground with an initial velocity of v_0 feet per second. Its height in t seconds is given by $s = v_0 t - 16t^2$ feet. What must its initial velocity be for the projectile to reach a maximum height of 1 mile?

36. An object thrown directly downward from the top of a cliff with an initial velocity of v_0 feet per second falls approximately $s = v_0 t - 16t^2$ feet in t seconds. If it strikes the ocean below in 3 seconds with a velocity of 140 feet per second, how high is the cliff?

37. A point moves along a horizontal coordinate line in such a way that its position at time t is specified by $s = t^3 - 3t^2 - 24t - 6$. Here s is measured in centimeters and t in seconds. When is the point slowing down; that is, when is its *speed* decreasing?

38. Explain why a point moving along a line is slowing down when its velocity and acceleration have opposite signs (see Problem 37).

39. Translate each of the following into the language of first, second, and third derivatives of distance with respect to time. For each part, sketch a plot of the car's position s against time t.

(a) The speed of the car is proportional to the distance it has traveled.

(b) The car is speeding up.

(c) I didn't say the car was slowing down; I said its rate of increase in speed was slowing down.

(d) The car's speed is increasing 10 miles per hour every minute.

(e) The car is slowing very gently to a stop.

(f) The car always travels the same distance in equal time intervals.

40. Translate each of the following into the language of derivates and sketch a plot of the appropriate function.

(a) Water is evaporating from the tank at a constant rate.

(b) Water is being poured into the tank at 3 gallons per minute but is also leaking out at $\frac{1}{2}$ gallon per minute.

(c) Since water is being poured into the conical tank at a constant rate, the water level is rising at a slower and slower rate.

(d) Inflation held steady this year but is expected to rise more and more rapidly in the years ahead.

(e) At present the price of oil is dropping, but this trend is expected to slow and then reverse directions in 2 years.

(f) David's temperature is still rising, but the penicillin seems to be taking effect.

41. Translate each of the following statements into mathematical language as in Example 5, and sketch a plot of the appropriate function.

(a) The cost of a car continues to increase and at a faster and faster rate.

(b) During the last 2 years, the United States has continued to cut its consumption of oil, but at a slower and slower rate.

(c) World population continues to grow, but at a slower and slower rate.

(d) The angle that the Leaning Tower of Pisa makes with the vertical is increasing more and more rapidly.

(e) Upper Midwest firm's profit growth slows.

(f) The XYZ Company has been losing money, but will soon turn this situation around.

42. Translate each statement from the following newspaper column into a statement about derivatives.

(a) In the United States, the ratio R of government debt to national income remained unchanged at around 28% up to 1981, but

(b) then it began to increase more and more sharply, reaching 36% during 1983.

EXPL **43.** Leibniz obtained a general formula for $D_x^n(uv)$, where u and v are both functions of x. See if you can find it. *Hint*: Begin by considering the cases $n = 1, n = 2$, and $n = 3$.

44. Use the formula of Problem 43 to find $D_x^4(x^4 \sin x)$.

GC **45.** Let $f(x) = x[\sin x - \cos(x/2)]$.

(a) Draw the graphs of $f(x), f'(x), f''(x)$, and $f'''(x)$ on $[0, 6]$ using the same axes.

(b) Evaluate $f'''(2.13)$.

GC **46.** Repeat Problem 45 for $f(x) = (x + 1)/(x^2 + 2)$.

Answers to Concepts Review: **1.** $f'''(x)$, $D_x^3 y$, $d^3 y/dx^3$ **2.** $ds/dt; |ds/dt|; d^2 s/dt^2$ **3.** $0; <0$ **4.** positive; negative

---

3.8
Implicit Differentiation

In the equation

$$y^3 + 7y = x^3$$

we cannot solve for y in terms of x. It still may be the case, however, that there is exactly one y corresponding to each x. For example, we may ask what y-values (if any) correspond to $x = 2$. To answer this equation, we must solve

$$y^3 + 7y = 8$$

Certainly, $y = 1$ is one solution, and it turns out that $y = 1$ is the *only* real solution. Given $x = 2$, the equation $y^3 + 7y = x^3$ determines a corresponding y-value. We say that the equation defines y as an **implicit** function of x. The graph of this equation, shown in Figure 1, certainly looks like the graph of a differentiable function. The new element is that we do not have an equation of the form $y = f(x)$. Based on the graph, we assume that y is some unknown function of x. If we denote this function by $y(x)$, we can write the equation as

$$[y(x)]^3 + 7y(x) = x^3$$

Even though we do not have a formula for $y(x)$, we can nevertheless get a relation between x, $y(x)$, and $y'(x)$, by differentiating both sides of the equation with respect to x. Remembering to apply the Chain Rule, we get

$$\frac{d}{dx}(y^3) + \frac{d}{dx}(7y) = \frac{d}{dx}x^3$$

$$3y^2 \frac{dy}{dx} + 7 \frac{dy}{dx} = 3x^2$$

$$\frac{dy}{dx}(3y^2 + 7) = 3x^2$$

$$\frac{dy}{dx} = \frac{3x^2}{3y^2 + 7}$$

Note that our expression for dy/dx involves both x and y, a fact that is often a nuisance. But if we wish only to find a slope at a point where we know both coordinates, no difficulty exists. At $(2, 1)$,

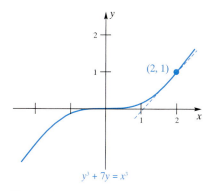

$y^3 + 7y = x^3$

Figure 1

33. If $s^2 t + t^3 = 1$, find ds/dt and dt/ds.

34. If $y = \sin(x^2) + 2x^3$, find dx/dy.

35. Sketch the graph of the circle $x^2 + 4x + y^2 + 3 = 0$ and then find equations of the two tangent lines that pass through the origin.

36. Find the equation of the **normal line** (line perpendicular to the tangent line) to the curve $8(x^2 + y^2)^2 = 100(x^2 - y^2)$ at $(3, 1)$.

37. Suppose that $xy + y^3 = 2$. Then implicit differentiation twice with respect to x yields in turn:
(a) $xy' + y + 3y^2 y' = 0$;
(b) $xy'' + y' + y' + 3y^2 y'' + 6y(y')^2 = 0$.
Solve (a) for y' and substitute in (b), and then solve for y''.

38. Find y'' if $x^3 - 4y^2 + 3 = 0$ (see Problem 37).

39. Find y'' at $(2, 1)$ if $2x^2 y - 4y^3 = 4$ (see Problem 37).

40. Use implicit differentiation twice to find y'' at $(3, 4)$ if $x^2 + y^2 = 25$.

41. Show that the normal line to $x^3 + y^3 = 3xy$ at $\left(\frac{3}{2}, \frac{3}{2}\right)$ passes through the origin.

42. Show that the hyperbolas $xy = 1$ and $x^2 - y^2 = 1$ intersect at right angles.

43. Show that the graphs of $2x^2 + y^2 = 6$ and $y^2 = 4x$ intersect at right angles.

44. Suppose that curves C_1 and C_2 intersect at (x_0, y_0) with slopes m_1 and m_2, respectively, as in Figure 4. Then (see Problem 40 of Section 2.3) the positive angle θ from C_1 (i.e., from the tangent line to C_1 at (x_0, y_0)) to C_2 satisfies

$$\tan \theta = \frac{m_2 - m_1}{1 + m_1 m_2}$$

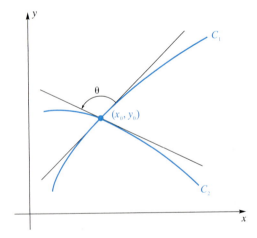

Figure 4

Find the angles from the circle $x^2 + y^2 = 1$ to the circle $(x - 1)^2 + y^2 = 1$ at the two points of intersection.

45. Find the angle from the line $y = 2x$ to the curve $x^2 - xy + 2y^2 = 28$ at their point of intersection in the first quadrant (see Problem 44).

46. A particle of mass m moves along the x-axis so that its position x and velocity $v = dx/dt$ satisfy

$$m(v^2 - v_0^2) = k(x_0^2 - x^2)$$

where v_0, x_0, and k are constants. Show by implicit differentiation that

$$m \frac{dv}{dt} = -kx$$

whenever $v \neq 0$.

47. The curve $x^2 - xy + y^2 = 16$ is an ellipse centered at the origin and with the line $y = x$ as its major axis. Find the equations of the tangent lines at the two points where the ellipse intersects the x-axis.

48. Find any points on the curve $x^2 y - xy^2 = 2$ where the tangent line is vertical, that is, where $dx/dy = 0$.

49. How high h must the light bulb in Figure 5 be if the point $(1.25, 0)$ is on the edge of the illuminated region?

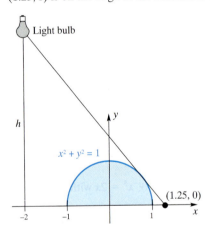

Figure 5

Answers to Concepts Review: **1.** $9/(x^3 - 3)$ **2.** $3y^2 \dfrac{dy}{dx}$

3. $x \cdot 2y \dfrac{dy}{dx} + y^2 + 3y^2 \dfrac{dy}{dx} - \dfrac{dy}{dx} = 3x^2$

4. $\dfrac{p}{q} x^{p/q - 1}; \frac{5}{3}(x^2 - 5x)^{2/3}(2x - 5)$

3.9
Related Rates

If a variable y depends on time t, then its derivative dy/dt is called a **time rate of change**. Of course, if y measures distance, then this time rate of change is also called velocity. We are interested in a wide variety of time rates: the rate at which water is flowing into a bucket, the rate at which the area of an oil spill is growing, the rate at which the value of a piece of real estate is increasing, and so on. If y is given explicitly in terms of t, the problem is simple; we just differentiate and then evaluate the derivative at the required time.

It may be that, in place of knowing y explicitly in terms of t, we know a relationship that connects y and another variable x and that we also know something about dx/dt. We may still be able to find dy/dt, since dy/dt and dx/dt are **related rates**. This will usually require implicit differentiation.

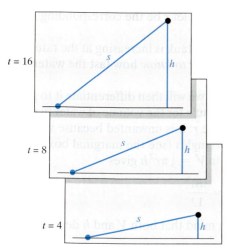

Figure 1

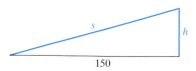

Figure 2

Two Simple Examples   In preparation for outlining a systematic procedure for solving related rate problems, we discuss two examples.

EXAMPLE 1   A small balloon is released at a point 150 feet away from an observer, who is on level ground. If the balloon goes straight up at a rate of 8 feet per second, how fast is the distance from the observer to the balloon increasing when the balloon is 50 feet high?

Solution   Let t denote the number of seconds after the balloon is released. Let h denote the height of the balloon and s its distance from the observer (see Figure 1). Both h and s are variables that depend on t; however, the base of the triangle (the distance from the observer to the point of release) remains unchanged as t increases. Figure 2 shows the key quantities in one simple diagram.

≈   Before going farther, we pick up a theme discussed earlier in the book, *estimating the answer*. Note that, initially, s changes hardly at all ($ds/dt \approx 0$), but eventually s changes about as fast as h changes ($ds/dt \approx dh/dt = 8$). An estimate for ds/dt when $h = 50$ might be about one-third to one-half of dh/dt, or 3. If we get an answer far from this value, we will know we have made a mistake. For example, answers such as 17 and even 7 are clearly wrong.

We continue with the exact solution. For emphasis, we ask and answer three fundamental questions.

(a) What is given? *Answer*: $dh/dt = 8$.

(b) What do we want to know? *Answer*: We want to know ds/dt at the instant when $h = 50$.

(c) How are s and h related?

The variables s and h change with time (they are implicit functions of t), but they are always related by the Pythagorean equation

$$s^2 = h^2 + (150)^2$$

If we differentiate implicitly with respect to t and use the Chain Rule, we obtain

$$2s \frac{ds}{dt} = 2h \frac{dh}{dt}$$

or

$$s \frac{ds}{dt} = h \frac{dh}{dt}$$

This relationship holds for all $t > 0$.

Now, and *not before now*, we turn to the specific instant when $h = 50$. From the Pythagorean Theorem, we see that, when $h = 50$,

$$s = \sqrt{(50)^2 + (150)^2} = 50 \sqrt{10}$$

Substituting in $s(ds/dt) = h(dh/dt)$ yields

$$50 \sqrt{10} \frac{ds}{dt} = 50(8)$$

or

$$\frac{ds}{dt} = \frac{8}{\sqrt{10}} \approx 2.53$$

At the instant when $h = 50$, the distance between the balloon and the observer is increasing at the rate of 2.53 feet per second.   ■

EXAMPLE 2   Water is pouring into a conical tank at the rate of 8 cubic feet per minute. If the height of the tank is 12 feet and the radius of its circular opening is 6 feet, how fast is the water level rising when the water is 4 feet deep?

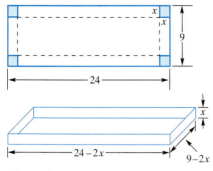

Figure 9

EXAMPLE 4   A rectangular box is to be made from a piece of cardboard 24 inches long and 9 inches wide by cutting out identical squares from the four corners and turning up the sides, as in Figure 9. Find the dimensions of the box of maximum volume. What is this volume?

Solution   Let x be the width of the square to be cut out and V the volume of the resulting box. Then

$$V = x(9 - 2x)(24 - 2x) = 216x - 66x^2 + 4x^3$$

Now x cannot be less than 0 nor more than 4.5. Thus, our problem is to maximize V on $[0, 4.5]$. The stationary points are found by setting dV/dx equal to 0 and solving the resulting equation:

$$\frac{dV}{dx} = 216 - 132x + 12x^2 = 12(18 - 11x + x^2) = 12(9 - x)(2 - x) = 0$$

This gives $x = 2$ or $x = 9$, but 9 is not in the interval $[0, 4.5]$. We see that there are only three critical points, 0, 2, and 4.5. At the end points 0 and 4.5, $V = 0$; at 2, $V = 200$. We conclude that the box has a maximum volume of 200 cubic inches if $x = 2$, that is, if the box is 20 inches long, 5 inches wide, and 2 inches deep. ■

Algebra and Geometry

Whenever possible, try to view a problem from both a geometric and an algebraic point of view. Example 4 is a good example for which this kind of thinking lends insight into the problem.

It is often helpful to plot the objective function. Plotting functions can be done easily with a graphing calculator or a CAS. Figure 10 shows a plot of the function $V(x) = 216x - 66x^2 + 4x^3$. When $x = 0$, $V(x)$ is equal to zero. In the context of folding the box, this means that when the width of the cut-out corner is zero there is nothing to fold up, so the volume is zero. Also, when $x = 4.5$, the cardboard gets folded in half, so there is no base to the box; this box will also have zero volume. Thus, $V(0) = 0$ and $V(4.5) = 0$. The greatest volume must be attained for some value of x between 0 and 4.5. The graph suggests that the maximum volume occurs when x is about 2; by using calculus, we can determine that the *exact* value of x that maximizes the volume of the box is $x = 2$.

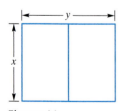

Figure 10

EXAMPLE 5   A farmer has 100 meters of wire fence with which he plans to build two identical adjacent pens, as shown in Figure 11. What are the dimensions of the enclosure that has maximum area?

Solution   Let x be the width and y the length of the total enclosure, both in meters. Because there are 100 meters of fence, $3x + 2y = 100$; that is,

$$y = 50 - \tfrac{3}{2}x$$

The total area A is given by

$$A = xy = 50x - \tfrac{3}{2}x^2$$

Since there must be three sides of length x, we see that $0 \le x \le \frac{100}{3}$. Thus, our problem is to maximize A on $\left[0, \frac{100}{3}\right]$. Now

$$\frac{dA}{dx} = 50 - 3x$$

When we set $50 - 3x$ equal to 0 and solve, we get $x = \frac{50}{3}$ as a stationary point. Thus, there are three critical points: $0, \frac{50}{3}$, and $\frac{100}{3}$. The two end points 0 and $\frac{100}{3}$ give $A = 0$, while $x = \frac{50}{3}$ yields $A \approx 416.67$. The desired dimensions are $x = \frac{50}{3} \approx 16.67$ meters and $y = 50 - \frac{3}{2}\left(\frac{50}{3}\right) = 25$ meters.

Is this answer sensible? Yes. We should expect to use more of the given fence in the y-direction than the x-direction because the former is fenced only twice, whereas the latter is fenced three times. ■

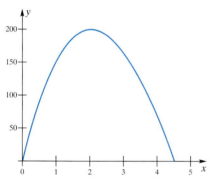

Figure 11

Our next example illustrates a problem faced by a firm that delivers its products by truck. The firm has found that as the truck's speed increases the operating cost (gasoline, oil, and so on) increases, whereas the driver cost goes down. What is the most economical speed at which a truck should be driven?

EXAMPLE 6   The operating cost for a certain truck is estimated to be $(30 + v/2)$ cents per mile when it is driven at a speed of v miles per hour. The driver is paid \$14 per hour. What speed will minimize the cost of making a delivery to a city k miles away? Assume that the law restricts the speed to $40 \leq v \leq 60$.

Solution   Let C be the total cost in cents of driving the truck k miles. Then

$$C = \text{driver cost} + \text{operating cost}$$

$$= \frac{k}{v}(1400) + k\left(30 + \frac{v}{2}\right) = 1400kv^{-1} + \left(\frac{k}{2}\right)v + 30k$$

Thus,

$$\frac{dC}{dv} = -1400kv^{-2} + \frac{k}{2} + 0$$

Setting dC/dv equal to 0 gives

$$\frac{1400k}{v^2} = \frac{k}{2}$$

$$v^2 = 2800$$

$$v \approx 53$$

A speed of 53 miles per hour would appear to be optimum, but we must evaluate C at the three critical points 40, 53, and 60 to be sure.

$$\text{At } v = 40, \quad C = k\left(\frac{1400}{40}\right) + k(30 + 20) = 85k$$

$$\text{At } v = 53, \quad C = k\left(\frac{1400}{53}\right) + k\left(30 + \frac{53}{2}\right) = 82.9k$$

$$\text{At } v = 60, \quad C = k\left(\frac{1400}{60}\right) + k(30 + 30) = 83.3k$$

We conclude that a speed of 53 miles per hour is best.

In many situations involving cost, the total cost is proportional to some other variable. In this case, the cost to operate the truck is proportional to the distance k that the truck must travel to make the delivery. What if we were to minimize the cost per mile, rather than the total cost? Would the optimum speed still be about 53 miles per hour? Since k is a constant (i.e., it doesn't depend on the velocity v), the answer is yes. The cost per mile is then

$$\text{Cost per mile} = \frac{C}{k} = \frac{1400}{v} + \frac{v}{2} + 30$$

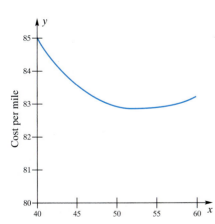

Figure 12 shows a graph of the cost per mile.   ■

Figure 12

We will have a good deal more to say about applied maxima and minima problems in Section 4.4.

Concepts Review

1. A _____ function on a _____ interval will always have both a maximum value and a minimum value on that interval.

2. The term _____ value denotes either a maximum or a minimum value.

3. A function can attain an extreme value only at a critical point. Critical points are of three types: _____, _____, and _____.

4. A stationary point for f is a number c such that _____; a singular point for f is a number c such that _____.

Problem Set 4.1

In Problems 1–16, identify the critical points and find the maximum value and minimum value (see Examples 1, 2, and 3).

1. $f(x) = x^2 + 4x + 4; I = [-4, 0]$

2. $h(x) = x^2 + x; I = [-2, 2]$

3. $\Psi(x) = x^2 + 3x; I = [-2, 1]$

4. $G(x) = \frac{1}{5}(2x^3 + 3x^2 - 12x); I = [-3, 3]$

5. $f(x) = x^3 - 3x + 1; I = (-\frac{3}{2}, 3)$ *Hint*: Sketch the graph.

6. $f(x) = x^3 - 3x + 1; I = [-\frac{3}{2}, 3]$

7. $h(r) = \frac{1}{r}; I = [-1, 3]$

8. $g(x) = \frac{1}{1 + x^2}; I = [-3, 1]$

9. $g(x) = \frac{1}{1 + x^2}; I = (-\infty, \infty)$ *Hint*: Sketch the graph.

10. $f(x) = \frac{x}{1 + x^2}; I = [-1, 4]$

11. $r(\theta) = \sin\theta; I = \left[-\frac{\pi}{4}, \frac{\pi}{6}\right]$

12. $s(t) = \sin t - \cos t; I = [0, \pi]$

13. $a(x) = |x - 1|; I = [0, 3]$

14. $f(s) = |3s - 2|; I = [-1, 4]$

15. $g(x) = \sqrt[3]{x}; I = [-1, 27]$

16. $s(t) = t^{2/5}; I = [-1, 32]$

17. Find two nonnegative numbers whose sum is 10 and whose product is a maximum. *Hint*: If x is one number, $10 - x$ is the other.

18. What number exceeds its square by the maximum amount? Begin by convincing yourself that this number is on the interval [0, 1].

19. Erika has 200 feet of fence with which she plans to enclose a rectangular yard for her dog. If she wishes to enclose maximum area, what should the dimensions be?

20. Show that for a rectangle of given perimeter K the one of maximum area is a square.

21. Find the volume of the largest open box that can be made from a piece of cardboard 24 inches square by cutting equal squares from the corners and turning up the sides (see Example 4).

22. A piece of wire 16 inches long is cut into two pieces; one piece is bent to form a square and the other is bent to form a circle. Where should the cut be made so that the sum of the areas of the square and the circle is a minimum? A maximum? (Allow the possibility of no cut.)

23. A farmer has 80 feet of fence with which he plans to enclose a rectangular pen along one side of his 100-foot barn, as shown in Figure 13 (the side along the barn needs no fence). What are the dimensions of the pen that has maximum area?

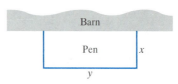

Figure 13

24. The farmer of Problem 23 decides to make three identical pens with his 80 feet of fence, as shown in Figure 14. What dimensions for the total enclosure make the area of the pens as large as possible?

Figure 14

25. Suppose that the farmer of Problem 23 has 180 feet of fence and wants the pen to adjoin to the whole side of the 100-foot barn, as shown in Figure 15. What should the dimensions be for maximum area? Note that $0 \le x \le 40$ in this case.

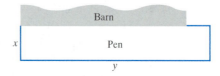

Figure 15

26. Suppose that the Farmer of Problem 23 decides to use his 80 feet of fence to make a rectangular pen to fit a 20-foot by 40-foot corner, as shown in Figure 16 (all the corner must be used and does not require fence). What dimensions give the pen a maximum area? *Hint*: Begin by deciding on the allowable values for x.

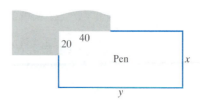

Figure 16

27. A rectangle has two corners on the x-axis and the other two on the parabola $y = 12 - x^2$, with $y \geq 0$ (Figure 17). What are the dimensions of the rectangle of this type with maximum area?

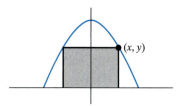

Figure 17

28. A rectangle is to be inscribed in a semicircle of radius r, as shown in Figure 18. What are the dimensions of the rectangle if its area is to be maximized?

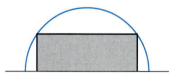

Figure 18

29. A metal rain gutter is to have 3–inch sides and a 3–inch horizontal bottom, the sides making an equal angle θ with the bottom (Figure 19). What should θ be in order to maximize the carrying capacity of the gutter? *Note:* $0 \leq \theta \leq \pi/2$.

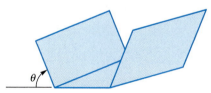

Figure 19

30. A huge conical tank is to be made from a circular piece of sheet metal of radius 10 meters by cutting out a sector with vertex angle θ and then welding together the straight edges of the remaining piece (Figure 20). Find θ so that the resulting cone has the largest possible volume.

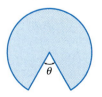

Figure 20

31. The operating cost of a certain truck is $25 + x/4$ cents per mile if the truck travels at x miles per hour. In addition, the driver gets $12 per hour. What is the most economical speed at which to operate the truck on a 400-mile run if the highway speed is required to be between 40 and 55 miles per hour?

32. Redo Problem 31 assuming that the operating cost is $40 + 0.05x^{3/2}$ cents per mile.

33. Find the points P and Q on the curve $y = x^2/4$, $0 \leq x \leq 2\sqrt{3}$, that are closest to and farthest from the point $(0, 4)$. *Hint:* The algebra is simpler if you consider the square of the required distance rather than the distance itself.

34. A humidifier uses a rotating disk of radius r, which is partially submerged in water. The most evaporation occurs when the exposed wetted region (shown as the upper shaded region in Figure 21) is maximized. Show that this happens when h (the distance from the center to the water) is equal to $r/\sqrt{1 + \pi^2}$.

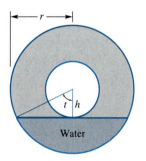

Figure 21

35. A covered box is to be made from a rectangular sheet of cardboard measuring 5 feet by 8 feet. This is done by cutting out the shaded regions of Figure 22 and then folding on the dotted lines. What are the dimensions x, y, and z that maximize the volume?

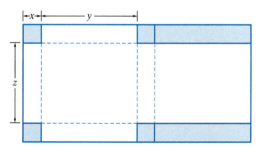

Figure 22

GC **36.** Identify the critical points and find the extreme values on $[-1, 5]$ for each function.
(a) $f(x) = x^3 - 6x^2 + x + 2$ (b) $g(x) = |f(x)|$

GC **37.** Follow the instructions of Problem 36 for $f(x) = \cos x + x \sin x + 2$.

Answers to Concepts Review: **1.** continuous; closed   **2.** extreme   **3.** end points; stationary points; singular points   **4.** $f'(c) = 0; f'(c)$ does not exist.

4.2
Monotonicity and Concavity

Figure 1

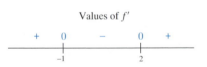

Figure 2

Consider the graph in Figure 1. No one will be surprised when we say that f is decreasing to the left of c and increasing to the right of c. But to make sure that we agree on terminology, we give precise definitions.

Definition

Let f be defined on an interval I (open, closed, or neither). We say that:

(i) f is increasing on I if, for every pair of numbers x_1 and x_2 in I,
$$x_1 < x_2 \Rightarrow f(x_1) < f(x_2)$$

(ii) f is decreasing on I if, for every pair of numbers x_1 and x_2 in I,
$$x_1 < x_2 \Rightarrow f(x_1) > f(x_2)$$

(iii) f is strictly monotonic on I if it is either increasing on I or decreasing on I.

How shall we decide where a function is increasing? Someone might suggest that we draw its graph and look at it. But a graph is usually drawn by plotting a few points and connecting those points with a smooth curve. Who can be sure that the graph does not wiggle between the plotted points? Even computer algebra systems and graphing calculators plot by connecting points. We need a better procedure.

The First Derivative and Monotonicity   Recall that the first derivative $f'(x)$ gives us the slope of the tangent line to the graph of f at the point x. Then, if $f'(x) > 0$, the tangent line is rising to the right (see Figure 2). Similarly, if $f'(x) < 0$, the tangent line is falling to the right. These facts make the following theorem intuitively clear. We postpone a rigorous proof until Section 4.7.

Theorem A   Monotonicity Theorem

Let f be continuous on an interval I and differentiable at every interior point of I.
(i) If $f'(x) > 0$ for all x interior to I, then f is increasing on I.
(ii) If $f'(x) < 0$ for all x interior to I, then f is decreasing on I.

This theorem usually allows us to determine precisely where a differentiable function increases and where it decreases. It is a matter of solving two inequalities.

EXAMPLE 1   If $f(x) = 2x^3 - 3x^2 - 12x + 7$, find where f is increasing and where it is decreasing.

Solution   We begin by finding the derivative of f.
$$f'(x) = 6x^2 - 6x - 12 = 6(x + 1)(x - 2)$$

We need to determine where
$$(x + 1)(x - 2) > 0$$

and also where
$$(x + 1)(x - 2) < 0$$

This problem was discussed in great detail in Section 1.3, a section worth reviewing now. The split points are -1 and 2; they split the x-axis into three intervals: $(-\infty, -1)$, $(-1, 2)$, and $(2, \infty)$. Using the test points -2, 0, and 3, we conclude that $f'(x) > 0$ on the first and last of these intervals and that $f'(x) < 0$ on the middle interval (Figure 3). Thus, by Theorem A, f is increasing on $(-\infty, -1]$ and $[2, \infty)$; it is decreasing on $[-1, 2]$. Note that the theorem allows us to include the end points of these intervals, even though $f'(x) = 0$ at those points. The graph of f is shown in Figure 4.

Values of f'

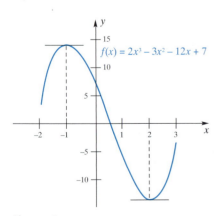

Figure 3

Figure 4

EXAMPLE 2   Determine where $g(x) = x/(1 + x^2)$ is increasing and where it is decreasing.

Solution

$$g'(x) = \frac{(1 + x^2) - x(2x)}{(1 + x^2)^2} = \frac{1 - x^2}{(1 + x^2)^2} = \frac{(1 - x)(1 + x)}{(1 + x^2)^2}$$

Since the denominator is always positive, $g'(x)$ has the same sign as the numerator $(1 - x)(1 + x)$. The split points, -1 and 1, determine the three intervals $(-\infty, -1)$, $(-1, 1)$, and $(1, \infty)$. When we test them, we find that $g'(x) < 0$ on the first and last of these intervals and that $g'(x) > 0$ on the middle one (Figure 5). We conclude from Theorem A that g is decreasing on $(-\infty, -1]$ and $[1, \infty)$ and that it is increasing on $[-1, 1]$. We postpone graphing g until later, but if you want to see the graph, turn to Figure 11 and Example 4.   ∎

The Second Derivative and Concavity   A function may be increasing and still have a very wiggly graph (Figure 6). To analyze wiggles, we need to study how the tangent line turns as we move from left to right along the graph. If the tangent line turns steadily in the counterclockwise direction, we say that the graph is *concave up*; if the tangent turns in the clockwise direction, the graph is *concave down*. Both definitions are better stated in terms of functions and their derivatives.

Definition

Let f be differentiable on an open interval I. We say that f (as well as its graph) is **concave up** on I if f' is increasing on I, and we say that f is **concave down** on I if f' is decreasing on I.

The diagrams in Figure 7 will help to clarify these notions. Note that a curve that is concave *up* is shaped like a *cup*.

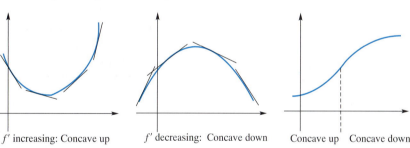

| f' increasing: Concave up | f' decreasing: Concave down | Concave up   Concave down |

Figure 7

In view of Theorem A, we have a simple criterion for deciding where a curve is concave up and where it is concave down. We simply keep in mind that the second derivative of f is the first derivative of f'. Thus, f' is increasing if f'' is positive; it is decreasing if f'' is negative.

Theorem B   Concavity Theorem

Let f be twice differentiable on the open interval I.
(i) If $f''(x) > 0$ for all x in I, then f is concave up on I.
(ii) If $f''(x) < 0$ for all x in I, then f is concave down on I.

For most functions, this theorem reduces the problem of determining concavity to the problem of solving inequalities. We are experts at this.

EXAMPLE 3   Where is $f(x) = \frac{1}{3}x^3 - x^2 - 3x + 4$ increasing, decreasing, concave up, and concave down?

Solution

$$f'(x) = x^2 - 2x - 3 = (x + 1)(x - 3)$$
$$f''(x) = 2x - 2 = 2(x - 1)$$

Values of g'

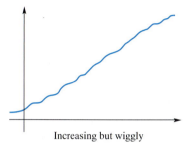

Figure 5

Increasing but wiggly

Figure 6

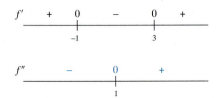

Figure 8

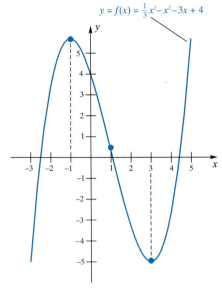

$y = f(x) = \frac{1}{3}x^3 - x^2 - 3x + 4$

Figure 9

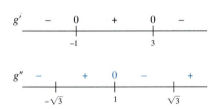

Figure 10

By solving the inequalities $(x + 1)(x - 3) > 0$ and its opposite, $(x + 1)(x - 3) < 0$, we conclude that f is increasing on $(-\infty, -1]$ and $[3, \infty)$ and decreasing on $[-1, 3]$ (Figure 8). Similarly, solving $2(x - 1) > 0$ and $2(x - 1) < 0$ shows that f is concave up on $(1, \infty)$ and concave down on $(-\infty, 1)$. The graph of f is shown in Figure 9. ∎

EXAMPLE 4   Where is $g(x) = x/(1 + x^2)$ concave up and where is it concave down? Sketch the graph of g.

Solution   We began our study of this function in Example 2. There we learned that g is decreasing on $(-\infty, -1]$ and $[1, \infty)$ and increasing on $[-1, 1]$. To analyze concavity, we calculate g''.

$$g'(x) = \frac{1 - x^2}{(1 + x^2)^2}$$

$$g''(x) = \frac{(1 + x^2)^2(-2x) - (1 - x^2)(2)(1 + x^2)(2x)}{(1 + x^2)^4}$$

$$= \frac{(1 + x^2)[(1 + x^2)(-2x) - (1 - x^2)(4x)]}{(1 + x^2)^4}$$

$$= \frac{2x^3 - 6x}{(1 + x^2)^3} = \frac{2x(x^2 - 3)}{(1 + x^2)^3}$$

Since the denominator is always positive, we need only solve $x(x^2 - 3) > 0$ and its opposite. The split points are $-\sqrt{3}, 0,$ and $\sqrt{3}$. These three split points determine four intervals. After testing them (Figure 10), we conclude that g is concave up on $(-\sqrt{3}, 0)$ and $(\sqrt{3}, \infty)$ and that it is concave down on $(-\infty, -\sqrt{3})$ and $(0, \sqrt{3})$.

To sketch the graph of g, we make use of all the information obtained so far, plus the fact that g is an odd function whose graph is symmetric with respect to the origin (Figure 11).

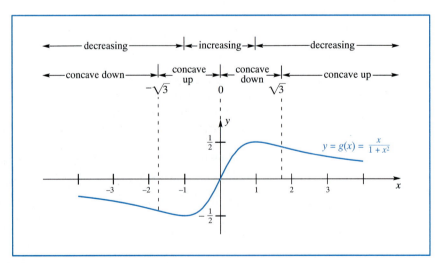

Figure 11   ∎

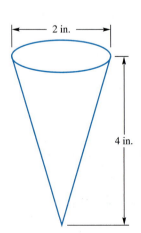

Figure 12

EXAMPLE 5   Suppose that water is poured into the conical container, as shown in Figure 12, at the constant rate of $\frac{1}{2}$ cubic inch per second. Determine the height h as a function of time t and plot $h(t)$ from time $t = 0$ until the time that the container is full.

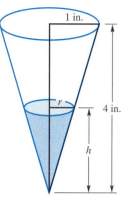

Figure 13

Solution Before we solve this problem, let's think about what the graph will look like. At first, the height will increase rapidly, since it takes very little water to fill the bottom of the cone. As the container fills up, the height will increase less rapidly. What do these statements say about the function $h(t)$, its derivative $h'(t)$, and its second derivative $h''(t)$? Since the water is steadily pouring in, the height will always increase, so $h'(t)$ will be positive. The height will increase more slowly as the water level rises. Thus, the function $h'(t)$ is decreasing so $h''(t)$ is negative. The graph of $h(t)$ is therefore increasing (because $h'(t)$ is positive) and concave down (because $h''(t)$ is negative).

Now, once we have an intuitive idea about what the graph should look like (increasing and concave down), let's solve the problem analytically. The volume of a right circular cone is $V = \frac{1}{3}\pi r^2 h$, where V, r, and h are all functions of time. Since water flows into the container at a rate of $\frac{1}{2}$ cubic inch per second, the function V is $V = \frac{1}{2}t$, where t is measured in seconds. The functions h and r are related; notice the similar triangles in Figure 13. Using properties of similar triangles, we have

$$\frac{r}{h} = \frac{1}{4}$$

Thus, $r = h/4$. The volume of the water inside the cone is thus

$$V = \frac{1}{3}\pi r^2 h = \frac{\pi}{3}\left(\frac{h}{4}\right)^2 h = \frac{\pi}{48}h^3$$

On the other hand, the volume is $V = \frac{1}{2}t$. Equating these two expressions for V gives

$$\frac{1}{2}t = \frac{\pi}{48}h^3$$

When $h = 4$, we have $t = \frac{2\pi}{48}4^3 = \frac{8}{3}\pi \approx 8.4$; thus, it takes about 8.4 seconds to fill the container. Now solve for h in the above equation relating h and t to obtain

$$h = \sqrt[3]{\frac{24}{\pi}t}$$

The first and second derivatives of h are

$$h'(t) = D_t\sqrt[3]{\frac{24}{\pi}t} = \frac{8}{\pi}\left(\frac{24}{\pi}t\right)^{-2/3} = \frac{2}{\sqrt[3]{9\pi t^2}}$$

which is positive, and

$$h''(t) = D_t\frac{2}{\sqrt[3]{9\pi t^2}} = -\frac{4}{3\sqrt[3]{9\pi t^5}}$$

which is negative. The graph of $h(t)$ is shown in Figure 14. As expected, the graph of h is increasing and concave down. ∎

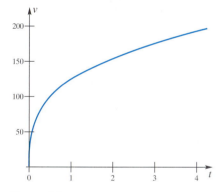

Figure 14

Inflection Points Let f be continuous at c. We call $(c, f(c))$ an **inflection point** of the graph of f if f is concave up on one side of c and concave down on the other side. The graph in Figure 15 indicates a number of possibilities.

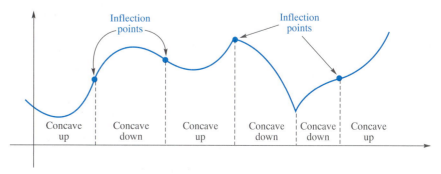

Figure 15

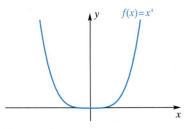

Figure 16

As you might guess, *points where $f''(x) = 0$ or where $f''(x)$ does not exist are the candidates for points of inflection.* We use the word *candidate* deliberately. Just as a candidate for political office may fail to be elected, so, for example, may a point where $f''(x) = 0$ fail to be a point of inflection. Consider $f(x) = x^4$, which has the graph shown in Figure 16. It is true that $f''(0) = 0$; yet the origin is not a point of inflection. Therefore, in searching for inflection points, we begin by identifying those points where $f''(x) = 0$ (and where $f''(x)$ does not exist). Then we check to see if they really are inflection points.

Look back at the graph in Example 4. You will see that it has three inflection points. They are $(-\sqrt{3}, -\sqrt{3}/4)$, $(0, 0)$, and $(\sqrt{3}, \sqrt{3}/4)$.

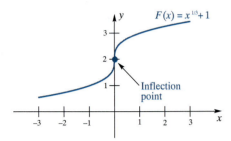

Figure 17

EXAMPLE 6   Find all points of inflection of $F(x) = x^{1/3} + 2$.

Solution

$$F'(x) = \frac{1}{3x^{2/3}}, \qquad F''(x) = \frac{-2}{9x^{5/3}}$$

The second derivative, $F''(x)$, is never 0; however, it fails to exist at $x = 0$. The point $(0, 2)$ is an inflection point since $F''(x) > 0$ for $x < 0$ and $F''(x) < 0$ for $x > 0$. The graph is sketched in Figure 17. ∎

Concepts Review

1. If $f'(x) > 0$ everywhere, then f is _____ everywhere; if $f''(x) > 0$ everywhere, then f is _____ everywhere.

2. If _____ and _____ on an open interval I, then f is both increasing and concave down on I.

3. A point on the graph of a continuous function where the concavity changes is called _____.

4. In trying to locate the inflection points for the graph of a function f, we should look at numbers c, where either _____ or _____.

Problem Set 4.2

In Problems 1–10, use the Monotonicity Theorem to find where the given function is increasing and where it is decreasing.

1. $f(x) = 3x + 3$
2. $g(x) = (x + 1)(x - 2)$

3. $h(t) = t^2 + 2t - 3$
4. $f(x) = x^3 - 1$

5. $G(x) = 2x^3 - 9x^2 + 12x$
6. $f(t) = t^3 + 3t^2 - 12$

7. $h(z) = \dfrac{z^4}{4} - \dfrac{4z^3}{6}$
8. $f(x) = \dfrac{x - 1}{x^2}$

9. $H(t) = \sin t, 0 \le t \le 2\pi$

10. $R(\theta) = \cos^2 \theta, 0 \le \theta \le 2\pi$

In Problems 11–18, use the Concavity Theorem to determine where the given function is concave up and where it is concave down. Also find all inflection points.

11. $f(x) = (x - 1)^2$
12. $G(w) = w^2 - 1$

13. $T(t) = 3t^3 - 18t$
14. $f(z) = z^2 - \dfrac{1}{z^2}$

15. $q(x) = x^4 - 6x^3 - 24x^2 + 3x + 1$

16. $f(x) = x^4 + 8x^3 - 2$
17. $F(x) = 2x^2 + \cos^2 x$

18. $G(x) = 24x^2 + 12 \sin^2 x$

In Problems 19–28, determine where the graph of the given function is increasing, decreasing, concave up, and concave down. Then sketch the graph (see Example 4).

19. $f(x) = x^3 - 12x + 1$

20. $g(x) = 4x^3 - 3x^2 - 6x + 12$

21. $g(x) = 3x^4 - 4x^3 + 2$
22. $F(x) = x^6 - 3x^4$

23. $G(x) = 3x^5 - 5x^3 + 1$
24. $H(x) = \dfrac{x^2}{x^2 + 1}$

25. $f(x) = \sqrt{\sin x}$ on $[0, \pi]$
26. $g(x) = x\sqrt{x - 2}$

27. $f(x) = x^{2/3}(1 - x)$
28. $g(x) = 8x^{1/3} + x^{4/3}$

In Problems 29–34, sketch the graph of a continuous function f on $[0, 6]$ that satisfies all the stated conditions.

29. $f(0) = 1; f(6) = 3$; increasing and concave down on $(0, 6)$

30. $f(0) = 8; f(6) = -2$; decreasing on $(0, 6)$; inflection point at the ordered pair $(2, 3)$, concave up on $(2, 6)$

31. $f(0) = 3; f(3) = 0; f(6) = 4;$
$f'(x) < 0$ on $(0,3); f'(x) > 0$ on $(3,6);$
$f''(x) > 0$ on $(0,5); f''(x) < 0$ on $(5,6)$

32. $f(0) = 3; f(2) = 2; f(6) = 0;$
$f'(x) < 0$ on $(0,2) \cup (2,6); f'(2) = 0;$
$f''(x) < 0$ on $(0,1) \cup (2,6); f''(x) > 0$ on $(1,2)$

33. $f(0) = f(4) = 1; f(2) = 2; f(6) = 0;$
$f'(x) > 0$ on $(0,2); f'(x) < 0$ on $(2,4) \cup (4,6);$
$f'(2) = f'(4) = 0; f''(x) > 0$ on $(0,1) \cup (3,4);$
$f''(x) < 0$ on $(1,3) \cup (4,6)$

34. $f(0) = f(3) = 3; f(2) = 4; f(4) = 2; f(6) = 0;$
$f'(x) > 0$ on $(0,2); f'(x) < 0$ on $(2,4) \cup (4,5);$
$f'(2) = f'(4) = 0; f'(x) = -1$ on $(5,6);$
$f''(x) < 0$ on $(0,3) \cup (4,5); f''(x) > 0$ on $(3,4)$

35. Prove that a quadratic function has no point of inflection.

36. Prove that a cubic function has exactly one point of inflection.

37. Prove that, if $f'(x)$ exists and is continuous on an interval I and if $f'(x) \neq 0$ at all interior points of I, then either f is increasing throughout I or decreasing throughout I. *Hint:* Use the Intermediate Value Theorem to show that there cannot be two points x_1 and x_2 of I where f' has opposite signs.

38. Suppose that f is a function whose derivative is $f'(x) = (x^2 - x + 1)/(x^2 + 1)$. Use Problem 37 to prove that f is increasing everywhere.

39. Use the Monotonicity Theorem to prove each statement if $0 < x < y$.

(a) $x^2 < y^2$ (b) $\sqrt{x} < \sqrt{y}$ (c) $\dfrac{1}{x} > \dfrac{1}{y}$

40. What conditions on a, b, and c will make $f(x) = ax^3 + bx^2 + cx + d$ always increasing?

41. Determine a and b so that $f(x) = a\sqrt{x} + b/\sqrt{x}$ has the point $(4, 13)$ as an inflection point.

42. Suppose that the cubic function $f(x)$ has three real zeros, $r_1, r_2,$ and r_3. Show that its inflection point has x-coordinate $(r_1 + r_2 + r_3)/3$. *Hint:* $f(x) = a(x - r_1)(x - r_2)(x - r_3)$.

43. Suppose that $f'(x) > 0$ and $g'(x) > 0$ for all x. What simple additional conditions (if any) are needed to guarantee that:

(a) $f(x) + g(x)$ is increasing for all x;
(b) $f(x) \cdot g(x)$ is increasing for all x;
(c) $f(g(x))$ is increasing for all x?

44. Suppose that $f''(x) > 0$ and $g''(x) > 0$ for all x. What simple additional conditions (if any) are needed to guarantee that:

(a) $f(x) + g(x)$ is concave up for all x;
(b) $f(x) \cdot g(x)$ is concave up for all x;
(c) $f(g(x))$ is concave up for all x?

`GC` *Use a graphing calculator or a computer to do Problems 45–48.*

45. Let $f(x) = \sin x + \cos(x/2)$ on the interval $I = (-2, 7)$.
(a) Draw the graph of f on I.
(b) Use this graph to estimate where $f'(x) < 0$ on I.
(c) Use this graph to estimate where $f''(x) < 0$ on I.
(d) Plot the graph of f' to confirm your answer to part (b).
(e) Plot the graph of f'' to confirm your answer to part (c).

46. Repeat Problem 45 for $f(x) = x \cos^2(x/3)$ on $(0, 10)$.

47. Let $f'(x) = x^3 - 5x^2 + 2$ on $I = [-2, 4]$. Where on I is f increasing?

48. Let $f''(x) = x^4 - 5x^3 + 4x^2 + 4$ on $I = [-2, 3]$. Where on I is f concave down?

49. Coffee is poured into the cup shown in Figure 18 at the rate of 2 cubic inches per second. The top diameter is 3.5 inches, the bottom diameter is 3 inches, and the height is 5 inches. This cup holds about 23 fluid ounces. Determine the height h as a function of time t, and sketch the plot from time $t = 0$ until the time that the cup is full.

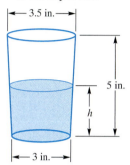

Figure 18

50. Water is being pumped at a constant rate of 5 gallons per minute into a cylindrical tank, as shown in Figure 19. The tank has diameter 3 feet and length 9.5 feet. The volume of the tank is $\pi r^2 l = \pi \times 1.5^2 \times 9.5 \approx 67.152$ cubic feet ≈ 500 gallons. Without doing any calculations sketch a graph of the height h of the water as a function of time t (see Example 5). Where is h concave up? Concave down?

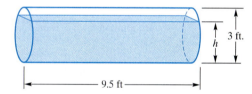

Figure 19

51. A liquid is poured into the container shown in Figure 20 at the rate of 3 cubic inches per second. The container holds about 24 cubic inches. Sketch a graph of the height h of the liquid as a function of time t. In your graph, pay special attention to the concavity of h.

52. A 20-gallon barrel, as shown in Figure 21, leaks at the constant rate of 0.1 gallon per day. Sketch a plot of the height h of the water as a function of time t, assuming that the barrel is full at time $t = 0$. In your graph, pay special attention to the concavity of h.

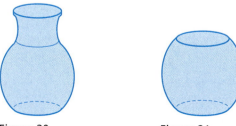

Figure 20   Figure 21

Answers to Concepts Review: **1.** increasing; concave up **2.** $f'(x) > 0; f''(x) < 0$ **3.** an inflection point **4.** $f''(c) = 0; f''(c)$ does not exist.

4.3
Local Maxima
and Minima

We recall from Section 4.1 that the maximum value (if it exists) of a function f on a set S is the largest value that f attains on the whole set S. It is sometimes referred to as the **global maximum value**, or the *absolute maximum value* of f. Thus, for the function f with domain $S = [a, b]$ whose graph is sketched in Figure 1, $f(a)$ is the global maximum value. But what about $f(c)$? It may not be king of the country, but at least it is chief of its own locality. We call it a **local maximum value**, or a *relative maximum value*. Of course, a global maximum value is automatically a local maximum value. Figure 2 illustrates a number of possibilities. Note that the global maximum value (if it exists) is simply the largest of the local maximum values. Similarly, the global minimum value is the smallest of the local minimum values.

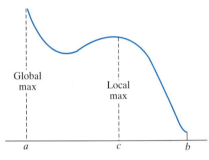

Figure 1

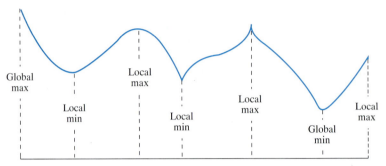

Figure 2

Here is the formal definition of local maxima and local minima. Recall that the symbol $\cap$ denotes the intersection (common part) of two sets.

Definition

Let S, the domain of f, contain the point c. We say that:

(i) $f(c)$ is a **local maximum value** of f if there is an interval (a, b) containing c such that $f(c)$ is the maximum value of f on $(a, b) \cap S$;

(ii) $f(c)$ is a **local minimum value** of f if there is an interval (a, b) containing c such that $f(c)$ is the minimum value of f on $(a, b) \cap S$;

(iii) $f(c)$ is a **local extreme value** of f if it is either a local maximum or a local minimum value.

Where Do Local Extreme Values Occur?

The Critical Point Theorem (Theorem 4.1B) holds with the phrase *extreme value* replaced by *local extreme value*; the proof is essentially the same. Thus, the critical points (end points, stationary points, and singular points) are the candidates for points where local extrema may occur. We say *candidates* because we are not claiming that there must be a local extremum at every critical point. The left graph in Figure 3 makes this clear. However, if the

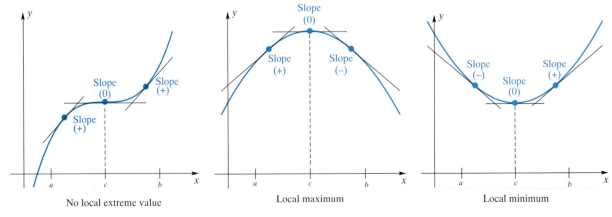

No local extreme value   Local maximum   Local minimum

Figure 3

derivative is positive on one side of the critical point and negative on the other, then we have a local extremum, as shown in the middle and right graphs of Figure 3.

Theorem A First Derivative Test

Let f be continuous on an open interval (a, b) that contains a critical point c.

(i) If $f'(x) > 0$ for all x in (a, c) and $f'(x) < 0$ for all x in (c, b), then $f(c)$ is a local maximum value of f.

(ii) If $f'(x) < 0$ for all x in (a, c) and $f'(x) > 0$ for all x in (c, b), then $f(c)$ is a local minimum value of f.

(iii) If $f'(x)$ has the same sign on both sides of c, then $f(c)$ is not a local extreme value of f.

Proof of (i) Since $f'(x) > 0$ for all x in (a, c), f is increasing on $(a, c]$ by the Monotonicity Theorem. Again, since $f'(x) < 0$ for all x in (c, b), f is decreasing on $[c, b)$. Thus, $f(x) < f(c)$ for all x in (a, b), except of course at $x = c$. We conclude that $f(c)$ is a local maximum.

The proofs of (ii) and (iii) are similar. ♦

EXAMPLE 1 Find the local extreme values of the function $f(x) = x^2 - 6x + 5$ on $(-\infty, \infty)$.

Solution The polynomial function f is continuous everywhere, and its derivative, $f'(x) = 2x - 6$, exists for all x. Thus, the only critical point for f is the single solution of $f'(x) = 0$; that is, $x = 3$.

Since $f'(x) = 2(x - 3) < 0$ for $x < 3$, f is decreasing on $(-\infty, 3]$; and because $2(x - 3) > 0$ for $x > 3$, f is increasing on $[3, \infty)$. Therefore, by the First Derivative Test $f(3) = -4$ is a local minimum value of f. Since 3 is the only critical point, there are no other extreme values. The graph of f is shown in Figure 4. Note that $f(3)$ is actually the (global) minimum value in this case. ■

EXAMPLE 2 Find the local extreme values of $f(x) = \frac{1}{3}x^3 - x^2 - 3x + 4$ on $(-\infty, \infty)$.

Solution Since $f'(x) = x^2 - 2x - 3 = (x + 1)(x - 3)$, the only critical points of f are -1 and 3. When we use the test points -2, 0, and 4, we learn that $(x + 1)(x - 3) > 0$ on $(-\infty, -1)$ and $(3, \infty)$ and $(x + 1)(x - 3) < 0$ on $(-1, 3)$. By the First Derivative Test, we conclude that $f(-1) = \frac{17}{3}$ is a local maximum value and that $f(3) = -5$ is a local minimum value (Figure 5). ■

EXAMPLE 3 Find the local extreme values of $f(x) = (\sin x)^{2/3}$ on $(-\pi/6, 2\pi/3)$.

Solution

$$f'(x) = \frac{2 \cos x}{3(\sin x)^{1/3}}, \qquad x \neq 0$$

The points 0 and $\pi/2$ are critical points, since $f'(0)$ does not exist and $f'(\pi/2) = 0$. Now $f'(x) < 0$ on $(-\pi/6, 0)$ and on $(\pi/2, 2\pi/3)$, while $f'(x) > 0$ on $(0, \pi/2)$. By the First Derivative Test, we conclude that $f(0) = 0$ is a local minimum value and that $f(\pi/2) = 1$ is a local maximum value. The graph of f is shown in Figure 6. ■

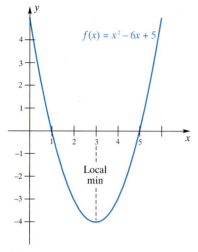

Figure 4

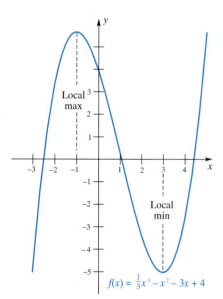

Figure 5

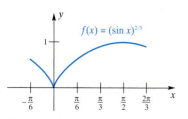

Figure 6

The Second Derivative Test   There is another test for local maxima and minima that is sometimes easier to apply than the First Derivative Test. It involves evaluating the second derivative at the stationary points. It does not apply to singular points.

Theorem B   Second Derivative Test

Let f' and f'' exist at every point in an open interval (a, b) containing c, and suppose that $f'(c) = 0$.
(i) If $f''(c) < 0$, $f(c)$ is a local maximum value of f.
(ii) If $f''(c) > 0$, $f(c)$ is a local minimum value of f.

Proof of (i)   It is tempting to say that, since $f''(c) < 0$, f is concave downward near c and to claim that this proves (i). However, to be sure that f is concave downward in a neighborhood of c, we need $f''(x) < 0$ in that neighborhood (not just at c), and nothing in our hypothesis guarantees that. We must be a bit more careful.

By definition and hypothesis,

$$f''(c) = \lim_{x \to c} \frac{f'(x) - f'(c)}{x - c} = \lim_{x \to c} \frac{f'(x) - 0}{x - c} < 0$$

so we can conclude that there is a (possibly small) interval (α, β) around c where

$$\frac{f'(x)}{x - c} < 0, \qquad x \neq c$$

But this inequality implies that $f'(x) > 0$ for $\alpha < x < c$ and $f'(x) < 0$ for $c < x < \beta$. Thus, by the First Derivative Test, $f(c)$ is a local maximum value. The proof of (ii) is similar. ♦

EXAMPLE 4   For $f(x) = x^2 - 6x + 5$, use the Second Derivative Test to identify local extrema.

Solution   This is the function of Example 1. Note that

$$f'(x) = 2x - 6 = 2(x - 3)$$
$$f''(x) = 2$$

Thus, $f'(3) = 0$ and $f''(3) > 0$. Therefore, by the Second Derivative Test, $f(3)$ is a local minimum value.   ∎

EXAMPLE 5   For $f(x) = \frac{1}{3}x^3 - x^2 - 3x + 4$, use the Second Derivative Test to identify local extrema.

Solution   This is the function of Example 2.

$$f'(x) = x^2 - 2x - 3 = (x + 1)(x - 3)$$
$$f''(x) = 2x - 2$$

The critical points are -1 and 3 ($f'(-1) = f'(3) = 0$). Since $f''(-1) = -4$ and $f''(3) = 4$, we conclude by the Second Derivative Test that $f(-1)$ is a local maximum value and that $f(3)$ is a local minimum value.   ∎

Unfortunately, the Second Derivative Test sometimes fails, since $f''(x)$ may be 0 at a stationary point. For both $f(x) = x^3$ and $f(x) = x^4$, $f'(0) = 0$ and $f''(0) = 0$ (see Figure 7). The first does not have a local maximum or minimum value at 0; the second has a local minimum there. This shows that if $f''(x) = 0$ at a stationary point we are unable to draw a conclusion about maxima or minima without more information.

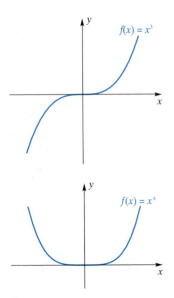

$f(x) = x^3$

$f(x) = x^4$

Figure 7

Knowing just a function's derivative can tell us a lot about the function itself, as the next example shows.

EXAMPLE 6   Figure 8 shows a plot of $y = f'(x)$. Find all local extrema, and points of inflection of f on the interval $[-1, 3]$. Given that $f(1) = 0$, sketch the graph of $y = f(x)$.

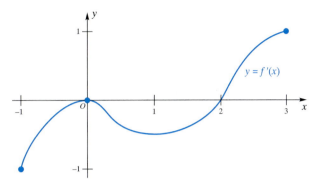

Figure 8

Solution   The derivative is negative on the intervals $(-1, 0)$ and $(0, 2)$ and positive on the interval $(2, 3)$. Thus, f is decreasing on $(-1, 0)$ and on $(0, 2)$ so there is a local maximum at the left end point $x = -1$. Since $f'(x)$ is positive on $(2, 3)$, f is increasing on $(2, 3)$ so there is a local maximum at the right end point $x = 3$. Since f is decreasing on $(0, 2)$ and increasing on $(2, 3)$, there is a relative minimum at $x = 2$. Figure 9 summarizes this information.

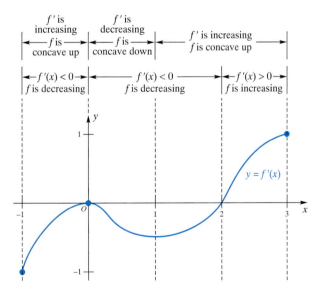

Figure 9

$f(-1)$	Local maximum
$f(2)$	Local minimum
$f(3)$	Local maximum
$(0, f(0))$	Inflection point
$(1, f(1))$	Inflection point

Inflection points for f occur when the concavity of f changes. Since f' is increasing on $(-1, 0)$ and on $(1, 3)$, f is concave up on $(-1, 0)$ and on $(1, 3)$. Since f' is decreasing on $(0, 1)$, f is concave down on $(0, 1)$. Thus, f changes concavity at $x = 0$ and at $x = 1$. The inflection points are therefore $(0, f(0))$ and $(1, f(1))$.

The information given above, together with the fact that $f(1) = 0$, can be used to sketch the graph of $y = f(x)$. (The sketch cannot be too precise because we still have limited information about f.) A sketch is shown in Figure 10.

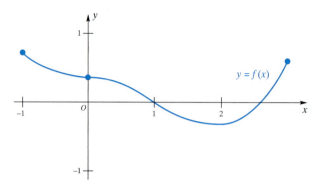

Figure 10

Concepts Review

1. If f is continuous at c, $f'(x) > 0$ near to c on the left, and $f'(x) < 0$ near to c on the right, then $f(c)$ is a local _____ value for f.

2. If $f'(x) = (x + 2)(x - 1)$, then $f(-2)$ is a local _____ value for f and $f(1)$ is a local _____ value for f.

3. If $f'(c) = 0$ and $f''(c) < 0$, we expect to find a local _____ value for f at c.

4. If $f(x) = x^3$, then $f(0)$ is neither a _____ nor a _____, even though $f''(0) = $ _____.

Problem Set 4.3

In Problems 1–6, identify the critical points. Then use (a) the First Derivative Test and (if possible) (b) the Second Derivative Test to decide which of the critical points give a local maximum and which give a local minimum.

1. $f(x) = x^3 - 6x^2 + 4$

2. $f(x) = x^3 - 12x + \pi$

3. $f(\theta) = \sin 2\theta, 0 < \theta < \dfrac{\pi}{4}$

4. $f(x) = \frac{1}{2}x + \sin x, 0 < x < 2\pi$

5. $\Psi(\theta) = \sin^2 \theta, -\pi/2 < \theta < \pi/2$

6. $r(z) = z^4 + 4$

In Problems 7–16, find the critical points and use the test of your choice to decide which critical points give a local maximum value and which give a local minimum value. What are these local maximum and minimum values?

7. $f(x) = x^3 - 3x$

8. $g(x) = x^4 + x^2 + 3$

9. $H(x) = x^4 - 2x^3$

10. $f(x) = (x - 2)^5$

11. $g(t) = \pi - (t - 2)^{2/3}$

12. $r(s) = 3s + s^{2/5}$

13. $f(t) = t - \dfrac{1}{t}, t \neq 0$

14. $f(x) = \dfrac{x^2}{\sqrt{x^2 + 4}}$

15. $\Lambda(\theta) = \dfrac{\cos \theta}{1 + \sin \theta}, 0 < \theta < 2\pi$

16. $g(\theta) = |\sin \theta|, 0 < \theta < 2\pi$

17. Find the (global) maximum and minimum values of $F(x) = 6\sqrt{x} - 4x$ on $[0, 4]$.

18. Do Problem 17 on the interval $[0, \infty)$.

19. Find (if possible) the maximum and minimum values of

$$f(x) = \frac{64}{\sin x} + \frac{27}{\cos x}$$

on $(0, \pi/2)$.

20. Find (if possible) the maximum and minimum values of

$$f(x) = x^2 + \frac{1}{x^2}$$

on $(0, \infty)$.

$\boxed{\text{C}}$ **21.** Find the minimum value of

$$g(x) = x^2 + \frac{16x^2}{(8 - x)^2}, \qquad x > 8$$

22. Consider $f(x) = Ax^2 + Bx + C$ with $A > 0$. Show that $f(x) \geq 0$ for all x if and only if $B^2 - 4AC \leq 0$.

In Problems 23–27, the first derivative f' is given. Find all values of x that make the function f (a) a local minimum and (b) a local maximum.

23. $f'(x) = x^3(1 - x)^2$

24. $f'(x) = -(x - 1)(x - 2)(x - 3)(x - 4)$

25. $f'(x) = (x - 1)^2(x - 2)^2(x - 3)(x - 4)$

26. $f'(x) = (x - 1)^2(x - 2)^2(x - 3)^2(x - 4)^2$

27. $f'(x) = 2(x + 2)(x + 1)^2(x - 2)^4(x - 3)^3$

28. What conclusions can you draw about f from the information that $f'(c) = f''(c) = 0$ and $f'''(c) > 0$?

29. Let f be a continuous function and let f' have the graph shown in Figure 11. Try to sketch a graph for f and answer the following questions.

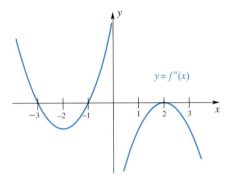

Figure 11

(a) Where is f increasing? Decreasing?
(b) Where is f concave up? Concave down?
(c) Where does f attain a local maximum? A local minimum?
(d) Where are there inflection points for f?

30. Repeat Problem 29 for Figure 12.

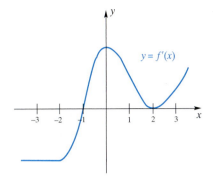

Figure 12

CAS *For each function in Problems 31–34, use your CAS to find the first and second derivatives, and plot the function and its derivative over the indicated interval. Find all local and global extrema. Use the Second Derivative Test to test whether you have a local maximum or a local minimum.*

31. $f(x) = x^5 - 5x^3 + 4$ on $[-2, 2.5]$
32. $f(x) = \sin x - \cos(x/2)$ on $[-2\pi, 2\pi]$
33. $f(x) = (x^3 - x)\tan x$ on $[-\pi/2, \pi/2]$
34. $f(x) = x^2/(x^4 + x^2 + 1)$ on $[-5, 5]$

Answers to Concepts Review: **1.** maximum   **2.** maximum; minimum   **3.** maximum   **4.** local maximum; local minimum; 0

4.4
More Max–Min Problems

The problems that we studied in Section 4.1 usually assumed that the set on which we wanted to maximize or minimize a function was a *closed* interval. However, the intervals that arise in practice are not always closed; they are sometimes open or even half-open, half-closed. We can still handle these problems if we correctly apply the theory developed in Section 4.3. Keep in mind that maximum (minimum) with no qualifying adjective means global maximum (minimum).

Extrema on Open Intervals

We give two examples to illustrate appropriate procedures for intervals that are open or half-open.

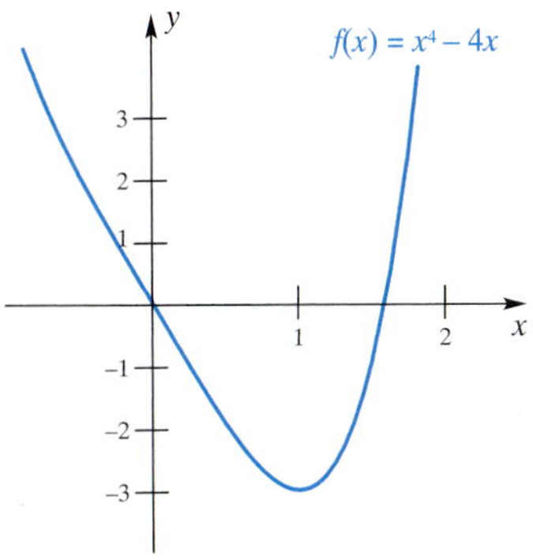

EXAMPLE 1   Find (if any exist) the minimum and maximum values of $f(x) = x^4 - 4x$ on $(-\infty, \infty)$.

Solution

$$f'(x) = 4x^3 - 4 = 4(x^3 - 1) = 4(x - 1)(x^2 + x + 1)$$

Since $x^2 + x + 1 = 0$ has no real solutions (quadratic formula), there is only one critical point, $x = 1$. For $x < 1$, $f'(x) < 0$, whereas for $x > 1$, $f'(x) > 0$. We conclude that $f(1) = -3$ is a local minimum value for f; and since f is decreasing on the left of 1 and increasing on the right of 1, it must actually be the minimum value of f.

The facts stated above imply that f cannot have a maximum value. The graph of f is shown in Figure 1.   ∎

EXAMPLE 2   Find (if any exist) the maximum and minimum values of

$$G(p) = \frac{1}{p(1 - p)} \text{ on } (0, 1).$$

Figure 1

Solution

$$G'(p) = \frac{2p - 1}{p^2(1 - p)^2}$$

The only critical point is $p = 1/2$. For every value of p in the interval $(0, 1)$ the denominator is positive; thus, the numerator determines the sign. If p is in the interval $(0, 1/2)$, then the numerator is negative; hence, $G'(p) < 0$. Similarly, if p is in the interval $(1/2, 1)$, $G'(p) > 0$. Thus, by the First Derivative Test, $G(1/2) = 4$ is a local minimum. Since there are no end points or singular points to check, $G(1/2)$ is a global minimum. The graph of $y = G(p)$ is shown in Figure 2.

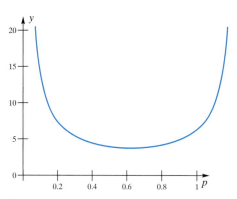

Figure 2

Practical Problems   Each of the following examples is different; yet there are common elements in the procedures that we use to solve them. Near the end of the section, we suggest a set of steps to use in solving any max–min problem.

EXAMPLE 3   A handbill is to contain 50 square inches of printed matter, with 4-inch margins at top and bottom and 2-inch margins on each side. What dimensions for the handbill would use the least paper?

Solution   Let x be the width and y the height of the handbill (Figure 3). Its area is

$$A = xy$$

We wish to minimize A.

As it stands, A is expressed in terms of two variables, a situation we do not know how to handle. However, we will find an equation connecting x and y so that one of these variables can be eliminated in the expression for A. The dimensions of the printed matter are $x - 4$ and $y - 8$, and its area is 50 square inches; so $(x - 4)(y - 8) = 50$. When we solve this equation for y, we obtain

$$y = \frac{50}{x - 4} + 8$$

Substituting this expression for y in $A = xy$ gives A in terms of x:

$$A = \frac{50x}{x - 4} + 8x$$

The allowable values for x are $4 < x < \infty$; we want to minimize A on the open interval $(4, \infty)$.

Now

$$\frac{dA}{dx} = \frac{(x - 4)50 - 50x}{(x - 4)^2} + 8 = \frac{8x^2 - 64x - 72}{(x - 4)^2} = \frac{8(x + 1)(x - 9)}{(x - 4)^2}$$

The only critical points are obtained by solving $\frac{dA}{dx} = 0$; this yields $x = 9$ and $x = -1$. We reject $x = -1$ because it is not in the interval $(4, \infty)$. Since $dA/dx < 0$ for x in $(4, 9)$ and $dA/dx > 0$ for x in $(9, \infty)$, we conclude that A attains its mini-

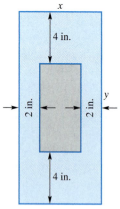

Figure 3

Common Sense

It would be hard to make any preliminary estimates in Example 3. However, common sense tells us that the handbill's height should be larger than its width. Why? Because we should capitalize on the narrow margins along the side.

mum value at $x = 9$. This value of x makes $y = 18$ (found by substituting in the equation connecting x and y). So the dimensions for the handbill that will use the least amount of paper are 9 inches by 18 inches. ∎

EXAMPLE 4   Andy, who is in a rowboat 2 miles from the nearest point B on a straight shoreline, notices smoke billowing from his house, which is 6 miles down the shoreline from B. He figures he can row at 3 miles per hour and run at 5 miles per hour. How should he proceed in order to get to his house in the least time?

Solution   We interpret the problem to mean that we are to determine the x in Figure 4 that will make Andy's traveling time a minimum. It is clear that we should restrict x to the closed interval $[0, 6]$.

The distance AD is $\sqrt{x^2 + 4}$ miles and the time to row it is $\sqrt{x^2 + 4}/3$ hours. The distance DC is $6 - x$ miles and the time to run it is $(6 - x)/5$ hours. Thus, the total time T in hours is

$$T = \frac{\sqrt{x^2 + 4}}{3} + \frac{6 - x}{5}$$

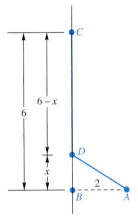

Figure 4

We wish to minimize T on $[0, 6]$.

This time there are three critical points, the end points 0 and 6 and a stationary point obtained by setting dT/dx equal to 0.

$$\frac{dT}{dx} = \frac{1}{3} \cdot \frac{1}{2}(x^2 + 4)^{-1/2}(2x) - \frac{1}{5}$$

$$= \frac{x}{3\sqrt{x^2 + 4}} - \frac{1}{5}$$

$$= \frac{5x - 3\sqrt{x^2 + 4}}{15\sqrt{x^2 + 4}}$$

When we set dT/dx equal to zero and solve, we obtain in successive steps,

$$\frac{5x - 3\sqrt{x^2 + 4}}{15\sqrt{x^2 + 4}} = 0$$

$$5x - 3\sqrt{x^2 + 4} = 0$$

$$5x = 3\sqrt{x^2 + 4}$$

$$25x^2 = 9(x^2 + 4)$$

$$16x^2 = 36$$

$$x^2 = \tfrac{36}{16}$$

$$x = \tfrac{3}{2}$$

x	T
0	1.87
1.5	1.73
6	2.11

Figure 5

Since the domain for T is a closed interval, T has a minimum (Theorem 4.1A), and this minimum must occur at a critical point (Theorem 4.1B). The critical points are thus $x = 0$, $x = 1.5$, and $x = 6$. Figure 5 shows the value of T at each critical point. Since the minimum occurs at $x = 1.5$, we conclude that Andy should row to a point 1.5 miles down the shoreline and then run the rest of the way. It will take him about 1.73 hours, or 104 minutes. For a similar problem in which one of the end points produces the minimum time, see Problem 15. ∎

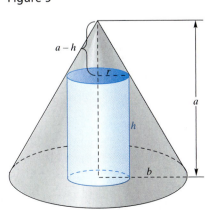

Figure 6

EXAMPLE 5   Find the dimensions of the right circular cylinder of greatest volume that can be inscribed in a given right circular cone.

Solution   Let a be the altitude and b the radius of the base of the given cone (both constants). Denote by h, r, and V the altitude, radius, and volume, respectively, of an inscribed cylinder (see Figure 6). Before proceeding, let's apply some intuition. If the cylinder's radius is close to the radius of the cone's base, then the cylinder's volume would be close to zero. Now, imagine inscribed cylinders with increasing height, but decreasing radius. Initially, the volumes would increase from zero, but then they would decrease to zero as the cylinders' heights get close to the cone's height. Intuitively, the volume should peak for some cylinder.

The volume of the inscribed cylinder is

$$V = \pi r^2 h$$

From similar triangles,

$$\frac{a - h}{r} = \frac{a}{b}$$

which gives

$$h = a - \frac{a}{b} r$$

When we substitute this expression for h in the formula for V, we obtain

$$V = \pi r^2 \left(a - \frac{a}{b} r \right) = \pi a r^2 - \pi \frac{a}{b} r^3$$

We wish to maximize V for r in the interval $[0, b]$. (Someone is sure to argue, and with good reason, that the appropriate interval is $(0, b)$. Actually, the answer is the same either way, though we have to use the First Derivative Test if we do the problem using $(0, b)$ as the domain.)

Now

$$\frac{dV}{dr} = 2\pi a r - 3\pi \frac{a}{b} r^2 = \pi a r \left(2 - \frac{3}{b} r \right)$$

This yields the stationary points $r = 0$ and $r = 2b/3$, giving us three critical points on $[0, b]$ to consider: 0, $2b/3$, and b. As expected, $r = 0$ and $r = b$ both give a volume of 0. Thus, $r = 2b/3$ has to give the maximum volume. When we substitute this value for r in the equation connecting r and h, we find that $h = a/3$. In other words, the inscribed cylinder has greatest volume when its radius is two-thirds the radius of the cone's base and its height is one-third the altitude of the cone.   ∎

A Summary of the Method

Based on the examples above, we suggest a step-by-step method that can be applied in many max–min problems. Do not follow it slavishly; common sense may sometimes suggest an alternative approach or omission of some steps.

Step 1: Draw a picture for the problem and assign appropriate variables to key quantities.

Step 2: Write a formula for the quantity Q to be maximized (minimized) in terms of those variables.

Step 3: Use the conditions of the problem to eliminate all but one of these variables, and thereby express Q as a function of a single variable, such as x.

Step 4: Determine the set of possible values for x, usually an interval.

Step 5: Find the critical points (end points, stationary points, singular points). Frequently, the key critical points are the stationary points where $dQ/dx = 0$.

Step 6: Use the theory of this chapter to decide which critical point gives the maximum (minimum).

Least Squares (Optional)

There are a number of physical, economic, and social phenomena in which one variable is proportional to another. For example, Newton's Second Law says that the force F on an object of mass m is proportional to its acceleration a $(F = ma)$. Hooke's Law says that the force exerted by a spring is proportional to the distance it is stretched $(F = kx)$. (Hooke's Law is often given as $F = -kx$, with the minus sign indicating that the force is in the direction oppo-

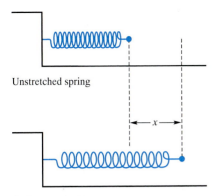

Unstretched spring

Spring stretched by amount x

Figure 7

Distance Stretched, x (meters)	Force y Exerted by Spring (newtons)
0.005	8
0.010	17
0.015	22
0.020	32
0.025	36

Figure 8

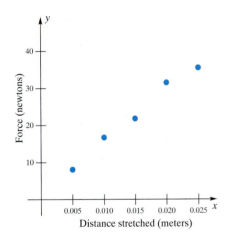

Figure 9

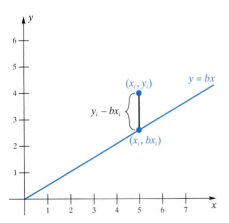

Figure 10

site the stretch. For now, we will ignore the sign of the force.) Manufacturing costs are proportional to the number of units produced. The number of traffic accidents is proportional to the volume of traffic. These are *models*, and in an experiment we will rarely find that the observed data fit the model exactly.

Suppose that we observe the force exerted by a spring when it is stretched by x centimeters (Figure 7). For example, when we stretch the spring by 0.5 centimeter, we observe a force of 8 newtons, when we stretch the spring by 1.0 centimeter, we observe a force of 17 newtons, and so on. Figure 8 shows additional observations, and Figure 9 shows a plot of the ordered pairs (x_i, y_i), where x_i is the distance stretched and y_i is the force exerted on the spring. A plot of ordered pairs like this is called a **scatter plot**.

Let's generalize the problem to one in which we are given n points (x_1, y_1), $(x_2, y_2),\ldots,(x_n, y_n)$. Our goal is to find a line through the origin that *best fits* these points. Before proceeding, we must introduce sigma (Σ) notation.

The symbol $\sum_{i=1}^{n} a_i$ means the sum the numbers $a_1, a_2,\ldots, a_n$. For example,

$$\sum_{i=1}^{3} i^2 = 1^2 + 2^2 + 3^2 = 14$$

and

$$\sum_{i=1}^{n} x_i y_i = x_1 y_1 + x_2 y_2 + \cdots + x_n y_n$$

In this case, we multiply x_i and y_i first and then sum. There will be more on sigma notation in Section 5.3.

To find the line that best fits these n points, we must be specific about how we measure the fit. Our *best-fit* line through the origin is defined to be the one that minimizes the sum of the squared vertical distances between (x_i, y_i) and the line $y = bx$. If (x_i, y_i) is a point, then (x_i, bx_i) is the point on the line $y = bx$ that is directly above or below (x_i, y_i). The vertical distance between (x_i, y_i) and (x_i, bx_i) is therefore $y_i - bx_i$. (See Figure 10.) The squared distance is thus $(y_i - bx_i)^2$. The problem is to find the value of b that minimizes the sum of these squared differences. If we define

$$S = \sum_{i=1}^{n} (y_i - bx_i)^2$$

then we must find the value of b that *minimizes S*. This is a minimization problem like the ones encountered before. Keep in mind, however, that the ordered pairs $(x_i, y_i), i = 1, 2,\ldots, n$ are *fixed*; the variable in this problem is b.

We proceed as before by finding dS/db, setting the result equal to 0, and solving for b. Since the derivative is a linear operator, we have

$$\frac{dS}{db} = \frac{d}{db} \sum_{i=1}^{n} (y_i - bx_i)^2$$

$$= \sum_{i=1}^{n} \frac{d}{db}(y_i - bx_i)^2$$

$$= \sum_{i=1}^{n} 2(y_i - bx_i)\left(\frac{d}{db}(y_i - bx_i)\right)$$

$$= -2 \sum_{i=1}^{n} x_i(y_i - bx_i)$$

Setting this result equal to zero and solving yields

$$0 = -2 \sum_{i=1}^{n} x_i (y_i - bx_i)$$

$$0 = \sum_{i=1}^{n} x_i y_i - b \sum_{i=1}^{n} x_i^2$$

$$b \sum_{i=1}^{n} x_i^2 = \sum_{i=1}^{n} x_i y_i$$

$$b = \frac{\sum_{i=1}^{n} x_i y_i}{\sum_{i=1}^{n} x_i^2}$$

To see that this yields a minimum value for S, we note that

$$\frac{d^2 S}{db^2} = 2 \sum_{i=1}^{n} x_i^2$$

which is always positive. There are no end points to check. Thus, by the Second Derivative Test, we conclude that the line $y = bx$, with $b = \sum_{i=1}^{n} x_i y_i / \sum_{i=1}^{n} x_1^2$, is the best-fit line, in the sense of minimizing S. The line $y = bx$ is called the **least-squares line through the origin**.

EXAMPLE 6   Find the least-squares line through the origin for the spring data in Figure 8.

Solution

$$b = \frac{0.005 \cdot 8 + 0.010 \cdot 17 + 0.015 \cdot 22 + 0.020 \cdot 32 + 0.025 \cdot 36}{0.005^2 + 0.010^2 + 0.015^2 + 0.020^2 + 0.025^2}$$

$$\approx 1512.7$$

The least-squares line through the origin is therefore $y = 1512.7x$ and is shown in Figure 11. The estimate of the spring constant is therefore $k = 1512.7$.   ■

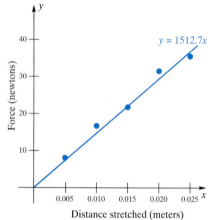

Distance stretched (meters)

Figure 11

Lot Size, x	Labor hours, y
10	21
14	25
6	13
7	14
10	18
13	29

Figure 12

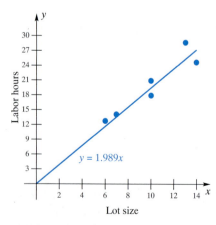

Lot size

Figure 13

EXAMPLE 7   The XYZ Company manufactures bookcases. The sizes of the customers' orders (called *lot sizes*) vary from one order to the next. One customer may order 10 bookcases, the next 14 bookcases, and the next 6. The number of labor hours required to produce the bookcases should be proportional to the size of the order, but there is always some variability. XYZ has obtained the data, shown in Figure 12, on lot size x and the labor hours y that were required. Find the least-squares line through the origin and use it to predict the number of labor hours required for a lot size of 11 bookcases.

Solution   Figure 13 suggests that the points fall near a straight line through the origin. (The line should go through the origin because an order size of 0 will require 0 labor hours.) The least-squares line has slope

$$b = \frac{10 \cdot 21 + 14 \cdot 25 + 6 \cdot 13 + 7 \cdot 14 + 10 \cdot 18 + 13 \cdot 29}{10^2 + 14^2 + 6^2 + 7^2 + 10^2 + 13^2}$$

$$= \frac{1293}{650} \approx 1.989$$

The equation of the least-squares line through the origin is therefore $y = 1.989x$ (see Figure 13). The prediction for the labor hours needed for an order of 11 bookcases is

$$y = 1.989 \times 11 \approx 21.9$$ ■

The assumption that the line pass through the origin is reasonable when one variable is proportional to another, but it is not reasonable in many other cases. For example, gas mileage is related to engine size, and the number of games won by a National Hockey League team is related to the number of goals allowed. Although in both of these examples the relationship is linear (at least approximately), in neither case does the line go through the origin. For a given set of data (x_1, y_1), $(x_2, y_2),\dots,(x_n, y_n)$, the line $y = a + bx$ that minimizes

$$S = \sum_{i=1}^{n} \left[y_i - \left(a + bx_i \right) \right]^2$$

is called the **least-squares line**. This minimization problem involves two variables, a and b. The formulas for the slope b and the intercept a that minimize S are

$$b = \frac{\displaystyle\sum_{i=1}^{n} x_i y_i - \frac{1}{n}\left(\sum_{i=1}^{n} x_i\right)\left(\sum_{i=1}^{n} y_i\right)}{\displaystyle\sum_{i=1}^{n} x_i^2 - \frac{1}{n}\left(\sum_{i=1}^{n} x_i\right)^2}$$

and

$$a = \frac{1}{n}\sum_{i=1}^{n} y_i - \frac{b}{n}\sum_{i=1}^{n} x_i$$

Since the minimization of functions of two variables is not covered until Chapter 15, the derivation will have to wait until then. The method of least-squares is used extensively by statisticians. You can study this method further in most courses on statistics or linear algebra.

Concepts Review

1. If a rectangle of area 100 has length x and width y, then the allowable values for x are _____.

2. The perimeter P of the rectangle of Question 1 expressed in terms of x is given by $P =$ _____.

3. If the rectangle of Question 1 is partitioned down the middle in both directions, the total length of fence L required to

enclose and partition it can be expressed in terms of x by $L =$ _____.

4. If f is continuous on $(0,\infty)$ and if $f'(x) < 0$ for $x < c$ and $f'(x) > 0$ for $x > c$, then f has a _____ at $x = c$. It has no _____ on $(0,\infty)$.

Problem Set 4.4

1. Find two numbers whose product is −16 and the sum of whose squares is a minimum.

2. For what number does the principal square root exceed eight times the number by the largest amount?

3. For what number does the principal fourth root exceed twice the number by the largest amount?

4. Find two numbers whose product is −12 and the sum of whose squares is a minimum.

5. Find the points on the parabola $y = x^2$ that are closest to the point $(0, 5)$. *Hint*: Minimize the square of the distance between (x, y) and $(0, 5)$.

6. Find the points on the parabola $x = 2y^2$ that are closest to the point $(10, 0)$. *Hint*: Minimize the square of the distance between (x, y) and $(10, 0)$.

7. A farmer wishes to fence off two identical adjoining rectangular pens, each with 900 square feet of area, as shown in Figure 14. What are x and y so that the least amount of fence is required?

Figure 14

8. A farmer wishes to fence off three identical adjoining rectangular pens (see Figure 15), each with 300 square feet of area. What should the width and length of each pen be so that the least amount of fence is required?

Figure 15

9. Suppose that the outer boundary of the pens in Problem 8 requires heavy fence that costs $3 per foot, but that the two internal partitions require fence costing only $2 per foot. What dimensions x and y will produce the least expensive cost for the pens?

10. Solve Problem 8 assuming that the area of each pen is 900 square feet. Study the solution to this problem and to Problem 8 and make a conjecture about the ratio of x/y in all problems of this type. Try to prove your conjecture.

☐ **11.** An object thrown from the edge of a 42-foot cliff follows the path given by $y = -\dfrac{2x^2}{25} + x + 42$ (Figure 16). An observer stands 2.6656 feet from the bottom of the cliff.

(a) Find the position of the object when it is closest to the observer.

(b) Find the position of the object when it is farthest from the observer.

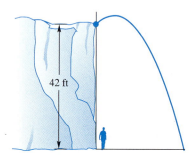

42 ft

Figure 16

12. The illumination at a point is inversely proportional to the square of the distance of the point from the light source and directly proportional to the intensity of the light source. If two light sources are s feet apart and their intensities are I_1 and I_2, respectively, at what point between them will the sum of their illuminations be a minimum?

☐ **13.** A small island is 2 miles from the nearest point P on the straight shoreline of a large lake. If a woman on the island can row a boat 3 miles per hour and can walk 4 miles per hour, where should the boat be landed in order to arrive at a town 10 miles down the shore from P in the least time? See Example 4.

☐ **14.** In Problem 13, suppose that the woman will be picked up by a car that will average 50 miles per hour when she gets to the shore. Then where should she land?

☐ **15.** In Problem 13, suppose that the woman uses a motorboat that goes 20 miles per hour. Then where should she land?

16. A powerhouse is located on one bank of a straight river that is w feet wide. A factory is situated on the opposite bank of the river, L feet downstream from the point A directly opposite the powerhouse. What is the most economical path for a cable connecting the powerhouse to the factory if it costs a dollars per foot to lay the cable under water and b dollars per foot on land $(a > b)$?

17. At 7:00 A.M. one ship was 60 miles due east from a second ship. If the first ship sailed west at 20 miles per hour and the second ship sailed southeast at 30 miles per hour, when were they closest together?

18. Find the equation of the line that is tangent to the ellipse $b^2 x^2 + a^2 y^2 = a^2 b^2$ in the first quadrant and forms with the coordinate axes the triangle with smallest possible area (a and b are positive constants).

19. Find the greatest volume that a right circular cylinder can have if it is inscribed in a sphere of radius r.

20. Show that the rectangle with maximum perimeter that can be inscribed in a circle is a square.

21. What are the dimensions of the right circular cylinder with greatest curved surface area that can be inscribed in a sphere of radius r?

22. A right circular cone is to be inscribed in another right circular cone of given volume, with the same axis and with the vertex of the inner cone touching the base of the outer cone. What must be the ratio of their altitudes for the inscribed cone to have maximum volume?

23. A wire of length 100 centimeters is cut into two pieces; one is bent to form a square, and the other is bent to form an equilateral triangle. Where should the cut be made if (a) the sum of the two areas is to be a minimum; (b) a maximum? (Allow the possibility of no cut.)

24. A closed box in the form of a rectangular parallelepiped with a square base is to have a given volume. If the material used in the bottom costs 20% more per square inch than the material in the sides, and the material in the top costs 50% more per square inch than that of the sides, find the most economical proportions for the box.

25. An observatory is to be in the form of a right circular cylinder surmounted by a hemispherical dome. If the hemispherical dome costs twice as much per square foot as the cylindrical wall, what are the most economical proportions for a given volume?

26. A weight connected to a spring moves along the x-axis so that its x-coordinate at time t is

$$x = \sin 2t + \sqrt{3} \cos 2t$$

What is the farthest that the weight gets from the origin?

27. A flower bed will be in the shape of a sector of a circle (a pie-shaped region) of radius r and vertex angle θ. Find r and θ if its area is a constant A and the perimeter is a minimum.

28. A fence h feet high runs parallel to a tall building and w feet from it (Figure 17). Find the length of the shortest ladder that will reach from the ground across the top of the fence to the wall of the building.

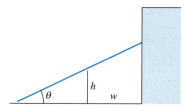

Figure 17

29. (Snell's Law). Fermat's Principle in optics says that light travels from point A to point B along the path that requires least time. Suppose that light travels in one medium at velocity c_1 and in a second medium at velocity c_2. If A is in medium 1 and B in medium 2 and the x-axis separates the two media, as shown in Figure 18, show that

$$\frac{\sin \theta_1}{c_1} = \frac{\sin \theta_2}{c_2}$$

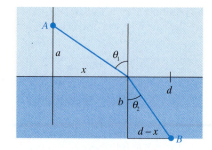

Figure 18

30. Light from A is reflected to B by a plane mirror. Use Fermat's Principle (Problem 29) to show that the angle of incidence is equal to the angle of reflection.

☑ **31.** One end of a 27-foot ladder rests on the ground and the other end rests on the top of an 8-foot wall. As the bottom end is pushed along the ground toward the wall, the top end extends beyond the wall. Find the maximum horizontal overhang of the top end.

32. I have enough pure silver to coat 1 square meter of surface area. I plan to coat a sphere and a cube. What dimensions should they be if the total volume of the silvered solids is to be a maximum? A minimum? (Allow the possibility of all the silver going onto one solid.)

33. One corner of a long narrow strip of paper is folded over so that it just touches the opposite side, as shown in Figure 19. With parts labeled as indicated, determine x in order to:

(a) maximize the area of triangle A;

(b) minimize the area of triangle B;

(c) minimize the length z.

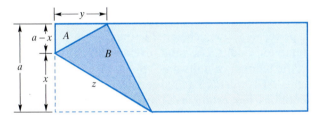

Figure 19

34. Determine θ so that the area of the symmetric cross shown in Figure 20 is maximized. Then find this maximum area.

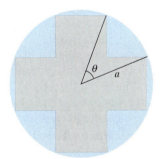

Figure 20

CAS **35.** A clock has hour and minute hands of lengths h and m, respectively, with $h \le m$. We wish to study this clock at times between 12:00 and 12:30. Let θ, ϕ, and L be as in Figure 21 and note that θ increases at a constant rate. By the Law of Cosines, $L = L(\theta) = \left(h^2 + m^2 - 2hm\cos\theta\right)^{1/2}$, and so

$$L'(\theta) = hm\left(h^2 + m^2 - 2hm\cos\theta\right)^{-1/2}\sin\theta$$

(a) For $h = 3$ and $m = 5$, determine L', L, and ϕ at the instant when L' is largest.

(b) Rework part (a) when $h = 5$ and $m = 13$.

(c) Based on parts (a) and (b), make conjectures about the values of L', L, and ϕ at the instant when the tips of the hands are separating most rapidly.

(d) Try to prove your conjectures.

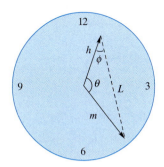

Figure 21

☑ C **36.** An object thrown from the edge of a 100-foot cliff follows the path given by $y = -\dfrac{x^2}{10} + x + 100$. An observer stands 2 feet from the bottom of the cliff.

(a) Find the position of the object when it is closest to the observer.

(b) Find the position of the object when it is farthest from the observer.

☑ CAS **37.** The earth's position in the solar system at time t can be described approximately by $P\left(93\cos(2\pi t), 93\sin(2\pi t)\right)$, where the sun is at the origin and distances are measured in millions of miles. Suppose that an asteroid has position $Q\left(60\cos\left[2\pi(1.51t - 1)\right], 120\sin\left[2\pi(1.51t - 1)\right]\right)$. When, over the time period $[0, 20]$ (i.e., over the next 20 years), does the asteroid come closest to the earth? How close does it come?

C **38.** Brass is produced in long rolls of a thin sheet. To monitor the quality, inspectors select at random a piece of the sheet, measure its area, and count the number of surface imperfections on that piece. The area varies from piece to piece. The following table gives data on the area (in square feet) of the selected piece and the number of surface imperfections found on that piece.

Piece	Area in Square Feet	Number of Surface Imperfections
1	1.0	3
2	4.0	12
3	3.6	9
4	1.5	5
5	3.0	8

(a) Make a scatter plot with area on the horizontal axis and number of surface imperfections on the vertical axis.

(b) Does it look like a line through the origin would be a good model for these data? Explain.

(c) Find the equation of the least-squares line through the origin.

(d) Use the result of part (c) to predict how many surface imperfections there would be on a sheet with area 2.0 square feet.

C **39.** Suppose that every customer order taken by the XYZ Company (see Example 8) requires exacty 5 hours of labor for handling the paperwork; this length of time is *fixed* and does not vary from lot to lot. The total number of hours y required to manufacture and sell a lot of size x would then be

$$y = (\text{number of hours to produce a lot of size } x) + 5$$

Some data on XYZ's brass bookcases are given in the following table.

Order	Lot Size x	Total Labor Hours y
1	11	38
2	16	52
3	8	29
4	7	25
5	10	38

(a) From the description of the problem, the least-squares line should have 5 as its y-intercept. Find a formula for the value of the slope b that minimizes the sum of squares

$$S = \sum_{i=1}^{n} \left[y_i - (5 + bx_i) \right]^2$$

(b) Use this formula to estimate the slope b.

(c) Use your least-squares line to predict the total number of labor hours to produce a lot consisting of 15 brass bookcases.

Answers to Concepts Review: **1.** $0 < x < \infty$ **2.** $2x + 200/x$ **3.** $3x + 300/x$ **4.** minimum; maximum

4.5
Economic Applications

Each discipline has its own language. This is certainly true of economics, which has a highly developed special vocabulary. Once we learn this vocabulary, we will discover that many of the problems of economics are just ordinary calculus problems dressed in new clothes.

Consider a typical company, the ABC Company. For simplicity, assume that ABC produces and markets a single product; it might be television sets, car batteries, or bars of soap. If it sells x units of the product in a fixed period of time (e.g., a year), it will be able to charge a **price**, $p(x)$, for each unit. In other words, $p(x)$ is the price required to attract a demand for x units. The **total revenue** that ABC can expect is given by $R(x) = xp(x)$, the number of units times the price per unit.

To produce and market x units, ABC will have a total cost, $C(x)$. This is normally the sum of a **fixed cost** (office utilities, real estate taxes, and so on) plus a **variable cost**, which depends directly on the number of units produced.

The key concept for a company is the **total profit**, $P(x)$. It is just the difference between revenue and cost; that is,

$$P(x) = R(x) - C(x) = xp(x) - C(x)$$

Generally, a company seeks to maximize its total profit.

There is a feature that tends to distinguish problems in economics from those in the physical sciences. In most cases, ABC's product will be in discrete units (you can't make or sell 8.23 television sets or π car batteries). Thus, the functions $R(x)$, $C(x)$, and $P(x)$ are usually defined only for $x = 0, 1, 2, \ldots$ and, consequently, their graphs consist of discrete points (Figure 1). In order to make the tools of calculus available, we connect these points with a smooth curve (Figure 2), thereby pretending that R, C, and P are nice differentiable functions. This illustrates an aspect of *mathematical modeling* that is almost always necessary, especially in economics. To model a real-world problem, we must make simplifying assumptions. This means that the answers we get are only approximations of the answers that we seek—one of the reasons economics is a less than perfect science. A well-known statistician once said: No model is accurate, but many models are useful.

Discrete versus Continuous

Most problems in the social sciences are properly viewed as discrete in nature. Moreover, the digital computer is a fast, accurate tool for handling discrete quantities. A natural question arises: Why not study discrete problems using discrete tools rather than by first modeling them with continuous curves? For this reason, most colleges now offer courses in discrete mathematics. However, because of its beauty and power, and because continuous models can be used to model (approximately at least) discrete relationships, calculus continues to enjoy popularity as a tool for analyzing social science as well as science problems.

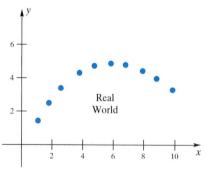

Figure 1

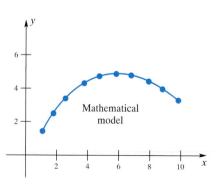

Figure 2

A related problem for an economist is how to obtain formulas for the functions $C(x)$ and $p(x)$. In a simple case, $C(x)$ might have the form

$$C(x) = 10{,}000 + 50x$$

If so, $10,000 is the **fixed cost** and $50x$ is the **variable cost**, based on a $50 direct cost for each unit produced. Perhaps a more typical situation is

$$C_1(x) = 10{,}000 + 45x + 100\sqrt{x}$$

In this case the **average variable cost** per unit is

$$\frac{45x + 100\sqrt{x}}{x} = 45 + \frac{100}{\sqrt{x}}$$

an amount that decreases as x increases (efficiency of size or, in economists' terms, economy of scale). The cost functions $C(x)$ and $C_1(x)$ are graphed together in Figure 3.

Selecting appropriate functions to model cost and price is a nontrivial task. Occasionally, they can be inferred from basic assumptions. In other cases, a careful study of the history of the firm will suggest reasonable choices. Sometimes, we must simply make intelligent guesses.

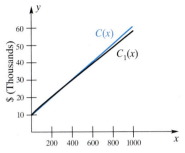

Figure 3

Use of the Word *Marginal*

Suppose that ABC knows its cost function $C(x)$ and that it has tentatively planned to produce 2000 units this year. President Hornblower would like to determine the additional cost per unit if ABC increased production slightly. Would it, for example, be less than the additional revenue per unit? If so, it would make good economic sense to increase production.

If the cost function is the one shown in Figure 4, President Hornblower is asking for the value of $\Delta C/\Delta x$ when $\Delta x = 1$. But we expect that this will be very close to the value of

$$\lim_{\Delta x \to 0} \frac{\Delta C}{\Delta x}$$

when $x = 2000$. This limit is called the **marginal cost**. We mathematicians recognize it as dC/dx, the derivative of C with respect to x.

In a similar vein, we define **marginal price** as dp/dx, **marginal revenue** as dR/dx, and **marginal profit** as dP/dx.

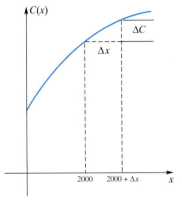

Figure 4

Examples

We now illustrate how to solve a wide variety of economic problems.

EXAMPLE 1 Suppose that $C(x) = 8300 + 3.25x + 40\sqrt[3]{x}$ dollars. Find the average cost per unit and the marginal cost, and then evaluate them when $x = 1000$.

Solution

$$\text{Average cost:} \quad \frac{C(x)}{x} = \frac{8300 + 3.25x + 40x^{1/3}}{x}$$

$$\text{Marginal cost:} \quad \frac{dC}{dx} = 3.25 + \frac{40}{3}x^{-2/3}$$

At $x = 1000$, these have the values 11.95 and 3.38, respectively. This means that it costs, on the average, $11.95 per unit to produce the first 1000 units; to produce one additional unit beyond 1000 costs only about $3.38. ∎

EXAMPLE 2 A company estimates that it can sell 1000 units per week if it sets the unit price at $3.00, but that its weekly sales will rise by 100 units for each $0.10 decrease in price. If x is the number of units sold each week ($x \geq 1000$), find:

(a) the price function, $p(x)$;
(b) the number of units and the corresponding price that will maximize weekly revenue;
(c) the maximum weekly revenue.

Economic Vocabulary

Because economics tends to be a study of discrete phenomena, your economics professor may define marginal cost at x as the cost of producing one additional unit, that is, as

$$C(x + 1) - C(x)$$

In the mathematical model, this number will be very close in value to dC/dx, and since the latter is a principal concept in calculus, we choose to take it as the definition of marginal cost. Similar statements hold for marginal revenue and marginal profit.

Solution

(a) We are given that

$$x = 1000 + \frac{3.00 - p(x)}{0.10}(100)$$

or, equivalently,

$$p(x) = 3.00 - (0.10)\frac{x - 1000}{100} = 4 - 0.001x$$

(b)
$$R(x) = xp(x) = 4x - 0.001x^2$$

$$\frac{dR}{dx} = 4 - 0.002x$$

The only critical points are the end point 1000 and the stationary point 2000, obtained by setting $dR/dx = 0$. The First Derivative Test ($R'(x) > 0$ for $1000 \le x < 2000$ and $R'(x) < 0$ for $x > 2000$) shows that $x = 2000$ gives the maximum revenue. This corresponds to a unit price $p(2000) = \$2.00$.

(c) The maximum weekly revenue is $R(2000) = \$4000$.   ∎

EXAMPLE 3   In manufacturing and selling x units of a certain commodity, the price function p and the cost function C (in dollars) are given by

$$p(x) = 5.00 - 0.002x$$

$$C(x) = 3.00 + 1.10x$$

Find expressions for the marginal revenue, marginal cost, and marginal profit. Determine the production level that will produce the maximum total profit.

Solution

$$R(x) = xp(x) = 5.00x - 0.002x^2$$

$$P(x) = R(x) - C(x) = -3.00 + 3.90x - 0.002x^2$$

Thus, we have the following derivatives:

Marginal revenue: $\dfrac{dR}{dx} = 5 - 0.004x$

Marginal cost: $\dfrac{dC}{dx} = 1.1$

Marginal profit: $\dfrac{dP}{dx} = \dfrac{dR}{dx} - \dfrac{dC}{dx} = 3.9 - 0.004x$

To maximize profit, we set $dP/dx = 0$ and solve. This gives $x = 975$ as the only critical number to consider. It does provide a maximum, as may be checked by the First Derivative Test. The maximum profit is $P(975) = \$1898.25$.   ∎

Note that at $x = 975$ both the marginal revenue and the marginal cost are \$1.10. In general, a company should expect to be at a maximum profit level when the cost of producing an additional unit equals the revenue from that unit.

The statement just made assumes that the cost function and the revenue function are nice, differentiable functions and that end points are not significant. In some situations, the cost function may take large jumps, as when a new employee or a new piece of equipment is added; also, a manufacturing plant may have a maximum capacity, thereby introducing an important end point. We illustrate these possibilities in the next example.

EXAMPLE 4   The XYZ Company manufactures wicker chairs. With its present machines, it has a maximum yearly output of 500 units. If it makes x chairs, it can set a price of $p(x) = 200 - 0.15x$ dollars each and will have a total yearly cost of $C(x) = 4000 + 6x - 0.001x^2$ dollars. The company has the opportunity to buy a new machine. With the new machine, the company can make up to 750 chairs per year. The cost function will then be

$$C(x) = \begin{cases} 4000 + 6x - 0.001x^2 & \text{if } 0 \le x \le 500 \\ 6000 + 6x - 0.003x^2 & \text{if } 500 < x \le 750 \end{cases}$$

What production level maximizes total yearly profit under these circumstances?

Solution The cost function results in the profit function

$$P(x) = \begin{cases} -4000 + 194x - 0.149x^2 & \text{if } 0 \le x \le 500 \\ -6000 + 194x - 0.147x^2 & \text{if } 500 < x \le 750 \end{cases}$$

On the interval $0 < x < 500$,

$$\frac{dP}{dx} = 194 - 0.298x$$

so there are no stationary points in $(0, 500)$. On the interval $500 < x < 750$,

$$\frac{dP}{dx} = 194 - 0.294x$$

which gives the stationary point 660. There are four critical points: 0, 500, 660, and 750 (see Figure 5). The corresponding values of P are -4000, 55,750, 58,007, and 56,813. We conclude that a production level of 660 units gives the maximum profit of $58,007. ■

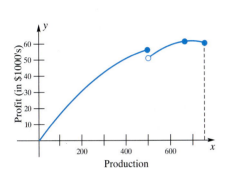

Figure 5

Notice that there are no critical points on the interval $(0, 500)$, so without the new machine a production level of 500 (at the right end point) would yield the maximum profit.

Concepts Review

1. To use calculus to model economic problems that are usually discrete in nature, we must transform these discrete problems into _____ ones.

2. Total revenue is $R(x) = xp(x)$, where x denotes _____ and $p(x)$ denotes _____ .

3. Total cost $C(x)$ usually consists of two kinds of cost: _____ cost, consisting of the cost of utilities, real estate taxes, and so on, and _____ cost, which depends on the number of units produced.

4. In economics, dR/dx is called _____ and dC/dx is called _____ .

Problem Set 4.5

1. It costs ACME $10 to make and sell one item. If the items sell at y dollars each and the number sold is $n = 100/(y - 10) + 20(100 - y)$, what sale price will bring the maximum profit?

2. The fixed monthly cost of operating a plant that makes Zbars is $7000, while the cost of manufacturing each unit is $100. Write an expression for $C(x)$, the total cost of making x Zbars in a month.

3. The manufacturer of Zbars estimates that 100 units per month can be sold if the unit price is $250 and that sales will increase by 10 units for each $5 decrease in price. Write an expression for the price $p(n)$ and the revenue $R(n)$ if n units are sold in one month, $n \ge 100$ (see Example 2).

4. Use the information in Problems 2 and 3 to write an expression for the total monthly profit $P(n)$, $n \ge 100$.

5. Sketch the graph of $P(n)$ of Problem 4, and from it estimate the value of n that maximizes P. Find this n exactly by the methods of calculus.

C 6. The total cost of producing and selling x units of Xbars per month is $C(x) = 100 + 3.002x - 0.0001x^2$. If the production level is 1600 units per month, find the average cost, $C(x)/x$, of each unit and the marginal cost.

7. The total cost of producing and selling n units of a certain commodity per week is $C(n) = 1000 + n^2/1200$. Find the average cost, $C(n)/n$, of each unit and the marginal cost at a production level of 800 units per week.

8. The total cost of producing and selling $100x$ units of a particular commodity per week is

$$C(x) = 1000 + 33x - 9x^2 + x^3$$

Find (a) the level of production at which the marginal cost is a minimum, and (b) the minimum marginal cost.

9. A price function, p, is defined by

$$p(x) = 20 + 4x - \frac{x^2}{3}$$

where $x \geq 0$ is the number of units.
(a) Find the total revenue function and the marginal revenue function.
(b) On what interval is the total revenue increasing?
(c) For what number x is the marginal revenue a maximum?

C **10.** For the price function defined by

$$p(x) = (182 - x/36)^{1/2}$$

find the number of units x_1 that makes the total revenue a maximum and state the maximum possible revenue. What is the marginal revenue when the optimum number of units, x_1, is sold?

11. For the price function given by

$$p(x) = 800/(x + 3) - 3$$

find the number of units x_1 that makes the total revenue a maximum and state the maximum possible revenue. What is the marginal revenue when the optimum number of units, x_1, is sold?

12. A riverboat company offers a fraternal organization a Fourth of July excursion with the understanding that there will be at least 400 passengers. The price of each ticket will be $12.00, and the company agrees to discount the price by $0.20 for each 10 passengers in excess of 400. Write an expression for the price function $p(x)$ and find the number x_1 of passengers that makes the total revenue a maximum.

13. A merchant finds that he can sell 4000 yards of a particular fabric each month if he prices it at $6.00 per yard, and that his monthly sales will increase by 250 yards for each $0.15 reduction in the price per yard. Write an expression for $p(x)$ and find the price per yard that would bring maximum revenue.

14. A manufacturer estimates that she can sell 500 articles per week if her unit price is $20.00, and that her weekly sales will rise by 50 with each $0.50 reduction in price. The cost of producing and selling x articles a week is

$$C(x) = 4200 + 5.10x + 0.0001x^2$$

Find each of the following.
(a) The price function
(b) The level of weekly production for maximum profit

(c) The price per article at the optimum level of production
(d) The marginal price at that level of production

15. The monthly overhead of a manufacturer of a certain commodity is $6000, and the cost of material is $1.00 per unit. If not more than 4500 units are manufactured per month, labor cost is $0.40 per unit; but for each unit over 4500, the manufacturer must pay time-and-a-half for labor. The manufacturer can sell 4000 units per month at $7.00 per unit and estimates that monthly sales will rise by 100 for each $0.10 reduction in price. Find (a) the total cost function, (b) the price function, and (c) the number of units that should be produced each month for maximum profit.

C **16.** The ZEE Company makes zingos, which it markets at a price $p(x) = 10 - 0.001x$ dollars, where x is the number produced each month. Its total monthly cost is $C(x) = 200 + 4x - 0.01x^2$. At peak production it can make 300 units. What is its maximum monthly profit and what level of production gives this profit?

C **17.** If the company of Problem 16 expands its facilities so that it can produce up to 450 units each month, its monthly cost function takes the form $C(x) = 800 + 3x - 0.01x^2$ for $300 < x \leq 450$. Find the production level that maximizes monthly profit and evaluate this profit. Sketch the graph of the monthly profit function $P(x)$ on $0 \leq x \leq 450$. (See Example 4.)

18. To be successful, a retail store must control its inventory. Overstocking results in excessive interest costs, extra warehouse rental costs, and the danger of obsolescence. Too small an inventory involves more paperwork in reordering, extra delivery charges, and the greater likelihood of running out of stock. Carvers Appliance Outlet estimates that it costs $20 to hold a microwave oven in stock for a year. To reorder a lot of ovens costs $200 plus $3 for each oven. What lot size will result in the smallest inventory cost? Assume that Carvers sells 1000 ovens per year and that ordering lots of size x means that, on the average, $x/2$ ovens will be in stock.

19. Suppose that a store expects to sell N units of a particular item per year, that it costs A dollars to keep one unit in stock for a year, and that to reorder a lot of size x costs $(F + Bx)$ dollars. Show that $x = \sqrt{2FN/A}$ is the lot size that minimizes inventory cost. See Problem 18.

20. If expected sales of an item quadruple, what will happen to the ideal lot size? See Problem 19.

Answers to Concepts Review: **1.** continuous **2.** number of units sold; price per unit **3.** fixed; variable **4.** marginal revenue; marginal cost

---

4.6
Sophisticated Graphing

Our treatment of graphing in Section 1.7 was elementary. We proposed plotting enough points so that the essential features of the graph were clear. We mentioned that symmetries of the graph could reduce the effort involved. We suggested that one should be alert to possible asymptotes. But if the equation to be graphed is complicated or if we want a very accurate graph, the techniques of Chapter 1 are inadequate.

Calculus provides a powerful tool for analyzing the fine structure of a graph, especially in identifying those points where the character of the graph changes. We can locate local maximum points, local minimum points, and inflection points; we can determine precisely where the graph is increasing or where it is concave up. Inclusion of all these ideas in our graphing procedure is the program for this section.

Polynomial Functions A polynomial of degree 1 or 2 is easy to graph by hand; one of degree 50 could be next to impossible. If the degree is of modest size, such as 3 to 6, we can use the tools of calculus to great advantage.

EXAMPLE 1 Sketch the graph of $f(x) = \dfrac{3x^5 - 20x^3}{32}$.

Solution Since $f(-x) = -f(x)$, f is an odd function, and therefore its graph is symmetric with respect to the origin. Setting $f(x) = 0$, we find the x-intercepts to be 0 and $\pm \sqrt{20/3} \approx \pm 2.6$. We can go this far without calculus.

When we differentiate f, we obtain

$$f'(x) = \frac{15x^4 - 60x^2}{32} = \frac{15x^2(x - 2)(x + 2)}{32}$$

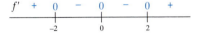

Figure 1

Thus, the critical points are $-2, 0$, and 2; we quickly discover (Figure 1) that $f'(x) > 0$ on $(-\infty, -2)$ and $(2, \infty)$ and that $f'(x) < 0$ on $(-2, 0)$ and $(0, 2)$. These facts tell us where f is increasing and where it is decreasing; they also confirm that $f(-2) = 2$ is a local maximum value and that $f(2) = -2$ is a local minimum value.

Differentiating again, we get

$$f''(x) = \frac{60x^3 - 120x}{32} = \frac{15x(x - \sqrt{2})(x + \sqrt{2})}{8}$$

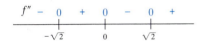

Figure 2

By studying the sign of $f''(x)$ (Figure 2), we deduce that f is concave upward on $(-\sqrt{2}, 0)$ and $(\sqrt{2}, \infty)$ and concave downward on $(-\infty, -\sqrt{2})$ and $(0, \sqrt{2})$. Thus, there are three points of inflection: $(-\sqrt{2}, 7\sqrt{2}/8) \approx (-1.4, 1.2)$, $(0, 0)$, and $(\sqrt{2}, -7\sqrt{2}/8) \approx (1.4, -1.2)$.

Much of this information is collected at the top of Figure 3, which we use to sketch the graph directly below it.

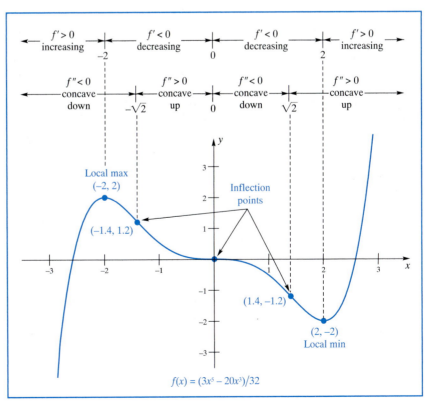

Figure 3

Rational Functions A rational function, being the quotient of two polynomial functions, is considerably more complicated to graph than a polynomial. In particular, we can expect dramatic behavior near where the denominator would be zero.

EXAMPLE 2 Sketch the graph of $f(x) = \dfrac{x^2 - 2x + 4}{x - 2}$.

Solution This function is neither even nor odd, so we do not have any of the usual symmetries. There are no x-intercepts, since the solutions to $x^2 - 2x + 4 = 0$ are not real numbers. The y-intercept is -2. We anticipate a vertical asymptote at $x = 2$. In fact,

$$\lim_{x \to 2^-} \frac{x^2 - 2x + 4}{x - 2} = -\infty \qquad \lim_{x \to 2^+} \frac{x^2 - 2x + 4}{x - 2} = \infty$$

Differentiation twice gives

$$f'(x) = \frac{x(x - 4)}{(x - 2)^2} \qquad f''(x) = \frac{8}{(x - 2)^3}$$

The stationary points are therefore $x = 0$ and $x = 4$.

Thus, $f'(x) > 0$ on $(-\infty, 0) \cup (4, \infty)$ and $f'(x) < 0$ on $(0, 2) \cup (2, 4)$. (Remember, $f'(x)$ does not exist when $x = 2$.) Also, $f''(x) > 0$ on $(2, \infty)$ and $f''(x) < 0$ on $(-\infty, 2)$. Since $f''(x)$ is never 0, there are no inflection points. On the other hand, $f(0) = -2$ and $f(4) = 6$ give local maximum and minimum values, respectively.

It is a good idea to check on the behavior of $f(x)$ for large $|x|$. Since

$$f(x) = \frac{x^2 - 2x + 4}{x - 2} = x + \frac{4}{x - 2}$$

the graph of $y = f(x)$ gets closer and closer to the line $y = x$ as $|x|$ gets larger and larger. We call the line $y = x$ an **oblique asymptote** for the graph of f (see Problem 43 of Section 2.8).

With all this information, we are able to sketch a rather accurate graph (Figure 4).

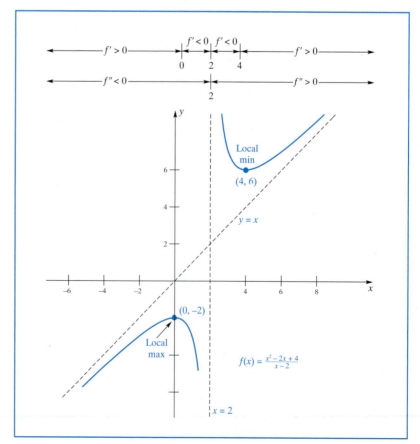

Figure 4

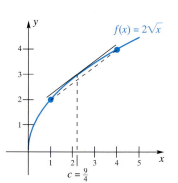

Figure 4

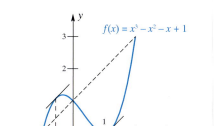

Figure 5

38
Sketch

39
$h(1) =$

40
Show t

41.
$a \neq 0,$

42.
$cx^2 + c$
flection

EXPL CA$
parame
flection
which ti

43.

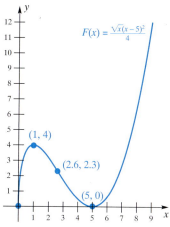

Figure 5

45.

47.
fies the

(a) $g(1$

(b) $g'($

(c) g is

(d) $f(x$

Sketch a

48.
that $H($
have a r
tion at x

49.
tinuous
sketch st
(a) $F'(x$

(b) $F''($

(c) $F''($

GC **50.**
each of t
mine the
tion poir
to at lea
$-5 \leq y$

(a) $f(x)$

(b) $f(x)$

(c) $f(x)$

(d) $f(x)$

GC **51.**
ing calcu
functions
cated at t

Functions Involving Roots There is an endless variety of functions involving roots. Here is one example.

EXAMPLE 3 Analyze the function

$$F(x) = \frac{\sqrt{x}(x-5)^2}{4}$$

and sketch its graph.

Solution The domain of F is $[0, \infty)$ and the range is $[0, \infty)$, so the graph of F is confined to the first quadrant and the positive coordinate axes. The x-intercepts are 0 and 5; the y-intercept is 0.
From

$$F'(x) = \frac{5(x-1)(x-5)}{8\sqrt{x}}, \qquad x > 0$$

we find the stationary points 1 and 5. Since $F'(x) > 0$ on $(0, 1)$ and $(5, \infty)$, while $F'(x) < 0$ on $(1, 5)$, we conclude that $F(1) = 4$ is a local maximum value and $F(5) = 0$ is a local minimum value.
So far, it has been clear sailing. But on calculating the second derivative, we obtain

$$F''(x) = \frac{5(3x^2 - 6x - 5)}{16x^{3/2}}, \qquad x > 0$$

which is quite complicated. However, $3x^2 - 6x - 5 = 0$ has one solution in $(0, \infty)$, namely $1 + 2\sqrt{6}/3 \approx 2.6$.
Using the test points 1 and 3, we conclude that $f''(x) < 0$ on $(0, 2.6)$ and $f''(x) > 0$ on $(2.6, \infty)$. It then follows that the point $(2.6, 2.3)$ is an inflection point.
As x grows large, $F(x)$ grows without bound and much faster than any linear function; there are no asymptotes. The graph is sketched in Figure 5. ■

Summary of the Method In graphing functions, there is no substitute for common sense. However, the following procedure will be helpful in most cases.

Step 1: Precalculus analysis.

(a) Check the *domain* and *range* of the function to see if any regions of the plane are excluded.

(b) Test for *symmetry* with respect to the y-axis and the origin. (Is the function even or odd?)

(c) Find the *intercepts*.

Step 2: Calculus analysis.

(a) Use the first derivative to find the critical points and to find out where the graph is *increasing* and *decreasing*.

(b) Test the critical points for *local maxima* and *minima*.

(c) Use the second derivative to find out where the graph is *concave upward* and *concave downward* and to locate *inflection points*.

(d) Find the *asymptotes*.

Step 3: Plot a few points (including all critical points and inflection points).

Step 4: Sketch the graph.

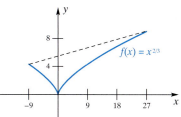

Figure 6

Con

1
$f(-x)$
spect

2
the gr

Prol

In Pro
above

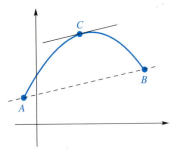

Figure 1

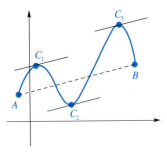

Figure 2

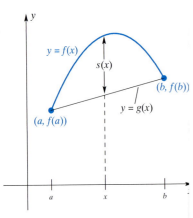

Figure 3

1

1

1

1

1

1

2

2

2

2

2

2

C **2**

2

prope
(a) f
(c) f
(e) f'

2
prope
(a) f

The Key to a Proof

The key to this proof is that c is the value at which $f'(c) = \dfrac{f(b) - f(a)}{b - a}$ and $s'(c) = 0$. Many proofs have one or two key ideas; if you understand the key, you will understand the proof.

4.
The Mean Valu
Theorer

Reflection and Refraction of Light

I. Preparation

Law of Reflection The law of reflection states that the angle of incidence θ_1 of a beam of light is equal to the angle of reflection θ_2; that is, $\theta_1 = \theta_2$ or, equivalently, $\alpha = \beta$ (see Figure 1). Furthermore, if the speed of light in the medium is constant, then the reflection path from a point P to a point Q consists of two straight line segments meeting at the reflection point R, as in Figure 1.

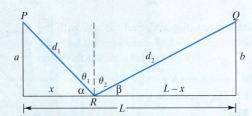

Figure 1
Geometry of reflection

Exercise 1 Given two points P and Q, we want to determine the location of the point of reflection on the interface (say, x measured horizontally from one of the points). As in Figure 1, assume that the two points are separated from each other by a total horizontal distance L and that their vertical distances from the reflecting interface are respectively a and b. Use the law of reflection to derive an expression for x in terms of a, b, and L. **Check:** For $a = 50$, $b = 25$, and $L = 150$, the correct answer for x is 100.

II. Using Technology

Snell's Law The law of refraction (or Snell's Law) states that, for an interface separating two media, the angle of incidence θ_1 is related to the angle of refraction θ_2 by the relation

$$\frac{\sin \theta_1}{c_1} = \frac{\sin \theta_2}{c_2} \qquad (1)$$

(see Figure 2). Here c_1 and c_2 are the respective speeds of light above and below the interface.

In contrast to the reflection case, where the reflection point can be found explicitly, the location of the re-

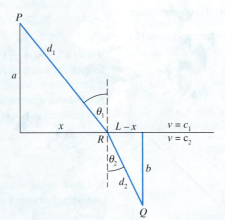

Figure 2
Geometry of refraction

fracting point x involves a fourth-degree polynomial:

$$0 = (c_2^2 - c_1^2)x^4 - 2L(c_2^2 - c_1^2)x^3$$
$$+ [L^2(c_2^2 - c_1^2) + c_2^2 b^2 - c_1^2 a^2]x^2$$
$$+ 2Lc_1^2 a^2 x - L^2 c_1^2 a^2 \qquad (2)$$

Exercise 2 Show that, for the values $a = b = 1$, $L = 4$, $c_1 = 1$, and $c_2 = 1/2$, the above equation for the location of the refraction point reduces to

$$3x^4 - 24x^3 + 51x^2 - 32x + 64 = 0$$

This latter equation actually has *two* real roots:

(a) Plot a graph that shows the approximate locations of the two roots.

(b) Explain why only one of the roots is relevant to the refraction problem.

(c) Find the relevant root to two significant figures, and explain how you know that your x value is this accurate.

Exercise 3 Suppose we want to find a c such that $f(c) = 0$, and we have an x_0 that is close to c. Think of x_0 as an initial guess for the root c. We can then improve on our initial guess x_0 to get a better approximation.

(a) Find the equation of the line that is tangent to the graph of $y = f(x)$ at the point $(x_0, f(x_0))$. [In Section 3.10, we called this the linear approximation to $f(x)$ at x_0.]

(b) Show that this line intersects the x-axis at the point

$$x_1 = x_0 - \frac{f(x_0)}{f'(x_0)}$$

(c) Turn in a graph showing in the same window the graphs of $y = f(x)$ and of the linear approximation to $f(x)$ at x_0.

Exercise 4
(a) Use your approximation to the root of the fourth-degree polynomial obtained in Exercise 2 as the x_0 value above and compute the improved x_1 from Exercise 3. Evaluate $f(x_0)$ and $f(x_1)$.

(b) Repeat the above process to get an x_2 that is an improvement of x_1. Evaluate $f(x_0)$, $f(x_1)$, and $f(x_2)$.

III. Reflection

Exercise 5 A boat with a spotlight is used to help underwater divers to repair a rupture in an underwater pipe (Figure 3). The rupture in the pipe is 6 feet below the surface of the water, and the spotlight on the boat is 5 feet above the surface of the water. If the boat can come within 20 feet of the pipe, at what angle should they point the spotlight to illuminate the rupture on the pipe? For air, $c_1 = 1$, and for water, $c_2 = 0.67$.

Exercise 6 Derive equation (2). *Hint*: Begin with equation (1) and substitute expressions for $\sin \theta_1$ and $\sin \theta_2$. Use a CAS to simplify the resulting expression. In Mathematica, you can use the **Collect** command by entering **Collect** [*expression, x*] to collect like powers of x. The **collect** command in Maple works similarly.

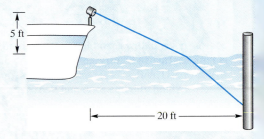

Figure 3

An Optimization Problem

I. Preparation

Often, the hardest part about applied max–min problems is setting them up. Once you find the objective function, it is usually not too difficult to find the optimum.

Exercise 1 A 6-foot-wide hallway makes a right-angle turn. What is the length of the longest thin rod that can be carried around the corner, assuming you cannot tilt the rod?

The longest rod will just barely touch the inside corner of the turn and the two outside walls. Label the points A, B, C, D, and E as in Figure 1. Let t denote the measure of the angle $\angle ABD$.
(a) Show that t is also the measure of $\angle ACE$.
(b) Use some trigonometry to find the lengths a and b.

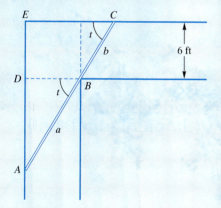

Figure 1

II. Using Technology

Exercise 2 The rod described in Exercise 1 has length

$$L = a + b = \frac{6}{\cos t} + \frac{6}{\sin t}$$

$$= 6 \sec t + 6 \csc t$$

Find the derivative dL/dt, set it equal to zero, and solve for t.

Exercise 3 To see whether we have found a minimum or a maximum, graph L as a function of t. You should also evaluate the second derivative of L at the optimum value of t. Does the t that you found yield the *smallest* rod or the *longest* rod that can be carried around the corner? Explain. *Hint*: If t is very close to zero, but still positive, will the rod be long or short? Will it fit around the corner? What if t is close to, but just less than, $\pi/4$ (i.e., 90°)?

Exercise 4 Now, let's change the dimensions of the hallway. Suppose that one corridor is 6.2 feet wide and the other is 8.6 feet wide. Find the length of the longest thin rod that just fits around the corner.

Exercise 5 For this problem, let's suppose that the corridors do not meet at a right angle, but instead meet at an angle of 105° as shown in Figure 2. Both

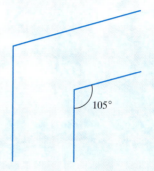

Figure 2

corridors are 6 feet wide. Again, find the length of the longest thin rod that just fits around the corner.

III. Reflection

Exercise 6 Finally, suppose that the ceiling is 9.7 feet above the floor, the corridors meet at a right angle, and the widths of the corridors are 6.2 feet and 8.6 feet. Assuming that you can tilt the thin rod, what is the length of the longest rod you can carry around the corner? *Hint*: This is not a calculus problem; use your answer to Exercise 4.

Now we pose an important question. Is every antiderivative of $f(x) = 4x^3$ of the form $F(x) = x^4 + C$? The answer is yes. This follows from Theorem 4.7B, which says that if two functions have the same derivative they must differ by a constant.

Our conclusion is this. If a function f has an antiderivative, it will have a whole family of them, and each member of this family can be obtained from one of them by the addition of an appropriate constant. We call this family of functions the **general antiderivative** of f. After we get used to this notion, we will often omit the adjective *general*.

EXAMPLE 2   Find the general antiderivative of $f(x) = x^2$ on $(-\infty, \infty)$.

Solution   The function $F(x) = x^3$ will not do since its derivative is $3x^2$. But this suggests $F(x) = \frac{1}{3}x^3$, which satisfies $F'(x) = \frac{1}{3} \cdot 3x^2 = x^2$. However, the general antiderivative is $\frac{1}{3}x^3 + C$.   ■

Notation for Antiderivatives

Since we used the symbol D_x for the operation of taking a derivative, it would be natural to use A_x for the operation of finding the antiderivative. Thus,

$$A_x(x^2) = \tfrac{1}{3}x^3 + C$$

This is the notation used by several authors and it was, in fact, used in earlier editions of this book. However, Leibniz's original notation continues to enjoy overwhelming popularity, and we therefore choose to follow him. Rather than A_x, Leibniz used the symbol $\int \ldots dx$. He wrote

$$\int x^2 \, dx = \tfrac{1}{3}x^3 + C$$

and

$$\int 4x^3 \, dx = x^4 + C$$

We will postpone explaining why Leibniz chose to use the elongated s, $\int$, and the dx until later. For the moment, simply think of $\int \ldots dx$ as indicating the antiderivative with respect to x, just as D_x indicates the derivative with respect to x. Note that

$$D_x \int f(x) \, dx = f(x) \quad \text{and} \quad \int D_x f(x) \, dx = f(x) + C$$

Theorem A   Power Rule

If r is any rational number except -1, then

$$\int x^r \, dx = \frac{x^{r+1}}{r+1} + C$$

Proof   To establish any result of the form

$$\int f(x) \, dx = F(x) + C$$

all we have to do is show that

$$D_x[F(x) + C] = f(x)$$

In this case,

$$D_x\left[\frac{x^{r+1}}{r+1} + C\right] = \frac{1}{r+1}(r+1)x^r = x^r \quad \blacklozenge$$

We make two comments about Theorem A. First, it is meant to include the case $r = 0$; that is,

$$\int 1 \, dx = x + C$$

Second, since no interval I is specified, the conclusion is understood to be valid only on intervals on which x^r is defined. In particular, we must exclude any interval containing the origin if $r < 0$.

Following Leibniz, we shall often use the term **indefinite integral** in place of antiderivative. To antidifferentiate is also to **integrate**. In the symbol $\int f(x)\,dx$, $\int$ is called the **integral sign** and $f(x)$ is called the **integrand**. Thus, we integrate the integrand and thereby evaluate the indefinite integral. Perhaps Leibniz used the adjective *indefinite* to suggest that the indefinite integral always involves an arbitrary constant.

EXAMPLE 3   Find the general antiderivative of $f(x) = x^{4/3}$.

Solution

$$\int x^{4/3}\,dx = \frac{x^{7/3}}{\frac{7}{3}} + C = \tfrac{3}{7}x^{7/3} + C \qquad\blacksquare$$

Note that *to integrate a power of x we increase the exponent by 1 and divide by the new exponent.*

Theorem B

$$\int \sin x\,dx = -\cos x + C \qquad \int \cos x\,dx = \sin x + C$$

Proof   Simply note that $D_x(-\cos x) = \sin x$ and $D_x(\sin x) = \cos x$. ◆

The Indefinite Integral Is Linear   Recall from Chapter 3 that D_x is a linear operator. This means two things.

1. $D_x[kf(x)] = kD_x f(x)$
2. $D_x[f(x) + g(x)] = D_x f(x) + D_x g(x)$

From these two properties, a third follows automatically.

3. $D_x[f(x) - g(x)] = D_x f(x) - D_x g(x)$

It turns out that $\int \ldots dx$ also has these properties of a linear operator.

Theorem C   Indefinite Integral Is a Linear Operator

Let f and g have antiderivatives (indefinite integrals) and let k be a constant. Then:
(i) $\int kf(x)\,dx = k\int f(x)\,dx$;
(ii) $\int[f(x) + g(x)]\,dx = \int f(x)\,dx + \int g(x)\,dx$; and consequently
(iii) $\int[f(x) - g(x)]\,dx = \int f(x)\,dx - \int g(x)\,dx$.

Proof   To show (i) and (ii), we simply differentiate the right side and observe that we get the integrand of the left side.

$$D_x\left[k\int f(x)\,dx\right] = kD_x\int f(x)\,dx = kf(x)$$

$$D_x\left[\int f(x)\,dx + \int g(x)\,dx\right] = D_x\int f(x)\,dx + D_x\int g(x)\,dx$$

$$= f(x) + g(x)$$

Property (iii) follows from (i) and (ii). ◆

EXAMPLE 4   Using the linearity of $\int$, evaluate

(a) $\int(3x^2 + 4x)\,dx$ (b) $\int(u^{3/2} - 3u + 14)\,du$ (c) $\int(1/t^2 + \sqrt{t})\,dt$

Solution

(a)
$$\int (3x^2 + 4x)\,dx = \int 3x^2\,dx + \int 4x\,dx$$

$$= 3\int x^2\,dx + 4\int x\,dx$$

$$= 3\left(\frac{x^3}{3} + C_1\right) + 4\left(\frac{x^2}{2} + C_2\right)$$

$$= x^3 + 2x^2 + (3C_1 + 4C_2)$$

$$= x^3 + 2x^2 + C$$

Two arbitrary constants C_1 and C_2 appeared, but they were combined into one constant, C, a practice we consistently follow.

(b) Note the use of the variable u rather than x. This is fine as long as the corresponding differential symbol is du, since we then have a complete change of notation.

$$\int (u^{3/2} - 3u + 14)\,du = \int u^{3/2}\,du - 3\int u\,du + 14\int 1\,du$$

$$= \tfrac{2}{5}u^{5/2} - \tfrac{3}{2}u^2 + 14u + C$$

(c)
$$\int \left(\frac{1}{t^2} + \sqrt{t}\right)dt = \int (t^{-2} + t^{1/2})\,dt = \int t^{-2}\,dt + \int t^{1/2}\,dt$$

$$= \frac{t^{-1}}{-1} + \frac{t^{3/2}}{\frac{3}{2}} + C = -\frac{1}{t} + \frac{2}{3}t^{3/2} + C \qquad \blacksquare$$

Generalized Power Rule   Recall the Chain Rule as applied to a power of a function. If $u = g(x)$ is a differentiable function and r is a rational number $(r \neq -1)$, then

$$D_x\left[\frac{u^{r+1}}{r+1}\right] = u^r \cdot D_x u$$

or, in functional notation,

$$D_x\left(\frac{[g(x)]^{r+1}}{r+1}\right) = [g(x)]^r \cdot g'(x)$$

From this we obtain an important rule for indefinite integrals.

Theorem D   Generalized Power Rule

Let g be a differentiable function and r a rational number different from -1. Then

$$\int [g(x)]^r\, g'(x)\,dx = \frac{[g(x)]^{r+1}}{r+1} + C$$

To apply Theorem D, we must be able to recognize the functions g and g' in the integrand.

EXAMPLE 5   Evaluate (a) $\int (x^4 + 3x)^{30}(4x^3 + 3)\,dx$ and (b) $\int \sin^{10} x \cos x \, dx$.

Solution
(a) Let $g(x) = x^4 + 3x$; then $g'(x) = 4x^3 + 3$. Thus, by Theorem D,

$$\int (x^4 + 3x)^{30}(4x^3 + 3)\,dx = \int [g(x)]^{30} g'(x)\,dx = \frac{[g(x)]^{31}}{31} + C$$

$$= \frac{(x^4 + 3x)^{31}}{31} + C$$

(b) Let $g(x) = \sin x$, then $g'(x) = \cos x$. Thus,

$$\int \sin^{10} x \cos x \, dx = \int [g(x)]^{10} g'(x) \, dx = \frac{[g(x)]^{11}}{11} + C$$

$$= \frac{\sin^{11} x}{11} + C \qquad \blacksquare$$

Example 5 shows why Leibniz used the differential dx in his notation $\int \ldots dx$. If we let $u = g(x)$, then $du = g'(x) \, dx$. The conclusion of Theorem D is therefore

$$\int u^r \, du = \frac{u^{r+1}}{r+1} + C, \qquad r \neq -1$$

which is the ordinary power rule with u as the variable. Thus, the generalized power rule is just the ordinary power rule applied to functions. But, in applying it, we must always make sure that we have du to go with u^r. The following examples illustrate what we mean.

EXAMPLE 6 Evaluate

(a) $\int (x^3 + 6x)^5 (6x^2 + 12) \, dx$ \qquad (b) $\int (x^2 + 4)^{10} x \, dx$, and

(c) $\int (x^2/2 + 3)^2 x^2 \, dx$.

Solution

(a) Let $u = x^3 + 6x$; then $du = (3x^2 + 6) \, dx$. Thus, $(6x^2 + 12) \, dx = 2(3x^2 + 6) \, dx$
$= 2 \, du$, and so

$$\int (x^3 + 6x)^5 (6x^2 + 12) \, dx = \int u^5 2 \, du$$

$$= 2 \int u^5 \, du$$

$$= 2 \left[\frac{u^6}{6} + C \right]$$

$$= \frac{u^6}{3} + 2C$$

$$= \frac{(x^3 + 6x)^6}{3} + K$$

Two things should be noted about our solution. First, the fact that $(6x^2 + 12) \, dx$ is $2du$ instead of du caused no trouble; the factor 2 could be moved in front of the integral sign by linearity. Second, we wound up with an arbitrary constant of $2C$. This is still an arbitrary constant; we called it K.

(b) Let $u = x^2 + 4$; then $du = 2x \, dx$. Thus,

$$\int (x^2 + 4)^{10} x \, dx = \int (x^2 + 4)^{10} \cdot \frac{1}{2} \cdot 2x \, dx$$

$$= \frac{1}{2} \int u^{10} \, du$$

$$= \frac{1}{2} \left(\frac{u^{11}}{11} + C \right)$$

$$= \frac{(x^2 + 4)^{11}}{22} + K$$

(c) Let $u = x^2/2 + 3$; then $du = x \, dx$. The method illustrated in parts (a) and (b) fails because $x^2 \, dx = x(x \, dx) = x \, du$, and the x cannot be passed in front of the integral sign. (That can be done only with a constant factor.) However, we can expand the integrand by ordinary algebra and then use the Power Rule.

$$\int \left(\frac{x^2}{2} + 3\right)^2 x^2 \, dx = \int \left(\frac{x^4}{4} + 3x^2 + 9\right)x^2 \, dx$$

$$= \int \left(\frac{x^6}{4} + 3x^4 + 9x^2\right) dx$$

$$= \frac{x^7}{28} + \frac{3x^5}{5} + 3x^3 + C \qquad ∎$$

Concepts Review

1. The Power Rule for derivatives says that $d(x^r)/dx =$ _____ . The Power Rule for integrals says that $\int x^r \, dx =$ _____ .

2. The Generalized Power Rule for derivatives says that $d[f(x)]^r/dx =$ _____ . The Generalized Power Rule for integrals says that $\int$ _____ $dx = [f(x)]^{r+1}/(r + 1) + C, r \neq -1$.

3. $\int (x^4 + 3x^2 + 1)^8(4x^3 + 6x) \, dx =$ _____ .

4. By linearity, $\int [c_1 f(x) + c_2 g(x)] \, dx =$ _____ .

Problem Set 5.1

Find the general antiderivative F(x) + C for each of the following.

1. $f(x) = 5$

2. $f(x) = x - 4$

3. $f(x) = x^2 + \pi$

4. $f(x) = 3x^2 + \sqrt{3}$

5. $f(x) = x^{5/4}$

6. $f(x) = 3x^{2/3}$

7. $f(x) = 1/\sqrt[3]{x^2}$

8. $f(x) = 7x^{-3/4}$

9. $f(x) = x^2 - x$

10. $f(x) = 3x^2 - \pi x$

11. $f(x) = 4x^5 - x^3$

12. $f(x) = x^{100} + x^{99}$

13. $f(x) = 27x^7 + 3x^5 - 45x^3 + \sqrt{2}x$

14. $f(x) = x^2(x^3 + 5x^2 - 3x + \sqrt{3})$

15. $f(x) = \dfrac{3}{x^2} - \dfrac{2}{x^3}$

16. $f(x) = \dfrac{\sqrt{2x}}{x} + \dfrac{3}{x^5}$

17. $f(x) = \dfrac{4x^6 + 3x^4}{x^3}$

18. $f(x) = \dfrac{x^6 - x}{x^3}$

In Problems 19–25, evaluate the indicated integrals.

19. $\displaystyle\int (x^2 + x) \, dx$

20. $\displaystyle\int (x^3 + \sqrt{x}) \, dx$

21. $\displaystyle\int (x + 1)^2 \, dx$

22. $\displaystyle\int (z + \sqrt{2}z)^2 \, dz$

23. $\displaystyle\int \frac{(z^2 + 1)^2}{\sqrt{z}} \, dz$

24. $\displaystyle\int \frac{s(s + 1)^2}{\sqrt{s}} \, ds$

25. $\displaystyle\int (\sin \theta - \cos \theta) \, d\theta$

26. $\displaystyle\int (t^2 - 2 \cos t) \, dt$

In Problems 27–32, use the methods of Examples 5 and 6 to evaluate the indefinite integrals.

27. $\displaystyle\int (\sqrt{2}x + 1)^3 \sqrt{2} \, dx$

28. $\displaystyle\int (\pi x^3 + 1)^4 3\pi x^2 \, dx$

29. $\displaystyle\int (5x^2 + 1)(5x^3 + 3x - 8)^6 \, dx$

30. $\displaystyle\int (5x^2 + 1)\sqrt{5x^3 + 3x - 2} \, dx$

31. $\displaystyle\int 3t \sqrt[3]{2t^2 - 11} \, dt$

32. $\displaystyle\int \frac{3y}{\sqrt{2y^2 + 5}} \, dy$

In Problems 33–38 f''(x) is given. Find f(x) by antidifferentiating twice. Note that in this case your answer should involve two arbitrary constants, one from each antidifferentiation. For example, if $f''(x) = x$, then $f'(x) = x^2/2 + C_1$ and $f(x) = x^3/6 + C_1 x + C_2$. The constants C_1 and C_2 cannot be combined because $C_1 x$ is not a constant.

33. $f''(x) = 3x + 1$

34. $f''(x) = -2x + 3$

35. $f''(x) = \sqrt{x}$

36. $f''(x) = x^{4/3}$

37. $f''(x) = \dfrac{x^4 + 1}{x^3}$

38. $f''(x) = 2\sqrt[3]{x} + 1$

39. Prove the formula

$$\int [f(x)g'(x) + g(x)f'(x)] \, dx = f(x)g(x) + C$$

Hint: See the first sentence in the proof of Theorem A.

40. Prove the formula

$$\int \frac{g(x)f'(x) - f(x)g'(x)}{g^2(x)} \, dx = \frac{f(x)}{g(x)} + C$$

41. Use the formula from Problem 39 to find

$$\int \left[\frac{x^2}{2\sqrt{x-1}} + 2x\sqrt{x-1}\right] dx$$

42. Use the formula from Problem 39 to find

$$\int \left[\frac{-x^3}{(2x+5)^{3/2}} + \frac{3x^2}{\sqrt{2x+5}}\right] dx$$

43. Find $\int f''(x) \, dx$ if $f(x) = x\sqrt{x^3 + 1}$.

44. Prove the formula

$$\int \frac{2g(x)f'(x) - f(x)g'(x)}{2[g(x)]^{3/2}} = \frac{f(x)}{\sqrt{g(x)}} + C$$

45. Prove the formula

$$\int f^{m-1}(x)g^{n-1}(x)[nf(x)g'(x) + mg(x)f'(x)] dx$$

$$= f^m(x)g^n(x) + C$$

46. Find the indefinite integral

$$\int \sin^3[(x^2 + 1)^4] \cos[(x^2 + 1)^4](x^2 + 1)^3 x \, dx$$

Hint: Let $u = \sin(x^2 + 1)^4$.

47. Find $\int |x| \, dx$.   **48.** Find $\int \sin^2 x \, dx$.

CAS **49.** Some software packages can evaluate indefinite integrals. Use your software on each of the following.

(a) $\int 6 \sin(3(x - 2)) \, dx$ (b) $\int \sin^3(x/6) \, dx$

(c) $\int (x^2 \cos 2x + x \sin 2x) \, dx$

EXPL CAS **50.** Let $F_0(x) = x \sin x$ and $F_{n+1}(x) = \int F_n(x) \, dx$.

(a) Determine $F_1(x)$, $F_2(x)$, $F_3(x)$, and $F_4(x)$.

(b) On the basis of part (a), conjecture the form of $F_{16}(x)$.

Answers to Concepts Review: **1.** rx^{r-1}; $x^{r+1}/(r + 1) + C$, $r \neq -1$ **2.** $r[f(x)]^{r-1} f'(x)$; $[f(x)]^r f'(x)$
3. $(x^4 + 3x^2 + 1)^9/9 + C$ **4.** $c_1 \int f(x) \, dx + c_2 \int g(x) \, dx$

5.2
Introduction to Differential Equations

In the previous section, our task was to integrate (antidifferentiate) a function f to obtain a new function F. We wrote

$$\int f(x) \, dx = F(x) + C$$

and this was correct by definition provided $F'(x) = f(x)$. Now $F'(x) = f(x)$ in the language of derivatives is equivalent to $dF(x) = f(x) \, dx$ in differential language (Section 3.10). Thus, we may look on the boxed formula as saying that

$$\int dF(x) = F(x) + C$$

From this perspective, we integrate the differential of a function to obtain the function (plus a constant). This was Leibniz's viewpoint; adopting it will help us to solve *differential equations*.

What Is a Differential Equation?   To motivate our answer, we begin with a simple example.

EXAMPLE 1   Find the xy-equation of the curve that passes through $(-1, 2)$ and whose slope at any point on the curve is equal to twice the abscissa (x-coordinate) of that point.

Solution   The condition that must hold at each point (x, y) on the curve is

$$\frac{dy}{dx} = 2x$$

We are looking for a function $y = f(x)$ that satisfies this equation and the additional condition that $y = 2$ when $x = -1$. We suggest two ways of looking at this problem.

Method 1   When an equation has the form $dy/dx = g(x)$, we observe that y must be an antiderivative of $g(x)$; that is,

$$y = \int g(x) \, dx$$

In our case,

$$y = \int 2x \, dx = x^2 + C$$

EXAMPLE 3   Show that if r is a rational number different from -1, then

$$\int_a^b x^r \, dx = \frac{b^{r+1}}{r+1} - \frac{a^{r+1}}{r+1}$$

Solution $F(x) = x^{r+1}/(r+1)$ is an antiderivative of $f(x) = x^r$. Thus, by the Second Fundamental Theorem of Calculus,

$$\int_a^b x^r \, dx = F(b) - F(a) = \frac{b^{r+1}}{r+1} - \frac{a^{r+1}}{r+1}$$

If $r < 0$, we require that 0 not be in $[a, b]$. Why?   ■

It is convenient to introduce a special symbol for $F(b) - F(a)$. We write

$$F(b) - F(a) = \left[F(x)\right]_a^b$$

With this notation,

$$\int_2^5 x^2 \, dx = \left[\frac{x^3}{3}\right]_2^5 = \frac{125}{3} - \frac{8}{3} = \frac{117}{3} = 39$$

Figur

EXAMPLE 4   Evaluate $\int_{-1}^2 (4x - 6x^2) \, dx$

(a) using the Second Fundamental Theorem of Calculus directly and
(b) using linearity (Theorem 5.6D) first.

Solution

(a) $$\int_{-1}^2 (4x - 6x^2) \, dx = \left[2x^2 - 2x^3\right]_{-1}^2$$

$$= (8 - 16) - (2 + 2) = -12$$

(b) Using linearity first, we have

$$\int_{-1}^2 (4x - 6x^2) \, dx = 4\int_{-1}^2 x \, dx - 6\int_{-1}^2 x^2 \, dx$$

$$= 4\left[\frac{x^2}{2}\right]_{-1}^2 - 6\left[\frac{x^3}{3}\right]_{-1}^2$$

$$= 4\left(\frac{4}{2} - \frac{1}{2}\right) - 6\left(\frac{8}{3} + \frac{1}{3}\right)$$

$$= -12$$ ■

EXAMPLE 5   Evaluate $\int_1^8 (x^{1/3} + x^{4/3}) \, dx$.

Solution

$$\int_1^8 (x^{1/3} + x^{4/3}) \, dx = \left[\tfrac{3}{4} x^{4/3} + \tfrac{3}{7} x^{7/3}\right]_1^8$$

$$= (\tfrac{3}{4} \cdot 16 + \tfrac{3}{7} \cdot 128) - (\tfrac{3}{4} \cdot 1 + \tfrac{3}{7} \cdot 1)$$

$$= \tfrac{45}{4} + \tfrac{381}{7} \approx 65.68$$ ■

EXAMPLE 6   Find $D_x \int_0^x 3 \sin t \, dt$ in two ways.

Solution   The easy way is to apply the First Fundamental Theorem of Calculus.

$$D_x \int_0^x 3 \sin t \, dt = 3 \sin x$$

A second way to do this problem is to apply the Second Fundamental Theorem of Calculus to evaluate the integral from 0 to x; then apply the rules of derivatives.

$$\int_0^x 3 \sin t \, dt = \left[-3 \cos t\right]_0^x = -3 \cos x - (-3 \cos 0) = -3 \cos x + 3$$

Then

$$D_x \int_0^x 3 \sin t \, dt = D_x(-3 \cos x + 3) = 3 \sin x$$ ■

In terms of the symbol for the indefinite integral, we may write the conclusion of the Second Fundamental Theorem of Calculus as

$$\int_a^b f(x) \, dx = \left[\int f(x) \, dx \right]_a^b$$

The nontrivial part of applying the theorem is always to find the indefinite integral $\int f(x) \, dx$. The Generalized Power Rule from Section 5.1 can be applied in many cases to evaluate a definite integral. There are, however, many functions that do not have antiderivatives that can be expressed in terms of elementary functions.

EXAMPLE 7   Evaluate $\int_0^4 \sqrt{x^2 + x} \, (2x + 1) \, dx$.

Solution   Let $u = x^2 + x$; then $du = (2x + 1) \, dx$. Thus,

$$\int \sqrt{x^2 + x} \, (2x + 1) \, dx = \int u^{1/2} \, du = \tfrac{2}{3} u^{3/2} + C$$

$$= \tfrac{2}{3}(x^2 + x)^{3/2} + C$$

Therefore, by the Second Fundamental Theorem of Calculus,

$$\int_0^4 \sqrt{x^2 + x} \, (2x + 1) \, dx = \left[\tfrac{2}{3}(x^2 + x)^{3/2} + C \right]_0^4$$

$$= \left[\tfrac{2}{3}(20)^{3/2} + C \right] - [0 + C]$$

$$= \tfrac{2}{3}(20)^{3/2} \approx 59.63$$ ■

Note that the C of the indefinite integration cancels out, as it always will, in the definite integration. That is why in the statement of the Second Fundamental Theorem we could use the phrase *any antiderivative*. In particular, we may always choose $C = 0$ in applying the Second Fundamental Theorem.

Note also that the derivative of u is precisely $2x + 1$. This is what makes the substition work. If the expression in parentheses were $3x + 1$ rather than $2x + 1$, the Generalized Power Rule would not apply and we would have a much more difficult problem.

EXAMPLE 8   Evaluate $\int_0^{\pi/4} \sin^3 2x \cos 2x \, dx$.

Solution   Let $u = \sin 2x$; then $du = 2 \cos 2x \, dx$. Thus,

$$\int \sin^3 2x \cos 2x \, dx = \frac{1}{2} \int (\sin^3 2x)(2 \cos 2x) \, dx = \frac{1}{2} \int u^3 \, du$$

$$= \frac{1}{2} \frac{u^4}{4} + C = \frac{\sin^4 2x}{8} + C$$

Therefore, by the Second Fundamental Theorem of Calculus,

$$\int_0^{\pi/4} \sin^3 2x \cos 2x \, dx = \left[\frac{\sin^4 2x}{8} \right]_0^{\pi/4} = \frac{1}{8} - 0 = \frac{1}{8}$$ ■

EXAMPLE 9   Evaluate $\int_0^1 \left[x^2 + (x^2 + 1)^4 x \right] dx$.

Solution

$$\int_0^1 \left[x^2 + (x^2 + 1)^4 \right] dx = \int_0^1 x^2 \, dx + \int_0^1 (x^2 + 1)^4 x \, dx$$

The first integral is easy to do directly. To handle the second, we let $u = x^2 + 1$, so $du = 2x\,dx$, and write

$$\int (x^2 + 1)^4 x\,dx = \frac{1}{2} \int (x^2 + 1)^4 2x\,dx = \frac{1}{2} \int u^4\,du$$

$$= \frac{1}{2}\frac{u^5}{5} + C = \frac{(x^2 + 1)^5}{10} + C$$

Therefore,

$$\int_0^1 x^2\,dx + \int_0^1 (x^2 + 1)^4 x\,dx = \left[\frac{x^3}{3}\right]_0^1 + \left[\frac{(x^2 + 1)^5}{10}\right]_0^1$$

$$= \left(\frac{1}{3} - 0\right) + \left(\frac{32}{10} - \frac{1}{10}\right)$$

$$= \frac{1}{3} + \frac{31}{10} = \frac{103}{30} \qquad\blacksquare$$

EXAMPLE 10 Figure 1 shows the graph of a function f that has a continuous third derivative. The dashed lines are tangent to the graph of $y = f(x)$ at $(1, 1)$ and $(5, 1)$. Based on what is shown, tell, if possible, whether the following integrals are positive, negative, or zero.

(a) $\displaystyle\int_1^5 f(x)\,dx$
(b) $\displaystyle\int_1^5 f'(x)\,dx$

(c) $\displaystyle\int_1^5 f''(x)\,dx$
(d) $\displaystyle\int_1^5 f'''(x)\,dx$

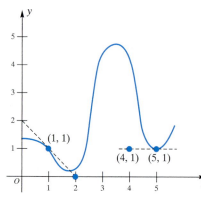

Figure 1

Solution

(a) The function f is positive for all x in the interval $[1, 5]$, and the graph indicates that there is some area above the x-axis. Thus, $\displaystyle\int_1^5 f(x)\,dx > 0$.

(b) By the Second Fundamental Theorem of Calculus,

$$\int_1^5 f'(x)\,dx = f(5) - f(1) = 1 - 1 = 0$$

(c) Again using the Second Fundamental Theorem of Calculus (this time with f' being an antiderivative of f''), we see that

$$\int_1^5 f''(x)\,dx = f'(5) - f'(1) = 0 - (-1) = 1$$

(d) The function f is concave up at $x = 5$, so $f''(5) > 0$, and it is concave down at $x = 1$, so $f''(1) < 0$. Thus,

$$\int_1^5 f'''(x)\,dx = f''(5) - f''(1) > 0 \qquad\blacksquare$$

This example illustrates the remarkable property that to evaluate a definite integral all we need to know are the values of an antiderivative at the end points a and b. For example, to evaluate $\displaystyle\int_1^5 f''(x)\,dx$, all we needed to know was $f'(5)$ and $f'(1)$; we did not need to know f' or f'' at any of the points in the open interval (a, b).

The Mean Value Theorem for Integrals By this time you realize that the Mean Value Theorem for Derivatives plays an important role in calculus. There is a theorem by the same name for integrals, which is also important. Geometrically, it says that there is some c in the interval $[a, b]$ such that the area of the rec-

Figure 2

tangle with height $f(c)$ and width $b - a$ is equal to the area under the curve. (See Figure 2.) In Figure 2 the area under the curve is equal to the area of the shaded rectangle.

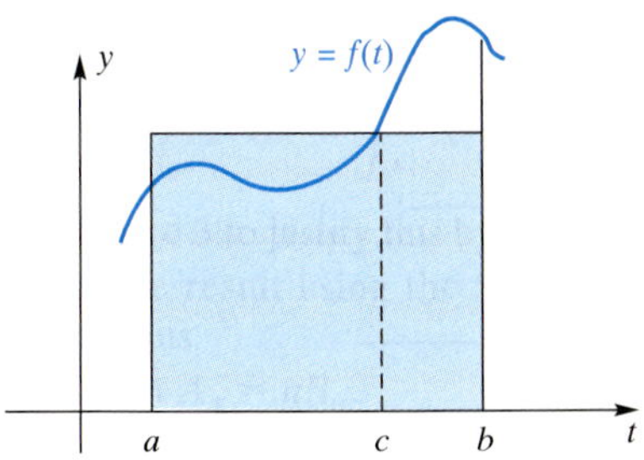

Theorem B   Mean Value Theorem for Integrals

If f is continuous on $[a, b]$, there is a number c between a and b such that

$$\int_a^b f(t)\,dt = f(c)(b - a)$$

Proof   Let

$$G(x) = \int_a^x f(t)\,dt, \qquad a \le x \le b$$

By the Mean Value Theorem for Derivatives applied to G, there is a point c in (a, b) such that

$$G(b) - G(a) = G'(c)(b - a)$$

That is,

$$\int_a^b f(t)\,dt - 0 = G'(c)(b - a)$$

But by the First Fundamental Theorem of Calculus, $G'(c) = f(c)$. The conclusion follows. ♦

≈ **Estimating Integrals**

Theorem B with its accompanying Figure 2 suggests a good way to estimate the value of a definite integral. The area of the region under a curve is equal to the area of a rectangle. One can make a good guess at this rectangle by simply "eyeballing" the region. In Figure 2, the area of the shaded part *above* the curve should match the area of the white part *below* the curve.

Note that if we solve for $f(c)$ in the conclusion of Theorem B we get

$$f(c) = \frac{\displaystyle\int_a^b f(t)\,dt}{b - a}$$

The number $\displaystyle\int_a^b f(x)\,dx / (b - a)$ is called the mean value, or **average value**, of f on $[a, b]$. To see why it has this name, consider a regular partition $P: a = x_0 < x_1 < x_2 < \cdots < x_n = b$ with $\Delta x = (b - a)/n$. The average of the n values $f(x_1), f(x_2), \ldots, f(x_n)$ is

$$\frac{f(x_1) + f(x_2) + \cdots f(x_n)}{n} = \sum_{i=1}^n f(x_i) \frac{1}{n}$$

$$= \frac{1}{b - a} \sum_{i=1}^n f(x_i) \frac{b - a}{n}$$

$$= \frac{1}{b - a} \sum_{i=1}^n f(x_i)\,\Delta x$$

The sum in the last expression is a Riemann sum for f on $[a, b]$ and therefore approaches $\displaystyle\int_a^b f(x)\,dx$ as $n \to \infty$. Thus, $\left(\displaystyle\int_a^b f(x)\,dx\right) \Big/ (b - a)$ appears as the natural extension of the familiar notion of average value.

Riemann Sums

I. Preparation

All continuous functions have antiderivatives, but not all have antiderivatives that can be expressed in terms of elementary functions, that is, polynomials, rational functions, root functions, trigonometric and inverse trigonometric functions, and logarithmic and exponential functions. When f is continuous and when you can find an antiderivative for f in terms of elementary functions, you can evaluate the definite integral

$$\int_a^b f(x)\,dx$$

by applying the Second Fundamental Theorem of Calculus; that is, if F is a function satisfying $F'(x) = f(x)$, then

$$\int_a^b f(x)\,dx = F(b) - F(a)$$

In this project, you will investigate the approximation of definite integrals using Riemann sums, so you should review the concept of the Riemann sum in Section 5.5. We will divide the interval $[a, b]$ into n subintervals of equal length, and we will evaluate the integrand at (1) the left end point, (2) the right end point, and (3) the midpoint of the interval $[x_{i-1}, x_i]$; we will call the resulting sum a left Riemann sum, a right Riemann sum, and a midpoint Riemann sum, respectively.

Exercise 1 Consider the definite integral $\int_0^5 (x^4 + 1)\,dx$.

Evaluate this integral using the Second Fundamental Theorem of Calculus.

II. Using Technology

Exercise 2 For the integral in Exercise 1 use a CAS to evaluate the left Riemann sum, that is, the Riemann sum obtained using the left end point of the each subinterval.

Exercise 3 Now, vary the value of n (the number of subintervals), run the program, and observe the effect on the Riemann sum. Choose $n = 5, 10, 20, 40, 80, 160, 320, 640, 1280$. For each value of n, make a note of the Riemann sum and the error:

error = |Riemann sum − exact value of integral|

(Notice that for this integrand you know the exact value of the definite integral, so you can compare the approxima-

tions with the exact value.) Copy and complete a table like the following:

n	Left Riemann Sum	Error for Left Riemann Sum
5		
10		
20		
40		
⋮		

Study the error column. What happens to the error as n is doubled? Is the error cut in half? Cut to one-third? Cut to one-fourth?

Exercise 4 Now use the right end point of each subinterval. Run the program as in Exercise 3, and complete a similar table. Now what happens to the error as n is doubled? Is the error cut in half? Cut to one-third? Cut to one-fourth?

Exercise 5 Now use the midpoint of each subinterval. Run the program as in Exercise 3, and complete a similar table. Now what happens to the error as n is doubled? Is the error cut in half? Cut to one-third? Cut to one-fourth?

Exercise 6 Next consider the definite integral

$$\int_0^{10} \sqrt{100 - x^2}\,dx$$

At this point in the course, you do not know how to find an antiderivative for $\sqrt{100 - x^2}$, but you can still find an exact value for this definite integral by using geometry. What is the exact value? Repeat Exercises 3 through 5 for this definite integral.

Exercise 7 Consider next the definite integral

$$\int_0^{\pi} \sqrt{\sin x}\,dx$$

For this integral you do not know the exact value. Based on what you learned from Exercises 3 through 5, how would you find an approximate value for this definite integral?

III. Reflection

Exercise 8 When you use a method to approximate a definite integral, the errors tend to be proportional to $1/n$, or $1/n^2$, or $1/n^3$, and so on, where n is the number of subintervals. On the basis of the tables you made in Exercises 3 through 5, try to detect this pattern for (1) the left Riemann sum, (2) the right Riemann sum, and (3) the midpoint Riemann sum. Explain your reasoning.

Accumulation Functions

I. Preparation

Some functions are defined as the area under a curve from a fixed point to a variable point. We have called such functions *accumulation functions*. One important function of this type is the sine integral function, which is defined to be

$$Si(x) = \int_0^x \frac{\sin t}{t}\,dt$$

Notice a few things about the integrand. First, the integrand is not defined for $t = 0$, but as you recall

$$\lim_{t \to 0} \frac{\sin t}{t} = 1$$

Second, notice that the integrand takes on both positive and negative values; thus, the integral from $t = 0$ to $t = x$ is really $A_{up} - A_{down}$.

II. Using Technology

Exercise 1 Use a graphing utility to graph $f(t) = \dfrac{\sin t}{t}$ over the interval $[0, 3\pi]$.

Exercise 2 Without using a graphing utility, try to sketch a graph of $Si(x)$ for x in the interval $[0, 3\pi]$.

Exercise 3 Now use a graphing utility to graph $Si(x)$. In Mathematica, the function is `SinIntegral` and in Maple the function is `Si`.

Exercise 4 Does

$$\lim_{x \to \infty} Si(x)$$

exist?

Exercise 5 What is

$$\frac{d}{dx} Si(x)?$$

Exercise 6 Show that $y = \dfrac{Si(x)}{x^2}$ is a solution to the differential equation $x^3 y' + 2x^2 y = \sin x$.

Exercise 7 Define

$$C(x) = \int_0^x \cos\left(\frac{\pi t^2}{2}\right) dt$$

and

$$S(x) = \int_0^x \sin\left(\frac{\pi t^2}{2}\right) dt$$

These are called *Fresnel integrals*. Find the name that your CAS uses for these functions and sketch both $C(x)$ and $S(x)$ over $[0, 4]$ in the same graph window.

III. Reflection

Exercise 8 Define your own continuous function f so that, to the best of your knowledge, $\int f(x)\,dx$ cannot be expressed in terms of elementary functions. Next, choose a value for a and define the accumulation function

$$F(x) = \int_a^x f(t)\,dt.$$ Use a CAS to graph your function F, and describe some of its the properties.

CHAPTER

6

Applications of the Integral

6.1
The Area of a Plane Region

The brief discussion of area in Section 5.4 served to motivate the definition of the definite integral. With the latter notion now firmly established, we use the definite integral to calculate areas of regions of more and more complicated shapes. As is our practice, we begin with simple cases.

A Region above the x-Axis Let $y = f(x)$ determine a curve in the xy-plane and suppose that f is continuous and nonnegative on the interval $a \le x \le b$ (as in Figure 1). Consider the region R bounded by the graphs of $y = f(x), x = a, x = b$, and $y = 0$. We refer to R as the region under $y = f(x)$ between $x = a$ and $x = b$. Its area $A(R)$ is given by

$$A(R) = \int_a^b f(x)\,dx$$

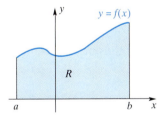

Figure 1

EXAMPLE 1 Find the area of the region R under $y = x^4 - 2x^3 + 2$ between $x = -1$ and $x = 2$.

$\boxed{\approx}$ **Solution** The graph of R is shown in Figure 2. A reasonable estimate for the area of R is its base times an average height, say $(3)(2) = 6$. The exact value is

$$A(R) = \int_{-1}^2 (x^4 - 2x^3 + 2)\,dx = \left[\frac{x^5}{5} - \frac{x^4}{2} + 2x\right]_{-1}^2$$

$$= \left(\frac{32}{5} - \frac{16}{2} + 4\right) - \left(-\frac{1}{5} - \frac{1}{2} - 2\right) = \frac{51}{10} = 5.1$$

273

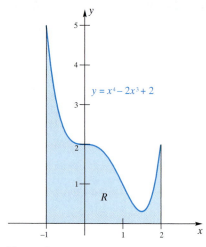

Figure 2

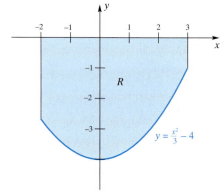

Figure 3

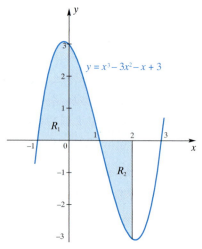

Figure 4

The calculated value 5.1 is close enough to our estimate, 6, to give us confidence in its correctness.   ∎

A Region Below the x-Axis   Area is a nonnegative number. If the graph of $y = f(x)$ is below the x-axis, then $\int_a^b f(x)\,dx$ is a negative number and therefore cannot be an area. However, it is just the negative of the area of the region bounded by $y = f(x)$, $x = a$, $x = b$, and $y = 0$.

EXAMPLE 2   Find the area of the region R bounded by $y = x^3/3 - 4$, the x-axis, $x = -2$, and $x = 3$.

≈ **Solution**   The region R is shown in Figure 3. Our preliminary estimate for its area is $(5)(3) = 15$. The exact value is

$$A(R) = -\int_{-2}^{3}\left(\frac{x^2}{3} - 4\right)dx = \int_{-2}^{3}\left(-\frac{x^2}{3} + 4\right)dx$$

$$= \left[-\frac{x^3}{9} + 4x\right]_{-2}^{3} = \left(-\frac{27}{9} + 12\right) - \left(\frac{8}{9} - 8\right) = \frac{145}{9} \approx 16.11$$

We are reassured by the nearness of 16.11 to our estimate.   ∎

EXAMPLE 3   Find the area of the region R bounded by $y = x^3 - 3x^2 - x + 3$, the segment of the x-axis between $x = -1$ and $x = 2$, and the line $x = 2$.

Solution   The region R is shaded in Figure 4. Note that part of it is above the x-axis and part is below. The areas of these two parts, R_1 and R_2, must be calculated separately. You can check that the curve crosses the x-axis at -1, 1, and 3. Thus,

$$A(R) = A(R_1) + A(R_2)$$

$$= \int_{-1}^{1}(x^3 - 3x^2 - x + 3)\,dx - \int_{1}^{2}(x^3 - 3x^2 - x + 3)\,dx$$

$$= \left[\frac{x^4}{4} - x^3 - \frac{x^2}{2} + 3x\right]_{-1}^{1} - \left[\frac{x^4}{4} - x^3 - \frac{x^2}{2} + 3x\right]_{1}^{2}$$

$$= 4 - \left(-\frac{7}{4}\right) = \frac{23}{4}$$

Notice that we could have written this area as one integral using the absolute value symbol.

$$A(R) = \int_{-1}^{2}\left|x^3 - 3x^2 - x + 3\right|\,dx$$

But this is no real simplification since, in order to evaluate this integral, we would have to split it into two parts, just as we did above.   ∎

A Helpful Way of Thinking   For simple regions of the type considered above, it is quite easy to write down the correct integral. When we consider more complicated regions (e.g., regions between two curves), the task of selecting the right integral is more difficult. However, there is a way of thinking that can be very helpful. It goes back to the definition of area and of the definite integral. Here it is in five steps.

Step 1:   Sketch the region.

Step 2:   Slice it into thin pieces (strips); label a typical piece.

Step 3:   Approximate the area of this typical piece as if it were a rectangle.

Step 4: Add up the approximations to the areas of the pieces.

Step 5: Take the limit as the width of the pieces approaches zero, thus getting a definite integral.

To illustrate, we consider yet another simple example.

EXAMPLE 4   Set up the integral for the area of the region under $y = 1 + \sqrt{x}$ between $x = 0$ and $x = 4$ (Figure 5).

Solution

1. Sketch

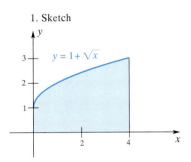

2. Slice

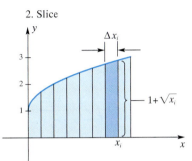

3. Approximate area of typical piece:

$$\Delta A_i \approx (1 + \sqrt{x_i})\, \Delta x_i$$

4. Add up: $A \approx \Sigma\, (1 + \sqrt{x_i})\, \Delta x_i$

5. Take limit: $A = \int_0^4 (1 + \sqrt{x})\, dx$

Figure 5

Once we understand this five-step procedure, we can abbreviate it to three: *slice, approximate, integrate.* Think of the word *integrate* as meaning to add the areas of the pieces and take the limit as the piece width tends to zero. In this process $\Sigma \ldots \Delta x$ transforms into $\int \ldots dx$ as we take the limit. Figure 6 gives the abbreviated form for the same problem.

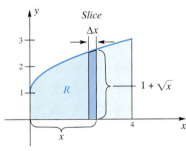

Approximate
$$\Delta A \approx (1 + \sqrt{x})\, \Delta x$$

Integrate
$$A = \int_0^4 (1 + \sqrt{x})\, dx$$

Figure 6

A Region Between Two Curves

Consider curves $y = f(x)$ and $y = g(x)$ with $g(x) \le f(x)$ on $a \le x \le b$. They determine the region shown in Figure 7. We use the *slice, approximate, integrate* method to find its area. Be sure to note that $f(x) - g(x)$ gives the correct height for the thin slice, even when the graph of g goes below the x-axis. For in this case $g(x)$ is negative; so subtracting $g(x)$ is the same as adding a positive number. You can check that $f(x) - g(x)$ also gives the correct height, even when both $f(x)$ and $g(x)$ are negative.

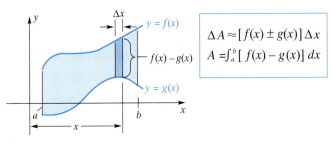

$$\Delta A \approx [f(x) \pm g(x)]\, \Delta x$$
$$A = \int_a^b [f(x) - g(x)]\, dx$$

Figure 7

EXAMPLE 5   Find the area of the region between the curves $y = x^4$ and $y = 2x - x^2$.

Solution   We start by finding where the two curves intersect. To do this, we need to solve $2x - x^2 = x^4$, a fourth-degree equation, which would usually be difficult to solve. However, in this case $x = 0$ and $x = 1$ are rather obvious solutions. Our sketch of the region, together with the appropriate approximation and the corresponding integral, is shown in Figure 8.

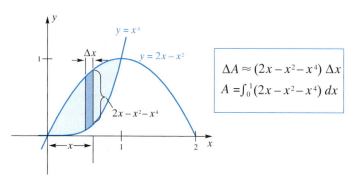

$$\Delta A \approx (2x - x^2 - x^4)\,\Delta x$$
$$A = \int_0^1 (2x - x^2 - x^4)\,dx$$

Figure 8

One job remains: to evaluate the integral.

$$\int_0^1 (2x - x^2 - x^4)\,dx = \left[x^2 - \frac{x^3}{3} - \frac{x^5}{5} \right]_0^1 = 1 - \frac{1}{3} - \frac{1}{5} = \frac{7}{15} \qquad \blacksquare$$

EXAMPLE 6

Horizontal Slicing   Find the area of the region between the parabola $y^2 = 4x$ and the line $4x - 3y = 4$.

Solution   We will need the points of intersection of these two curves. The y-coordinates of these points can be found by writing the second equation as $4x = 3y + 4$ and then equating the two expressions for $4x$.

$$y^2 = 3y + 4$$
$$y^2 - 3y - 4 = 0$$
$$(y - 4)(y + 1) = 0$$
$$y = 4, -1$$

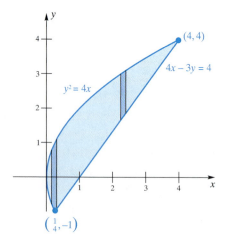

Figure 9

From this, we conclude that the points of intersection are $(4, 4)$ and $\left(\frac{1}{4}, -1\right)$. The region between the curves is shown in Figure 9.

Now imagine slicing this region vertically. We face a problem, because the lower boundary consists of two different curves. Slices at the extreme left extend from the lower branch of the parabola to its upper branch. For the rest of the region, slices extend from the line to the parabola. To solve the problem with vertical slices requires that we first split our region into two parts, set up an integral for each part, and then evaluate both integrals.

A far better approach is to slice the region horizontally as shown in Figure 10, thus using y rather than x as the integration variable. Note that horizontal slices always go from the parabola (at the left) to the line (at the right). The length of such a slice is the largest x-value $\left(x = \frac{1}{4}(3y + 4)\right)$ minus the smallest x-value $\left(x = \frac{1}{4}y^2\right)$.

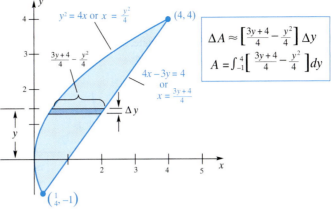

Figure 10

$$A = \int_{-1}^{4}\left[\frac{3y + 4 - y^2}{4}\right]dy = \frac{1}{4}\int_{-1}^{4}(3y + 4 - y^2)\,dy$$

$$= \frac{1}{4}\left[\frac{3y^2}{2} + 4y - \frac{y^3}{3}\right]_{-1}^{4}$$

$$= \frac{1}{4}\left[\left(24 + 16 - \frac{64}{3}\right) - \left(\frac{3}{2} - 4 + \frac{1}{3}\right)\right]$$

$$= \frac{125}{24} \approx 5.21$$

There are two items to note: (1) The integrand resulting from a horizontal slicing involves y, not x; and (2) to get the integrand, solve both equations for x and subtract the smaller x-value from the larger.   ■

Distance and Displacement   Consider an object moving along a straight line with velocity $v(t)$ at time t. If $v(t) \geq 0$, then $\int_{a}^{b} v(t)\,dt$ gives the distance traveled during the time interval $a \leq t \leq b$ (see Sections 5.4 and 5.6). However, if $v(t)$ is sometimes negative (which corresponds to the object moving in reverse), then

$$\int_{a}^{b} v(t)\,dt = s(b) - s(a)$$

measures the **displacement** of the object, that is, the directed distance from its starting position $s(a)$ to its ending position $s(b)$. To get the **total distance** that the object traveled during $a \leq t \leq b$, we must calculate $\int_{a}^{b} |v(t)|\,dt$, the area between the velocity curve and the t-axis. Problems 31 through 33 illustrate these ideas.

Concepts Review

1. Let R be the region between the curve $y = f(x)$ and the x-axis on the interval $[a, b]$. If $f(x) \geq 0$ for all x in $[a, b]$, then $A(R) =$ _____, but if $f(x) \leq 0$ for all x in $[a, b]$, then $A(R) =$ _____.

2. To find the area of the region between two curves, it is wise to think of the following three-word motto: _____.

3. Suppose that the curves $y = f(x)$ and $y = g(x)$ bound a region R on which $f(x) \leq g(x)$. Then the area of R is given by

$$A(R) = \int_{a}^{b} \underline{\hspace{1.5cm}}\,dx,$$

where a and b are determined by solving the equation _____.

4. If $p(y) \leq q(y)$ for all y in $[c, d]$, then the area $A(R)$ of the region R bounded by the curves $x = p(y)$ and $x = q(y)$ between $y = c$ and $y = d$ is given by $A(R) =$ _____.

6.2
Volumes of Solids: Slabs, Disks, Washers

That the definite integral can be used to calculate *areas* is not surprising; it was invented for that purpose. But uses of the integral go far beyond that application. Many quantities can be thought of as a result of slicing something into small pieces, approximating each piece, adding up, and taking the limit as the pieces shrink in size. This method of slice, approximate, and integrate can be used to find the *volumes* of solids provided that the volume of each slice is easy to approximate.

What is volume? We start with simple solids called *right cylinders*, four of which are shown in Figure 1. In each case the solid is generated by moving a plane region (the base) through a distance h in a direction perpendicular to that region. And in each case the volume of the solid is defined to be the area A of the base times the height h; that is,

$$V = A \cdot h$$

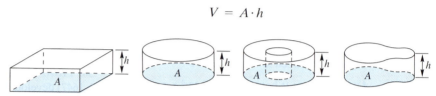

Figure 1

The Volume of a Coin

Consider an ordinary coin, say a quarter.

A quarter has radius about 1 centimeter and thickness about 0.2 centimeter. Its volume is the area of the base, $A = \pi(1^2)$, times the thickness $h = 0.2$; that is

$$V = (1\pi)(0.2) \approx 0.63$$

cubic centimeters.

Next consider a solid with the property that cross sections perpendicular to a given line have known area. In particular, suppose that the line is the x-axis and that the area of the cross section at x is $A(x), a \le x \le b$ (Figure 2). We partition the interval $[a, b]$ by inserting points $a = x_0 < x_1 < x_2 < \cdots < x_n = b$. We then pass planes through these points perpendicular to the x-axis, thus slicing the solid into thin **slabs** (Figure 3). The *volume* ΔV_i of a slab should be approximately the volume of a cylinder; that is,

$$\Delta V_i \approx A(\bar{x}_i)\, \Delta x_i$$

$\left(\text{Recall that } \bar{x}_i, \text{ called a } sample\ point, \text{ is any number in the interval } [x_{i-1}, x_i].\right)$

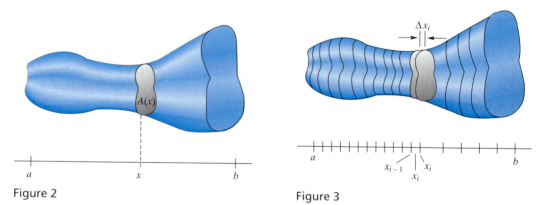

Figure 2   **Figure 3**

The "volume" V of the solid should be given approximately by the Riemann sum

$$V \approx \sum_{i=1}^{n} A(\bar{x}_i)\, \Delta x_i$$

When we let the norm of the partition approach zero, we obtain a definite integral; this integral is defined to be the **volume** of the solid.

$$V = \int_a^b A(x)\, dx$$

Rather than routinely applying the boxed formula to obtain volumes, we suggest that in each problem you go through the process that led to it. Just as for areas, we call this process *slice, approximate, integrate*. It is illustrated in the examples that follow.

Solids of Revolution: Method of Disks   When a plane region, lying entirely on one side of a fixed line in its plane, is revolved about that line, it generates a **solid of revolution**. The fixed line is called the **axis** of the solid of revolution.

As an illustration, if the region bounded by a semicircle and its diameter is revolved about that diameter, it sweeps out a spherical solid (Figure 4). If the region inside a right triangle is revolved about one of its legs, it generates a conical solid (Figure 5). When a circular region is revolved about a line in its plane that does not intersect the circle (Figure 6), it sweeps out a torus (doughnut). In each case, it is possible to represent the volume as a definite integral.

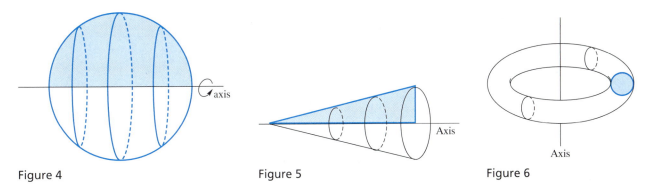

Figure 4   Figure 5   Figure 6

EXAMPLE 1   Find the volume of the solid of revolution obtained by revolving the plane region R bounded by $y = \sqrt{x}$, the x-axis, and the line $x = 4$ about the x-axis.

Solution   The region R, with a typical slice, is displayed as the left part of Figure 7. When revolved about the x-axis, this region generates a solid of revolution and the slice generates a disk, a thin coin-shaped object.

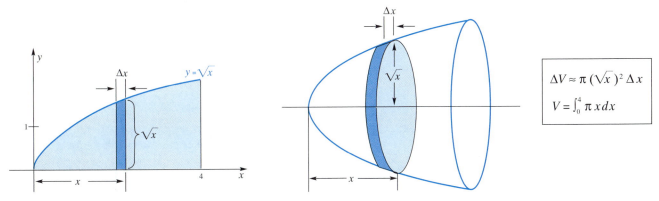

$$\Delta V \approx \pi \left(\sqrt{x} \right)^2 \Delta x$$
$$V = \int_0^4 \pi x \, dx$$

Figure 7

Recalling that the volume of a circular cylinder is $\pi r^2 h$, we approximate the volume ΔV of this disk with $\Delta V \approx \pi (\sqrt{x})^2 \Delta x$ and then integrate.

$$V = \pi \int_0^4 x \, dx = \pi \left[\frac{x^2}{2} \right]_0^4 = \pi \frac{16}{2} = 8\pi \approx 25.13 \qquad \blacksquare$$

EXAMPLE 2   Find the volume of the solid generated by revolving the region bounded by the curve $y = x^3$, the y-axis, and the line $y = 3$ about the y-axis (Figure 8).

Solution   Here we slice horizontally, which makes y the choice for the integration variable. Note that $y = x^3$ is equivalent to $x = \sqrt[3]{y}$ and $\Delta V \approx \pi (\sqrt[3]{y})^2 \Delta y$.

The volume is therefore

$$V = \pi \int_0^3 y^{2/3} \, dy = \pi \left[\frac{3}{5} y^{5/3} \right]_0^3 = \pi \frac{9 \sqrt[3]{9}}{5} \approx 11.76$$

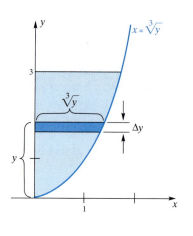

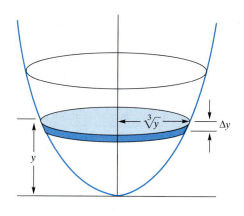

$$\Delta V \approx \pi (\sqrt[3]{y})^2 \Delta y$$

$$V = \int_0^3 \pi y^{2/3} \, dy$$

Figure 8

Method of Washers   Sometimes, slicing a solid of revolution results in disks with holes in the middle. We call them **washers**. See the diagram and accompanying volume formula shown in Figure 9.

EXAMPLE 3   Find the volume of the solid generated by revolving the region bounded by the parabolas $y = x^2$ and $y^2 = 8x$ about the x-axis.

Solution   The key words are still *slice, approximate, integrate* (see Figure 10).

$$V = \pi \int_0^2 (8x - x^4) \, dx = \pi \left[\frac{8x^2}{2} - \frac{x^5}{5} \right]_0^2 = \frac{48\pi}{5} \approx 30.16$$

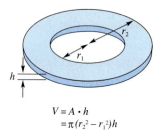

$$V = A \cdot h$$
$$= \pi (r_2^2 - r_1^2) h$$

Figure 9

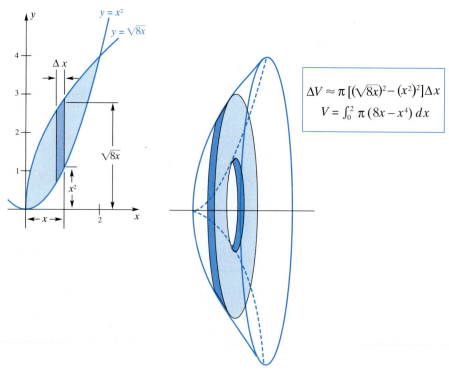

$$\Delta V \approx \pi [(\sqrt{8x})^2 - (x^2)^2] \Delta x$$

$$V = \int_0^2 \pi (8x - x^4) \, dx$$

Figure 10

EXAMPLE 4   The semicircular region bounded by the curve $x = \sqrt{4 - y^2}$ and the y-axis is revolved about the line $x = -1$. Set up the integral that represents its volume.

Solution   Here the outer radius of the washer is $1 + \sqrt{4 - y^2}$ and the inner radius is 1. Figure 11 exhibits the solution. The integral can be simplified.

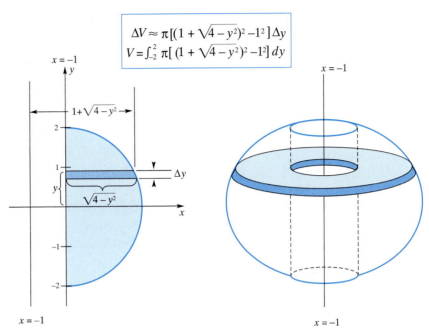

Figure 11

The part above the x-axis has the same volume as the part below it (which manifests itself in an even integrand). Thus, we may integrate from 0 to 2 and double the result. Also, the integrand simplifies.

$$V = \pi \int_{-2}^{2} \left[(1 + \sqrt{4 - y^2})^2 - 1^2 \right] dy$$

$$= 2\pi \int_{0}^{2} \left[2\sqrt{4 - y^2} + 4 - y^2 \right] dy$$

Now see Problem 35 for a way to evaluate this integral.   ■

Other Solids with Known Cross Sections   So far, our solids have had circular cross sections. However, the method for finding volume works just as well for solids whose cross sections are squares or triangles. In fact, all that is really needed is that the areas of the cross sections can be determined, since, in this case, we can also approximate the volume of the slice—a slab—with this cross section. The volume is then found by integrating.

EXAMPLE 5   Let the base of a solid be the first quadrant plane region bounded by $y = 1 - x^2/4$, the x-axis, and the y-axis. Suppose that cross sections perpendicular to the x-axis are squares. Find the volume of the solid.

Solution   When we slice this solid perpendicularly to the x-axis, we get thin square boxes (Figure 12), like slices of cheese.

$$V = \int_{0}^{2} \left(1 - \frac{x^2}{2} + \frac{x^4}{16} \right) dx = \left[x - \frac{x^3}{6} + \frac{x^5}{80} \right]_{0}^{2}$$

$$= 2 - \frac{8}{6} + \frac{32}{80} = \frac{16}{15} \approx 1.07$$

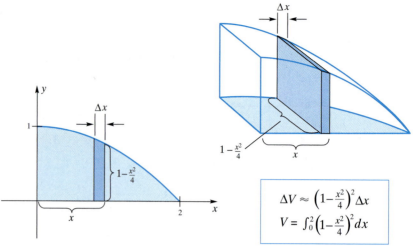

Figure 12

EXAMPLE 6 The base of a solid is the region between one arch of $y = \sin x$ and the x-axis. Each cross section perpendicular to the x-axis is an equilateral triangle sitting on this base. Find the volume of the solid.

Solution We need the fact that the area of an equilateral triangle of side u is $\sqrt{3}\, u^2/4$ (see Figure 13). We proceed as shown in Figure 14.

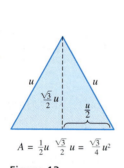

$$A = \tfrac{1}{2} u\, \tfrac{\sqrt{3}}{2} u = \tfrac{\sqrt{3}}{4} u^2$$

Figure 13

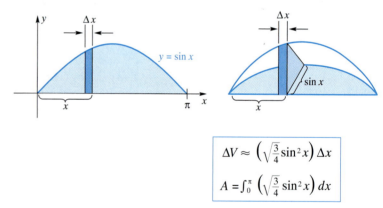

$$\Delta V \approx \left(\sqrt{\tfrac{3}{4}}\sin^2 x \right) \Delta x$$

$$A = \int_0^\pi \left(\sqrt{\tfrac{3}{4}}\sin^2 x \right) dx$$

Figure 14

To perform the indicated integration, we use the half-angle formula $\sin^2 x = (1 - \cos 2x)/2$.

$$V = \frac{\sqrt{3}}{4} \int_0^\pi \frac{1 - \cos 2x}{2}\, dx = \frac{\sqrt{3}}{8} \int_0^\pi (1 - \cos 2x)\, dx$$

$$= \frac{\sqrt{3}}{8} \left[\int_0^\pi 1\, dx - \frac{1}{2} \int_0^\pi \cos 2x \cdot 2\, dx \right]$$

$$= \frac{\sqrt{3}}{8} \left[x - \frac{1}{2}\sin 2x \right]_0^\pi = \frac{\sqrt{3}}{8} \pi \approx 0.68 \qquad \blacksquare$$

Concepts Review

1. The volume of a disk of radius r and thickness h is _____.

2. The volume of a washer of inner radius r, outer radius R, and thickness h is _____.

3. If the region R bounded by $y = x^2$, $y = 0$, and $x = 3$ is revolved about the x-axis, the disk at x will have volume $\Delta V \approx$ _____.

4. If the region R of Question 3 is revolved about the line $y = -2$, the washer at x will have volume $\Delta V \approx$ _____.

Problem Set 6.2

In Problems 1–4, find the volume of the solid generated when the indicated region is revolved about the specified axis; slice, approximate, integrate.

1. *x*-axis

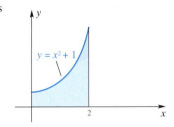

$y = x^2 + 1$

2

2. *x*-axis

$y = -x^2 + 4x$

3

3. (a) *x*-axis
   (b) *y*-axis

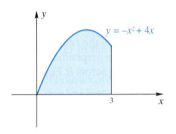

$y = 4 - x^2$

2

4. (a) *x*-axis
   (b) *y*-axis

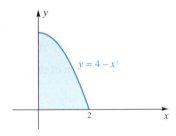

$y = 4 - 2x$

In Problems 5–10, sketch the region R bounded by the graphs of the given equations, and show a typical vertical slice. Then find the volume of the solid generated by revolving R about the x-axis.

5. $y = \dfrac{x^2}{\pi}, x = 4, y = 0$

6. $y = x^3, x = 3, y = 0$

7. $y = \dfrac{1}{x}, x = 2, x = 4, y = 0$

8. $y = x^{3/2}, y = 0$, between $x = 2$ and $x = 3$

9. $y = \sqrt{9 - x^2}, y = 0$, between $x = -2$ and $x = 3$

10. $y = x^{2/3}, y = 0$, between $x = 1$ and $x = 27$

In Problems 11–16, sketch the region R bounded by the graphs of the given equations and show a typical horizontal slice. Find the volume of the solid generated by revolving R about the y-axis.

11. $x = y^2, x = 0, y = 3$

12. $x = \dfrac{2}{y}, y = 2, y = 6, x = 0$

13. $x = 2\sqrt{y}, y = 4, x = 0$ **14.** $x = y^{2/3}, y = 27, x = 0$

15. $x = y^{3/2}, y = 9, x = 0$ **16.** $x = \sqrt{4 - y^2}, x = 0$

17. Find the volume of the solid generated by revolving about the *x*-axis the region bounded by the upper half of the ellipse

$$\frac{x^2}{a^2} + \frac{y^2}{b^2} = 1$$

and the *x*-axis, and thus find the volume of a *prolate spheroid*. Here *a* and *b* are positive constants, with $a > b$.

18. Find the volume of the solid generated by revolving about the *x*-axis the region bounded by the line $y = 6x$ and the parabola $y = 6x^2$.

19. Find the volume of the solid generated by revolving about the *x*-axis the region bounded by the line $x - 2y = 0$ and the parabola $y^2 = 4x$.

20. Find the volume of the solid generated by revolving about the *x*-axis the region in the first quadrant bounded by the circle $x^2 + y^2 = r^2$, the *x*-axis, and the line $x = r - h, 0 < h < r$, and thus find the volume of a *spherical segment* of height *h*, of a sphere of radius *r*.

21. Find the volume of the solid generated by revolving about the *y*-axis the region bounded by the line $y = 4x$ and the parabola $y = 4x^2$.

22. Find the volume of the solid generated by revolving about the line $y = 2$ the region in the first quadrant bounded by the parabolas $3x^2 - 16y + 48 = 0$ and $x^2 - 16y + 80 = 0$ and the *y*-axis.

23. The base of a solid is the region inside the circle $x^2 + y^2 = 4$. Find the volume of the solid if every cross section by a plane perpendicular to the *x*-axis is a square. *Hint:* See Examples 5 and 6.

24. Do Problem 23 assuming that every cross section by a plane perpendicular to the *x*-axis is an isosceles triangle with base on the *xy*-plane and altitude 4. *Hint:* To complete the evaluation, interpret $\displaystyle\int_{-2}^{2} \sqrt{4 - x^2}\, dx$ as the area of a semicircle.

25. The base of a solid is bounded by one arch of $y = \sqrt{\cos x}, -\pi/2 \le x \le \pi/2$, and the *x*-axis. Each cross section perpendicular to the *x*-axis is a square sitting on this base. Find the volume of the solid.

26. The base of a solid is the region bounded by $y = 1 - x^2$ and $y = 1 - x^4$. Cross sections of the solid that are perpendicular to the *x*-axis are squares. Find the volume of the solid.

Volume in an Elliptical Cylinder

I. Preparation

Exercise 1 We know that the equation $x^2 + y^2 = r^2$ represents a circle of radius r. Use this fact to explain why

$$A_{\text{circle}} = 4r \int_0^r \sqrt{1 - \frac{x^2}{r^2}} \, dx = \pi r^2 \qquad (1)$$

Similarly, the equation $\frac{x^2}{a^2} + \frac{y^2}{b^2} = 1$ represents an ellipse.

Justify both of the formulas

$$A_{\text{ellipse}} = 4b \int_0^a \sqrt{1 - \frac{x^2}{a^2}} \, dx \qquad (2)$$

$$A_{\text{ellipse}} = 4a \int_0^b \sqrt{1 - \frac{y^2}{b^2}} \, dy \qquad (3)$$

for the area of an ellipse.

Exercise 2 Make the substitution $u = x/a$ in the integral in (2), and use (1) to compute A_{ellipse} explicitly. Check your result in each of the following ways:

(a) Derive the result again using (3).

(b) Show that your result holds for the case when the ellipse is a circle of radius r.

II. Using Technology

Figure 1 shows an underground fuel tank such as might be found at a service station. Each cross section of the tank is an ellipse whose axes have lengths $2a$ feet and $2b$ feet. The length of the tank is L feet. The owner has only a crude way of measuring the contents: he lowers a stick down into the tank and measures how high the fuel level is on the stick. You are to help him to convert this measurement, call it h (in feet), into gallons of fuel in the tank. (See Figure 2.)

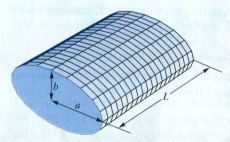

Figure 1

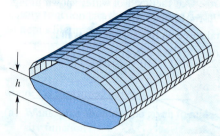

Figure 2

Exercise 3 Derive the following formula for the volume $V(h)$, where $0 \le h \le 2b$, of the fuel when the height is h:

$$V(h) = 2La \int_{-b}^{-b+h} \sqrt{1 - \frac{y^2}{b^2}} \, dy \qquad (4)$$

Exercise 4 Use your computing technology to evaluate the integrals in equations (1), (2), and (3). (Some software will not perform the integration with parameters r, a, or b. In this case, substitute some specific values for these variables and ask your software to evaluate the integral.)

Exercise 5 Suppose that the tank described above has length $L = 20$ feet, width $2a = 10$ feet, and height $2b = 5$ feet. Make a table with two columns containing, respectively, h and $V(h)$, for $h = 0$ to $h = 5$ in increments of 0.25 foot.

Exercise 6 The owner would like to put a mark on the stick as a warning to indicate when only 1500 gallons is left (1 cubic foot equals 7.48 gallons). Using the same tank dimensions as in Exercise 5, determine to the nearest hundredth of a foot the height h at which the stick should be marked. *Suggestions:* Your technology may have root-finding capability, you could use Newton's method, or you could use the zooming technique from earlier projects.

III. Reflection

Exercise 7 Explain why Exercise 4 implies that the total volume of the cylindrical tank is πabL, and use this result to check your output in Exercise 5 for the values $h = 0$, $h = 2.5$, and $h = 5$.

Exercise 8 For arbitrary tank dimensions a, b, and L, find $V'(h)$ and $V''(h)$. Over what intervals is $V(h)$ increasing? Over what intervals is $V(h)$ concave up? Concave down?

Arc Length

I. Preparation

This project involves approximating the length of the curve $y = f(x), a \leq x \leq b$.

Exercise 1 Consider the curve that represents the graph of $y = f(x) = \sqrt{2x + 1}, 0 \leq x \leq 4$. Let

$$x_i = i\frac{4}{n}, \qquad i = 0, 1, \ldots, n$$

$$y_i = f(x_i) = \sqrt{2x_i + 1}$$

The points $(x_i, y_i), i = 0, 1, \ldots, n$, lie on the graph of $y = f(x)$. Explain why the total length of the polygonal line segments connecting $(x_0, y_0), (x_1, y_1), \ldots, (x_n, y_n)$ is

$$\sum_{i=1}^{n} \sqrt{\left(\frac{4}{n}\right)^2 + \left(\sqrt{2x_i + 1} - \sqrt{2x_{i-1} + 1}\right)^2} \qquad (1)$$

Exercise 2 Use a calculator (not a CAS) to evaluate the above sum for $n = 1, 2$, and 4.

Exercise 3 Derive an expression similar to (1) for any given f, a, b, and n. *Hint*: First, define x_i appropriately.

II. Using Technology

Exercise 4 Use your technology and your answer to Exercise 3 to compute the length of the polygonal line for $f(x) = \sqrt{1 - x^2}, a = -1, b = 1$, and for each of $n = 2, 4, 8, 16, 32, 64, 128, 256, 512, 1024$. Make a table of n and the estimated arc lengths (i.e., the total lengths of the polygonal line segments).

Exercise 5 Using a geometrical argument, determine the arc length of the curve in Exercise 4. Add a column to your table that gives the error:

$$|(\text{exact arc length}) - (\text{estimated arc length})|$$

What happens when you double n? Does the error become half of what it was before? One-fourth of what it was before?

Exercise 6 Determine a definite integral that gives the length of the curve defined by the function $f(x) = \sqrt{1 - x^2}, -1 \leq x \leq 1$. Study this definite integral carefully. In what sense is this definite integral different from ones that you have seen before? Can your technology evaluate this integral?

Exercise 7 Consider the curve defined by the function $f(x) = x^{3/2}, 0 \leq x \leq 1$. Evaluate the definite integral that gives the arc length. Compute the length of the polygonal line for each of $n = 2, 4, 8, 16, 32, 64, 128, 256, 512, 1024$. Make a table of n, the estimated arc length (i.e., the total length of the polygonal line segments), and the error.

III. Reflection

Exercise 8 Suppose that we were to approximate the definite integral

$$\int_a^b \sqrt{1 + [f'(x)]^2} \, dx$$

with a Riemann sum, using the left end point of each interval. Assuming a regular partition of $[a, b]$ with n subintervals, determine this Riemann sum. Apply your Riemann sum formula to the definite integral in Exercise 7.

Exercise 9 In Exercises 1 through 7 we approximated the length of a curve by the total length of polygonal line segments. Now, let's take a different approach to approximating arc length. Let $a = x_0 < x_1 < \ldots < x_n = b$ be an equally spaced partition of the interval $[a, b]$.

(a) Referring to Figure 1, determine the coordinates of the point Q, assuming that PQ is tangent to the curve $y = f(x)$ at P.

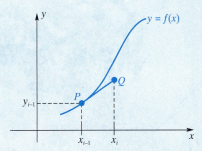

Figure 1

(b) Next, determine the length of the line segment PQ.

(c) Finally, sum up the lengths of these segments as i goes from 1 to n. (Figure 2 show these segments for the curve $y = x^{3/2}$, for $n = 2, 4, 8$.)

(d) What is the relationship between your answers for Exercises 8 and 9?

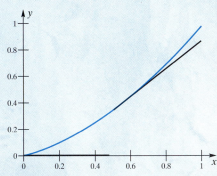
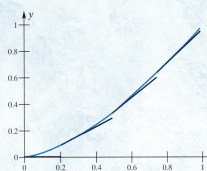
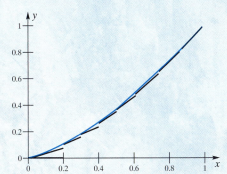

Figure 2

7

Transcendental Functions

7.1
The Natural Logarithm Function

The power of calculus, both that of derivatives and integrals, has already been amply demonstrated. Yet we have only scratched the surface of potential applications. To dig deeper, we need to expand the class of functions with which we can work. That is the object of this chapter.

We begin by asking you to notice a peculiar gap in our knowledge of derivatives.

$$D_x(x^3/3) = x^2$$
$$D_x(x^2/2) = x^1$$
$$D_x(x) = 1 = x^0$$
$$D_x(????) = x^{-1}$$
$$D_x(-x^{-1}) = x^{-2}$$
$$D_x(-x^{-2}/2) = x^{-3}$$

Is there a function whose derivative is $1/x$? Alternatively, is there an antiderivative $\int 1/x\,dx$? The First Fundamental Theorem of Calculus states that the accumulation function

$$F(x) = \int_a^x f(t)\,dt$$

is a function whose derivative is $f(x)$, provided that f is continuous on an interval I that contains a and x. In this sense, we can find an antiderivative of *any* continuous function. The existence of an antiderivative does not mean that the antiderivative can be expressed in terms of functions that we have studied so far. In this chapter we will introduce and study a number of new functions.

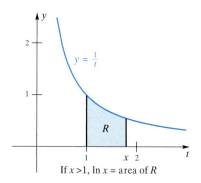

If $x > 1$, $\ln x = $ area of R

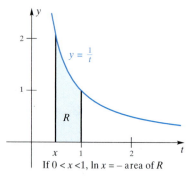

If $0 < x < 1$, $\ln x = -$ area of R

Figure 1

Our first new function is chosen to fill the gap noticed above. We call it the **natural logarithm function**, and it does have something to do with the logarithm studied in algebra, but this relationship will only appear later. For the time being, just accept the fact that we are going to define a *new* function and study its properties.

Definition   Natural Logarithm Function

The **natural logarithm function**, denoted by ln, is defined by

$$\ln x = \int_1^x \frac{1}{t}\, dt, \qquad x > 0$$

The domain of the natural logarithm function is the set of positive real numbers.

The diagrams in Figure 1 indicate the geometric meaning of $\ln x$. The natural logarithm (or natural log) function measures the area under the curve $y = 1/t$ between 1 and x if $x > 1$ and the negative of the area if $0 < x < 1$. The natural logarithm is an accumulation function because it accummulates area under the curve $y = 1/t$. Clearly, $\ln x$ is well defined for $x > 0$; $\ln x$ is not defined for $x \le 0$ because this definite integral does not exist over an interval that includes 0.

And what is the derivative of this new function? Just exactly what we want.

Derivative of the Natural Logarithm Function   From the First Fundamental Theorem of Calculus, we have

$$D_x \int_1^x \frac{1}{t}\, dt = D_x \ln x = \frac{1}{x}, \qquad x > 0$$

This can be combined with the Chain Rule. If $u = f(x) > 0$ and if f is differentiable, then

$$D_x \ln u = \frac{1}{u} D_x u$$

EXAMPLE 1   Find $D_x \ln \sqrt{x}$.

Solution   Let $u = \sqrt{x} = x^{1/2}$. Then

$$D_x \ln \sqrt{x} = \frac{1}{x^{1/2}} \cdot D_x(x^{1/2}) = \frac{1}{x^{1/2}} \cdot \frac{1}{2} x^{-1/2} = \frac{1}{2x} \qquad \blacksquare$$

EXAMPLE 2   Find $D_x \ln(x^2 - x - 2)$.

Solution   This problem makes sense, provided that $x^2 - x - 2 > 0$. Now $x^2 - x - 2 = (x - 2)(x + 1)$, which is positive provided that $x < -1$ or $x > 2$. Thus, the domain of $\ln(x^2 - x - 2)$ is $(-\infty, -1) \cup (2, \infty)$. On this domain,

$$D_x \ln(x^2 - x - 2) = \frac{1}{x^2 - x - 2} D_x(x^2 - x - 2) = \frac{2x - 1}{x^2 - x - 2} \qquad \blacksquare$$

EXAMPLE 3   Show that

$$D_x \ln|x| = \frac{1}{x}, \qquad x \ne 0$$

Solution   Two cases are to be considered. If $x > 0$, $|x| = x$, and

$$D_x \ln|x| = D_x \ln x = \frac{1}{x}$$

If $x < 0$, $|x| = -x$, and so

$$D_x \ln|x| = D_x \ln(-x) = \frac{1}{-x} D_x(-x) = \left(\frac{1}{-x}\right)(-1) = \frac{1}{x} \quad \blacksquare$$

We know that for every differentiation formula there is a corresponding integration formula. Thus, Example 3 implies that

$$\int \frac{1}{x} \, dx = \ln|x| + C, \qquad x \neq 0$$

or, with u replacing x,

$$\boxed{\int \frac{1}{u} \, du = \ln|u| + C, \qquad u \neq 0}$$

This fills the long-standing gap in the Power Rule: $\int u^r \, du = u^{r+1}/(r+1)$, from which we had to exclude the exponent $r = -1$.

EXAMPLE 4 Find $\displaystyle\int \frac{5}{2x + 7} \, dx$.

Solution Let $u = 2x + 7$, so $du = 2 \, dx$. Then

$$\int \frac{5}{2x + 7} \, dx = \frac{5}{2} \int \frac{1}{2x + 7} 2 \, dx = \frac{5}{2} \int \frac{1}{u} \, du$$

$$= \frac{5}{2} \ln|u| + C = \frac{5}{2} \ln|2x + 7| + C \quad \blacksquare$$

EXAMPLE 5 Evaluate $\displaystyle\int_{-1}^{3} \frac{x}{10 - x^2} \, dx$.

Solution Let $u = 10 - x^2$, so $du = -2x \, dx$. Then

$$\int \frac{x}{10 - x^2} \, dx = -\frac{1}{2} \int \frac{-2x}{10 - x^2} \, dx = -\frac{1}{2} \int \frac{1}{u} \, du$$

$$= -\frac{1}{2} \ln|u| + C = -\frac{1}{2} \ln|10 - x^2| + C$$

Thus, by the Second Fundamental Theorem of Calculus,

$$\int_{-1}^{3} \frac{x}{10 - x^2} \, dx = \left[-\frac{1}{2} \ln|10 - x^2| \right]_{-1}^{3}$$

$$= -\frac{1}{2} \ln 1 + \frac{1}{2} \ln 9 = \frac{1}{2} \ln 9$$

For the above calculation to be valid, $10 - x^2$ must never be 0 on the interval $[-1, 3]$. It is easy to see that this is true. $\blacksquare$

Properties of the Natural Logarithm
You should be reminded of common logarithms by the results of our next theorem.

Common Logarithms

Properties (ii) and (iii) for **common logarithms** (base 10 logarithms) were what motivated the invention of logarithms. John Napier (1550–1617) wanted to simplify the complicated calculations that arose in astronomy and navigation. To replace multiplication by addition and division by subtraction was his goal—exactly what (ii) and (iii) accomplish. For over 350 years, common logarithms were an essential aid in computation, but today we use calculators and computers for this purpose. However, natural logarithms retain their importance for other reasons, as you will see.

Theorem A

If a and b are positive numbers and r is any rational number, then

(i) $\ln 1 = 0$;

(ii) $\ln ab = \ln a + \ln b$;

(iii) $\ln \dfrac{a}{b} = \ln a - \ln b$;

(iv) $\ln a^r = r \ln a$.

Proof

(i) $\ln 1 = \displaystyle\int_1^1 \frac{1}{t}\, dt = 0.$

(ii) Since, for $x > 0$,

$$D_x \ln ax = \frac{1}{ax} \cdot a = \frac{1}{x}$$

and

$$D_x \ln x = \frac{1}{x}$$

it follows from the theorem about two functions with the same derivative (Theorem 4.7B) that

$$\ln ax = \ln x + C$$

To evaluate C, let $x = 1$, obtaining $\ln a = C$. Thus,

$$\ln ax = \ln x + \ln a$$

Finally, let $x = b$.

(iii) Replace a by $1/b$ in (ii) to obtain

$$\ln \frac{1}{b} + \ln b = \ln \left(\frac{1}{b} \cdot b\right) = \ln 1 = 0$$

Thus,

$$\ln \frac{1}{b} = -\ln b$$

Applying (ii) again, we get

$$\ln \frac{a}{b} = \ln \left(a \cdot \frac{1}{b}\right) = \ln a + \ln \frac{1}{b} = \ln a - \ln b$$

(iv) Since, for $x > 0$,

$$D_x\left(\ln x^r\right) = \frac{1}{x^r} \cdot rx^{r-1} = \frac{r}{x}$$

and

$$D_x(r \ln x) = r \cdot \frac{1}{x} = \frac{r}{x}$$

it follows by Theorem 4.8B, which we used in (ii), that

$$\ln x^r = r \ln x + C$$

Let $x = 1$, which gives $C = 0$. Thus,

$$\ln x^r = r \ln x$$

a result equivalent to (iv). ◆

EXAMPLE 6   Find dy/dx if $y = \ln \sqrt[3]{(x - 1)/x^2}$, $x > 1$.

Solution   Our task is easier if we first use the properties of natural logarithms to simplify y.

$$y = \ln \left(\frac{x - 1}{x^2}\right)^{1/3} = \frac{1}{3} \ln \left(\frac{x - 1}{x^2}\right)$$

$$= \frac{1}{3}\left[\ln(x - 1) - \ln x^2\right] = \frac{1}{3}\left[\ln(x - 1) - 2\ln x\right]$$

Thus,

$$\frac{dy}{dx} = \frac{1}{3}\left[\frac{1}{x-1} - \frac{2}{x}\right] = \frac{2-x}{3x(x-1)}$$ ■

Logarithmic Differentiation   The labor of differentiating expressions involving quotients, products, or powers can often be substantially reduced by first applying the natural logarithm function and using its properties. This method, called **logarithmic differentiation**, is illustrated in Example 7.

EXAMPLE 7   Differentiate $y = \dfrac{\sqrt{1-x^2}}{(x+1)^{2/3}}$.

Solution   First we take natural logarithms; then we differentiate implicitly with respect to x (recall Section 3.8).

$$\ln y = \frac{1}{2}\ln(1-x^2) - \frac{2}{3}\ln(x+1)$$

$$\frac{1}{y}\frac{dy}{dx} = \frac{-2x}{2(1-x^2)} - \frac{2}{3(x+1)} = \frac{-(x+2)}{3(1-x^2)}$$

Thus,

$$\frac{dy}{dx} = \frac{-y(x+2)}{3(1-x^2)} = \frac{-\sqrt{1-x^2}\,(x+2)}{3(x+1)^{2/3}(1-x^2)}$$

$$= \frac{-(x+2)}{3(x+1)^{2/3}(1-x^2)^{1/2}}$$ ■

Example 7 could have been done directly without first taking logarithms, and we suggest you try it. You should be able to make the two answers agree.

The Graph of the Natural Logarithm   The domain of $\ln x$ consists of the set of all positive real numbers, so the graph of $y = \ln x$ is in the right half-plane. Also, for $x > 0$,

$$D_x \ln x = \frac{1}{x} > 0$$

and

$$D_x^2 \ln x = -\frac{1}{x^2} < 0$$

The first formula tells us that the natural log function is continuous (why?) and rises as x increases; the second tells us that the graph is everywhere concave downward. In Problems 39 and 40, you are asked to show that

$$\lim_{x \to \infty} \ln x = \infty$$

and

$$\lim_{x \to 0^+} \ln x = -\infty$$

Finally, $\ln 1 = 0$. These facts imply that the graph of $y = \ln x$ is similar in shape to that shown in Figure 2.

If your calculator has an $\boxed{\text{ln}}$ button, values for the natural logarithm are at your fingertips. For example,

$$\ln 2 \approx 0.6931$$

$$\ln 3 \approx 1.0986$$

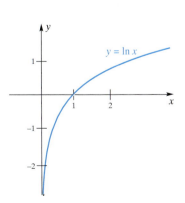

Figure 2

Concepts Review

1. The function ln is defined by $\ln x =$ _____. The domain of this function is _____ and its range is _____.

2. From the preceding definition, it follows that $D_x \ln x =$ _____ for $x > 0$.

3. More generally, for $x \neq 0$, $D_x \ln|x| =$ _____ and so $\int (1/x)\, dx =$ _____.

4. Some common properties of ln are $\ln(xy) =$ _____, $\ln(x/y) =$ _____, and $\ln(x^r) =$ _____.

Problem Set 7.1

1. Use the approximations $\ln 2 = 0.693$ and $\ln 3 = 1.099$ together with the properties stated in Theorem A to calculate approximations to each of the following. For example, $\ln 6 = \ln(2 \cdot 3) = \ln 2 + \ln 3 = 0.693 + 1.099 = 1.792$.

(a) $\ln 6$ (b) $\ln 1.5$ (c) $\ln 81$

(d) $\ln \sqrt{2}$ (e) $\ln\left(\frac{1}{36}\right)$ (f) $\ln 48$

[C] **2.** Use your calculator to make the computations in Problem 1 directly.

In Problems 3–14, find the indicated derivative (see Examples 1 and 2). Assume in each case that x is restricted so that ln is defined.

3. $D_x \ln(x^2 + 3x + \pi)$ **4.** $D_x \ln(3x^3 + 2x)$

5. $D_x \ln(x - 4)^3$ **6.** $D_x \ln \sqrt{3x - 2}$

7. $\dfrac{dy}{dx}$ if $y = 3 \ln x$ **8.** $\dfrac{dy}{dx}$ if $y = x^2 \ln x$

9. $\dfrac{dz}{dx}$ if $z = x^2 \ln x^2 + (\ln x)^3$

10. $\dfrac{dr}{dx}$ if $r = \dfrac{\ln x}{x^2 \ln x^2} + \left(\ln \dfrac{1}{x}\right)^3$

11. $g'(x)$ if $g(x) = \ln(x + \sqrt{x^2 + 1})$

12. $h'(x)$ if $h(x) = \ln(x + \sqrt{x^2 - 1})$

13. $f'(81)$ if $f(x) = \ln \sqrt[3]{x}$

14. $f'\left(\dfrac{\pi}{2}\right)$ if $f(x) = \ln(\cos x)$

In Problems 15–22, find the integrals (see Examples 4 and 5).

15. $\displaystyle\int \dfrac{1}{2x + 1}\, dx$ **16.** $\displaystyle\int \dfrac{1}{1 - 2x}\, dx$

17. $\displaystyle\int \dfrac{6v + 9}{3v^2 + 9v}\, dv$ **18.** $\displaystyle\int \dfrac{z}{2z^2 + 8}\, dz$

19. $\displaystyle\int \dfrac{2 \ln x}{x}\, dx$ **20.** $\displaystyle\int \dfrac{-1}{x(\ln x)^2}\, dx$

21. $\displaystyle\int_0^3 \dfrac{x^4}{2x^5 + \pi}\, dx$ **22.** $\displaystyle\int_0^1 \dfrac{t + 1}{2t^2 + 4t + 3}\, dt$

In Problems 23–26, use Theorem A to write the expressions as the logarithm of a single quantity.

23. $2 \ln(x + 1) - \ln x$ **24.** $\frac{1}{2} \ln(x - 9) + \frac{1}{2} \ln x$

25. $\ln(x - 2) - \ln(x + 2) + 2 \ln x$

26. $\ln(x^2 - 9) - 2 \ln(x - 3) - \ln(x + 3)$

In Problems 27–30, find dy/dx by logarithmic differentiation (see Example 7).

27. $y = \dfrac{x + 11}{\sqrt{x^3 - 4}}$

28. $y = (x^2 + 3x)(x - 2)(x^2 + 1)$

29. $y = \dfrac{\sqrt{x + 13}}{(x - 4)\sqrt[3]{2x + 1}}$

30. $y = \dfrac{(x^2 + 3)^{2/3}(3x + 2)^2}{\sqrt{x + 1}}$

In Problems 31–34, make use of the known graph of $y = \ln x$ to sketch the graphs of the equations.

31. $y = \ln|x|$ **32.** $y = \ln \sqrt{x}$

33. $y = \ln\left(\dfrac{1}{x}\right)$ **34.** $y = \ln(x - 2)$

35. Sketch the graph of $y = \ln \cos x + \ln \sec x$ on $(-\pi/2, \pi/2)$, but think before you begin.

36. Explain why $\displaystyle\lim_{x \to 0} \ln \dfrac{\sin x}{x} = 0$.

37. Find all local extreme values of $f(x) = 2x^2 \ln x - x^2$ on its domain.

38. The rate of transmission in a telegraph cable is observed to be proportional to $x^2 \ln(1/x)$, where x is the ratio of the radius of the core to the thickness of the insulation $(0 < x < 1)$. What value of x gives the maximum rate of transmission?

39. Use the fact that $\ln 4 > 1$ to show that $\ln 4^m > m$ for $m > 0$. Conclude that $\ln x$ can be made as large as desired by choosing x sufficiently large. What does this imply about $\lim_{x \to \infty} \ln x$?

40. Use the fact that $\ln x = -\ln(1/x)$ and Problem 39 to show that $\lim_{x \to 0^+} \ln x = -\infty$.

41. Solve for x: $\displaystyle\int_{1/3}^x \dfrac{1}{t}\, dt = 2\int_1^x \dfrac{1}{t}\, dt$

42. Prove.

(a) Since $1/t < 1/\sqrt{t}$ for $t > 1$, $\ln x < 2(\sqrt{x} - 1)$ for $x > 1$.

(b) $\displaystyle\lim_{x \to \infty} (\ln x)/x = 0$.

43. Calculate

$$\lim_{n \to \infty}\left[\dfrac{1}{n + 1} + \dfrac{1}{n + 2} + \cdots + \dfrac{1}{2n}\right]$$

by writing the expression in brackets as

$$\left[\frac{1}{1 + 1/n} + \frac{1}{1 + 2/n} + \cdots + \frac{1}{1 + n/n} \right] \frac{1}{n}$$

and recognizing the latter as a Riemann sum.

C **44.** A famous theorem (the Prime Number Theorem) says that the number of primes less than n for large n is approximately $n/(\ln n)$. About how many primes are there less than 1,000,000?

45. Find and simplify $f'(1)$.

(a) $f(x) = \ln\left(\dfrac{ax - b}{ax + b} \right)^c$, where $c = \dfrac{a^2 - b^2}{2ab}$.

(b) $f(x) = \displaystyle\int_1^u \cos^2 t \, dt$, where $u = \ln(x^2 + x - 1)$.

46. Evaluate

(a) $\displaystyle\int_0^{\pi/3} \tan x \, dx$ (b) $\displaystyle\int_{\pi/4}^{\pi/3} \sec x \csc x \, dx$

47. The region bounded by $y = (x^2 + 4)^{-1}$, $y = 0$, $x = 1$, and $x = 4$, is revolved about the y-axis, generating a solid. Find its volume.

48. Find the length of the curve $y = x^2/4 - \ln \sqrt{x}$, $1 \le x \le 2$.

49. By appealing to the graph of $y = 1/x$, show that

$$\frac{1}{2} + \frac{1}{3} + \cdots + \frac{1}{n} < \ln n < 1 + \frac{1}{2} + \frac{1}{3} + \cdots + \frac{1}{n-1}$$

50. Prove **Napier's Inequality**, which says that, for $0 < x < y$,

$$\frac{1}{y} < \frac{\ln y - \ln x}{y - x} < \frac{1}{x}$$

CAS **51.** Let $f(x) = \ln(1.5 + \sin x)$.

(a) Find the absolute extreme points on $[0, 3\pi]$.

(b) Find any inflection points on $[0, 3\pi]$.

(c) Evaluate $\displaystyle\int_0^{3\pi} \ln(1.5 + \sin x)\, dx$.

CAS **52.** Let $f(x) = \cos(\ln x)$.

(a) Find the absolute extreme points on $[0.1, 20]$.

(b) Find the absolute extreme points on $[0.01, 20]$.

(c) Evaluate $\displaystyle\int_{0.1}^{20} \cos(\ln x)\, dx$.

CAS **53.** Draw the graphs of $f(x) = x \ln(1/x)$ and $g(x) = x^2 \ln(1/x)$ on $(0, 1]$.

(a) Find the area of the region between these curves on $(0, 1]$.

(b) Find the absolute maximum value of $|f(x) - g(x)|$ on $(0, 1]$.

CAS **54.** Follow the directions of Problem 53 for $f(x) = x \ln x$ and $g(x) = \sqrt{x} \ln x$.

Answers to Concepts Review: **1.** $\displaystyle\int_1^x (1/t)\, dt; (0, \infty); (-\infty, \infty)$

2. $1/x$ **3.** $1/x; \ln|x| + C$ **4.** $\ln x + \ln y; \ln x - \ln y; r \ln x$

7.2
Inverse Functions and Their Derivatives

Our stated aim for this chapter is to expand the number of functions in our repertoire. One way to manufacture new functions is to take old ones and "reverse" them. When we do this for the natural logarithm function, we will be led to the natural exponential function, the subject of Section 7.3. In this section, we study the general problem of reversing (or inverting) a function. Here is the idea.

A function f takes a number x from its domain D and assigns to it a single value y from its range R. If we are lucky, as in the case of the two functions graphed in Figures 1 and 2, we can reverse f; that is, for any given y in R, we can unambiguously go back and find the x from which it came. This new function that takes y and assigns x to it is denoted by f^{-1}. Note that its domain is R and its range is D. It is called the **inverse** of f, or simply f-inverse. Here we are using the superscript -1 in a new way. The symbol f^{-1} does not denote $1/f$, as you might have expected. We, and all mathematicians, use it to name the inverse function.

Undoing Machir

We may view a fu
machine that acce
produces an outp
machine and the
hooked together
undo each other.

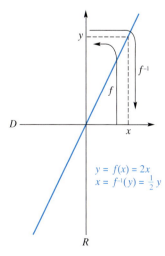

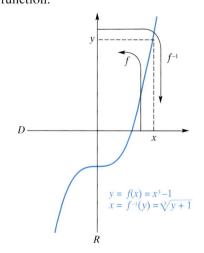

Figure 1   Figure 2

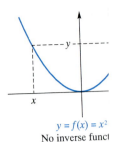

$y = f(x) = x^2$
No inverse func

Figure 3

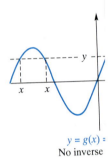

$y = g(x) =$
No inverse

Figure 4

(a) the area of R;
(b) the volume of the solid obtained when R is revolved about the x-axis.

47. Evaluate $\displaystyle\lim_{n\to\infty} \frac{e^{1/n} + e^{2/n} + \cdots + e^{n/n}}{n}$.

48. The **normal probability density function** with mean μ and standard deviation σ is defined by

$$y = f(x) = \frac{1}{\sigma\sqrt{2\pi}} \exp\left[-\frac{(x - \mu)^2}{2\sigma^2}\right]$$

Show that:
(a) its graph is symmetric about the line $x = \mu$;
(b) it has a maximum at $x = \mu$ and inflection points at $x = \mu \pm \sigma$.

GC *Use a graphing calculator or a CAS to do Problems 49–54.*

49. Evaluate.

(a) $\displaystyle\int_{-3}^{3} \exp(-1/x^2)\,dx$ (b) $\displaystyle\int_{0}^{8\pi} e^{-0.1x} \sin x \, dx$

50. Evaluate.

(a) $\displaystyle\lim_{x\to 0}(1 + x)^{1/x}$ (b) $\displaystyle\lim_{x\to 0}(1 + x)^{-1/x}$

51. Find the area of the region between the graphs of $y = f(x) = \exp(-x^2)$ and $y = f''(x)$ on $[-3, 3]$.

EXPL **52.** Draw the graphs of $y = x^p e^{-x}$ for various positive values of p using the same axes. Make conjectures about:

(a) $\displaystyle\lim_{x\to\infty} x^p e^{-x}$,

(b) the x-coordinate of the maximum point for $f(x) = x^p e^{-x}$.

53. Describe the behavior of $\ln(x^2 + e^{-x})$ for large negative x. For large positive x.

54. Draw the graphs of f and f', where $f(x) = 1/(1 + e^{1/x})$. Then determine each of the following:

(a) $\displaystyle\lim_{x\to 0^+} f(x)$ (b) $\displaystyle\lim_{x\to 0^-} f(x)$
(c) $\displaystyle\lim_{x\to\pm\infty} f(x)$ (d) $\displaystyle\lim_{x\to 0} f'(x)$
(e) The maximum and minimum values of f (if they exist).

Answers to Concepts Review: **1.** increasing; exp **2.** $\ln e = 1$; 2.72 **3.** x; x **4.** e^x; $e^x + C$

7.4
General Exponential and Logarithmic Functions

We defined $e^{\sqrt{2}}$, e^{π}, and all other irrational powers of e in the previous section. But what about $2^{\sqrt{2}}$, π^{π}, π^{e}, $\sqrt{2}^{\pi}$, and similar irrational powers of other numbers? In fact, we want to give meaning to a^x for $a > 0$ and x any real number. Now, if $r = p/q$ is a rational number, then $a^r = (\sqrt[q]{a})^p$. But we also know that

$$a^r = \exp(\ln a^r) = \exp(r \ln a) = e^{r \ln a}$$

This suggests the definition of the **exponential function to the base a.**

What is 2^π?

In algebra, 2^n is first defined for positive integers n. Thus, $2^1 = 2$ and $2^4 = 2 \cdot 2 \cdot 2 \cdot 2$. Next, we define 2^n for zero,

$$2^0 = 1$$

and for negative integers:

$$2^{-n} = 1/2^n \quad \text{if} \quad n > 0$$

This means that $2^{-3} = 1/2^3 = 1/8$. Finally, we used root functions to define 2^r for rational numbers r. Thus,

$$2^{7/3} = \sqrt[3]{2^7}$$

Calculus is required to extend the definition of 2^x to the set of real numbers. One way to define 2^π would be to say that it is the limit of the sequence

$$2^3, 2^{3.1}, 2^{3.14}, 2^{3.141}, \ldots$$

The definition we use is

$$2^\pi = e^{\pi \ln 2}$$

This definition involves calculus, because our definition of the natural log function involved the definite integral.

Definition

For $a > 0$ and any real number x,

$$a^x = e^{x \ln a}$$

Of course, this definition will be appropriate only if the usual properties of exponents are valid for it, a matter we take up shortly. To shore up our confidence in the definition, we use it to calculate 3^2 (with a little help from our calculator):

$$3^2 = e^{2 \ln 3} \approx e^{2(1.0986123)} \approx 9.000000$$

Your calculator may give a result that differs slightly from 9. Calculators use approximations for e^x and $\ln x$, and they round to a fixed number of decimal places (usually about 8).

Now we can fill a small gap in the properties of the natural logarithm left over from Section 7.1.

$$\ln(a^x) = \ln(e^{x \ln a}) = x \ln a$$

Thus, Property (iv) of Theorem 7.1A holds for all real x, not just rational x as claimed there. We will need this fact in the proof of Theorem A below.

Properties of a^x
Theorem A summarizes the familiar properties of exponents, which can all be proved now in a completely rigorous manner. Theorem B shows us how to differentiate and integrate a^x.

Theorem A   Properties of Exponents

If $a > 0, b > 0$, and x and y are real numbers, then

(i) $a^x a^y = a^{x+y}$;

(ii) $\dfrac{a^x}{a^y} = a^{x-y}$;

(iii) $\left(a^x\right)^y = a^{xy}$;

(iv) $(ab)^x = a^x b^x$;

(v) $\left(\dfrac{a}{b}\right)^x = \dfrac{a^x}{b^x}$.

Proof   We content ourselves with proving (ii) and (iii), leaving the others for you.

(ii) $\dfrac{a^x}{a^y} = e^{\ln(a^x/a^y)} = e^{\ln a^x - \ln a^y}$

$= e^{x \ln a - y \ln a} = e^{(x-y)\ln a} = a^{x-y}$

(iii) $\left(a^x\right)^y = e^{y \ln a^x} = e^{yx \ln a} = a^{yx} = a^{xy}$ ◆

Theorem B   Exponential Function Rules

$$D_x a^x = a^x \ln a$$

$$\int a^x \, dx = \left(\frac{1}{\ln a}\right)a^x + C, \qquad a \neq 1$$

Proof

$$D_x a^x = D_x\left(e^{x \ln a}\right) = e^{x \ln a} D_x(x \ln a)$$
$$= a^x \ln a$$

The integral formula follows immediately from the derivative formula. ◆

EXAMPLE 1   Find $D_x\left(3^{\sqrt{x}}\right)$.

Solution   We use the Chain Rule with $u = \sqrt{x}$.

$$D_x\left(3^{\sqrt{x}}\right) = 3^{\sqrt{x}} \ln 3 \cdot D_x \sqrt{x} = \frac{3^{\sqrt{x}} \ln 3}{2\sqrt{x}}$$ ■

EXAMPLE 2   Find dy/dx if $y = \left(x^4 + 2\right)^5 + 5^{x^4 + 2}$.

Solution

$$\frac{dy}{dx} = 5\left(x^4 + 2\right)^4 \cdot 4x^3 + 5^{x^4 + 2} \ln 5 \cdot 4x^3$$

$$= 4x^3\left[5\left(x^4 + 2\right)^4 + 5^{x^4 + 2} \ln 5\right]$$

$$= 20x^3\left[\left(x^4 + 2\right)^4 + 5^{x^4 + 1} \ln 5\right]$$ ■

EXAMPLE 3   Find $\displaystyle\int 2^{x^3} x^2 \, dx$.

Solution   Let $u = x^3$, so $du = 3x^2 \, dx$. Then

$$\int 2^{x^3} x^2 \, dx = \frac{1}{3} \int 2^{x^3}\left(3x^2 \, dx\right) = \frac{1}{3} \int 2^u \, du$$

$$= \frac{1}{3}\frac{2^u}{\ln 2} + C = \frac{2^{x^3}}{3 \ln 2} + C$$ ■

Why Other Bases?

Are bases other than e really needed? No. The formulas

$$a^x = e^{x \ln a}$$

and

$$\log_a x = \frac{\ln x}{\ln a}$$

allow us to turn any problem involving exponential functions or logarithmic functions with base a to corresponding functions with base e. This supports our terminology: *natural* exponential and *natural* logarithmic functions. It also explains the universal use of the latter functions in advanced work.

From a^x to [f(x)]

Note the increasin
the functions that
considered. The p
to x^x is one chain.
chain is $a^{f(x)}$ to $\left[f(\right.$
We now know how
derivatives of all t
Finding the deriva
these is best accon
logarithmic differe
technique introdu
and illustrated in I

The Function $\log_a$ Finally, we are ready to make a connection with the logarithms that you studied in algebra. We note that if $0 < a < 1$ then $f(x) = a^x$ is a decreasing function; if $a > 1$, it is an increasing function, as you may check by

Thus, by the Second Fundamental Theorem of Calculus,

$$\int_{1/2}^{1} \frac{5^{1/x}}{x^2}\,dx = \left[-\frac{5^{1/x}}{\ln 5} \right]_{1/2}^{1} = \frac{1}{\ln 5}(5^2 - 5)$$

$$= \frac{20}{\ln 5} \approx 12.43 \qquad \blacksquare$$

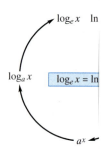

Figure 1

Concepts Review

1. In terms of e and $\ln$, $\pi^{\sqrt{3}} =$ _____. More generally $a^x =$ _____.

2. $\ln x = \log_a x$, where $a =$ _____.

3. $\log_a x$ can be expressed in terms of $\ln$ by $\log_a x =$ _____.

4. The derivative of the power function $f(x) = x^a$ is $f'(x) =$ _____; the derivative of the exponential function $g(x) = a^x$ is $g'(x) =$ _____.

Problem Set 7.4

In Problems 1–8, solve for x. Hint: $\log_a b = c \Leftrightarrow a^c = b$.

1. $\log_2 8 = x$

2. $\log_5 x = 2$

3. $\log_4 x = \frac{3}{2}$

4. $\log_x 64 = 4$

5. $2\log_9\left(\frac{x}{3}\right) = 1$

6. $\log_4\left(\frac{1}{2x}\right) = 3$

7. $\log_2(x + 3) - \log_2 x = 2$

8. $\log_5(x + 3) - \log_5 x = 1$

C *Use* $\log_a x = (\ln x)/(\ln a)$ *to calculate each of the logarithms in Problems 9–12.*

9. $\log_5 12$

10. $\log_7(0.11)$

11. $\log_{11}(8.12)^{1/5}$

12. $\log_{10}(8.57)^7$

C *In Problems 13–16, use natural logarithms to solve each of the exponential equations. Hint: To solve* $3^x = 11$, *take* $\ln$ *of both sides, obtaining* $x \ln 3 = \ln 11$; *then* $x = (\ln 11)/(\ln 3) \approx 2.1827$.

13. $2^x = 17$

14. $5^x = 13$

15. $5^{2s-3} = 4$

16. $12^{1/(\theta-1)} = 4$

Find the indicated derivative or integral (see Examples 1–4).

17. $D_x(6^{2x})$

18. $D_x(3^{2x^2-3x})$

19. $D_x \log_3 e^x$

20. $D_x \log_{10}(x^3 + 9)$

21. $D_z[3^z \ln(z + 5)]$

22. $D_\theta \sqrt{\log_{10}(3^{\theta^2 - \theta})}$

23. $\int x2^{x^2}\,dx$

24. $\int 10^{5x-1}\,dx$

25. $\int_{1}^{4} \frac{5^{\sqrt{x}}}{\sqrt{x}}\,dx$

26. $\int_{0}^{1} (10^{3x} + 10^{-3x})\,dx$

In Problems 27–32, find dy/dx. Note: You must distinguish among problems of the type a^x, x^a, *and* x^x *as in Examples 5–7.*

27. $y = 10^{(x^2)} + (x^2)^{10}$

28. $y = \sin^2 x + 2^{\sin x}$

29. $y = x^{\pi+1} + (\pi + 1)^x$

30. $y = 2^{(e^x)} + (2^e)^x$

31. $y = (x^2 + 1)^{\ln x}$

32. $y = (\ln x^2)^{2x+3}$

33. If $f(x) = x^{\sin x}$, find $f'(1)$

C **34.** Let $f(x) = \pi^x$ and $g(x) = x^\pi$. Which is larger, $f(e)$ or $g(e)$? $f'(e)$ or $g'(e)$?

35. How are $\log_{1/2} x$ and $\log_2 x$ related?

36. Sketch the graphs of $\log_{1/3} x$ and $\log_3 x$ using the same coordinate axes.

C **37.** The magnitude M of an earthquake on the **Richter scale** is

$$M = 0.67 \log_{10}(0.37E) + 1.46$$

where E is the energy of the earthquake in kilowatt-hours. Find the energy of an earthquake of magnitude 7. Of magnitude 8.

C **38.** The loudness of sound is measured in decibels in honor of Alexander Graham Bell (1847–1922), inventor of the telephone. If the variation in pressure is P pounds per square inch, then the loudness L in decibels is

$$L = 20 \log_{10}(121.3P)$$

Find the variation in pressure caused by a rock band at 115 decibels.

C **39.** In the equally tempered scale to which keyed instruments have been tuned since the days of J.S. Bach (1685–1750), the frequencies of successive notes C, C#, D, D#, E, F, F#, G, G#, A, A#, B, $\overline{C}$ form a geometric sequence (progression), with $\overline{C}$ having twice the frequency of C (C# is read C sharp). What is the ratio r between the frequencies of successive notes? If the frequency of A is 440, find the frequency of $\overline{C}$.

40. Prove that $\log_2 3$ is irrational. *Hint:* Use proof by contradiction.

GC **41.** You are suspicious that the xy-data that you collect lie on either an exponential curve $y = Ab^x$ or a power curve $y = Cx^d$. To check, you plot $\ln y$ against x on one graph and $\ln y$ against $\ln x$ on another graph. (Graphing calculators and CASs have options to make the vertical axis, or both the vertical and horizontal axes, a logarithmic scale.) Explain how these graphs will help you to come to a conclusion.

42. (An Amusement) Given the problem of finding y' if $y = x^x$, student A did the following:

$$\text{WRONG 1} \qquad \begin{aligned} y &= x^x \\ y' &= x \cdot x^{x-1} \cdot 1 \\ &= x^x \end{aligned} \qquad \left(\begin{matrix} \text{misapplying the} \\ \text{Power Rule} \end{matrix}\right)$$

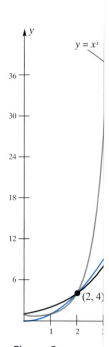

Figure 2

Student B did this:

$$y = x^x$$
WRONG 2 $\quad y' = x^x \cdot \ln x \cdot 1 \quad \begin{pmatrix} \text{misapplying the} \\ \text{Exponential} \\ \text{FunctionRule} \end{pmatrix}$
$$= x^x \ln x$$

The sum $x^x + x^x \ln x$ is correct (Example 5), so

$$\text{WRONG 1} + \text{WRONG 2} = \text{RIGHT}$$

Show that the same procedure yields a correct answer for $y = f(x)^{g(x)}$.

43. Convince yourself that $f(x) = (x^x)^x$ and $g(x) = x^{(x^x)}$ are not the same function. Then find $f'(x)$ and $g'(x)$. *Note:* When mathematicians write x^{x^x}, they mean $x^{(x^x)}$.

44. Consider $f(x) = \dfrac{a^x - 1}{a^x + 1}$ for fixed a, $a > 0$, $a \neq 1$. Show that f has an inverse and find a formula for $f^{-1}(x)$.

45. For fixed $a > 1$, let $f(x) = x^a / a^x$ on $[0, \infty)$. Show:
(a) $\lim\limits_{x \to \infty} f(x) = 0$ (study $\ln f(x)$);
(b) $f(x)$ is maximized at $x_0 = a/\ln a$;
(c) $x^a = a^x$ has two positive solutions if $a \neq e$ and only one such solution if $a = e$;
(d) $\pi^e < e^\pi$.

46. Let $f_u(x) = x^u e^{-x}$ for $x \geq 0$. Show that for any fixed $u > 0$:
(a) $f_u(x)$ attains its maximum at $x_0 = u$;

(b) $f_u(u) > f_u(u + 1)$ and $f_{u+1}(u + 1) > f_{u+1}(u)$ imply

$$\left(\frac{u+1}{u}\right)^u < e < \left(\frac{u+1}{u}\right)^{u+1}$$

(c) $\dfrac{u}{u+1} e < \left(\dfrac{u+1}{u}\right)^u < e$.

Conclude from part (c) that $\lim\limits_{u \to \infty} \left(1 + \dfrac{1}{u}\right)^u = e$.

GC **47.** Find $\lim\limits_{x \to 0^+} x^x$. Also find the coordinates of the minimum point for $f(x) = x^x$ on $[0, 4]$.

GC **48.** Draw the graphs of $y = x^3$ and $y = 3^x$ using the same axes and find all their intersection points.

CAS **49.** Evaluate $\displaystyle\int_0^{4\pi} x^{\sin x}\, dx$.

CAS **50.** Refer to Problem 43. Draw the graphs of f and g using the same axes. Then draw the graphs of f' and g' using the same axes.

Answers to Concepts Review: $\quad$ **1.** $e^{\sqrt{3} \ln \pi}$; $e^{x \ln a}$ **2.** e
3. $(\ln x)/(\ln a)$ **4.** ax^{a-1}; $a^x \ln a$

7.5
Exponential Growth and Decay

At the beginning of 1998, the world's population was about 5.9 billion. It is said that by the year 2020 it will reach 7.9 billion. How are such predictions made?

To treat the problem mathematically, let $y = f(t)$ denote the size of the population at time t, where t is the number of years after 1998. Actually, $f(t)$ is an integer, and its graph "jumps" when someone is born or someone dies. However, for a large population, these jumps are so small relative to the total population that we will not go far wrong if we pretend that f is a nice differentiable function.

It seems reasonable to suppose that the increase Δy in population (births minus deaths) during a short time period Δt is proportional to the size of the population at the beginning of the period and to the length of that period. Thus, $\Delta y = ky\,\Delta t$, or

$$\frac{\Delta y}{\Delta t} = ky$$

In its limiting form, this gives the differential equation

$$\boxed{\frac{dy}{dt} = ky}$$

If $k > 0$, the population is growing; if $k < 0$, it is shrinking. For world population, history indicates that k is about 0.0132 (assuming that t is measured in years), though some agencies report a different figure.

Solving the Differential Equation
We began our study of differential equations in Section 5.2, and you might refer to that section now. We want to solve $dy/dt = ky$ subject to the condition that $y = y_0$ when $t = 0$. Separating variables and integrating, we obtain

$$\frac{dy}{y} = k\, dt$$

$$\int \frac{dy}{y} = \int k\, dt$$

$$\ln y = kt + C$$

The condition $y = y_0$ at $t = 0$ gives $C = \ln y_0$. Thus,

$$\ln y - \ln y_0 = kt$$

or

$$\ln \frac{y}{y_0} = kt$$

Changing to exponential form yields

$$\frac{y}{y_0} = e^{kt}$$

or, finally,

$$\boxed{y = y_0 e^{kt}}$$

When $k > 0$, this type of growth is called **exponential growth**, and when $k < 0$, it is called **exponential decay**.

Returning to the problem of world population, we choose to measure time t in years after January 1, 1998, and y in billions of people. Thus, $y_0 = 5.9$ and, since $k = 0.0132$,

$$y = 5.9e^{0.0132t}$$

By the year 2020, when $t = 22$, we can predict that y will be about

$$y = 5.9e^{0.0132(22)} \approx 7.9 \text{ billion}$$

EXAMPLE 1   How long will it take world population to double under the assumptions above?

Solution   The question is equivalent to asking "In how many years after 1998 will the population reach 11.8 billion?" We need to solve

$$11.8 = 5.9e^{0.0132t}$$

$$2 = e^{0.0132t}$$

for t. Taking logarithms of both sides gives

$$\ln 2 = 0.0132t$$

$$t = \frac{\ln 2}{0.0132} \approx 53 \text{ years} \qquad \blacksquare$$

If world population will double in the first 53 years after 1998, it will double in any 53-year period; so, for example, it will quadruple in 106 years. More generally, if an exponentially growing quantity doubles from y_0 to $2y_0$ in an initial interval of length T, it will double in *any* interval of length T, since

$$\frac{y(t + T)}{y(t)} = \frac{y_0 e^{k(t+T)}}{y_0 e^{kt}} = \frac{y_0 e^{kT}}{y_0} = \frac{2y_0}{y_0} = 2$$

We call the number T the **doubling time**.

EXAMPLE 2   The number of bacteria in a rapidly growing culture was estimated to be 10,000 at noon and 40,000 after 2 hours. Predict how many bacteria there will be at 5 P.M.

Solution   We assume that the differential equation $dy/dt = ky$ is applicable, so $y = y_0 e^{kt}$. Now we have two conditions ($y_0 = 10,000$ and $y = 40,000$ at $t = 2$), from which we conclude that

$$40{,}000 = 10{,}000e^{k(2)}$$

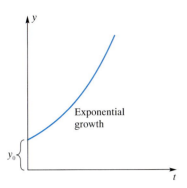

Figure 1

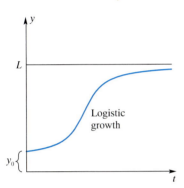

Figure 2

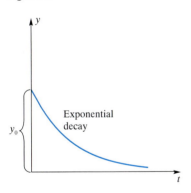

Figure 3

or

$$4 = e^{2k}$$

Taking logarithms yields

$$\ln 4 = 2k$$

or

$$k = \tfrac{1}{2}\ln 4 = \ln \sqrt{4} = \ln 2$$

Thus,

$$y = 10{,}000e^{(\ln 2)t}$$

and, at $t = 5$, this gives

$$y = 10{,}000e^{0.693(5)} \approx 320{,}000 \qquad \blacksquare$$

The exponential model $y = y_0 e^{kt}$, $k > 0$, for population growth is flawed since it projects faster and faster growth indefinitely far into the future (Figure 1). In most cases (including that of world population), the limited amount of space and resources will eventually force a slowing of the growth rate. This suggests another model for population growth, called the **logistic model**, in which we assume that the rate of growth is proportional both to the population size y and to the difference $L - y$, where L is the maximum population that can be supported. This leads to the differential equation

$$\frac{dy}{dt} = ky(L - y)$$

Note that for small y, $dy/dt \approx kLy$, which suggests exponential type growth. But as y nears L, growth is curtailed and dy/dt gets smaller and smaller, producing a growth curve like Figure 2. This model is explored in Problems 24, 25, and 35 of this section and again in Technology Project 8.2.

Radioactive Decay   Not everything grows; some things decrease over time. For example, radioactive elements *decay*, and they do it at a rate proportional to the amount present. Thus, their change rates also satisfy the differential equation

$$\frac{dy}{dt} = ky$$

but now with k negative. It is still true that $y = y_0 e^{kt}$ is the solution to this equation. A typical graph appears in Figure 3.

EXAMPLE 3   Carbon 14, an isotope of carbon, is radioactive and decays at a rate proportional to the amount present. Its **half-life** is 5730 years; that is, it takes 5730 years for a given amount of carbon 14 to decay to one-half its original size. If 10 grams was present originally, how much will be left after 2000 years?

Solution   The half-life of 5730 allows us to determine k, since it implies that

$$\frac{1}{2} = 1e^{k(5730)}$$

or, after taking logarithms,

$$-\ln 2 = 5730k$$

$$k = \frac{-\ln 2}{5730} \approx -0.000121$$

Thus,

$$y = 10e^{-0.000121t}$$

17. If \$375 is put in the bank today, what will it be worth at the end of 2 years if interest is 9.5% and is compounded as specified?

(a) Annually (b) Monthly

(c) Daily (d) Continuously

18. Do Problem 17 assuming that the interest rate is 14.4%.

19. How long does it take money to double in value for the specified interest rate?

(a) 12% compounded monthly

(b) 12% compounded continuously

20. Inflation between 1977 and 1981 ran at about 11.5% per year. On this basis, what would you expect a car that would have cost \$4000 in 1977 to cost in 1981?

21. Manhattan Island is said to have been bought by Peter Minuit in 1626 for \$24. Suppose that Minuit had instead put the \$24 in the bank at 6% interest compounded continuously. What would that \$24 have been worth in 2000? It would be interesting to compare this result with the actual value of Manhattan Island in 2000.

22. If Methuselah's parents had put \$100 in the bank for him at birth and he left it there, what would Methuselah have had at his death (969 years later) if interest was 8% compounded annually?

23. It will be shown later for small x that $\ln(1 + x) \approx x$. Use this fact to show that the doubling time for money invested at p percent compounded annually is about $70/p$ years.

24. The equation for logistic growth is

$$\frac{dy}{dt} = ky(L - y)$$

Show that this differential equation has the solution

$$y = \frac{Ly_0}{y_0 + (L - y_0)e^{-Lkt}}$$

Hint: $\dfrac{1}{y(L - y)} = \dfrac{1}{Ly} + \dfrac{1}{L(L - y)}.$

25. Sketch the graph of the solution in Problem 24 when $y_0 = 5$, $L = 16$, and $k = 0.00186$ (a *logistic model* for world population; see the discussion at the beginning of this section). Note that $\lim_{t \to \infty} y = 16$.

26. Find each of the following limits.

(a) $\lim_{x \to 0}(1 + x)^{1000}$ (b) $\lim_{x \to 0}(1)^{1/x}$

(c) $\lim_{x \to 0^+}(1 + \varepsilon)^{1/x}, \varepsilon > 0$ (d) $\lim_{x \to 0^-}(1 + \varepsilon)^{1/x}, \varepsilon > 0$

(e) $\lim_{x \to 0}(1 + x)^{1/x}$

27. Use the fact that $e = \lim_{h \to 0}(1 + h)^{1/h}$ to find each limit.

(a) $\lim_{x \to 0}(1 - x)^{1/x}$ *Hint:* $(1 - x)^{1/x} = \left[(1 - x)^{1/(-x)}\right]^{-1}$

(b) $\lim_{x \to 0}(1 + 3x)^{1/x}$ (c) $\lim_{n \to \infty}\left(\dfrac{n + 2}{n}\right)^n$

(d) $\lim_{n \to \infty}\left(\dfrac{n - 1}{n}\right)^{2n}$

28. Show that the differential equation

$$\frac{dy}{dt} = ay + b$$

has solution

$$y = \left(y_0 + \frac{b}{a}\right)e^{at} - \frac{b}{a}$$

Assume that $a \neq 0$.

29. Consider a country with a population of 10 million in 1985, a growth rate of 1.2% per year, and immigration from other countries of 60,000 per year. Use the differential equation of Problem 28 to model this situation and predict the population in 2010. Take $a = 0.012$.

30. Important news is said to diffuse through an adult population of fixed size L at a time rate proportional to the number of people who have not heard the news. Five days after a scandal in City Hall was reported, a poll showed that half the people had heard it. How long will it take for 99% of the people to hear it?

31. Assume that (1) world population continues to grow exponentially with growth constant $k = 0.0132$, (2) it takes $\frac{1}{2}$ acre of land to supply food for one person, and (3) there are 13,500,000 square miles of arable land in the world. How long will it be before the world reaches the maximum population? *Note:* There were 5.9 billion people in 1998 and 1 square mile is 640 acres.

GC **32.** The Census Bureau estimates that the growth rate k of the world population will decrease by roughly 0.0002 per year for the next few decades. In 1998, k was 0.0132.

(a) Express k as a function of time t, where t is measured in years since 1998.

(b) Find a differential equation that models the population y for this problem.

(c) Solve the differential equation with the additional condition that the population in 1998 ($t = 0$) was 5.9 billion.

(d) Graph the population y for the next 150 years.

(e) With this model, when will the population reach a maximum? When will the population drop below the 1998 level?

GC **33.** Repeat Exercise 32 under the assumption that k will decrease by 0.0001 per year.

EXPL **34.** Let E be a differentiable function satisfying $E(u + v) = E(u)E(v)$ for all u and v. Find a formula for $E(x)$. *Hint:* First find $E'(x)$.

GC **35.** Using the same axes, draw the graphs for $0 \leq t \leq 100$ of the following two models for the growth of world population (both described in this section).

(a) Exponential growth: $y = 5.9e^{0.0132t}$

(b) Logistic growth: $y = 94.4/(6 + 10e^{-0.030t})$

Compare what the two models predict for world population in 2010, 2040, and 2090. *Note:* Both models assume that world population was 5.9 billion in 1998 ($t = 0$).

GC **36.** Evaluate:

(a) $\lim_{x \to 0}(1 + x)^{1/x}$ (b) $\lim_{x \to 0}(1 - x)^{1/x}$

The limit in part (a) determines e. What special number does the limit in part (b) determine?

Answers to Concepts Review: **1.** $ky; ky(L - y)$ **2.** 8 **3.** half-life **4.** $(1 + h)^{1/h}$

7.6
First-Order Linear Differential Equations

We first solved differential equations in Section 5.2. There we developed the method of separation of variables for finding a solution. In the previous section we used the method of separation of variables to solve differential equations involving growth and decay.

Not all differential equations are *separable*. For example, in the differential equation

$$\frac{dy}{dx} = 2 - 3y$$

there is no way to separate the variables in such a way as to have dy and all expressions involving y on one side and dx and all expressions involving x on the other side. This equation can, however, be put in the form

$$\frac{dy}{dx} + P(x)y = Q(x)$$

where $P(x)$ and $Q(x)$ are functions of x only. A differential equation of this form is said to be a **first-order linear differential equation**. *First-order* refers to the fact that the only derivative is a first derivative. *Linear* refers to fact that the equation can be written in the form $D_x y + P(x)Iy = Q(x)$, where D_x is the derivative operator, and I is the identity operator (that is $Iy = y$). Both D_x and I are *linear operators*.

The family of all solutions of a differential solution is called the **general solution**. Many problems require that the solution satisfy the condition $y = b$ when $x = a$, where a and b are given. Such a condition is called an **initial condition**, and a function that satisfies the differential equation and the initial condition is called a **particular solution**.

Solving First-Order Linear Equations

To solve the first-order linear differential equation, we first multiply both sides by the **integrating factor**

$$e^{\int P(x)\,dx}$$

(The reason for this step will become clear shortly.) The differential equation is then

$$e^{\int P(x)\,dx}\frac{dy}{dx} + e^{\int P(x)\,dx}P(x)y = e^{\int P(x)\,dx}Q(x)$$

The left side is the derivative of the product $y \cdot e^{\int P(x)\,dx}$, so the equation takes the form

$$\frac{d}{dx}\left(y \cdot e^{\int P(x)\,dx}\right) = e^{\int P(x)\,dx}Q(x)$$

Integration of both sides yields

$$ye^{\int P(x)\,dx} = \int \left(Q(x)e^{\int P(x)\,dx}\right) dx$$

The general solution is thus

$$y = e^{-\int P(x)\,dx} \int \left(Q(x)e^{\int P(x)\,dx}\right) dx$$

It is not worth memorizing this final result; the process of getting there is easily recalled and that is what we illustrate.

EXAMPLE 1   Solve

$$\frac{dy}{dx} + \frac{2}{x}y = \frac{\sin 3x}{x^2}$$

Solution   Our integrating factor is

$$e^{\int P(x)\,dx} = e^{\int (2/x)\,dx} = e^{2\ln|x|} = e^{\ln x^2} = x^2$$

(We have taken the arbitrary constant from the integration $\int P(x)\,dx$ to be 0. The choice for the constant does not affect the answer. See Problems 27 and 28. Multiplying both sides of the original equation by x^2, we obtain

$$x^2\frac{dy}{dx} + 2xy = \sin 3x$$

The left side of this equation is the derivative of the product $x^2 y$. Thus,

$$\frac{d}{dx}(x^2 y) = \sin 3x$$

Integration of both members yields

$$x^2 y = \int \sin 3x \, dx = -\tfrac{1}{3}\cos 3x + C$$

or

$$y = \left(-\tfrac{1}{3}\cos 3x + C\right)x^{-2} \qquad \blacksquare$$

EXAMPLE 2   Find the particular solution of

$$\frac{dy}{dx} - 3y = xe^{3x}$$

that satisfies $y = 4$ when $x = 0$.

Solution   The appropriate integrating factor is

$$e^{\int (-3)\,dx} = e^{-3x}$$

Upon multiplication by this factor, our equation takes the form

$$\frac{d}{dx}\left(e^{-3x} y\right) = x$$

or

$$e^{-3x} y = \int x \, dx = \frac{1}{2} x^2 + C$$

Thus, the general solution is

$$y = \frac{1}{2} x^2 e^{3x} + Ce^{3x}$$

Substitution of $y = 4$ when $x = 0$ makes $C = 4$. The desired particular solution is

$$y = \frac{1}{2} x^2 e^{3x} + 4e^{3x} \qquad \blacksquare$$

Applications   We begin with a mixture problem, typical of many problems that arise in chemistry.

EXAMPLE 3   A tank initially contains 120 gallons of brine, holding 75 pounds of dissolved salt in solution. Salt water containing 1.2 pounds of salt per gallon is entering the tank at the rate of 2 gallons per minute and brine flows out at the same rate (Figure 1). If the mixture is kept uniform by constant stirring, find the amount of salt in the tank at the end of 1 hour.

Solution   Let y be the number of pounds of salt in the tank at the end of t minutes. From the brine flowing in, the tank gains 2.4 pounds of salt per minute; from that flowing out, it loses $\frac{2}{120} y$ pounds per minute. Thus,

$$\frac{dy}{dt} = 2.4 - \frac{1}{60} y$$

subject to the condition $y = 75$ when $t = 0$. The equivalent equation

$$\frac{dy}{dt} + \frac{1}{60} y = 2.4$$

has the integrating factor $e^{t/60}$, and so

$$\frac{d}{dt}\left[ye^{t/60}\right] = 2.4e^{t/60}$$

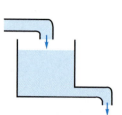

Figure 1

A General Principle

In flow problems such as Example 3, we apply a general principle. Let y measure the quantity of interest that is in the tank at time t. Then the rate of change of y with respect to time is the input rate minus the output rate; that is

$$\frac{dy}{dt} = \text{rate in} - \text{rate out}$$

We conclude that

$$ye^{t/60} = \int 2.4e^{t/60}\, dt = (60)(2.4)e^{t/60} + C$$

Substituting $y = 75$ when $t = 0$ yields $C = -69$, and so

$$y = e^{-t/60}\left[144e^{t/60} - 69\right] = 144 - 69e^{-t/60}$$

At the end of 1 hour ($t = 60$),

$$y = 144 - 69e^{-1} \approx 118.62 \text{ pounds}$$

Note that the limiting value for y as $t \to \infty$ is 144. This corresponds to the fact that the tank will ultimately take on the complexion of the brine entering the tank. One hundred twenty gallons of brine with a concentration of 1.2 pounds of salt per gallon will contain 144 pounds of salt.   ■

We turn next to an example from electricity. According to **Kirchhoff's Law**, a simple electrical circuit (Figure 2) containing a resistor with a resistance of R ohms and an inductor with an inductance of L henrys in series with a source of electromotive force (a battery or generator) that supplies a voltage of $E(t)$ volts at time t satisfies

$$L\frac{dI}{dt} + RI = E(t)$$

where I is the current measured in amperes. This is a linear equation, easily solved by the method of this section.

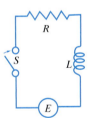

Figure 2

EXAMPLE 4   Consider a circuit (Figure 2) with $L = 2$ henrys, $R = 6$ ohms, and a battery supplying a constant voltage of 12 volts. If $I = 0$ at $t = 0$ (when the switch S is closed), find I at time t.

Solution   The differential equation is

$$2\frac{dI}{dt} + 6I = 12 \quad \text{or} \quad \frac{dI}{dt} + 3I = 6$$

Following our standard procedure (multiply by the integrating factor e^{3t}, integrate, and multiply by e^{-3t}), we obtain

$$I = e^{-3t}\left(2e^{3t} + C\right) = 2 + Ce^{-3t}$$

The initial condition, $I = 0$ at $t = 0$, gives $C = -2$; hence

$$I = 2 - 2e^{-3t}$$

As t increases, the current tends toward a current of 2 amps.   ■

Concepts Review

1. The general first-order linear differential equation has the form $dy/dx + P(x)y = Q(x)$. An integrating factor for this equation is _____.

2. Multiplying both sides of the first-order linear differential equation in Question 1 by its integrating factor makes the left side $\dfrac{d}{dx}($ _____ $)$.

3. The integrating factor for $dy/dx - (1/x)y = x$, where $x > 0$, is _____. When we multiply both sides by this factor, the equation takes the form _____. The general solution to this equation is $y = $ _____.

4. The solution to the differential equation in Question 1 satisfying $y(a) = b$ is called a _____ solution.

Problem Set 7.6

In Problems 1–14, solve each differential equation.

1. $\dfrac{dy}{dx} + y = e^{-x}$

2. $(x + 1)\dfrac{dy}{dx} + y = x^2 - 1$

3. $(1 - x^2)\dfrac{dy}{dx} + xy = ax,\ |x| < 1$

4. $y' + y \tan x = \sec x$

5. $\dfrac{dy}{dx} - \dfrac{y}{x} = xe^x$

6. $y' - ay = f(x)$

7. $\dfrac{dy}{dx} + \dfrac{y}{x} = \dfrac{1}{x}$

8. $y' + \dfrac{2y}{x + 1} = (x + 1)^3$

9. $y' + yf(x) = f(x)$

10. $\dfrac{dy}{dx} + 2y = x$

11. $\dfrac{dy}{dx} - \dfrac{y}{x} = 3x^3;\ y = 3$ when $x = 1$.

12. $y' = e^{2x} - 3y;\ y = 1$ when $x = 0$.

13. $xy' + (1 + x)y = e^{-x};\ y = 0$ when $x = 1$.

14. $\sin x\dfrac{dy}{dx} + 2y \cos x = \sin 2x;\ y = 2$ when $x = \dfrac{\pi}{6}$.

15. A tank contains 20 gallons of a solution, with 10 pounds of chemical A in the solution. At a certain instant, we begin pouring in a solution containing the same chemical in a concentration of 2 pounds per gallon. We pour at a rate of 3 gallons per minute while simultaneously draining off the resulting (well-stirred) solution at the same rate. Find the amount of chemical A in the tank after 20 minutes.

16. A tank initially contains 200 gallons of brine, with 50 pounds of salt in solution. Brine containing 2 pounds of salt per gallon is entering the tank at the rate of 4 gallons per minute and is flowing out at the same rate. If the mixture in the tank is kept uniform by constant stirring, find the amount of salt in the tank at the end of 40 minutes.

17. A tank initially contains 120 gallons of pure water. Brine with 1 pound of salt per gallon flows into the tank at 4 gallons per minute, and the well-stirred solution runs out at 6 gallons per minute. How much salt is in the tank after t minutes, $0 \le t \le 60$?

18. A tank initially contains 50 gallons of brine, with 30 pounds of salt in solution. Water runs into the tank at 3 gallons per minute and the well-stirred solution runs out at 2 gallons per minute. How long will it be until there are 25 pounds of salt in the tank?

19. Find the current I as a function of time for the circuit of Figure 3 if the switch S is closed and $I = 0$ at $t = 0$.

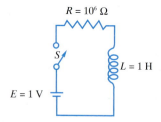

$R = 10^6\ \Omega$

S

$L = 1$ H

$E = 1$ V

Figure 3

20. Find I as a function of time for the circuit of Figure 4, assuming that the switch is closed and $I = 0$ at $t = 0$.

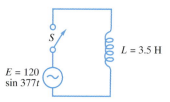

S

$L = 3.5$ H

$E = 120$ $\sin 377t$

Figure 4

21. Find I as a function of time for the circuit of Figure 5, assuming that the switch is closed and $I = 0$ at $t = 0$.

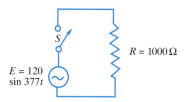

S

$R = 1000\ \Omega$

$E = 120$ $\sin 377t$

Figure 5

22. Suppose that tank 1 initially contains 100 gallons of solution, with 50 pounds of dissolved salt, and tank 2 contains 200 gallons, with 150 pounds of dissolved salt. Pure water flows into tank 1 at 2 gallons per minute, the well-mixed solution flows out and into tank 2 at the same rate, and finally, the solution in tank 2 drains away also at the same rate. Let $x(t)$ and $y(t)$ denote the amounts of salt in tanks 1 and 2, respectively, at time t. Find $y(t)$. *Hint*: First find $x(t)$ and use it in setting up the differential equation for tank 2.

23. A tank of capacity 100 gallons is initially full of pure alcohol. The flow rate of the drain pipe is 5 gallons per minute; the flow rate of the filler pipe can be adjusted to c gallons per minute. An unlimited amount of 25% alcohol solution can be brought in through the filler pipe. Our goal is to reduce the amount of alcohol in the tank so that it will contain 100 gallons of 50% solution. Let T be the number of minutes required to accomplish the desired change.

(a) Evaluate T if $c = 5$ and both pipes are opened.

(b) Evaluate T if $c = 5$ and we first drain away a sufficient amount of the pure alcohol and then close the drain and open the filler pipe.

(c) For what values of c (if any) would strategy (b) give a faster time than (a)?

(d) Suppose that $c = 4$. Determine the equation for T if we initially open both pipes and then close the drain.

EXPL **24.** The differential equation for a falling body near the earth's surface with air resistance proportional to the velocity v is $dv/dt = -g - av$, where $g = 32$ feet per second per second is the acceleration of gravity and $a > 0$ is the *drag coefficient*. Show each of the following:

(a) $v(t) = (v_0 - v_\infty)e^{-at} + v_\infty$, where $v_0 = v(0)$, and

$$v_\infty = -g/a = \lim_{t \to \infty} v(t)$$

the so-called terminal velocity.

(b) If $y(t)$ denotes the altitude, then

$$y(t) = y_0 + tv_\infty + (1/a)(v_0 - v_\infty)(1 - e^{-at})$$

25. A ball is thrown straight up from ground level with an initial velocity $v_0 = 120$ feet per second. Assuming a drag coefficient of $a = 0.05$, determine each of the following:

(a) the maximum altitude

(b) an equation for T, the time when the ball hits the ground

26. Megan bailed out of her plane at an altitude of 8000 feet, fell freely for 15 seconds, and then opened her parachute. Assume that the drag coefficients are $a = 0.10$ for free fall and $a = 1.6$ with the parachute. When did she land?

27. For the differential equation $\dfrac{dy}{dx} - \dfrac{y}{x} = x^2, x > 0$, the integrating factor is $e^{\int (-1/x)\,dx}$. The general antiderivative $\displaystyle\int \left(-\dfrac{1}{x}\right) dx$ is equal to $-\ln x + C$.

(a) Multiply both sides of the differential equation by $\exp\!\left(\displaystyle\int \left(-\dfrac{1}{x}\right) dx\right) = \exp(-\ln x + C)$, and show that $\exp(-\ln x + C)$ is an integrating factor for every value of C.

(b) Solve the resulting equation for y, and show that the solution agrees with the solution obtained when we assumed that $C = 0$.

28. Multiply both sides of the equation $\dfrac{dy}{dx} + P(x)y = Q(x)$ by the factor $e^{\int P(x)\,dx + C}$.

(a) Show that $e^{\int P(x)\,dx + C}$ is an integrating factor for every value of C.

(b) Solve the resulting equation for y, and show that it agrees with the general solution given before Example 1.

Answers to Concepts Review: **1.** $\exp\!\left(\int P(x)\,dx\right)$

2. $y \exp\!\left(\int P(x)\,dx\right)$ **3.** $1/x$; $\dfrac{d}{dx}\left(\dfrac{y}{x}\right) = 1$; $x^2 + Cx$ **4.** particular

---

7.7
The Inverse Trigonometric Functions and Their Derivatives

The six basic trigonometric functions (sine, cosine, tangent, cotangent, secant, and cosecant) were defined in Section 2.3, and we have used them occasionally in examples and problems. With respect to the notion of inverse, they are miserable functions, since for each y in their range there are infinitely many x's that correspond to it (Figure 1). Nonetheless, we are going to introduce a notion of inverse for them. That this is possible rests on a procedure called **restricting the domain**, which was discussed briefly in Section 7.2.

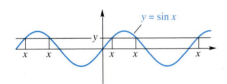

Figure 1

Inverse Sine and Inverse Cosine   In the case of sine and cosine, we restrict the domain, keeping the range as large as possible while insisting that the resulting function have an inverse. This can be done in many ways, but the agreed procedure is suggested by Figures 2 and 3. We show also the graph of the corresponding inverse function, obtained, as usual, by reflecting across the line $y = x$.

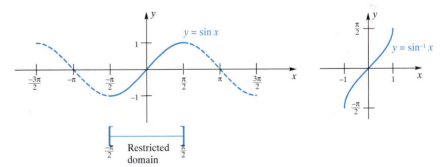

Figure 2

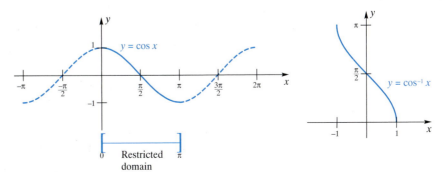

Figure 3

We formalize what we have shown in a definition.

Definition

To obtain inverses for sine and cosine, we restrict their domains to $[-\pi/2, \pi/2]$ and $[0, \pi]$, respectively. Thus,

$$x = \sin^{-1} y \quad \Leftrightarrow \quad y = \sin x \quad \text{and} \quad -\frac{\pi}{2} \leq x \leq \frac{\pi}{2}$$

$$x = \cos^{-1} y \quad \Leftrightarrow \quad y = \cos x \quad \text{and} \quad 0 \leq x \leq \pi$$

The symbol arcsin is often used for $\sin^{-1}$, and arccos is similarly used for $\cos^{-1}$. Think of arcsin as meaning "the arc whose sine is" or "the angle whose sine is" (Figure 4). We will use both forms throughout the rest of this book.

EXAMPLE 1 Calculate

(a) $\sin^{-1}(\sqrt{2}/2)$, (b) $\cos^{-1}(-\frac{1}{2})$,

(c) $\cos(\cos^{-1} 0.6)$, and (d) $\sin^{-1}(\sin 3\pi/2)$

Solution

(a) $\sin^{-1}\left(\frac{\sqrt{2}}{2}\right) = \frac{\pi}{4}$ (b) $\cos^{-1}\left(-\frac{1}{2}\right) = \frac{2\pi}{3}$

(c) $\cos(\cos^{-1} 0.6) = 0.6$ (d) $\sin^{-1}\left(\sin\frac{3\pi}{2}\right) = -\frac{\pi}{2}$

The only one of these that is tricky is (d). Note that it would be wrong to give $3\pi/2$ as the answer, since $\sin^{-1} y$ is always in the interval $[-\pi/2, \pi/2]$. Work the problem in steps, as follows.

$$\sin^{-1}\left(\sin\frac{3\pi}{2}\right) = \sin^{-1}(-1) = -\pi/2 \qquad \blacksquare$$

EXAMPLE 2 Calculate

(a) $\cos^{-1}(-0.61)$, (b) $\sin^{-1}(1.21)$, (c) $\sin^{-1}(\sin 4.13)$

Solution Use a calculator in radian mode. It has been programmed to give answers that are consistent with the definitions that we have given.

(a) $\cos^{-1}(-0.61) = 2.2268569$

(b) Your calculator should indicate an error, since $\sin^{-1}(1.21)$ does not exist.

(c) $\sin^{-1}(\sin 4.13) = -0.9884073$ $\blacksquare$

Figure 4

Another Way To Say It

$$\sin^{-1} y$$

is the number in the interval $[-\pi/2, \pi/2]$ whose sine is y.

$$\cos^{-1} y$$

is the number in the interval $[0, \pi]$ whose cosine is y.

$$\tan^{-1} y$$

is the number in the interval $(-\pi/2, \pi/2)$ whose tangent is y.

Inverse Tangent and Inverse Secant In Figure 5, we show the graph of the tangent function, its restricted domain, and the graph of $y = \tan^{-1} x$.

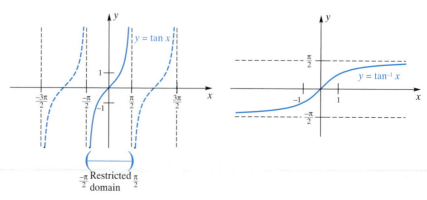

Figure 5

There is a standard way to restrict the domain of the cotangent function, that is, to $(0, \pi)$, so that it has an inverse. However, this function does not play a significant role in calculus.

To obtain an inverse for secant, we graph $y = \sec x$, restrict its domain appropriately, and then graph $y = \sec^{-1} x$ (Figure 6).

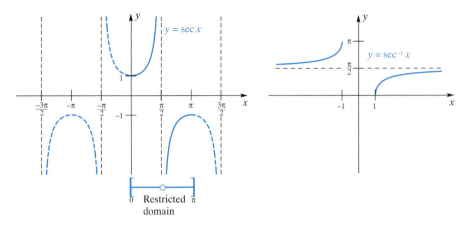

Figure 6

Definition

To obtain inverses for tangent and secant, we restrict their domains to $(-\pi/2, \pi/2)$ and $[0, \pi/2) \cup (\pi/2, \pi]$, respectively. Thus,

$$x = \tan^{-1} y \iff y = \tan x \quad \text{and} \quad -\frac{\pi}{2} < x < \frac{\pi}{2}$$

$$x = \sec^{-1} y \iff y = \sec x \quad \text{and} \quad 0 \le x \le \pi, x \ne \frac{\pi}{2}$$

Some authors restrict the domain of the secant in a different way. Thus, if you refer to another book, you must check that author's definition. We will have no need to define $\csc^{-1}$, though this can also be done.

EXAMPLE 3 Calculate

(a) $\tan^{-1}(1)$,
(c) $\tan^{-1}(\tan 5.236)$,
(e) $\sec^{-1}(2)$, and

(b) $\tan^{-1}(-\sqrt{3})$,
(d) $\sec^{-1}(-1)$,
(f) $\sec^{-1}(-1.32)$

Solution

(a) $\tan^{-1}(1) = \dfrac{\pi}{4}$ (b) $\tan^{-1}(-\sqrt{3}) = -\dfrac{\pi}{3}$

(c) $\tan^{-1}(\tan 5.236) = -1.0471853$

Most of us have trouble remembering our secants; moreover, most calculators fail to have a secant button. Therefore, we suggest that you remember that $\sec x = 1/\cos x$. From this, it follows that

$$\boxed{\sec^{-1} y = \cos^{-1}\left(\frac{1}{y}\right)}$$

and this allows us to use known facts about the cosine.

(d) $\sec^{-1}(-1) = \cos^{-1}(-1) = \pi$

(e) $\sec^{-1}(2) = \cos^{-1}\left(\dfrac{1}{2}\right) = \dfrac{\pi}{3}$

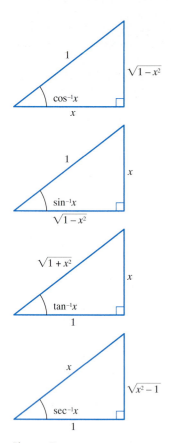

Figure 7

(f) $\sec^{-1}(-1.32) = \cos^{-1}\left(-\dfrac{1}{1.32}\right) = \cos^{-1}(0.7575758)$

$$= 2.4303875 \qquad \blacksquare$$

Four Useful Identities Theorem A gives some useful identities. You can recall them by reference to the triangles in Figure 7.

Theorem A

(i) $\sin(\cos^{-1} x) = \sqrt{1 - x^2}$

(ii) $\cos(\sin^{-1} x) = \sqrt{1 - x^2}$

(iii) $\sec(\tan^{-1} x) = \sqrt{1 + x^2}$

(iv) $\tan(\sec^{-1} x) = \begin{cases} \sqrt{x^2 - 1}, & \text{if } x \geq 1 \\ -\sqrt{x^2 - 1}, & \text{if } x \leq -1 \end{cases}$

Proof To prove (i), recall that $\sin^2 \theta + \cos^2 \theta = 1$. If $0 \leq \theta \leq \pi$, then

$$\sin \theta = \sqrt{1 - \cos^2 \theta}$$

Now apply this with $\theta = \cos^{-1} x$ and use the fact that $\cos(\cos^{-1} x) = x$ to get

$$\sin(\cos^{-1} x) = \sqrt{1 - \cos^2(\cos^{-1} x)} = \sqrt{1 - x^2}$$

Identity (ii) is proved in a completely similar manner. To prove (iii) and (iv), use the identity $\sec^2 \theta = 1 + \tan^2 \theta$ in place of $\sin^2 \theta + \cos^2 \theta = 1$. ◆

EXAMPLE 4 Calculate $\sin\left[2 \cos^{-1}\left(\tfrac{2}{3}\right)\right]$.

Solution Recall the double-angle identity $\sin 2\theta = 2 \sin \theta \cos \theta$. Thus,

$$\sin\left[2 \cos^{-1}\left(\frac{2}{3}\right)\right] = 2 \sin\left[\cos^{-1}\left(\frac{2}{3}\right)\right] \cos\left[\cos^{-1}\left(\frac{2}{3}\right)\right]$$

$$= 2 \cdot \sqrt{1 - \left(\frac{2}{3}\right)^2} \cdot \frac{2}{3} = \frac{4\sqrt{5}}{9} \qquad \blacksquare$$

Derivatives of Trigonometric Functions We learned in Section 3.4 the derivative formulas for the six trigonometric functions. They should be memorized.

$$
\begin{array}{ll}
D_x \sin x = \cos x & D_x \cos x = -\sin x \\
D_x \tan x = \sec^2 x & D_x \cot x = -\csc^2 x \\
D_x \sec x = \sec x \tan x & D_x \csc x = -\csc x \cot x
\end{array}
$$

We can combine the rules above with the Chain Rule. For example, if $u = f(x)$ is differentiable, then

$$D_x \sin u = \cos u \cdot D_x u$$

Inverse Trigonometric Functions From the Inverse Function Theorem (Theorem 7.2B), we conclude that $\sin^{-1}$, $\cos^{-1}$, $\tan^{-1}$, and $\sec^{-1}$ are differentiable. Our aim is to find formulas for their derivatives. We state the results and then show how they can be derived.

Theorem B   Derivatives of Four Inverse Trigonometric Functions

(i) $D_x \sin^{-1} x = \dfrac{1}{\sqrt{1 - x^2}}$, $-1 < x < 1$

(ii) $D_x \cos^{-1} x = \dfrac{-1}{\sqrt{1 - x^2}}$, $-1 < x < 1$

(iii) $D_x \tan^{-1} x = \dfrac{1}{1 + x^2}$

(iv) $D_x \sec^{-1} x = \dfrac{1}{|x|\sqrt{x^2 - 1}}$, $|x| > 1$

Proof   Our proofs follow the same pattern in each case. To prove (i), let $y = \sin^{-1} x$, so that

$$x = \sin y$$

Now differentiate both sides with respect to x, using the Chain Rule on the right-hand side. Then

$$1 = \cos y\, D_x y = \cos(\sin^{-1} x)\, D_x(\sin^{-1} x)$$
$$= \sqrt{1 - x^2}\, D_x(\sin^{-1} x)$$

At the last step, we used Theorem A(ii). We conclude that $D_x(\sin^{-1} x) = 1/\sqrt{1 - x^2}$.

Results (ii), (iii), and (iv) are proved similarly, but (iv) has a little twist. Let $y = \sec^{-1} x$, so

$$x = \sec y$$

Differentiating both sides with respect to x and using Theorem A(iv), we obtain

$$1 = \sec y \tan y\, D_x y$$
$$= \sec(\sec^{-1} x)\tan(\sec^{-1} x)D_x(\sec^{-1} x)$$
$$= \begin{cases} x\sqrt{x^2 - 1}\, D_x(\sec^{-1} x), & \text{if } x \geq 1 \\ x(-\sqrt{x^2 - 1})D_x(\sec^{-1} x), & \text{if } x \leq -1 \end{cases}$$
$$= |x|\sqrt{x^2 - 1}\, D_x(\sec^{-1} x)$$

The desired result follows immediately. ♦

EXAMPLE 5   Find $D_x \sin^{-1}(3x - 1)$.

Solution   We use Theorem B(i) and the Chain Rule.

$$D_x \sin^{-1}(3x - 1) = \frac{1}{\sqrt{1 - (3x - 1)^2}}\, D_x(3x - 1)$$
$$= \frac{3}{\sqrt{-9x^2 + 6x}} \qquad \blacksquare$$

Of course, every differentiation formula leads to an integration formula, a matter we will say much more about in the next chapter. In particular,

(i) $\displaystyle\int \frac{1}{\sqrt{1 - x^2}}\, dx = \sin^{-1} x + C$

(ii) $\displaystyle\int \frac{1}{1 + x^2}\, dx = \tan^{-1} x + C$

(iii) $\displaystyle\int \frac{2}{x\sqrt{x^2 - 1}}\, dx = \sec^{-1}|x| + C$

$D_x \sec^{-1} x$

Here is another way to derive the formula for the derivative of $\sec^{-1} x$.

$$D_x \sec^{-1} x = D_x \cos^{-1}\left(\frac{1}{x}\right)$$
$$= \frac{-1}{\sqrt{1 - 1/x^2}} \cdot \frac{-1}{x^2}$$
$$= \frac{1}{\sqrt{x^2 - 1}} \cdot \frac{\sqrt{x^2}}{x^2}$$
$$= \frac{1}{\sqrt{x^2 - 1}} \cdot \frac{|x|}{x^2}$$
$$= \frac{1}{|x|\sqrt{x^2 - 1}}$$

EXAMPLE 6　Evaluate $\int_0^{1/2} \dfrac{dx}{\sqrt{1-x^2}}$.

Solution

$$\int_0^{1/2} \frac{1}{\sqrt{1-x^2}}\, dx = [\sin^{-1} x]_0^{1/2} = \sin^{-1}\frac{1}{2} - \sin^{-1} 0$$

$$= \frac{\pi}{6} - 0 = \frac{\pi}{6} \qquad ■$$

EXAMPLE 7　A man standing on top of a vertical cliff is 200 feet above a lake. As he watches, a motorboat moves directly away from the foot of the cliff at a rate of 25 feet per second. How fast is the angle of depression of his line of sight changing when the boat is 150 feet from the foot of the cliff?

Solution　The essential details are shown in Figure 8. Note that θ, the angle of depression, is

$$\theta = \tan^{-1}\left(\frac{200}{x}\right)$$

Thus,

$$\frac{d\theta}{dt} = \frac{1}{1 + (200/x)^2} \cdot \frac{-200}{x^2} \cdot \frac{dx}{dt} = \frac{-200}{x^2 + 40{,}000} \cdot \frac{dx}{dt}$$

When we substitute $x = 150$ and $dx/dt = 25$, we obtain $d\theta/dt = -0.08$ radians per second. ■

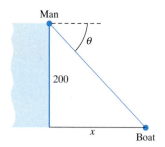

Figure 8

Concepts Review

1. To obtain an inverse for the sine function, we restrict its domain to _____. The resulting inverse function is denoted by $\sin^{-1}$ or by _____.

2. To obtain an inverse for the tangent function, we restrict the domain to _____. The resulting inverse function is denoted by $\tan^{-1}$ or by _____.

3. $D_x \sin(\arcsin x) = $ _____.

4. Since $D_x \arctan x = 1/(1 + x^2)$, it follows that $4\int_0^1 1/(1 + x^2)\, dx = $ _____.

Problem Set 7.7

In Problems 1–10, find the exact value without using a calculator.

1. $\arccos\left(\dfrac{\sqrt{2}}{2}\right)$

2. $\arcsin\left(-\dfrac{\sqrt{3}}{2}\right)$

3. $\sin^{-1}\left(-\dfrac{\sqrt{3}}{2}\right)$

4. $\sin^{-1}\left(-\dfrac{\sqrt{2}}{2}\right)$

5. $\arctan(\sqrt{3})$

6. $\operatorname{arcsec}(2)$

7. $\arcsin\left(-\tfrac{1}{2}\right)$

8. $\tan^{-1}\left(-\dfrac{\sqrt{3}}{3}\right)$

9. $\sin(\sin^{-1} 0.4567)$

10. $\cos(\sin^{-1} 0.56)$

C　*In Problems 11–18, approximate each value.*

11. $\sin^{-1}(0.1113)$

12. $\arccos(0.6341)$

13. $\cos(\operatorname{arccot} 3.212)$

14. $\sec(\arccos 0.5111)$

15. $\sec^{-1}(-2.222)$

16. $\tan^{-1}(-60.11)$

17. $\cos(\sin(\tan^{-1} 2.001))$

18. $\sin^2(\ln(\cos 0.5555))$

In Problems 19–24, express θ in terms of x using the inverse trigonometric functions $\sin^{-1}$, $\cos^{-1}$, $\tan^{-1}$, and $\sec^{-1}$.

19.

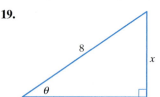

20.

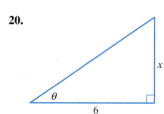

21.

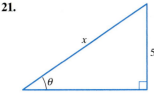

22.

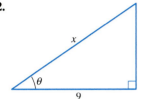

23.

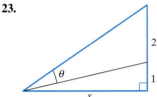

24.

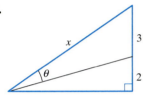

In Problems 25–28, find each value without using a calculator (see Example 4).

25. $\cos\left[2\sin^{-1}\left(-\frac{2}{3}\right)\right]$ **26.** $\tan\left[2\tan^{-1}\left(\frac{1}{3}\right)\right]$

27. $\sin\left[\cos^{-1}\left(\frac{3}{5}\right) + \cos^{-1}\left(\frac{5}{13}\right)\right]$

28. $\cos\left[\cos^{-1}\left(\frac{4}{5}\right) + \sin^{-1}\left(\frac{12}{13}\right)\right]$

In Problems 29–32, show that each equation is an identity.

29. $\tan\left(\sin^{-1}x\right) = \dfrac{x}{\sqrt{1 - x^2}}$

30. $\sin\left(\tan^{-1}x\right) = \dfrac{x}{\sqrt{1 + x^2}}$

31. $\cos\left(2\sin^{-1}x\right) = 1 - 2x^2$

32. $\tan\left(2\tan^{-1}x\right) = \dfrac{2x}{1 - x^2}$

33. Find each limit.

(a) $\lim\limits_{x \to \infty} \tan^{-1}x$ (b) $\lim\limits_{x \to -\infty} \tan^{-1}x$

34. Find each limit.

(a) $\lim\limits_{x \to \infty} \sec^{-1}x$ (b) $\lim\limits_{x \to -\infty} \sec^{-1}x$

35. Sketch the graph of $y = \cot^{-1}x$, assuming that it has been obtained by restricting the domain of the cotangent to $(0, \pi)$.

In Problems 36–47, find dy/dx.

36. $y = e^{\tan x}$ **37.** $y = \ln(\sec x + \tan x)$

38. $y = -\ln(\csc x + \cot x)$ **39.** $y = \sin^{-1}(2x^2)$

40. $y = \arccos(e^x)$ **41.** $y = x^3 \tan^{-1}(e^x)$

42. $y = e^x \arcsin x^2$ **43.** $y = \left(\tan^{-1}x\right)^3$

44. $y = \tan\left(\cos^{-1}x\right)$ **45.** $y = \sec^{-1}\left(x^3\right)$

46. $y = \left(\sec^{-1}x\right)^3$ **47.** $y = \left(1 + \sin^{-1}x\right)^3$

In Problems 48–58, evaluate each integral.

48. $\displaystyle\int x \sin(x^2)\,dx$ **49.** $\displaystyle\int \sin 2x \cos 2x\,dx$

50. $\displaystyle\int \tan x\,dx = \int \frac{\sin x}{\cos x}\,dx$ **51.** $\displaystyle\int_0^1 e^{2x}\cos(e^{2x})\,dx$

52. $\displaystyle\int_0^{\pi/2} \sin^2 x \cos x\,dx$ **53.** $\displaystyle\int_0^{\sqrt{2}/2} \frac{1}{\sqrt{1 - x^2}}\,dx$

54. $\displaystyle\int_{\sqrt{2}}^2 \frac{dx}{x\sqrt{x^2 - 1}}$ **55.** $\displaystyle\int_{-1}^1 \frac{1}{1 + x^2}\,dx$

56. $\displaystyle\int_0^{\pi/2} \frac{\sin\theta}{1 + \cos^2\theta}\,d\theta$ **57.** $\displaystyle\int \frac{1}{1 + 4x^2}\,dx$

58. $\displaystyle\int \frac{e^x}{1 + e^{2x}}\,dx$

C   **59.** A picture 5 feet in height is hung on a wall so that its bottom is 8 feet from the floor, as shown in Figure 9. A viewer with eye level at 5.4 feet stands b feet from the wall. Express θ, the vertical angle subtended by the picture at her eye, in terms of b, and then find θ if $b = 12.9$ feet.

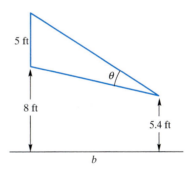

5 ft

8 ft

5.4 ft

b

Figure 9

60. Find formulas for $f^{-1}(x)$ for each of the following functions f, first indicating how you would restrict the domain so that f has an inverse. For example, if $f(x) = 3\sin 2x$ and we restrict the domain to $-\pi/4 \le x \le \pi/4$, then $f^{-1}(x) = \frac{1}{2}\sin^{-1}(x/3)$.

(a) $f(x) = 3\cos 2x$ (b) $f(x) = 2\sin 3x$

(c) $f(x) = \frac{1}{2}\tan x$ (d) $f(x) = \sin\dfrac{1}{x}$

61. By repeated use of the addition formula

$$\tan(x + y) = (\tan x + \tan y)/(1 - \tan x \tan y)$$

show that

$$\frac{\pi}{4} = 3\tan^{-1}\left(\frac{1}{4}\right) + \tan^{-1}\left(\frac{5}{99}\right)$$

62. Verify that

$$\frac{\pi}{4} = 4\tan^{-1}\left(\frac{1}{5}\right) - \tan^{-1}\left(\frac{1}{239}\right)$$

a result discovered by John Machin in 1706 and used by him to calculate the first 100 decimal places of π.

63. Without using calculus, find a formula for the area of the shaded region in Figure 10 in terms of a and b. Note that the center of the larger circle is on the rim of the smaller.

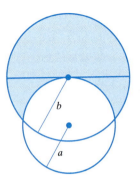

Figure 10

GC **64.** Draw the graphs of

$$y = \arcsin x \quad \text{and} \quad y = \arctan\left(x/\sqrt{1 - x^2}\right)$$

using the same axes. Make a conjecture. Prove it.

GC **65.** Draw the graph of $y = \pi/2 - \arcsin x$. Make a conjecture. Prove it.

GC **66.** Draw the graph of $y = \sin(\arcsin x)$ on $[-1, 1]$. Then draw the graph of $y = \arcsin(\sin x)$ on $[-2\pi, 2\pi]$. Explain the differences that you observe.

67. Show that

$$\int \frac{dx}{\sqrt{a^2 - x^2}} = \sin^{-1}\frac{x}{a} + C, \quad a > 0$$

by writing $a^2 - x^2 = a^2[1 - (x/a)^2]$ and making the substitution $u = x/a$.

68. Show the result in Problem 67 by differentiating the right side to get the integrand.

69. Show that

$$\int \frac{dx}{a^2 + x^2} = \frac{1}{a}\tan^{-1}\frac{x}{a} + C, \quad a \neq 0$$

70. Show that

$$\int \frac{dx}{x\sqrt{x^2 - a^2}} = \frac{1}{a}\sec^{-1}\frac{|x|}{a} + C, \quad a > 0$$

71. Show, by differentiating the right side, that

$$\int \sqrt{a^2 - x^2}\, dx = \frac{x}{2}\sqrt{a^2 - x^2} + \frac{a^2}{2}\sin^{-1}\frac{x}{a} + C, \quad a > 0$$

72. Use the result of Problem 71 to show that

$$\int_{-a}^{a} \sqrt{a^2 - x^2}\, dx = \frac{\pi a^2}{2}$$

Why is this result expected?

73. The lower edge of a wall hanging, 10 feet in height, is 2 feet above the observer's eye level. Find the ideal distance b to stand from the wall for viewing the hanging; that is, find b that maximizes the angle subtended at the viewer's eye. (See Problem 59.)

74. Express $d\theta/dt$ in terms of x, dx/dt, and the constants a and b.

(a)

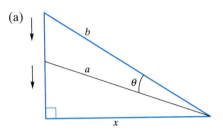

(b)

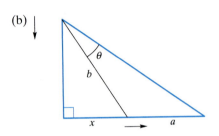

75. The structural steel work of a new office building is finished. Across the street, 60 feet from the ground floor of the freight elevator shaft in the building, a spectator is standing and watching the freight elevator ascend at a constant rate of 15 feet per second. How fast is the angle of elevation of the spectator's line of sight to the elevator increasing 6 seconds after his line of sight passes the horizontal?

76. An airplane is flying at a constant altitude of 2 miles and a constant speed of 600 miles per hour on a straight course that will take it directly over an observer on the ground. How fast is the angle of elevation of the observer's line of sight increasing when the distance from her to the plane is 3 miles? Give your result in radians per minute.

77. A revolving beacon light is located on an island and is 2 miles away from the nearest point P of the straight shoreline of the mainland. The beacon throws a spot of light that moves along the shoreline as the beacon revolves. If the speed of the spot of light on the shoreline is 5π miles per minute when the spot is 1 mile from P, how fast is the beacon revolving?

78. A man on a dock is pulling in a rope attached to a rowboat at a rate of 5 feet per second. If the man's hands are 8 feet higher than the point where the rope is attached to the boat, how fast is the angle of depression of the rope changing when there are still 17 feet of rope out?

C **79.** A visitor from outer space is approaching the earth (radius = 6376 kilometers) at 2 kilometers per second. How fast is the angle θ subtended by the earth at her eye increasing when she is 3000 kilometers from the surface?

Answers to Concepts Review: **1.** $[-\pi/2, \pi/2]$; arcsin **2.** $(-\pi/2, \pi/2)$; arctan **3.** 1 **4.** π

7.8
The Hyperbolic Functions and Their Inverses

In both mathematics and science, certain combinations of e^x and e^{-x} occur so often that they are given special names.

Definition   Hyperbolic Functions

The hyperbolic sine, hyperbolic cosine, and four related functions are defined by

$$\sinh x = \frac{e^x - e^{-x}}{2} \qquad \cosh x = \frac{e^x + e^{-x}}{2}$$

$$\tanh x = \frac{\sinh x}{\cosh x} \qquad \coth x = \frac{\cosh x}{\sinh x}$$

$$\operatorname{sech} x = \frac{1}{\cosh x} \qquad \operatorname{csch} x = \frac{1}{\sinh x}$$

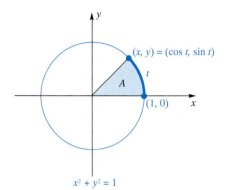

Figure 1

The terminology suggests that there must be some connection with the trigonometric functions; there is. First, the fundamental identity for the hyperbolic functions (reminiscent of $\cos^2 x + \sin^2 x = 1$ in trigonometry) is

$$\boxed{\cosh^2 x - \sinh^2 x = 1}$$

To verify it, we write

$$\cosh^2 x - \sinh^2 x = \frac{e^{2x} + 2 + e^{-2x}}{4} - \frac{e^{2x} - 2 + e^{-2x}}{4} = 1$$

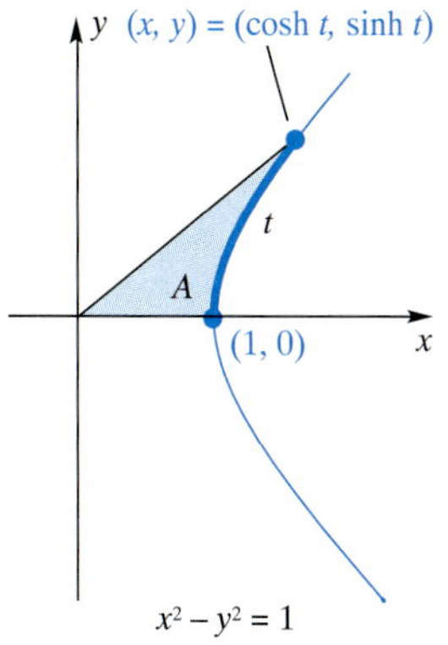

Second, recall that the trigonometric functions are intimately related to the unit circle (Figure 1), so much so that they are sometimes called the *circular functions*. In fact, the parametric equations $x = \cos t$, $y = \sin t$ describe the unit circle. In parallel fashion, the parametric equations $x = \cosh t$, $y = \sinh t$ describe the right branch of the unit hyperbola $x^2 - y^2 = 1$ (Figure 2). Moreover, in both cases the parameter t is related to the shaded area A by $t = 2A$, though this is not obvious in the second case (Problem 56).

Since $\sinh(-x) = -\sinh x$, sinh is an odd function; $\cosh(-x) = \cosh x$, so cosh is an even function. Correspondingly, the graph of $y = \sinh x$ is symmetric with respect to the origin and the graph of $y = \cosh x$ is symmetric with respect to the y-axis. Similarly, tanh is an odd function and sech is an even function. The graphs are shown in Figure 3.

Figure 2

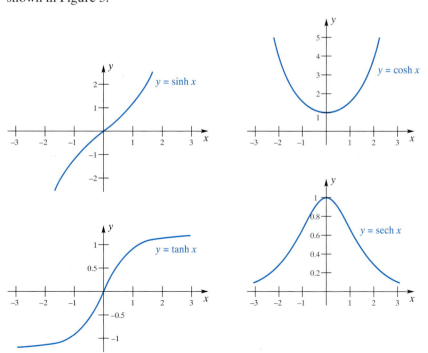

Figure 3

Derivatives of Hyperbolic Functions We can find $D_x \sinh x$ and $D_x \cosh x$ directly from the definitions.

$$D_x \sinh x = D_x\left(\frac{e^x - e^{-x}}{2}\right) = \frac{e^x + e^{-x}}{2} = \cosh x$$

and

$$D_x \cosh x = D_x\left(\frac{e^x + e^{-x}}{2}\right) = \frac{e^x - e^{-x}}{2} = \sinh x$$

Note that these facts confirm the character of the graphs that we drew. For example, since $D_x(\sinh x) = \cosh x > 0$, the graph of hyperbolic sine is always increasing. Similarly, $D_x^2(\cosh x) = \cosh x > 0$, which means that the graph of hyperbolic cosine is concave upward.

The derivatives of the other four hyperbolic functions follow from those for the first two, combined with the Quotient Rule. The results are summarized in Theorem A.

Theorem A Derivatives of Hyperbolic Functions

$$D_x \sinh x = \cosh x \qquad D_x \cosh x = \sinh x$$
$$D_x \tanh x = \text{sech}^2 x \qquad D_x \coth x = -\text{csch}^2 x$$
$$D_x \text{sech } x = -\text{sech } x \tanh x$$
$$D_x \text{csch } x = -\text{csch } x \coth x$$

Another way that the trigonometric and hyperbolic functions are connected concerns differential equations. The functions $\sin x$ and $\cos x$ are solutions to the second-order differential equation $y'' = -y$, and $\sinh x$ and $\cosh x$ are solutions to the differential equation $y'' = y$.

EXAMPLE 1 Find $D_x \tanh(\sin x)$.

Solution

$$D_x \tanh(\sin x) = \text{sech}^2(\sin x)D_x(\sin x)$$
$$= \cos x \cdot \text{sech}^2(\sin x) \qquad \blacksquare$$

EXAMPLE 2 Find $D_x \cosh^2(3x - 1)$.

Solution We apply the Chain Rule twice.

$$D_x \cosh^2(3x - 1) = 2\cosh(3x - 1)D_x \cosh(3x - 1)$$
$$= 2\cosh(3x - 1)\sinh(3x - 1)D_x(3x - 1)$$
$$= 6\cosh(3x - 1)\sinh(3x - 1) \qquad \blacksquare$$

EXAMPLE 3 Find $\int \tanh x \, dx$.

Solution Let $u = \cosh x$, so $du = \sinh x \, dx$.

$$\int \tanh x \, dx = \int \frac{\sinh x}{\cosh x} dx = \int \frac{1}{u} du$$

$$= \ln|u| + C = \ln|\cosh x| + C = \ln(\cosh x) + C$$

We could drop the absolute value signs because $\cosh x > 0$. $\qquad \blacksquare$

Inverse Hyperbolic Functions   Since hyperbolic sine and hyperbolic tangent have positive derivatives, they are increasing functions and automatically have inverses. To obtain inverses for hyperbolic cosine and hyperbolic secant, we restrict their domains to $x \geq 0$. Thus,

$$x = \sinh^{-1} y \quad \Leftrightarrow \quad y = \sinh x$$

$$x = \cosh^{-1} y \quad \Leftrightarrow \quad y = \cosh x \quad \text{and} \quad x \geq 0$$

$$x = \tanh^{-1} y \quad \Leftrightarrow \quad y = \tanh x$$

$$x = \operatorname{sech}^{-1} y \quad \Leftrightarrow \quad y = \operatorname{sech} x \quad \text{and} \quad x \geq 0$$

Since the hyperbolic functions are defined in terms of e^x and e^{-x}, it is not too surprising that the inverse hyperbolic functions can be expressed in terms of the natural logarithm. For example, consider $y = \cosh x$ for $x \geq 0$; that is, consider

$$y = \frac{e^x + e^{-x}}{2}, \quad x \geq 0$$

Our goal is to solve this equation for x, which will give $\cosh^{-1} y$. Multiplying both members by $2e^x$, we get $2ye^x = e^{2x} + 1$, or

$$\left(e^x\right)^2 - 2ye^x + 1 = 0, \quad x \geq 0$$

If we solve this quadratic equation in e^x, we obtain

$$e^x = \frac{2y + \sqrt{(2y)^2 - 4}}{2} = y + \sqrt{y^2 - 1}$$

The Quadratic Formula gives two solutions, the one given above and $\left(2y - \sqrt{(2y)^2 - 4}\right)/2$. This latter solution is extraneous because it is less than 1, whereas e^x is greater than 1 for all $x > 0$. Thus, $x = \ln\left(y + \sqrt{y^2 - 1}\right)$, so

$$x = \cosh^{-1} y = \ln\left(y + \sqrt{y^2 - 1}\right)$$

Similar arguments apply to each of the inverse hyperbolic functions. We obtain the following results (note that the roles of x and y have been interchanged). Figure 3 suggests the necessary domain restrictions. Graphs of the inverse hyperbolic functions are shown in Figure 4.

$$\sinh^{-1} x = \ln\left(x + \sqrt{x^2 + 1}\right)$$

$$\cosh^{-1} x = \ln\left(x + \sqrt{x^2 - 1}\right), \quad x \geq 1$$

$$\tanh^{-1} x = \frac{1}{2} \ln \frac{1 + x}{1 - x}, \quad -1 < x < 1$$

$$\operatorname{sech}^{-1} x = \ln\left(\frac{1 + \sqrt{1 - x^2}}{x}\right), \quad 0 < x \leq 1$$

Each of these functions is differentiable. In fact,

$$D_x \sinh^{-1} x = \frac{1}{\sqrt{x^2 + 1}}$$

$$D_x \cosh^{-1} x = \frac{1}{\sqrt{x^2 - 1}}, \quad x > 1$$

$$D_x \tanh^{-1} x = \frac{1}{1 - x^2}, \quad -1 < x < 1$$

$$D_x \operatorname{sech}^{-1} x = \frac{-1}{x\sqrt{1 - x^2}}, \quad 0 < x < 1$$

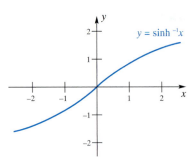

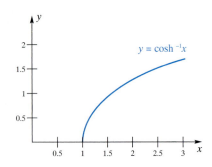

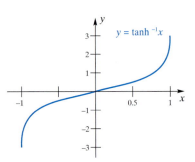

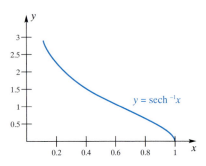

Figure 4

An Inverted Catenary

EXAMPLE 4 Show that $D_x \sinh^{-1} x = 1/\sqrt{x^2 + 1}$ by two different methods.

Solution

Method 1 Let $y = \sinh^{-1} x$, so

$$x = \sinh y$$

Now differentiate both sides with respect to x.

$$1 = (\cosh y)D_x y$$

Thus,

$$D_x y = D_x(\sinh^{-1} x) = \frac{1}{\cosh y} = \frac{1}{\sqrt{1 + \sinh^2 y}} = \frac{1}{\sqrt{1 + x^2}}$$

Method 2 Use the logarithmic expression for $\sinh^{-1} x$.

$$D_x(\sinh^{-1} x) = D_x \ln(x + \sqrt{x^2 + 1})$$

$$= \frac{1}{x + \sqrt{x^2 + 1}} D_x(x + \sqrt{x^2 + 1})$$

$$= \frac{1}{x + \sqrt{x^2 + 1}} \left(1 + \frac{x}{\sqrt{x^2 + 1}}\right)$$

$$= \frac{1}{\sqrt{x^2 + 1}}　\blacksquare$$

Applications: The Catenary　If a homogeneous flexible cable or chain is suspended between two fixed points at the same height, it forms a curve called a **catenary** (Figure 5). Furthermore (see Problem 53), a catenary can be placed in a coordinate system so that its equation takes the form

$$y = a \cosh \frac{x}{a}$$

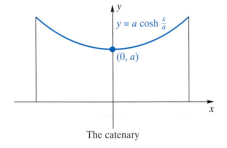

The catenary

Figure 5

EXAMPLE 5　Find the length of the catenary $y = a \cosh(x/a)$ between $x = -a$ and $x = a$.

Solution   The desired length (see Section 6.4) is given by

$$\int_{-a}^{a} \sqrt{1 + \left(\frac{dy}{dx}\right)^2}\, dx = \int_{-a}^{a} \sqrt{1 + \sinh^2\left(\frac{x}{a}\right)}\, dx$$

$$= \int_{-a}^{a} \sqrt{\cosh^2\left(\frac{x}{a}\right)}\, dx$$

$$= 2\int_{0}^{a} \cosh\left(\frac{x}{a}\right) dx$$

$$= 2a\int_{0}^{a} \cosh\left(\frac{x}{a}\right)\left(\frac{1}{a}\, dx\right)$$

$$= \left[2a\sinh\frac{x}{a}\right]_{0}^{a}$$

$$= 2a\sinh 1 \approx 2.35a \qquad \blacksquare$$

Concepts Review

1. sinh and cosh are defined by $\sinh x =$ _____ and $\cosh x =$ _____ .

2. In *hyperbolic* trigonometry, the identity corresponding to $\sin^2 x + \cos^2 x = 1$ is _____ .

3. Because of the identity in Question 2, the graph of the parametric equations $x = \cosh t$, $y = \sinh t$ is _____ .

4. The graph of $y = a\cosh(x/a)$ is a curve called a _____ ; this curve is important as a model for _____ .

Problem Set 7.8

In Problems 1–12, verify that the given equations are identities.

1. $e^x = \cosh x + \sinh x$ **2.** $e^{2x} = \cosh 2x + \sinh 2x$

3. $e^{-x} = \cosh x - \sinh x$

4. $e^{-2x} = \cosh 2x - \sinh 2x$

5. $\sinh(x + y) = \sinh x \cosh y + \cosh x \sinh y$

6. $\sinh(x - y) = \sinh x \cosh y - \cosh x \sinh y$

7. $\cosh(x + y) = \cosh x \cosh y + \sinh x \sinh y$

8. $\cosh(x - y) = \cosh x \cosh y - \sinh x \sinh y$

9. $\tanh(x + y) = \dfrac{\tanh x + \tanh y}{1 + \tanh x \tanh y}$

10. $\tanh(x - y) = \dfrac{\tanh x - \tanh y}{1 - \tanh x \tanh y}$

11. $\sinh 2x = 2\sinh x \cosh x$

12. $\cosh 2x = \cosh^2 x + \sinh^2 x$

In Problems 13–36, find $D_x y$.

13. $y = \sinh^2 x$ **14.** $y = \cosh^2 x$

15. $y = 5\sinh^2 x$ **16.** $y = \cosh^3 x$

17. $y = \cosh(3x + 1)$ **18.** $y = \sinh(x^2 + x)$

19. $y = \ln(\sinh x)$ **20.** $y = \ln(\coth x)$

21. $y = x^2 \cosh x$ **22.** $y = x^{-2}\sinh x$

23. $y = \cosh 3x \sinh x$ **24.** $y = \sinh x \cosh 4x$

25. $y = \tanh x \sinh 2x$ **26.** $y = \coth 4x \sinh x$

27. $y = \sinh^{-1}(x^2)$ **28.** $y = \cosh^{-1}(x^3)$

29. $y = \tanh^{-1}(2x - 3)$ **30.** $y = \coth^{-1}(x^5)$

31. $y = x\cosh^{-1}(3x)$ **32.** $y = x^2 \sinh^{-1}(x^5)$

33. $y = \ln(\cosh^{-1} x)$ **34.** $y = \cosh^{-1}(\cos x)$

35. $y = \tanh(\cot x)$ **36.** $y = \coth^{-1}(\tanh x)$

37. Find the area of the region bounded by $y = \cosh 2x$, $y = 0$, $x = 0$, and $x = \ln 3$.

In Problems 38–48, evaluate each integral.

38. $\displaystyle\int \sinh(3x + 2)\, dx$ **39.** $\displaystyle\int x\cosh(\pi x^2 + 5)\, dx$

40. $\displaystyle\int \frac{\cosh \sqrt{z}}{\sqrt{z}}\, dz$ **41.** $\displaystyle\int \frac{\sinh(2z^{1/4})}{\sqrt[4]{z^3}}\, dz$

42. $\displaystyle\int e^x \sinh e^x\, dx$ **43.** $\displaystyle\int \cos x \sinh(\sin x)\, dx$

44. $\displaystyle\int \tanh x \ln(\cosh x)\, dx$

45. $\displaystyle\int x\coth x^2 \ln(\sinh x^2)\, dx$

46. Find the area of the region bounded by $y = \cosh 2x$, $y = 0$, $x = -\ln 5$, and $x = \ln 5$.

47. Find the area of the region bounded by $y = \sinh x$, $y = 0$, and $x = \ln 2$.

48. Find the area of the region bounded by $y = \tanh x$, $y = 0$, $x = -8$, and $x = 8$.

49. The region bounded by $y = \cosh x$, $y = 0$, $x = 0$, and $x = 1$ is revolved about the x-axis. Find the volume of the resulting solid. *Hint*: $\cosh^2 x = (1 + \cosh 2x)/2$.

50. The region bounded by $y = \sinh x$, $y = 0$, $x = 0$, and $x = \ln 10$ is revolved about the x-axis. Find the volume of the resulting solid.

51. The curve $y = \cosh x$, $0 \le x \le 1$, is revolved about the x-axis. Find the area of the resulting surface.

52. The curve $y = \sinh x$, $0 \le x \le 1$, is revolved about the x-axis. Find the area of the resulting surface.

53. To derive the equation of a hanging cable (catenary), we consider the section AP from the lowest point A to a general point $P(x, y)$ (see Figure 6) and imagine the rest of the cable to have been removed.

The forces acting on the cable are:

1. H = horizontal tension pulling at A;
2. T = tangential tension pulling at P;
3. $W = \delta s$ = weight of s feet of cable of density δ pounds per foot.

To be in equilibrium, the horizontal and vertical components of T must just balance H and W, respectively. Thus, $T \cos \phi = H$ and $T \sin \phi = W = \delta s$, and so

$$\frac{T \sin \phi}{T \cos \phi} = \tan \phi = \frac{\delta s}{H}$$

But since $\tan \phi = dy/dx$, we get

$$\frac{dy}{dx} = \frac{\delta s}{H}$$

and therefore

$$\frac{d^2 y}{dx^2} = \frac{\delta}{H} \frac{ds}{dx} = \frac{\delta}{H} \sqrt{1 + \left(\frac{dy}{dx}\right)^2}$$

Now show that $y = a \cosh(x/a) + C$ satisfies this differential equation with $a = H/\delta$.

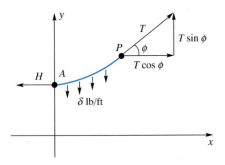

Figure 6

C **54.** Call the graph of $y = b - a \cosh(x/a)$ an inverted catenary and imagine it to be an arch sitting on the x-axis. Show that if the width of this arch along the x-axis is $2a$ then each of the following is true.

(a) $b = a \cosh 1 \approx 1.54308a$.

(b) The height of the arch is approximately $0.54308a$.

(c) The height of an arch of width 48 is approximately 13.

C **55.** A farmer built a large hayshed of length 100 feet and width 48 feet. A cross section has the shape of an inverted catenary (see Problem 54) with equation $y = 37 - 24 \cosh(x/24)$.

(a) Draw a picture of this shed.

(b) Find the volume of the shed.

(c) Find the surface area of the roof of the shed.

56. Show that $A = t/2$, where A denotes the area in Figure 2 of this section. *Hint*: At some point you will need to use Formula 44 from the back of the book.

57. Demonstrate for r any real number:

(a) $(\sinh x + \cosh x)^r = \sinh rx + \cosh rx$

(b) $(\cosh x - \sinh x)^r = \cosh rx - \sinh rx$

(c) $(\cos x + i \sin x)^r = \cos rx + i \sin rx$

(d) $(\cos x - i \sin x)^r = \cos rx - i \sin rx$

58. The **gudermannian** of t is defined by

$$gd(t) = \tan^{-1}(\sinh t)$$

Show that

(a) gd is odd and increasing with an inflection point at the origin;

(b) $gd(t) = \sin^{-1}(\tanh t) = \int_0^t \operatorname{sech} u \, du$

59. Show that the area under the curve $y = \cosh t$, $0 \le t \le x$, is numerically equal to its arc length.

60. Find the equation of the Gateway Arch in St. Louis, Missouri, given that it is an inverted catenary. Assume that it stands on the x-axis, that it is symmetric with respect to the y-axis, and that it is 630 feet wide at the base and 630 feet high at the center.

GC **61.** Draw the graphs of $y = \sinh x$, $y = \ln(x + \sqrt{x^2 + 1})$, and $y = x$ using the same axes and scaled so that $-3 \le x \le 3$ and $-3 \le y \le 3$. What does this demonstrate?

CAS **62.** Refer to Problem 58. Derive a formula for $gd^{-1}(x)$. Draw its graph and also that of $gd(x)$ using the same axes and thereby confirm your formula.

Answers to Concepts Review: **1.** $(e^x - e^{-x})/2$; $(e^x + e^{-x})/2$ **2.** $\cosh^2 x - \sinh^2 x = 1$ **3.** the graph of $x^2 - y^2 = 1$ (a hyperbola)   **4.** catenary; a hanging cable

7.9 Chapter Review

Concepts Test

Respond with true or false to each of the following assertions. Be prepared to justify your answer.

1. $\ln |x|$ is defined for all real x.

2. The graph of $y = \ln x$ has no inflection points.

3. $\int_1^{e^3} \frac{1}{t}\, dt = 3$

4. The graph of an invertible function $y = f(x)$ is intersected exactly once by every horizontal line.

5. The domain of $\ln^{-1}$ is the set of all real numbers.

6. $\ln x / \ln y = \ln x - \ln y$.

7. $(\ln x)^4 = 4 \ln x$.

8. $\ln(2e^{x+1}) - \ln(2e^x) = 1$ for all x in $\mathbb{R}$.

9. The functions $f(x) = 4 + e^x$ and $g(x) = \ln(x - 4)$ are inverses of each other.

10. $\exp x + \exp y = \exp(x + y)$.

11. If $x > y$, then $\ln x > \ln y$.

12. If $a \ln x < b \ln x$, then $a < b$.

13. If $a < b$, then $ae^x < be^x$.

14. If $a < b$, then $e^a < e^b$.

15. $\lim_{x \to 0^+} (\ln \sin x - \ln x) = 0$.

16. $\pi^{\sqrt{2}} = e^{\sqrt{2} \ln \pi}$

17. $\dfrac{d}{dx}(\ln \pi) = \dfrac{1}{\pi}$

18. $\int \dfrac{1}{x}\, dx = \ln 3|x| + C$

19. $D_x(x^e) = ex^{e-1}$

20. If $f(x) \cdot \exp[g(x)] = 0$ for $x = x_0$, then $f(x_0) = 0$.

21. $D_x(x^x) = x^x \ln x$.

22. $y = \tan x + \sec x$ is a solution of $2y' - y^2 = 1$.

23. An integrating factor for $y' + \dfrac{4}{x} y = e^x$ is x^4.

24. $\sin(\arcsin x) = x$ for all real numbers x.

25. $\arcsin(\sin x) = x$ for all real numbers x.

26. If $a < b$, then $\sinh a < \sinh b$.

27. If $a < b$, then $\cosh a < \cosh b$.

28. $\cosh x \le e^{|x|}$.

29. $|\sinh x| \le e^{|x|}/2$.

30. $\tan^{-1} x = \sin^{-1} x / \cos^{-1} x$.

31. $\cosh(\ln 3) = \frac{5}{6}$.

32. $\lim_{x \to 0} \ln\left(\dfrac{\sin x}{x}\right) = 1$.

33. $\lim_{x \to -\infty} \tan^{-1} x = -\dfrac{\pi}{2}$

34. $\sin^{-1}(\cosh x)$ is defined for all real x.

35. $f(x) = \tanh x$ is an odd function.

36. Both $y = \sinh x$ and $y = \cosh x$ satisfy the differential equation $y'' + y = 0$.

37. $\ln(3^{100}) > 100$.

38. $\ln(2x^2 - 18) - \ln(x - 3) - \ln(x + 3) = \ln 2$ for all x in $\mathbb{R}$.

39. If y is growing exponentially and if y triples between $t = 0$ and $t = t_1$, then y will also triple between $t = 2t_1$ and $t = 3t_1$.

40. The time necessary for $x(t) = Ce^{-kt}$ to drop to half its value is $\dfrac{\ln 2}{\ln k}$.

41. If $y'(t) = ky(t)$ and $z'(t) = kz(t)$, then $(y(t) + z(t))' = k(y(t) + z(t))$.

42. If $y_1(t)$ and $y_2(t)$ both satisfy $y'(t) = ky(t) + C$, then so does $(y_1(t) + y_2(t))$.

43. $\lim_{h \to 0}(1 - h)^{-1/h} = e^{-1}$.

44. It is to a saver's advantage to have money invested at 5% compounded continuously rather than 6% compounded monthly.

45. If $D_x(a^x) = a^x$ with $a > 0$, then $a = e$.

Sample Test Problems

In Problems 1–24, differentiate each function.

1. $\ln \dfrac{x^4}{2}$

2. $\sin^2(x^3)$

3. $e^{x^2 - 4x}$

4. $\log_{10}(x^5 - 1)$

5. $\tan(\ln e^x)$

6. $e^{\ln \cot x}$

7. $2 \tanh \sqrt{x}$

8. $\tanh^{-1}(\sin x)$

9. $\sinh^{-1}(\tan x)$

10. $2 \sin^{-1}\sqrt{3x}$

11. $\sec^{-1} e^x$

12. $\ln \sin^2\left(\dfrac{x}{2}\right)$

13. $3 \ln(e^{5x} + 1)$

14. $\ln(2x^3 - 4x + 5)$

15. $\cos e^{\sqrt{x}}$

16. $\ln(\tanh x)$

17. $2 \cos^{-1}\sqrt{x}$

18. $4^{3x} + (3x)^4$

19. $2 \csc e^{\ln \sqrt{x}}$

20. $(\log_{10} 2x)^{2/3}$

21. $4 \tan 5x \sec 5x$

22. $x \tan^{-1} \dfrac{x^2}{2}$

23. x^{1+x}

24. $(1 + x^2)^e$

In Problems 25–34, find the antiderivative of each function and verify your result by differentiation.

25. e^{3x-1}

26. $6 \cot 3x$

27. $e^x \sin e^x$

28. $\dfrac{6x + 3}{x^2 + x - 5}$

29. $\dfrac{e^{x+2}}{e^{x+3} + 1}$

30. $4x \cos x^2$

31. $\dfrac{4}{\sqrt{1-4x^2}}$

32. $\dfrac{\cos x}{1+\sin^2 x}$

33. $\dfrac{-1}{x+x(\ln x)^2}$

34. $\text{sech}^2(x-3)$

In Problems 35 and 36, find the intervals on which f is increasing and the intervals on which f is decreasing. Find where the graph of f is concave upward and where it is concave downward. Find any extreme values and points of inflection. Then sketch the graph of f.

35. $f(x)=\sin x+\cos x,\ \dfrac{-\pi}{2}\le x\le\dfrac{\pi}{2}$

36. $f(x)=\dfrac{x^2}{e^x},\ -\infty<x<\infty$

37. Let $f(x)=x^5+2x^3+4x,\ -\infty<x<\infty$.
(a) Prove that f has an inverse $g=f^{-1}$.
(b) Evaluate $g(7)=f^{-1}(7)$.
(c) Evaluate $g'(7)$.

38. A certain radioactive substance has a half-life of 10 years. How long will it take for 100 grams to decay to 1 gram?

39. If \$100 is put in the bank today at 12% interest, how much will it be worth at the end of 1 year if interest is compounded as indicated?
(a) annually
(b) monthly
(c) daily
(d) continuously

40. An airplane is flying horizontally at an altitude of 500 feet with a speed of 300 feet per second directly away from a searchlight on the ground. The searchlight is kept directed at the plane. At what rate is the angle between the light beam and the ground changing when this angle is $30°$?

41. Find the equation of the tangent line to $y=(\cos x)^{\sin x}$ at $(0,1)$.

42. A town grew exponentially from 10,000 in 1990 to 14,000 in 2000. Assuming that the same type of growth continues, what will the population be in 2010?

In Problems 43–47, solve each differential equations.

43. $\dfrac{dy}{dx}+\dfrac{y}{x}=0$

44. $\dfrac{dy}{dx}-\dfrac{x^2-2y}{x}=0$

45. $\dfrac{dy}{dx}+2x(y-1)=0;\ y=3$ when $x=0$.

46. $\dfrac{dy}{dx}-ay=e^{ax}$

47. $\dfrac{dy}{dx}-2y=e^x$

48. Suppose that glucose is infused into the bloodstream of a patient at the rate of 3 grams per minute, but that the patient's body converts and removes glucose from its blood at a rate proportional to the amount present (with constant of proportionality 0.02). Let $Q(t)$ be the amount present at time t, with $Q(0)=120$.
(a) Write the differential equation for Q.
(b) Solve this differential equation.
(c) Determine what happens to Q in the long run.

7.10 Additional Problems

EXPL *Here we explore compound interest. We saw in Section 7.5 that an amount of money A_0 in the bank at $100r\%$ interest per year compounded continuously grows to $A(t)=A_0e^{rt}$, when t is measured in years. We investigate other financial scenarios in Problems 1–5.*

1. Find the value of \$1000 at the end of 1 year when the interest is compounded continuously at 5%. This is called the **future value**.

2. Suppose that after 1 year you have \$1000 in the bank. If the interest was compounded continuously at 5%, how much money did you put in the bank one year ago? This is called the **present value**.

3. In many cases, as you earn money you deposit it in a bank account at regular time intervals. The money in the bank generally earns interest that is compounded continuously. Suppose that you deposit \$100 every 30 days and it is earning interest of 5% compounded continuously. After 60 days you would have $100e^{0.05\cdot(60/365)}+100e^{0.05\cdot(30/365)}$.
(a) How much money would you have after 360 days?
(b) Can you create a compact formula that gives this amount?

4. An **annuity** is money that you can take out of the bank at regular intervals. Often you want to calculate the amount of money that you must deposit at the start to be able to remove a given amount of money for a set number of months. Suppose that you want to remove \$144 every 30 days for the next 3,600 days (≈ 10 years) from your account, which is earning 5% interest. Assume that at the end of that time you will have no money left in the bank.

(a) Show that if interest is compounded every 30 days then the deposit A_0 needed would be
$$A_0=144\,\frac{1-[1+0.05\cdot(30/365)]^{-10\cdot(365/30)}}{0.05\cdot(30/365)}$$
(b) Find a formula for A_0 if the payments are made every 30 days, but the interest is compounded at 5% continuously.

5. The *rule of 70* gives a quick estimate of how long it takes your money to double when it is earning interest compounded continuously. The rule states that if it takes T years for money in the bank to double, then $T\approx 70/(\text{interest rate in percent})$.
(a) Show that T satisfies the equation $A_0e^{rT}=2A_0$, where r is the continuous rate of interest
(b) Show that $T\approx 70/(100r)$.
(c) Estimate how many years it would take money earning interest at 7% to double.

EXPL *Besides providing an easy way to differentiate products, the logarithmic differentiation also provides a measure of the **relative** or **fractional rate of change**, defined as y'/y. We explore this concept in Problems 6–10.*

6. Show that the relative rate of change of e^{kt} as a function of t is k.

7. Show that the relative rate of change of any polynomial approaches zero as the independent variable approaches infinity.

8. Prove that if the relative rate of change is a positive constant then the function must represent exponential growth.

9. Prove that if the relative rate of change is a negative constant then the function must represent exponential decay.

10. The **condition number** for a function provides a measure of how errors in the input are reflected as errors in the output. Suppose that the input to $f(x)$ at $x = 1.00$ is only known to ± 0.01. How is this margin of error reflected in the computation of $f(x)$?

(a) Show that the relative error in $f(x)$ due to a relative error in the input of size Δx is

$$\frac{f(x + \Delta x) - f(x)}{f(x)} = \frac{f(x + \Delta x) - f(x)}{\Delta x \cdot f(x)} \cdot \frac{(x + \Delta x) - x}{x} \cdot x$$

(b) By taking limits of the above expression, justify why one could say that if $\rho_f = $ *relative error in* $f(x)$, and $\rho_x = $ *relative error in x* then

$$\rho_f \approx x \cdot \frac{d \ln(f(x))}{dx} \cdot \rho_x$$

The quantity $x \cdot \dfrac{d \ln(f(x))}{dx}$ is called the *condition number* of the computation.

(c) Find the condition number for computing e^x.

Our graphing experience so far has been restricted to using standard (linear) coordinate spacings. When working with exponential and logarithmic functions it may be more instructive to use logarithmic and log–log scales. We explore these techniques in Problems 11 and 12.

[GC] **11.** On a single set of axes, use your calculator to draw the graphs of $y = 2^x$, $y = 3^x$, and $y = 4^x$ over the interval $0 < x < 4$. Do the same for the inverse functions $y = \log_2 x$, $y = \log_3 x$, and $y = \log_4 x$. If we use a computer graphing program that permits the use of semilog axes (a logarithmic scale on the y-axis and a normal scale on the x-axis) to graph the functions $y = 2^x$, $y = 3^x$, and $y = 4^x$ over the region $-5 < x < 5$ (Figure 1), we get three lines.

(a) Identify each of the curves in Figure 1.

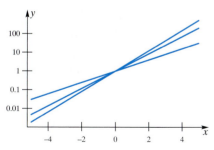

Figure 1

(b) Noting that, if $y = Cb^x$ then $\ln y = \ln C + x \ln b$, explain why all the curves in Figure 1 are lines through the point $(0, 1)$.

(c) Based on the semilog plot given by Figure 2, determine the C and b in the equation $y = Cb^x$.

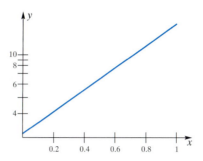

Figure 2

12. If we use log scaling for the x-axis as well as the y-axis (called a log–log plot) and graph several power functions, we will also get lines. Using the result that, upon taking logs, $y = Cx^r$ becomes $\log y = \log C + r \log x$, identify the equations that are graphed in Figure 3.

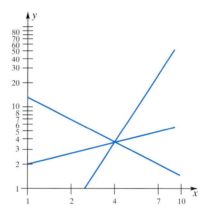

Figure 3

13. In 1895 an equation was derived by Korteweg and de Vries for a finite amplitude water wave that moves unchanged in form down a channel. Such waves, called **solitons**, play a very important role in fiber-optical signal propagation. If you move with the wave, the shape satisfies the nonlinear ordinary differential equation $\eta'(x) = \eta(x)\eta'(x) + \eta'''(x)$. This equation can be integrated once and then multiplied by $\eta'(x)$ and integrated again to give $3\eta^2 = \eta^3 + 3(\eta')^2$.

(a) Verify that the shape of the wave is given by the sech^2 function by showing by substitution that $\eta = 3\,\text{sech}^2(x/2)$ satisfies the equation.

[GC] (b) Plot the function $\eta = 3\,\text{sech}^2(x/2)$ for the domain $(-5, 5)$ and verify that it does look like a water wave.

Special Functions

I. Preparation

We have seen several functions that are defined by integrals. Technology Project 5.2 showed us the *sine integral* and the *Fresnel integrals*. In this chapter we defined the natural logarithm as an integral.

In many applications we have a continuous function f that does not possess an elementary function antiderivative and yet its antiderivative is needed. How can we proceed in this case? The First Fundamental Theorem of Calculus tells us that the function F defined by

$$F(x) = \int_a^x f(t)\, dt \qquad (1)$$

is a well-defined antiderivative. This is true whether or not an elementary antiderivative exists for f. The lower limit a is chosen as some fixed value compatible with the particular historical tradition for the specific application.

Exercise 1 We defined the natural logarithm function as

$$\ln x = \int_1^x \frac{1}{t}\, dt \qquad (2)$$

For the natural logarithm function, identify f, F and a as used in equation (1).

Exercise 2 From equation (2), evaluate $\dfrac{d}{dx}\ln x$ and $\ln 1$.

Another useful special function is the error function, abbreviated erf, which is defined by

$$\operatorname{erf}(x) = \frac{2}{\sqrt{\pi}} \int_0^x e^{-t^2}\, dt \qquad (3)$$

Exercise 3 For the error function, identify f, F, and a as used in equation (1).

Exercise 4 Use equation (3) to evaluate $\dfrac{d}{dx}\operatorname{erf}(x)$ and $\operatorname{erf}(0)$.

II. Using Technology

Exercise 5 A special function arising in probability is

$$P(x) = \frac{1}{\sqrt{2\pi}} \int_0^x e^{-t^2/2}\, dt$$

Use your CAS to evaluate the integral above. What is the result?

Exercise 6 If your system gave an answer in terms of the error function, make the appropriate change of variable to verify that the answer is correct.

Exercise 7 Another special function arising in probability is

$$Q(x) = \frac{1}{2} + \frac{1}{\sqrt{2\pi}} \int_0^x e^{-t^2/2}\, dt$$

This function has domain $(-\infty, \infty)$, and it gives the area under the curve $y = (1/\sqrt{2\pi})e^{-t^2/2}$ to the left of x. Plot $Q(x)$ and $Q'(x)$.

Exercise 8 Estimate the following limits

$$\lim_{x \to \infty} \operatorname{erf}(x)$$

$$\lim_{x \to \infty} Q(x)$$

$$\lim_{x \to -\infty} Q(x)$$

by evaluating erf(2), erf(3), erf(4), erf(5), erf(6), and so on. *Hint*: The exact answers are simple integers.

III. Reflection

Exercise 9 Explain why $Q(x)$ defined in Exercise 7 has an inverse. ($Q^{-1}(y)$ gives the x-value that makes the region under the curve $y = (1/\sqrt{2\pi})e^{-t^2/2}$ to the left of x have an area of y.)

Exercise 10

(a) What are the domain and range of $Q^{-1}(x)$?

(b) Sketch by hand the graph of $Q^{-1}(x)$.

(c) Find $(Q^{-1})'(0)$.

Exercise 11 Define

$$G(x) = \int_{-x}^x \frac{\cos t}{1 + t^2}\, dt$$

Derive as many properties of this function as you can. (For example, its derivative, $G(0)$, symmetry, limit as $x \to \infty$, existence of an inverse, and so on.)

The bible for special functions is *Handbook of Mathematical Functions with Formulas, Graphs, and Mathematical Tables*, edited by Milton Abramowitz and Irene A. Stegun. This large book has over 20 chapters discussing the special functions common in scientific applications, many of which are defined by integrals. It is cited a *vast* number of times in the scientific literature and you may want to consider buying your own copy of the paperback edition.

Population Growth and Least Squares

I. Preparation

A population that grows exponentially satisfies $P = P_0 e^{kt}$, where P_0 is the population size at time $t = 0$ and k is the rate of growth. If we take the logarithm of both sides, we obtain

$$\ln P = \ln P_0 + kt$$

This means that the graph of $y = \ln P$ against t is a line with slope k and intercept $\ln P_0$. This fact can be used to approximate the growth constant k and the initial population size P_0.

Exercise 1 The following table shows a population size for $t = 1, 2, 3, 4$. Use a calculator to compute the natural logarithm (to two decimal places) of each population size and fill in the table. Then plot the points $(t, \ln P)$.

t	P	$y = \ln P$
1	75	
2	141	
3	264	
4	523	

Exercise 2 Consider the data in Exercise 1.

(a) Sketch a line that comes close to fitting these four points.

(b) Approximate the slope and the intercept.

(c) Use your approximation from part (b) to find P in terms of t.

(d) Use your answer to (c) to predict the population size for $t = 5$.

II. Using Technology

Exercise 3 In Section 4.4 we stated that the least-squares line through the points (x_i, y_i) has slope

$$b = \frac{\sum_{i=1}^{n} x_i y_i - \frac{1}{n}\left(\sum_{i=1}^{n} x_i\right)\left(\sum_{i=1}^{n} y_i\right)}{\sum_{i=1}^{n} x_i^2 - \frac{1}{n}\left(\sum_{i=1}^{n} x_i\right)^2}$$

and intercept

$$a = \frac{1}{n}\sum_{i=1}^{n} y_i - \frac{b}{n}\sum_{i=1}^{n} x_i$$

Use these results to find the least-squares line for the data in Exercise 1, and plot the least-squares line on the scatter plot.

Use your least-squares line to express P as a function of t, and predict the population size for time $t = 5$.

Exercise 4 The following table shows the size of the U.S. population P (in millions) from 1790 to 1990.

Year	t	P	$y = \ln P$	Year	t	P	$y = \ln P$
1790	1	3.9		1900	12	76.0	
1800	2	5.3		1910	13	92.0	
1810	3	7.2		1920	14	106	
1820	4	9.6		1930	15	123	
1830	5	12.8		1940	16	132	
1840	6	17.1		1950	17	151	
1850	7	23.2		1960	18	179	
1860	8	31.4		1970	19	203	
1870	9	39.8		1980	20	226	
1880	10	50.2		1990	21	249	
1890	11	62.9					

Compute $y = \ln P$ and plot the ordered pairs (t, y). Find the least-squares line and plot it on the scatter plot. Use your linear regression line to express P as a function of t, and use this relationship to predict the U.S. population size for the year 2010.

III. Reflection

Exercise 5 For the U.S. population data, graph the scatter plot of (t, P) and, in the same graph window, plot the curve that predicts population size P as a function of t. Discuss how well the model fits the data, especially for recent years.

Exercise 6 The scatter plot of (t, y) suggests that the slope changed in the early twentieth century. Use this scatter plot to estimate when this change occurred. Then use data from that point onward to predict the U.S. population size for the year 2010. Compare your answer with the answer to Exercise 4.

8

Techniques of Integration

8.1
Integration by Substitution

Our repertoire of functions now includes all the elementary functions. These are the constant functions, the power functions, the logarithmic and exponential functions, the trigonometric and inverse trigonometric functions, and all functions obtained from them by addition, subtraction, multiplication, division, and composition. Thus,

$$f(x) = \frac{e^x + e^{-x}}{2} = \cosh x$$

$$g(x) = \left(1 + \cos^4 x\right)^{1/2}$$

$$h(x) = \frac{3^{x^2 - 2x}}{\ln(x^2 + 1)} - \sin\left[\cos(\cosh x)\right]$$

are elementary functions.

Differentiation of an elementary function is straightforward, requiring only a systematic use of the rules that we have learned. And the result is always an elementary function. Integration (antidifferentiation) is a far different matter. It involves a few techniques and a large bag of tricks; what is worse, it does not always yield an elementary function. For example, it is known that the antiderivatives of e^{-x^2} and $(\sin x)/x$ are not elementary functions.

The two principal techniques for integration are *substitution* (Sections 8.1–8.3) and *integration by parts* (Section 8.4). The method of substitution was introduced in Sections 5.1 and 5.8; we have used it occasionally in the intervening chapters. Here, we review the method and apply it in a wide variety of situations.

Finally in Section 8.5, we discuss the problem of integrating a rational function, that is, a quotient of two polynomial functions. We learn that, in theory, we can always perform such an integration and the result will be an elementary function.

Standard Forms Effective use of the method of substitution depends on the ready availability of a list of known integrals. One such list (but too long to memorize) appears inside the back cover of this book. The short list shown below is so useful that we think that every calculus student should memorize it.

Standard Integral Forms

Constants, Powers

1. $\displaystyle\int k\,du = ku + C$

2. $\displaystyle\int u^r\,du = \begin{cases} \dfrac{u^{r+1}}{r+1} + C & r \neq -1 \\ \ln|u| + C & r = -1 \end{cases}$

Exponentials

3. $\displaystyle\int e^u\,du = e^u + C$

4. $\displaystyle\int a^u\,du = \dfrac{a^u}{\ln a} + C, a \neq 1, a > 0$

Trigonometric Functions

5. $\displaystyle\int \sin u\,du = -\cos u + C$

6. $\displaystyle\int \cos u\,du = \sin u + C$

7. $\displaystyle\int \sec^2 u\,du = \tan u + C$

8. $\displaystyle\int \csc^2 u\,du = -\cot u + C$

9. $\displaystyle\int \sec u \tan u\,du = \sec u + C$

10. $\displaystyle\int \csc u \cot u\,du = -\csc u + C$

11. $\displaystyle\int \tan u\,du = -\ln|\cos u| + C$

12. $\displaystyle\int \cot u\,du = \ln|\sin u| + C$

Algebraic Functions

13. $\displaystyle\int \dfrac{du}{\sqrt{a^2 - u^2}} = \sin^{-1}\left(\dfrac{u}{a}\right) + C$

14. $\displaystyle\int \dfrac{du}{a^2 + u^2} = \dfrac{1}{a}\tan^{-1}\left(\dfrac{u}{a}\right) + C$

15. $\displaystyle\int \dfrac{du}{u\sqrt{u^2 - a^2}} = \dfrac{1}{a}\sec^{-1}\left(\dfrac{|u|}{a}\right) + C = \dfrac{1}{a}\cos^{-1}\left(\dfrac{a}{|u|}\right) + C$

16. $\displaystyle\int \sinh u\,du = \cosh u + C$

17. $\displaystyle\int \cosh u\,du = \sinh u + C$

Substitution in Indefinite Integrals Suppose that you face an indefinite integration. If it is a standard form, simply write the answer. If not, look for a substitution that will change it to a standard form. If the first substitution that you try does not work, try another. Skill at this, like most worthwhile activities, depends on practice.

The method of substitution was given in Theorem 5.8A and is restated here for easy reference.

Theorem A Substitution in Indefinite Integrals

Let g be a differentiable function and suppose that F is an antiderivative of f. Then, if $u = g(x)$,

$$\int f(g(x))g'(x)\,dx = \int f(u)\,du = F(u) + C = F(g(x)) + C$$

EXAMPLE 1 Find $\displaystyle\int \dfrac{x}{\cos^2(x^2)}\,dx$.

Solution Stare at this integral for a few moments. Since $1/\cos^2 x = \sec^2 x$, you may be reminded of the standard form $\int \sec^2 u\,du$. Let $u = x^2$, $du = 2x\,dx$. Then

$$\int \frac{x}{\cos^2(x^2)}\,dx = \frac{1}{2}\int \frac{1}{\cos^2(x^2)}\cdot 2x\,dx = \frac{1}{2}\int \sec^2 u\,du$$

$$= \frac{1}{2}\tan u + C = \frac{1}{2}\tan^2(x^2) + C \qquad \blacksquare$$

EXAMPLE 2   Find $\displaystyle\int \frac{3}{\sqrt{5-9x^2}}\,dx$.

Solution   Think of $\displaystyle\int \frac{du}{\sqrt{a^2-u^2}}$. Let $u = 3x$, so $du = 3\,dx$. Then,

$$\int \frac{3}{\sqrt{5-9x^2}}\,dx = \int \frac{1}{\sqrt{5-u^2}}\,du = \sin^{-1}\!\left(\frac{u}{\sqrt{5}}\right) + C$$

$$= \sin^{-1}\!\left(\frac{3x}{\sqrt{5}}\right) + C \qquad \blacksquare$$

EXAMPLE 3   Find $\displaystyle\int \frac{6e^{1/x}}{x^2}\,dx$.

Solution   Think of $\int e^u\,du$. Let $u = 1/x$, so $du = (-1/x^2)\,dx$. Then,

$$\int \frac{6e^{1/x}}{x^2}\,dx = -6\int e^{1/x}\!\left(\frac{-1}{x^2}\,dx\right) = -6\int e^u\,du$$

$$= -6e^u + C = -6e^{1/x} + C \qquad \blacksquare$$

EXAMPLE 4   Find $\displaystyle\int \frac{e^x}{4+9e^{2x}}\,dx$.

Solution   Think of $\displaystyle\int \frac{1}{a^2+u^2}\,du$. Let $u = 3e^x$, so $du = 3e^x\,dx$. Then,

$$\int \frac{e^x}{4+9e^{2x}}\,dx = \frac{1}{3}\int \frac{1}{4+9e^{2x}}\,(3e^x\,dx) = \frac{1}{3}\int \frac{1}{4+u^2}\,du$$

$$= \frac{1}{3}\cdot\frac{1}{2}\tan^{-1}\!\left(\frac{u}{2}\right) + C = \frac{1}{6}\tan^{-1}\!\left(\frac{3e^x}{2}\right) + C \qquad \blacksquare$$

No law says that you have to write out the u-substitution. If you can do the substitution mentally, that is fine. Here are two illustrations.

EXAMPLE 5   Find $\displaystyle\int x^3\sqrt{x^4+11}\,dx$.

Solution   Mentally, substitute $u = x^4 + 11$.

$$\int x^3\sqrt{x^4+11}\,dx = \frac{1}{4}\int (x^4+11)^{1/2}(4x^3\,dx)$$

$$= \frac{1}{6}(x^4+11)^{3/2} + C$$

EXAMPLE 6   Find $\displaystyle\int \frac{a^{\tan t}}{\cos^2 t}\,dt$.

Solution   Mentally, substitute $u = \tan t$.

$$\int \frac{a^{\tan t}}{\cos^2 t}\,dt = \int a^{\tan t}(\sec^2 t\,dt) = \frac{a^{\tan t}}{\ln a} + C \qquad \blacksquare$$

Manipulating the Integrand   Before you make a substitution, you may find it helpful to rewrite the integrand in a more convenient form. Integrals with quadratic expressions in the denominator can often be reduced to standard forms by *completing the square*. Recall that $x^2 + bx$ becomes a perfect square by the addition of $(b/2)^2$.

EXAMPLE 7   Find $\displaystyle\int \frac{7}{x^2 - 6x + 25}\, dx$

Solution

$$\int \frac{7}{x^2 - 6x + 25}\, dx = \int \frac{7}{x^2 - 6x + 9 + 16}\, dx$$

$$= 7 \int \frac{1}{(x - 3)^2 + 4^2}\, dx$$

$$= \frac{7}{4} \tan^{-1}\left(\frac{x - 3}{4}\right) + C$$

We made the mental substitution $u = x - 3$ at the final stage.   ■

When the integrand is the quotient of two polynomials (that is, a rational function) and the numerator is of equal or greater degree than the denominator, always *divide the denominator into the numerator first*.

EXAMPLE 8   Find $\displaystyle\int \frac{x^2 - x}{x + 1}\, dx$.

Solution   By long division (Figure 1),

$$\frac{x^2 - x}{x + 1} = x - 2 + \frac{2}{x + 1}$$

Hence,

$$\int \frac{x^2 - x}{x + 1}\, dx = \int (x - 2)\, dx + 2\int \frac{1}{x + 1}\, dx$$

$$= \frac{x^2}{2} - 2x + 2\int \frac{1}{x + 1}\, dx$$

$$= \frac{x^2}{2} - 2x + 2\ln|x + 1| + C$$ ■

$$
\begin{array}{r}
x - 2 \\
x + 1\,\overline{)\,x^2 - x} \\
\underline{x^2 + x} \\
-2x \\
\underline{-2x - 2} \\
2
\end{array}
$$

Figure 1

Substitution in Definite Integrals   This topic was covered in Section 5.8. It is just like substitution in indefinite integrals, but we must remember to make the appropriate change in the limits of integration.

EXAMPLE 9   Evaluate $\displaystyle\int_2^5 t\sqrt{t^2 - 4}\, dt$.

Solution   Let $u = t^2 - 4$, so $du = 2t\, dt$; note that when $t = 2, u = 0$, and when $t = 5, u = 21$. Thus,

$$\int_2^5 t\sqrt{t^2 - 4}\, dt = \frac{1}{2} \int_2^5 (t^2 - 4)^{1/2}(2t\, dt)$$

$$= \frac{1}{2} \int_0^{21} u^{1/2}\, du$$

$$= \left[\frac{1}{3} u^{3/2}\right]_0^{21} = \frac{1}{3}(21)^{3/2} \approx 32.08$$ ■

Use of Tables of Integrals   Our list of standard forms is very short (17 formulas); the list inside the back cover of this book is longer (113 formulas) and potentially more useful. Notice that the integrals listed there are grouped according to type. We illustrate the use of this list.

EXAMPLE 10   Find $\displaystyle\int \sqrt{6x - x^2}\, dx$ and $\displaystyle\int_0^{\pi/2} (\cos x)\sqrt{6 \sin x - \sin^2 x}\, dx$.

Solution   We use Formula 102 with $a = 3$.

$$\int \sqrt{6x - x^2}\, dx = \frac{x - 3}{2} \sqrt{6x - x^2} + \frac{9}{2} \sin^{-1}\left(\frac{x - 3}{3}\right) + C$$

In the second integral, let $u = \sin x$, so $du = \cos x\, dx$. Then apply Formula 102 again.

$$\int_0^{\pi/2} \cos x \sqrt{6 \sin x - \sin^2 x}\, dx = \int_0^1 \sqrt{6u - u^2}\, du$$

$$= \left[\frac{u - 3}{2}\sqrt{6u - u^2} + \frac{9}{2}\sin^{-1}\left(\frac{u - 3}{3}\right)\right]_0^1$$

$$= -\sqrt{5} + \frac{9}{2}\sin^{-1}\left(\frac{-2}{3}\right) - \frac{9}{2}\sin^{-1}(-1)$$

$$\approx 1.55 \qquad\blacksquare$$

Much more extensive tables of integrals may be found in most libraries. One of the better known is *Standard Mathematical Tables*, published by the Chemical Rubber Company.

Concepts Review

1. The substitution $u = 1 + x^3$ transforms $\displaystyle\int 3x^2(1 + x^3)^5\, dx$ to _____.

2. The substitution $u =$ _____ transforms $\displaystyle\int e^x/(4 + e^{2x})\, dx$ to $\displaystyle\int 1/(4 + u^2)\, du$.

3. The substitution $u = 1 + \sin x$ transforms $\displaystyle\int_0^{\pi/2} (1 + \sin x)^3 \cos x\, dx$ to _____.

4. To evaluate the integral $\displaystyle\int 1/(x^2 + 2x + 2)\, dx$, we should _____ in the denominator.

Problem Set 8.1

In Problems 1–54, perform the indicated integrations.

1. $\displaystyle\int (x - 2)^5\, dx$

2. $\displaystyle\int \sqrt{3x}\, dx$

3. $\displaystyle\int_0^2 x(x^2 + 1)^5\, dx$

4. $\displaystyle\int_0^1 x\sqrt{1 - x^2}\, dx$

5. $\displaystyle\int \frac{dx}{x^2 + 4}$

6. $\displaystyle\int \frac{e^x}{2 + e^x}\, dx$

7. $\displaystyle\int \frac{x}{x^2 + 4}\, dx$

8. $\displaystyle\int \frac{2t^2}{2t^2 + 1}\, dt$

9. $\displaystyle\int 6z\sqrt{4 + z^2}\, dz$

10. $\displaystyle\int \frac{5}{\sqrt{2t + 1}}\, dt$

11. $\displaystyle\int \frac{\tan z}{\cos^2 z}\, dz$

12. $\displaystyle\int e^{\cos z} \sin z\, dz$

13. $\displaystyle\int \frac{\sin \sqrt{t}}{\sqrt{t}}\, dt$

14. $\displaystyle\int \frac{2x\, dx}{\sqrt{1 - x^4}}$

15. $\displaystyle\int_0^{\pi/4} \frac{\cos x}{1 + \sin^2 x}\, dx$

16. $\displaystyle\int_0^{3/4} \frac{\sin\sqrt{1 - x}}{\sqrt{1 - x}}\, dx$

17. $\displaystyle\int \frac{3x^2 + 2x}{x + 1}\, dx$

18. $\displaystyle\int \frac{x^3 + 7x}{x - 1}\, dx$

19. $\displaystyle\int \frac{\sin(\ln 4x^2)}{x}\, dx$

20. $\displaystyle\int \frac{\sec^2(\ln x)}{2x}\, dx$

21. $\displaystyle\int \frac{6e^x}{\sqrt{1 - e^{2x}}}\, dx$

22. $\displaystyle\int \frac{x}{x^4 + 4}\, dx$

23. $\displaystyle\int \frac{3e^{2x}}{\sqrt{1 - e^{2x}}}\, dx$

24. $\displaystyle\int \frac{x^3}{x^4 + 4}\, dx$

25. $\displaystyle\int_0^1 t\, 3^{t^2}\, dt$

26. $\displaystyle\int_0^{\pi/6} 2^{\cos x} \sin x\, dx$

27. $\displaystyle\int \frac{\sin x - \cos x}{\sin x}\, dx$

28. $\displaystyle\int \frac{\sin(4t - 1)}{1 - \sin^2(4t - 1)}\, dt$

29. $\displaystyle\int e^x \sec e^x\, dx$

30. $\displaystyle\int e^x \sec^2(e^x)\, dx$

31. $\displaystyle\int \frac{\sec^3 x + e^{\sin x}}{\sec x}\, dx$

32. $\displaystyle\int \frac{(6t - 1)\sin\sqrt{3t^2 - t - 1}}{\sqrt{3t^2 - t - 1}}\, dt$

33. $\displaystyle\int \frac{t^2 \cos(t^3 - 2)}{\sin^2(t^3 - 2)}\, dt$

34. $\displaystyle\int \frac{1 + \cos 2x}{\sin^2 2x}\, dx$

35. $\displaystyle\int \frac{t^2 \cos^2(t^3 - 2)}{\sin^2(t^3 - 2)}\, dt$

36. $\displaystyle\int \frac{\csc^2 2t}{\sqrt{1 + \cot 2t}}\, dt$

37. $\displaystyle\int \frac{e^{\tan^{-1} 2t}}{1 + 4t^2}\, dt$

38. $\displaystyle\int (t + 1)e^{-t^2 - 2t - 5}\, dt$

39. $\displaystyle\int \frac{y}{\sqrt{16 - 9y^4}}\, dy$

40. $\displaystyle\int \cosh 3x\, dx$

41. $\displaystyle\int x^2 \sinh x^3\, dx$

42. $\displaystyle\int \frac{5}{\sqrt{9 - 4x^2}}\, dx$

43. $\displaystyle\int \frac{e^{3t}}{\sqrt{4 - e^{6t}}}\, dt$

44. $\displaystyle\int \frac{dt}{2t\sqrt{4t^2 - 1}}$

45. $\displaystyle\int_0^{\pi/2} \frac{\sin x}{16 + \cos^2 x}\, dx$

46. $\displaystyle\int_0^1 \frac{e^{2x} - e^{-2x}}{e^{2x} + e^{-2x}}\, dx$

47. $\displaystyle\int \frac{1}{x^2 + 2x + 5}\, dx$

48. $\displaystyle\int \frac{1}{x^2 - 4x + 9}\, dx$

49. $\displaystyle\int \frac{dx}{9x^2 + 18x + 10}$

50. $\displaystyle\int \frac{dx}{\sqrt{16 + 6x - x^2}}$

51. $\displaystyle\int \frac{x + 1}{9x^2 + 18x + 10}\, dx$

52. $\displaystyle\int \frac{3 - x}{\sqrt{16 + 6x - x^2}}\, dx$

53. $\displaystyle\int \frac{dt}{t\sqrt{2t^2 - 9}}$

54. $\displaystyle\int \frac{\tan x}{\sqrt{\sec^2 x - 4}}\, dx$

In Problems 55–66, use the table of integrals on the inside back cover, perhaps combined with a substitution, to perform the indicated integrations.

55. $\displaystyle\int x\sqrt{3x + 2}\, dx$

56. $\displaystyle\int 2t\sqrt{3 - 4t}\, dt$

57. $\displaystyle\int \frac{dx}{9 - 16x^2}$

58. $\displaystyle\int \frac{dx}{5x^2 - 11}$

59. $\displaystyle\int x^2\sqrt{9 - 2x^2}\, dx$

60. $\displaystyle\int \frac{\sqrt{16 - 3t^2}}{t}\, dt$

61. $\displaystyle\int \frac{dx}{\sqrt{5 + 3x^2}}$

62. $\displaystyle\int t^2\sqrt{3 + 5t^2}\, dt$

63. $\displaystyle\int \frac{dt}{\sqrt{t^2 + 2t - 3}}$

64. $\displaystyle\int \frac{\sqrt{x^2 + 2x - 3}}{x + 1}\, dx$

65. $\displaystyle\int \frac{\sin t \cos t}{\sqrt{3 \sin t + 5}}\, dt$

66. $\displaystyle\int \frac{\sin x}{\cos x\sqrt{5 - 4\cos x}}\, dx$

67. Find the length of the curve $y = \ln(\cos x)$ between $x = 0$ and $x = \pi/4$.

68. Establish the identity

$$\sec x = \frac{\sin x}{\cos x} + \frac{\cos x}{1 + \sin x}$$

and then use it to derive the formula

$$\int \sec x\, dx = \ln|\sec x + \tan x| + C$$

69. Evaluate $\displaystyle\int_0^{2\pi} \frac{x|\sin x|}{1 + \cos^2 x}\, dx$. *Hint*: Make the substitution $u = x - \pi$ in the definite integral and then use symmetry properties.

70. Let R be the region bounded by $y = \sin x$ and $y = \cos x$ between $x = -\pi/4$ and $x = 3\pi/4$. Find the volume of the solid obtained when R is revolved about $x = -\pi/4$. *Hint*: Use cylindrical shells to write a single integral, make the substitution $u = x - \pi/4$, and apply symmetry properties.

Answers to Concepts Review: **1.** $\displaystyle\int u^5\, du$ **2.** e^x **3.** $\displaystyle\int_1^2 u^3\, du$

4. complete the square

8.2
Some Trigonometric Integrals

When we combine the method of substitution with a clever use of trigonometric identities, we can integrate a wide variety of trigonometric forms. We consider three commonly encountered types.

1. $\displaystyle\int \sin^n x \, dx$ and $\displaystyle\int \cos^n x \, dx$

2. $\displaystyle\int \sin^m x \cos^n x \, dx$

3. $\displaystyle\int \sin mx \cos nx \, dx, \quad \int \sin mx \sin nx \, dx, \quad \int \cos mx \cos nx \, dx$

Type 1 $\left(\int\sin^n x \, dx, \ \int\cos^n x \, dx \right)$ Consider first the case where n is an odd positive integer. After taking out either the factor $\sin x$ or $\cos x$, use the identity $\sin^2 x + \cos^2 x = 1$.

EXAMPLE 1

***n* Odd**   Find $\displaystyle\int \sin^5 x \, dx$.

Solution

$$\int \sin^5 x \, dx = \int \sin^4 x \sin x \, dx$$

$$= \int \left(1 - \cos^2 x\right)^2 \sin x \, dx$$

$$= \int \left(1 - 2\cos^2 x + \cos^4 x\right) \sin x \, dx$$

$$= -\int \left(1 - 2\cos^2 x + \cos^4 x\right)(-\sin x \, dx)$$

$$= -\cos x + \tfrac{2}{3}\cos^3 x - \tfrac{1}{5}\cos^5 x + C \qquad\blacksquare$$

EXAMPLE 2

***n* Even**   Find $\displaystyle\int \sin^2 x \, dx$ and $\displaystyle\int \cos^4 x \, dx$.

Solution   Here we make use of half-angle identities.

$$\int \sin^2 x \, dx = \int \frac{1 - \cos 2x}{2} \, dx$$

$$= \frac{1}{2} \int dx - \frac{1}{4} \int (\cos 2x)(2 \, dx)$$

$$= \frac{1}{2} x - \frac{1}{4} \sin 2x + C$$

$$\int \cos^4 x \, dx = \int \left(\frac{1 + \cos 2x}{2} \right)^2 dx$$

$$= \frac{1}{4} \int \left(1 + 2\cos 2x + \cos^2 2x\right) dx$$

$$= \frac{1}{4} \int dx + \frac{1}{4} \int (\cos 2x)(2) \, dx + \frac{1}{8} \int (1 + \cos 4x) \, dx$$

$$= \frac{3}{8} \int dx + \frac{1}{4} \int \cos 2x(2 \, dx) + \frac{1}{32} \int \cos 4x(4 \, dx)$$

$$= \frac{3}{8} x + \frac{1}{4} \sin 2x + \frac{1}{32} \sin 4x + C \qquad\blacksquare$$

Useful Identities

Some trigonometric identities needed in this section are the following.

Pythagorean Identities

$$\sin^2 x + \cos^2 x = 1$$
$$1 + \tan^2 x = \sec^2 x$$
$$1 + \cot^2 x = \csc^2 x$$

Half-Angle Identities

$$\sin^2 x = \frac{1 - \cos 2x}{2}$$

$$\cos^2 x = \frac{1 + \cos 2x}{2}$$

Type 2 ($\int \sin^m x \cos^n x \, dx$)   If either m or n is an odd positive integer and the other exponent is any number, we factor out $\sin x$ or $\cos x$ and use the identity $\sin^2 x + \cos^2 x = 1$.

EXAMPLE 3

m or n Odd   Find $\int \sin^3 x \cos^{-4} x \, dx$.

Solution

$$\int \sin^3 x \cos^{-4} x \, dx = \int (1 - \cos^2 x)(\cos^{-4} x)(\sin x) \, dx$$

$$= -\int (\cos^{-4} x - \cos^{-4} x)(-\sin x \, dx)$$

$$= -\left[\frac{(\cos x)^{-3}}{-3} - \frac{(\cos x)^{-1}}{-1} \right] + C$$

$$= \frac{1}{3} \sec^3 x - \sec x + C \qquad \blacksquare$$

If both m and n are even positive integers, we use half-angle identities to reduce the degree of the integrand. Example 4 gives an illustration.

EXAMPLE 4

Both m and n Even   Find $\int \sin^2 x \cos^4 x \, dx$.

Solution

$$\int \sin^2 x \cos^4 x \, dx$$

$$= \int \left(\frac{1 - \cos 2x}{2} \right) \left(\frac{1 + \cos 2x}{2} \right)^2 dx$$

$$= \frac{1}{8} \int (1 + \cos 2x - \cos^2 2x - \cos^3 2x) \, dx$$

$$= \frac{1}{8} \int \left[1 + \cos 2x - \frac{1}{2}(1 + \cos 4x) - (1 - \sin^2 2x) \cos 2x \right] dx$$

$$= \frac{1}{8} \int \left[\frac{1}{2} - \frac{1}{2} \cos 4x + \sin^2 2x \cos 2x \right] dx$$

$$= \frac{1}{8} \left[\int \frac{1}{2} \, dx - \frac{1}{8} \int \cos 4x (4 \, dx) + \frac{1}{2} \int \sin^2 2x (2 \cos 2x \, dx) \right]$$

$$= \frac{1}{8} \left[\frac{1}{2} x - \frac{1}{8} \sin 4x + \frac{1}{6} \sin^3 2x \right] + C \qquad \blacksquare$$

Type 3 ($\int \sin mx \cos nx \, dx$, $\int \sin mx \sin nx \, dx$, $\int \cos mx \cos nx \, dx$)   Integrals of this type occur in alternating current theory, heat-transfer problems, and in many other applications. To handle these integrals, we use the product identities.

1. $\sin mx \cos nx = \dfrac{1}{2} \big[\sin(m + n)x + \sin(m - n)x \big]$

2. $\sin mx \sin nx = -\dfrac{1}{2} \big[\cos(m + n)x - \cos(m - n)x \big]$

3. $\cos mx \cos nx = \dfrac{1}{2} \big[\cos(m + n)x + \cos(m - n)x \big]$

EXAMPLE 5   Find $\displaystyle\int \sin 2x \cos 3x \, dx$.

Solution   Apply product identity 1.

$$\int \sin 2x \cos 3x \, dx = \frac{1}{2} \int \left[\sin 5x + \sin(-x)\right] dx$$

$$= \frac{1}{10} \int \sin 5x (5 \, dx) - \frac{1}{2} \int \sin x \, dx$$

$$= -\frac{1}{10} \cos 5x + \frac{1}{2} \cos x + C \qquad \blacksquare$$

EXAMPLE 6   If m and n are positive integers, show that

$$\int_{-\pi}^{\pi} \sin mx \sin nx \, dx = \begin{cases} 0 & \text{if } n \neq m \\ \pi & \text{if } n = m \end{cases}$$

Solution   Apply product identity 2. If $m \neq n$, then

$$\int_{-\pi}^{\pi} \sin mx \sin nx \, dx = -\frac{1}{2} \int_{-\pi}^{\pi} \left[\cos(m+n)x - \cos(m-n)x\right] dx$$

$$= -\frac{1}{2} \left[\frac{1}{m+n} \sin(m+n)x - \frac{1}{m-n} \sin(m-n)x\right]_{-\pi}^{\pi}$$

$$= 0$$

If $m = n$,

$$\int_{-\pi}^{\pi} \sin mx \sin nx \, dx = -\frac{1}{2} \int_{-\pi}^{\pi} \left[\cos 2mx - 1\right] dx$$

$$= -\frac{1}{2} \left[\frac{1}{2m} \sin 2mx - x\right]_{-\pi}^{\pi}$$

$$= -\frac{1}{2} \left[-2\pi\right] = \pi \qquad \blacksquare$$

EXAMPLE 7   If m and n are positive integers, find

$$\int_{-L}^{L} \sin \frac{m\pi x}{L} \sin \frac{n\pi x}{L} \, dx$$

Solution   Let $u = \pi x/L$, $du = \pi \, dx/L$. If $x = -L$, then $u = -\pi$, and if $x = L$, then $u = \pi$. Thus,

$$\int_{-L}^{L} \sin \frac{m\pi x}{L} \sin \frac{n\pi x}{L} \, dx = \frac{L}{\pi} \int_{-\pi}^{\pi} \sin mu \sin nu \, du$$

$$= \begin{cases} \dfrac{L}{\pi} \cdot 0 & \text{if } n \neq m \\[2mm] \dfrac{L}{\pi} \cdot \pi & \text{if } n = m \end{cases}$$

$$= \begin{cases} 0 & \text{if } n \neq m \\ L & \text{if } n = m \end{cases}$$

Here we have used the result of Example 6. $\qquad \blacksquare$

A number of times in this book we have suggested that you should view things from both an algebraic and a geometric point of view. So far this section has been entirely algebraic, but with definite integrals such as those in Examples 6 and 7, we have an opportunity to view things geometrically.

Figure 1 shows graphs of $y = \sin(3x)\sin(2x)$ and $y = \sin(3\pi x/10)\sin(2\pi x/10)$. The graphs suggest that the areas above and below the x-axis are the same, leaving $A_{\text{up}} - A_{\text{down}} = 0$. Example 7 confirms this.

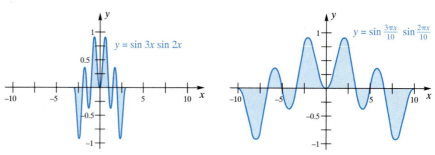

Figure 1

Figure 2 shows graphs of $y = \sin 2x \sin 2x = \sin^2 2x$, $-\pi \leq x \leq \pi$, and $y = \sin(2\pi x/10)\sin(2\pi x/10) = \sin^2(2\pi x/10)$, $-10 \leq x \leq 10$. These two graphs look the same, except the one on the right has been stretched horizontally by the factor $10/\pi$. Does it then make sense that the area will increase by this same factor? That would make the shaded area in the figure on the right equal to $10/\pi$ times the shaded area in the figure on the left; that is, the area on the right should be $(10/\pi) \cdot \pi = 10$, which corresponds to the result of Example 7 with $L = 10$.

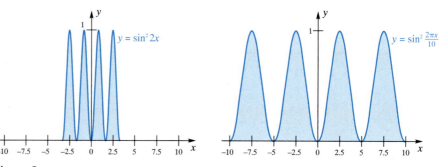

Figure 2

Concepts Review

1. To handle $\int \cos^2 x \, dx$, we first rewrite it as _____.

2. To handle $\int \cos^3 x \, dx$, we first rewrite it as _____.

3. To handle $\int \sin^2 x \cos^3 x \, dx$, we first rewrite it as _____.

4. To handle $\int_{-\pi}^{\pi} \cos mx \cos nx \, dx$, where $m \neq n$, we use the trigonometric identity _____.

Problem Set 8.2

In Problems 1–16, perform the indicated integrations.

1. $\int \sin^2 x \, dx$

2. $\int \sin^4 6x \, dx$

3. $\int \sin^3 x \, dx$

4. $\int \cos^3 x \, dx$

5. $\int_0^{\pi/2} \cos^5 \theta \, d\theta$

6. $\int_0^{\pi/2} \sin^6 \theta \, d\theta$

7. $\int \sin^5 4x \cos^2 4x \, dx$

8. $\int (\sin^3 2t)\sqrt{\cos 2t} \, dt$

9. $\int \cos^3 3\theta \sin^{-2} 3\theta \, d\theta$

10. $\int \sin^{1/2} 2z \cos^3 2z \, dz$

11. $\int \sin^4 3t \cos^4 3t \, dt$

12. $\int \cos^6 \theta \sin^2 \theta \, d\theta$

13. $\int \sin 4y \cos 5y \, dy$

14. $\int \cos y \cos 4y \, dy$

15. $\int \sin^4\left(\frac{w}{2}\right)\cos^2\left(\frac{w}{2}\right) dw$

16. $\int \sin 3t \sin t \, dt$

Integrals of the form $\int \tan^n x \, dx$ can be evaluated by factoring out $\tan^2 x = \sec^2 x - 1$, and integrals of the form $\int \cot^n x \, dx$ can be evaluated by factoring out $\cot^2 x = \csc^2 x - 1$. Use this method to evaluate the integrals in Problems 17–22.

17. $\int \tan^4 x \, dx$

18. $\int \cot^4 x \, dx$

19. $\int \tan^3 x \, dx$

20. $\int \cot^3 2t \, dt$

21. $\int \tan^5\left(\frac{\theta}{2}\right) d\theta$

22. $\int \cot^5 2t \, dt$

When n is even, integrals of the form $\int \tan^m x \sec^n x \, dx$ can be evaluated by factoring out $\sec^2 x = 1 + \tan^2 x$ and using the fact that $D_x \tan x = \sec^2 x$. When m is odd, integrals of this form can be evaluated by factoring out $\tan x \sec x$ and using the fact that $D_x \sec x = \sec x \tan x$. Use this method to evaluate the integrals in Problems 23–26.

23. $\int \tan^{-3} x \sec^4 x \, dx$ **24.** $\int \tan^{-3/2} x \sec^4 x \, dx$

25. $\int \tan^3 x \sec^2 x \, dx$ **26.** $\int \tan^3 x \sec^{-1/2} x \, dx$

27. Find $\int_{-\pi}^{\pi} \cos mx \cos nx \, dx$, $m \neq n$; m, n integers

28. Find $\int_{-L}^{L} \cos \frac{m\pi x}{L} \cos \frac{n\pi x}{L} \, dx$, $m \neq n$, m, n integers

29. The region bounded by $y = x + \sin x$, $y = 0$, $x = \pi$, is revolved about the x-axis. Find the volume of the resulting solid.

30. The region bounded by $y = \sin^2(x^2)$, $y = 0$, and $x = \sqrt{\pi}/2$ is revolved about the y-axis. Find the volume of the resulting solid.

31. Let $f(x) = \sum_{n=1}^{N} a_n \sin(nx)$. Use Example 6 to show each of the following.

(a) $\frac{1}{\pi} \int_{-\pi}^{\pi} f(x) \sin(mx) \, dx = \begin{cases} a_m & \text{if } m \leq N \\ 0 & \text{if } m > N \end{cases}$

(b) $\frac{1}{\pi} \int_{-\pi}^{\pi} f^2(x) \, dx = \sum_{n=1}^{N} a_n^2$

Note: Integrals of this type occur in a subject called *Fourier series*, which has applications to heat, vibrating strings, and other physical phenomena.

32. Show that

$$\lim_{n \to \infty} \cos \frac{x}{2} \cos \frac{x}{4} \cos \frac{x}{8} \cdots \cos \frac{x}{2^n} = \frac{\sin x}{x}$$

by completing the following steps.

(a) $\cos \frac{x}{2} \cos \frac{x}{4} \cdots \cos \frac{x}{2^n}$

$$= \left[\cos \frac{1}{2^n} x + \cos \frac{3}{2^n} x + \cdots + \cos \frac{2^n - 1}{2^n} x \right] \frac{1}{2^{n-1}}$$

(See Problem 46 of Section 2.3.)

(b) Recognize a Riemann sum leading to a definite integral.

(c) Evaluate the definite integral.

33. Use the result of Problem 32 to obtain the famous formula of François Viète (1540–1603):

$$\frac{2}{\pi} = \frac{\sqrt{2}}{2} \cdot \frac{\sqrt{2 + \sqrt{2}}}{2} \cdot \frac{\sqrt{2 + \sqrt{2 + \sqrt{2}}}}{2} \cdots$$

34. The shaded region (Figure 3) between one arch of $y = \sin x$, $0 \leq x \leq \pi$, and the line $y = k$, $0 \leq k \leq 1$, is revolved about the line $y = k$, generating a solid S. Determine k so that S has

(a) minimum volume and   (b) maximum volume.

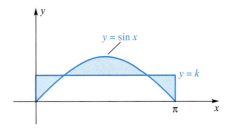

Figure 3

8.3
Rationalizing Substitutions

Radicals in an integrand are always troublesome and we usually try to get rid of them. Often an appropriate substitution will rationalize the integrand.

Integrands Involving $\sqrt[n]{ax + b}$ If $\sqrt[n]{ax + b}$ appears in an integral, the substitution $u = \sqrt[n]{ax + b}$ will eliminate the radical.

EXAMPLE 1   Find $\int \frac{dx}{x - \sqrt{x}}$.

Solution   Let $u = \sqrt{x}$, so $u^2 = x$ and $2u \, du = dx$. Then

$$\int \frac{dx}{x - \sqrt{x}} = \int \frac{2u}{u^2 - u} \, du = 2 \int \frac{1}{u - 1} \, du$$

$$= 2 \ln|u - 1| + C = 2 \ln|\sqrt{x} - 1| + C \qquad \blacksquare$$

EXAMPLE 2   Find $\int x\sqrt[3]{x-4}\,dx$.

Solution   Let $u = \sqrt[3]{x-4}$, so $u^3 = x - 4$ and $3u^2\,du = dx$. Then

$$\int x\sqrt[3]{x-4}\,dx = \int (u^3 + 4)u \cdot (3u^2\,du) = 3\int (u^6 + 4u^3)\,du$$

$$= 3\left[\frac{u^7}{7} + u^4\right] + C = \frac{3}{7}(x-4)^{7/3} + 3(x-4)^{4/3} + C \quad \blacksquare$$

EXAMPLE 3   Find $\int x\sqrt[5]{(x+1)^2}\,dx$.

Solution   Let $u = (x+1)^{1/5}$, so $u^5 = x + 1$ and $5u^4\,du = dx$. Then

$$\int x(x+1)^{2/5}\,dx = \int (u^5 - 1)u^2 \cdot 5u^4\,du$$

$$= 5\int (u^{11} - u^6)\,du = \tfrac{5}{12}u^{12} - \tfrac{5}{7}u^7 + C$$

$$= \tfrac{5}{12}(x+1)^{12/5} - \tfrac{5}{7}(x+1)^{7/5} + C \quad \blacksquare$$

Integrands Involving $\sqrt{a^2 - x^2}$, $\sqrt{a^2 + x^2}$, and $\sqrt{x^2 - a^2}$ To rationalize these three expressions, we may assume that a is positive and make the following trigonometric substitutions.

Radical	Substitution	Restriction on t
1. $\sqrt{a^2 - x^2}$	$x = a\sin t$	$-\pi/2 \le t \le \pi/2$
2. $\sqrt{a^2 + x^2}$	$x = a\tan t$	$-\pi/2 < t < \pi/2$
3. $\sqrt{x^2 - a^2}$	$x = a\sec t$	$0 \le t \le \pi, t \ne \pi/2$

Now note the simplifications that these substitutions achieve.

1. $\sqrt{a^2 - x^2} = \sqrt{a^2 - a^2\sin^2 t} = \sqrt{a^2\cos^2 t} = |a\cos t| = a\cos t$
2. $\sqrt{a^2 + x^2} = \sqrt{a^2 + a^2\tan^2 t} = \sqrt{a^2\sec^2 t} = |a\sec t| = a\sec t$
3. $\sqrt{x^2 - a^2} = \sqrt{a^2\sec^2 t - a^2} = \sqrt{a^2\tan^2 t} = |a\tan t| = \pm a\tan t$

The restrictions on t allowed us to remove the absolute value signs in the first two cases, but they also achieved something else. These restrictions are exactly the ones we introduced in Section 7.7 in order to make sine, tangent, and secant invertible. This means that we can solve the substitution equations for t in each case, and this will allow us to write our final answers in the following examples in terms of x.

EXAMPLE 4   Find $\int \sqrt{a^2 - x^2}\,dx$.

Solution   We make the substitution

$$x = a\sin t, \qquad -\frac{\pi}{2} \le t \le \frac{\pi}{2}$$

Then $dx = a\cos t\,dt$ and $\sqrt{a^2 - x^2} = a\cos t$. Thus,

$$\int \sqrt{a^2 - x^2}\, dx = \int a \cos t \cdot a \cos t\, dt = a^2 \int \cos^2 t\, dt$$

$$= \frac{a^2}{2} \int (1 + \cos 2t)\, dt$$

$$= \frac{a^2}{2} \left(t + \frac{1}{2} \sin 2t \right) + C$$

$$= \frac{a^2}{2} (t + \sin t \cos t) + C$$

Now, $x = a \sin t$ is equivalent to $x/a = \sin t$ and, since t was restricted to make the sine function invertible,

$$t = \sin^{-1}\left(\frac{x}{a} \right)$$

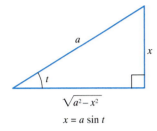

$x = a \sin t$

Figure 1

Using the right triangle in Figure 1 (as we did in Section 7.7), we see that

$$\cos t = \cos \left[\sin^{-1}\left(\frac{x}{a} \right) \right] = \sqrt{1 - \frac{x^2}{a^2}} = \frac{1}{a} \sqrt{a^2 - x^2}$$

Thus,

$$\int \sqrt{a^2 - x^2}\, dx = \frac{a^2}{2} \sin^{-1}\left(\frac{x}{a} \right) + \frac{x}{2} \sqrt{a^2 - x^2} + C \qquad \blacksquare$$

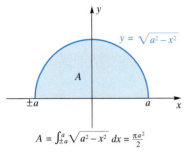

$$A = \int_{-a}^{a} \sqrt{a^2 - x^2}\, dx = \frac{\pi a^2}{2}$$

Figure 2

The result in Example 4 allows us to calculate the following definite integral, which represents the area of a semicircle (Figure 2). Thus, calculus confirms a result that we already know.

$$\int_{-a}^{a} \sqrt{a^2 - x^2}\, dx = \left[\frac{a^2}{2} \sin^{-1}\left(\frac{x}{a} \right) + \frac{x}{2} \sqrt{a^2 - x^2} \right]_{-a}^{a} = \frac{a^2}{2} \left[\frac{\pi}{2} + \frac{\pi}{2} \right] = \frac{\pi a^2}{2}$$

EXAMPLE 5   Find $\displaystyle\int \frac{dx}{\sqrt{9 + x^2}}$.

Solution   Let $x = 3 \tan t, -\pi/2 < t < \pi/2$. Then $dx = 3 \sec^2 t\, dt$ and $\sqrt{9 + x^2} = 3 \sec t$.

$$\int \frac{dx}{\sqrt{9 + x^2}} = \int \frac{3 \sec^2 t}{3 \sec t}\, dt = \int \sec t\, dt$$

$$= \ln|\sec t + \tan t| + C$$

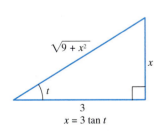

$x = 3 \tan t$

Figure 3

The last step, the integration of $\sec t$, was handled in Problem 68 of Section 8.1. Now $\tan t = x/3$, which suggests the triangle in Figure 3, from which we conclude that $\sec t = \sqrt{9 + x^2}/3$. Thus,

$$\int \frac{dx}{\sqrt{9 + x^2}} = \ln \left| \frac{\sqrt{9 + x^2} + x}{3} \right| + C$$

$$= \ln|\sqrt{9 + x^2} + x| - \ln 3 + C$$

$$= \ln|\sqrt{9 + x^2} + x| + K \qquad \blacksquare$$

EXAMPLE 6   Calculate $\displaystyle\int_{2}^{4} \frac{\sqrt{x^2 - 4}}{x}\, dx$.

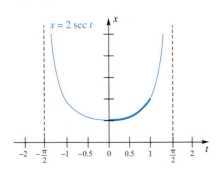

$x = 2 \sec t$

Figure 4

Solution   Let $x = 2 \sec t$, where $0 \le t < \pi/2$. Note that the restriction of t to this interval is acceptable, since x is in the interval $2 \le x \le 4$ (see Figure 4). This is important because it allows us to remove the absolute value sign that normally appears when we simplify $\sqrt{x^2 - a^2}$. In our case,

$$\sqrt{x^2 - 4} = \sqrt{4\sec^2 t - 4} = \sqrt{4\tan^2 t} = 2|\tan t| = 2\tan t$$

We now use the theorem on substituting in a definite integral (which requires changing the limits of integration) to write

$$\int_2^4 \frac{\sqrt{x^2 - 4}}{x}\, dx = \int_0^{\pi/3} \frac{2\tan t}{2\sec t} 2\sec t \tan t\, dt$$

$$= \int_0^{\pi/3} 2\tan^2 t\, dt = 2\int_0^{\pi/3} \left(\sec^2 t - 1\right) dt$$

$$= 2[\tan t - t]_0^{\pi/3} = 2\sqrt{3} - \frac{2\pi}{3} \approx 1.37 \qquad \blacksquare$$

Completing the Square   When a quadratic expression of the type $x^2 + Bx + C$ appears under a radical, completing the square will prepare it for a trigonometric substitution.

EXAMPLE 7   Find $\displaystyle\int \frac{dx}{\sqrt{x^2 + 2x + 26}}$ and $\displaystyle\int \frac{2x}{\sqrt{x^2 + 2x + 26}}\, dx$.

Solution $x^2 + 2x + 26 = x^2 + 2x + 1 + 25 = (x + 1)^2 + 25$. Let $u = x + 1$ and $du = dx$. Then

$$\int \frac{dx}{\sqrt{x^2 + 2x + 26}} = \int \frac{du}{\sqrt{u^2 + 25}}$$

Next let $u = 5\tan t$, $-\pi/2 < t < \pi/2$. Then $du = 5\sec^2 t\, dt$ and $\sqrt{u^2 + 25} = \sqrt{25(\tan^2 t + 1)} = 5\sec t$, so

$$\int \frac{du}{\sqrt{u^2 + 25}} = \int \frac{5\sec^2 t\, dt}{5\sec t} = \int \sec t\, dt$$

$$= \ln|\sec t + \tan t| + C$$

$$= \ln\left|\frac{\sqrt{u^2 + 25}}{5} + \frac{u}{5}\right| + C \qquad \text{(by Figure 5)}$$

$$= \ln\left|\sqrt{u^2 + 25} + u\right| - \ln 5 + C$$

$$= \ln\left|\sqrt{x^2 + 2x + 26} + x + 1\right| + K$$

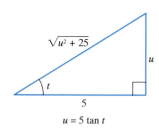

$u = 5\tan t$

Figure 5

To handle the second integral, we write

$$\int \frac{2x}{\sqrt{x^2 + 2x + 26}}\, dx = \int \frac{2x + 2}{\sqrt{x^2 + 2x + 26}}\, dx - 2\int \frac{1}{\sqrt{x^2 + 2x + 26}}\, dx$$

The first of the integrals on the right is handled by the substitution $u = x^2 + 2x + 26$; the second was just done. We obtain

$$\int \frac{2x}{\sqrt{x^2 + 2x + 26}}\, dx$$

$$= 2\sqrt{x^2 + 2x + 26} - 2\ln\left|\sqrt{x^2 + 2x + 26} + x + 1\right| + K \qquad \blacksquare$$

Concepts Review

1. To handle $\int x\sqrt{x-3}\,dx$, make the substitution $u = $ _____.

2. To handle an integral involving $\sqrt{4-x^2}$, make the substitution $x = $ _____.

3. To handle an integral involving $\sqrt{4+x^2}$, make the substitution $x = $ _____.

4. To handle an integral involving $\sqrt{x^2-4}$, make the substitution $x = $ _____.

Problem Set 8.3

In Problems 1–16, perform the indicated integrations.

1. $\displaystyle\int x\sqrt{x+1}\,dx$

2. $\displaystyle\int x\sqrt[3]{x+\pi}\,dx$

3. $\displaystyle\int \frac{t\,dt}{\sqrt{3t+4}}$

4. $\displaystyle\int \frac{x^2+3x}{\sqrt{x+4}}\,dx$

5. $\displaystyle\int_1^2 \frac{dt}{\sqrt{t}+e}$

6. $\displaystyle\int_0^1 \frac{\sqrt{t}}{t+1}\,dt$

7. $\displaystyle\int t(3t+2)^{3/2}\,dt$

8. $\displaystyle\int x(1-x)^{2/3}\,dx$

9. $\displaystyle\int \frac{\sqrt{4-x^2}}{x}\,dx$

10. $\displaystyle\int \frac{x^2\,dx}{\sqrt{16-x^2}}$

11. $\displaystyle\int \frac{dx}{(x^2+4)^{3/2}}$

12. $\displaystyle\int_2^3 \frac{dt}{t^2\sqrt{t^2-1}}$

13. $\displaystyle\int_{-2}^{-3} \frac{\sqrt{t^2-1}}{t^3}\,dt$

14. $\displaystyle\int \frac{t}{\sqrt{1-t^2}}\,dt$

15. $\displaystyle\int \frac{2z-3}{\sqrt{1-z^2}}\,dz$

16. $\displaystyle\int_0^\pi \frac{\pi x-1}{\sqrt{x^2+\pi^2}}\,dx$

In Problems 17–26, use the method of completing the square, along with a trigonometric substitution if needed, to evaluate each integral.

17. $\displaystyle\int \frac{dx}{\sqrt{x^2+2x+5}}$

18. $\displaystyle\int \frac{dx}{\sqrt{x^2+4x+5}}$

19. $\displaystyle\int \frac{3x}{\sqrt{x^2+2x+5}}\,dx$

20. $\displaystyle\int \frac{2x-1}{\sqrt{x^2+4x+5}}\,dx$

21. $\displaystyle\int \sqrt{5-4x-x^2}\,dx$

22. $\displaystyle\int \frac{dx}{\sqrt{16+6x-x^2}}$

23. $\displaystyle\int \frac{dx}{\sqrt{4x-x^2}}$

24. $\displaystyle\int \frac{x}{\sqrt{4x-x^2}}\,dx$

25. $\displaystyle\int \frac{2x+1}{x^2+2x+2}\,dx$

26. $\displaystyle\int \frac{2x-1}{x^2-6x+18}\,dx$

27. The region bounded by $y = 1/(x^2+2x+5)$, $y = 0$, $x = 0$, and $x = 1$, is revolved about the x-axis. Find the volume of the resulting solid.

28. The region of Problem 27 is revolved about the y-axis. Find the volume of the resulting solid.

29. Find $\displaystyle\int \frac{x\,dx}{x^2+9}$ by

(a) an algebraic substitution and

(b) a trigonometric substitution. Then reconcile your answers.

30. Find $\displaystyle\int_0^3 \frac{x^3\,dx}{\sqrt{9+x^2}}$ by making the substitutions

$$u = \sqrt{9+x^2}, \quad u^2 = 9+x^2, \quad 2u\,du = 2x\,dx$$

31. Find $\displaystyle\int \frac{\sqrt{4-x^2}}{x}\,dx$ by

(a) the substitution $u = \sqrt{4-x^2}$ and

(b) a trigonometric substitution. Then reconcile your answers.

Hint: $\displaystyle\int \csc x\,dx = \ln|\csc x - \cot x| + C.$

32. Two circles of radius b intersect as shown in Figure 6 with their centers $2a$ apart ($0 \le a \le b$). Find the area of the region of their overlap.

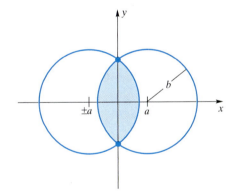

Figure 6

33. Hippocrates of Chios (ca. 430 B.C.) showed that the two shaded regions in Figure 7 have the same area (he squared the lune). Note that C is the center of the lower arc of the lune. Show Hippocrates' result

(a) using calculus and   (b) without calculus.

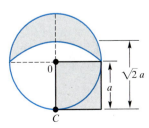

Figure 7

34. Generalize the idea in Problem 33 by finding a formula for the area of the shaded lune shown in Figure 8.

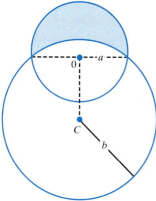

Figure 8

35. Starting at $(a, 0)$, an object is pulled along by a string of length a with the pulling end moving along the positive y-axis. The path of the object is a curve called a **tractrix** and has the property that the string is always tangent to the curve (see Figure 9). Set up a differential equation for the curve and solve it.

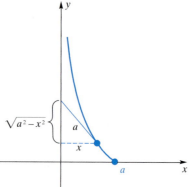

Figure 9

Answers to Concepts Review:　**1.** $\sqrt{x-3}$ **2.** $2\sin t$ **3.** $2\tan t$
4. $2\sec t$

8.4
Integration by Parts

If integration by substitution fails, it may be possible to use a double substitution, better known as *integration by parts*. This method is based on the integration of the formula for the derivative of a product of two functions.

Let $u = u(x)$ and $v = v(x)$. Then

$$D_x\big[u(x)v(x)\big] = u(x)v'(x) + v(x)u'(x)$$

or

$$u(x)v'(x) = D_x\big[u(x)v(x)\big] - v(x)u'(x)$$

By integrating both members of this equation, we obtain

$$\int u(x)v'(x)\,dx = u(x)v(x) - \int v(x)u'(x)\,dx$$

Since $dv = v'(x)\,dx$ and $du = u'(x)\,dx$, the preceding equation is usually written symbolically as follows:

Integration by Parts: Indefinite Integrals

$$\int u\,dv = uv - \int v\,du$$

The corresponding formula for definite integrals is

$$\int_a^b u(x)v'(x)\,dx = \big[u(x)v(x)\big]_a^b - \int_a^b v(x)u'(x)\,dx$$

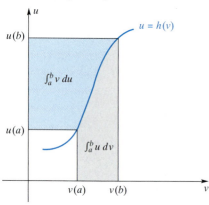

A Geometric Interpretation
of
Integration by Parts

$\int_a^b u\,dv = u(b)v(b) - u(a)v(a) - \int_a^b v\,du$

Figure 1

Figure 1 illustrates a geometric interpretation of integration by parts. We abbreviate this as follows:

Integration by Parts: Definite Integrals

$$\int_a^b u\,dv = \big[uv\big]_a^b - \int_a^b v\,du$$

These formulas allow us to shift the problem of integrating $u\,dv$ to that of integrating $v\,du$. Success depends on the proper choice of u and dv, which comes with practice.

EXAMPLE 1   Find $\int x \cos x \, dx$.

Solution   We wish to write $x \cos x \, dx$ as $u \, dv$. One possibility is to let $u = x$ and $dv = \cos x \, dx$. Then $du = dx$ and $v = \int \cos x \, dx = \sin x$ (we can omit the arbitrary constant at this stage). Here is a summary of this double substitution in a convenient format.

$$u = x \qquad\qquad dv = \cos x \, dx$$

$$du = dx \qquad\qquad v = \sin x$$

The formula for integration by parts gives

$$\int \underset{u}{\underbrace{x}} \, \underset{dv}{\underbrace{\cos x \, dx}} \; = \; \underset{u}{\underbrace{x}} \, \underset{v}{\underbrace{\sin x}} - \int \underset{v}{\underbrace{\sin x}} \, \underset{du}{\underbrace{dx}}$$

$$= x \sin x + \cos x + C$$

We were successful on our first try. Another substitution would be

$$u = \cos x \qquad dv = x \, dx$$

$$du = -\sin x \, dx \qquad v = \frac{x^2}{2}$$

This time the formula for integration by parts gives

$$\int \underset{u}{\underbrace{\cos x}} \, \underset{dv}{\underbrace{x \, dx}} \; = \; \underset{u}{\underbrace{(\cos x)}} \underset{v}{\underbrace{\frac{x^2}{2}}} - \int \underset{v}{\underbrace{\frac{x^2}{2}}} \underset{du}{\underbrace{(-\sin x \, dx)}}$$

which is correct but not helpful. The new integral on the right-hand side is more complicated than the original one. Thus, we see the importance of a wise choice for u and dv. ∎

EXAMPLE 2   Find $\int_1^2 \ln x \, dx$.

Solution   We make the following substitutions:

$$u = \ln x \qquad dv = dx$$

$$du = \left(\frac{1}{x}\right) dx \qquad v = x$$

Then

$$\int_1^2 \ln x \, dx = \left[x \ln x\right]_1^2 - \int_1^2 x \frac{1}{x} \, dx$$

$$= 2 \ln 2 - \int_1^2 dx$$

$$= 2 \ln 2 - 1 \approx 0.386$$ ∎

EXAMPLE 3   Find $\int \arcsin x \, dx$.

Solution   We make the substitutions

$$u = \arcsin x \qquad\qquad dv = dx$$

$$du = \frac{1}{\sqrt{1 - x^2}} \, dx \qquad v = x$$

Then

$$\int \arcsin x \, dx = x \arcsin x - \int \frac{x}{\sqrt{1 - x^2}} \, dx$$

$$= x \arcsin x + \frac{1}{2} \int (1 - x^2)^{-1/2}(-2x \, dx)$$

$$= x \arcsin x + \frac{1}{2} \cdot 2(1 - x^2)^{1/2} + C$$

$$= x \arcsin x + \sqrt{1 - x^2} + C$$

Repeated Integration by Parts   Sometimes it is necessary to apply integration by parts several times.

EXAMPLE 4   Find $\int x^2 \sin x \, dx$.

Solution   Let

$$u = x^2 \qquad\qquad dv = \sin x \, dx$$
$$du = 2x \, dx \qquad\quad v = -\cos x$$

Then

$$\int x^2 \sin x \, dx = -x^2 \cos x + 2 \int x \cos x \, dx$$

We have improved our situation (the exponent on x has gone from 2 to 1), which suggests reapplying integration by parts to the integral on the right. Actually, we did this integration in Example 1, so we will make use of the result obtained there.

$$\int x^2 \sin x \, dx = -x^2 \cos x + 2(x \sin x + \cos x + C)$$

$$= -x^2 \cos x + 2x \sin x + 2 \cos x + K$$

EXAMPLE 5   Find $\int e^x \sin x \, dx$.

Solution   Take $u = e^x$ and $dv = \sin x \, dx$. Then $du = e^x \, dx$ and $v = -\cos x$. Thus,

$$\int e^x \sin x \, dx = -e^x \cos x + \int e^x \cos x \, dx$$

which does not seem to have improved things—but does not leave us any worse off either. So, let's not give up and try integration by parts again. In the integral on the right, let $u = e^x$ and $dv = \cos x \, dx$, so $du = e^x \, dx$ and $v = \sin x$. Then

$$\int e^x \cos x \, dx = e^x \sin x - \int e^x \sin x \, dx$$

When we substitute this in our first result, we get

$$\int e^x \sin x \, dx = -e^x \cos x + e^x \sin x - \int e^x \sin x \, dx$$

By moving the last term to the left side and combining terms, we obtain

$$2 \int e^x \sin x \, dx = e^x (\sin x - \cos x) + C$$

from which

$$\int e^x \sin x \, dx = \frac{1}{2} e^x (\sin x - \cos x) + K$$

The fact that the integral we wanted to find reappeared on the right side is what made Example 5 work.

Reduction Formulas   A formula of the form

$$\int f^n(x)\,dx = g(x) + \int f^k(x)\,dx,$$

where $k < n$, is called a **reduction formula** (the exponent on f is reduced). Such formulas can often be obtained via integration by parts.

EXAMPLE 6   Derive a reduction formula for $\int \sin^n x\,dx$.

Solution   Let $u = \sin^{n-1} x$ and $dv = \sin x\,dx$. Then

$$du = (n-1)\sin^{n-2} x \cos x\,dx \quad \text{and} \quad v = -\cos x$$

from which

$$\int \sin^n x\,dx = -\sin^{n-1} x \cos x + (n-1)\int \sin^{n-2} x \cos^2 x\,dx$$

If we replace $\cos^2 x$ by $1 - \sin^2 x$ in the last integral, we obtain

$$\int \sin^n x\,dx = -\sin^{n-1} x \cos x + (n-1)\int \sin^{n-2} x\,dx - (n-1)\int \sin^n x\,dx$$

After combining the first and last integrals above and solving for $\int \sin^n x\,dx$, we get the reduction formula (valid for $n \geq 2$)

$$\int \sin^n x\,dx = \frac{-\sin^{n-1} x \cos x}{n} + \frac{n-1}{n}\int \sin^{n-2} x\,dx \qquad ■$$

EXAMPLE 7   Use the reduction formula above to evaluate $\int_0^{\pi/2} \sin^8 x\,dx$.

Solution   Note first that

$$\int_0^{\pi/2} \sin^n x\,dx = \left[\frac{-\sin^{n-1} x \cos x}{n} \right]_0^{\pi/2} + \frac{n-1}{n}\int_0^{\pi/2} \sin^{n-2} x\,dx$$

$$= 0 + \frac{n-1}{n}\int_0^{\pi/2} \sin^{n-2} x\,dx$$

Thus,

$$\int_0^{\pi/2} \sin^8 x\,dx = \frac{7}{8}\int_0^{\pi/2} \sin^6 x\,dx$$

$$= \frac{7}{8}\cdot\frac{5}{6}\int_0^{\pi/2} \sin^4 x\,dx$$

$$= \frac{7}{8}\cdot\frac{5}{6}\cdot\frac{3}{4}\int_0^{\pi/2} \sin^2 x\,dx$$

$$= \frac{7}{8}\cdot\frac{5}{6}\cdot\frac{3}{4}\cdot\frac{1}{2}\int_0^{\pi/2} 1\,dx$$

$$= \frac{7}{8}\cdot\frac{5}{6}\cdot\frac{3}{4}\cdot\frac{1}{2}\cdot\frac{\pi}{2} = \frac{35}{256}\pi \qquad ■$$

The general formula for $\int_0^{\pi/2} \sin^n x\,dx$ can be found in a similar way (Formula 113 at the back of the book).

Concepts Review

1. The integration-by-parts formula says that $\int u \, dv =$ _____.

2. To apply this formula to $\int x \sin x \, dx$, let $u =$ _____ and $dv =$ _____.

3. Applying the integration-by-parts formula yields the value _____ for $\int_0^{\pi/2} x \sin x \, dx$.

4. A formula that expresses $\int f^n(x) \, dx$ in terms of $\int f^k(x) \, dx$, where $k < n$, is called a _____ formula.

Problem Set 8.4

In Problems 1–40, use integration by parts to evaluate each integral.

1. $\int xe^x \, dx$

2. $\int xe^{3x} \, dx$

3. $\int te^{5t+\pi} \, dt$

4. $\int (t + 7)e^{2t+3} \, dt$

5. $\int x \cos x \, dx$

6. $\int x \sin 2x \, dx$

7. $\int (t - 3)\cos(t - 3) \, dt$

8. $\int (x - \pi)\sin x \, dx$

9. $\int t\sqrt{t + 1} \, dt$

10. $\int t\sqrt[3]{2t + 7} \, dt$

11. $\int \ln 3x \, dx$

12. $\int \ln(7x^5) \, dx$

13. $\int \arctan x \, dx$

14. $\int \arctan 5x \, dx$

15. $\int \frac{\ln x}{x^2} \, dx$

16. $\int_2^3 \frac{\ln 2x^5}{x^2} \, dx$

17. $\int_1^e \sqrt{t} \ln t \, dt$

18. $\int_5^1 \sqrt{2x} \ln x^3 \, dx$

19. $\int z^3 \ln z \, dz$

20. $\int t \arctan t \, dt$

21. $\int \arctan(1/t) \, dt$

22. $\int t^5 \ln(t^7) \, dt$

23. $\int x \cos^2 x \sin x \, dx$

24. $\int x \sin^3 \pi x \cos \pi x \, dx$

25. $\int_{\pi/6}^{\pi/2} x \csc^2 x \, dx$

26. $\int_{\pi/4}^{\pi/2} \csc^2 x \, dx$

27. $\int_{\pi/6}^{\pi/4} x \sec^2 x \, dx$

28. $\int \sec^{-1} \sqrt{x} \, dx$

29. $\int x^5 \sqrt{x^3 + 4} \, dx$

30. $\int x^{13}\sqrt{x^7 + 1} \, dx$

31. $\int \frac{t^7}{(7 - 3t^4)^{3/2}} \, dt$

32. $\int x^3 \sqrt{4 - x^2} \, dx$

33. $\int \frac{z^7}{(4 - z^4)^2} \, dz$

34. $\int x \cosh x \, dx$

35. $\int x \sinh x \, dx$

36. $\int \sec^3 x \, dx$

37. $\int \frac{\ln x}{\sqrt{x}} \, dx$

38. $\int x(3x + 10)^{49} \, dx$

39. $\int x \, 2^x \, dx$

40. $\int z \, a^z \, dz$

In Problems 41–52, apply integration by parts twice to evaluate each integral (see Examples 4 and 5).

41. $\int x^2 e^x \, dx$

42. $\int x^5 e^{x^2} \, dx$

43. $\int \ln^2 z \, dz$

44. $\int \ln^2 x^{20} \, dx$

45. $\int e^t \cos t \, dt$

46. $\int e^{at} \sin t \, dt$

47. $\int x^2 \cos x \, dx$

48. $\int r^2 \sin r \, dr$

49. $\int \sin(\ln x) \, dx$

50. $\int \cos(\ln x) \, dx$

51. $\int (\ln x)^3 \, dx$ Hint: Use Problem 43.

52. $\int (\ln x)^4 \, dx$ Hint: Use Problems 43 and 51.

In Problems 53–58, use integration by parts to derive the given formula.

53. $\int \sin(x)\sin(3x) \, dx$
$$= -\tfrac{3}{8}\sin x \cos 3x + \tfrac{1}{8}\cos x \sin 3x + C$$

54. $\int \cos(5x)\sin(7x) \, dx$
$$= -\tfrac{7}{24}\cos 5x \cos 7x - \tfrac{5}{24}\sin 5x \sin 7x + C$$

55. $\int e^{\alpha z} \sin \beta z \, dz = \dfrac{e^{\alpha z}(\alpha \sin \beta z - \beta \cos \beta z)}{\alpha^2 + \beta^2} + C$

56. $\int e^{\alpha z} \cos \beta z \, dz = \dfrac{e^{\alpha z}(\alpha \cos \beta z + \beta \sin \beta z)}{\alpha^2 + \beta^2} + C$

57. $\int x^\alpha \ln x \, dx = \dfrac{x^{\alpha+1}}{\alpha + 1} \ln x - \dfrac{x^{\alpha+1}}{(\alpha + 1)^2} + C, \alpha \ne -1$

58. $\int x^\alpha (\ln x)^2 \, dx = \dfrac{x^{\alpha+1}}{\alpha + 1}(\ln x)^2$
$$- 2\dfrac{x^{\alpha+1}}{(\alpha + 1)^2} \ln x + 2\dfrac{x^{\alpha+1}}{(\alpha + 1)^3} + C, \alpha \ne -1$$

In Problems 59–65 derive the given reduction formula using integration by parts.

59. $\int x^\alpha e^{\beta x} \, dx = \dfrac{x^\alpha e^{\beta x}}{\beta} - \dfrac{\alpha}{\beta}\int x^{\alpha-1} e^{\beta x} \, dx$

60. $\displaystyle\int x^\alpha \sin \beta x \, dx = -\frac{x^\alpha \cos \beta x}{\beta} + \frac{\alpha}{\beta} \int x^{\alpha-1} \cos \beta x \, dx$

61. $\displaystyle\int x^\alpha \cos \beta x \, dx = \frac{x^\alpha \sin \beta x}{\beta} - \frac{\alpha}{\beta} \int x^{\alpha-1} \sin \beta x \, dx$

62. $\displaystyle\int (\ln x)^\alpha \, dx = x(\ln x)^\alpha - \alpha \int (\ln x)^{\alpha-1} \, dx$

63. $\displaystyle\int (a^2 - x^2)^\alpha \, dx = x(a^2 - x^2)^\alpha + 2\alpha \int x^2(a^2 - x^2)^{\alpha-1} \, dx$

64. $\displaystyle\int \cos^\alpha x \, dx = \frac{\cos^{\alpha-1} x \sin x}{\alpha} + \frac{\alpha - 1}{\alpha} \int \cos^{\alpha-2} x \, dx$

65. $\displaystyle\int \cos^\alpha \beta x \, dx$
$$= \frac{\cos^{\alpha-1} \beta x \sin \beta x}{\alpha\beta} + \frac{\alpha - 1}{\alpha} \int \cos^{\alpha-2} \beta x \, dx$$

66. Use Problem 59 to derive

$$\int x^4 e^{3x} \, dx = \tfrac{1}{3}x^4 e^{3x} - \tfrac{4}{9}x^3 e^{3x} + \tfrac{4}{9}x^2 e^{3x} - \tfrac{8}{27}xe^{3x} + \tfrac{8}{81}e^{3x} + C$$

67. Use Problems 60 and 61 to derive $\displaystyle\int x^4 \cos 3x \, dx =$
$\tfrac{1}{3}x^4 \sin 3x + \tfrac{4}{9}x^3 \cos 3x - \tfrac{4}{9}x^2 \sin 3x + \tfrac{8}{81} \sin 3x - \tfrac{8}{27}x \cos 3x + C.$

68. Use Problem 65 to derive $\displaystyle\int \cos^6 3x \, dx =$
$\tfrac{1}{18} \sin 3x \cos^5 3x + \tfrac{5}{72} \sin 3x \cos^3 3x + \tfrac{5}{48} \sin 3x \cos 3x + \tfrac{5}{16}x + C.$

69. Find the area of the region bounded by the curve $y = \ln x$, the x-axis, and the line $x = e$.

70. Find the volume of the solid generated by revolving the region of Problem 69 about the x-axis.

71. Find the area of the region bounded by the curves $y = 3e^{-x/3}$, $y = 0$, $x = 0$, and $x = 9$. Make a sketch.

72. Find the volume of the solid generated by revolving the region described in Problem 71 about the x-axis.

73. Find the area of the region bounded by the graphs of $y = x \sin x$ and $y = x \cos x$ from $x = 0$ to $x = \pi/4$.

74. Find the volume of the solid obtained by revolving the region under the graph of $y = \sin(x/2)$ from $x = 0$ to $x = 2\pi$ about the y-axis.

75. Find the centroid (see Section 6.6) of the region bounded by $y = \ln x^2$ and the x-axis from $x = 1$ to $x = e$.

76. Evaluate the integral $\displaystyle\int \cot x \csc^2 x \, dx$ by parts in two

different ways:
(a) By differentiating $\cot x$ (b) By differentiating $\csc x$
(c) Show that the two results are equivalent up to a constant.

77. If $p(x)$ is a polynomial of degree n and $G_1, G_2, \ldots, G_{n+1}$, are successive antiderivatives of g, then, by repeated integration by parts,

$$\int p(x)g(x) \, dx = p(x)G_1(x) - p'(x)G_2(x) + p''(x)G_3(x) - \cdots$$
$$+ (-1)^n p^{(n)}(x)G_{n+1}(x) + C$$

Use this result to find each of the following:

(a) $\displaystyle\int (x^3 - 2x)e^x \, dx$ (b) $\displaystyle\int (x^2 - 3x + 1)\sin x \, dx$

78. Establish the identity

$$2 \sec^3 x \, dx = \sec x \, dx + d(\sec x \tan x)$$

and use it to derive a formula for $\int \sec^3 x \, dx$.

79. Evaluate $\displaystyle\int_0^{2\pi} \sin^n x \, dx$. *Hint:* First rewrite this expression in terms of $\displaystyle\int_0^{\pi/2} \sin^n x \, dx$ and then use Formula 113 from the back of the book.

80. The graph of $y = x \sin x$ for $x \geq 0$ is sketched in Figure 2.
(a) Find a formula for the area of the nth arch.
(b) The second arch is revolved about the y-axis. Find the volume of the resulting solid.

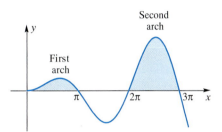

Figure 2

81. The Fourier coefficient $a_n = \dfrac{1}{\pi} \displaystyle\int_{-\pi}^{\pi} f(x) \sin nx \, dx$ plays an important role in applied mathematics. Show that if $f'(x)$ is continuous on $[-\pi, \pi]$, then $\displaystyle\lim_{n\to\infty} a_n = 0$. *Hint:* Integration by parts.

82. Let $G_n = \sqrt[n]{(n+1)(n+2)\cdots(n+n)}$. Show that $\displaystyle\lim_{n\to\infty}(G_n/n) = 4/e$. *Hint:* Consider $\ln(G_n/n)$, recognize it as a Riemann sum, and use Example 2.

83. Find the error in the following "proof" that $0 = 1$. In $\int(1/t) \, dt$, set $u = 1/t$ and $dv = dt$. Then $du = -t^{-2} \, dt$ and $uv = 1$. Integration by parts gives

$$\int (1/t) \, dt = 1 - \int (-1/t) \, dt$$

or $0 = 1$.

84. Suppose that you want to compute the integral

$$\int e^{5x}(4 \cos 7x + 6 \sin 7x) \, dx$$

and you know from experience that the result will be of the form $e^{5x}(C_1 \cos 7x + C_2 \sin 7x) + C_3$. Compute C_1 and C_2 by differentiating the result and setting it equal to the integrand.

Many surprising theoretical results can be derived through the use of integration by parts. In all cases, one starts with an integral. We explore two of these results here.

85. Show that

$$\int_a^b f(x) \, dx = xf(x)\Big|_a^b - \int_a^b xf'(x) \, dx$$
$$= (x - a)f(x)\Big|_a^b - \int_a^b (x - a)f'(x) \, dx$$

86. Using Problem 85 and replacing f by f', show that

$$\begin{aligned} f(b) - f(a) &= \int_a^b f'(x)\,dx \\ &= f'(b)(b-a) - \int_a^b (x-a)f''(x)\,dx \\ &= f'(a)(b-a) - \int_a^b (x-b)f''(x)\,dx \end{aligned}$$

87. Show that

$$f(t) = f(a) + \sum_{i=1}^{n} \frac{f^{(i)}(a)}{i!}(t-a)^i + \int_a^t \frac{(t-a)^n}{n!} f^{(n+1)}(x)\,dx,$$

provided that f can be differentiated $n+1$ times.

88. The *Beta function*, which is important in many branches of mathematics, is defined as

$$B(\alpha, \beta) = \int_0^1 x^{\alpha}(1-x)^{\beta}\,dx,$$

with the condition that $\alpha > 0$ and $\beta > 0$.

(a) Show by a change of variables that

$$B(\alpha, \beta) = \int_0^1 x^{\beta}(1-x)^{\alpha}dx = B(\beta, \alpha)$$

(b) Integrate by parts to show that

$$B(\alpha, \beta) = \frac{\alpha}{\beta+1}B(\alpha-1, \beta+1) = \frac{\beta}{\alpha+1}B(\alpha+1, \beta-1)$$

(c) Assume now that $\alpha = n$ and $\beta = m$ and that n and m are positive integers. By using the result in part (b) repeatedly, show that

$$B(n, m) = \frac{n!m!}{(n+m+1)!}$$

This result is even valid in the case where n and m are not integers, provided that we can give meaning to $n!$, $m!$, and $(n+m+1)!$.

89. Suppose that $f(t)$ has the property that $f'(a) = f'(b) = 0$ and that $f(t)$ has two continuous derivatives. Use integration by parts to prove that $\int_a^b f''(t)f(t)\,dt \le 0$. *Hint*: Use integration by parts by differentiating $f(t)$ and integrating $f''(t)$. This result has many applications in the field of applied mathematics and partial differential equations.

90. Derive the formula

$$\int_0^x \left(\int_0^t f(z)\,dz \right) dt = \int_0^x f(t)(x-t)\,dt$$

using integration by parts.

91. Generalize the formula given in Problem 90 to one for an n-fold iterated integral

$$\int_0^x \int_0^{t_1} \cdots \int_0^{t_{n-1}} f(t_n)\,dt_n \ldots dt_1 = \frac{1}{(n-1)!}\int_0^x f(t_1)(x-t_1)^{n-1}\,dt_1$$

92. If $P_n(x)$ is a polynomial of degree n, show that

$$\int e^x P_n(x)\,dx = e^x \sum_{j=0}^{n} (-1)^j \frac{d^j P_n(x)}{dx^j}$$

93. Use the result from Problem 92 to evaluate

$$\int (3x^4 + 2x^2)e^x\,dx$$

Answers to Concepts Review: **1.** $uv - \int v\,du$ **2.** x; $\sin x\,dx$ **3.** 1 **4.** reduction

8.5 Integration of Rational Functions

A **rational function** is by definition the quotient of two polynomial functions. Examples are

$$f(x) = \frac{2}{(x+1)^3}, \quad g(x) = \frac{2x+2}{x^2-4x+8}, \quad h(x) = \frac{x^5+2x^3-x+1}{x^3+5x}$$

Of these, f and g are **proper rational functions**, meaning that the degree of the numerator is less than that of the denominator. An improper (not proper) rational function can always be written as a sum of a polynomial function and a proper rational function. Thus, for example,

$$h(x) = \frac{x^5+2x^3-x+1}{x^3+5x} = x^2 - 3 + \frac{14x+1}{x^3+5x}$$

a result obtained by long division (Figure 1). Since polynomials are easy to integrate, the problem of integrating rational functions is really that of integrating proper rational functions. But can we always integrate proper rational functions? In theory, the answer is yes, though the practical details may overwhelm us. Consider first the integrals of f and g above.

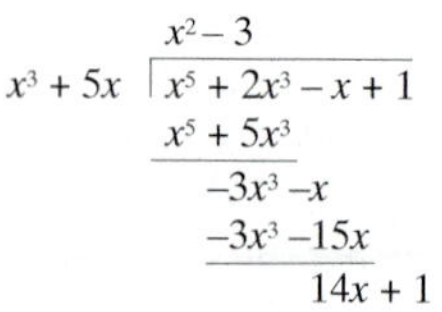

x² – 3
x³ + 5x | x⁵ + 2x³ – x + 1
x⁵ + 5x³
–3x³ –x
–3x³ –15x
14x + 1

Figure 1

EXAMPLE 1   Find $\displaystyle\int \frac{2}{(x+1)^3}\,dx$.

Solution   Think of the substitution $u = x + 1$.

$$\int \frac{2}{(x+1)^3}\,dx = 2\int (x+1)^{-3}\,dx = \frac{2(x+1)^{-2}}{-2} + C$$

$$= \frac{-1}{(x+1)^2} + C \qquad\blacksquare$$

EXAMPLE 2   Find $\displaystyle\int \frac{2x+2}{x^2-4x+8}\,dx$.

Solution   Think first of the substitution $u = x^2 - 4x + 8$ for which $du = (2x - 4)\,dx$. Then write the given integral as a sum of two integrals.

$$\int \frac{2x+2}{x^2-4x+8}\,dx = \int \frac{2x-4}{x^2-4x+8}\,dx + \int \frac{6}{x^2-4x+8}\,dx$$

$$= \ln\left|x^2 - 4x + 8\right| + 6\int \frac{1}{x^2-4x+8}\,dx$$

In the second integral, complete the square.

$$\int \frac{1}{x^2-4x+8}\,dx = \int \frac{1}{x^2-4x+4+4}\,dx = \int \frac{1}{(x-2)^2+4}\,dx$$

$$= \int \frac{1}{(x-2)^2+4}\,dx = \frac{1}{2}\tan^{-1}\left(\frac{x-2}{2}\right) + C$$

We conclude that

$$\int \frac{2x+2}{x^2-4x+8}\,dx = \ln\left|x^2 - 4x + 8\right| + 3\tan^{-1}\left(\frac{x-2}{2}\right) + K \qquad\blacksquare$$

It is a remarkable fact that any proper rational function can be written as a sum of *simple* proper rational functions like those illustrated in Examples 1 and 2. We must be more precise.

Partial Fraction Decomposition (Linear Factors)   To add fractions is a standard algebraic exercise: find a common denominator and add. For example,

$$\frac{2}{x-1} + \frac{3}{x+1} = \frac{2(x+1)+3(x-1)}{(x-1)(x+1)} = \frac{5x-1}{(x-1)(x+1)} = \frac{5x-1}{x^2-1}$$

It is the reverse process of decomposing a fraction into a sum of simpler fractions that interests us now. We focus on the denominator and consider cases.

EXAMPLE 3

Distinct Linear Factors   Decompose $(3x-1)/(x^2-x-6)$ and then find its indefinite integral.

Solution   Since the denominator factors as $(x+2)(x-3)$, it seems reasonable to hope for a decomposition of the following form:

(1) $$\frac{3x-1}{(x+2)(x-3)} = \frac{A}{x+2} + \frac{B}{x-3}$$

Our job is, of course, to determine A and B so that (1) is an identity, a task that we find easier after we have multiplied both sides by $(x+2)(x-3)$. We obtain

(2) $$3x - 1 = A(x-3) + B(x+2)$$

or, equivalently,

(3) $3x - 1 = (A + B)x + (-3A + 2B)$

However, (3) is an identity if and only if coefficients of like powers of x on both sides are equal; that is,

$$A + B = 3$$

$$-3A + 2B = -1$$

By solving this pair of equations for A and B, we obtain $A = \frac{7}{5}$, $B = \frac{8}{5}$. Consequently,

$$\frac{3x - 1}{x^2 - x - 6} = \frac{3x - 1}{(x + 2)(x - 3)} = \frac{\frac{7}{5}}{x + 2} + \frac{\frac{8}{5}}{x - 3}$$

and

$$\int \frac{3x - 1}{x^2 - x - 6} \, dx = \frac{7}{5} \int \frac{1}{x + 2} \, dx + \frac{8}{5} \int \frac{1}{x - 3} \, dx$$

$$= \frac{7}{5} \ln|x + 2| + \frac{8}{5} \ln|x - 3| + C \qquad \blacksquare$$

If there was anything difficult about this process, it was the determination of A and B. We found their values by "brute force," but there is an easier way. In (2), which we wish to be an identity (that is, true for *every* value of x) substitute the convenient values $x = 3$ and $x = -2$, obtaining

$$8 = A \cdot 0 + B \cdot 5$$

$$-7 = A \cdot (-5) = B \cdot 0$$

This immediately gives $B = \frac{8}{5}$ and $A = \frac{7}{5}$.

You have just witnessed an odd, but correct, mathematical maneuver. Equation (1) turns out to be an identity (true for all x except -2 and 3) if and only if the essentially equivalent equation (2) is true precisely at -2 and 3. Ask yourself why this is so. Ultimately it depends on the fact that the two sides of equation (2), both linear polynomials, are identical if they have the same values at any two points.

EXAMPLE 4

Distinct Linear Factors   Find $\displaystyle\int \frac{5x + 3}{x^3 - 2x^2 - 3x} \, dx$.

Solution   Since the denominator factors as $x(x + 1)(x - 3)$, we write

$$\frac{5x + 3}{x(x + 1)(x - 3)} = \frac{A}{x} + \frac{B}{x + 1} + \frac{C}{x - 3}$$

and seek to determine A, B, and C. Clearing of fractions gives

$$5x + 3 = A(x + 1)(x - 3) + Bx(x - 3) + Cx(x + 1)$$

Substitution of the values $x = 0$, $x = -1$, and $x = 3$ results in

$$3 = A(-3)$$

$$-2 = B(4)$$

$$18 = C(12)$$

or $A = -1$, $B = -\frac{1}{2}$, $C = \frac{3}{2}$. Thus,

$$\int \frac{5x + 3}{x^3 - 2x^2 - 3x} \, dx = -\int \frac{1}{x} \, dx - \frac{1}{2} \int \frac{1}{x + 1} \, dx + \frac{3}{2} \int \frac{1}{x - 3} \, dx$$

$$= -\ln|x| - \frac{1}{2} \ln|x + 1| + \frac{3}{2} \ln|x - 3| + C \qquad \blacksquare$$

Solve This D.E.

"often, there is little resemblance between a differential equation and its solution. Who would suppose that an expression as simple as

$$\frac{dy}{dx} = \frac{1}{a^2 - x^2}$$

could be transformed into

$$y = \frac{1}{2a} \log_e \left(\frac{a + x}{a - x} \right) + C$$

This resembles the transformation of a chrysalis into a butterfly."

Silvanus P. Thompson

The method of partial fractions makes this an easy transformation. Do you see how it is done?

EXAMPLE 5

Repeated Linear Factors   Find $\displaystyle\int \frac{x}{(x-3)^2}\, dx$.

Solution   Now the decomposition takes the form

$$\frac{x}{(x-3)^2} = \frac{A}{x-3} + \frac{B}{(x-3)^2}$$

with A and B to be determined. After clearing the fractions, we get

$$x = A(x-3) + B$$

If we now substitute the convenient value $x = 3$ and any other value, such as $x = 0$, we obtain $B = 3$ and $A = 1$. Thus,

$$\int \frac{x}{(x-3)^2}\, dx = \int \frac{1}{x-3}\, dx + 3\int \frac{1}{(x-3)^2}\, dx$$

$$= \ln|x-3| - \frac{3}{x-3} + C \qquad \blacksquare$$

EXAMPLE 6

Some Distinct, Some Repeated Linear Factors   Find $\displaystyle\int \frac{3x^2 - 8x + 13}{(x+3)(x-1)^2}\, dx$

Solution   We decompose the integrand in the following way:

$$\frac{3x^2 - 8x + 13}{(x+3)(x-1)^2} = \frac{A}{x+3} + \frac{B}{x-1} + \frac{C}{(x-1)^2}$$

Clearing the fractions changes this to

$$3x^2 - 8x + 13 = A(x-1)^2 + B(x+3)(x-1) + C(x+3)$$

Substitution of $x = 1$, $x = -3$, and $x = 0$ yields $C = 2$, $A = 4$, and $B = -1$. Thus,

$$\int \frac{3x^2 - 8x + 13}{(x+3)(x-1)^2}\, dx = 4\int \frac{dx}{x+3} - \int \frac{dx}{x-1} + 2\int \frac{dx}{(x-1)^2}$$

$$= 4\ln|x+3| - \ln|x-1| - \frac{2}{x-1} + C \qquad \blacksquare$$

Be sure to note the inclusion of the two fractions $B/(x-1)$ and $C/(x-1)^2$ in the decomposition above. The general rule for decomposing fractions with repeated linear factors in the denominator is this: For each factor $(ax+b)^k$ of the denominator, there are k terms in the partial fraction decomposition:

$$\frac{A_1}{ax+b} + \frac{A_2}{(ax+b)^2} + \frac{A_3}{(ax+b)^3} + \cdots + \frac{A_k}{(ax+b)^k}$$

Partial Fraction Decomposition (Quadratic Factors)   In factoring the denominator of a fraction, we may well get some quadratic factors (such as $x^2 + 1$), that cannot be factored into linear factors without introducing complex numbers.

EXAMPLE 7

A Single Quadratic Factor   Decompose $\displaystyle\frac{6x^2 - 3x + 1}{(4x+1)(x^2+1)}$ and then find its indefinite integral.

Solution   The best we can hope for is a decomposition of the form

$$\frac{6x^2 - 3x + 1}{(4x+1)(x^2+1)} = \frac{A}{4x+1} + \frac{Bx+C}{x^2+1}$$

To determine the constants A, B, and C, we multiply both sides by $(4x + 1)(x^2 + 1)$ and obtain

$$6x^2 - 3x + 1 = A(x^2 + 1) + (Bx + C)(4x + 1)$$

Substitution of $x = -\frac{1}{4}$, $x = 0$, and $x = 1$ yields

$$\frac{6}{16} + \frac{3}{4} + 1 = A\left(\frac{17}{16}\right) \qquad \Rightarrow A = 2$$
$$1 = 2 + C \qquad \Rightarrow C = -1$$
$$4 = 4 + (B - 1)5 \Rightarrow B = 1$$

Thus,

$$\int \frac{6x^2 - 3x + 1}{(4x + 1)(x^2 + 1)}\, dx = \int \frac{2}{4x + 1}\, dx + \int \frac{x - 1}{x^2 + 1}\, dx$$

$$= \frac{1}{2} \int \frac{4\, dx}{4x + 1} + \frac{1}{2} \int \frac{2x\, dx}{x^2 + 1} - \int \frac{dx}{x^2 + 1}$$

$$= \frac{1}{2} \ln|4x + 1| + \frac{1}{2} \ln(x^2 + 1) - \tan^{-1} x + C \qquad \blacksquare$$

EXAMPLE 8

A Repeated Quadratic Factor Find $\displaystyle \int \frac{6x^2 - 15x + 22}{(x + 3)(x^2 + 2)^2}\, dx$.

Solution Here the appropriate decomposition is

$$\frac{6x^2 - 15x + 22}{(x + 3)(x^2 + 2)^2} = \frac{A}{x + 3} + \frac{Bx + C}{x^2 + 2} + \frac{Dx + E}{(x^2 + 2)^2}$$

After considerable work, we discover that $A = 1$, $B = -1$, $C = 3$, $D = -5$, and $E = 0$. Thus,

$$\int \frac{6x^2 - 15x + 22}{(x + 3)(x^2 + 2)^2}\, dx$$

$$= \int \frac{dx}{x + 3} - \int \frac{x - 3}{x^2 + 2}\, dx - 5 \int \frac{x}{(x^2 + 2)^2}\, dx$$

$$= \int \frac{dx}{x + 3} - \frac{1}{2} \int \frac{2x}{x^2 + 2}\, dx + 3 \int \frac{dx}{x^2 + 2} - \frac{5}{2} \int \frac{2x\, dx}{(x^2 + 2)^2}$$

$$= \ln|x + 3| - \frac{1}{2} \ln(x^2 + 2) + \frac{3}{\sqrt{2}} \tan^{-1}\left(\frac{x}{\sqrt{2}}\right) + \frac{5}{2(x^2 + 2)} + C \qquad \blacksquare$$

Summary To decompose a rational function $f(x) = p(x)/q(x)$ into partial fractions, proceed as follows:

Step 1: If $f(x)$ is improper, that is, if $p(x)$ is of degree at least that of $q(x)$, divide $p(x)$ by $q(x)$, obtaining

$$f(x) = \text{a polynomial} + \frac{N(x)}{D(x)}$$

Step 2: Factor $D(x)$ into a product of linear and irreducible quadratic factors with real coefficients. By a theorem of algebra, this is always (theoretically) possible.

Step 3: For each factor of the form $(ax + b)^k$, expect the decomposition to have the terms

$$\frac{A_1}{(ax + b)} + \frac{A_2}{(ax + b)^2} + \cdots + \frac{A_k}{(ax + b)^k}$$

Step 4: For each factor of the form $(ax^2 + bx + c)^m$, expect the decomposition to have the terms

$$\frac{B_1 x + C_1}{ax^2 + bx + c} + \frac{B_2 x + C_2}{(ax^2 + bx + c)^2} + \cdots + \frac{B_m x + C_m}{(ax^2 + bx + c)^m}$$

Step 5: Set $N(x)/D(x)$ equal to the sum of all the terms found in Steps 3 and 4. The number of constants to be determined should equal the degree of the denominator, $D(x)$.

Step 6: Multiply both sides of the equation found in Step 5 by $D(x)$ and solve for the unknown constants. This can be done by either of two methods: (1) Equate coefficients of like-degree terms or (2) assign convenient values to the variable x.

Concepts Review

1. If the degree of the polynomial $p(x)$ is less than the degree of $q(x)$, then $f(x) = p(x)/q(x)$ is called a _____ rational function.

2. To integrate the improper rational function $f(x) = (x^2 + 4)/(x + 1)$, we first rewrite it as $f(x) =$ _____.

3. If $(x - 1)(x + 1) + 3x + x^2 = ax^2 + bx + c$, then $a =$ _____, $b =$ _____, and $c =$ _____.

4. $(3x + 1)/[(x - 1)^2(x^2 + 1)]$ can be decomposed in the form _____.

Problem Set 8.5

In Problems 1–40, use the method of partial fraction decomposition to perform the required integration.

1. $\displaystyle\int \frac{1}{x(x + 1)}\, dx$

2. $\displaystyle\int \frac{2}{x^2 + 3x}\, dx$

3. $\displaystyle\int \frac{3}{x^2 - 1}\, dx$

4. $\displaystyle\int \frac{5x}{2x^3 + 6x^2}\, dx$

5. $\displaystyle\int \frac{x - 11}{x^2 + 3x - 4}\, dx$

6. $\displaystyle\int \frac{x - 7}{x^2 - x - 12}\, dx$

7. $\displaystyle\int \frac{3x - 13}{x^2 + 3x - 10}\, dx$

8. $\displaystyle\int \frac{x + \pi}{x^2 - 3\pi x + 2\pi^2}\, dx$

9. $\displaystyle\int \frac{2x + 21}{2x^2 + 9x - 5}\, dx$

10. $\displaystyle\int \frac{2x^2 - x - 20}{x^2 + x - 6}\, dx$

11. $\displaystyle\int \frac{17x - 3}{3x^2 + x - 2}\, dx$

12. $\displaystyle\int \frac{5 - x}{x^2 - x(\pi + 4) + 4\pi}\, dx$

13. $\displaystyle\int \frac{2x^2 + x - 4}{x^3 - x^2 - 2x}\, dx$

14. $\displaystyle\int \frac{7x^2 + 2x - 3}{(2x - 1)(3x + 2)(x - 3)}\, dx$

15. $\displaystyle\int \frac{6x^2 + 22x - 23}{(2x - 1)(x^2 + x - 6)}\, dx$

16. $\displaystyle\int \frac{x^3 - 6x^2 + 11x - 6}{4x^3 - 28x^2 + 56x - 32}\, dx$

17. $\displaystyle\int \frac{x^3}{x^2 + x - 2}\, dx$

18. $\displaystyle\int \frac{x^3 + x^2}{x^2 + 5x + 6}\, dx$

19. $\displaystyle\int \frac{x^4 + 8x^2 + 8}{x^3 - 4x}\, dx$

20. $\displaystyle\int \frac{x^6 + 4x^3 + 4}{x^3 - 4x^2}\, dx$

21. $\displaystyle\int \frac{x + 1}{(x - 3)^2}\, dx$

22. $\displaystyle\int \frac{5x + 7}{x^2 + 4x + 4}\, dx$

23. $\displaystyle\int \frac{3x + 2}{x^3 + 3x^2 + 3x + 1}\, dx$

24. $\displaystyle\int \frac{x^6}{(x - 2)^2(1 - x)^5}\, dx$

25. $\displaystyle\int \frac{3x^2 - 21x + 32}{x^3 - 8x^2 + 16x}\, dx$

26. $\displaystyle\int \frac{x^2 + 19x + 10}{2x^4 + 5x^3}\, dx$

27. $\displaystyle\int \frac{2x^2 + x - 8}{x^3 + 4x}\, dx$

28. $\displaystyle\int \frac{3x + 2}{x(x + 2)^2 + 16x}\, dx$

29. $\displaystyle\int \frac{2x^2 - 3x - 36}{(2x - 1)(x^2 + 9)}\, dx$

30. $\displaystyle\int \frac{1}{x^4 - 16}\, dx$

31. $\displaystyle\int \frac{1}{(x - 1)^2(x + 4)^2}\, dx$

32. $\displaystyle\int \frac{x^3 - 8x^2 - 1}{(x + 3)(x^2 - 4x + 5)}\, dx$

33. $\displaystyle\int \frac{(\sin^3 t - 8\sin^2 t - 1)\cos t}{(\sin t + 3)(\sin^2 t - 4\sin t + 5)}\, dt$

34. $\displaystyle\int \frac{\cos t}{\sin^4 t - 16}\, dt$

35. $\displaystyle\int \frac{x^3 - 4x}{(x^2 + 1)^2}\, dx$

36. $\displaystyle\int \frac{(\sin t)(4\cos^2 t - 1)}{(\cos t)(1 + 2\cos^2 t + \cos^4 t)}\, dt$

37. $\displaystyle\int \frac{2x^3 + 5x^2 + 16x}{x^5 + 8x^3 + 16x}\, dx$

38. $\displaystyle\int_4^6 \frac{x - 17}{x^2 + x - 12}\, dx$

39. $\displaystyle\int_0^{\pi/4} \frac{\cos\theta}{(1 - \sin^2\theta)(\sin^2\theta + 1)^2}\, d\theta$

40. $\displaystyle\int_1^5 \frac{3x + 13}{x^2 + 4x + 3}\, dx$

41. The Law of Mass Action in chemistry results in the differential equation

$$\frac{dx}{dt} = k(a - x)(b - x), \qquad k > 0, \quad a > 0, \quad b > 0$$

where x is the amount of a substance at time t resulting from the reaction of two others. Assume that $x = 0$ when $t = 0$.

(a) Solve this differential equation in the case $b > a$.

(b) Show that $x \to a$ as $t \to \infty$ (if $b > a$).

(c) Suppose that $a = 2$ and $b = 4$ and that 1 gram of the substance is formed in 20 minutes. How much will be present in 1 hour?

(d) Solve the differential equation if $a = b$.

42. In many population growth problems, there is an upper limit beyond which the population cannot grow. Let us suppose that the earth will not support a population of more than 16 billion and that there were 2 billion people in 1925 and 4 billion people in 1975. Then, if y is the population t years after 1925, an appropriate model is the differential equation

$$\frac{dy}{dt} = ky(16 - y)$$

(a) Solve this differential equation.

(b) Find the population in 2015.

(c) When will the population be 9 billion?

This model, called the logistic model, was discussed in Section 7.5.

43. Do Problem 42 assuming that the upper limit for the population is 10 billion.

44. The differential equation

$$\frac{dy}{dt} = k(y - m)(M - y)$$

with $k > 0$ and $0 \leq m < y_0 < M$ is used to model some growth problems. Solve the equation and find $\lim\limits_{t \to \infty} y$.

45. As a model for the production of trypsin from trypsinogen in digestion, biochemists have proposed the model

$$\frac{dy}{dt} = k(A - y)(B + y)$$

where $k > 0$, A is the initial amount of trypsinogen, and B is the original amount of trypsin. Solve this differential equation.

46. Evaluate

$$\int_{\pi/6}^{\pi/2} \frac{\cos x}{\sin x (\sin^2 x + 1)^2} \, dx$$

Answers to Concepts Review: **1.** proper   **2.** $x - 1 + \dfrac{5}{x + 1}$

3. $2; 3; -1$ **4.** $\dfrac{A}{x - 1} + \dfrac{B}{(x - 1)^2} + \dfrac{Cx + D}{x^2 + 1}$

8.6 Chapter Review

Concepts Test

Respond with true or false to each of the following assertions. Be prepared to justify your answer.

1. To evaluate $\displaystyle\int x \sin(x^2) \, dx$, make the substitution $u = x^2$.

2. To evaluate $\displaystyle\int \frac{x}{9 + x^4} \, dx$, make the substitution $u = x^2$.

3. To evaluate $\displaystyle\int \frac{x^3}{9 + x^4} \, dx$, make the substitution $u = x^2$.

4. To evaluate $\displaystyle\int \frac{2x - 3}{x^2 - 3x + 5} \, dx$, begin by completing the square of the denominator.

5. To evaluate $\displaystyle\int \frac{3}{x^2 - 3x + 5} \, dx$, begin by completing the square of the denominator.

6. To evaluate $\displaystyle\int \frac{1}{\sqrt{4 - 5x^2}} \, dx$, make the substitution $u = \sqrt{5}\, x$.

7. To evaluate $\displaystyle\int \frac{t + 2}{t^3 - 9t} \, dt$, make a partial fraction decomposition.

8. To evaluate $\displaystyle\int \frac{t^4}{t^2 - 1} \, dt$, use integration by parts.

9. To evaluate $\displaystyle\int \sin^6 x \cos^2 x \, dx$, use half-angle formulas.

10. To evaluate $\displaystyle\int \frac{e^x}{1 + e^x} \, dx$, use integration by parts.

11. To evaluate $\displaystyle\int \frac{x + 2}{\sqrt{-x^2 - 4x}} \, dx$, use a trigonometric substitution.

12. To evaluate $\displaystyle\int x^2 \sqrt[3]{3 - 2x} \, dx$, let $u = \sqrt[3]{3 - 2x}$.

13. To evaluate $\displaystyle\int \sin^2 x \cos^5 x \, dx$, rewrite the integrand as $\sin^2 x (1 - \sin^2 x)^2 \cos x$.

14. To evaluate $\displaystyle\int \frac{1}{x^2 \sqrt{9 - x^2}} \, dx$, make a trigonometric substitution.

15. To evaluate $\displaystyle\int x^2 \ln x \, dx$, use integration by parts.

16. To evaluate $\displaystyle\int \sin 2x \cos 4x \, dx$, use half-angle formulas.

17. $\dfrac{x^2}{x^2 - 1}$ can be expressed in the form $\dfrac{A}{x - 1} + \dfrac{B}{x + 1}$.

18. $\dfrac{x^2 + 2}{x(x^2 - 1)}$ can be expressed in the form $\dfrac{A}{x} + \dfrac{B}{x - 1} + \dfrac{C}{x + 1}$.

19. $\dfrac{x^2 + 2}{x(x^2 + 1)}$ can be expressed in the form $\dfrac{A}{x} + \dfrac{Bx + C}{x^2 + 1}$.

20. $\dfrac{x + 2}{x^2(x^2 - 1)}$ can be expressed in the form

$\dfrac{A}{x^2} + \dfrac{B}{x - 1} + \dfrac{C}{x + 1}$.

21. To complete the square of $ax^2 + bx$, add $(b/2)^2$.

22. Any polynomial with real coefficients can be factored into a product of linear polynomials with real coefficients.

23. Two polynomials in x have the same values for all x if and only if the coefficients of like-degree terms are identical.

Sample Test Problems

In Problems 1–42, evaluate each integral.

1. $\displaystyle\int_0^4 \frac{t}{\sqrt{9 + t^2}}\, dt$

2. $\displaystyle\int \cot^2 (2\theta)\, d\theta$

3. $\displaystyle\int_0^{\pi/2} e^{\cos x} \sin x\, dx$

4. $\displaystyle\int_0^{\pi/4} x \sin 2x\, dx$

5. $\displaystyle\int \frac{y^3 + y}{y + 1}\, dy$

6. $\displaystyle\int \sin^3 (2t)\, dt$

7. $\displaystyle\int \frac{y - 2}{y^2 - 4y + 2}\, dy$

8. $\displaystyle\int_0^{3/2} \frac{dy}{\sqrt{2y + 1}}$

9. $\displaystyle\int \frac{e^{2t}}{e^t - 2}\, dt$

10. $\displaystyle\int \frac{\sin x + \cos x}{\tan x}\, dx$

11. $\displaystyle\int \frac{dx}{\sqrt{16 + 4x - 2x^2}}$

12. $\displaystyle\int x^2 e^x\, dx$

13. $\displaystyle\int \frac{dy}{\sqrt{2 + 3y^2}}$

14. $\displaystyle\int \frac{w^3}{1 - w^2}\, dw$

15. $\displaystyle\int \frac{\tan x}{\ln|\cos x|}\, dx$

16. $\displaystyle\int \frac{3\, dt}{t^3 - 1}$

17. $\displaystyle\int \sinh x\, dx$

18. $\displaystyle\int \frac{(\ln y)^5}{y}\, dy$

19. $\displaystyle\int x \cot^2 x\, dx$

20. $\displaystyle\int \frac{\sin \sqrt{x}}{\sqrt{x}}\, dx$

21. $\displaystyle\int \frac{\ln t^2}{t}\, dt$

22. $\displaystyle\int \ln(y^2 + 9)\, dy$

23. $\displaystyle\int e^{t/3} \sin 3t\, dt$

24. $\displaystyle\int \frac{t + 9}{t^3 + 9t}\, dt$

25. $\displaystyle\int \sin \frac{3x}{2} \cos \frac{x}{2}\, dx$

26. $\displaystyle\int \cos^4 \left(\frac{x}{2}\right) dx$

27. $\displaystyle\int \tan^3 2x \sec 2x\, dx$

28. $\displaystyle\int \frac{\sqrt{x}}{1 + \sqrt{x}}\, dx$

29. $\displaystyle\int \tan^{3/2} x \sec^4 x\, dx$

30. $\displaystyle\int \frac{dt}{t(t^{1/6} + 1)}$

31. $\displaystyle\int \frac{e^{2y}\, dy}{\sqrt{9 - e^{2y}}}$

32. $\displaystyle\int \cos^5 x \sqrt{\sin x}\, dx$

33. $\displaystyle\int e^{\ln(3\cos x)}\, dx$

34. $\displaystyle\int \frac{\sqrt{9 - y^2}}{y}\, dy$

35. $\displaystyle\int \frac{e^{4x}}{1 + e^{8x}}\, dx$

36. $\displaystyle\int \frac{\sqrt{x^2 + a^2}}{x^4}\, dx$

37. $\displaystyle\int \frac{w}{\sqrt{w + 5}}\, dw$

38. $\displaystyle\int \frac{\sin t\, dt}{\sqrt{1 + \cos t}}$

39. $\displaystyle\int \frac{\sin y \cos y}{9 + \cos^4 y}\, dy$

40. $\displaystyle\int \frac{dx}{\sqrt{1 - 6x - x^2}}$

41. $\displaystyle\int \frac{4x^2 + 3x + 6}{x^2(x^2 + 3)}\, dx$

42. $\displaystyle\int \frac{dx}{(16 + x^2)^{3/2}}$

43. Express the partial fraction decomposition of each rational function without computing the exact coefficients. For example,

$$\frac{3x + 1}{(x - 1)^2} = \frac{A}{(x - 1)} + \frac{B}{(x - 1)^2}$$

(a) $\displaystyle\frac{3 - 4x^2}{(2x + 1)^3}$

(b) $\displaystyle\frac{7x - 41}{(x - 1)^2(2 - x)^3}$

(c) $\displaystyle\frac{3x + 1}{(x^2 + x + 10)^2}$

(d) $\displaystyle\frac{(x + 1)^2}{(x^2 - x + 10)^2(1 - x^2)^2}$

(e) $\displaystyle\frac{x^5}{(x + 3)^4(x^2 + 2x + 10)^2}$

(f) $\displaystyle\frac{(3x^2 + 2x - 1)^2}{(2x^2 + x + 10)^3}$

44. Find the volume of the solid generated by revolving the region under the graph of

$$y = \frac{1}{\sqrt{3x - x^2}}$$

from $x = 1$ to $x = 2$ about

(a) the x-axis;   (b) the y-axis.

45. Find the length of the curve $y = x^2/16$ from $x = 0$ to $x = 4$.

46. The region under the curve

$$y = \frac{1}{x^2 + 5x + 6}$$

from $x = 0$ to $x = 3$ is rotated about the x-axis. Compute the volume of the solid that is generated.

47. If the curve given in Problem 46 is rotated about the y-axis, find the volume of the solid.

48. Find the volume of the solid created by revolving the region bounded by the x-axis and the curve $y = 4x\sqrt{2 - x}$ about the y-axis.

49. Find the volume when the area created by the x-axis, y-axis, the curve $y = 2(e^x - 1)$ and the curve $x = \ln 3$ is revolved about the line $x = \ln 3$.

50. Find the are of the region bounded by the x-axis, the curve $y = 18/(x^2\sqrt{x^2 + 9})$, and the lines $x = \sqrt{3}$ and $x = 3\sqrt{3}$.

51. Find the are of the region bounded by the curve $s = t/(t - 1)^2$, $s = 0$, $t = -6$, and $t = 0$.

≈ **52.** Find the volume of the solid generated by revolving the region

$$\left\{ (x, y) : -3 \le x \le -1, \frac{6}{x\sqrt{x + 4}} \le y \le 0 \right\}$$

about the x-axis. Make a sketch.

≈ **53.** Find the length of the segment of the curve $y = \ln(\sin x)$ from $x = \pi/6$ to $x = \pi/3$.

Integration Using a Computer Algebra System

I. Preparation

Mathematica and Maple are widely used computer algebra systems (CASs). We assume that you have one of these, or a TI-89 or TI-92 calculator, available to help evaluate definite and indefinite integrals. We will investigate how these systems can be used, as well as some of their pitfalls.

Exercise 1   Study the syntax for doing definite and indefinite integration on your CAS. In Mathematica, the indefinite integral of $f(x)$ is obtained with `Integrate [f[x], x]` and the definite integral $\int_a^b f(x)\,dx$ with `Integrate [f[x],{x,a,b}]`. The corresponding forms in Maple are `int(f(x), x)` and `int(f(x), x=a..b)`. The TI-89 and TI-92 can evaluate integrals such as $\int x^2\,dx$ and $\int_1^2 x^2\,dx$ by entering

$$\boxed{\text{2nd}} \quad \int(\text{x}^2,\text{x}) \quad \text{and} \quad \boxed{\text{2nd}} \quad \int(\text{x}^2,\text{x},1,2)$$

respectively. Both Mathematica (using the `Apart` command) and Maple (using the `convert` command) are able to compute the partial decomposition of a rational function.

Exercise 2   Evaluate the following integrals by hand.

(a) $\displaystyle\int (3x + 10)^{50}\,dx$ (b) $\displaystyle\int x(3x + 10)^{50}\,dx$

Hint: Use integration by parts for part (b).

(c) $\displaystyle\int x(3x^2 + 10)^{50}\,dx$ (d) $\displaystyle\int x^2(3x^3 + 10)^{50}\,dx$

Exercise 3   Decompose $\dfrac{x}{x^2 - 4}$ by hand.

II. Using Technology

Exercise 4   Use a CAS to evaluate each indefinite integral in Exercise 2. In which cases (if any) does the CAS's answer differ from the one that you obtained by hand in Exercise 2? Which answer is simpler? (In some cases, CASs will multiply out and integrate term by term, even when a simpler method works.)

Exercise 5   Use your CAS to find the partial fraction decomposition of each function. After you obtain the partial fraction decomposition, evaluate the indefinite integral of each expression.

(a) $\dfrac{x}{x^2 - 4}$ (b) $\dfrac{x^2}{x^2 - 4}$

(c) $\dfrac{x^3}{x^2 - 4}$ (d) $\dfrac{x^4}{x^2 - 4}$

(e) $\dfrac{x^3 + 3x^2 - 2x + 6}{706x^3 - 435x^2 - 273x^4 - 118x + 120}$

(f) $\dfrac{3x^3 + 5x^2 + 7x + 16}{x^4 + 8x^3 + 48x^2 + 119x + 230}$

Exercise 6   Evaluate by hand the indefinite integral

$$\int \sin^2 x \cos^3 x\,dx$$

Next use your CAS to evaluate the integral. Maple gives

$$\int \sin^2 x \cos^3 x\,dx$$

$$= -\frac{1}{5}\sin x \cos^4 x + \frac{1}{15}\cos^2 x \sin x + \frac{2}{15}\sin x$$

whereas Mathematica gives

$$\int \sin^2 x \cos^3 x\,dx = \frac{1}{240}(30\sin x - 5\sin 3x - 3\sin 5x)$$

Learn the commands `combine` and `expand` in Maple or `TrigReduce` and `TrigExpand` in Mathematica. By using the proper commands, transform one answer into the other.

Exercise 7   A CAS may omit absolute value signs, causing incorrect results. Consider

$$\int \sqrt{9x^2 + x^4}\,dx$$

(a) Explain why $\sqrt{9x^2 + x^4} = |x|\sqrt{9 + x^2}$.

(b) Explain why

$$\int_{-1}^{1} \sqrt{9x^2 + x^4}\,dx = 2\int_0^1 \sqrt{9x^2 + x^4}\,dx$$

(c) Compute $\displaystyle\int_0^1 \sqrt{9x^2 + x^4}\,dx$ by hand.

(d) Use your CAS to evaluate $\displaystyle\int \sqrt{9x^2 + x^4}\,dx$,

$$\int_{-1}^{1} \sqrt{9x^2 + x^4}\,dx, \text{ and } \int_0^1 \sqrt{9x^2 + x^4}\,dx.$$

III. Reflection

Exercise 8   If you received contradictory results in Exercise 7, determine which results are correct and explain why.

Exercise 9   Use your CAS to find partial fraction decompositions of $\dfrac{x^k}{x^2 - 4}$ for $k = 0, 1, 2, \ldots, 10$. Identify a pattern and use the pattern to predict the partial fraction decompositions for $\dfrac{x^{11}}{x^2 - 4}$ and $\dfrac{x^{12}}{x^2 - 4}$. Find an expression for the partial fraction decomposition of $\dfrac{x^n}{x^2 - 4}$. *Hint*: You should have a two-part answer; one when n is even and one when n is odd. *Bonus*: Prove, using mathematical induction, that your conjecture is correct.

The Logistic Differential Equation

I. Preparation

In Chapter 7 we used the differential equation $y' = ky$ to model the growth of a population. This model says that the growth rate is proportional to the size y of the population. This differential equation has solution $y = y_0 e^{kt}$, where y_0 is the initial population size. This model, which leads to *exponential growth*, is reasonable until the size of the population becomes so large that factors such as space and resources affect the growth of the population. Pierre Verhulst (1804–1849) introduced in 1838 the model

$$(1) \qquad y' = ay - by^2$$

Typically, the value of b is small, so when the population size y is small, the rate of growth is roughly proportional to y. However, when the population becomes larger, the $-by^2$ term becomes significant, and it restricts the rate of growth and forces a limiting population size L.

It is a remarkable fact that we can determine the limiting population size L without solving the differential equation. If there is a number L to which the population converges as $t \to \infty$ (that is, if $\lim_{t \to \infty} y(t) = L$), then

$$(2) \qquad \lim_{t \to \infty} y'(t) = 0$$

(If the population settles down to nearly the value L, then the rate of change will be near zero.)

Exercise 1 In the equation $y' = ay - by^2$, take the limit as $t \to \infty$ on both sides (remember, y is a function of time t) and use the result in (2). Then solve for the limiting population L.

Exercise 2 Show that the model in (1) can be written in the form

$$y' = ky(L - y)$$

This shows that the rate of growth of the population is proportional to both the population size y and the "room" that is left for the population to grow (that is, the difference between the limiting population size L and the actual population size y).

II. Using Technology

Exercise 3 Solve $y' = ky(L - y)$ with the initial condition $y(0) = y_0$. *Hint:* Separate variables and integrate.

Exercise 4 Set $L = 1000$, $k = 0.0005$, and $y_0 = 200$ and plot the solution to the logistic equation. Holding L and k constant, vary the value of y_0. Plot the resulting solutions. Be sure to include some values of y_0 below and above L; you should also try $y_0 = 0$ and $y_0 = L$. Describe the effect of y_0 on the solution.

Exercise 5 Set $k = 0.0005$ and $y_0 = 200$. Vary the value of L and plot the solutions. Describe the effect of L on the solution.

Exercise 6 Set $L = 1000$ and $y_0 = 200$. Vary the value of k and plot the solutions. Describe the effect of k on the solution.

III. Reflection

An **autonomous equation** is a first-order differential equation of the form

$$y' = F(y)$$

In other words, the time variable t does not appear explicitly in the differential equation. For example, $y' = 2y$ and $y' = 2y(3 - y)$ are autonomous, whereas $y' = t^2 y$ is not. A **stationary point** of an autonomous differential equation is a constant c with the property that the constant function $y(t) = c$ is a solution to the differential equation.

Exercise 7 Explain why all roots of $F(y) = 0$ are stationary points.

Exercise 8 Find all stationary points of the logistic equation $y' = 0.002y (500 - y)$. Explain the significance of the stationary points as they relate to the population size.

Exercise 9 Find all stationary points of the equations

$$y' = 2y(3 - y)(4 - y)(5 - y)$$

and

$$y' = y^3 - 6y^2 + 8y.$$

L'Hôpital's Rule In 1696, Guillaume François Antoine de l'Hôpital published the first textbook on differential calculus; it included the following rule, which he had learned from his teacher Johann Bernoulli.

Theorem A   L'Hôpital's Rule for forms of type 0/0

Suppose that $\lim_{x \to u} f(x) = \lim_{x \to u} g(x) = 0$. If $\lim_{x \to u} \left[f'(x)/g'(x) \right]$ exists in either the finite or infinite sense (i.e., if this limit is a finite number or $-\infty$ or $+\infty$), then

$$\lim_{x \to u} \frac{f(x)}{g(x)} = \lim_{x \to u} \frac{f'(x)}{g'(x)}$$

Here u may stand for any of the symbols a, a^-, a^+, $-\infty$, or $+\infty$.

Geometric Interpretation of l'Hôpital's Rule

Study the diagrams below. They should make l'Hôpital's Rule seem quite reasonable. (See Problems 36–40.)

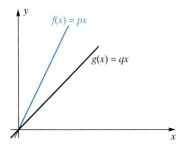

$$\lim_{x \to 0} \frac{f(x)}{g(x)} = \lim_{x \to 0} \frac{px}{qx} = \frac{p}{q} = \lim_{x \to 0} \frac{f'(x)}{g'(x)}$$

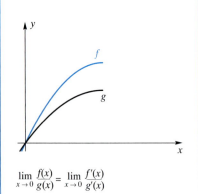

$$\lim_{x \to 0} \frac{f(x)}{g(x)} = \lim_{x \to 0} \frac{f'(x)}{g'(x)}$$

Before attempting to prove this theorem, we illustrate it. Note that l'Hôpital's Rule allows us to replace one limit by another, which may be simpler and, in particular, may not have the 0/0 form.

EXAMPLE 1   Use l'Hôpital's Rule to show that

$$\lim_{x \to 0} \frac{\sin x}{x} = 1 \quad \text{and} \quad \lim_{x \to 0} \frac{1 - \cos x}{x} = 0$$

Solution   We worked pretty hard to demonstrate these two facts in Section 2.7. After noting that trying to evaluate both limits by substitution leads to the form 0/0, we can now establish the desired results in two lines (but see Problem 25). By l'Hôpital's Rule,

$$\lim_{x \to 0} \frac{\sin x}{x} = \lim_{x \to 0} \frac{D_x \sin x}{D_x x} = \lim_{x \to 0} \frac{\cos x}{1} = 1$$

$$\lim_{x \to 0} \frac{1 - \cos x}{x} = \lim_{x \to 0} \frac{D_x(1 - \cos x)}{D_x x} = \lim_{x \to 0} \frac{\sin x}{1} = 0 \qquad \blacksquare$$

EXAMPLE 2   Find $\lim_{x \to 3} \dfrac{x^2 - 9}{x^2 - x - 6}$ and $\lim_{x \to 2^+} \dfrac{x^2 + 3x - 10}{x^2 - 4x + 4}$.

Solution   Both limits have the 0/0 form, so, by l'Hôpital's Rule,

$$\lim_{x \to 3} \frac{x^2 - 9}{x^2 - x - 6} = \lim_{x \to 3} \frac{2x}{2x - 1} = \frac{6}{5}$$

$$\lim_{x \to 2^+} \frac{x^2 + 3x - 10}{x^2 - 4x + 4} = \lim_{x \to 2^+} \frac{2x + 3}{2x - 4} = \infty$$

The first of these limits was handled at the beginning of this section by factoring and simplifying. Of course, we get the same answer either way. $\qquad \blacksquare$

EXAMPLE 3   Find $\lim_{x \to 0} \dfrac{\tan 2x}{\ln(1 + x)}$.

Solution   Both numerator and denominator have limit 0. Hence,

$$\lim_{x \to 0} \frac{\tan 2x}{\ln(1 + x)} = \lim_{x \to 0} \frac{2 \sec^2 2x}{1/(1 + x)} = \frac{2}{1} = 2 \qquad \blacksquare$$

Sometimes $\lim f'(x)/g'(x)$ also has the indeterminate form 0/0. Then we may apply l'Hôpital's Rule again, as we now illustrate. Each application of l'Hôpital's Rule is flagged with the symbol Ⓛ.

EXAMPLE 4   Find $\lim\limits_{x \to 0} \dfrac{\sin x - x}{x^3}$.

Solution   By l'Hôpital's Rule applied three times in succession,

$$\lim_{x \to 0} \frac{\sin x - x}{x^3} \;\overset{\text{(L)}}{=}\; \lim_{x \to 0} \frac{\cos x - 1}{3x^2}$$

$$\overset{\text{(L)}}{=}\; \lim_{x \to 0} \frac{-\sin x}{6x}$$

$$\overset{\text{(L)}}{=}\; \lim_{x \to 0} \frac{-\cos x}{6}$$

$$= -\frac{1}{6}$$ ∎

Just because we have an elegant rule does not mean that we should use it indiscriminately. In particular, we must always make sure that it applies; that is, we must make sure that the limit has the indeterminate form 0/0. Otherwise we will be led into all kinds of errors, as we now illustrate.

EXAMPLE 5   Find $\lim\limits_{x \to 0} \dfrac{1 - \cos x}{x^2 + 3x}$.

Solution

$$\lim_{x \to 0} \frac{1 - \cos x}{x^2 + 3x} \;\overset{\text{(L)}}{=}\; \lim_{x \to 0} \frac{\sin x}{2x + 3} \;\overset{\text{(L)}}{=}\; \lim_{x \to 0} \frac{\cos x}{2} = \frac{1}{2} \qquad \text{WRONG}$$

The first application of l'Hôpital's Rule was correct; the second was not since, at that stage, the limit did not have the 0/0 form. Here is what we should have done.

$$\lim_{x \to 0} \frac{1 - \cos x}{x^2 + 3x} \;\overset{\text{(L)}}{=}\; \lim_{x \to 0} \frac{\sin x}{2x + 3} = 0 \qquad \text{RIGHT}$$

We stop differentiating as soon as either the numerator or denominator has a nonzero limit.   ∎

Even if the conditions of l'Hôpital's Rule hold, an application of l'Hôpital's Rule may not help us; witness the following example.

EXAMPLE 6   Find $\lim\limits_{x \to \infty} \dfrac{e^{-x}}{x^{-1}}$.

Solution   Since the numerator and denominator both tend to 0, the limit is indeterminate of the form 0/0. Thus, the conditions of Theorem A hold. We may apply l'Hôpital's Rule indefinitely.

$$\lim_{x \to \infty} \frac{e^{-x}}{x^{-1}} \;\overset{\text{(L)}}{=}\; \lim_{x \to \infty} \frac{e^{-x}}{x^{-2}} \;\overset{\text{(L)}}{=}\; \lim_{x \to \infty} \frac{e^{-x}}{2x^{-3}} = \cdots$$

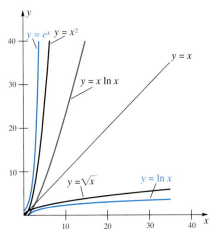

Figure 2

A similar argument works for any $a > 0$. Let m denote the greatest integer less than a. Then $m + 1$ applications of l'Hôpital's Rule give

$$\lim_{x \to \infty} \frac{x^a}{e^x} \overset{L}{=} \lim_{x \to \infty} \frac{ax^{a-1}}{e^x} \overset{L}{=} \lim_{x \to \infty} \frac{a(a-1)x^{a-2}}{e^x} \overset{L}{=} \cdots \overset{L}{=} \lim_{x \to \infty} \frac{a(a-1)\cdots(a-m)}{x^{m+1-a}e^x} = 0$$

∎

EXAMPLE 3   Show that, if a is any positive real number, $\displaystyle\lim_{x \to \infty} \frac{\ln x}{x^a} = 0$.

Solution   Both $\ln x$ and x^a tend to ∞ as $x \to \infty$. Hence, by one application of l'Hôpital's Rule,

$$\lim_{x \to \infty} \frac{\ln x}{x^a} \overset{L}{=} \lim_{x \to \infty} \frac{1/x}{ax^{a-1}} = \lim_{x \to \infty} \frac{1}{ax^a} = 0$$

∎

Examples 2 and 3 say something that is worth remembering: *for sufficiently large x, e^x grows faster as x increases than any constant power of x, whereas $\ln x$ grows more slowly than any constant power of x.* For example, when x is sufficiently large, e^x grows faster than x^{100} and $\ln x$ grows more slowly than $\sqrt[100]{x}$. The chart in the margin and Figure 2 offer additional illustration.

EXAMPLE 4   Find $\displaystyle\lim_{x \to 0^+} \frac{\ln x}{\cot x}$.

Solution   As $x \to 0^+$, $\ln x \to -\infty$ and $\cot x \to \infty$, so l'Hôpital's Rule applies.

$$\lim_{x \to 0^+} \frac{\ln x}{\cot x} \overset{L}{=} \lim_{x \to 0^+} \left[\frac{1/x}{-\csc^2 x} \right]$$

This is still indeterminate as it stands, but rather than apply l'Hôpital's Rule again (which only makes things worse), we rewrite the bracketed expression as

$$\frac{1/x}{-\csc^2 x} = -\frac{\sin^2 x}{x} = -\sin x \frac{\sin x}{x}$$

Thus,

$$\lim_{x \to 0^+} \frac{\ln x}{\cot x} = \lim_{x \to 0^+} \left[-\sin x \frac{\sin x}{x} \right] = 0 \cdot 1 = 0$$

∎

The Indeterminate Forms $0 \cdot \infty$ and $\infty - \infty$

Suppose that $A(x) \to 0$, but $B(x) \to \infty$. What is going to happen to the product $A(x)B(x)$? Two competing forces are at work, tending to pull the product in opposite directions. Which will win this battle, A or B or neither? It depends on whether one is stronger (i.e., doing its job at a faster rate) or whether they are evenly matched. L'Hôpital's Rule will help us to decide, but only after we have transformed the problem to a 0/0 or ∞/∞ form.

EXAMPLE 5   Find $\lim\limits_{x \to \pi/2} (\tan x \cdot \ln \sin x)$.

Solution   Since $\lim\limits_{x \to \pi/2} \ln \sin x = 0$ and $\lim\limits_{x \to \pi/2} |\tan x| = \infty$, this is a $0 \cdot \infty$ indeterminate form. We can rewrite it as a $0/0$ form by the simple device of changing $\tan x$ to $1/\cot x$. Thus,

$$\lim_{x \to \pi/2} (\tan x \cdot \ln \sin x) = \lim_{x \to \pi/2} \frac{\ln \sin x}{\cot x}$$

$$\overset{\text{(L)}}{=} \lim_{x \to \pi/2} \frac{\frac{1}{\sin x} \cdot \cos x}{-\csc^2 x}$$

$$= \lim_{x \to \pi/2} (-\cos x \cdot \sin x) = 0 \qquad \blacksquare$$

EXAMPLE 6   Find $\lim\limits_{x \to 1^+} \left(\dfrac{x}{x - 1} - \dfrac{1}{\ln x} \right)$.

Solution   The first term is growing without bound; so is the second. We say that the limit is an $\infty - \infty$ indeterminate form. L'Hôpital's Rule will determine the result, but only after we rewrite the problem in a form for which the rule applies. In this case, the two fractions must be combined, a procedure that changes the problem to a $0/0$ form. Two applications of l'Hôpital's Rule yield

$$\lim_{x \to 1^+} \left(\frac{x}{x-1} - \frac{1}{\ln x} \right) = \lim_{x \to 1^+} \frac{x \ln x - x + 1}{(x-1)\ln x} \overset{\text{(L)}}{=} \lim_{x \to 1^+} \frac{x \cdot 1/x + \ln x - 1}{(x-1)(1/x) + \ln x}$$

$$= \lim_{x \to 1^+} \frac{x \ln x}{x - 1 + x \ln x} \overset{\text{(L)}}{=} \lim_{x \to 1^+} \frac{1 + \ln x}{2 + \ln x} = \frac{1}{2} \qquad \blacksquare$$

The Indeterminate Forms 0^0, ∞^0, 1^∞   We turn now to three indeterminate forms of exponential type. Here the trick is to consider not the original expression, but rather its logarithm. Usually, l'Hôpital's Rule will apply to the logarithm.

EXAMPLE 7   Find $\lim\limits_{x \to 0^+} (x + 1)^{\cot x}$.

Solution   This takes the indeterminate form 1^∞. Let $y = (x + 1)^{\cot x}$, so

$$\ln y = \cot x \ln (x + 1) = \frac{\ln(x + 1)}{\tan x}$$

Using l'Hôpital's Rule for $0/0$ forms, we obtain

$$\lim_{x \to 0^+} \ln y = \lim_{x \to 0^+} \frac{\ln(x + 1)}{\tan x} \overset{\text{(L)}}{=} \lim_{x \to 0^+} \frac{\frac{1}{x + 1}}{\sec^2 x} = 1$$

Now $y = e^{\ln y}$, and since the exponential function $f(x) = e^x$ is continuous,

$$\lim_{x \to 0^+} y = \lim_{x \to 0^+} \exp(\ln y) = \exp \left(\lim_{x \to 0^+} \ln y \right) = \exp 1 = e \qquad \blacksquare$$

EXAMPLE 8   Find $\lim\limits_{x \to \pi/2^-} (\tan x)^{\cos x}$.

Solution   This has the indeterminate form ∞^0. Let $y = (\tan x)^{\cos x}$, so

$$\ln y = \cos x \cdot \ln \tan x = \frac{\ln \tan x}{\sec x}$$

Then

$$\lim_{x \to \pi/2^-} \ln y = \lim_{x \to \pi/2^-} \frac{\ln \tan x}{\sec x} \overset{L}{=} \lim_{x \to \pi/2^-} \frac{\frac{1}{\tan x} \cdot \sec^2 x}{\sec x \tan x}$$

$$= \lim_{x \to \pi/2^-} \frac{\sec x}{\tan^2 x} = \lim_{x \to \pi/2^-} \frac{\cos x}{\sin^2 x} = 0$$

Therefore,

$$\lim_{x \to \pi/2^-} y = e^0 = 1 \qquad \blacksquare$$

Summary   We have classified certain limit problems as indeterminate forms, using the seven symbols $0/0, \infty/\infty, 0 \cdot \infty, \infty - \infty, 0^0, \infty^0$, and 1^∞. Each involves a competition of opposing forces, which means that the result is not obvious. However, with the help of l'Hôpital's Rule, which applies directly only to the $0/0$ and ∞/∞ forms, we can usually determine the limit.

There are many other possibilities symbolized by, for example, $0/\infty, \infty/0$, $\infty + \infty, \infty \cdot \infty, 0^\infty$, and ∞^∞. Why don't we call these indeterminate forms? Because, in each of these cases, the forces work together, not in competition.

EXAMPLE 9   Find $\lim\limits_{x \to 0^+} (\sin x)^{\cot x}$.

Solution   We might call this a 0^∞ form, but it is not indeterminate. Note that $\sin x$ is approaching zero, and raising it to the exponent $\cot x$, an increasingly large number, serves only to make it approach zero faster. Thus,

$$\lim_{x \to 0^+} (\sin x)^{\cot x} = 0 \qquad \blacksquare$$

Concepts Review

1. If $\lim\limits_{x \to a} f(x) = \lim\limits_{x \to a} g(x) = \infty$, then l'Hôpital's Rule says that $\lim\limits_{x \to a} f(x)/g(x) = \lim\limits_{x \to a}$ _____ .

2. If $\lim\limits_{x \to a} f(x) = 0$ and $\lim\limits_{x \to a} g(x) = \infty$, then $\lim\limits_{x \to a} f(x)g(x)$ is an indeterminate form. To apply l'Hôpital's Rule, we may rewrite this latter limit as _____ .

3. Seven indeterminate forms are discussed in this book. They are symbolized by $0/0, \infty/\infty, 0 \cdot \infty$, and _____ .

4. e^x grows faster than any power of x, but _____ grows more slowly than any power of x.

Problem Set 9.2

Find each limit in Problems 1–40. Be sure you have an indeterminate form before applying l'Hôpital's Rule.

1. $\lim\limits_{x \to \infty} \dfrac{\ln x^{10000}}{x}$

2. $\lim\limits_{x \to \infty} \dfrac{(\ln x)^2}{2^x}$

3. $\lim\limits_{x \to \infty} \dfrac{x^{10000}}{e^x}$

4. $\lim\limits_{x \to \infty} \dfrac{3x}{\ln(100x + e^x)}$

5. $\lim\limits_{x \to \pi/2} \dfrac{3 \sec x + 5}{\tan x}$

6. $\lim\limits_{x \to 0^+} \dfrac{\ln \sin^2 x}{3 \ln \tan x}$

7. $\lim\limits_{x \to \infty} \dfrac{\ln(\ln x^{1000})}{\ln x}$

8. $\lim\limits_{x \to (1/2)^-} \dfrac{\ln(4 - 8x)^2}{\tan \pi x}$

9. $\lim\limits_{x \to 0^+} \dfrac{\cot x}{\sqrt{-\ln x}}$

10. $\lim\limits_{x \to 0} \dfrac{2 \csc^2 x}{\cot^2 x}$

11. $\lim\limits_{x \to 0} (x \ln x^{1000})$

12. $\lim\limits_{x \to 0} 3x^2 \csc^2 x$

13. $\lim\limits_{x \to 0} (\csc^2 x - \cot^2 x)$

14. $\lim\limits_{x \to \pi/2} (\tan x - \sec x)$

15. $\lim\limits_{x \to 0^+} (3x)^{x^2}$

16. $\lim\limits_{x \to 0} (\cos x)^{\csc x}$

17. $\lim\limits_{x \to (\pi/2)^-} (5 \cos x)^{\tan x}$

18. $\lim\limits_{x \to 0} \left(\csc^2 x - \dfrac{1}{x^2} \right)^2$

19. $\lim\limits_{x \to 0} (x + e^{x/3})^{3/x}$

20. $\lim\limits_{x \to (\pi/2)^-} (\cos 2x)^{x - \pi/2}$

21. $\lim\limits_{x \to \pi/2} (\sin x)^{\cos x}$

22. $\lim\limits_{x \to \infty} x^x$

23. $\lim\limits_{x \to \infty} x^{1/x}$

24. $\lim\limits_{x \to 0} (\cos x)^{1/x^2}$

25. $\lim\limits_{x \to 0^+} (\tan x)^{2/x}$

26. $\lim\limits_{x \to -\infty} (e^{-x} - x)$

27. $\lim\limits_{x \to 0^+} (\sin x)^x$

28. $\lim\limits_{x \to 0} (\cos x - \sin x)^{1/x}$

29. $\lim\limits_{x \to 0} \left(\csc x - \dfrac{1}{x} \right)$

30. $\lim\limits_{x \to \infty} \left(1 + \dfrac{1}{x} \right)^x$

31. $\lim\limits_{x \to 0^+} (1 + 2e^x)^{1/x}$

32. $\lim\limits_{x \to 1} \left(\dfrac{1}{x - 1} - \dfrac{x}{\ln x} \right)$

33. $\lim\limits_{x \to 0} (\cos x)^{1/x}$

34. $\lim\limits_{x \to 0^+} (x^{1/2} \ln x)$

35. $\lim\limits_{x \to \infty} e^{\cos x}$

36. $\lim\limits_{x \to \infty} [\ln(x + 1) - \ln(x - 1)]$

37. $\lim\limits_{x \to 0^+} \dfrac{x}{\ln x}$

38. $\lim\limits_{x \to 0^+} (\ln x \cot x)$

39. $\lim\limits_{x \to \infty} \dfrac{\displaystyle \int_1^x \sqrt{1 + e^{-t}} \, dt}{x}$

40. $\lim\limits_{x \to 1^+} \dfrac{\displaystyle \int_1^x \sin t \, dt}{x - 1}$

41. Find each limit. *Hint:* Transform to problems involving a continuous variable x.

(a) $\lim\limits_{n \to \infty} \sqrt[n]{a}$

(b) $\lim\limits_{n \to \infty} \sqrt[n]{n}$

(c) $\lim\limits_{n \to \infty} n(\sqrt[n]{a} - 1)$

(d) $\lim\limits_{n \to \infty} n(\sqrt[n]{n} - 1)$

42. Find each limit.

(a) $\lim\limits_{x \to 0^+} x^x$

(b) $\lim\limits_{x \to 0^+} (x^x)^x$

(c) $\lim\limits_{x \to 0^+} x^{(x^x)}$

(d) $\lim\limits_{x \to 0^+} ((x^x)^x)^x$

(e) $\lim\limits_{x \to 0^+} x^{(x^{(x^x)})}$

[GC] **43.** Graph $y = x^{1/x}$ for $x > 0$. Show what happens for very small x and very large x. Indicate the maximum value.

44. Find each limit.

(a) $\lim\limits_{x \to 0^+} (1^x + 2^x)^{1/x}$

(b) $\lim\limits_{x \to 0^-} (1^x + 2^x)^{1/x}$

(c) $\lim\limits_{x \to \infty} (1^x + 2^x)^{1/x}$

(d) $\lim\limits_{x \to -\infty} (1^x + 2^x)^{1/x}$

45. For $k \geq 0$, find

$$\lim\limits_{n \to \infty} \dfrac{1^k + 2^k + \cdots + n^k}{n^{k+1}}$$

Hint: Though this has the ∞/∞ form, l'Hôpital's Rule is not helpful. Think of another often-used technique.

46. Let $c_1, c_2, \ldots, c_n$ be positive constants with $\sum\limits_{i=1}^{n} c_i = 1$, and let $x_1, x_2, \ldots, x_n$ be positive numbers. Take natural logarithms and then use l'Hôpital's Rule to show that

$$\lim\limits_{t \to 0^+} \left(\sum\limits_{i=1}^{n} c_i x_i^t \right)^{1/t} = x_1^{c_1} x_2^{c_2} \cdots x_n^{c_n} = \prod\limits_{i=1}^{n} x_i^{c_i}$$

Here $\prod$ means product; that is, $\prod\limits_{i=1}^{n} a_i$ means $a_1 \cdot a_2 \cdot \cdots \cdot a_n$. In particular, if a, b, x, and y are positive and $a + b = 1$, then

$$\lim\limits_{t \to 0^+} (ax^t + by^t)^{1/t} = x^a y^b$$

[GC] **47.** Verify the last statement in Problem 46 by calculating each of the following.

(a) $\lim\limits_{t \to 0^+} (\tfrac{1}{2} 2^t + \tfrac{1}{2} 5^t)^{1/t}$

(b) $\lim\limits_{t \to 0^+} (\tfrac{1}{5} 2^t + \tfrac{4}{5} 5^t)^{1/t}$

(c) $\lim\limits_{t \to 0^+} (\tfrac{1}{10} 2^t + \tfrac{9}{10} 5^t)^{1/t}$

[GC] **48.** Consider $f(x) = n^2 x e^{-nx}$.

(a) Graph $f(x)$ for $n = 1, 2, 3, 4, 5, 6$ on $[0, 1]$ in the same graph window.

(b) For $x > 0$, find $\lim\limits_{n \to \infty} f(x)$.

(c) Evaluate $\displaystyle \int_0^1 f(x) \, dx$ for $n = 1, 2, 3, 4, 5, 6$.

(d) Guess at $\lim\limits_{n \to \infty} \displaystyle \int_0^1 f(x) \, dx$. Then justify your answer rigorously.

[GC] **49.** Find the absolute maximum and minimum points (if they exist) for $f(x) = (x^{25} + x^3 + 2^x) e^{-x}$ on $[0, \infty)$.

Answers to Concepts Review: **1.** $f'(x)/g'(x)$
2. $\lim\limits_{x \to a} f(x)/[1/g(x)]$ or $\lim\limits_{x \to a} g(x)/[1/f(x)]$ **3.** $\infty - \infty, 0^0, \infty^0, 1^\infty$
4. $\ln x$

9.3
Improper Integrals: Infinite Limits of Integration

In the definition of $\int_a^b f(x)\,dx$, it was assumed that the interval $[a, b]$ was finite. However, in many applications in physics, economics, and probability we wish to allow a or b (or both) to be ∞ or $-\infty$. We must therefore find a way to give meaning to symbols like

$$\int_0^\infty \frac{1}{1 + x^2}\,dx, \qquad \int_{-\infty}^{-1} xe^{-x^2}\,dx, \qquad \int_{-\infty}^\infty x^2 e^{-x^2}\,dx$$

These integrals are called **improper integrals** with infinite limits.

One Infinite Limit   The graph of $f(x) = e^{-x}$ on $[0, \infty)$ is shown in Figure 1. The integral $\int_0^b e^{-x}\,dx$ makes perfectly good sense no matter how large we make b; in fact, we can evaluate this integral explicitly.

$$\int_0^b e^{-x}\,dx = -\int_0^b e^{-x}(-dx) = \left[-e^{-x}\right]_0^b = 1 - e^{-b}$$

Now $\lim_{b \to \infty}\left(1 - e^{-b}\right) = 1$, so it seems natural to define

$$\int_0^\infty e^{-x}\,dx = 1$$

Here is the general definition.

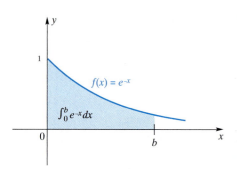

Figure 1

Definition

$$\int_{-\infty}^b f(x)\,dx = \lim_{a \to -\infty} \int_a^b f(x)\,dx$$

$$\int_a^\infty f(x)\,dx = \lim_{b \to \infty} \int_a^b f(x)\,dx$$

If the limits on the right exist and have finite values, then we say that the corresponding improper integrals **converge** and have those values. Otherwise, the integrals are said to **diverge**.

EXAMPLE 1   Find, if possible, $\int_{-\infty}^{-1} xe^{-x^2}\,dx$.

Solution

$$\int_a^{-1} xe^{-x^2}\,dx = -\frac{1}{2}\int_a^{-1} e^{-x^2}(-2x\,dx) = \left[-\frac{1}{2}e^{-x^2}\right]_a^{-1}$$

$$= -\frac{1}{2}e^{-1} + \frac{1}{2}e^{-a^2}$$

Thus,

$$\int_{-\infty}^{-1} xe^{-x^2}\,dx = \lim_{a \to -\infty}\left[-\frac{1}{2}e^{-1} + \frac{1}{2}e^{-a^2}\right] = -\frac{1}{2e}$$

We say the integral converges and has value $-1/2e$.   ∎

EXAMPLE 2 Find, if possible,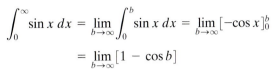

Solution

$$\int_0^\infty \sin x \, dx = \lim_{b \to \infty} \int_0^b \sin x \, dx = \lim_{b \to \infty} \left[-\cos x \right]_0^b$$

$$= \lim_{b \to \infty} \left[1 - \cos b \right]$$

The latter limit does not exist; we conclude that the given integral diverges. Think about the geometric meaning of $\int_0^\infty \sin x \, dx$ to support this result (Figure 2). ∎

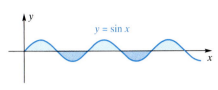

Figure 2

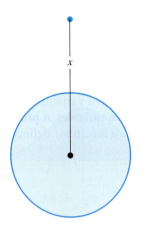

Figure 3

EXAMPLE 3 According to Newton's Inverse Square Law, the force exerted by the earth on a space capsule is $-k/x^2$, where x is the distance (in miles, for instance) from the capsule to the center of the earth (Figure 3). The force $F(x)$ required to lift the capsule is therefore $F(x) = k/x^2$. How much work is done in propelling a 1000-pound capsule out of the earth's gravitational field?

Solution We can evaluate k by noting that at $x = 3960$ miles (the radius of the earth) $F = 1000$ pounds. This yields $k = 1000(3960)^2 \approx 1.568 \times 10^{10}$. The work done in mile-pounds is therefore

$$1.568 \times 10^{10} \int_{3960}^\infty \frac{1}{x^2} \, dx = \lim_{b \to \infty} 1.568 \times 10^{10} \left[-\frac{1}{x} \right]_{3960}^b$$

$$= \lim_{b \to \infty} 1.568 \times 10^{10} \left[-\frac{1}{b} + \frac{1}{3960} \right]$$

$$= \frac{1.568 \times 10^{10}}{3960} \approx 3.96 \times 10^6 \qquad ∎$$

Both Limits Infinite We can now give a definition for $\int_{-\infty}^\infty f(x) \, dx$.

Definition

If both $\int_{-\infty}^0 f(x) \, dx$ and $\int_0^\infty f(x) \, dx$ converge, then $\int_{-\infty}^\infty f(x) \, dx$ is said to converge and have value

$$\int_{-\infty}^\infty f(x) \, dx = \int_{-\infty}^0 f(x) \, dx + \int_0^\infty f(x) \, dx$$

Otherwise, $\int_{-\infty}^\infty f(x) \, dx$ diverges.

EXAMPLE 4 Evaluate 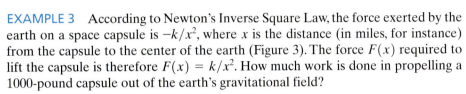 or state that it diverges.

Solution

$$\int_0^\infty \frac{1}{1 + x^2} \, dx = \lim_{b \to \infty} \int_0^b \frac{1}{1 + x^2} \, dx$$

$$= \lim_{b \to \infty} \left[\tan^{-1} x \right]_0^b$$

$$= \lim_{b \to \infty} \left[\tan^{-1} b - \tan^{-1} 0 \right] = \frac{\pi}{2}$$

Since the integrand is an even function,

29. Suppose that a company expects its annual profits t years from now to be $f(t)$ dollars and that interest is considered to be compounded continuously at an annual rate r. Then the present value of all future profits can be shown to be

$$FP = \int_0^\infty e^{-rt} f(t)\, dt$$

Find FP if $r = 0.08$ and $f(t) = 100{,}000$.

30. Do Problem 29 assuming that $f(t) = 100{,}000 + 1000t$.

31. A continuous random variable X has a **uniform distribution** if it has a probability density function of the form

$$f(x) = \begin{cases} \dfrac{1}{b - a} & \text{if } a < x < b \\ 0 & \text{if } x \le a \text{ or } x \ge b \end{cases}$$

(a) Show that $\displaystyle\int_{-\infty}^\infty f(x)\, dx = 1$.

(b) Find the mean μ and variance σ^2 of the uniform distribution.

(c) If $a = 0$ and $b = 10$, find the probability that X is less than 2.

32. A random variable X has a **Weibull distribution** if it has probability density function

$$f(x) = \begin{cases} \dfrac{\beta}{\theta}\left(\dfrac{x}{\theta}\right)^{\beta - 1} e^{-(x/\theta)^\beta} & \text{if } x > 0 \\ 0 & \text{if } x \le 0 \end{cases}$$

(a) Show that $\displaystyle\int_{-\infty}^\infty f(x)\, dx = 1$. (Assume $\beta > 1$.)

(b) If $\theta = 3$ and $\beta = 2$, find the mean μ and the variance σ^2.

(c) If the lifetime of a computer monitor is a random variable X that has a Weibull distribution with $\theta = 3$ and $\beta = 2$ (where age is measured in years) find the probability that a monitor fails before two years.

33. In probability theory, *waiting times* tend to be continuous random variables that have an **exponential distribution** with probability density function

$$f(x) = \begin{cases} \alpha e^{-\alpha x} & \text{if } x > 0 \\ 0 & \text{if } x \le 0 \end{cases}$$

(a) Show that $\displaystyle\int_{-\infty}^\infty f(x)\, dx = 1$.

(b) Find the mean μ and variance σ^2 of the exponential distribution.

34. In electromagnetic theory, the magnetic potential u at a point on the axis of a circular coil is given by

$$u = Ar \int_a^\infty \frac{dx}{(r^2 + x^2)^{3/2}}$$

where A, r, and a are constants. Evaluate u.

35. There is a subtlety in the definition of $\displaystyle\int_{-\infty}^\infty f(x)\, dx$ illustrated by the following: Show that

(a) $\displaystyle\int_{-\infty}^\infty \sin x\, dx$ diverges and (b) $\displaystyle\lim_{a \to \infty} \int_{-a}^a \sin x\, dx = 0$.

36. Consider an infinitely long wire coinciding with the positive x-axis and having mass density $\delta(x) = (1 + x^2)^{-1}, 0 \le x < \infty$.

(a) Calculate the total mass of the wire (Example 4).

(b) Show that this wire does not have a center of mass.

37. Give an example of a region in the first quadrant that gives a solid of finite volume when revolved about the x-axis, but gives a solid of infinite volume when revolved about the y-axis.

38. Let f be a nonnegative continuous function defined on $0 \le x < \infty$ with $\displaystyle\int_0^\infty f(x)\, dx < \infty$. Show that

(a) if $\displaystyle\lim_{x \to \infty} f(x)$ exists it must be 0;

(b) it is possible that $\displaystyle\lim_{x \to \infty} f(x)$ does not exist.

CAS 39. We can use a computer to approximate $\displaystyle\int_1^\infty f(x)\, dx$ by taking b very large in $\displaystyle\int_1^b f(x)\, dx$ *provided* we know that the first integral converges. Calculate $\displaystyle\int_1^{100} (1/x^p)\, dx$ for $p = 2, 1.1, 1.01, 1$, and 0.99. Note that this gives no hint that the integral $\displaystyle\int_1^\infty (1/x^p)\, dx$ converges for $p > 1$ and diverges for $p \le 1$.

CAS 40. Calculate $\displaystyle\int_0^a \frac{1}{\pi}(1 + x^2)^{-1}\, dx$ for $a = 10, 50$, and 100.

CAS 41. Calculate $\displaystyle\int_0^a \frac{1}{\sqrt{2\pi}} \exp(-x^2/2)\, dx$ for $a = 1, 2, 3$, and 4.

Answers to Concepts Review: **1.** converge **2.** $\displaystyle\lim_{b \to \infty} \int_0^b \cos x\, dx$ **3.** $\displaystyle\int_{-\infty}^0 f(x)\, dx; \int_0^\infty f(x)\, dx$ **4.** $p > 1$

9.4
Improper Integrals: Infinite Integrands

Considering the many complicated integrations that we have done, here is one that looks simple enough but is incorrect.

$$\int_{-2}^1 \frac{1}{x^2}\, dx = \left[-\frac{1}{x} \right]_{-2}^1 = -1 - \frac{1}{2} = -\frac{3}{2} \qquad \text{WRONG}$$

One glance at Figure 1 tells us that something is terribly wrong. The value of the integral (if there is one) has to be a positive number. (Why?)

Where is our mistake? To answer, we refer back to Section 5.5. Recall that for a function to be integrable in the standard (or proper) sense it must be bounded. Our function, $f(x) = 1/x^2$, is not bounded, so it is not integrable in the proper sense.

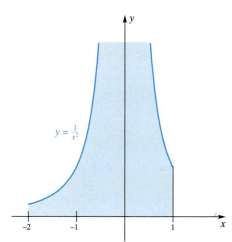

Figure 1

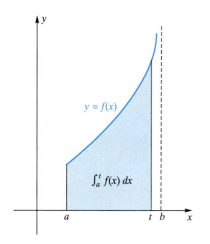

Figure 2

We say that $\int_{-2}^{1} x^{-2}\,dx$ is an improper integral with an infinite integrand (*unbounded integrand* is a more accurate but less colorful term).

Until now, we have carefully avoided infinite integrands in all our examples and problems. We could continue to do this, but this would be to avoid a kind of integral that has important applications. Our task for this section is to define and analyze this new kind of integral.

Integrands That Are Infinite at an End Point

We give the definition for the case where f tends to infinity at the right end point of the interval of integration. There is a completely analogous definition for the case where f tends to infinity at the left end point.

Definition

Let f be continuous on the half-open interval $[a,\ b)$ and suppose that $\lim_{x \to b^-} |f(x)| = \infty$. Then

$$\int_a^b f(x)\,dx = \lim_{t \to b^-} \int_a^t f(x)\,dx$$

provided that this limit exists and is finite, in which case we say that the integral converges. Otherwise, we say that the integral diverges.

Note the geometric interpretation shown in Figure 2.

EXAMPLE 1   Evaluate, if possible, the improper integral $\int_0^2 \dfrac{dx}{\sqrt{4 - x^2}}$.

Solution   Note that the integrand tends to infinity at 2.

$$\int_0^2 \frac{dx}{\sqrt{4 - x^2}} = \lim_{t \to 2^-} \int_0^t \frac{dx}{\sqrt{4 - x^2}} = \lim_{t \to 2^-} \left[\sin^{-1}\left(\frac{x}{2}\right) \right]_0^t$$

$$= \lim_{t \to 2^-} \left[\sin^{-1}\left(\frac{t}{2}\right) - \sin^{-1}\left(\frac{0}{2}\right) \right] = \frac{\pi}{2} \qquad \blacksquare$$

EXAMPLE 2   Evaluate, if possible, $\int_0^{16} \dfrac{1}{\sqrt[4]{x}}\,dx$.

Solution

$$\int_0^{16} x^{-1/4}\,dx = \lim_{t \to 0^+} \int_t^{16} x^{-1/4}\,dx = \lim_{t \to 0^+} \left[\frac{4}{3} x^{3/4} \right]_t^{16}$$

$$= \lim_{t \to 0^+} \left[\frac{32}{3} - \frac{4}{3} t^{3/4} \right] = \frac{32}{3} \qquad \blacksquare$$

EXAMPLE 3   Evaluate, if possible, $\int_0^1 \dfrac{1}{x}\,dx$.

Solution

$$\int_0^1 \frac{1}{x}\,dx = \lim_{t \to 0^+} \int_t^1 \frac{1}{x}\,dx = \lim_{t \to 0^+} \left[\ln x \right]_t^1$$

$$= \lim_{t \to 0^+} [-\ln t] = \infty$$

We conclude that the integral diverges. $\qquad \blacksquare$

Two Key Examples

From Example 6 of Section 9.3, we learned that

$$\int_1^\infty \frac{1}{x^p}\,dx$$

converges if and only if $p > 1$. From Example 4 of the present section, we learn that

$$\int_0^1 \frac{1}{x^p}\,dx$$

converges if and only if $p < 1$. The first has an infinite limit of integration; the second has an infinite integrand. If you feel at home with these two integrals, you should also be at ease with any other improper integrals that you may meet.

EXAMPLE 4   Show that $\displaystyle\int_0^1 \frac{1}{x^p}\,dx$ converges if $p < 1$, but diverges if $p \geq 1$.

Solution   Example 3 took care of the case $p = 1$. If $p \neq 1$,

$$\int_0^1 \frac{1}{x^p}\,dx = \lim_{t \to 0^+} \int_t^1 x^{-p}\,dx = \lim_{t \to 0^+} \left[\frac{x^{-p+1}}{-p+1} \right]_t^1$$

$$= \lim_{t \to 0^+} \left[\frac{1}{1-p} - \frac{1}{1-p} \cdot \frac{1}{t^{p-1}} \right] = \begin{cases} \dfrac{1}{1-p} & \text{if } p < 1 \\ \infty & \text{if } p > 1 \end{cases}$$ ∎

EXAMPLE 5   Sketch the graph of the hypocycloid of four cusps, $x^{2/3} + y^{2/3} = 1$, and find its perimeter.

≈ ***Solution***   The graph is shown in Figure 3. To find the perimeter, it is enough to find the length L of the first quadrant portion and quadruple it. We estimate L to be a bit more than $\sqrt{2} \approx 1.4$. Its exact value (see Section 6.4) is

$$L = \int_0^1 \sqrt{1 + (y')^2}\,dx$$

By implicit differentiation of $x^{2/3} + y^{2/3} = 1$, we obtain

$$\frac{2}{3}x^{-1/3} + \frac{2}{3}y^{-1/3}y' = 0$$

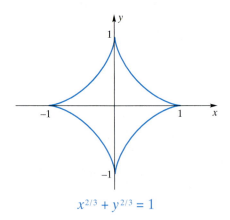

$x^{2/3} + y^{2/3} = 1$

Figure 3

or

$$y' = -\frac{y^{1/3}}{x^{1/3}}$$

Thus,

$$1 + (y')^2 = 1 + \frac{y^{2/3}}{x^{2/3}} = 1 + \frac{1 - x^{2/3}}{x^{2/3}} = \frac{1}{x^{2/3}}$$

and so

$$L = \int_0^1 \sqrt{1 + (y')^2}\,dx = \int_0^1 \frac{1}{x^{1/3}}\,dx$$

The value of this improper integral can be read from the solution to Example 4; it is $L = 1/(1 - \frac{1}{3}) = \frac{3}{2}$. We conclude that the hypocycloid has perimeter $4L = 6$. ∎

Integrands That Are Infinite at an Interior Point   The integral $\displaystyle\int_{-2}^1 1/x^2\,dx$

of our introduction has an integrand that tends to infinity at $x = 0$, an interior point of the interval $[-2, 1]$. Here is the appropriate definition to give meaning to such an integral.

Definition

Let f be continuous on $[a, b]$ except at a number c, where $a < c < b$, and suppose that $\lim_{x \to c}|f(x)| = \infty$. Then we define

$$\int_a^b f(x)\,dx = \int_a^c f(x)\,dx + \int_c^b f(x)\,dx$$

provided both integrals on the right converge. Otherwise, we say that $\displaystyle\int_a^b f(x)\,dx$ diverges.

EXAMPLE 6   Show that $\int_{-2}^{1} 1/x^2\, dx$ diverges.

Solution

$$\int_{-2}^{1} \frac{1}{x^2}\, dx = \int_{-2}^{0} \frac{1}{x^2}\, dx + \int_{0}^{1} \frac{1}{x^2}\, dx$$

The second of the integrals on the right diverges by Example 4. This is enough to give the conclusion.   ∎

EXAMPLE 7   Evaluate, if possible, the improper integral $\int_{0}^{3} \dfrac{dx}{(x-1)^{2/3}}$.

Solution   The integrand tends to infinity at $x = 1$ (see Figure 4). Thus,

$$\int_{0}^{3} \frac{dx}{(x-1)^{2/3}} = \int_{0}^{1} \frac{dx}{(x-1)^{2/3}} + \int_{1}^{3} \frac{dx}{(x-1)^{2/3}}$$

$$= \lim_{t \to 1^-} \int_{0}^{t} \frac{dx}{(x-1)^{2/3}} + \lim_{s \to 1^+} \int_{s}^{3} \frac{dx}{(x-1)^{2/3}}$$

$$= \lim_{t \to 1^-} \left[3(x-1)^{1/3}\right]_{0}^{t} + \lim_{s \to 1^+} \left[3(x-1)^{1/3}\right]_{s}^{3}$$

$$= 3 \lim_{t \to 1^-} \left[(t-1)^{1/3} + 1\right] + 3 \lim_{s \to 1^+} \left[2^{1/3} - (s-1)^{1/3}\right]$$

$$= 3 + 3\left(2^{1/3}\right) \approx 6.78$$ ∎

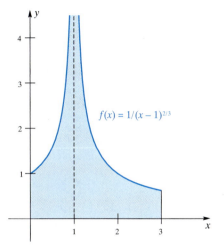

Figure 4

Concepts Review

1. The integral $\int_{0}^{1}(1/\sqrt{x})\, dx$ does not exist in the proper sense because the function $f(x) = 1/\sqrt{x}$ is _____ on the interval $(0, 1]$.

2. Considered as an improper integral,
$$\int_{0}^{1}(1/\sqrt{x})\, dx = \lim_{a \to 0^+} \int_{a}^{1} x^{-1/2}\, dx = \underline{\qquad}.$$

3. The improper integral $\int_{0}^{4}(1/\sqrt{4-x})\, dx$ is defined by _____.

4. The improper integral $\int_{0}^{1}(1/x^p)\, dx$ converges if and only if _____.

Problem Set 9.4

In Problems 1–32, evaluate each improper integral or show that it diverges.

1. $\displaystyle\int_{1}^{3} \frac{dx}{(x-1)^{1/3}}$

2. $\displaystyle\int_{1}^{3} \frac{dx}{(x-1)^{4/3}}$

3. $\displaystyle\int_{3}^{10} \frac{dx}{\sqrt{x-3}}$

4. $\displaystyle\int_{0}^{9} \frac{dx}{\sqrt{9-x}}$

5. $\displaystyle\int_{0}^{1} \frac{dx}{\sqrt{1-x^2}}$

6. $\displaystyle\int_{100}^{\infty} \frac{x}{\sqrt{1+x^2}}\, dx$

7. $\displaystyle\int_{-1}^{3} \frac{1}{x^3}\, dx$

8. $\displaystyle\int_{5}^{-5} \frac{1}{x^{2/3}}\, dx$

9. $\displaystyle\int_{-1}^{128} x^{-5/7}\, dx$

10. $\displaystyle\int_{0}^{1} \frac{x}{\sqrt[3]{1-x^2}}\, dx$

11. $\displaystyle\int_{0}^{4} \frac{dx}{(2-3x)^{1/3}}$

12. $\displaystyle\int_{\sqrt{5}}^{\sqrt{8}} \frac{x}{(16-2x^2)^{2/3}}\, dx$

13. $\displaystyle\int_{0}^{-4} \frac{x}{16-2x^2}\, dx$

14. $\displaystyle\int_{0}^{3} \frac{x}{\sqrt{9-x^2}}\, dx$

15. $\displaystyle\int_{-2}^{-1} \frac{dx}{(x+1)^{4/3}}$

16. $\displaystyle\int_{0}^{3} \frac{dx}{x^2+x-2}$

17. $\displaystyle\int_{0}^{3} \frac{dx}{x^3-x^2-x+1}$

18. $\displaystyle\int_{0}^{27} \frac{x^{1/3}}{x^{2/3}-9}\, dx$

19. $\displaystyle\int_{0}^{\pi/4} \tan 2x\, dx$

20. $\displaystyle\int_{0}^{\pi/2} \csc x\, dx$

21. $\displaystyle\int_{0}^{\pi/2} \frac{\sin x}{1-\cos x}\, dx$

22. $\displaystyle\int_{0}^{\pi/2} \frac{\cos x}{\sqrt[3]{\sin x}}\, dx$

23. $\displaystyle\int_{0}^{\pi/2} \tan^2 x \sec^2 x\, dx$

24. $\displaystyle\int_{0}^{\pi/4} \frac{\sec^2 x}{(\tan x - 1)^2}\, dx$

25. $\displaystyle\int_{0}^{\pi} \frac{dx}{\cos x - 1}$

26. $\displaystyle\int_{-3}^{-1} \frac{dx}{x\sqrt{\ln(-x)}}$

27. $\displaystyle\int_{0}^{\ln 3} \frac{e^x\, dx}{\sqrt{e^x-1}}$

28. $\displaystyle\int_{2}^{4} \frac{dx}{\sqrt{4x-x^2}}$

29. $\displaystyle\int_{1}^{e} \frac{dx}{x\ln x}$

30. $\displaystyle\int_{1}^{10} \frac{dx}{x\ln^{100} x}$

31. $\displaystyle\int_{2c}^{4c} \frac{dx}{\sqrt{x^2 - 4c^2}}$

32. $\displaystyle\int_{c}^{2c} \frac{x\,dx}{\sqrt{x^2 + xc - 2c^2}}, c > 0$

33. It is often possible to change an improper integral into a proper one by using integration by parts. Consider $\displaystyle\lim_{c \to 0^+} \int_{c}^{1} \frac{dx}{\sqrt{x}\,(1 + x)}$. Use integration by parts on the interval $[c, 1]$ where $c > 0$ to show that

$$\int_{c}^{1} \frac{dx}{\sqrt{x}\,(1 + x)} = 1 - \frac{2\sqrt{c}}{c + 1} + 2\int_{c}^{1} \frac{\sqrt{x}\,dx}{(1 + x)^2}$$

and thus conclude that upon taking the limit as $c \to 0$ an improper integral can be turned into a proper integral.

34. Use integration by parts and the technique of Problem 33 to transform the improper integral $\displaystyle\int_{0}^{1} \frac{dx}{\sqrt{x}(1 + x)}$ into a proper integral.

35. If $f(x)$ tends to infinity at both a and b, then we define

$$\int_{a}^{b} f(x)\,dx = \int_{a}^{c} f(x)\,dx + \int_{c}^{b} f(x)\,dx,$$

where c is any point between a and b, provided of course that both the latter integrals converge. Otherwise, we say that the given integral diverges. Use this to evaluate $\displaystyle\int_{-3}^{3} \frac{x}{\sqrt{9 - x^2}}\,dx$ or show that it diverges.

36. Evaluate $\displaystyle\int_{-3}^{3} \frac{x}{9 - x^2}\,dx$ or show that it diverges. See Problem 35.

37. Evaluate $\displaystyle\int_{-4}^{4} \frac{1}{16 - x^2}\,dx$ or show that it diverges. See Problem 35.

38. Evaluate $\displaystyle\int_{-1}^{1} \frac{1}{x\sqrt{-\ln|x|}}\,dx$ or show that it diverges.

39. If $\displaystyle\lim_{x \to 0^+} f(x) = \infty$, we define

$$\int_{0}^{\infty} f(x)\,dx = \lim_{c \to 0^+} \int_{c}^{1} f(x)\,dx + \lim_{b \to \infty} \int_{1}^{b} f(x)\,dx$$

provided both limits exist. Otherwise, we say that $\displaystyle\int_{0}^{\infty} f(x)\,dx$ diverges. Show that $\displaystyle\int_{0}^{\infty} \frac{1}{x^p}\,dx$ diverges for all p.

40. Suppose that f is continuous on $[0, \infty)$ except at $x = 1$, where $\displaystyle\lim_{x \to 1}|f(x)| = \infty$. How would you define $\displaystyle\int_{0}^{\infty} f(x)\,dx$?

41. Find the area of the region between the curves $y = (x - 8)^{-2/3}$ and $y = 0$ for $0 \le x < 8$.

42. Find the area of the region between the curves $y = 1/x$ and $y = 1/(x^3 + x)$ for $0 < x \le 1$.

43. Let R be the region in the first quadrant below the curve $y = x^{-2/3}$ and to the left of $x = 1$.

(a) Show that the area of R is finite by finding its value.

(b) Show that the volume of the solid generated by revolving R about the x-axis is infinite.

44. Find b so that $\displaystyle\int_{0}^{b} \ln x\,dx = 0$.

45. Is $\displaystyle\int_{0}^{1} \frac{\sin x}{x}\,dx$ an improper integral? Explain.

[EXPL] **46. Comparison Test** If $0 \le f(x) \le g(x)$ on $[a, \infty)$, it can be shown that the convergence of $\displaystyle\int_{a}^{\infty} g(x)\,dx$ implies the convergence of $\displaystyle\int_{a}^{\infty} f(x)\,dx$, and the divergence of $\displaystyle\int_{a}^{\infty} f(x)\,dx$ implies the divergence of $\displaystyle\int_{a}^{\infty} g(x)\,dx$. Use this to show that $\displaystyle\int_{1}^{\infty} \frac{1}{x^4(1 + x^4)}\,dx$ converges.

Hint: On $[1, \infty)$, $1/[x^4(1 + x^4)] \le 1/x^4$.

47. Use the Comparison Test of Problem 46 to show that $\displaystyle\int_{1}^{\infty} e^{-x^2}\,dx$ converges. *Hint*: $e^{-x^2} \le e^{-x}$ on $[1, \infty)$.

48. Use the Comparison Test of Problem 46 to show that $\displaystyle\int_{2}^{\infty} \frac{1}{\sqrt{x + 2} - 1}\,dx$ diverges.

49. Use the Comparison Test of Problem 46 to determine whether $\displaystyle\int_{1}^{\infty} \frac{1}{x^2 \ln(x + 1)}\,dx$ converges or diverges.

50. Formulate a comparison test for improper integrals with infinite integrands.

51. (a) Use Example 2 of Section 9.2 to show that for any positive number n there is a number M such that

$$0 < \frac{x^{n-1}}{e^x} \le \frac{1}{x^2} \quad \text{for } x \ge M$$

(b) Use part (a) and Problem 46 to show that $\displaystyle\int_{1}^{\infty} x^{n-1} e^{-x}\,dx$ converges.

52. Using Problem 50, prove that $\displaystyle\int_{0}^{1} x^{n-1} e^{-x}\,dx$ converges for $n > 0$.

[EXPL] **53. Gamma Function** Let $\Gamma(n) = \displaystyle\int_{0}^{\infty} x^{n-1} e^{-x}\,dx, n > 0$. This integral converges by Problems 51 and 52. Show each of the following (note that the Gamma Function is defined for any positive real number n):

(a) $\Gamma(1) = 1$ (b) $\Gamma(n + 1) = n\Gamma(n)$

(c) $\Gamma(n + 1) = n!$, if n is a positive integer.

CAS **54.** Evaluate $\int_0^\infty x^{n-1} e^{-x} \, dx$ for $n = 1, 2, 3, 4,$ and $5,$ thereby confirming Problem 53(c).

55. By interpreting each of the following integrals as an area and then calculating this area by a y-integration, evaluate:

(a) $\int_0^1 \sqrt{\dfrac{1-x}{x}} \, dx$ 　　　 (b) $\int_{-1}^1 \sqrt{\dfrac{1+x}{1-x}} \, dx$

EXPL **56.** Suppose that $0 < p < q$ and $\int_0^\infty \dfrac{1}{x^p + x^q} \, dx$ converges. What can you say about p and q?

Answers to Concepts Review: 　 **1.** unbounded **2.** 2

3. $\lim\limits_{b \to 4^-} \int_0^b \left(1/\sqrt{4-x}\right) dx$ **4.** $p < 1$

9.5 Chapter Review

Concepts Test

Respond with true or false to each of the following assertions. Be prepared to justify your answer.

1. $\lim\limits_{x \to \infty} \dfrac{x^{100}}{e^x} = 0$

2. $\lim\limits_{x \to \infty} \dfrac{x^{1/10}}{\ln x} = \infty$

3. $\lim\limits_{x \to \infty} \dfrac{1000x^4 + 1000}{0.001x^4 + 1} = \infty$

4. $\lim\limits_{x \to \infty} xe^{-1/x} = 0$

5. If $\lim\limits_{x \to a} f(x) = \lim\limits_{x \to a} g(x) = \infty$, then $\lim\limits_{x \to a} \dfrac{f(x)}{g(x)} = 1$.

6. If $\lim\limits_{x \to a} f(x) = 1$ and $\lim\limits_{x \to a} g(x) = \infty$, then $\lim\limits_{x \to a} [f(x)]^{g(x)} = 1$.

7. If $\lim\limits_{x \to a} f(x) = 1$, then $\lim\limits_{n \to \infty} \left\{\lim\limits_{x \to a} [f(x)]^n\right\} = 1$.

8. If $\lim\limits_{x \to a} f(x) = 0$ and $\lim\limits_{x \to a} g(x) = \infty$, then $\lim\limits_{x \to a} [f(x)]^{g(x)} = 0$. (Assume $f(x) \geq 0$ for $x \neq a$.)

9. If $\lim\limits_{x \to a} f(x) = -1$ and $\lim\limits_{x \to a} g(x) = \infty$, then $\lim\limits_{x \to a} [f(x)g(x)] = -\infty$.

10. If $\lim\limits_{x \to a} f(x) = 0$ and $\lim\limits_{x \to a} g(x) = \infty$, then $\lim\limits_{x \to a} [f(x)g(x)] = 0$.

11. If $\lim\limits_{x \to \infty} \dfrac{f(x)}{g(x)} = 3$, then $\lim\limits_{x \to \infty} [f(x) - 3g(x)] = 0$.

12. If $\lim\limits_{x \to a} f(x) = 2$ and $\lim\limits_{x \to a} g(x) = 0$, then $\lim\limits_{x \to a} \dfrac{f(x)}{|g(x)|} = \infty$. (Assume $g(x) \neq 0$ for $x \neq a$.)

13. If $\lim\limits_{x \to \infty} \ln f(x) = 2$, then $\lim\limits_{x \to \infty} f(x) = e^2$.

14. If $f(x) \neq 0$ for $x \neq a$ and $\lim\limits_{x \to a} f(x) = 0$, then $\lim\limits_{x \to a} [1 + f(x)]^{1/f(x)} = e$.

15. If $p(x)$ is a polynomial, then $\lim\limits_{x \to \infty} \dfrac{p(x)}{e^x} = 0$.

16. If $p(x)$ is a polynomial, then $\lim\limits_{x \to 0} \dfrac{p(x)}{e^x} = p(0)$.

17. If $f(x)$ and $g(x)$ are both differentiable and $\lim\limits_{x \to 0} \dfrac{f'(x)}{g'(x)} = L$, then $\lim\limits_{x \to 0} \dfrac{f(x)}{g(x)} = L$.

18. $\int_0^1 \dfrac{1}{x^{1.001}} \, dx$ converges.

19. $\int_0^\infty \dfrac{1}{x^p} \, dx$ diverges for all $p > 0$.

20. If f is continuous on $[0, \infty)$ and $\lim\limits_{x \to \infty} f(x) = 0$, then $\int_0^\infty f(x) \, dx$ converges.

21. If f is an even function and $\int_0^\infty f(x) \, dx$ converges, then $\int_{-\infty}^\infty f(x) \, dx$ converges.

22. If $\lim\limits_{b \to \infty} \int_{-b}^b f(x) \, dx$ exists and is finite, then $\int_{-\infty}^\infty f(x) \, dx$ converges.

23. If f' is continuous on $[0, \infty)$ and $\lim\limits_{x \to \infty} f(x) = 0$, then $\int_0^\infty f'(x) \, dx$ converges.

24. If $0 \leq f(x) \leq e^{-x}$ on $[0, \infty)$, then $\int_0^\infty f(x) \, dx$ converges.

25. $\int_0^{\pi/4} \dfrac{\tan x}{x} \, dx$ is an improper integral.

Sample Test Problems

Find each limit in Problems 1–18.

1. $\lim\limits_{x \to 0} \dfrac{4x}{\tan x}$

2. $\lim\limits_{x \to 0} \dfrac{\tan 2x}{\sin 3x}$

3. $\lim\limits_{x \to 0} \dfrac{\sin x - \tan x}{\frac{1}{3}x^2}$

4. $\lim\limits_{x \to 0} \dfrac{\cos x}{x^2}$

5. $\lim\limits_{x \to 0} 2x \cot x$

6. $\lim\limits_{x \to 1^-} \dfrac{\ln(1-x)}{\cot \pi x}$

7. $\lim\limits_{t \to \infty} \dfrac{\ln t}{t^2}$

8. $\lim\limits_{x \to \infty} \dfrac{2x^3}{\ln x}$

9. $\lim\limits_{x \to 0^+} (\sin x)^{1/x}$

10. $\lim\limits_{x \to 0^+} x \ln x$

11. $\lim\limits_{x \to 0^+} x^x$

12. $\lim\limits_{x \to 0} (1 + \sin x)^{2/x}$

13. $\lim\limits_{x \to 0^+} \sqrt{x} \ln x$

14. $\lim\limits_{t \to \infty} t^{1/t}$

15. $\lim\limits_{x \to 0^+} \left(\dfrac{1}{\sin x} - \dfrac{1}{x}\right)$

16. $\lim\limits_{x \to \pi/2} \dfrac{\tan 3x}{\tan x}$

17. $\lim_{x \to \pi/2} (\sin x)^{\tan x}$

18. $\lim_{x \to \pi/2} \left(x \tan x - \dfrac{\pi}{2} \sec x \right)$

In Problems 19–38, evaluate the given improper integral or show that it diverges.

19. $\displaystyle\int_0^\infty \dfrac{dx}{(x+1)^2}$

20. $\displaystyle\int_0^\infty \dfrac{dx}{1+x^2}$

21. $\displaystyle\int_{-\infty}^1 e^{2x} \, dx$

22. $\displaystyle\int_{-1}^1 \dfrac{dx}{1-x}$

23. $\displaystyle\int_0^\infty \dfrac{dx}{x+1}$

24. $\displaystyle\int_{1/2}^2 \dfrac{dx}{x(\ln x)^{1/5}}$

25. $\displaystyle\int_1^\infty \dfrac{dx}{x^2 + x^4}$

26. $\displaystyle\int_{-\infty}^1 \dfrac{dx}{(2-x)^2}$

27. $\displaystyle\int_{-2}^0 \dfrac{dx}{2x+3}$

28. $\displaystyle\int_1^4 \dfrac{dx}{\sqrt{x-1}}$

29. $\displaystyle\int_2^\infty \dfrac{dx}{x(\ln x)^2}$

30. $\displaystyle\int_0^\infty \dfrac{dx}{e^{x/2}}$

31. $\displaystyle\int_3^5 \dfrac{dx}{(4-x)^{2/3}}$

32. $\displaystyle\int_2^\infty x e^{-x^2} \, dx$

33. $\displaystyle\int_{-\infty}^\infty \dfrac{x}{x^2+1} \, dx$

34. $\displaystyle\int_{-\infty}^\infty \dfrac{x}{1+x^4} \, dx$

35. $\displaystyle\int_0^\infty \dfrac{e^x}{e^{2x}+1} \, dx$

36. $\displaystyle\int_{-\infty}^\infty x^2 e^{-x^3} \, dx$

37. $\displaystyle\int_{-3}^3 \dfrac{x}{\sqrt{9-x^2}} \, dx$

38. $\displaystyle\int_{\pi/3}^{\pi/2} \dfrac{\tan x}{(\ln \cos x)^2} \, dx$

39. For what values of p does the integral $\displaystyle\int_1^\infty \dfrac{1}{x^p} \, dx$ converge and for what values does it diverge?

40. For what values of p does the integral $\displaystyle\int_0^1 \dfrac{1}{x^p} \, dx$ converge and for what values does it diverge?

In Problems 41–44, use a comparison test (see Problem 46 of Section 9.4) to decide whether each of the following converges or diverges.

41. $\displaystyle\int_1^\infty \dfrac{dx}{\sqrt{x^6 + x}}$

42. $\displaystyle\int_1^\infty \dfrac{\ln x}{e^{2x}} \, dx$

43. $\displaystyle\int_3^\infty \dfrac{\ln x}{x} \, dx$

44. $\displaystyle\int_1^\infty \dfrac{\ln x}{x^3} \, dx$

9.6 Additional Problems

1. One can transform an improper integral into a proper integral by changing the variable of integration (Sections 5.8 and 8.1). If you use a change of variable given by $u = g(x)$ in the integral $\displaystyle\int_a^b f(x) \, dx$, then $g(x)$ must be a differentiable function for all x such that $a < x < b$.

(a) Show that the integral $\displaystyle\int_1^c \dfrac{1}{1+x^2} \, dx$ using the substitution $u = 1/x$ is transformed into $\displaystyle\int_{1/c}^1 \dfrac{1}{1+u^2} \, du$.

(b) Show that under the transformation given in part (a) the improper integral $\displaystyle\int_1^\infty \dfrac{1}{1+x^2} \, dx$ is equal to $\displaystyle\int_0^1 \dfrac{1}{1+u^2} \, du$, and evaluate the integral.

2. Use an appropriate change of variable to convert the improper integral $\displaystyle\int_1^\infty \dfrac{x}{x^3+1} \, dx$ into the proper integral $\displaystyle\int_0^1 \dfrac{1}{u^3+1} \, du$.

3. Define a probability density function by $f(x) = C|x|e^{-kx^2}$ for positive k.

(a) Find the constant C that makes $f(x)$ a probability density function.

(b) Using the value of C found in part (a), find the mean $\mu = \displaystyle\int_{-\infty}^\infty x f(x) \, dx$.

4. Sketch the graph of the normal probability density function

$$f(x) = \dfrac{1}{\sigma \sqrt{2\pi}} e^{-(x-\mu)^2/2\sigma^2}$$

and show, using calculus, that σ is the distance from the mean μ to the x-coordinate of one of the inflection points.

5. The **Pareto** probability density function has the form

$$f(x) = \begin{cases} \dfrac{CM^k}{x^{k+1}} & \text{if } x \geq M \\ 0 & \text{if } x < M \end{cases}$$

where k and M are positive constants.

(a) Find the value of C that makes $f(x)$ a probability density function.

(b) For the value of C found in part (a), find the value of the mean μ. Is the mean finite for all positive k? If not, how does the mean depend on k?

(c) For the value of C found in part (a), find the variance σ^2. How does the variance depend on k?

6. The **gamma** probability density function is

$$f(x) = \begin{cases} Cx^{\alpha-1} e^{-\beta x} & \text{if } x > 0 \\ 0 & \text{if } x \leq 0 \end{cases}$$

where α and β are positive constants. (Both the gamma and the Weibull distributions are used to model lifetimes of people, animals, and equipment.)

(a) Remembering that the gamma function is defined by

$$\Gamma(\alpha) = \int_0^\infty x^{\alpha-1} e^{-x} \, dx \text{ for } \alpha > 0, \text{ find the value of } C, \text{ de-}$$

pending on both α and β, that makes $f(x)$ a probability density function.

(b) For the value of C found in part (a), find the value of the mean μ.

(c) For the value of C found in part (a), find the variance σ^2.

7. The function

$$f(x) = \begin{cases} e^{-1/x} & \text{if } x > 0 \\ 0 & \text{if } x \le 0 \end{cases}$$

has an infinite number of derivatives. Denote $\dfrac{d^n}{dx^n} f(x)$ by $f^{(n)}(x)$.

(a) Sketch the graphs of $f(x)$, $f^{(1)}(x)$, $f^{(2)}(x)$, and $f^{(3)}(x)$ for $-2 < x < 5$.

(b) Compute $\lim_{x \to 0^+} f^{(n)}(x)$, for $n = 0, 1, 2, 3$ and show that $f^{(n)}(x)$ is a continuous function of x for these values of n.

(c) Give a plausible argument for the conclusion that $f^{(n)}(x)$ is a continuous function for all positive integral values of n.

EXPL **8.** The **Laplace transform**, named after the French mathematician Pierre-Simon de Laplace (1749–1827), of a function $f(x)$ is given by $L\{f(t)\}(s) = \displaystyle\int_0^\infty f(t) e^{-st} \, dt$. Laplace transforms are useful for solving differential equations.

(a) Show that the Laplace transform of t^α is given by $\Gamma(\alpha+1)/s^{\alpha+1}$ and is defined for $s > 0$.

(b) Show that the Laplace transform of $e^{\alpha t}$ is given by $1/(s-\alpha)$ and is defined for $s > \alpha$.

(c) Show that the Laplace transform of $\sin(\alpha t)$ is given by $\alpha/(s^2 + \alpha^2)$ and is defined for $s > 0$.

(d) Show that the Laplace transform of $\cos(\alpha t)$ is given by $s/(s^2 + \alpha^2)$ and is defined for $s > 0$.

TECHNOLOGY PROJECT 9.1

Probability Density Functions

I. Preparation

Exercise 1 Suppose f is an even function; that is, $f(-x) = f(x)$. Prove that

$$\int_{-a}^a f(x) \, dx = 2 \int_0^a f(x) \, dx$$

Exercise 2 Suppose g is an odd function; that is, $g(-x) = -g(x)$. Prove that

$$\int_{-a}^a g(x) \, dx = 0$$

II. Using Technology

Two important probability density functions are the normal

$$(1) \quad f(x) = \frac{1}{\sigma\sqrt{2\pi}} e^{-(x-\mu)^2/2\sigma^2}$$

and the Cauchy

$$g(x) = \frac{1}{\pi(1+x^2)}$$

Exercise 3 Plot the normal density with $\mu = 0$ and $\sigma = 2$ and the Cauchy density on the same graph. Then plot the functions $xf(x)$ and $xg(x)$, again on the same axes. Describe how these graphs are similar and how they are different.

Exercise 4 Vary μ and σ in (1) and plot the density function. Explain the effect of μ and σ on the shape and location of the probability density function.

Exercise 5 Find the mean of the normal distribution. (*Hint*: In the integral make the substitution $u = (x - \mu)/\sqrt{2}\,\sigma$. Factor constants outside the integral. Then use your technology to evaluate the integral.) Does the mean of the Cauchy distribution exist? Explain.

Exercise 6 Find the variance of the normal distribution. Does the variance of the Cauchy distribution exist? Explain.

The **cumulative distribution function** is related to the probability density function as follows:

$$F(x) = \int_{-\infty}^x f(t) \, dt$$

Thus, the cumulative distribution function accumulates probability much like an accumulation function accumulates area under a curve.

Exercise 7 Plot the cumulative distribution functions for the normal distribution (with $\mu = 0$ and $\sigma = 2$) and the Cauchy distribution.

III. Reflection

Exercise 8 Explain why it is not always true that

$$\int_{-\infty}^\infty f(x) \, dx = 0$$

when f is an odd function.

The Normal Distribution

I. Preparation

The probability density function gives probability as area under a curve. For example, if the density

$$f(x) = \frac{1}{2\sqrt{2\pi}} e^{-(x-70)^2/8}$$

is a model for the height in inches of male college students, then the probability that the height of a male college student chosen at random will exceed 72 inches is

$$\int_{72}^{\infty} \frac{1}{2\sqrt{2\pi}} e^{-(x-70)^2/8} \, dx$$

Exercise 1 Make the substitution $u = (x - 70)/2$ in this integral. What is the resulting integral? (Be sure to adjust the limits.)

Exercise 2 Write an integral that gives the probability that a randomly selected male college student is between 65 and 70 inches tall. Make the substitution $u = (x - 70)/2$ in this integral. What is the resulting integral?

II. Using Technology

Exercise 3 Use technology to evaluate the integrals in Exercises 1 and 2.

Exercise 4 For the normal distribution with mean $\mu = 0$ and variance $\sigma^2 = 1$ (this is called the *standard* normal distribution), find the probabilities for the intervals $(-1, 1)$, $(-2, 2)$, $(-3, 3)$, and $(-4, 4)$.

Exercise 5 It is often of interest to find the value of L such that the interval $(-L, L)$ has probability $1 - \alpha$, where α is typically a number close to, but not equal to, zero. Find the appropriate value of L for $\alpha = 0.05$ and for $\alpha = 0.10$. *Hint*: Zero in on L.

III. Reflection

Exercise 6 Nearly every statistics book tabulates the cumulative normal distribution function for the normal distribution (see Technology Project 9.1), but they do it only for the normal distribution with mean $\mu = 0$ and $\sigma = 1$. Explain why this is sufficient to get probabilities from *any* normal distribution.

10

Infinite Series

10.1
Infinite Sequences

In simple language, a sequence

$$a_1, a_2, a_3, a_4, \ldots$$

is an ordered arrangement of real numbers, one for each positive integer. More formally, an **infinite sequence** is a function whose domain is the set of positive integers and whose range is a set of real numbers. We may denote a sequence by $a_1, a_2, a_3, \ldots$, by $\{a_n\}_{n=1}^{\infty}$, or simply by $\{a_n\}$. Occasionally, we will extend the notion slightly by allowing the domain to consist of all integers greater than or equal to a specified integer, as in $b_0, b_1, b_2, \ldots$ and $c_8, c_9, c_{10}, \ldots$, which are also denoted by $\{b_n\}_{n=0}^{\infty}$ and $\{c_n\}_{n=8}^{\infty}$, respectively.

A sequence may be specified by giving enough initial terms to establish a pattern, as in

$$1, 4, 7, 10, 13, \ldots$$

by an **explicit formula** for the nth term, as in

$$a_n = 3n - 2, \qquad n \geq 1$$

or by a **recursion formula**

$$a_1 = 1, \qquad a_n = a_{n-1} + 3, \qquad n \geq 2$$

Note that each of our three illustrations describes the same sequence. Here are four more explicit formulas and the first few terms of the sequences that they generate.

Patterns

Someone is sure to argue that there are many different sequences that begin

$$1, 4, 7, 10, 13$$

and we agree. For example, the formula

$$3n - 2 + (n - 1) \cdot (n - 2) \cdots$$
$$(n - 5)$$

generates those five numbers. Who but an expert would think of this formula? When we ask you to look for a pattern, we mean a simple and obvious pattern.

429

(1) $a_n = 1 - \dfrac{1}{n}$, $\qquad n \geq 1$: $\quad 0, \dfrac{1}{2}, \dfrac{2}{3}, \dfrac{3}{4}, \dfrac{4}{5}, \ldots$

(2) $b_n = 1 + (-1)^n \dfrac{1}{n}$, $\qquad n \geq 1$: $\quad 0, \dfrac{3}{2}, \dfrac{2}{3}, \dfrac{5}{4}, \dfrac{4}{5}, \dfrac{7}{6}, \dfrac{6}{7}, \ldots$

(3) $c_n = (-1)^n + \dfrac{1}{n}$, $\qquad n \geq 1$: $\quad 0, \dfrac{3}{2}, \dfrac{-2}{3}, \dfrac{5}{4}, \dfrac{-4}{5}, \dfrac{7}{6}, \dfrac{-6}{7}, \ldots$

(4) $d_n = 0.999$, $\qquad n \geq 1$: $\quad 0.999, 0.999, 0.999, 0.999, \ldots$

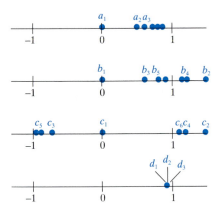

Figure 1

Convergence Consider the four sequences just defined. Each has values that pile up near 1 (see the diagrams in Figure 1). But do they all *converge* to 1? The correct response is that sequences $\{a_n\}$ and $\{b_n\}$ converge to 1, but $\{c_n\}$ and $\{d_n\}$ do not.

For a sequence to converge to 1 means first that values of the sequence should get close to 1. But they must do more than get close; they must *remain* close, for all n beyond a certain value. This rules out sequence $\{c_n\}$. And close means arbitrarily close, that is, within *any* specified nonzero distance from 1, which rules out sequence $\{d_n\}$. While sequence $\{d_n\}$ does not converge to 1, it is correct to say that it converges to 0.999. Sequence $\{c_n\}$ does not converge at all; we say it diverges.

Here is the formal definition; it should sound vaguely familiar.

Definition

The sequence $\{a_n\}$ is said to **converge** to L, and we write

$$\lim_{n \to \infty} a_n = L$$

if for each positive number ε there is a corresponding positive number N such that

$$n \geq N \Rightarrow |a_n - L| < \varepsilon$$

A sequence that fails to converge to any finite number L is said to **diverge**, or to be divergent.

To see a relationship with limits at infinity (Section 2.8), consider graphing $a_n = 1 - 1/n$ and $a(x) = 1 - 1/x$. The only difference is that in the sequence case the domain is restricted to the positive integers. In the first case, we write $\lim_{n \to \infty} a_n = 1$; in the second, $\lim_{x \to \infty} a(x) = 1$. Note the interpretations of ε and N in the diagrams in Figure 2.

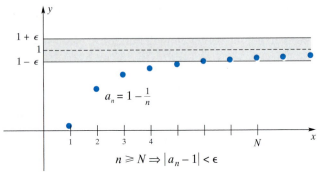

Figure 2

EXAMPLE 1 Show that if p is a positive integer then

$$\lim_{n \to \infty} \frac{1}{n^p} = 0$$

Solution This is almost obvious from earlier work, but we can give a formal demonstration. Let an arbitrary $\varepsilon > 0$ be given. Choose N to be any number greater than $\sqrt[p]{1/\varepsilon}$. Then $n \geq N$ implies that

$$|a_n - L| = \left| \frac{1}{n^p} - 0 \right| = \frac{1}{n^p} \leq \frac{1}{N^p} < \frac{1}{\left(\sqrt[p]{1/\varepsilon} \right)^p} = \varepsilon \qquad \blacksquare$$

All the familiar limit theorems hold for convergent sequences. We state them without proof.

Theorem A Properties of Limits of Sequences

Let $\{a_n\}$ and $\{b_n\}$ be convergent sequences and k a constant. Then:

1. $\displaystyle\lim_{n\to\infty} k = k$;

2. $\displaystyle\lim_{n\to\infty} ka_n = k \lim_{n\to\infty} a_n$;

3. $\displaystyle\lim_{n\to\infty} (a_n \pm b_n) = \lim_{n\to\infty} a_n \pm \lim_{n\to\infty} b_n$;

4. $\displaystyle\lim_{n\to\infty} (a_n \cdot b_n) = \lim_{n\to\infty} a_n \cdot \lim_{n\to\infty} b_n$;

5. $\displaystyle\lim_{n\to\infty} \frac{a_n}{b_n} = \frac{\displaystyle\lim_{n\to\infty} a_n}{\displaystyle\lim_{n\to\infty} b_n}$, provided that $\displaystyle\lim_{n\to\infty} b_n \neq 0$.

EXAMPLE 2 Find $\displaystyle\lim_{n\to\infty} \frac{3n^2}{7n^2 + 1}$.

Solution To decide what is happening to a quotient of two polynomials in n as n gets large, it is wise to divide numerator and denominator by the largest power of n that occurs in the denominator. This justifies our first step below; the others are justified by appealing to statements from Theorem A as indicated by the circled numbers.

$$\lim_{n\to\infty} \frac{3n^2}{7n^2 + 1} = \lim_{n\to\infty} \frac{3}{7 + (1/n^2)}$$

$$\overset{\text{\textcircled{5}}}{=} \frac{\displaystyle\lim_{n\to\infty} 3}{\displaystyle\lim_{n\to\infty} [7 + (1/n^2)]}$$

$$\overset{\text{\textcircled{3}}}{=} \frac{\displaystyle\lim_{n\to\infty} 3}{\displaystyle\lim_{n\to\infty} 7 + \lim_{n\to\infty} 1/n^2}$$

$$\overset{\text{\textcircled{1}}}{=} \frac{3}{7 + \displaystyle\lim_{n\to\infty} 1/n^2} = \frac{3}{7 + 0} = \frac{3}{7}$$

By this time, the limit theorems are so familiar that we will normally jump directly from the first step to the final result. $\qquad \blacksquare$

EXAMPLE 3 Does the sequence $\{(\ln n)/e^n\}$ converge and, if so, to what number?

Solution Here and in many sequence problems, it is convenient to use the following almost obvious fact (see Figure 2).

$$\boxed{\text{If } \lim_{x\to\infty} f(x) = L, \text{ then } \lim_{n\to\infty} f(n) = L.}$$

This is convenient because we can apply l'Hôpital's Rule to the continuous variable problem. In particular, by l'Hôpital's Rule,

$$\lim_{x\to\infty} \frac{\ln x}{e^x} = \lim_{x\to\infty} \frac{1/x}{e^x} = 0$$

Thus,

$$\lim_{n\to\infty} \frac{\ln n}{e^n} = 0$$

That is, $\{(\ln n)/e^n\}$ converges to 0.   ■

Here is another theorem that we have seen before in a slightly different guise (Theorem 2.6C).

Theorem B   Squeeze Theorem

Suppose that $\{a_n\}$ and $\{c_n\}$ both converge to L and that $a_n \le b_n \le c_n$ for $n \ge K$ (K a fixed integer). Then $\{b_n\}$ also converges to L.

EXAMPLE 4   Show that $\displaystyle\lim_{n\to\infty} \frac{\sin^3 n}{n} = 0$.

Solution   For $n \ge 1$, $-1/n \le (\sin^3 n)/n \le 1/n$. Since $\displaystyle\lim_{n\to\infty} (-1/n) = 0$ and $\displaystyle\lim_{n\to\infty} (1/n) = 0$, the result follows by the Squeeze Theorem.   ■

For sequences of variable sign, it is helpful to have the following result.

Theorem C

If $\displaystyle\lim_{n\to\infty} |a_n| = 0$, then $\displaystyle\lim_{n\to\infty} a_n = 0$.

Proof   Since $-|a_n| \le a_n \le |a_n|$, the result follows from the Squeeze Theorem.   ◆

What happens to the numbers in the sequence $\{0.999^n\}$ as $n \to \infty$? We suggest that you calculate 0.999^n for $n = 10, 100, 1000$, and $10,000$ on your calculator to make a good guess. Then note the following example.

EXAMPLE 5   Show that if $-1 < r < 1$ then $\displaystyle\lim_{n\to\infty} r^n = 0$.

Solution   If $r = 0$, the result is trivial, so suppose otherwise. Then $1/|r| > 1$, and so $1/|r| = 1 + p$ for some number $p > 0$. By the Binomial Formula,

$$\frac{1}{|r|^n} = (1 + p)^n = 1 + pn + (\text{positive terms}) \ge pn$$

Thus,

$$0 \le |r|^n \le \frac{1}{pn}$$

Since $\displaystyle\lim_{n\to\infty} (1/pn) = (1/p)\lim_{n\to\infty} (1/n) = 0$, it follows from the Squeeze Theorem that $\displaystyle\lim_{n\to\infty} |r|^n = 0$ or, equivalently, $\displaystyle\lim_{n\to\infty} |r^n| = 0$. By Theorem C, $\displaystyle\lim_{n\to\infty} r^n = 0$.   ■

What if $r > 1$; for example, $r = 1.5$? Then r^n will march off toward ∞. In this case, we write

$$\lim_{n\to\infty} r^n = \infty, \qquad r > 1$$

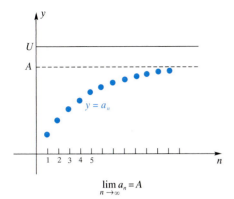

$$\lim_{n \to \infty} a_n = A$$

Figure 3

However, we say that the sequence $\{r^n\}$ diverges. To converge, a sequence must approach a *finite* limit. The sequence $\{r^n\}$ also diverges when $r \leq -1$.

Monotonic Sequences   Consider now an arbitrary **nondecreasing sequence** $\{a_n\}$, by which we mean $a_n \leq a_{n+1}, n \geq 1$. One example is the sequence $a_n = n^2$; another is $a_n = 1 - 1/n$. If you think about it a little, you may convince yourself that such a sequence can do one of only two things. Either it marches off to infinity or, if it cannot do that because it is bounded above, then it must approach a lid (see Figure 3). Here is the formal statement of this very important result.

> **Theorem D**   Monotonic Sequence Theorem
>
> If U is an upper bound for a nondecreasing sequence $\{a_n\}$, then the sequence converges to a limit A that is less than or equal to U. Similarly, if L is a lower bound for a nonincreasing sequence $\{b_n\}$, then the sequence $\{b_n\}$ converges to a limit B that is greater than or equal to L.

The expression **monotonic sequence** is used to describe either a nondecreasing or nonincreasing sequence; hence the name for this theorem.

Theorem D describes a very deep property of the real number system. It is equivalent to the *completeness property* of the real numbers, which in simple language says that the real line has no "holes" in it (see Problems 47 and 48). It is this property that distinguishes the real number line from the rational number line (which is full of holes). A great deal more could be said about this topic; we hope Theorem D appeals to your intuition and that you will accept it on faith until you take a more advanced course.

We make one more comment about Theorem D. It is not necessary that the sequences $\{a_n\}$ and $\{b_n\}$ be monotonic initially, only that they be monotonic from some point on, that is, for $n \geq K$. In fact, *the convergence or divergence of a sequence does not depend on the character of the initial terms, but rather on what is true for large n.*

EXAMPLE 6   Show that the sequence $b_n = n^2/2^n$ converges by using Theorem D.

Solution   The first few terms of this sequence are

$$\frac{1}{2}, 1, \frac{9}{8}, 1, \frac{25}{32}, \frac{36}{64}, \frac{49}{128}, \ldots$$

For $n \geq 3$ the sequence appears to be decreasing $(b_n > b_{n+1})$, a fact that we now establish. Each of the following inequalities is equivalent to the others.

$$\frac{n^2}{2^n} > \frac{(n+1)^2}{2^{n+1}}$$

$$n^2 > \frac{(n+1)^2}{2}$$

$$2n^2 > n^2 + 2n + 1$$

$$n^2 - 2n > 1$$

$$n(n-2) > 1$$

The last inequality is clearly true for $n \geq 3$. Since the sequence is decreasing (a stronger condition than nonincreasing) and is bounded below by zero, the Monotonic Sequence Theorem guarantees that it has a limit.

It would be easy using l'Hôpital's Rule to show that the limit is zero.   ■

Consider the partial sums

$$S_1 = \frac{1}{2}$$

$$S_2 = \frac{1}{2} + \frac{1}{4} = \frac{3}{4}$$

$$S_3 = \frac{1}{2} + \frac{1}{4} + \frac{1}{8} = \frac{7}{8}$$

$$\vdots$$

$$S_n = \frac{1}{2} + \frac{1}{4} + \frac{1}{8} + \cdots + \frac{1}{2^n} = 1 - \frac{1}{2^n}$$

Clearly, these partial sums get increasingly close to 1. In fact,

$$\lim_{n \to \infty} S_n = \lim_{n \to \infty} \left(1 - \frac{1}{2^n}\right) = 1$$

The infinite sum is then defined to be the limit of the partial sum S_n.

More generally, consider the **infinite series**

$$a_1 + a_2 + a_3 + a_4 + \cdots$$

which is also denoted by $\sum_{k=1}^{\infty} a_k$ or $\sum a_k$. Then S_n, the **nth partial sum**, is given by

$$S_n = a_1 + a_2 + a_3 + \cdots + a_n = \sum_{k=1}^{n} a_k$$

We make the following formal definition.

Definition

The infinite series $\sum_{k=1}^{\infty} a_k$ **converges** and has **sum** S if the sequence of partial sums $\{S_n\}$ converges to S. If $\{S_n\}$ diverges, then the series **diverges**. A divergent series has no sum.

Geometric Series   A series of the form

$$\sum_{k=1}^{\infty} ar^{k-1} = a + ar + ar^2 + ar^3 + \cdots$$

where $a \neq 0$, is called a **geometric series**.

EXAMPLE 1   Show that a geometric series converges with sum $S = a/(1 - r)$ if $|r| < 1$, but diverges if $|r| \geq 1$.

Solution   Let $S_n = a + ar + ar^2 + \cdots + ar^{n-1}$. If $r = 1$, $S_n = na$, which grows without bound, and so $\{S_n\}$ diverges. If $r \neq 1$, we may write

$$S_n - rS_n = \left(a + ar + \cdots + ar^{n-1}\right) - \left(ar + ar^2 + \cdots + ar^n\right) = a - ar^n$$

and so

$$S_n = \frac{a - ar^n}{1 - r} = \frac{a}{1 - r} - \frac{a}{1 - r}r^n$$

If $|r| < 1$, then $\lim_{n \to \infty} r^n = 0$ (Section 10.1, Example 5), and thus

$$S = \lim_{n \to \infty} S_n = \frac{a}{1 - r}$$

If $|r| > 1$ or $r = -1$, the sequence $\{r^n\}$ diverges, and consequently so does $\{S_n\}$. ∎

EXAMPLE 2 Use the result of Example 1 to find the sum of the following two geometric series.

(a) $\dfrac{4}{3} + \dfrac{4}{9} + \dfrac{4}{27} + \dfrac{4}{81} + \cdots$

(b) $0.515151\ldots = \dfrac{51}{100} + \dfrac{51}{10,000} + \dfrac{51}{1,000,000} + \cdots$

Solution

(a) $S = \dfrac{a}{1-r} = \dfrac{\frac{4}{3}}{1 - \frac{1}{3}} = \dfrac{\frac{4}{3}}{\frac{2}{3}} = 2$ (b) $S = \dfrac{\frac{51}{100}}{1 - \frac{1}{100}} = \dfrac{\frac{51}{100}}{\frac{99}{100}} = \dfrac{51}{99} = \dfrac{17}{33}$

Incidentally, the procedure in part (b) suggests how to show that any repeating decimal represents a rational number. ∎

EXAMPLE 3 The diagram in Figure 2 represents an equilateral triangle containing infinitely many circles, tangent to the triangle and to neighboring circles, and reaching into the corners. What fraction of the area of the triangle is occupied by the circles?

Solution Suppose for convenience that the large circle has radius 1, which makes the triangle have sides of length $2\sqrt{3}$. Concentrate attention on the vertical stack of circles. With a bit of geometric reasoning (the center of the large circle is two-thirds of the way from the upper vertex to the base), we see that the radii of these circles are $1, \frac{1}{3}, \frac{1}{9}, \ldots$ and conclude that the vertical stack has area

$$\pi\left[1^2 + \left(\frac{1}{3}\right)^2 + \left(\frac{1}{9}\right)^2 + \left(\frac{1}{27}\right)^2 + \cdots\right]$$

$$= \pi\left[1 + \frac{1}{9} + \frac{1}{81} + \frac{1}{729} + \cdots\right] = \pi\left[\frac{1}{1 - \frac{1}{9}}\right] = \frac{9\pi}{8}$$

The total area of all the circles is three times this number minus twice the area of the big circle, that is, $27\pi/8 - 2\pi$, or $11\pi/8$. Since the triangle has area $3\sqrt{3}$, the fraction of this area occupied by the circles is

$$\frac{11\pi}{24\sqrt{3}} \approx 0.83$$ ∎

A General Test for Divergence Consider the geometric series $a + ar + ar^2 + \cdots + ar^{n-1} + \cdots$ once more. Its nth term a_n is given by $a_n = ar^{n-1}$. Example 1 shows that a geometric series converges *if and only if* $\lim_{n\to\infty} a_n = 0$.

Could this possibly be true of all series? The answer is no, although half of the statement (the "only-if" half) is correct. This leads to an important divergence test for series.

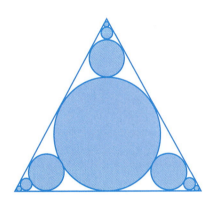

Figure 2

Logic

Consider these two statements:

1. If $\sum_{k=1}^{\infty} a_n$ converges, then $\lim_{n\to\infty} a_n = 0$.

2. If $\lim_{n\to\infty} a_n = 0$, then $\sum_{n=1}^{\infty} a_n$ converges.

The first statement is true for any sequence $\{a_n\}$; the second is not. This provides another example of a true statement (the first) whose *converse* is false.

Recall that the *contrapositive* of a statement is true whenever the statement is true. The contrapositive of the first statement is

3. If $\lim_{n\to\infty} a_n \neq 0$, then $\sum_{n=1}^{\infty} a_n$ diverges.

Theorem A nth-Term Test for Divergence

If the series $\sum_{n=1}^{\infty} a_n$ converges, then $\lim_{n\to\infty} a_n = 0$. Equivalently, if $\lim_{n\to\infty} a_n \neq 0$ or if $\lim_{n\to\infty} a_n$ does not exist, then the series diverges.

Proof Let S_n be the nth partial sum and $S = \lim_{n\to\infty} S_n$. Note that $a_n = S_n - S_{n-1}$. Since $\lim_{n\to\infty} S_{n-1} = \lim_{n\to\infty} S_n = S$, it follows that

$$\lim_{n\to\infty} a_n = \lim_{n\to\infty} S_n - \lim_{n\to\infty} S_{n-1} = S - S = 0 \; ♦$$

EXAMPLE 4 Show that $\displaystyle\sum_{n=1}^{\infty} \frac{n^3}{3n^3 + 2n^2}$ diverges.

Solution

$$\lim_{n\to\infty} a_n = \lim_{n\to\infty} \frac{n^3}{3n^3 + 2n^2} = \lim_{n\to\infty} \frac{1}{3 + 2/n} = \frac{1}{3}$$

Thus, by the nth-Term Test, the series diverges. ∎

The Harmonic Series Students invariably want to turn Theorem A around and make it say that $a_n \to 0$ implies convergence of Σa_n. The **harmonic series**

$$\sum_{n=1}^{\infty} \frac{1}{n} = 1 + \frac{1}{2} + \frac{1}{3} + \cdots + \frac{1}{n} + \cdots$$

shows that this is false. Clearly, $\lim_{n\to\infty} a_n = \lim_{n\to\infty} (1/n) = 0$. However, the series diverges, as we now show.

EXAMPLE 5 Show that the harmonic series diverges.

Solution We show that S_n grows without bound. Imagine n to be large and write

$$S_n = 1 + \frac{1}{2} + \frac{1}{3} + \frac{1}{4} + \frac{1}{5} + \cdots + \frac{1}{n}$$

$$= 1 + \frac{1}{2} + \left(\frac{1}{3} + \frac{1}{4}\right) + \left(\frac{1}{5} + \frac{1}{6} + \frac{1}{7} + \frac{1}{8}\right) + \left(\frac{1}{9} + \cdots + \frac{1}{16}\right) + \cdots + \frac{1}{n}$$

$$> 1 + \frac{1}{2} + \frac{2}{4} + \frac{4}{8} + \frac{8}{16} + \cdots + \frac{1}{n}$$

$$= 1 + \frac{1}{2} + \frac{1}{2} + \frac{1}{2} + \frac{1}{2} + \cdots + \frac{1}{n}$$

It is clear that by taking n sufficiently large we can introduce as many $\frac{1}{2}$'s into the last expression as we wish. Thus, S_n grows without bound, and so $\{S_n\}$ diverges. Hence, the harmonic series diverges. ∎

Collapsing Series A geometric series is one of the few series where we can actually give an explicit formula for S_n; a **collapsing series** is another (see Example 2 of Section 5.3).

EXAMPLE 6 Show that the following series converges and find its sum.

$$\sum_{k=1}^{\infty} \frac{1}{(k + 2)(k + 3)}$$

Solution Use a partial fraction decomposition to write

$$\frac{1}{(k + 2)(k + 3)} = \frac{1}{k + 2} - \frac{1}{k + 3}$$

Then,

$$S_n = \sum_{k=1}^{n} \left(\frac{1}{k + 2} - \frac{1}{k + 3}\right) = \left(\frac{1}{3} - \frac{1}{4}\right) + \left(\frac{1}{4} - \frac{1}{5}\right) + \cdots + \left(\frac{1}{n + 2} - \frac{1}{n + 3}\right)$$

$$= \frac{1}{3} - \frac{1}{n + 3}$$

Therefore,

$$\lim_{n\to\infty} S_n = \tfrac{1}{3}$$

The series converges and has sum $\frac{1}{3}$. ∎

Properties of Convergent Series   Convergent series behave much like finite sums; what you expect to be true often *is* true.

A Note on Terminology

This theorem introduces a subtle shift in terminology. The symbol $\sum_{k=1}^{\infty} a_k$ is now being used both for the infinite series $a_1 + a_2 + \cdots$ and for the sum of this series, which is a number.

Theorem B   Linearity of Convergent Series

If $\displaystyle\sum_{k=1}^{\infty} a_k$ and $\displaystyle\sum_{k=1}^{\infty} b_k$ both converge and c is a constant, then $\displaystyle\sum_{k=1}^{\infty} ca_k$ and $\displaystyle\sum_{k=1}^{\infty} (a_k + b_k)$ also converge, and

(i) $\displaystyle\sum_{k=1}^{\infty} ca_k = c \sum_{k=1}^{\infty} a_k$;

(ii) $\displaystyle\sum_{k=1}^{\infty} (a_k + b_k) = \sum_{k=1}^{\infty} a_k + \sum_{k=1}^{\infty} b_k$.

Proof   By hypothesis, $\displaystyle\lim_{n\to\infty} \sum_{k=1}^{n} a_k$ and $\displaystyle\lim_{n\to\infty} \sum_{k=1}^{n} b_k$ both exist. Thus, use the properties of sums with finitely many terms and the properties of limits of sequences.

(i) $\displaystyle\sum_{k=1}^{\infty} ca_k = \lim_{n\to\infty} \sum_{k=1}^{n} ca_k = \lim_{n\to\infty} c \sum_{k=1}^{n} a_k$

$\displaystyle\qquad\quad = c \lim_{n\to\infty} \sum_{k=1}^{n} a_k = c \sum_{k=1}^{\infty} a_k$

(ii) $\displaystyle\sum_{k=1}^{\infty} (a_k + b_k) = \lim_{n\to\infty} \sum_{k=1}^{n} (a_k + b_k) = \lim_{n\to\infty} \left[\sum_{k=1}^{n} a_k + \sum_{k=1}^{n} b_k \right]$

$\displaystyle\qquad\qquad = \lim_{n\to\infty} \sum_{k=1}^{n} a_k + \lim_{n\to\infty} \sum_{k=1}^{n} b_k = \sum_{k=1}^{\infty} a_k + \sum_{k=1}^{\infty} b_k$ ♦

EXAMPLE 7   Calculate $\displaystyle\sum_{k=1}^{\infty} \left[3\left(\tfrac{1}{8}\right)^k - 5\left(\tfrac{1}{3}\right)^k \right]$.

Solution   By Theorem B and Example 1,

$$\sum_{k=1}^{\infty} \left[3\left(\frac{1}{8}\right)^k - 5\left(\frac{1}{3}\right)^k \right] = 3 \sum_{k=1}^{\infty} \left(\frac{1}{8}\right)^k - 5 \sum_{k=1}^{\infty} \left(\frac{1}{3}\right)^k$$

$$= 3 \frac{\frac{1}{8}}{1 - \frac{1}{8}} - 5 \frac{\frac{1}{3}}{1 - \frac{1}{3}} = \frac{3}{7} - \frac{5}{2} = -\frac{29}{14} \qquad\blacksquare$$

Theorem C

If $\displaystyle\sum_{k=1}^{\infty} a_k$ diverges and $c \neq 0$, then $\displaystyle\sum_{k=1}^{\infty} ca_k$ diverges.

We leave the proof of this theorem to you (Problem 35). It implies, for example, that

$$\sum_{k=1}^{\infty} \frac{1}{3k} = \sum_{k=1}^{\infty} \frac{1}{3} \cdot \frac{1}{k}$$

diverges, since we know that the harmonic series diverges.

The associative law of addition allows us to group terms in a *finite* sum in any way that we please. For example,

$$2 + 7 + 3 + 4 + 5 = (2 + 7) + (3 + 4) + 5 = 2 + (7 + 3) + (4 + 5)$$

But sometimes we lose sight of the definition of an *infinite* series as the limit of a sequence of partial sums, and we let our intuition guide us into a paradox. For example, the series

$$1 - 1 + 1 - 1 + \cdots + (-1)^{n+1} + \cdots$$

has partial sums

$$S_1 = 1$$
$$S_2 = 1 - 1 = 0$$
$$S_3 = 1 - 1 + 1 = 1$$
$$S_4 = 1 - 1 + 1 - 1 = 0$$
$$\vdots$$

The sequence of partial sums, $1, 0, 1, 0, 1, \ldots$, diverges; thus the series $1 - 1 + 1 - 1 + \cdots$ diverges. We might, however, view the series as

$$(1 - 1) + (1 - 1) + \cdots$$

and claim that the sum is 0. Alternatively, we might view the series as

$$1 - (1 - 1) - (1 - 1) - \cdots$$

and claim that the sum is 1. The sum of the series cannot be equal to both 0 and 1. It turns out that grouping of terms in a series is acceptable provided that the series is convergent; in such a case we can group terms in any way that we wish.

Theorem D   Grouping Terms in an Infinite Series

The terms of a convergent series can be grouped in any way (provided that the order of the terms is maintained), and the new series will converge with the same sum as the original series.

Proof   Let Σa_n be the original convergent series and let $\{S_n\}$ be its sequence of partial sums. If Σb_m is a series formed by grouping the terms of Σa_n and if $\{T_m\}$ is its sequence of partial sums, then each T_m is one of the S_n's. For example, T_4 might be

$$T_4 = a_1 + (a_2 + a_3) + (a_4 + a_5 + a_6) + (a_7 + a_8)$$

in which case $T_4 = S_8$. Thus, $\{T_m\}$ is a "subsequence" of $\{S_n\}$. A moment's thought should convince you that if $S_n \to S$ then $T_m \to S$. ♦

Concepts Review

1. An expression of the form $a_1 + a_2 + a_3 + \cdots$ is called _____.

2. A series $a_1 + a_2 + \cdots$ is said to converge if the sequence $\{S_n\}$ converges, where $S_n = $ _____.

3. The geometric series $a + ar + ar^2 + \cdots$ converges if _____; in this case the sum of the series is _____.

4. If $\lim\limits_{n \to \infty} a_n \neq 0$, we can be sure that the series $\sum\limits_{n=1}^{\infty} a_n$ _____.

Problem Set 10.2

In Problems 1–14, indicate whether the given series converges or diverges. If it converges, find its sum. Hint: It may help you to write out the first few terms of the series.

1. $\sum\limits_{k=1}^{\infty} \left(\frac{1}{7}\right)^k$

2. $\sum\limits_{k=1}^{\infty} \left(-\frac{1}{4}\right)^{-k-2}$

3. $\sum\limits_{k=0}^{\infty} \left[2\left(\frac{1}{4}\right)^k + 3\left(-\frac{1}{5}\right)^k\right]$

4. $\sum\limits_{k=1}^{\infty} \left[5\left(\frac{1}{2}\right)^k - 3\left(\frac{1}{7}\right)^{k+1}\right]$

5. $\sum\limits_{k=1}^{\infty} \frac{k-5}{k+2}$

6. $\sum\limits_{k=1}^{\infty} \left(\frac{9}{8}\right)^k$

7. $\sum\limits_{k=2}^{\infty} \left(\frac{1}{k} - \frac{1}{k-1}\right)$ Hint: Example 6.

8. $\sum\limits_{k=1}^{\infty} \frac{3}{k}$

9. $\sum\limits_{k=1}^{\infty} \frac{k!}{100^k}$

10. $\sum\limits_{k=1}^{\infty} \frac{2}{(k+2)k}$

11. $\sum\limits_{k=1}^{\infty} \left(\frac{e}{\pi}\right)^{k+1}$

12. $\sum\limits_{k=1}^{\infty} \frac{4^{k+1}}{7^{k-1}}$

13. $\sum\limits_{k=2}^{\infty} \left(\frac{3}{(k-1)^2} - \frac{3}{k^2}\right)$

14. $\sum\limits_{k=6}^{\infty} \frac{2}{k-5}$

In Problems 15–20, write the given decimal as an infinite series, then find the sum of the series, and finally use the result to write the decimal as a ratio of two integers (see Example 2).

15. $0.22222\ldots$

16. $0.21212121\ldots$

17. $0.013013013\ldots$

18. $0.125125125\ldots$

19. $0.49999\ldots$

20. $0.36717171\ldots$

21. Evaluate $\displaystyle\sum_{k=0}^{\infty} r(1-r)^k, 0 < r < 2$.

22. Evaluate $\displaystyle\sum_{k=0}^{\infty} (-1)^k x^k, -1 < x < 1$.

23. Show that $\displaystyle\sum_{k=1}^{\infty} \ln\frac{k}{k+1}$ diverges. *Hint*: Obtain a formula for S_n.

24. Show that $\displaystyle\sum_{k=2}^{\infty} \ln\left(1 - \frac{1}{k^2}\right) = -\ln 2$.

25. A ball is dropped from a height of 100 feet. Each time it hits the floor, it rebounds to $\frac{2}{3}$ its previous height. Find the total distance it travels before coming to rest.

26. Three people, A, B, and C, divide an apple as follows. First they divide it into fourths, each taking a quarter. Then they divide the leftover quarter into fourths, each taking a quarter, and so on. Show that each gets a third of the apple.

27. Suppose that the government pumps an extra $1 billion into the economy. Assume that each business and individual saves 25% of its income and spends the rest, so of the initial $1 billion 75% is respent by individuals and businesses. Of that amount, 75% is spent, and so forth. What is the total increase in spending due to the government action? (This is called the *multiplier effect* in economics.)

28. Do Problem 27 assuming that only 10% of the income is saved at each stage.

29. Assume that square $ABCD$ (Figure 3) has sides of length 1 and that $E, F, G,$ and H are midpoints of the sides. If the indicated pattern is continued indefinitely, what will be the area of the painted region?

30. If the pattern shown in Figure 4 is continued indefinitely, what fraction of the original square will eventually be painted?

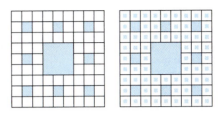
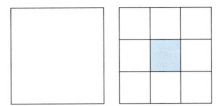

Figure 4

31. Each triangle in the descending chain (Figure 5) has its vertices at the midpoints of the sides of the next larger one. If the indicated pattern of painting is continued indefinitely, what fraction of the original triangle will be painted? Does the original triangle need to be equilateral for this to be true?

Figure 5

32. Circles are inscribed in the triangles of Problem 31 as indicated in Figure 6. If the original triangle is equilateral, what fraction of the area is eventually painted?

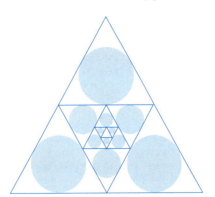

Figure 6

33. In another version of Zeno's paradox, Achilles can run ten times as fast as the tortoise, but the tortoise has a 100–yard headstart. Achilles cannot catch the tortoise, says Zeno, because

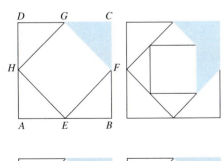

Figure 3

when Achilles runs 100 yards the tortoise will have moved 10 yards ahead, when Achilles runs another 10 yards, the tortoise will have moved 1 yard ahead, and so on. Convince Zeno that Achilles will catch the tortoise and tell him exactly how many yards Achilles will have to run to do it.

34. Tom and Joel are good runners, both able to run at a constant speed of 10 miles per hour. Their amazing dog Trot can do even better; he runs at 20 miles per hour. Starting from towns 60 miles apart, Tom and Joel run toward each other while Trot runs back and forth between them. How far does Trot run by the time the boys meet? Assume that Trot started with Tom running toward Joel and that he is able to make instant turnarounds. Solve the problem two ways.
(a) Use a geometric series.
(b) Find a shorter way to do the problem.

35. Prove: If $\sum_{k=1}^{\infty} a_k$ diverges, so does $\sum_{k=1}^{\infty} ca_k$ for $c \neq 0$.

36. Use Problem 35 to conclude that $\frac{1}{2} + \frac{1}{4} + \frac{1}{6} + \frac{1}{8} + \cdots$ diverges.

37. Suppose that one has an unlimited supply of identical blocks each 1 unit long.
(a) Convince yourself that they may be stacked as in Figure 7 without toppling. *Hint*: Consider centers of mass.
(b) How far can one make the top block protrude to the right of the bottom block using this method of stacking?

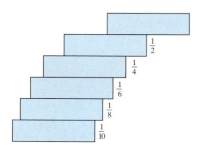

Figure 7

38. How large must N be in order for $S_N = \sum_{k=1}^{N} (1/k)$ just to exceed 4? *Note*: Computer calculations show that for S_N to exceed 20, $N = 272,400,600$, and for S_N to exceed 100, $N \approx 1.5 \times 10^{43}$.

39. Prove that if $\sum a_n$ diverges and $\sum b_n$ converges then $\sum(a_n + b_n)$ diverges.

40. Show that it is possible for $\sum a_n$ and $\sum b_n$ both to diverge and yet for $\sum(a_n + b_n)$ to converge.

41. By looking at the region in Figure 8 first vertically and then horizontally, conclude that

$$1 + \frac{1}{2} + \frac{1}{4} + \frac{1}{8} + \cdots = \frac{1}{2} + \frac{2}{4} + \frac{3}{8} + \frac{4}{16} + \cdots$$

and use this fact to calculate:
(a) $\sum_{k=1}^{\infty} \frac{k}{2^k}$

(b) $\bar{x}$, the horizontal coordinate of the centroid of the region.

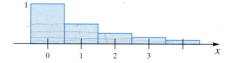

Figure 8

42. Let r be a fixed number with $|r| < 1$. Then it can be shown that $\sum_{k=1}^{\infty} kr^k$ converges, say with sum S. Use the properties of $\sum$ to show that

$$(1 - r)S = \sum_{k=1}^{\infty} r^k$$

and then obtain a formula for S, thus generalizing Problem 41a.

43. Many drugs are eliminated from the body in an exponential manner. Thus, if a drug is given in dosages of size C at time intervals of length t, the amount A_n of the drug in the body just after the $(n + 1)$st dose is

$$A_n = C + Ce^{-kt} + Ce^{-2kt} + \cdots + Ce^{-nkt}$$

where k is a positive constant that depends on the type of drug.
(a) Derive a formula for A, the amount of drug in the body just after a dose if a person has been on the drug for a very long time (assume an infinitely long time).
(b) Evaluate A if it is known that one-half of a dose is eliminated from the body in 6 hours and doses of size 2 milligrams are given every 12 hours.

44. Find the sum of the series

$$\sum_{k=1}^{\infty} \frac{2^k}{(2^{k+1} - 1)(2^k - 1)}$$

45. Evaluate $\sum_{k=1}^{\infty} \frac{1}{f_k f_{k+2}}$ where $\{f_k\}$ is the Fibonacci sequence introduced in Problem 52 of Section 10.1. *Hint*: First show that

$$\frac{1}{f_k f_{k+2}} = \frac{1}{f_k f_{k+1}} - \frac{1}{f_{k+1} f_{k+2}}$$

Answers to Concepts Review: **1.** an infinite series
2. $a_1 + a_2 + a_3 + \cdots + a_n$ **3.** $|r| < 1$; $a/(1 - r)$ **4.** diverges

---

10.3
Positive Series:
The Integral Test

We introduced some important ideas in Section 10.2, but we illustrated them mainly for two very special types of series: geometric series and collapsing series. For these series we can give exact formulas for the partial sums S_n, something that we can rarely do for most other types of series. Our task now is to begin a study of very general infinite series.

There are always two important questions to ask about a series.

1. Does the series converge?

2. If it converges, what is its sum?

How shall we answer these questions? Someone may suggest that we use a computer. To answer the first question, simply add up more and more terms of the series, watching the numbers you get as partial sums. If these numbers seem to settle down on a fixed number S, the series converges. And in this case, S is the sum of the series, answering the second question. This response is plain wrong for question 1 and only partially adequate for question 2. Let us see why.

Consider the harmonic series

$$1 + \tfrac{1}{2} + \tfrac{1}{3} + \tfrac{1}{4} + \cdots$$

introduced in Section 10.2 and discussed in Example 5 and Problem 38 of that section. We know that this series diverges, but a computer would not help us to discover this fact. The partial sums S_n of this series grow without bound, but they grow so slowly that it takes over 272 million terms for S_n to reach 20 and over 10^{43} terms for S_n to reach 100. Because of the inherent limitation in the number of digits that it can handle, a computer would eventually give repeated values for S_n, suggesting wrongly that the S_n's were converging. What is true for the harmonic series is true for any slowly diverging series. We state it emphatically: A computer is no substitute for mathematical tests of convergence and divergence, a subject to which we now turn.

In this and the next section, we restrict our attention to series with positive (or at least nonnegative) terms. With this restriction, we will be able to give some remarkably simple convergence tests. Tests for series with terms of arbitrary sign are presented in Section 10.5.

Bounded Partial Sums Our first result flows directly from the Monotonic Sequence Theorem (Theorem 10.1D).

Theorem A Bounded Sum Test

A series $\sum a_k$ of nonnegative terms converges if and only if its partial sums are bounded above.

Proof As usual, let $S_n = a_1 + a_2 + \cdots + a_n$. Since $a_k \geq 0$, $S_{n+1} \geq S_n$; that is, $\{S_n\}$ is a nondecreasing sequence. Thus, by Theorem 10.1D, the sequence $\{S_n\}$ will converge provided that there is a number U such that $S_n \leq U$ for all n. Otherwise, the S_n's will grow without bound, in which case $\{S_n\}$ diverges. ◆

EXAMPLE 1 Show that the series $\dfrac{1}{1!} + \dfrac{1}{2!} + \dfrac{1}{3!} + \cdots$ converges.

Solution We aim to show that the partial sums S_n are bounded above. Note first that

$$n! = 1 \cdot 2 \cdot 3 \cdots n \geq 1 \cdot 2 \cdot 2 \cdots 2 = 2^{n-1}$$

and so $1/n! \leq 1/2^{n-1}$. Thus,

$$S_n = \frac{1}{1!} + \frac{1}{2!} + \frac{1}{3!} + \cdots + \frac{1}{n!}$$

$$\leq 1 + \frac{1}{2} + \frac{1}{4} + \cdots + \frac{1}{2^{n-1}}$$

These latter terms come from a geometric series with $r = \tfrac{1}{2}$. They can be added by a formula in Example 1 of Section 10.2. We obtain

$$S_n \leq \frac{1 - \left(\frac{1}{2}\right)^n}{1 - \frac{1}{2}} = 2\left[1 - \left(\frac{1}{2}\right)^n\right] < 2$$

Thus, by the Bounded Sum Test, the given series converges. The argument also shows that its sum S is at most 2. Later we will show that $S = e - 1 \approx 1.71828$. ■

Series and Improper Integrals The behavior of $\displaystyle\sum_{k=1}^{\infty} f(k)$ and $\displaystyle\int_{1}^{\infty} f(x)\, dx$ with respect to convergence is similar and gives a very powerful test.

Theorem B Integral Test

Let f be a continuous, positive, nonincreasing function on the interval $[1, \infty)$ and suppose that $a_k = f(k)$ for all positive integers k. Then the infinite series

$$\sum_{k=1}^{\infty} a_k$$

converges if and only if the improper integral

$$\int_1^{\infty} f(x)\, dx$$

converges.

We remark that the integer 1 may be replaced by any positive integer M throughout this theorem (see Example 4).

Proof The diagrams in Figure 1 indicate how we may interpret the partial sums of the series $\sum a_k$ as areas and thereby relate the series to a corresponding integral. Note that the area of each rectangle is equal to its height, since the width is 1 in each case. From these diagrams, we easily see that

$$\sum_{k=2}^{n} a_k \leq \int_1^n f(x)\, dx \leq \sum_{k=1}^{n-1} a_k$$

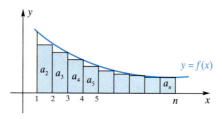 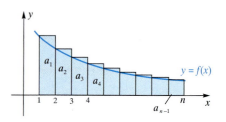

Figure 1

Now suppose that $\int_1^{\infty} f(x)\, dx$ converges. Then, by the left inequality above,

$$S_n = a_1 + \sum_{k=2}^{n} a_k \leq a_1 + \int_1^n f(x)\, dx \leq a_1 + \int_1^{\infty} f(x)\, dx$$

Therefore, by the Bounded Sum Test, $\sum_{k=1}^{\infty} a_k$ converges.

On the other hand, suppose that $\sum_{k=1}^{\infty} a_k$ converges. Then, by the right inequality above, if $t \leq n$,

$$\int_1^t f(x)\, dx \leq \int_1^n f(x)\, dx \leq \sum_{k=1}^{n-1} a_k \leq \sum_{k=1}^{\infty} a_k$$

Since $\int_1^t f(x)\, dx$ increases with t and is bounded above, $\lim_{t \to \infty} \int_1^t f(x)\, dx$ must exist; that is, $\int_1^{\infty} f(x)\, dx$ converges. ◆

The conclusion to Theorem B is often stated this way. *The series $\sum_{k=1}^{\infty} f(k)$ and the improper integral $\int_1^{\infty} f(x)\, dx$ converge or diverge together.* You should see that this is equivalent to our statement.

EXAMPLE 2

***p*-Series Test**   The series

$$\sum_{k=1}^{\infty} \frac{1}{k^p} = 1 + \frac{1}{2^p} + \frac{1}{3^p} + \frac{1}{4^p} + \cdots$$

where p is a constant, is called a ***p*-series**. Show each of the following:

(a) The p-series converges if $p > 1$.
(b) The p-series diverges if $p \leq 1$.

Solution   If $p \geq 0$, the function $f(x) = 1/x^p$ is continuous, positive, and nonincreasing on $[1, \infty)$ and $f(k) = 1/k^p$. Thus, by the Integral Test, $\sum(1/k^p)$ converges if and only if $\displaystyle\lim_{t \to \infty} \int_1^t x^{-p}\, dx$ exists (as a finite number).

If $p \neq 1$,

$$\int_1^t x^{-p}\, dx = \left[\frac{x^{1-p}}{1-p} \right]_1^t = \frac{t^{1-p} - 1}{1 - p}$$

If $p = 1$,

$$\int_1^t x^{-1}\, dx = [\ln x]_1^t = \ln t$$

Since $\displaystyle\lim_{t\to\infty} t^{1-p} = 0$ if $p > 1$ and $\displaystyle\lim_{t\to\infty} t^{1-p} = \infty$ if $p < 1$ and since $\displaystyle\lim_{t\to\infty} \ln t = \infty$, we conclude that the p-series converges if $p > 1$ and diverges if $0 \leq p \leq 1$.

We still have the case $p < 0$ to consider. In this case, the nth term of $\sum(1/k^p)$, that is, $1/n^p$, does not even tend toward 0. Thus, by the nth-Term Test, the series diverges.

Note that the case $p = 1$ gives the harmonic series, which was treated in Section 10.2. Our results here and there are consistent. The harmonic series diverges. ∎

The Tail of a Series

The beginning of a series plays no role in its convergence or divergence. Only the tail is important (the tail really does wag the dog). By the *tail* of a series, we mean

$$a_N + a_{N+1} + a_{N+2} + \cdots$$

where N denotes an arbitrarily large number. Hence, in testing for convergence or divergence of a series, we can ignore the beginning terms or even change them. Clearly, however, the sum of a series does depend on all its terms, including the initial ones.

EXAMPLE 3   Does $\displaystyle\sum_{k=4}^{\infty} \frac{1}{k^{1.001}}$ converge or diverge?

Solution   By the p-Series Test, $\displaystyle\sum_{k=1}^{\infty} (1/k^{1.001})$ converges. *The insertion or removal of a finite number of terms in a series cannot affect its convergence or divergence (though it may affect the sum).* Thus, the given series converges. ∎

EXAMPLE 4   Determine whether $\displaystyle\sum_{k=2}^{\infty} \frac{1}{k \ln k}$ converges or diverges.

Solution   The hypotheses of the Integral Test are satisfied for $f(x) = 1/(x \ln x)$ on $[2, \infty)$. That the interval is $[2, \infty)$ rather than $[1, \infty)$ is inconsequential, as we noted right after Theorem B. Now,

$$\int_2^\infty \frac{1}{x \ln x}\, dx = \lim_{t\to\infty} \int_2^t \frac{1}{\ln x}\left(\frac{1}{x}\, dx\right) = \lim_{t\to\infty} [\ln \ln x]_2^t = \infty$$

Thus, $\sum 1/(k \ln k)$ diverges. ∎

EXAMPLE 5   By means of an improper integral, find a good upper bound for the error in using the sum of the first five terms of the convergent series

$$\sum_{n=1}^{\infty} \frac{n}{e^{n^2}}$$

to approximate the sum of the series.

Solution   The error E is

$$E = \sum_{n=6}^{\infty} \frac{n}{e^{n^2}}$$

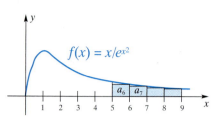

Figure 2

The function $f(x) = x/e^{x^2}$ is continuous, positive, and nonincreasing on $[5, \infty)$ (see Figure 2). Thus,

$$E = \sum_{n=6}^{\infty} \frac{n}{e^{n^2}} < \int_5^{\infty} x e^{-x^2} \, dx$$

$$= \lim_{t \to \infty} \left(-\frac{1}{2} \right) \int_5^t e^{-x^2}(-2x \, dx)$$

$$= \lim_{t \to \infty} \left(-\frac{1}{2} \right) \left[e^{-x^2} \right]_5^t = \frac{1}{2} e^{-25} \approx 6.94 \times 10^{-12} \quad \blacksquare$$

Concepts Review

1. A series of nonnegative terms converges if and only if its partial sums are _____.

2. The Integral Test relates the convergence of $\sum_{k=1}^{\infty} a_k$ and $\int_1^{\infty} f(x) \, dx$, assuming $a_k =$ _____ and f is _____, _____, and _____ on $[1, \infty)$.

3. The insertion or removal of a finite number of terms in a series does not affect its _____, although it may affect its sum.

4. The p-series $\sum_{k=1}^{\infty} (1/k^p)$ converges if and only if _____.

Problem Set 10.3

Use the Integral Test to decide the convergence or divergence of each of the following series.

1. $\displaystyle\sum_{k=0}^{\infty} \frac{1}{k + 3}$

2. $\displaystyle\sum_{k=1}^{\infty} \frac{3}{2k - 3}$

3. $\displaystyle\sum_{k=0}^{\infty} \frac{k}{k^2 + 3}$

4. $\displaystyle\sum_{k=1}^{\infty} \frac{3}{2k^2 + 1}$

5. $\displaystyle\sum_{k=1}^{\infty} \frac{-2}{\sqrt{k + 2}}$

6. $\displaystyle\sum_{k=100}^{\infty} \frac{3}{(k + 2)^2}$

7. $\displaystyle\sum_{k=2}^{\infty} \frac{7}{4k + 2}$

8. $\displaystyle\sum_{k=1}^{\infty} \frac{k^2}{e^k}$

9. $\displaystyle\sum_{k=1}^{\infty} \frac{3}{(4 + 3k)^{7/6}}$

10. $\displaystyle\sum_{k=1}^{\infty} \frac{1000k^2}{1 + k^3}$

11. $\displaystyle\sum_{k=1}^{\infty} k e^{-3k^2}$

12. $\displaystyle\sum_{k=5}^{\infty} \frac{1000}{k(\ln k)^2}$

In Problems 13–22, use any test developed so far, including any from Section 10.2, to decide about the convergence or divergence of the series. Give a reason for your conclusion.

13. $\displaystyle\sum_{k=1}^{\infty} \frac{k^2 + 1}{k^2 + 5}$

14. $\displaystyle\sum_{k=1}^{\infty} \left(\frac{3}{\pi} \right)^k$

15. $\displaystyle\sum_{k=1}^{\infty} \left[\left(\frac{1}{2} \right)^k + \frac{k - 1}{2k + 1} \right]$

16. $\displaystyle\sum_{k=1}^{\infty} \left(\frac{1}{k^2} + \frac{1}{2^k} \right)$

17. $\displaystyle\sum_{k=1}^{\infty} \sin \left(\frac{k\pi}{2} \right)$

18. $\displaystyle\sum_{k=1}^{\infty} k \sin \frac{1}{k}$

19. $\displaystyle\sum_{k=1}^{\infty} k^2 e^{-k^3}$

20. $\displaystyle\sum_{k=1}^{\infty} \left(\frac{1}{k} - \frac{1}{k + 1} \right)$

21. $\displaystyle\sum_{k=1}^{\infty} \frac{\tan^{-1} k}{1 + k^2}$

22. $\displaystyle\sum_{k=1}^{\infty} \frac{1}{1 + 4k^2}$

In Problems 23–26, estimate the error that is made by approximating the sum of the given series by the sum of the first five terms (see Example 5).

23. $\displaystyle\sum_{k=1}^{\infty} \frac{k}{e^k}$

24. $\displaystyle\sum_{k=1}^{\infty} \frac{1}{k\sqrt{k}}$

25. $\displaystyle\sum_{k=1}^{\infty} \frac{1}{1 + k^2}$

26. $\displaystyle\sum_{k=1}^{\infty} \frac{1}{k(k + 1)} = \sum_{k=1}^{\infty} \left(\frac{1}{k} - \frac{1}{k + 1} \right)$

27. For what values of p does $\displaystyle\sum_{n=2}^{\infty} 1/[n(\ln n)^p]$ converge? Explain.

28. Does $\displaystyle\sum_{n=3}^{\infty} 1/[n \cdot \ln n \cdot \ln(\ln n)]$ converge or diverge? Explain.

29. Use diagrams, as in Figure 1, to show that

$$\ln(n + 1) < 1 + \frac{1}{2} + \frac{1}{3} + \cdots + \frac{1}{n} < 1 + \ln n$$

Hint: $\displaystyle\int_1^n (1/x) \, dx = \ln n$.

30. Using Problem 29, show that the sequence

$$B_n = 1 + \frac{1}{2} + \frac{1}{3} + \cdots + \frac{1}{n} - \ln(n + 1)$$

is increasing and bounded above by 1.

31. Use the result of Problem 29 to prove that $\lim_{n \to \infty} B_n$ exists. (The limit, denoted γ, is called **Euler's constant** and is approximately 0.5772. It is currently not known whether γ is rational or irrational. It is known, however, that *if* γ is rational then the denominator in its reduced fraction is at least $10^{244,663}$.)

32. Use Problem 29 to get good upper and lower bounds for the sum of the first 10 million terms of the harmonic series.

33. From Problem 31, we infer that

$$1 + \frac{1}{2} + \frac{1}{3} + \cdots + \frac{1}{n} \approx \gamma + \ln(n+1)$$

Use this to estimate the number of terms of the harmonic series that are needed to get a sum greater than 20 and compare with the result reported in Problem 38 of Section 10.2.

34. Now that we have shown the existence of Euler's constant the hard way (Problems 29–31), we will solve a much more general problem the easy way and watch γ appear out of thin air, so to speak. Let f be continuous and decreasing on $[1, \infty)$ and let

$$B_n = f(1) + f(2) + \cdots + f(n) - \int_1^{n+1} f(x)\,dx$$

Note that B_n is the area of the shaded region in Figure 3.
(a) Why is it obvious that B_n increases with n?
(b) Show that $B_n \le f(1)$. *Hint*: Simply shift all the little shaded pieces leftward into the outlined rectangle.
(c) Conclude that $\lim_{n \to \infty} B_n$ exists.
(d) How do we get γ out of this?

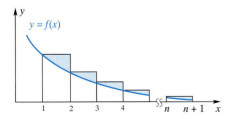

Figure 3

35. Let f be continuous, increasing, and concave down on $[1, \infty)$ as in Figure 4. Furthermore, let A_n be the area of the shaded region. Show that A_n is increasing with n, that $A_n \le T$ where T is the area of the outlined triangle, and thus that $\lim_{n \to \infty} A_n$ exists.

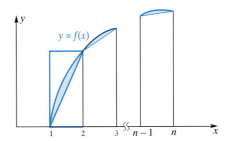

Figure 4

36. Specialize f of Problem 35 to $f(x) = \ln x$.

(a) Show that

$$A_n = \int_1^n \ln x\,dx - \left[\frac{\ln 1 + \ln 2}{2} + \cdots + \frac{\ln(n-1) + \ln n}{2} \right]$$

$$= n \ln n - n + 1 - \ln n! + \ln \sqrt{n}$$

$$= 1 + \ln \frac{(n/e)^n \sqrt{n}}{n!}$$

(b) Conclude from part (a) and Problem 35 that

$$k = \lim_{n \to \infty} \frac{n!}{(n/e)^n \sqrt{n}}$$

exists. It can be shown that $k = \sqrt{2\pi}$.

(c) This means that $n! \approx \sqrt{2\pi n}(n/e)^n$, which is called **Stirling's Formula**. Use it to approximate 15! and compare it with the value that your calculator gives for 15!

10.4
Positive Series: Other Tests

We have completely analyzed the convergence and divergence of two series, the geometric series and the *p*-series.

$$\sum_{n=1}^{\infty} r^n \quad \text{converges if } -1 < r < 1, \text{ diverges otherwise}$$

$$\sum_{n=1}^{\infty} \frac{1}{n^p} \quad \text{converges if } p > 1, \text{ diverges otherwise}$$

In the first we have found what the series converges to, provided that it converges; in the second, we have not. These series provide standards, or models, against which we can measure other series. Keep in mind that we are still considering series whose terms are positive (or at least nonnegative).

Comparing One Series with Another A series with terms less than the corresponding terms of a convergent series ought to converge; a series with terms greater than the corresponding terms of a divergent series ought to diverge. What ought to be true is true.

Theorem A Ordinary Comparison Test

Suppose that $0 \le a_n \le b_n$ for $n \ge N$.

(i) If Σb_n converges, so does Σa_n.
(ii) If Σa_n diverges, so does Σb_n.

Proof We suppose that $N = 1$; the case $N > 1$ is only slightly harder. To prove (i), let $S_n = a_1 + a_2 + \cdots + a_n$ and note that $\{S_n\}$ is a nondecreasing sequence. If Σb_n converges, for instance, with sum B, then

$$S_n \le b_1 + b_2 + \cdots + b_n \le \sum_{n=1}^{\infty} b_n = B$$

By the Bounded Sum Test (Theorem 10.3A), Σa_n converges.

Property (ii) follows from (i); for if Σb_n converged, then Σa_n would have to converge. ◆

EXAMPLE 1 Does $\displaystyle\sum \frac{n}{5n^2 - 4}$ converge or diverge?

Solution A good guess would be that it diverges, since the nth term behaves like $1/5n$ for large n. In fact,

$$\frac{n}{5n^2 - 4} > \frac{n}{5n^2} = \frac{1}{5} \cdot \frac{1}{n}$$

We know that $\displaystyle\sum \frac{1}{5} \cdot \frac{1}{n}$ diverges since it is one-fifth of the harmonic series (Theorem 10.2C). Thus, by the Ordinary Comparison Test, the given series also diverges. ∎

EXAMPLE 2 Does $\displaystyle\sum \frac{n}{2^n(n + 1)}$ converge or diverge?

Solution A good guess would be that it converges, since the nth term behaves like $(1/2)^n$ for large n. To substantiate our guess, we note that

$$\frac{n}{2^n(n + 1)} = \left(\frac{1}{2}\right)^n \frac{n}{n + 1} < \left(\frac{1}{2}\right)^n$$

Since $\Sigma \left(\frac{1}{2}\right)^n$ converges (it is a geometric series with $r = \frac{1}{2}$), we conclude that the given series converges. ∎

If there is a problem in applying the Ordinary Comparison Test, it is in finding exactly the right known series with which to compare the series to be tested. Suppose that we wish to determine the convergence or divergence of

$$\sum_{n=3}^{\infty} \frac{1}{(n - 2)^2} = \sum_{n=3}^{\infty} \frac{1}{n^2 - 4n + 4}$$

We suspect convergence, so our inclination is to compare $1/(n - 2)^2$ with $1/n^2$, but, unfortunately,

$$\frac{1}{(n - 2)^2} > \frac{1}{n^2}$$

which gives no test at all (the inequality goes the wrong way for what we want). After some experimenting, we discover that

$$\frac{1}{(n - 2)^2} \le \frac{9}{n^2}$$

for $n \ge 3$; since $\Sigma 9/n^2$ converges, so does $\Sigma 1/(n - 2)^2$.

Can we avoid these contortions with inequalities? Our intuition tells us that Σa_n and Σb_n converge or diverge together, provided that a_n and b_n are approximately the same size for large n (give or take a multiplicative constant). This is the essential content of our next theorem.

Theorem B   Limit Comparison Test

Suppose that $a_n \geq 0, b_n > 0$, and

$$\lim_{n \to \infty} \frac{a_n}{b_n} = L$$

If $0 < L < \infty$, then Σa_n and Σb_n converge or diverge together. If $L = 0$ and Σb_n converges, then Σa_n converges.

Proof   Begin by taking $\varepsilon = L/2$ in the definition of limit of a sequence (Section 10.1). There is a number N such that $n \geq N \Rightarrow |(a_n/b_n) - L| < L/2$; that is,

$$-\frac{L}{2} < \frac{a_n}{b_n} - L < \frac{L}{2}$$

This inequality is equivalent (by adding L throughout) to

$$\frac{L}{2} < \frac{a_n}{b_n} < \frac{3L}{2}$$

Hence, for $n \leq N$,

$$b_n < \frac{2}{L} a_n \quad \text{and} \quad a_n < \frac{3L}{2} b_n$$

These two inequalities, together with the Ordinary Comparison Test, show that Σa_n and Σb_n converge or diverge together. We leave the proof of the final statement of the theorem to the reader (Problem 37). ◆

EXAMPLE 3   Determine the convergence or divergence of each series.

(a) $\displaystyle\sum_{n=1}^{\infty} \frac{3n - 2}{n^3 - 2n^2 + 11}$ (b) $\displaystyle\sum_{n=1}^{\infty} \frac{1}{\sqrt{n^2 + 19n}}$

Solution   We apply the Limit Comparison Test, but we still must decide to what we should compare the nth term. We see what the nth term is like for large n by looking at the largest-degree terms in the numerator and denominator. In the first case, the nth term is like $3/n^2$; in the second, it is like $1/n$.

(a) $\displaystyle\lim_{n \to \infty} \frac{a_n}{b_n} = \lim_{n \to \infty} \frac{(3n - 2)/(n^3 - 2n^2 + 11)}{3/n^2} = \lim_{n \to \infty} \frac{3n^3 - 2n^2}{3n^3 - 6n^2 + 33} = 1$

(b) $\displaystyle\lim_{n \to \infty} \frac{a_n}{b_n} = \lim_{n \to \infty} \frac{1/\sqrt{n^2 + 19n}}{1/n} = \lim_{n \to \infty} \sqrt{\frac{n^2}{n^2 + 19n}} = 1$

Since $\Sigma 3/n^2$ converges and $\Sigma 1/n$ diverges, we conclude that the series in (a) converges and the series in (b) diverges. ∎

EXAMPLE 4   Does $\displaystyle\sum_{n=1}^{\infty} \frac{\ln n}{n^2}$ converge or diverge?

Solution   To what shall we compare $(\ln n)/n^2$? If we try $1/n^2$, we get

$$\lim_{n \to \infty} \frac{a_n}{b_n} = \lim_{n \to \infty} \frac{\ln n}{n^2} \div \frac{1}{n^2} = \lim_{n \to \infty} \ln n = \infty$$

The test fails because its conditions are not satisfied. On the other hand, if we use $1/n$, we get

$$\lim_{n\to\infty} \frac{a_n}{b_n} = \lim_{n\to\infty} \frac{\ln n}{n^2} \div \frac{1}{n} = \lim_{n\to\infty} \frac{\ln n}{n} = 0$$

Again, the test fails. Possibly something between $1/n^2$ and $1/n$ will work, such as $1/n^{3/2}$.

$$\lim_{n\to\infty} \frac{a_n}{b_n} = \lim_{n\to\infty} \frac{\ln n}{n^2} \div \frac{1}{n^{3/2}} = \lim_{n\to\infty} \frac{\ln n}{\sqrt{n}} = 0$$

(The last equality follows from l'Hôpital's Rule.) We conclude from the second part of the Limit Comparison Test that $\sum(\ln n)/n^2$ converges (since $\sum 1/n^{3/2}$ converges). ■

Comparing a Series with Itself Getting useful results from the comparison tests requires insight or perseverance. We must choose wisely among known series to find one that is just right for comparison with the series that we wish to test. Wouldn't it be nice if we could somehow compare a series with itself and thereby determine convergence or divergence? Roughly speaking, this is what we do in the Ratio Test.

Theorem C Ratio Test

Let $\sum a_n$ be a series of positive terms and suppose that

$$\lim_{n\to\infty} \frac{a_{n+1}}{a_n} = \rho$$

(i) If $\rho < 1$, the series converges.
(ii) If $\rho > 1$ or if $\lim_{n\to\infty} a_{n+1}/a_n = \infty$, the series diverges.
(iii) If $\rho = 1$, the test is inconclusive.

Proof Here is what is behind the Ratio Test. Since $\lim_{n\to\infty} a_{n+1}/a_n = \rho$, $a_{n+1} \approx \rho a_n$; that is, the series behaves like a geometric series with ratio ρ. A geometric series converges when its ratio is less than 1 and diverges when its ratio is greater than 1. Tying down this argument is the task before us.

(i) Since $\rho < 1$, we may choose a number r such that $\rho < r < 1$, for example, $r = (\rho + 1)/2$. Next choose N so large that $n \geq N$ implies that $a_{n+1}/a_n < r$. (This can be done since $\lim a_{n+1}/a_n = \rho < r$.)
Then,

$$a_{N+1} < ra_N$$
$$a_{N+2} < ra_{N+1} < r^2 a_N$$
$$a_{N+3} < ra_{N+2} < r^3 a_N$$
$$\vdots$$

Since $ra_N + r^2 a_N + r^3 a_N + \cdots$ is a geometric series with $0 < r < 1$, it converges. By the Ordinary Comparison Test, $\sum_{n=N+1}^{\infty} a_n$ converges, and hence so does $\sum_{n=1}^{\infty} a_n$.

(ii) Since $\rho > 1$, there is a number N such that $a_{n+1}/a_n > 1$ for all $n \geq N$. Thus,

$$a_{N+1} > a_N$$
$$a_{N+2} > a_{N+1} > a_N$$
$$\vdots$$

Hence, $a_n > a_N > 0$ for all $n > N$, which means that $\lim_{n\to\infty} a_n$ cannot be zero. By the nth-Term Test for Divergence, $\sum a_n$ diverges.

(iii) We know that $\sum 1/n$ diverges, whereas $\sum 1/n^2$ converges. For the first series,

$$\lim_{n \to \infty} \frac{a_{n+1}}{a_n} = \lim_{n \to \infty} \frac{1}{n+1} \div \frac{1}{n} = \lim_{n \to \infty} \frac{n}{n+1} = 1$$

For the second series,

$$\lim_{n \to \infty} \frac{a_{n+1}}{a_n} = \lim_{n \to \infty} \frac{1}{(n+1)^2} \div \frac{1}{n^2} = \lim_{n \to \infty} \frac{n^2}{(n+1)^2} = 1$$

Thus, the Ratio Test does not distinguish between convergence and divergence when $\rho = 1$. ◆

The Ratio Test will always be inconclusive for any series whose nth term is a rational expression in n, since in this case $\rho = 1$ (the cases $a_n = 1/n$ and $a_n = 1/n^2$ were treated above). However, for a series whose nth term involves $n!$ or r^n, the Ratio Test usually works beautifully.

EXAMPLE 5   Test for convergence or divergence: $\displaystyle\sum_{n=1}^{\infty} \frac{2^n}{n!}$.

Solution

$$\rho = \lim_{n \to \infty} \frac{a_{n+1}}{a_n} = \lim_{n \to \infty} \frac{2^{n+1}}{(n+1)!} \frac{n!}{2^n} = \lim_{n \to \infty} \frac{2}{n+1} = 0$$

We conclude by the Ratio Test that the series converges.   ■

EXAMPLE 6   Test for convergence or divergence: $\displaystyle\sum_{n=1}^{\infty} \frac{2^n}{n^{20}}$.

Solution

$$\rho = \lim_{n \to \infty} \frac{a_{n+1}}{a_n} = \lim_{n \to \infty} \frac{2^{n+1}}{(n+1)^{20}} \frac{n^{20}}{2^n}$$

$$= \lim_{n \to \infty} \left(\frac{n}{n+1} \right)^{20} \cdot 2 = 2$$

We conclude that the given series diverges.   ■

EXAMPLE 7   Test for convergence or divergence: $\displaystyle\sum_{n=1}^{\infty} \frac{n!}{n^n}$.

Solution   We will need the fact that

$$\lim_{n \to \infty} \left(1 + \frac{1}{n} \right)^n = \lim_{h \to 0} (1 + h)^{1/h} = e$$

which follows from Theorem 7.5A. Taking this as known, we may write

$$\rho = \lim_{n \to \infty} \frac{a_{n+1}}{a_n} = \lim_{n \to \infty} \frac{(n+1)!}{(n+1)^{n+1}} \frac{n^n}{n!} = \lim_{n \to \infty} \left(\frac{n}{n+1} \right)^n$$

$$= \lim_{n \to \infty} \frac{1}{((n+1)/n)^n} = \lim_{n \to \infty} \frac{1}{(1 + 1/n)^n} = \frac{1}{e} < 1$$

Therefore, the given series converges.   ■

Summary   To test a series Σa_n of positive terms for convergence or divergence, look carefully at a_n.

1. If $\lim_{n \to \infty} a_n \neq 0$, conclude from the nth-Term Test that the series diverges.

2. If a_n involves $n!$, r^n, or n^n, try the Ratio Test.

3. If a_n involves only constant powers of n, try the Limit Comparison Test. In particular, if a_n is a rational expression in n, use this test with b_n as the quotient of the leading terms from the numerator and denominator.

4. If the tests above do not work, try the Ordinary Comparison Test, the Integral Test, or the Bounded Sum Test.

5. Some series require a clever manipulation or a neat trick to determine convergence or divergence.

Concepts Review

1. The Ordinary Comparison Test says that if _____ and if Σb_k converges, then Σa_k also converges.

2. Assume that $a_k \geq 0$ and $b_k > 0$. The Limit Comparison Test says that if $0 < $ _____ $< \infty$ then Σa_k and Σb_k converge or diverge together.

3. Let $\rho = \lim\limits_{n \to \infty} \dfrac{a_{n+1}}{a_n}$. The Ratio Test says that a series Σa_k of positive terms converges if _____, diverges if _____, and may do either if _____.

4. $\Sigma(3^k/k!)$ is an obvious candidate for the _____ Test, whereas $\Sigma k/(k^3 - k - 1)$ is an obvious candidate for the _____ Test.

Problem Set 10.4

In Problems 1–4, use the Limit Comparison Test to determine convergence or divergence.

1. $\sum\limits_{n=1}^{\infty} \dfrac{n}{n^2 + 2n + 3}$

2. $\sum\limits_{n=1}^{\infty} \dfrac{3n + 1}{n^3 - 4}$

3. $\sum\limits_{n=1}^{\infty} \dfrac{1}{n\sqrt{n + 1}}$

4. $\sum\limits_{n=1}^{\infty} \dfrac{\sqrt{2n + 1}}{n^2}$

In Problems 5–10, use the Ratio Test to determine convergence or divergence.

5. $\sum\limits_{n=1}^{\infty} \dfrac{8^n}{n!}$

6. $\sum\limits_{n=1}^{\infty} \dfrac{5^n}{n^5}$

7. $\sum\limits_{n=1}^{\infty} \dfrac{n!}{n^{100}}$

8. $\sum\limits_{n=1}^{\infty} n(\tfrac{1}{3})^n$

9. $\sum\limits_{n=1}^{\infty} \dfrac{n^3}{(2n)!}$

10. $\sum\limits_{k=1}^{\infty} \dfrac{3^k + k}{k!}$

In Problems 11–34, determine convergence or divergence for each of the series. Indicate the test you use.

11. $\sum\limits_{n=1}^{\infty} \dfrac{n}{n + 200}$

12. $\sum\limits_{n=1}^{\infty} \dfrac{n!}{5 + n}$

13. $\sum\limits_{n=1}^{\infty} \dfrac{n + 3}{n^2\sqrt{n}}$

14. $\sum\limits_{n=1}^{\infty} \dfrac{\sqrt{n + 1}}{n^2 + 1}$

15. $\sum\limits_{n=1}^{\infty} \dfrac{n^2}{n!}$

16. $\sum\limits_{n=1}^{\infty} \dfrac{\ln n}{2^n}$

17. $\sum\limits_{n=1}^{\infty} \dfrac{4n^3 + 3n}{n^5 - 4n^2 + 1}$

18. $\sum\limits_{n=1}^{\infty} \dfrac{n^2 + 1}{3^n}$

19. $\dfrac{1}{1 \cdot 2} + \dfrac{1}{2 \cdot 3} + \dfrac{1}{3 \cdot 4} + \dfrac{1}{4 \cdot 5} + \cdots$

Hint: $a_n = \dfrac{1}{n(n + 1)}$.

20. $\dfrac{1}{2^2} + \dfrac{2}{3^2} + \dfrac{3}{4^2} + \dfrac{4}{5^2} + \cdots$

21. $\dfrac{2}{1 \cdot 3 \cdot 4} + \dfrac{3}{2 \cdot 4 \cdot 5} + \dfrac{4}{3 \cdot 5 \cdot 6} + \dfrac{5}{4 \cdot 6 \cdot 7} + \cdots$

22. $\dfrac{1}{1^2 + 1} + \dfrac{2}{2^2 + 1} + \dfrac{3}{3^2 + 1} + \dfrac{4}{4^2 + 1} + \cdots$

23. $\dfrac{1}{3} + \dfrac{2}{3^2} + \dfrac{3}{3^3} + \dfrac{4}{3^4} + \cdots$

24. $3 + \dfrac{3^2}{2!} + \dfrac{3^3}{3!} + \dfrac{3^4}{4!} + \cdots$

25. $1 + \dfrac{1}{2\sqrt{2}} + \dfrac{1}{3\sqrt{3}} + \dfrac{1}{4\sqrt{4}} + \cdots$

26. $\dfrac{\ln 2}{2^2} + \dfrac{\ln 3}{3^2} + \dfrac{\ln 4}{4^2} + \dfrac{\ln 5}{5^2} + \cdots$

27. $\sum\limits_{n=1}^{\infty} \dfrac{1}{2 + \sin^2 n}$

28. $\sum\limits_{n=1}^{\infty} \dfrac{5}{3^n + 1}$

29. $\sum\limits_{n=1}^{\infty} \dfrac{4 + \cos n}{n^3}$

30. $\sum\limits_{n=1}^{\infty} \dfrac{5^{2n}}{n!}$

31. $\sum\limits_{n=1}^{\infty} \dfrac{n^n}{(2n)!}$

32. $\sum\limits_{n=2}^{\infty} \left(1 - \dfrac{1}{n}\right)^n$

33. $\sum\limits_{n=1}^{\infty} \dfrac{4^n + n}{n!}$

34. $\sum\limits_{n=1}^{\infty} \dfrac{n}{2 + n5^n}$

35. Let $a_n > 0$ and suppose that Σa_n converges. Prove that Σa_n^2 converges.

36. Prove that $\lim\limits_{n \to \infty} (n!/n^n) = 0$ by considering the series $\Sigma n!/n^n$. Hint: Example 7, followed by nth-Term Test.

37. Prove that if $a_n \geq 0, b_n > 0, \lim\limits_{n \to \infty} a_n/b_n = 0$, and Σb_n converges then Σa_n converges.

38. Prove that if $a_n \geq 0, b_n > 0, \lim\limits_{n \to \infty} a_n/b_n = \infty$, and Σb_n diverges then Σa_n diverges.

39. Suppose that $\lim\limits_{n \to \infty} na_n = 1$. Prove that Σa_n diverges.

40. Prove that if Σa_n is a convergent series of positive terms then $\Sigma \ln(1 + a_n)$ converges.

41. Root Test Prove that if $a_n > 0$ and $\lim\limits_{n \to \infty} (a_n)^{1/n} = R$ then Σa_n converges if $R < 1$ and diverges if $R > 1$.

42. Test for convergence or divergence using the Root Test.

(a) $\displaystyle\sum_{n=2}^{\infty}\left(\frac{1}{\ln n}\right)^{n}$

(b) $\displaystyle\sum_{n=1}^{\infty}\left(\frac{n}{3n+2}\right)^{n}$

(c) $\displaystyle\sum_{n=1}^{\infty}\left(\frac{1}{2}+\frac{1}{n}\right)^{n}$

43. Test for convergence or divergence. In some cases, a clever manipulation using the properties of logarithms will simplify the problem.

(a) $\displaystyle\sum_{n=1}^{\infty}\ln\left(1+\frac{1}{n}\right)$

(b) $\displaystyle\sum_{n=1}^{\infty}\ln\left[\frac{(n+1)^{2}}{n(n+2)}\right]$

(c) $\displaystyle\sum_{n=2}^{\infty}\frac{1}{(\ln n)^{\ln n}}$

(d) $\displaystyle\sum_{n=3}^{\infty}\frac{1}{[\ln(\ln n)]^{\ln n}}$

(e) $\displaystyle\sum_{n=2}^{\infty}\frac{1}{(\ln n)^{4}}$

(f) $\displaystyle\sum_{n=1}^{\infty}\left[\frac{\ln n}{n}\right]^{2}$

[EXPL] **44.** Let $p(n)$ and $q(n)$ be polynomials in n with nonnegative coefficients. Give simple conditions that determine the convergence or divergence of $\displaystyle\sum_{n=1}^{\infty}\frac{p(n)}{q(n)}$.

[EXPL] **45.** Give conditions on p that determine the convergence or divergence of $\displaystyle\sum_{n=1}^{\infty}\frac{1}{n^{p}}\left(1+\frac{1}{2^{p}}+\frac{1}{3^{p}}+\cdots+\frac{1}{n^{p}}\right)$.

46. Test for convergence or divergence.

(a) $\displaystyle\sum_{n=1}^{\infty}\sin^{2}\left(\frac{1}{n}\right)$

(b) $\displaystyle\sum_{n=1}^{\infty}\tan\left(\frac{1}{n}\right)$

(c) $\displaystyle\sum_{n=1}^{\infty}\sqrt{n}\left[1-\cos\left(\frac{1}{n}\right)\right]$

Answers to Concepts Review: **1.** $0\le a_{k}\le b_{k}$ **2.** $\displaystyle\lim_{k\to\infty}\left(a_{k}/b_{k}\right)$ **3.** $\rho<1;\rho>1;\rho=1$ **4.** Ratio; Limit Comparison

10.5
Alternating Series, Absolute Convergence, and Conditional Convergence

In the last two sections, we considered series of nonnegative terms. Now we remove that restriction, allowing some terms to be negative. In particular, we study **alternating series**, that is, series of the form

$$a_{1}-a_{2}+a_{3}-a_{4}+\cdots$$

where $a_{n}>0$ for all n. An important example is the **alternating harmonic series**

$$1-\tfrac{1}{2}+\tfrac{1}{3}-\tfrac{1}{4}+\cdots$$

We have seen that the harmonic series diverges; we shall soon see that the *alternating* harmonic series converges.

A Convergence Test Let us suppose that the sequence $\{a_{n}\}$ is decreasing; that is, $a_{n+1}<a_{n}$ for all n. Also, let S_{n} have its usual meaning. Thus, for the alternating series $a_{1}-a_{2}+a_{3}-a_{4}+\cdots$, we have

$$S_{1}=a_{1}$$
$$S_{2}=a_{1}-a_{2}=S_{1}-a_{2}$$
$$S_{3}=a_{1}-a_{2}+a_{3}=S_{2}+a_{3}$$
$$S_{4}=a_{1}-a_{2}+a_{3}-a_{4}=S_{3}-a_{4}$$

and so on. A geometric interpretation of these partial sums is shown in Figure 1. Note that the even-numbered terms $S_{2},S_{4},S_{6},\ldots$ are increasing and bounded above and hence must converge to a limit, call it S'. Similarly, the odd-numbered terms $S_{1},S_{3},S_{5},\ldots$ are decreasing and bounded below. They also converge, say to S''.

Both S' and S'' are between S_{n} and S_{n+1} for all n (see Figure 2), and so

$$\left|S''-S'\right|\le\left|S_{n+1}-S_{n}\right|=a_{n+1}$$

Thus, the condition $a_{n+1}\to0$ as $n\to\infty$ will guarantee that $S'=S''$ and, consequently, the convergence of the series to their common value, which we call S. Finally, we note that, since S is between S_{n} and S_{n+1},

$$\left|S-S_{n}\right|\le\left|S_{n+1}-S_{n}\right|=a_{n+1}$$

That is, the error made by using S_{n} as an approximation to the sum S of the whole series is not more than the magnitude of the first neglected term. We have proved the following theorem.

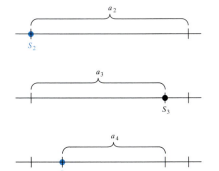

Figure 1

Figure 2

Theorem A Alternating Series Test

Let

$$a_{1}-a_{2}+a_{3}-a_{4}+\cdots \qquad \text{(continued on next page)}$$

be an alternating series with $a_n > a_{n+1} > 0$. If $\lim_{n \to \infty} a_n = 0$, then the series converges. Moreover, the error made by using the sum S_n of the first n terms to approximate the sum S of the series is not more than a_{n+1}.

EXAMPLE 1 Show that the alternating harmonic series

$$1 - \tfrac{1}{2} + \tfrac{1}{3} - \tfrac{1}{4} + \cdots$$

converges. How many terms of this series would we need to take in order to get a partial sum S_n within 0.01 of the sum S of the whole series?

Solution The alternating harmonic series satisfies the hypotheses of Theorem A and so converges. We want $|S - S_n| \leq 0.01$, and this will hold if $a_{n+1} \leq 0.01$. Since $a_{n+1} = 1/(n + 1)$, we require $1/(n + 1) \leq 0.01$, which is satisfied if $n \geq 99$. Thus, we need to take 99 terms to make sure that we have the desired accuracy. This gives you an idea of how slowly the alternating harmonic series converges. (See Problem 45 for a clever way to find the exact sum of this series.) ∎

EXAMPLE 2 Show that

$$\frac{1}{1!} - \frac{1}{2!} + \frac{1}{3!} - \frac{1}{4!} + \cdots$$

converges. Calculate S_5 and estimate the error made by using this as a value for the sum of the whole series.

Solution The Alternating Series Test (Theorem A) applies and guarantees convergence.

$$S_5 = 1 - \frac{1}{2} + \frac{1}{6} - \frac{1}{24} + \frac{1}{120} \approx 0.6333$$

$$|S - S_5| \leq a_6 = \frac{1}{6!} \approx 0.0014$$

∎

EXAMPLE 3 Show that $\sum_{n=1}^{\infty} (-1)^{n-1} \dfrac{n^2}{2^n}$ converges.

Solution To get a feeling for this series, we write the first few terms:

$$\tfrac{1}{2} - 1 + \tfrac{9}{8} - 1 + \tfrac{25}{32} - \tfrac{36}{64} + \cdots$$

The series is alternating and $\lim_{n \to \infty} n^2/2^n = 0$ (l'Hôpital's Rule), but unfortunately the terms are not decreasing initially. However, they do appear to be decreasing after the first two terms; this is good enough, since what happens at the beginning of a series never affects convergence or divergence. To show that the sequence $\{n^2/2^n\}$ is decreasing from the third term on, consider the function

$$f(x) = \frac{x^2}{2^x}$$

Note that if $x \geq 3$ the derivative

$$f'(x) = \frac{2x \cdot 2^x - x^2 2^x \ln 2}{2^{2x}} = \frac{x 2^x (2 - x \ln 2)}{2^{2x}}$$

$$\approx \frac{x(2 - 0.69x)}{2^x} < 0$$

Thus, f is decreasing on $[3, \infty)$, and so $\{n^2/2^n\}$ is decreasing for $n \geq 3$. For a different demonstration of this last fact, see Example 6 of Section 10.1. ∎

Absolute Convergence Does a series such as

$$1 + \tfrac{1}{4} - \tfrac{1}{9} + \tfrac{1}{16} + \tfrac{1}{25} - \tfrac{1}{36} + \cdots$$

in which there is a pattern of two positive terms followed by one negative term, converge or diverge? The Alternating Series Test does not apply. However, since the corresponding series of all positive terms

$$1 + \tfrac{1}{4} + \tfrac{1}{9} + \tfrac{1}{16} + \tfrac{1}{25} + \tfrac{1}{36} + \cdots$$

converges (*p*-series with $p = 2$), it seems plausible to think that the same series with some terms negative should converge (even better). This is the content of our next theorem.

Theorem B   Absolute Convergence Test

If $\Sigma |u_n|$ converges, then Σu_n converges.

Proof   We use a trick. Let $v_n = u_n + |u_n|$, so

$$u_n = v_n - |u_n|$$

Now $0 \le v_n \le 2|u_n|$, and so Σv_n converges by the Ordinary Comparison Test. It follows from the Linearity Theorem (Theorem 10.2B) that $\Sigma u_n = \Sigma(v_n - |u_n|)$ converges. ♦

A series Σu_n is said to **converge absolutely** if $\Sigma |u_n|$ converges. Theorem B asserts that absolute convergence implies convergence. All our tests for convergence of series of positive terms are automatically tests for the absolute convergence of a series in which some terms are negative. In particular, this is true of the Ratio Test, which we now restate.

Theorem C   Absolute Ratio Test

Let Σu_n be a series of nonzero terms and suppose that

$$\lim_{n \to \infty} \frac{|u_{n+1}|}{|u_n|} = \rho$$

(i)   If $\rho < 1$, the series converges absolutely (hence converges).
(ii)   If $\rho > 1$, the series diverges.
(iii)   If $\rho = 1$, the test is inconclusive.

Proof   Proofs of (i) and (iii) are direct results of the Ratio Test. For (ii), we could conclude from the original Ratio Test, that $\Sigma |u_n|$ diverges, but here we are claiming more, that Σu_n diverges. Since

$$\lim_{n \to \infty} \frac{|u_{n+1}|}{|u_n|} > 1$$

it follows that for n sufficiently large, say $n \ge N$, $|u_{n+1}| > |u_n|$. This, in turn, implies that $|u_n| > |u_N| > 0$ for all $n \ge N$, and so $\lim_{n \to \infty} u_n$ cannot be 0. We conclude by the nth-Term Test that Σu_n diverges. ♦

EXAMPLE 4   Show that $\displaystyle\sum_{n=1}^{\infty} (-1)^{n+1} \frac{3^n}{n!}$ converges absolutely.

Solution

$$\rho = \lim_{n \to \infty} \frac{|u_{n+1}|}{|u_n|} = \lim_{n \to \infty} \frac{3^{n+1}}{(n+1)!} \div \frac{3^n}{n!}$$

$$= \lim_{n \to \infty} \frac{3}{n+1} = 0$$

We conclude from the Absolute Ratio Test that the series converges absolutely (and therefore converges). ■

EXAMPLE 5 Test for the convergence or divergence of $\sum_{n=1}^{\infty} \dfrac{\cos(n!)}{n^2}$.

Solution If you write out the first 100 terms of this series, you will discover that the signs of the terms vary in a rather random way. The series is in fact a difficult one to analyze directly. However,

$$\left|\frac{\cos(n!)}{n^2}\right| \le \frac{1}{n^2}$$

and so the series converges absolutely by the Ordinary Comparison Test. We conclude from the Absolute Convergence Test (Theorem B) that the series converges. ■

Conditional Convergence A common error is to try to turn Theorem B around. It does *not* say that convergence implies absolute convergence. That is clearly false; witness the alternating harmonic series. We know that

$$1 - \tfrac{1}{2} + \tfrac{1}{3} - \tfrac{1}{4} + \cdots$$

converges, but that

$$1 + \tfrac{1}{2} + \tfrac{1}{3} + \tfrac{1}{4} + \cdots$$

diverges. A series $\sum u_n$ is called **conditionally convergent** if $\sum u_n$ converges but $\sum |u_n|$ diverges. The alternating harmonic series is the premier example of a conditionally convergent series, but there are many others.

EXAMPLE 6 Show that $\sum_{n=1}^{\infty} (-1)^{n+1} \dfrac{1}{\sqrt{n}}$ is conditionally convergent.

Solution $\sum_{n=1}^{\infty} (-1)^{n+1}\left[1/\sqrt{n}\right]$ converges by the Alternating Series Test. However, $\sum_{n=1}^{\infty} 1/\sqrt{n}$ diverges, since it is a p-series with $p = \tfrac{1}{2}$. ■

Absolutely convergent series behave much better than do conditionally convergent ones. Here is a nice theorem about absolutely convergent series. It is spectacularly false for conditionally convergent series (see Problems 35–38). The proof is difficult, so we do not include it here.

Theorem D Rearrangement Theorem

The terms of an absolutely convergent series can be rearranged without affecting either the convergence or the sum of the series.

For example, the series

$$1 + \tfrac{1}{4} - \tfrac{1}{9} + \tfrac{1}{16} + \tfrac{1}{25} - \tfrac{1}{36} + \tfrac{1}{49} + \tfrac{1}{64} - \tfrac{1}{81} + \cdots$$

converges absolutely. The rearrangement

$$1 + \tfrac{1}{4} + \tfrac{1}{16} - \tfrac{1}{9} + \tfrac{1}{25} + \tfrac{1}{49} + \tfrac{1}{64} - \tfrac{1}{36} + \cdots$$

converges and has the same sum as the original series.

Concepts Review

1. If $a_n \geq 0$ for all n, the alternating series $a_1 - a_2 + a_3 - \cdots$ will converge provided that the terms are decreasing in size and _____.

2. If $\Sigma|u_k|$ converges, we say that the series Σu_k converges _____; if Σu_k converges, but $\Sigma|u_k|$ diverges, we say that Σu_k converges _____.

3. The premier example of a conditionally convergent series is _____.

4. The terms of an absolutely convergent series may be _____ at will without affecting its convergence or its sum.

Problem Set 10.5

In Problems 1–6, show that each alternating series converges and then estimate the error made by using the partial sum S_9 as an approximation to the sum S of the series (see Examples 1–3).

1. $\displaystyle\sum_{n=1}^{\infty} (-1)^{n+1} \frac{2}{3n+1}$

2. $\displaystyle\sum_{n=1}^{\infty} (-1)^{n+1} \frac{1}{\sqrt{n}}$

3. $\displaystyle\sum_{n=1}^{\infty} (-1)^{n+1} \frac{1}{\ln(n+1)}$

4. $\displaystyle\sum_{n=1}^{\infty} (-1)^{n+1} \frac{n}{n^2+1}$

5. $\displaystyle\sum_{n=1}^{\infty} (-1)^{n+1} \frac{\ln n}{n}$

6. $\displaystyle\sum_{n=1}^{\infty} (-1)^{n+1} \frac{\ln n}{\sqrt{n}}$

In Problems 7–12, show that each series converges absolutely.

7. $\displaystyle\sum_{n=1}^{\infty} \left(-\tfrac{3}{4}\right)^n$

8. $\displaystyle\sum_{n=1}^{\infty} (-1)^n \frac{1}{n\sqrt{n}}$

9. $\displaystyle\sum_{n=1}^{\infty} (-1)^{n+1} \frac{n}{2^n}$

10. $\displaystyle\sum_{n=1}^{\infty} (-1)^{n+1} \frac{n^2}{e^n}$

11. $\displaystyle\sum_{n=1}^{\infty} (-1)^{n+1} \frac{1}{n(n+1)}$

12. $\displaystyle\sum_{n=1}^{\infty} (-1)^{n+1} \frac{2^n}{n!}$

In Problems 13–30, classify each series as absolutely convergent, conditionally convergent, or divergent.

13. $\displaystyle\sum_{n=1}^{\infty} (-1)^{n+1} \frac{1}{5n}$

14. $\displaystyle\sum_{n=1}^{\infty} (-1)^{n+1} \frac{1}{5n^{1.1}}$

15. $\displaystyle\sum_{n=1}^{\infty} (-1)^{n+1} \frac{n}{10n+1}$

16. $\displaystyle\sum_{n=1}^{\infty} (-1)^{n+1} \frac{n}{10n^{1.1}+1}$

17. $\displaystyle\sum_{n=2}^{\infty} (-1)^n \frac{1}{n \ln n}$

18. $\displaystyle\sum_{n=1}^{\infty} (-1)^{n+1} \frac{1}{n(1+\sqrt{n})}$

19. $\displaystyle\sum_{n=1}^{\infty} (-1)^{n+1} \frac{n^4}{2^n}$

20. $\displaystyle\sum_{n=2}^{\infty} (-1)^n \frac{1}{\sqrt{n^2-1}}$

21. $\displaystyle\sum_{n=1}^{\infty} (-1)^{n+1} \frac{n}{n^2+1}$

22. $\displaystyle\sum_{n=1}^{\infty} (-1)^{n+1} \frac{n-1}{n}$

23. $\displaystyle\sum_{n=1}^{\infty} \frac{\cos n\pi}{n}$

24. $\displaystyle\sum_{n=1}^{\infty} \frac{\sin(n\pi/2)}{n^2}$

25. $\displaystyle\sum_{n=1}^{\infty} (-1)^n \frac{\sin n}{n\sqrt{n}}$

26. $\displaystyle\sum_{n=1}^{\infty} n \sin\left(\frac{1}{n}\right)$

27. $\displaystyle\sum_{n=1}^{\infty} (-1)^{n+1} \frac{1}{\sqrt{n(n+1)}}$

28. $\displaystyle\sum_{n=1}^{\infty} \frac{(-1)^{n+1}}{\sqrt{n+1}+\sqrt{n}}$

29. $\displaystyle\sum_{n=1}^{\infty} \frac{(-3)^{n+1}}{n^2}$

30. $\displaystyle\sum_{n=1}^{\infty} (-1)^{n+1} \sin\frac{\pi}{n}$

31. Prove that if Σa_n diverges, so does $\Sigma|a_n|$.

32. Give an example of two series Σa_n and Σb_n, both convergent, such that $\Sigma a_n b_n$ diverges.

33. Show that the positive terms of the alternating harmonic series form a divergent series. Show the same for the negative terms.

34. Show that the results in Problem 33 hold for any conditionally convergent series.

35. Show that the alternating harmonic series

$$1 - \tfrac{1}{2} + \tfrac{1}{3} - \tfrac{1}{4} + \tfrac{1}{5} - \tfrac{1}{6} + \cdots$$

(whose sum is actually $\ln 2 \approx 0.69$) can be rearranged to converge to 1.3 by using the following steps.

(a) Take enough of the positive terms $1 + \tfrac{1}{3} + \tfrac{1}{5} + \cdots$ to just exceed 1.3.

(b) Now add enough of the negative terms $-\tfrac{1}{2} - \tfrac{1}{4} - \tfrac{1}{6} - \cdots$ so that the partial sum S_n falls just below 1.3.

(c) Add just enough more positive terms to again exceed 1.3, and so on.

[C] **36.** Use your calculator to help you to find the first 20 terms of the series described in Problem 35. Calculate S_{20}.

37. Explain why a conditionally convergent series can be rearranged to converge to any given number.

38. Show that a conditionally convergent series can be rearranged so as to diverge.

39. Show that $\lim_{n\to\infty} a_n = 0$ is not sufficient to guarantee the convergence of the alternating series $\Sigma(-1)^{n+1} a_n$. *Hint:* Alternate the terms of $\Sigma 1/n$ and $\Sigma -1/n^2$.

40. Discuss the convergence or divergence of

$$\frac{1}{\sqrt{2}-1} - \frac{1}{\sqrt{2}+1} + \frac{1}{\sqrt{3}-1} - \frac{1}{\sqrt{3}+1}$$

$$+ \frac{1}{\sqrt{4}-1} - \frac{1}{\sqrt{4}+1} + \cdots$$

41. Prove that if $\sum_{k=1}^{\infty} a_k^2$ and $\sum_{k=1}^{\infty} b_k^2$ both converge then $\sum_{k=1}^{\infty} a_k b_k$ converges absolutely. *Hint:* First show that $2|a_k b_k| \le a_k^2 + b_k^2$.

42. Sketch the graph of $y = (\sin x)/x$ and then show that $\int_0^\infty (\sin x)/x \, dx$ converges.

43. Show that $\int_0^\infty |\sin x|/x \, dx$ diverges.

44. Show that the graph of $y = x \sin \dfrac{\pi}{x}$ on $(0, 1]$ has infinite length.

45. Note that

$$1 - \frac{1}{2} + \frac{1}{3} - \frac{1}{4} + \cdots - \frac{1}{2n}$$

$$= 1 + \frac{1}{2} + \frac{1}{3} + \cdots + \frac{1}{2n} - \left(1 + \frac{1}{2} + \frac{1}{3} + \cdots + \frac{1}{n}\right)$$

$$= \frac{1}{n+1} + \frac{1}{n+2} + \cdots + \frac{1}{2n}$$

Recognize the latter expression as a Riemann sum and use it to find the sum of the alternating harmonic series.

Answers to Concepts Review: **1.** $\lim\limits_{n \to \infty} a_n = 0$ **2.** absolutely; conditionally **3.** the alternating harmonic series **4.** rearranged

---

10.6
Power Series

So far we have been studying what might be called *series of constants*, that is, series of the form $\sum u_n$, where each u_n is a number. Now we consider *series of functions*, series of the form $\sum u_n(x)$. A typical example of such a series is

$$\sum_{n=1}^{\infty} \frac{\sin nx}{n^2} = \frac{\sin x}{1} + \frac{\sin 2x}{4} + \frac{\sin 3x}{9} + \cdots$$

Of course, as soon as we substitute a value for x (such as $x = 2.1$), we are back to familiar territory; we have a series of constants.

There are two important questions to ask about a series of functions.

1. For what x's does the series converge?
2. To what function does it converge; that is, what is the sum $S(x)$ of the series?

The general situation is a proper subject for an advanced calculus course. However, even in elementary calculus, we can learn a good deal about the special case of a power series. A **power series in x** has the form

$$\sum_{n=0}^{\infty} a_n x^n = a_0 + a_1 x + a_2 x^2 + \cdots$$

(Here we interpret $a_0 x^0$ to be a_0 even if $x = 0$.) We can immediately answer our two questions for one such power series.

EXAMPLE 1   For what x's does the power series

$$\sum_{n=0}^{\infty} a x^n = a + ax + ax^2 + ax^3 + \cdots$$

converge and what is its sum? Assume that $a \ne 0$.

Solution   We actually studied this series in Section 10.2 (with r in place of x) and called it a geometric series. It converges for $-1 < x < 1$ and has sum $S(x)$ given by

$$S(x) = \frac{a}{1-x}, \qquad -1 < x < 1 \qquad \blacksquare$$

Fourier Series

The series of sine functions mentioned in the introduction is an example of a *Fourier series*, named after Jean Baptiste Joseph Fourier (1768–1830). Fourier series are of immense importance in the study of wave phenomena, since they allow us to represent a complicated wave as a sum of its fundamental components (called the pure tones in the case of sound waves). It is a large field, which we leave to other authors and other books.

Convergence set

Figure 1

Convergence set

Figure 2

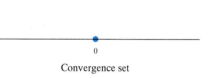

Convergence set

Figure 3

Convergence set

Figure 4

The Convergence Set

We call the set on which a power series converges its **convergence set**. What kind of set can be a convergence set? Example 1 suggests that it can be an open interval (see Figure 1). Are there other possibilities?

EXAMPLE 2 What is the convergence set for

$$\sum_{n=0}^{\infty} \frac{x^n}{(n+1)2^n} = 1 + \frac{1}{2}\frac{x}{2} + \frac{1}{3}\frac{x^2}{2^2} + \frac{1}{4}\frac{x^3}{2^3} + \cdots$$

Solution Note that some of the terms may be negative (if x is negative). Let's test for absolute convergence using the Absolute Ratio Test (Theorem 10.5C).

$$\rho = \lim_{n \to \infty} \left| \frac{x^{n+1}}{(n+2)2^{n+1}} \div \frac{x^n}{(n+1)2^n} \right| = \lim_{n \to \infty} \frac{|x|}{2} \cdot \frac{n+1}{n+2} = \frac{|x|}{2}$$

The series converges absolutely (hence converges) when $\rho = |x|/2 < 1$ and diverges when $|x|/2 > 1$. Consequently, it converges when $|x| < 2$ and diverges when $|x| > 2$.

If $x = 2$ or $x = -2$, the Ratio Test fails. However, when $x = 2$, the series is the harmonic series, which diverges; and when $x = -2$, it is the alternating harmonic series, which converges. We conclude that the convergence set for the given series is the interval $-2 \le x < 2$ (Figure 2). ∎

EXAMPLE 3 Find the convergence set for $\displaystyle\sum_{n=0}^{\infty} \frac{x^n}{n!}$.

Solution

$$\rho = \lim_{n \to \infty} \left| \frac{x^{n+1}}{(n+1)!} \div \frac{x^n}{n!} \right| = \lim_{n \to \infty} \frac{|x|}{n+1} = 0$$

We conclude from the Absolute Ratio Test that the series converges for all x (Figure 3). ∎

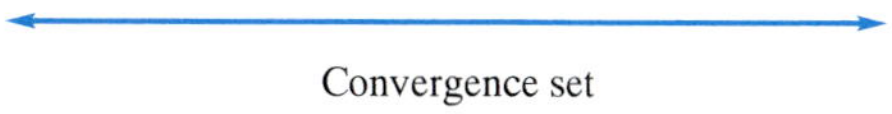

EXAMPLE 4 Find the convergence set for $\displaystyle\sum_{n=0}^{\infty} n!x^n$.

Solution

$$\rho = \lim_{n \to \infty} \left| \frac{(n+1)!x^{n+1}}{n!x^n} \right| = \lim_{n \to \infty} (n+1)|x| = \begin{cases} 0 & \text{if } x = 0 \\ \infty & \text{if } x \ne 0 \end{cases}$$

We conclude that the series converges only at $x = 0$ (Figure 4). ∎

In each of our examples, the convergence set was an interval (a degenerate interval in the last example). This will always be the case. For example, it is impossible for a power series to have a convergence set consisting of two disconnected parts (like $[0, 1] \cup [2, 3]$). Our next theorem tells the whole story.

Theorem A

The convergence set for a power series $\sum a_n x^n$ is always an interval of one of the following three types:

(i) The single point $x = 0$.
(ii) An interval $(-R, R)$, plus possibly one or both end points.
(iii) The whole real line.

In (i), (ii), and (iii), the series is said to have **radius of convergence** 0, R, and ∞, respectively.

Proof Suppose that the series converges at $x = x_1 \neq 0$. Then $\lim\limits_{n\to\infty} a_n x_1^n = 0$, and so there is certainly a number N such that $|a_n x_1^n| < 1$ for $n \geq N$. Then, for any x for which $|x| < |x_1|$,

$$|a_n x^n| = |a_n x_1^n| \left|\frac{x}{x_1}\right|^n < \left|\frac{x}{x_1}\right|^n$$

this holding for $n \geq N$. Now $\sum |x/x_1|^n$ converges, since it is a geometric series with ratio less than 1. Thus, by the Ordinary Comparison Test (Theorem 10.4A), $\sum |a_n x^n|$ converges. We have shown that if a power series converges at x_1 it converges (absolutely) for all x such that $|x| < |x_1|$.

On the other hand, suppose that a power series diverges at x_2. Then it must diverge for all x for which $|x| > |x_2|$. For if it converged at x_1 such that $|x_1| > |x_2|$, then, by what we have already shown, it would converge at x_2, contrary to hypothesis.

These two paragraphs together eliminate all possible types of convergence sets except the three types mentioned in the theorem. ♦

Actually we have proved slightly more than we have claimed in Theorem A, and it is worth stating this as another theorem.

Theorem B

A power series $\sum a_n x^n$ converges absolutely on the interior of its interval of convergence.

Of course, it might even converge absolutely at the end points of the interval of convergence, but of that we cannot be sure; witness Example 2.

Power Series in $x - a$ A series of the form

$$\sum a_n (x - a)^n = a_0 + a_1(x - a) + a_2(x - a)^2 + \cdots$$

is called a **power series in $x - a$**. All that we have said about power series in x applies equally well for series in $x - a$. In particular, its convergence set is always one of the following kinds of intervals:

1. The single point $x = a$.
2. An interval $(a - R, a + R)$, plus possibly one or both end points (Figure 5).
3. The whole real line.

Convergence set

Figure 5

EXAMPLE 5 Find the convergence set for $\sum\limits_{n=0}^{\infty} \dfrac{(x - 1)^n}{(n + 1)^2}$.

Solution We apply the Absolute Ratio Test.

$$\rho = \lim_{n\to\infty} \left| \frac{(x - 1)^{n+1}}{(n + 2)^2} \div \frac{(x - 1)^n}{(n + 1)^2} \right| = \lim_{n\to\infty} |x - 1| \frac{(n + 1)^2}{(n + 2)^2}$$
$$= |x - 1|$$

Thus, the series converges if $|x - 1| < 1$, that is, if $0 < x < 2$; it diverges if $|x - 1| > 1$. It also converges (even absolutely) at both of the end points 0 and 2, as we see by substitution of these values. The convergence set is the closed interval $[0, 2]$ (Figure 6). ■

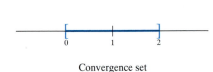

Convergence set

Figure 6

EXAMPLE 6 Determine the convergence set for

$$\frac{(x + 2)^2 \ln 2}{2 \cdot 9} + \frac{(x + 2)^3 \ln 3}{3 \cdot 27} + \frac{(x + 2)^4 \ln 4}{4 \cdot 81} + \cdots$$

Solution The nth term is $u_n = \dfrac{(x+2)^n \ln n}{n \cdot 3^n}$, $n \geq 2$. Thus,

$$\rho = \lim_{n \to \infty} \left| \frac{(x+2)^{n+1} \ln(n+1)}{(n+1)3^{n+1}} \cdot \frac{n3^n}{(x+2)^n \ln n} \right|$$

$$= \frac{|x+2|}{3} \lim_{n \to \infty} \frac{n}{n+1} \frac{\ln(n+1)}{\ln n} = \frac{|x+2|}{3}$$

We know that the series converges when $\rho < 1$, that is, when $|x+2| < 3$ or, equivalently, $-5 < x < 1$, but we must check the end points -5 and 1.

At $x = -5$,

$$u_n = \frac{(-3)^n \ln n}{n3^n} = (-1)^n \frac{\ln n}{n}$$

and $\Sigma(-1)^n (\ln n)/n$ converges by the Alternating Series Test.

At $x = 1$, $u_n = (\ln n)/n$ and $\Sigma(\ln n)/n$ diverges by comparison with the harmonic series.

We conclude that the given series converges on the interval $-5 \leq x < 1$. ∎

Concepts Review

1. A series of the form $a_0 + a_1 x + a_2 x^2 + \cdots$ is called a _____.

2. Rather than asking whether a power series converges, we should ask _____.

3. A power series always converges on a(n) _____, which may or may not include its _____.

4. The series $5 + x + x^2 + x^3 + \cdots$ converges on the interval _____.

Problem Set 10.6

In Problems 1–20, find the convergence set of the given power series. Hint: *First find a formula for the nth term; then use the Absolute Ratio Test.*

1. $\dfrac{x}{1 \cdot 2} - \dfrac{x^2}{2 \cdot 3} + \dfrac{x^3}{3 \cdot 4} - \dfrac{x^4}{4 \cdot 5} + \dfrac{x^5}{5 \cdot 6} - \cdots$

2. $1 + x + \dfrac{x^2}{2!} + \dfrac{x^3}{3!} + \dfrac{x^4}{4!} + \cdots$

3. $x - \dfrac{x^3}{3!} + \dfrac{x^5}{5!} - \dfrac{x^7}{7!} + \dfrac{x^9}{9!} - \cdots$

4. $1 - \dfrac{x^2}{2!} + \dfrac{x^4}{4!} - \dfrac{x^6}{6!} + \dfrac{x^8}{8!} - \dfrac{x^{10}}{10!} + \cdots$

5. $x + 2x^2 + 3x^3 + 4x^4 + \cdots$

6. $x + 2^2 x^2 + 3^2 x^3 + 4^2 x^4 + \cdots$

7. $1 - x + \dfrac{x^2}{2} - \dfrac{x^3}{3} + \dfrac{x^4}{4} - \cdots$

8. $1 + x + \dfrac{x^2}{\sqrt{2}} + \dfrac{x^3}{\sqrt{3}} + \dfrac{x^4}{\sqrt{4}} + \dfrac{x^5}{\sqrt{5}} + \cdots$

9. $1 - \dfrac{x}{1 \cdot 3} + \dfrac{x^2}{2 \cdot 4} - \dfrac{x^3}{3 \cdot 5} + \dfrac{x^4}{4 \cdot 6} - \cdots$

10. $\dfrac{x}{2^2 - 1} + \dfrac{x^2}{3^2 - 1} + \dfrac{x^3}{4^2 - 1} + \dfrac{x^4}{5^2 - 1} + \cdots$

11. $1 - \dfrac{x}{2} + \dfrac{x^2}{2^2} - \dfrac{x^3}{2^3} + \dfrac{x^4}{2^4} - \cdots$

12. $1 + 2x + 2^2 x^2 + 2^3 x^3 + 2^4 x^4 + \cdots$

13. $1 + 2x + \dfrac{2^2 x^2}{2!} + \dfrac{2^3 x^3}{3!} + \dfrac{2^4 x^4}{4!} + \cdots$

14. $\dfrac{x}{2} + \dfrac{2x^2}{3} + \dfrac{3x^3}{4} + \dfrac{4x^4}{5} + \dfrac{5x^5}{6} + \cdots$

15. $\dfrac{x-1}{1} + \dfrac{(x-1)^2}{2} + \dfrac{(x-1)^3}{3} + \dfrac{(x-1)^4}{4} + \cdots$

16. $1 + (x+2) + \dfrac{(x+2)^2}{2!} + \dfrac{(x+2)^3}{3!} + \cdots$

17. $1 + \dfrac{x+1}{2} + \dfrac{(x+1)^2}{2^2} + \dfrac{(x+1)^3}{2^3} + \cdots$

18. $\dfrac{x-2}{1^2} + \dfrac{(x-2)^2}{2^2} + \dfrac{(x-2)^3}{3^2} + \dfrac{(x-2)^4}{4^2} + \cdots$

19. $\dfrac{x+5}{1 \cdot 2} + \dfrac{(x+5)^2}{2 \cdot 3} + \dfrac{(x+5)^3}{3 \cdot 4} + \dfrac{(x+5)^4}{4 \cdot 5} + \cdots$

20. $(x+3) - 2(x+3)^2 + 3(x+3)^3 - 4(x+3)^4 + \cdots$

21. From Example 3, we know that $\Sigma x^n/n!$ converges for all x. Why can we conclude that $\lim\limits_{n \to \infty} x^n/n! = 0$ for all x?

22. Let k be an arbitrary number and $-1 < x < 1$. Prove that

$$\lim_{n \to \infty} \frac{k(k-1)(k-2) \cdots (k-n)}{n!} x^n = 0$$

Hint: See Problem 21.

23. Find the radius of convergence of

$$\sum_{n=1}^{\infty} \frac{1 \cdot 2 \cdot 3 \cdots n}{1 \cdot 3 \cdot 5 \cdots (2n - 1)} x^{2n+1}$$

24. Find the radius of convergence of

$$\sum_{n=0}^{\infty} \frac{(pn)!}{(n!)^p} x^n$$

where p is a positive integer.

25. Find the sum $S(x)$ of $\sum_{n=0}^{\infty} (x - 3)^n$. What is the convergence set?

26. Suppose that $\sum_{n=0}^{\infty} a_n(x - 3)^n$ converges at $x = -1$. Why can you conclude that it converges at $x = 6$? Can you be sure that it converges at $x = 7$? Explain.

27. Find the convergence set for each series.

(a) $\sum_{n=1}^{\infty} \frac{(3x + 1)^n}{n \cdot 2^n}$ (b) $\sum_{n=1}^{\infty} (-1)^n \frac{(2x - 3)^n}{4^n \sqrt{n}}$

28. Refer to Problem 52 of Section 10.1, where the Fibonacci sequence $f_1, f_2, f_3, \ldots$ was defined. Find the radius of convergence of $\sum_{n=1}^{\infty} f_n x^n$.

29. Suppose that $a_{n+3} = a_n$ and let $S(x) = \sum_{n=0}^{\infty} a_n x^n$. Show that the series converges for $|x| < 1$ and give a formula for $S(x)$.

30. Follow the directions of Problem 29 for the case where $a_{n+p} = a_n$ for some fixed positive integer p.

Answers to Concepts Review: **1.** power series **2.** where it converges **3.** interval; end points **4.** $(-1, 1)$

10.7
Operations
on Power Series

We know from the previous section that the convergence set of a power series $\Sigma a_n x^n$ is an interval I. This interval is the domain for a new function $S(x)$, the sum of the series. The most obvious question to ask about $S(x)$ is whether we can give a simple formula for it. We have done this for one series, a geometric series.

$$\sum_{n=0}^{\infty} ax^n = \frac{a}{1 - x}, \quad -1 < x < 1$$

Actually, there is little reason to hope that the sum of an arbitrarily given power series will be one of the elementary functions studied earlier in this book, though we will make a little progress in that direction in this section and more in Section 10.8.

A better question to ask now is whether we can say anything about the properties of $S(x)$. For example, is it differentiable? Is it integrable? The answer to both questions is yes.

Term-by-Term Differentiation and Integration
Think of a power series as a polynomial with infinitely many terms. It behaves like a polynomial under both integration and differentiation; these operations can be performed term by term, as follows.

Theorem A

Suppose that $S(x)$ is the sum of a power series on an interval I; that is,

$$S(x) = \sum_{n=0}^{\infty} a_n x^n = a_0 + a_1 x + a_2 x^2 + a_3 x^3 + \cdots$$

Then, if x is interior to I,

(i) $\displaystyle S'(x) = \sum_{n=0}^{\infty} D_x(a_n x^n) = \sum_{n=1}^{\infty} n a_n x^{n-1}$

$\qquad = a_1 + 2a_2 x + 3a_3 x^2 + \cdots$

(ii) $\displaystyle \int_0^x S(t)\, dt = \sum_{n=0}^{\infty} \int_0^x a_n t^n\, dt = \sum_{n=0}^{\infty} \frac{a_n}{n + 1} x^{n+1}$

$\qquad = a_0 x + \frac{1}{2} a_1 x^2 + \frac{1}{3} a_2 x^3 + \frac{1}{4} a_3 x^4 + \cdots$

The theorem entails several things. It asserts that S is both differentiable and integrable, it shows how the derivative and integral may be calculated, and it implies that the radius of convergence of both the differentiated and integrated series is the same as for the original series (though it says nothing about the end points of the interval of convergence). The theorem is hard to prove. We leave the proof to more advanced books.

A nice consequence of Theorem A is that we can apply it to a power series with a known sum formula to obtain sum formulas for other series.

EXAMPLE 1   Apply Theorem A to the geometric series

$$\frac{1}{1-x} = 1 + x + x^2 + x^3 + \cdots, \qquad -1 < x < 1$$

to obtain formulas for two new series.

Solution   Differentiating term by term yields

$$\frac{1}{(1-x)^2} = 1 + 2x + 3x^2 + 4x^3 + \cdots, \qquad -1 < x < 1$$

Integrating term by term gives

$$\int_0^x \frac{1}{1-t}\, dt = \int_0^x 1\, dt + \int_0^x t\, dt + \int_0^x t^2\, dt + \cdots$$

That is,

$$-\ln(1-x) = x + \frac{x^2}{2} + \frac{x^3}{3} + \cdots, \qquad -1 < x < 1$$

If we replace x by $-x$ in the latter and multiply both sides by -1, we obtain

$$\boxed{\ln(1+x) = x - \frac{x^2}{2} + \frac{x^3}{3} - \frac{x^4}{4} + \cdots, \qquad -1 < x < 1}$$

From Problem 45 of Section 10.5, we learn that this result is valid at the end point $x = 1$ (also see the note in the margin).   ∎

EXAMPLE 2   Find the power series representation for $\tan^{-1} x$.

Solution   Recall that

$$\tan^{-1} x = \int_0^x \frac{1}{1+t^2}\, dt$$

From the geometric series for $1/(1-x)$ with x replaced by $-t^2$, we get

$$\frac{1}{1+t^2} = 1 - t^2 + t^4 - t^6 + \cdots, \qquad -1 < t < 1$$

Thus,

$$\tan^{-1} x = \int_0^x \left(1 - t^2 + t^4 - t^6 + \cdots\right) dt$$

That is,

$$\boxed{\tan^{-1} x = x - \frac{x^3}{3} + \frac{x^5}{5} - \frac{x^7}{7} + \cdots, \qquad -1 < x < 1}$$

(By the note in the margin, this also holds at $x = \pm 1$.)   ∎

EXAMPLE 3   Find a formula for the sum of the series

$$S(x) = 1 + x + \frac{x^2}{2!} + \frac{x^3}{3!} + \cdots$$

Solution   We saw earlier (Section 10.6, Example 3) that this series converges for all x. Differentiating term by term, we obtain

---

An End Point Result

The question of what is true at an end point of the interval of convergence of a power series is tricky. One result is due to Norway's greatest mathematician, Niels Henrik Abel (1802–1829). Suppose that

$$f(x) = \sum_{n=0}^{\infty} a_n x^n$$

for $|x| < R$. If f is continuous at an end point (R or $-R$) and if the series converges there, then the formula also holds at that end point.

$$S'(x) = 1 + x + \frac{x^2}{2!} + \frac{x^3}{3!} + \cdots$$

That is, $S'(x) = S(x)$ for all x. Furthermore, $S(0) = 1$. This differential equation has the unique solution $S(x) = e^x$ (see Section 7.5). Thus,

$$\boxed{e^x = 1 + x + \frac{x^2}{2!} + \frac{x^3}{3!} + \cdots}$$ ■

EXAMPLE 4   Obtain the power series representation for e^{-x^2}.

Solution   Simply substitute $-x^2$ for x in the series for e^x.

$$e^{-x^2} = 1 - x^2 + \frac{x^4}{2!} - \frac{x^6}{3!} + \cdots$$ ■

Algebraic Operations   Convergent power series can be added and subtracted term by term (Theorem 10.2B). In that sense they behave like polynomials. Convergent power series can also be multiplied and divided in a manner suggested by the multiplication and "long" division of polynomials.

EXAMPLE 5   Multiply and divide the power series for $\ln(1 + x)$ by that for e^x.

Solution   We refer to Examples 1 and 3 for the required series. The key to multiplication is to first find the constant term, then the x-term, then the x^2-term, and so on. We arrange our work as follows.

$$0 + x - \frac{x^2}{2} + \frac{x^3}{3} - \frac{x^4}{4} + \cdots$$

$$1 + x + \frac{x^2}{2!} + \frac{x^3}{3!} + \frac{x^4}{4!} + \cdots$$

$$\overline{\rule{12cm}{0.4pt}}$$

$$0 + (0 + 1)x + \left(0 + 1 - \frac{1}{2}\right)x^2 + \left(0 + \frac{1}{2!} - \frac{1}{2} + \frac{1}{3}\right)x^3$$

$$+ \left(0 + \frac{1}{3!} - \frac{1}{2!2} + \frac{1}{3} - \frac{1}{4}\right)x^4 + \cdots$$

$$= 0 + x + \frac{1}{2}x^2 + \frac{1}{3}x^3 + 0 \cdot x^4 + \cdots$$

Here is how division is done.

$$
\begin{array}{r}
x - \frac{3}{2}x^2 + \frac{4}{3}x^3 - \quad x^4 + \cdots \\
1 + x + \frac{1}{2}x^2 + \frac{1}{6}x^3 + \cdots \overline{\smash{\big)}\, x - \frac{1}{2}x^2 + \frac{1}{3}x^3 - \frac{1}{4}x^4 + \cdots} \\
\underline{x + x^2 + \frac{1}{2}x^3 + \frac{1}{6}x^4 + \cdots} \\
-\frac{3}{2}x^2 - \frac{1}{6}x^3 - \frac{5}{12}x^4 + \cdots \\
\underline{-\frac{3}{2}x^2 - \frac{3}{2}x^3 - \frac{3}{4}x^4 + \cdots} \\
\frac{4}{3}x^3 + \frac{1}{3}x^4 + \cdots \\
\underline{\frac{4}{3}x^3 + \frac{4}{3}x^4 + \cdots} \\
-x^4 + \cdots
\end{array}
$$ ■

The real question relative to Example 5 is whether the two series that we have obtained converge to $\left[\ln(1 + x)\right]e^x$ and $\left[\ln(1 + x)\right]/e^x$, respectively. Our next theorem, stated without proof, answers this question.

Theorem B

Let $f(x) = \sum a_n x^n$ and $g(x) = \sum b_n x^n$, with both of these series converging at least for $|x| < r$. If the operations of addition, subtraction, and multiplication are performed on these series as if they were polynomials, the resulting series will converge for $|x| < r$ and represent $f(x) + g(x)$, $f(x) - g(x)$, and $f(x) \cdot g(x)$, respectively. If $b_0 \neq 0$, the corresponding result holds for division, but we can guarantee its validity only for $|x|$ sufficiently small.

We mention that the operation of substituting one power series in another is also legitimate for $|x|$ sufficiently small, provided that the constant term of the substituted series is zero. Here is an illustration.

EXAMPLE 6 Find the power series for $e^{\tan^{-1} x}$ through terms of degree 4.

Solution Since

$$e^u = 1 + u + \frac{u^2}{2!} + \frac{u^3}{3!} + \frac{u^4}{4!} + \cdots$$

$$e^{\tan^{-1} x} = 1 + \tan^{-1} x + \frac{\left(\tan^{-1} x\right)^2}{2!} + \frac{\left(\tan^{-1} x\right)^3}{3!} + \frac{\left(\tan^{-1} x\right)^4}{4!} + \cdots$$

Now substitute the series for $\tan^{-1} x$ from Example 2 and combine like terms.

$$e^{\tan^{-1} x} = 1 + \left(x - \frac{x^3}{3} + \cdots\right) + \frac{\left(x - \frac{x^3}{3} + \cdots\right)^2}{2!} + \frac{\left(x - \frac{x^3}{3} + \cdots\right)^3}{3!}$$

$$+ \frac{\left(x - \frac{x^3}{3} + \cdots\right)^4}{4!} + \cdots$$

$$= 1 + \left(x - \frac{x^3}{3} + \cdots\right) + \frac{\left(x^2 - \frac{2}{3} x^4 + \cdots\right)}{2} + \frac{\left(x^3 + \cdots\right)}{6}$$

$$+ \frac{\left(x^4 + \cdots\right)}{24} + \cdots$$

$$= 1 + x + \frac{x^2}{2} - \frac{x^3}{6} - \frac{7x^4}{24} + \cdots \qquad \blacksquare$$

S. Ramanujan (1887–1920)

One of the most remarkable people of the early 20th century was the Indian mathematician Srinivasa Ramanujan. Largely self-educated, Ramanujan left at his death a number of notebooks in which he had recorded his discoveries. These notebooks are only now being thoroughly studied. In them are many strange and wonderful formulas, some for the sums of infinite series. Here is one.

$$\frac{1}{\pi} = \frac{\sqrt{8}}{9801} \sum_{n=0}^{\infty} \frac{(4n)! [1103 + 26{,}390n]}{(n!)^4 (396)^{4n}}$$

Formulas like this were used in 1989 to calculate the decimal expansion of π to over 1 billion places. (See Problem 35.)

Power Series in $x - a$ We have stated the theorems of this section for power series in x, but with obvious modifications they are equally valid for power series in $x - a$.

Concepts Review

1. A power series may be differentiated or _____ term by term on the _____ of its interval of convergence.

2. The first five terms in the power series expansion for $\ln(1 - x)$ are _____.

3. The first four terms in the power series expansion for $\exp(x^2)$ are _____.

4. The first five terms in the power series expansion for $\exp(x^2) - \ln(1 - x)$ are _____.

Problem Set 10.7

In Problems 1–10, find the power series representation for f(x) and specify the radius of convergence. Each is somehow related to a geometric series (see Examples 1 and 2).

1. $f(x) = \dfrac{1}{1 + x}$

2. $f(x) = \dfrac{1}{(1 + x)^2}$ *Hint*: Differentiate Problem 1.

3. $f(x) = \dfrac{1}{(1 - x)^3}$ **4.** $f(x) = \dfrac{x}{(1 + x)^2}$

5. $f(x) = \dfrac{1}{2 - 3x} = \dfrac{\frac{1}{2}}{1 - \frac{3}{2}x}$ **6.** $f(x) = \dfrac{1}{3 + 2x}$

7. $f(x) = \dfrac{x^2}{1 - x^4}$ **8.** $f(x) = \dfrac{x^3}{2 - x^3}$

9. $f(x) = \displaystyle\int_0^x \ln(1 + t)\, dt$ **10.** $f(x) = \displaystyle\int_0^x \tan^{-1} t\, dt$

11. Obtain the power series in x for $\ln[(1 + x)/(1 - x)]$ and specify its radius of convergence. *Hint*:

$$\ln[(1 + x)/(1 - x)] = \ln(1 + x) - \ln(1 - x)$$

12. Show that any positive number M can be represented by $(1 + x)/(1 - x)$, where x lies within the interval of convergence of the series of Problem 11. Hence conclude that the natural logarithm of any positive number can be found by means of this series. Find $\ln 8$ this way to three decimal places.

In Problems 13–16, use the result of Example 3 to find the power series in x for the given functions.

13. $f(x) = e^{-x}$ **14.** $f(x) = xe^{x^2}$

15. $f(x) = e^x + e^{-x}$ **16.** $f(x) = e^{2x} - 1 - 2x$

In Problems 17–24, use the methods of Example 5 to find power series in x for each function f.

17. $f(x) = e^{-x} \cdot \dfrac{1}{1 - x}$ **18.** $f(x) = e^x \tan^{-1} x$

19. $f(x) = \dfrac{\tan^{-1} x}{e^x}$ **20.** $f(x) = \dfrac{e^x}{1 + \ln(1 + x)}$

21. $f(x) = (\tan^{-1} x)(1 + x^2 + x^4)$

22. $f(x) = \dfrac{\tan^{-1} x}{1 + x^2 + x^4}$

23. $f(x) = \displaystyle\int_0^x \dfrac{e^t}{1 + t}\, dt$ **24.** $f(x) = \displaystyle\int_0^x \dfrac{\tan^{-1} t}{t}\, dt$

25. Find the sum of each of the following series by recognizing how it is related to something familiar.
(a) $x - x^2 + x^3 - x^4 + x^5 - \cdots$

(b) $\dfrac{1}{2!} + \dfrac{x}{3!} + \dfrac{x^2}{4!} + \dfrac{x^3}{5!} + \cdots$

(c) $2x + \dfrac{4x^2}{2} + \dfrac{8x^3}{3} + \dfrac{16x^4}{4} + \cdots$

26. Follow the directions of Problem 25.
(a) $1 + x^2 + x^4 + x^6 + x^8 + \cdots$
(b) $\cos x + \cos^2 x + \cos^3 x + \cos^4 x + \cdots$
(c) $\dfrac{x^2}{2} + \dfrac{x^4}{4} + \dfrac{x^6}{6} + \dfrac{x^8}{8} + \cdots$

27. Find the sum of $\displaystyle\sum_{n=1}^{\infty} nx^n$.

28. Find the sum of $\displaystyle\sum_{n=1}^{\infty} n(n + 1)x^n$.

29. Use the method of substitution (Example 6) to find power series through terms of degree 3.
(a) $\tan^{-1}(e^x - 1)$ (b) $e^{e^x - 1}$

30. Suppose that $f(x) = \displaystyle\sum_{n=0}^{\infty} a_n x^n = \sum_{n=0}^{\infty} b_n x^n$ for $|x| < R$. Show that $a_n = b_n$ for all n. *Hint*: Let $x = 0$; then differentiate and let $x = 0$ again. Continue.

31. Find the power series representation of $x/(x^2 - 3x + 2)$. *Hint*: Use partial fractions.

32. Let $y = y(x) = x - \dfrac{x^3}{3!} + \dfrac{x^5}{5!} - \dfrac{x^7}{7!} + \cdots$. Show that y satisfies the differential equation $y'' + y = 0$ with the conditions $y(0) = 0$ and $y'(0) = 1$. From this, guess at a simple formula for y.

33. Let $\{f_n\}$ be the Fibonacci sequence defined by

$$f_0 = 0, \qquad f_1 = 1, \qquad f_{n+2} = f_{n+1} + f_n$$

(See Problem 52 of Section 10.1 and Problem 28 of Section 10.6.) If $F(x) = \displaystyle\sum_{n=0}^{\infty} f_n x^n$, show that

$$F(x) - xF(x) - x^2 F(x) = x$$

and then use this fact to obtain a simple formula for $F(x)$.

34. Let $y = y(x) = \displaystyle\sum_{n=0}^{\infty} \dfrac{f_n}{n!} x^n$, where f_n is as in Problem 33. Show that y satisfies the differential equation $y'' - y' - y = 0$.

C **35.** Did you ever wonder how people find the decimal expansion of π to a large number of places? One method depends on the following identity (see Problem 62 of Section 7.7).

$$\pi = 16 \tan^{-1}\left(\tfrac{1}{5}\right) - 4 \tan^{-1}\left(\tfrac{1}{239}\right)$$

Find the first 6 digits of π using this identity and the series for $\tan^{-1} x$. (You will need terms through $x^9/9$ for $\tan^{-1}\left(\tfrac{1}{5}\right)$, but only the first term for $\tan^{-1}(1/239)$.) In 1706, John Machin used this method to calculate the first 100 digits of π, while in 1973, Jean Guilloud and Martine Bouyer found the first 1 million digits using the related identity

$$\pi = 48 \tan^{-1}\left(\tfrac{1}{18}\right) + 32 \tan^{-1}\left(\tfrac{1}{57}\right) - 20 \tan^{-1}\left(\tfrac{1}{239}\right)$$

In 1983, π was calculated to over 16 million digits by a somewhat different method. Of course, computers were used in these recent calculations.

36. The number e is readily calculated to as many digits as desired using the rapidly converging series

$$e = 1 + 1 + \frac{1}{2!} + \frac{1}{3!} + \frac{1}{4!} + \cdots$$

This series can also be used to show that e is irrational. Do so by completing the following argument. Suppose that $e = p/q$, where p and q are positive integers. Choose $n > q$ and let

$$M = n!\left(e - 1 - 1 - \frac{1}{2!} - \frac{1}{3!} - \cdots - \frac{1}{n!}\right)$$

Now M is a positive integer. (Why?) Also,

$$M = n!\left[\frac{1}{(n+1)!} + \frac{1}{(n+2)!} + \frac{1}{(n+3)!} + \cdots\right]$$

$$= \frac{1}{n+1} + \frac{1}{(n+1)(n+2)} + \frac{1}{(n+1)(n+2)(n+3)} + \cdots$$

$$< \frac{1}{n+1} + \frac{1}{(n+1)^2} + \frac{1}{(n+1)^3} + \cdots$$

$$= \frac{1}{n}$$

which gives a contradiction (to what?).

Answers to Concepts Review: **1.** integrated; interior
2. $-x - \frac{1}{2}x^2 - \frac{1}{3}x^3 - \frac{1}{4}x^4 - \frac{1}{5}x^5$ **3.** $1 + x^2 + \frac{1}{2}x^4 + \frac{1}{6}x^6$
4. $1 + x + \frac{3}{2}x^2 + \frac{1}{3}x^3 + \frac{3}{4}x^4$

10.8
Taylor and Maclaurin Series

The major question still dangling is this: Given a function f (e.g., $\sin x$ or $\ln(\cos^2 x)$), can we represent it as a power series in x or, more generally, in $x - a$? More precisely, can we find numbers $c_0, c_1, c_2, c_3, \ldots$ such that

$$f(x) = c_0 + c_1(x - a) + c_2(x - a)^2 + c_3(x - a)^3 + \cdots$$

on some interval around a?

Suppose that such a representation exists. Then, by the theorem on differentiating series (Theorem 10.7A),

$$f'(x) = c_1 + 2c_2(x - a) + 3c_3(x - a)^2 + 4c_4(x - a)^3 + \cdots$$

$$f''(x) = 2!c_2 + 3!c_3(x - a) + 4 \cdot 3c_4(x - a)^2 + \cdots$$

$$f'''(x) = 3!c_3 + 4!c_4(x - a) + 5 \cdot 4 \cdot 3c_5(x - a)^2 + \cdots$$

$$\vdots$$

When we substitute $x = a$ and solve for c_n, we get

$$c_0 = f(a)$$

$$c_1 = f'(a)$$

$$c_2 = \frac{f''(a)}{2!}$$

$$c_3 = \frac{f'''(a)}{3!}$$

and, more generally,

$$c_n = \frac{f^{(n)}(a)}{n!}$$

(To make this valid for $n = 0$, we define $f^{(0)}(a)$ to mean $f(a)$ and $0!$ to be 1.) Thus, the coefficients c_n are determined by the function f. This also shows that a function f cannot be represented by two different power series in $x - a$, an important point that we have glossed over until now. We summarize in the following theorem.

Theorem A   Uniqueness Theorem

Suppose that f satisfies

$$f(x) = c_0 + c_1(x - a) + c_2(x - a)^2 + c_3(x - a)^3 + \cdots$$

for all x in some interval around a. Then

$$c_n = \frac{f^{(n)}(a)}{n!}$$

Thus, a function cannot be represented by more than one power series in $x - a$. The power series representation of a function in $x - a$ is called its **Taylor series** after the English mathematician Brook Taylor (1685–1731). If $a = 0$, the corresponding series is called the **Maclaurin series** after the Scottish mathematician Colin Maclaurin (1698–1746).

Convergence of Taylor Series   But the existence question remains. Given a function f, can we represent it in a power series in $x - a$ (which must necessarily be the Taylor series)? The next two theorems give the answer.

Theorem B   Taylor's Formula with Remainder

Let f be a function whose $(n + 1)$st derivative $f^{(n+1)}(x)$ exists for each x in an open interval I containing a. Then, for each x in I,

$$f(x) = f(a) - f'(a)(x - a) + \frac{f''(a)}{2!}(x - a)^2 + \cdots$$

$$+ \frac{f^{(n)}(a)}{n!}(x - a)^n + R_n(x)$$

where the remainder (or error) $R_n(x)$ is given by the formula

$$R_n(x) = \frac{f^{(n+1)}(c)}{(n + 1)!}(x - a)^{n+1}$$

and c is some point between x and a.

Proof   We will prove the theorem for the special case of $n = 4$; the proof for an arbitrary n follows the same structure and is left as an exercise. (See Problem 37.) First define the function $R_4(x)$ on I by

$$R_4(x) = f(x) - f(a) - f'(a)(x - a) - \frac{f''(a)}{2!}(x - a)^2$$

$$- \frac{f'''(a)}{3!}(x - a)^3 - \frac{f^{(4)}(a)}{4!}(x - a)^4$$

Now think of x and a as constants, and define a new function g on I by

$$g(t) = f(x) - f(t) - f'(t)(x - t) - \frac{f''(t)(x - t)^2}{2!} - \frac{f'''(t)(x - t)^3}{3!}$$

$$- \frac{f^{(4)}(t)(x - t)^4}{4!} - R_4(x)\frac{(x - t)^5}{(x - a)^5}$$

Clearly, $g(x) = 0$ (remember, x is considered fixed) and

$$g(a) = f(x) - f(a) - f'(a)(x - a) - \frac{f''(a)(x - a)^2}{2!} - \frac{f'''(a)(x - a)^3}{3!}$$

$$- \frac{f^{(4)}(a)(x - a)^4}{4!} - R_4(x)\frac{(x - a)^5}{(x - a)^5}$$

$$= R_4(x) - R_4(x)$$

$$= 0$$

Since a and x are points in I with the property that $g(a) = g(x) = 0$, we can apply the Mean Value Theorem for Derivatives. There exists, therefore, a real number c between a and x such that $g'(c) = 0$. To obtain the derivative of g, we must repeatedly apply the product rule.

$$g'(t) = 0 - f'(t) - \left[f'(t)(-1) + (x - t)f''(t)\right] - \frac{1}{2!}\left[f''(t)2(x - t)(-1) + (x - t)^2 f'''(t)\right]$$

$$- \frac{1}{3!}\left[f'''(t)3(x - t)^2(-1) + (x - t)^3 f^{(4)}(t)\right]$$

$$- \frac{1}{4!}\left[f^{(4)}(t)4(x - t)^3(-1) + (x - t)^4 f^{(5)}(t)\right] - R_4(x)\frac{5(x - t)^4(-1)}{(x - a)^5}$$

$$= -\frac{1}{4!}(x - t)^4 f^{(5)}(t) + 5R_4(x)\frac{(x - t)^4}{(x - a)^5}$$

Thus, by the Mean Value Theorem for Derivatives, there is some c between x and a such that,

$$0 = g'(c) = -\frac{1}{4!}(x - c)^4 f^{(5)}(c) + 5R_4(x)\frac{(x - c)^4}{(x - a)^5}$$

This leads to

$$\frac{1}{4!}(x - c)^4 f^{(5)}(c) = 5R_4(x)\frac{(x - c)^4}{(x - a)^5}$$

$$R_4(x) = \frac{f^{(5)}(c)}{5!}(x - a)^5 \quad \blacklozenge$$

This theorem tells us what the error can be when we approximate a function with a finite number of terms of its Taylor series. In the next chapter, we will further exploit the relationship given in Theorem B.

We now—finally—answer the question about whether a function f can be represented by a power series in $x - a$.

Theorem C Taylor's Theorem

Let f be a function with derivatives of all orders in some interval $(a - r, a + r)$. The Taylor series

$$f(a) + f'(a)(x - a) + \frac{f''(a)}{2!}(x - a)^2 + \frac{f'''(a)}{3!}(x - a)^3 + \cdots$$

represents the function f on the interval $(a - r, a + r)$ if and only if

$$\lim_{n \to \infty} R_n(x) = 0$$

where $R_n(x)$ is the remainder in Taylor's Formula,

$$R_n(x) = \frac{f^{(n+1)}(c)}{(n + 1)!}(x - a)^{n+1}$$

and c is some point in $(a - r, a + r)$.

Proof We need only recall Taylor's Formula with Remainder (Theorem B),

$$f(x) = f(a) + f'(a)(x - a) + \cdots + \frac{f^{(n)}(a)}{n!}(x - a)^n + R_n(x)$$

and the result follows. ◆

Note that if $a = 0$, we get the Maclaurin series

$$f(0) + f'(0)x + \frac{f''(0)}{2!}x^2 + \frac{f'''(0)}{3!}x^3 + \cdots$$

Warning

Here is a fact that surprises many students. It is possible that the Taylor series for $f(x)$ converges on an interval but does not represent $f(x)$ there. This is shown by example in Problem 40. Of course,

$$\lim_{n \to \infty} R_n(x) \neq 0$$

in this example.

EXAMPLE 1 Find the Maclaurin series for $\sin x$ and prove that it represents $\sin x$ for all x.

Solution

$$f(x) = \sin x \qquad\qquad f(0) = 0$$
$$f'(x) = \cos x \qquad\qquad f'(0) = 1$$
$$f''(x) = -\sin x \qquad\qquad f''(0) = 0$$
$$f'''(x) = -\cos x \qquad\qquad f'''(0) = -1$$
$$f^{(4)}(x) = \sin x \qquad\qquad f^{(4)}(0) = 0$$
$$\vdots \qquad\qquad\qquad \vdots$$

Thus,

$$\sin x = x - \frac{x^3}{3!} + \frac{x^5}{5!} - \frac{x^7}{7!} + \cdots$$

and this is valid for all x, provided we can show that

$$\lim_{n\to\infty} R_n(x) = \lim_{n\to\infty} \frac{f^{(n+1)}(c)}{(n+1)!} x^{n+1} = 0$$

Now, $\left|f^{(n+1)}(x)\right| = |\cos x|$ or $\left|f^{(n+1)}(x)\right| = |\sin x|$, and so

$$\left|R_n(x)\right| \le \frac{|x|^{n+1}}{(n+1)!}$$

But $\lim_{n\to\infty} x^n/n! = 0$ for all x, since $x^n/n!$ is the nth term of a convergent series (see Example 3 and Problem 21 of Section 10.6). As a consequence, we see that $\lim_{n\to\infty} R_n(x) = 0$. ∎

EXAMPLE 2 Find the Maclaurin series for $\cos x$ and show that it represents $\cos x$ for all x.

Solution We could proceed as in Example 1. However, it is easier to get the result by differentiating the series of that example (a valid procedure according to Theorem 10.7A). We obtain

$$\cos x = 1 - \frac{x^2}{2!} + \frac{x^4}{4!} - \frac{x^6}{6!} + \cdots$$
∎

EXAMPLE 3 Find the Maclaurin series for $f(x) = \cosh x$ in two different ways, and show that it represents $\cosh x$ for all x.

Solution

Method 1. This is the direct method.

$$f(x) = \cosh x \qquad\qquad f(0) = 1$$
$$f'(x) = \sinh x \qquad\qquad f'(0) = 0$$
$$f''(x) = \cosh x \qquad\qquad f''(0) = 1$$
$$f'''(x) = \sinh x \qquad\qquad f'''(0) = 0$$
$$\vdots \qquad\qquad\qquad \vdots$$

Thus,

$$\cosh x = 1 + \frac{x^2}{2!} + \frac{x^4}{4!} + \frac{x^6}{6!} + \cdots$$

provided we can show that $\lim_{n\to\infty} R_n(x) = 0$ for all x.

Now let B be an arbitrary number and suppose that $|x| \leq B$. Then

$$|\cosh x| = \left| \frac{e^x + e^{-x}}{2} \right| \leq \frac{e^x}{2} + \frac{e^{-x}}{2} \leq \frac{e^B}{2} + \frac{e^B}{2} = e^B$$

By similar reasoning, $|\sinh x| \leq e^B$. Since $f^{(n+1)}(x)$ is either $\cosh x$ or $\sinh x$, we conclude that

$$|R_n(x)| = \left| \frac{f^{(n+1)}(c)x^{n+1}}{(n+1)!} \right| \leq \frac{e^B |x|^{n+1}}{(n+1)!}$$

The latter expression tends to zero as $n \to \infty$, just as in Example 1.

Method 2. We use the fact that $\cosh x = (e^x + e^{-x})/2$. From Example 3 of Section 10.7,

$$e^x = 1 + x + \frac{x^2}{2!} + \frac{x^3}{3!} + \frac{x^4}{4!} + \cdots$$

$$e^{-x} = 1 - x + \frac{x^2}{2!} - \frac{x^3}{3!} + \frac{x^4}{4!} - \cdots$$

The previously obtained result follows by adding these two series and dividing by 2. ∎

EXAMPLE 4 Find the Maclaurin series for $\sinh x$ and show that it represents $\sinh x$ for all x.

Solution We do both jobs at once when we differentiate the series for $\cosh x$ (Example 3) term by term and use Theorem 10.7A.

$$\sinh x = x + \frac{x^3}{3!} + \frac{x^5}{5!} + \frac{x^7}{7!} + \cdots$$

∎

The Binomial Series We are all familiar with the Binomial Formula. For a positive integer p,

$$(1 + x)^p = 1 + \binom{p}{1} x + \binom{p}{2} x^2 + \cdots + \binom{p}{p} x^p$$

where

$$\binom{p}{k} = \frac{p!}{k!(p-k)!} = \frac{p(p-1)(p-2) \cdots (p-k+1)}{k!}$$

Note that if we redefine $\binom{p}{k}$ to be

$$\binom{p}{k} = \frac{p(p-1)(p-2) \cdots (p-k+1)}{k!}$$

then $\binom{p}{k}$ makes sense for *any* real number p, provided that k is a positive integer. Of course, if p is a positive integer, then our new definition reduces to $p!/[k!(p-k)!]$.

Theorem D Binomial Series

For any real number p and for $|x| < 1$,

$$(1 + x)^p = 1 + \binom{p}{1} x + \binom{p}{2} x^2 + \binom{p}{3} x^3 + \cdots$$

Partial Proof   Let $f(x) = (1 + x)^p$. Then

$$f(x) = (1 + x)^p \qquad\qquad f(0) = 1$$

$$f'(x) = p(1 + x)^{p-1} \qquad\qquad f'(0) = p$$

$$f''(x) = p(p - 1)(1 + x)^{p-2} \qquad\qquad f''(0) = p(p - 1)$$

$$f'''(x) = p(p - 1)(p - 2)(1 + x)^{p-3} \qquad f'''(0) = p(p - 1)(p - 2)$$

$$\vdots \qquad\qquad\qquad\qquad\qquad \vdots$$

Thus, the Maclaurin series for $(1 + x)^p$ is as indicated in the theorem. To show that it represents $(1 + x)^p$, we need to show that $\lim_{n\to\infty} R_n(x) = 0$. This unfortunately, is difficult, and we leave it for more advanced courses. (See Problem 38 for a completely different way to prove Theorem D.) ◆

If p is a positive integer, $\binom{p}{k} = 0$ for $k > p$, and so the Binomial Series collapses to a series with finitely many terms, the usual Binomial Formula.

EXAMPLE 5   Represent $(1 - x)^{-2}$ in a Maclaurin series for $-1 < x < 1$.

Solution   By Theorem D,

$$(1 + x)^{-2} = 1 + (-2)x + \frac{(-2)(-3)}{2!} x^2 + \frac{(-2)(-3)(-4)}{3!} x^3 + \cdots$$

$$= 1 - 2x + 3x^2 - 4x^3 + \cdots$$

Thus,

$$(1 - x)^{-2} = 1 + 2x + 3x^2 + 4x^3 + \cdots$$

Naturally, this agrees with a result we obtained by a different method in Example 1 of Section 10.7.   ■

EXAMPLE 6   Represent $\sqrt{1 + x}$ in a Maclaurin series and use it to approximate $\sqrt{1.1}$ to five decimal places.

Solution   By Theorem D,

$$(1 + x)^{1/2} = 1 + \frac{1}{2} x + \frac{\left(\frac{1}{2}\right)\left(-\frac{1}{2}\right)}{2!} x^2 + \frac{\left(\frac{1}{2}\right)\left(-\frac{1}{2}\right)\left(-\frac{3}{2}\right)}{3!} x^3$$

$$+ \frac{\left(\frac{1}{2}\right)\left(-\frac{1}{2}\right)\left(-\frac{3}{2}\right)\left(-\frac{5}{2}\right)}{4!} x^4 + \cdots$$

$$= 1 + \frac{1}{2} x - \frac{1}{8} x^2 + \frac{1}{16} x^3 - \frac{5}{128} x^4 + \cdots$$

Thus,

$$\sqrt{1.1} = 1 + \frac{0.1}{2} - \frac{0.01}{8} + \frac{0.001}{16} - \frac{5(0.0001)}{128} + \cdots$$

$$\approx 1.04881 \qquad\qquad\qquad ■$$

EXAMPLE 7   Compute $\displaystyle\int_0^{0.4} \sqrt{1 + x^4}\, dx$ to five decimal places.

Solution   From Example 6,

$$\sqrt{1 + x^4} = 1 + \frac{1}{2} x^4 - \frac{1}{8} x^8 + \frac{1}{16} x^{12} - \frac{5}{128} x^{16} + \cdots$$

Thus,

$$\int_0^{0.4} \sqrt{1 + x^4}\, dx = \left[x + \frac{x^5}{10} - \frac{x^9}{72} + \frac{x^{13}}{208} + \cdots \right]_0^{0.4} \approx 0.40102 \qquad ■$$

Summary   We conclude our discussion of series with a list of the important Maclaurin series we have found. These series will be useful in doing the problem set, but, what is more significant, they find application throughout mathematics and science.

Important Maclaurin Series

1. $\dfrac{1}{1 - x} = 1 + x + x^2 + x^3 + x^4 + \cdots$ $\qquad\qquad -1 < x < 1$

2. $\ln(1 + x) = x - \dfrac{x^2}{2} + \dfrac{x^3}{3} - \dfrac{x^4}{4} + \dfrac{x^5}{5} - \cdots$ $\qquad -1 < x \le 1$

3. $\tan^{-1} x = x - \dfrac{x^3}{3} + \dfrac{x^5}{5} - \dfrac{x^7}{7} + \dfrac{x^9}{9} + \cdots$ $\qquad -1 \le x \le 1$

4. $e^x = 1 + x + \dfrac{x^2}{2!} + \dfrac{x^3}{3!} + \dfrac{x^4}{4!} + \cdots$

5. $\sin x = x - \dfrac{x^3}{3!} + \dfrac{x^5}{5!} - \dfrac{x^7}{7!} + \dfrac{x^9}{9!} - \cdots$

6. $\cos x = 1 - \dfrac{x^2}{2!} + \dfrac{x^4}{4!} - \dfrac{x^6}{6!} + \dfrac{x^8}{8!} - \cdots$

7. $\sinh x = x + \dfrac{x^3}{3!} + \dfrac{x^5}{5!} + \dfrac{x^7}{7!} + \dfrac{x^9}{9!} + \cdots$

8. $\cosh x = 1 + \dfrac{x^2}{2!} + \dfrac{x^4}{4!} + \dfrac{x^6}{6!} + \dfrac{x^8}{8!} + \cdots$

9. $(1 + x)^p = 1 + \dbinom{p}{1}x + \dbinom{p}{2}x^2 + \dbinom{p}{3}x^3 + \dbinom{p}{4}x^4 + \cdots$ $\quad -1 < x < 1$

Concepts Review

1. If a function $f(x)$ is represented by the power series $\sum c_k x^k$, then $c_k = $ _____.

2. The Taylor series for a function will represent the function for those x for which the remainder $R_n(x)$ in Taylor's Formula satisfies _____.

3. The Maclaurin series for $\sin x$ represents $\sin x$ for _____ $< x < $ _____.

4. The first four terms in the Maclaurin series for $(1 + x)^{1/3}$ are _____.

Problem Set 10.8

In Problems 1–18, find the terms through x^5 in the Maclaurin series for $f(x)$. Hint: It may be easiest to use known Maclaurin series and then perform multiplications, divisions, and so on. For example, $\tan x = (\sin x)/(\cos x)$.

1. $f(x) = \tan x$

2. $f(x) = \tanh x$

3. $f(x) = e^x \sin x$

4. $f(x) = e^{-x} \cos x$

5. $f(x) = (\cos x)\ln(1 + x)$

6. $f(x) = (\sin x)\sqrt{1 + x}$

7. $f(x) = e^x + x + \sin x$

8. $f(x) = \dfrac{\cos x - 1 + x^2/2}{x^4}$

9. $f(x) = \dfrac{1}{1 - x} \cosh x$

10. $f(x) = \dfrac{1}{1 + x}\ln\left(\dfrac{1}{1 + x}\right) = \dfrac{-\ln(1 + x)}{1 + x}$

11. $f(x) = \dfrac{1}{1 + x + x^2}$

12. $f(x) = \dfrac{1}{1 - \sin x}$

13. $f(x) = \sin^3 x$

14. $f(x) = x(\sin 2x + \sin 3x)$

15. $f(x) = x\sec(x^2) + \sin x$

16. $f(x) = \dfrac{\cos x}{\sqrt{1 + x}}$

17. $f(x) = (1 + x)^{3/2}$

18. $f(x) = \left(1 - x^2\right)^{2/3}$

In Problems 19–24, find the Taylor series in $x - a$ through $(x - a)^3$.

19. $e^x, a = 1$

20. $\sin x, a = \dfrac{\pi}{6}$

21. $\cos x, a = \dfrac{\pi}{3}$

22. $\tan x, a = \dfrac{\pi}{4}$

23. $1 + x^2 + x^3, a = 1$

24. $2 - x + 3x^2 - x^3, a = -1$

25. Let $f(x) = \sum a_n x^n$ be an even function $(f(-x) = f(x))$ for x in $(-R, R)$. Prove that $a_n = 0$ if n is odd. *Hint:* Use the Uniqueness Theorem.

26. State and prove a theorem analogous to that in Problem 25 for odd functions.

27. Recall that

$$\sin^{-1} x = \int_0^x \frac{1}{\sqrt{1 - t^2}}\, dt$$

Find the first four nonzero terms in the Maclaurin series for $\sin^{-1} x$.

28. Given that

$$\sinh^{-1} x = \int_0^x \frac{1}{\sqrt{1 + t^2}}\, dt$$

find the first four nonzero terms in the Maclaurin series for $\sinh^{-1} x$.

C 29. Calculate, accurate to four decimal places,

$$\int_0^1 \cos(x^2)\, dx$$

C 30. Calculate, accurate to five decimal places,

$$\int_0^{0.5} \sin \sqrt{x}\, dx$$

31. By writing $1/x = 1/[1 - (1 - x)]$ and using the known expansion of $1/(1 - x)$, find the Taylor series for $1/x$ in powers of $x - 1$.

32. Let $f(x) = (1 + x)^{1/2} + (1 - x)^{1/2}$. Find the Maclaurin series for f and use it to find $f^{(4)}(0)$ and $f^{(51)}(0)$.

33. In each case, find the Maclaurin series for $f(x)$ by use of known series and then use it to calculate $f^{(4)}(0)$.

(a) $f(x) = e^{x + x^2}$

(b) $f(x) = e^{\sin x}$

(c) $f(x) = \displaystyle\int_0^x \frac{e^{t^2} - 1}{t^2}\, dt$

(d) $f(x) = e^{\cos x} = e \cdot e^{\cos x - 1}$

(e) $f(x) = \ln(\cos^2 x)$

34. One can sometimes find a Maclaurin series by the *method of equating coefficients*. For example, let

$$\tan x = \frac{\sin x}{\cos x} = a_0 + a_1 x + a_2 x^2 + \cdots$$

Then multiply by $\cos x$ and replace $\sin x$ and $\cos x$ by their series to obtain

$$x - \frac{x^3}{6} + \cdots = \left(a_0 + a_1 x + a_2 x^2 + \cdots\right)\left(1 - \frac{x^2}{2} + \cdots\right)$$

$$= a_0 + a_1 x + \left(a_2 - \frac{a_0}{2}\right)x^2 + \left(a_3 - \frac{a_1}{2}\right)x^3 + \cdots$$

Thus,

$$a_0 = 0, \quad a_1 = 1, \quad a_2 - \frac{a_0}{2} = 0, \quad a_3 - \frac{a_1}{2} = -\frac{1}{6}, \quad \cdots$$

so

$$a_0 = 0, \quad a_1 = 1, \quad a_2 = 0, \quad a_3 = \tfrac{1}{3}, \quad \cdots$$

and therefore

$$\tan x = 0 + x + 0 + \tfrac{1}{3}x^3 + \cdots$$

which agrees with Problem 1. Use this method to find the terms through x^4 in the series for $\sec x$.

35. Use the method of Problem 34 to find the terms through x^5 in the Maclaurin series for $\tanh x$.

36. Use the method of Problem 34 to find the terms through x^4 in the series for $\operatorname{sech} x$.

37. Prove Theorem B for

(a) the special case of $n = 3$, and

(b) an arbitrary n.

38. Prove Theorem D as follows: Let

$$f(x) = 1 + \sum_{n=1}^{\infty} \binom{p}{n} x^n.$$

(a) Show that the series converges for $|x| < 1$.

(b) Show that $(1 + x)f'(x) = pf(x)$ and $f(0) = 1$.

(c) Solve this differential equation to get $f(x) = (1 + x)^p$.

39. Let

$$f(t) = \begin{cases} 0 & t < 0 \\ t^4 & t \geq 0 \end{cases}$$

Explain why $f(t)$ cannot be represented by a Maclaurin series. Also show that, if $g(t)$ gives the distance traveled by a car that is stationary for $t < 0$ and moving ahead for $t \geq 0$, $g(t)$ cannot be represented by a Maclaurin series.

40. Let

$$f(x) = \begin{cases} e^{-1/x^2} & x \neq 0 \\ 0 & x = 0 \end{cases}$$

(a) Show that $f'(0) = 0$ by using the definition of the derivative.

(b) Show that $f''(0) = 0$.

(c) Assuming the known fact that $f^{(n)}(0) = 0$ for all n, find the Maclaurin series for $f(x)$.

(d) Does the Maclaurin series represent $f(x)$?

(e) When $a = 0$, the formula in Theorem B is called **Maclaurin's Formula**. What is the remainder in Maclaurin's Formula for $f(x)$?

This shows that a Maclaurin series may exist and yet not represent the given function (the remainder does not tend to 0 as $n \to \infty$).

CAS *Use a CAS to find the first four nonzero terms in the Maclaurin series for each of the following. Check Problems 43–48 to see that you get the same answers using the results of Section 10.7.*

41. $\sin x$

42. $\exp x$

43. $3 \sin x - 2 \exp x$

44. $\exp(x^2)$

45. $\sin(\exp x - 1)$

46. $\exp(\sin x)$

47. $(\sin x)(\exp x)$

48. $(\sin x)/(\exp x)$

Answers to Concepts Review: **1.** $f^{(k)}(0)/k!$ **2.** $\lim_{n \to \infty} R_n(x) = 0$

3. $-\infty, \infty$ **4.** $1 + \tfrac{1}{3}x - \tfrac{1}{9}x^2 + \tfrac{5}{81}x^3$

10.9 Chapter Review

Concepts Test

Respond with true or false to each of the following assertions. Be prepared to justify your answer.

1. If $0 \le a_n \le b_n$ for all n in $\mathbb{N}$ and $\lim\limits_{n \to \infty} b_n$ exists, then $\lim\limits_{n \to \infty} a_n$ exists.

2. For every positive integer n, it is true that $n! \le n^n \le (2n - 1)!$.

3. If $\lim\limits_{n \to \infty} a_n = L$, then $\lim\limits_{n \to \infty} a_{3n+4} = L$.

4. If $\lim\limits_{n \to \infty} a_{2n} = L$ and $\lim\limits_{n \to \infty} a_{3n} = L$, then $\lim\limits_{n \to \infty} a_n = L$.

5. If $\lim\limits_{n \to \infty} a_{mn} = L$ for every positive integer $m \ge 2$, then $\lim\limits_{n \to \infty} a_n = L$.

6. If $\lim\limits_{n \to \infty} a_{2n} = L$ and $\lim\limits_{n \to \infty} a_{2n+1} = L$, then $\lim\limits_{n \to \infty} a_n = L$.

7. If $\lim\limits_{n \to \infty} (a_n - a_{n+1}) = 0$, then $\lim\limits_{n \to \infty} a_n$ exists and is finite.

8. If $\{a_n\}$ and $\{b_n\}$ both diverge, then $\{a_n + b_n\}$ diverges.

9. If $\{a_n\}$ converges, then $\{a_n/n\}$ converges to 0.

10. If $\sum\limits_{n=1}^{\infty} a_n$ converges, so does $\sum\limits_{n=1}^{\infty} a_n^2$.

11. If $0 < a_{n+1} < a_n$ for all n in $\mathbb{N}$ and $\lim\limits_{n \to \infty} a_n = 0$, then $\sum\limits_{n=1}^{\infty} (-1)^{n+1} a_n$ converges and has sum S satisfying $0 < S < a_1$.

12. $\sum\limits_{n=1}^{\infty} \left(\frac{1}{n}\right)^n$ converges and has sum S satisfying $1 < S < 2$.

13. If a series $\sum a_n$ diverges, then its sequence of partial sums is unbounded.

14. If $0 \le a_n \le b_n$ for all n in $\mathbb{N}$ and if $\sum\limits_{n=1}^{\infty} b_n$ diverges, then $\sum\limits_{n=1}^{\infty} a_n$ diverges.

15. The Ratio Test will not help in determining the convergence or divergence of $\sum\limits_{n=1}^{\infty} \dfrac{2n + 3}{3n^4 + 2n^3 + 3n + 1}$.

16. If $a_n > 0$ for all n in $\mathbb{N}$ and $\sum\limits_{n=1}^{\infty} a_n$ converges, then $\lim\limits_{n \to \infty} (a_{n+1}/a_n) < 1$.

17. $\sum\limits_{n=1}^{\infty} \left(1 - \frac{1}{n}\right)^n$ converges.

18. $\sum\limits_{n=1}^{\infty} \dfrac{1}{\ln(n^4 + 1)}$ converges.

19. $\sum\limits_{n=2}^{\infty} \dfrac{n + 1}{(n \ln n)^2}$ converges.

20. $\sum\limits_{n=1}^{\infty} \dfrac{\sin^2(n\pi/2)}{n}$ converges.

21. If $0 \le a_{n+100} \le b_n$ for all n in $\mathbb{N}$ and $\sum\limits_{n=1}^{\infty} b_n$ converges, then $\sum\limits_{n=1}^{\infty} a_n$ converges.

22. If, for some $c > 0$, $ca_n \ge 1/n$ for all n in $\mathbb{N}$, then $\sum\limits_{n=1}^{\infty} a_n$ diverges.

23. $\frac{1}{3} + \left(\frac{1}{3}\right)^2 + \left(\frac{1}{3}\right)^3 + \cdots + \left(\frac{1}{3}\right)^{1000} < \frac{1}{2}$.

24. If $\sum\limits_{n=1}^{\infty} a_n$ converges, then $\sum\limits_{n=1}^{\infty} (-1)^n a_n$ converges.

25. If $b_n \le a_n \le 0$ for all n in $\mathbb{N}$ and $\sum\limits_{n=1}^{\infty} b_n$ converges, then $\sum\limits_{n=1}^{\infty} a_n$ converges.

26. If $0 \le a_n$ for all n in $\mathbb{N}$ and $\sum\limits_{n=1}^{\infty} a_n$ converges, then $\sum\limits_{n=1}^{\infty} (-1)^n a_n$ converges.

27. $\left| \sum\limits_{n=1}^{\infty} (-1)^{n+1} \dfrac{1}{n} - \sum\limits_{n=1}^{99} (-1)^{n+1} \dfrac{1}{n} \right| < 0.01$.

28. If $\sum\limits_{n=1}^{\infty} a_n$ diverges, then $\sum\limits_{n=1}^{\infty} |a_n|$ diverges.

29. If the power series $\sum\limits_{n=0}^{\infty} a_n(x - 3)^n$ converges at $x = -1.1$, it also converges at $x = 7$.

30. If $\sum\limits_{n=0}^{\infty} a_n x^n$ converges at $x = -2$, it also converges at $x = 2$.

31. If $f(x) = \sum\limits_{n=0}^{\infty} a_n x^n$ and the series converges at $x = 1.5$, then $\int_0^1 f(x)\, dx = \sum\limits_{n=0}^{\infty} a_n/(n + 1)$.

32. Every power series converges for at least two values of the variable.

33. If $f(0), f'(0), f''(0), \ldots$ all exist, then the Maclaurin series for $f(x)$ converges to $f(x)$ in a neighborhood of $x = 0$.

34. The function $f(x) = 1 + x + x^2 + x^3 + \cdots$ satisfies the differential equation $y' = y^2$ on the interval $(-1, 1)$.

35. The function $f(x) = \sum\limits_{n=0}^{\infty} (-1)^n x^n/n!$ satisfies the differential equation $y' + y = 0$ on the whole real line.

Sample Test Problems

In Problems 1–8, determine whether the given sequence converges or diverges and, if it converges, find $\lim\limits_{n \to \infty} a_n$.

1. $a_n = \dfrac{9n}{\sqrt{9n^2 + 1}}$

2. $a_n = \dfrac{\ln n}{\sqrt{n}}$

3. $a_n = \left(1 + \dfrac{4}{n}\right)^n$

4. $a_n = \dfrac{n!}{3^n}$

5. $a_n = \sqrt[n]{n}$

6. $a_n = \dfrac{1}{\sqrt[3]{n}} + \dfrac{1}{\sqrt[4]{3}}$

7. $a_n = \dfrac{\sin^2 n}{\sqrt{n}}$ **8.** $a_n = \cos\left(\dfrac{n\pi}{6}\right)$

In Problems 9–18, determine whether the given series converges or diverges and, if it converges, find its sum.

9. $\displaystyle\sum_{k=1}^{\infty}\left(\dfrac{1}{\sqrt{k}} - \dfrac{1}{\sqrt{k+1}}\right)$ **10.** $\displaystyle\sum_{k=1}^{\infty}\left(\dfrac{1}{k} - \dfrac{1}{k+2}\right)$

11. $\ln\frac{1}{2} + \ln\frac{2}{3} + \ln\frac{3}{4} + \cdots$ **12.** $\displaystyle\sum_{k=0}^{\infty}\cos k\pi$

13. $\displaystyle\sum_{k=0}^{\infty}e^{-2k}$ **14.** $\displaystyle\sum_{k=0}^{\infty}\left(\dfrac{3}{2^k} + \dfrac{4}{3^k}\right)$

15. $0.91919191\ldots = \displaystyle\sum_{k=1}^{\infty}91\left(\dfrac{1}{100}\right)^k$

16. $\displaystyle\sum_{k=1}^{\infty}\left(\dfrac{1}{\ln 2}\right)^k$ **17.** $1 - \dfrac{2^2}{2!} + \dfrac{2^4}{4!} - \dfrac{2^6}{6!} + \cdots$

18. $1 - \dfrac{1}{1!} + \dfrac{1}{2!} - \dfrac{1}{3!} + \dfrac{1}{4!} - \cdots$

In Problems 19–32, indicate whether the given series converges or diverges and give a reason for your conclusion.

19. $\displaystyle\sum_{n=1}^{\infty}\dfrac{n}{1+n^2}$ **20.** $\displaystyle\sum_{n=1}^{\infty}\dfrac{n+5}{1+n^3}$

21. $\displaystyle\sum_{n=1}^{\infty}(-1)^{n+1}\dfrac{1}{\sqrt[3]{n}}$ **22.** $\displaystyle\sum_{n=1}^{\infty}(-1)^{n+1}\dfrac{1}{\sqrt[3]{3}}$

23. $\displaystyle\sum_{n=1}^{\infty}\dfrac{2^n+3^n}{4^n}$ **24.** $\displaystyle\sum_{n=1}^{\infty}\dfrac{n}{e^{n^2}}$

25. $\displaystyle\sum_{n=1}^{\infty}(-1)^{n+1}\dfrac{n+1}{10n+12}$ **26.** $\displaystyle\sum_{n=1}^{\infty}\dfrac{\sqrt{n}}{n^2+7}$

27. $\displaystyle\sum_{n=1}^{\infty}\dfrac{n^2}{n!}$ **28.** $\displaystyle\sum_{n=1}^{\infty}\dfrac{n^3 3^n}{(n+1)!}$

29. $\displaystyle\sum_{n=1}^{\infty}\dfrac{2^n n!}{(n+2)!}$ **30.** $\displaystyle\sum_{n=2}^{\infty}\left(1-\dfrac{1}{n}\right)^n$

31. $\displaystyle\sum_{n=1}^{\infty}n^2\left(\tfrac{2}{3}\right)^n$ **32.** $\displaystyle\sum_{n=1}^{\infty}\dfrac{(-1)^n}{1+\ln n}$

In Problems 33–36, state whether the given series is absolutely convergent, conditionally convergent, or divergent.

33. $\displaystyle\sum_{n=1}^{\infty}(-1)^n\dfrac{1}{3n-1}$ **34.** $\displaystyle\sum_{n=1}^{\infty}\dfrac{(-1)^n n^3}{2^n}$

35. $\displaystyle\sum_{n=1}^{\infty}(-1)^n\dfrac{3^n}{2^{n+8}}$ **36.** $\displaystyle\sum_{n=2}^{\infty}\dfrac{(-1)^n\sqrt[n]{n}}{\ln n}$

In Problems 37–42, find the convergence set for the power series.

37. $\displaystyle\sum_{n=0}^{\infty}\dfrac{x^n}{n^3+1}$ **38.** $\displaystyle\sum_{n=0}^{\infty}\dfrac{(-2)^{n+1}x^n}{2n+3}$

39. $\displaystyle\sum_{n=0}^{\infty}\dfrac{(-1)^n(x-4)^n}{n+1}$ **40.** $\displaystyle\sum_{n=0}^{\infty}\dfrac{3^n x^{3n}}{(3n)!}$

41. $\displaystyle\sum_{n=0}^{\infty}\dfrac{(x-3)^n}{2^n+1}$ **42.** $\displaystyle\sum_{n=0}^{\infty}\dfrac{n!(x+1)^n}{3^n}$

43. By differentiating the geometric series

$$\dfrac{1}{1+x} = 1 - x + x^2 - x^3 + x^4 - \cdots, \qquad |x| < 1,$$

find a power series that represents $1/(1+x)^2$. What is its interval of convergence?

44. Find a power series that represents $1/(1+x)^3$ on the interval $(-1, 1)$.

45. Find the Maclaurin series for $\sin^2 x$. For what values of x does the series represent the function?

46. Find the first five terms of the Taylor series for e^x based at the point $x = 2$.

47. Write the Maclaurin series for $f(x) = \sin x + \cos x$. For what values of x does it represent f?

$\boxed{\text{C}}$ **48.** Write the Maclaurin series for $f(x) = \cos x^2$ and use it to approximate

$$\int_0^1 \cos x^2 \, dx$$

How many terms of the series are needed to compute the value of this integral correct to four decimal places?

$\boxed{\text{C}}$ **49.** Calculate the following integral correct to five decimal places.

$$\int_0^{0.2}\dfrac{e^x-1}{x}\, dx$$

50. How many terms do we have to take in the convergent series

$$1 - \dfrac{1}{\sqrt{2}} + \dfrac{1}{\sqrt{3}} - \dfrac{1}{\sqrt{4}} + \dfrac{1}{\sqrt{5}} - \dfrac{1}{\sqrt{6}} + \cdots$$

to be sure that we have approximated its sum to within 0.001?

51. Give a good bound for the maximum error made in approximating $\cos x$ by $1 - x^2/2$ for $-0.1 \le x \le 0.1$.

52. Use the simplest method you can think of to find the first three nonzero terms of the Maclaurin series for each of the following:

(a) $\dfrac{1}{1-x^3}$ (b) $\sqrt{1+x^2}$

(c) $e^{-x} - 1 + x$ (d) $x\sec x$

(e) $e^{-x}\sin x$ (f) $\dfrac{1}{1+\sin x}$

Using Infinite Series to Approximate π

I. Preparation

Exercise 1   Derive the relationships

$$\pi = 4 \tan^{-1} 1$$

$$\pi = 12 \tan^{-1}\left(\frac{1}{4}\right) + 4 \tan^{-1}\left(\frac{5}{99}\right)$$

$$\pi = 16 \tan^{-1}\left(\frac{1}{5}\right) - 4 \tan^{-1}\left(\frac{1}{239}\right)$$

Hint: Repeatedly apply the addition formula

$$\tan(x + y) = \frac{\tan x + \tan y}{1 - \tan x \tan y}$$

II. Using Technology

Exercise 2   Use Taylor series to find infinite series for

(a) $\tan^{-1} 1$

(b) $\tan^{-1}\left(\frac{1}{4}\right)$

(c) $\tan^{-1}\left(\frac{5}{99}\right)$

(d) $\tan^{-1}\left(\frac{1}{5}\right)$

(e) $\tan^{-1}\left(\frac{1}{239}\right)$

Numerically evaluate the fifth term of each series and the fifth partial sum. Next evaluate the tenth term and the tenth partial sum. Finally, evaluate the twentieth term and the twentieth partial sum. Which of these five series converges "fastest"? If you had to estimate π using infinite series for the inverse tangent function along with the expressions from Exercise 1, which one would be most efficient?

Exercise 3   Use your CAS to evaluate the first five terms and the five partial sums of the series

$$\sum_{n=0}^{\infty} \frac{(4n)!}{(n!)^4} \frac{(1103 + 26{,}390n)}{396^{4n}}$$

Using these results evaluate the fraction

(1)
$$\frac{9801}{\sqrt{8} \displaystyle\sum_{n=0}^{N} \frac{(4n)!}{(n!)^4} \frac{(1103 + 26{,}390n)}{396^{4n}}}$$

for $N = 0, 1, 2, 3, 4, 5$. What do you notice?

III. Reflection

Exercise 4   You have undoubtedly noticed that the expression in (1) appears to converge to π as N increases without bound. This result was established by the Indian mathematician Srinivasa Ramanujan (1887–1920). Estimate how large N must be in order to approximate π to 500 decimal places, and explain why you think that your value of N is sufficient.

Euler's Derivation of $\dfrac{1}{1^2} + \dfrac{1}{2^2} + \dfrac{1}{3^2} + \cdots = \dfrac{\pi^2}{6}$

I. Preparation

Exercise 1 Let P be a polynomial of degree n with the property that $a_1, a_2, \ldots, a_n$ are distinct zeros of P (i.e., $P(a_i) = 0$ for $i = 1, 2, \ldots, n$) and $P(0) = 1$. Show that P must be of the form

$$(1) \qquad P(x) = \left(1 - \frac{x}{a_1}\right)\left(1 - \frac{x}{a_2}\right) \cdots \left(1 - \frac{x}{a_n}\right)$$

Exercise 2 Derive power series for $\sin x$ and $\dfrac{\sin x}{x}$. For what values of x is $\dfrac{\sin x}{x}$ equal to zero?

II. Using Technology

Exercise 3 Use your technology to obtain the first n terms, $n = 1, 2, 3, 4, 5$, of the Maclaurin series for the function $f(x) = \dfrac{\sin x}{x}$. Notice that each of these functions is a polynomial. Plot f along with each of these polynomials. Comment on the closeness of these polynomials to the function f.

Exercise 4 In Example 6 of Section 10.7 we saw that we can substitute one power series into another to find a power series for a composite function. As you might expect, the calculations can become quite messy; fortunately, we now have computers to do much of the algebra for us. Use technology to find the power series for $\tan(\sin x)$ through terms of degree 6.

Exercise 5 Although power series are not polynomials, it is often helpful to think of them as such. Leonhard Euler (1707–1783) naively treated power series as polynomials of infinite degree and derived a number of remarkable formulas. (Of course, a polynomial of infinite degree is not a

polynomial at all.) In this exercise you will see how he derived one of his results:

$$(2) \qquad \frac{1}{1^2} + \frac{1}{2^2} + \frac{1}{3^2} + \cdots = \frac{\pi^2}{6}$$

Euler treated the power series for $\dfrac{\sin x}{x}$ as a big polynomial. He then applied the result from Exercises 1 and 2 to obtain

$$1 - \frac{x^2}{3!} + \frac{x^4}{5!} - \frac{x^6}{7!} + \frac{x^8}{9!} - \cdots$$

$$= \left(1 - \frac{x}{\pi}\right)\left(1 - \frac{x}{-\pi}\right)\left(1 - \frac{x}{2\pi}\right)$$

$$\left(1 - \frac{x}{-2\pi}\right)\left(1 - \frac{x}{3\pi}\right)\left(1 - \frac{x}{-3\pi}\right)\cdots$$

Explain how this result follows from Exercises 1 and 2. Now multiply the expressions $\left(1 - \dfrac{x}{\pi}\right)$ and $\left(1 - \dfrac{x}{-\pi}\right)$, then multiply $\left(1 - \dfrac{x}{2\pi}\right)$ and $\left(1 - \dfrac{x}{-2\pi}\right)$, and so on. Next, use your technology to multiply out as much of the right side as necessary until you recognize a pattern. Finally, equate the coefficients of x^2 on the left and right. You should end up with the relationship in (2).

III. Reflection

Exercise 6 Continue with Exercise 5 by equating the coefficients of x^4. You should end up with a value for

$$\frac{1}{1^4} + \frac{1}{2^4} + \frac{1}{3^4} + \cdots$$

We must point out that even though these formulas were obtained by treating power series as polynomials of infinite degree, they are still true. Since Euler's death, these results have been established through more rigorous methods.

Numerical Methods, Approximations

11.1 The Taylor Approximation to a Function

So far in this book we have emphasized what might be called *exact* methods. There have, however, been some exceptions, such as when we write $\frac{1}{3} \approx 0.333$ or $\pi \approx 3.1416$. Most notable was our discussion of differentials in Section 3.10, where we used the differential dy to approximate the actual change Δy in $y = f(x)$ when x changed by an amount Δx. That example illustrates the kind of approximation methods we want to highlight in this chapter.

Two factors contribute to the importance of approximation methods. First is the fact that many of the mathematical entities that occur in applications cannot be calculated by exact methods. We mention, for example, the integrals $\int_0^b \sin(x^2)\, dx$, which is used extensively in optics, and $\int_a^b e^{-x^2}\, dx$, which plays a key role in statistics. Second, the invention and now widespread availability of computers and calculators have made approximate numerical methods practical. In fact, it is often easier to calculate something approximately using a calculator (and get an answer to desired accuracy) than to use exact methods, even when exact methods are available.

The Taylor Polynomial of Order 1 In Section 3.10 we emphasized that a function f can be approximated near a point a by its tangent line through the point $(a, f(a))$ (see Figure 1). We called such a line the linear approximation to f near a and we found it to be

$$P_1(x) = f(a) + f'(a)(x - a)$$

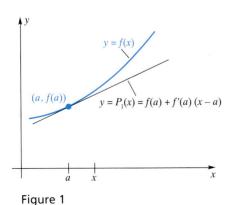

Figure 1

$$\underbrace{}_{\substack{\text{first} \\ \text{error}}} \qquad \underbrace{}_{\substack{\text{second} \\ \text{error}}}$$

$$\cos(0.2) \approx 1 - \frac{1}{2!}(0.2)^2 + \frac{1}{4!}(0.2)^4 \approx 0.980067$$

This example illustrates the two kinds of errors that occur in approximation processes. First, there is the **error of the method**. In this case, we approximated $\cos x$ by a fourth-degree polynomial instead of evaluating the *exact* sum of the series. Second, there is the **error of calculation**. This includes errors due to rounding, as when we replaced the unending decimal $0.9800666\ldots$ by 0.9800667 in the last term above.

Now notice a sad fact of the numerical analyst's life. We can reduce the error of the method by using Maclaurin polynomials of higher order. But using polynomials of higher order means more calculations, which potentially increases the error of calculation. To be a good numerical analyst is to know how to compromise between these two types of error. Unfortunately this is more of an art than a science. However, we can say something definite about the first type of error, the subject to which we now turn.

The Error in the Method   In Chapter 10 we gave a formula for the error of approximating a function by its Taylor polynomial. Taylor's Formula with Remainder is

$$f(x) = f(a) + f'(a)(x - a) + \frac{f''(a)}{2!}(x - a)^2 + \cdots$$

$$+ \frac{f^{(n)}(a)}{n!}(x - a)^n + R_n(x)$$

$$= P_n(x) + R_n(x)$$

The error, or remainder, $R_n(x)$ is given by

$$R_n(x) = \frac{f^{(n+1)}(c)}{(n + 1)!}(x - a)^{n+1}$$

where c is some real number between a and x. This formula for the error is due to the French–Italian mathematician Joseph Louis Lagrange (1736–1813) and is often called the Lagrange error bound for Taylor polynomials. When $a = 0$, Taylor's Formula is called **Maclaurin's Formula**.

One problem that you might foresee at this point is that we do not know what c is; all we know is that it is some real number between a and x. For most problems we must settle for a bound on the remainder using the known bounds on c. The next example illustrates this point.

EXAMPLE 4   Approximate $e^{0.8}$ with an error of less than 0.001.

Solution   For $f(x) = e^x$, Maclaurin's Formula gives the remainder

$$R_n(x) = \frac{f^{(n+1)}(c)}{(n + 1)!}x^{n+1} = \frac{e^c}{(n + 1)!}x^{n+1}$$

and so

$$R_n(0.8) = \frac{e^c}{(n + 1)!}(0.8)^{n+1}$$

where $0 < c < 0.8$. Our goal is to choose n large enough so $|R_n(0.8)| < 0.001$. Now, $e^c < e^{0.8} < 3$ and $(0.8)^{n+1} < (1)^{n+1}$, and so

$$|R_n(0.8)| < \frac{3(1)^{n+1}}{(n + 1)!} = \frac{3}{(n + 1)!}$$

It is easy to check that $3/(n + 1)! < 0.001$ when $n \geq 6$, and so we can obtain the desired accuracy by using the Maclaurin polynomial of order 6:

$$e^{0.8} \approx 1 + (0.8) + \frac{(0.8)^2}{2!} + \frac{(0.8)^3}{3!} + \frac{(0.8)^4}{4!} + \frac{(0.8)^5}{5!} + \frac{(0.8)^6}{6!}$$

Our calculator gives 2.2254948 for this sum.

Can we be sure that this value is within 0.001 of the true result? Certainly the error of the method is less than 0.001. But could the error of calculation have distorted our answer? Possibly so; however, so few calculations are involved that we feel confident in reporting an answer of 2.2255 accurate within 0.001.   ■

Useful Tools for Bounding $|R_n|$ The precise value of R_n is almost never obtainable, since we do not know c, only that c lies on a certain interval. Our task is therefore to find the maximum possible value of $|R_n|$ for c in the given interval. To do this exactly is often difficult, so we usually content ourselves with getting a "good" upper bound for $|R_n|$. This involves a sensible use of inequalities. Our chief tools are the triangle inequality $|a \pm b| \le |a| + |b|$ and the fact that a fraction gets larger when we make its numerator larger or its denominator smaller.

EXAMPLE 5   If c is known to be in $[2,4]$, give a good bound for the maximum value of

$$\left| \frac{c^2 - \sin c}{c} \right|$$

Solution

$$\left| \frac{c^2 - \sin c}{c} \right| = \frac{|c^2 - \sin c|}{|c|} \le \frac{|c^2| + |\sin c|}{|c|} \le \frac{4^2 + 1}{2} = 8.5$$

A different and better bound is obtained as follows:

$$\left| \frac{c^2 - \sin c}{c} \right| = \left| c - \frac{\sin c}{c} \right| \le |c| + \left| \frac{\sin c}{c} \right| \le 4 + \frac{1}{2} = 4.5$$ ■

EXAMPLE 6   Use a Taylor polynomial of order 2 to approximate $\cos 62°$ and then give a bound for the error of the approximation.

Solution   Since $62°$ is near $60°$ (whose cosine and sine are known), we use radian measure and the Taylor polynomial based at $a = \pi/3$.

$$f(x) = \cos x \qquad f\!\left(\frac{\pi}{3}\right) = \frac{1}{2}$$

$$f'(x) = -\sin x \qquad f'\!\left(\frac{\pi}{3}\right) = -\frac{\sqrt{3}}{2}$$

$$f''(x) = -\cos x \qquad f''\!\left(\frac{\pi}{3}\right) = -\frac{1}{2}$$

$$f'''(x) = \sin x \qquad f'''(c) = \sin c$$

Now

$$62° = \frac{\pi}{3} + \frac{\pi}{90} \text{ radians}$$

Thus,

$$\cos x = \frac{1}{2} - \frac{\sqrt{3}}{2}\left(x - \frac{\pi}{3}\right) - \frac{1}{4}\left(x - \frac{\pi}{3}\right)^2 + R_2(x)$$

and

$$\cos\left(\frac{\pi}{3} + \frac{\pi}{90}\right) = \frac{1}{2} - \frac{\sqrt{3}}{2}\left(\frac{\pi}{90}\right) - \frac{1}{4}\left(\frac{\pi}{90}\right)^2 + R_2\left(\frac{\pi}{3} + \frac{\pi}{90}\right)$$

$$\approx 0.4694654 + R_2$$

and

$$|R_2| = \left| \frac{\sin c}{3!} \left(\frac{\pi}{90} \right)^3 \right| < \frac{1}{6} \left(\frac{\pi}{90} \right)^3 \approx 0.0000071$$

Again the number of calculations is small, so we feel safe in reporting $\cos 62° = 0.4694654$ with an error of less than 0.0000071. ■

The Error of Calculation

In all our examples so far, we have assumed that the error of calculation is small enough so that it can be ignored. We will ordinarily make that assumption in this book, since our problems will always involve a small number of calculations. We feel obligated, however, to warn you that when computers are used to do thousands or millions of operations, these errors of calculation may well accumulate and distort an answer.

There are two sources of calculation errors that may be significant even in using a calculator. Consider calculating

$$a + b_1 + b_2 + b_3 + \cdots + b_m$$

where a is very much larger than any of the b's; for example, $a = 10,000,000$ and $b_i = 0.4, i = 1, 2, \ldots, m$. If we use eight-digit floating-point arithmetic and proceed from left to right, first adding b_1 to a, then adding b_2 to the result, and so on, we will simply get $10,000,000$ at each stage. Yet a sum of just 25 of the b's ought to affect the seventh digit of the overall sum. The moral here is that in adding a large number of small terms to one or more large ones it is wise to find the sum of the small terms first.

A more likely source of calculation error is due to the loss of significant digits in a subtraction of nearly equal numbers. For example, subtracting 0.823421 from 0.823445, each with six significant digits, results in 0.000024, which has only two significant digits. That this can cause trouble is easily illustrated by calculating a numerical approximation to a derivative.

Consider calculating $f'(2)$ for $f(x) = x^4$ by using the difference quotient

$$f'(2) \approx \frac{f(2 + h) - f(2)}{h} = \frac{(2 + 10^{-n})^4 - 2^4}{10^{-n}}$$

Theoretically, as n increases (and $h = 10^{-n}$ correspondingly decreases) the result should get closer and closer to the correct value, 32. But note what happens on one eight-digit calculator when n gets too large.

n	$(2 + 10^{-n})^4 - 2^4$	$[(2 + 10^{-n})^4 - 2^4]10^n$
2	0.32240801	32.240801
3	0.03202401	32.024010
4	0.00320024	32.002400
5	0.00032000	32.000000
6	0.00003200	32.000000
7	0.00000320	32.000000
8	0.00000032	32.000000
9	0.00000003	30.000000
10	0.00000000	0.000000
⋮	⋮	⋮

Problems like this arise even if we use 16-digit or 32-digit floating-point arithmetic. Regardless of the number of significant digits used in the calculations, the difference quotient in the preceding table will be 0 for sufficiently large n. Numerical analysts must be aware of such calculation errors.

Concepts Review

1. If $P_2(x)$ is the Taylor polynomial of order 2 based at 1 for $f(x)$, then $P_2(1) =$ _____ , $P_2'(1) =$ _____ , and $P_2''(1) =$ _____ .

2. The coefficient of x^6 in the Maclaurin polynomial of order 9 for $f(x)$ is _____ .

3. The two types of errors that arise in approximation theory are called _____ and _____ .

4. Calculation errors in using Taylor's Formula tend to _____ as n increases, whereas errors of the method tend to _____ as n increases.

Problem Set 11.1

[C] In Problems 1–8, find the Maclaurin polynomial of order 4 for *f(x)* and use it to approximate *f(0.12)*.

1. $f(x) = e^{2x}$

2. $f(x) = e^{-3x}$

3. $f(x) = \sin 2x$

4. $f(x) = \tan x$

5. $f(x) = \ln(1 + x)$

6. $f(x) = \sqrt{1 + x}$

7. $f(x) = \tan^{-1} x$

8. $f(x) = \sinh x$

[C] In Problems 9–14, find the Taylor polynomial of order 3 based at a for the given function.

9. $e^x; a = 1$

10. $\sin x; a = \dfrac{\pi}{4}$

11. $\tan x; a = \dfrac{\pi}{6}$

12. $\sec x; a = \dfrac{\pi}{4}$

13. $\cot^{-1} x; a = 1$

14. $\sqrt{x}; a = 2$

15. Find the Taylor polynomial of order 3 based at 1 for $f(x) = x^3 - 2x^2 + 3x + 5$ and show that it is an exact representation of $f(x)$.

16. Find the Taylor polynomial of order 4 based at 2 for $f(x) = x^4$ and show that it represents $f(x)$ exactly.

17. Find the Maclaurin polynomial of order n for $f(x) = 1/(1 - x)$. Then use it with $n = 4$ to approximate each of the following.
(a) $f(0.1)$ (b) $f(0.5)$ (c) $f(0.9)$ (d) $f(2)$

[C] **18.** Find the Maclaurin polynomial of order n (n odd) for $\sin x$. Then use it with $n = 5$ to approximate each of the following. (This example should convince you that the Maclaurin approximation can be exceedingly poor if x is far from zero.) Compare your answers with those given by your calculator. What conclusion do you draw?
(a) $\sin(0.1)$ (b) $\sin(0.5)$ (c) $\sin(1)$ (d) $\sin(10)$

19. Use a Maclaurin polynomial to obtain the approximation $A \approx r^2 t^3/12$ for the area of the shaded region in Figure 5. First express A exactly, then approximate.

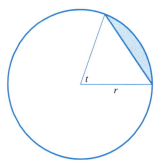

Figure 5

20. If an object of rest mass m_0 has velocity v, then (according to the theory of relativity) its mass m is given by $m = m_0/\sqrt{1 - v^2/c^2}$, where c is the velocity of light. Explain how physicists get the approximation

$$m \approx m_0 + \frac{m_0}{2}\left(\frac{v}{c}\right)^2$$

21. If money is invested at interest rate r compounded monthly, it will double in n years, where n satisfies

$$\left(1 + \frac{r}{12}\right)^{12n} = 2$$

(a) Show that

$$n = \ln 2 \left[\frac{1}{12 \ln(1 + r/12)}\right]$$

(b) Use the Maclaurin polynomial of order 2 for $\ln(1 + x)$ and a partial fraction decomposition to obtain the approximation

$$n \approx \frac{0.693}{r} + 0.029$$

[C] (c) Some people use the *Rule of 72, $n \approx 72/(100r)$*, to approximate n. Fill in the table to compare the values obtained from these three formulas.

r	n (Exact)	n (Approximation)	n (Rule of 72)
0.05			
0.10			
0.15			
0.20			

22. The author of a biology text claimed that the smallest positive solution to $x = 1 - e^{-(1+k)x}$ is approximately $x = 2k$, provided k is very small. Show how she reached this conclusion and check on it for $k = 0.01$.

[CAS] *Use a computer algebra system to work Problems 23 and 24.*

23. For each of the following, draw using the same axes the graphs of the Maclaurin polynomials of orders 1, 2, 3, and 4.
(a) $\sin(e^x)$ (b) $(\sin x)/(2 + \sin x)$

24. Follow the directions of Problem 23.
(a) $\exp(-x^2)$ (b) $\sin(\ln(1 + x))$

In Problems 25–32, find a good bound for the maximum value of the given expression, given that c is in the stated interval. Answers may vary depending on the technique used. (See Example 5.)

25. $|e^{2c} + e^{-2c}|; [0, 3]$

26. $|\tan c + \sec c|; \left[0, \dfrac{\pi}{4}\right]$

27. $\left|\dfrac{4c}{\sin c}\right|; \left[\dfrac{\pi}{4}, \dfrac{\pi}{2}\right]$

28. $\left|\dfrac{4c}{c+4}\right|; [0, 1]$

29. $\left|\dfrac{e^c}{c+5}\right|; [-2, 4]$

30. $\left|\dfrac{\cos c}{c+2}\right|; \left[0, \dfrac{\pi}{4}\right]$

31. $\left|\dfrac{c^2 + \sin c}{10 \ln c}\right|; [2, 4]$

32. $\left|\dfrac{c^2 - c}{\cos c}\right|; \left[0, \dfrac{\pi}{4}\right]$

In Problems 33–36, find a formula for $R_6(x)$, the remainder for the Taylor polynomial of order 6 based at a. Then obtain a good bound for $|R_6(0.5)|$. See Examples 4 and 6.

33. $\ln(2 + x); a = 0$

34. $e^{-x}; a = 1$

35. $\sin x; a = \pi/4$

36. $\dfrac{1}{x-3}; a = 1$

In the formula for the remainder

$$R_n(x) = \frac{f^{(n+1)}(c)}{(n+1)!}(x-a)^{n+1}$$

there is a value of c for which $R_n(x)$ is the exact value of the remainder. Sometimes it is useful to know the minimum as well as the maximum estimate for $R_n(x)$. In Problems 37 and 38 we explore this situation.

C 37. Consider the Maclaurin polynomial of order 3 for e^x. Estimate the minimum and maximum value of the error made in calculating $e^{-0.1}$ using that polynomial. Compare the estimated maximum and minimum errors with the actual error at that point.

C 38. Consider the Taylor polynomial of order 3 for $\sin x$ about the point $a = \dfrac{\pi}{4}$. Estimate the minimum and maximum value of the error made in calculating $\sin\dfrac{\pi}{8}$ using that polynomial. Compare the estimated maximum and minimum errors with the actual error at that point.

39. Determine the order n of the Maclaurin polynomial for e^x that is required to approximate e to five decimal places, that is, so that $|R_n(1)| \leq 0.000005$ (see Example 4).

40. Find the third-order Maclaurin polynomial for $(1 + x)^{3/2}$ and bound the error $R_3(x)$ if $-0.1 \leq x \leq 0$.

41. Find the third-order Maclaurin polynomial for $(1 + x)^{-1/2}$ and bound the error $R_3(x)$ if $-0.05 \leq x \leq 0.05$.

42. Find the fourth-order Maclaurin polynomial for $\ln[(1+x)/(1-x)]$ and bound the error $R_4(x)$ for $-0.5 \leq x \leq 0.5$.

43. Note that the fourth-order Maclaurin polynomial for $\sin x$ is really of third degree since the coefficient of x^4 is 0. Thus,

$$\sin x = x - \frac{x^3}{6} + R_4(x)$$

Show that if $0 \leq x \leq 0.5, |R_4(x)| \leq 0.0002605$. Use this result to approximate $\displaystyle\int_0^{0.5} \sin x \, dx$ and give a bound for the error.

44. In analogy with Problem 43,

$$\cos x = 1 - \frac{x^2}{2} + \frac{x^4}{24} + R_5(x)$$

If $0 \leq x \leq 1$, give a good bound for $|R_5(x)|$. Then use your result to approximate $\displaystyle\int_0^1 \cos x \, dx$ and give a bound for the error.

45. Expand $x^4 - 3x^3 + 2x^2 + x - 2$ in a Taylor polynomial of order 4 based at 1 and show that $R_4(x) = 0$ for all x.

46. Let $f(x)$ be a function that possesses at least n derivatives at $x = a$ and let $P_n(x)$ be the Taylor polynomial of order n based at a. Show that

$$P_n(a) = f(a), \quad P'_n(a) = f'(a), \quad P''_n(a) = f''(a),$$

$$\dots, \quad P_n^{(n)}(a) = f^{(n)}(a)$$

C 47. Calculate $\sin 43° = \sin 43\pi/180$ by using the Taylor polynomial of order 3 based at $\pi/4$ for $\sin x$. Then obtain a good bound for the error made. See Example 6.

C 48. Calculate $\cos 63°$ by the method illustrated in Example 6. Choose n large enough so that $|R_n| \leq 0.0005$.

C 49. Show that if x is in $[0, \pi/2]$ the error in using

$$\sin x \approx x - \frac{x^3}{3!} + \frac{x^5}{5!} - \frac{x^7}{7!} + \frac{x^9}{9!}$$

is less than 5×10^{-5} and, therefore, that this formula is good enough to build a four-place sine table.

50. Use Maclaurin's Formula rather than l'Hôpital's Rule to find:

(a) $\displaystyle\lim_{x \to 0} \frac{\sin x - x + x^3/6}{x^5}$

(b) $\displaystyle\lim_{x \to 0} \frac{\cos x - 1 + x^2/2 - x^4/24}{x^6}$

EXPL 51. Let $g(x) = p(x) + x^{n+1} f(x)$, where $p(x)$ is a polynomial of degree at most n and f has derivatives through order n. Show that $p(x)$ is the Maclaurin polynomial of order n for g.

EXPL 52. Recall that the Second Derivative Test for Local Extrema (Section 4.3) does not apply when $f''(c) = 0$. Prove the following generalization, which may help determine a maximum or a minimum when $f''(c) = 0$. Suppose that

$$f'(c) = f''(c) = f'''(c) = \dots = f^{(n)}(c) = 0$$

where n is odd and $f^{(n+1)}(x)$ is continuous near c.

(i) If $f^{(n+1)}(c) < 0, f(c)$ is a local maximum value.

(ii) If $f^{(n+1)}(c) > 0, f(c)$ is a local minimum value.

Test this result on $f(x) = x^4$.

EXPL 53. Many other polynomial approximations to functions exist besides Taylor and Maclaurin polynomials. Here we consider the *Lagrange interpolating polynomials* as a specific example.

(a) Show that the polynomial

$$L_{5,1}(x) = \frac{(x - x_2)(x - x_3)(x - x_4)(x - x_5)}{(x_1 - x_2)(x_1 - x_3)(x_1 - x_4)(x_1 - x_5)}$$

is of degree 4 and has the property that $L_{5,1}(x_1) = 1$ while $L_{5,1}(x_j) = 0$ for $j = 2, 3, 4, 5$.

(b) Using $L_{5,1}(x)$ as a model, construct fourth-degree polynomials $L_{5,i}(x)$ that are 1 at x_i and 0 at x_j for $j \neq i$, where $i = 2, 3, 4, 5$.

(c) Consider the polynomial

$$L_5(x) \equiv L_{5,1}(x)y_1 + L_{5,2}(x)y_2 + L_{5,3}(x)y_3$$
$$+ L_{5,4}(x)y_4 + L_{5,5}(x)y_5$$

and show that L_5 is a polynomial of degree less than or equal to 4 that takes the value y_i at $x = x_i, i = 1, \dots, 5$. Such a polynomial is called the *Lagrange interpolating polynomial* that goes through (i.e., interpolates) the points $(x_i, y_i), i = 1, \ldots, 5$ where the x_i are all distinct.

(d) Construct the second-degree Lagrange interpolating polynomial that goes through the points $(1, 2), (2, 2.5)$ and $(0, 0)$.

54. Plot the points $(1, 2), (2, 3), (3, 4), (4, 5), (5, 6)$. Construct the Lagrange interpolating polynomial that goes through these points and show that, after some algebra, the answer reduces to $x + 1$.

It can be shown that the error for the Lagrange interpolating polynomial $P_n(x)$ of degree n for the function $f(x)$ is given by

$$R_n(x) = \frac{(x - x_1)(x - x_2)\cdots(x - x_{n+1})}{(n + 1)!} f^{(n+1)}(\alpha)$$

where α is some point contained in the interval that includes all the points $x_1, x_2, \ldots, x_{n+1}, x$. Use this error formula in Problems 55–57.

55. Given that $\ln 1 = 0, \ln 3 = 1.099$, and $\ln 5 = 1.609$, write a second-degree polynomial that interpolates these values. Use this interpolation to compute an approximation for $\ln 2$. Use the error formula to give an estimate for the maximum error. Compare

your estimate for the error to the actual error in your approximation to $\ln 2$.

56. Given that $e^0 = 1, e^{0.2} = 1.221$, and $e^{0.3} = 1.350$, write a second-degree polynomial that interpolates these values. Use this interpolation to compute an approximation for $e^{0.25}$. Use the error formula to give an estimate for the maximum and minimum error. Compare your estimate for the error to the actual error in your approximation to $e^{0.25}$.

57. Construct a second-degree Maclaurin polynomial for e^x. Also, construct a second-order interpolating polynomial using the values $e^0 = 1, e^{0.2} = 1.221$, and $e^{0.3} = 1.350$. Using the expressions for the error in the Maclaurin polynomial and the error in the interpolating polynomial, compute an estimate for the maximum error for $0 \le x \le 0.3$. Compare your answers with the actual error at $x = 0.1$.

Answers to Concepts Review: **1.** $f(1); f'(1); f''(1)$ **2.** $f^{(6)}(0)/6!$
3. error of the method; error of calculation **4.** increase; decrease

---

11.2
Numerical Integration

We know that if f is continuous on a closed interval $[a, b]$ then the definite integral $\int_a^b f(x)\,dx$ must exist. Existence is one thing; evaluation is a very different matter.

There are many definite integrals that cannot be evaluated by the methods that we have learned, that is, by use of the Second Fundamental Theorem of Calculus. For example, the indefinite integrals of such integrands as

$$e^{-x^2}, \qquad \sin(x^2), \qquad \sqrt{1 - x^4}, \qquad \frac{\sin x}{x}$$

cannot be expressed algebraically in terms of elementary functions, that is, in terms of functions studied in a first calculus course (see the introductory remarks in Section 8.1). Even when elementary indefinite integrals can be found, it is often advantageous to use the approximation methods of this section, since they lead to efficient algorithms that can be directly programmed on a calculator or computer. In Section 5.5 we saw how Riemann sums can be used to approximate a definite integral. Here we present two additional methods: the Trapezoidal Rule and the Parabolic Rule.

The Trapezoidal Rule Consider the graph of $y = f(x)$ on $[a, b]$; it might look something like the curve in Figure 1. Partition the interval $[a, b]$ into n subintervals, each of length $h = (b - a)/n$, by means of points $a = x_0 < x_1 < x_2 < \cdots < x_n = b$. Join the pairs of points $(x_{i-1}, f(x_{i-1}))$ and $(x_i, f(x_i))$ by line segments, as shown in the figure, thus forming n trapezoids.

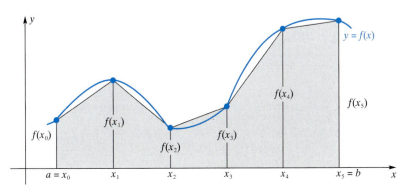

Figure 1

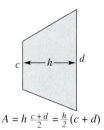

$A = h \frac{c+d}{2} = \frac{h}{2}(c + d)$

Figure 2

Recalling the area formula shown in Figure 2, we can write the area of the *i*th trapezoid as

$$A_i = \frac{h}{2}[f(x_{i-1}) + f(x_i)]$$

More accurately, we should say *signed* area, since A_i will be negative for a subinterval where f is negative. The definite integral $\int_a^b f(x)\,dx$ is approximately equal to $A_1 + A_2 + \cdots + A_n$, that is, to

$$\frac{h}{2}[f(x_0) + f(x_1)] + \frac{h}{2}[f(x_1) + f(x_2)] + \cdots + \frac{h}{2}[f(x_{n-1}) + f(x_n)]$$

This simplifies to the **Trapezoidal Rule**:

Trapezoidal Rule

$$\int_a^b f(x)\,dx \approx \frac{h}{2}[f(x_0) + 2f(x_1) + 2f(x_2) + \cdots + 2f(x_{n-1}) + f(x_n)]$$

$$= \frac{h}{2}\left[f(x_0) + 2\sum_{i=1}^{n-1} f(x_i) + f(x_n)\right]$$

We illustrate this rule first for a definite integral where we know its exact value.

EXAMPLE 1 Use the Trapezoidal Rule with $n = 8$ to approximate

$$\int_1^3 x^4\,dx$$

Solution Since $n = 8$, $h = (3 - 1)/8 = 0.25$, and

$$\begin{aligned}
x_0 &= 1.00 & f(x_0) &= (1.00)^4 = 1.0000 \\
x_1 &= 1.25 & f(x_1) &= (1.25)^4 \approx 2.4414 \\
x_2 &= 1.50 & f(x_2) &= (1.50)^4 = 5.0625 \\
x_3 &= 1.75 & f(x_3) &= (1.75)^4 \approx 9.3789 \\
x_4 &= 2.00 & f(x_4) &= (2.00)^4 = 16.0000 \\
x_5 &= 2.25 & f(x_5) &= (2.25)^4 = 25.6289 \\
x_6 &= 2.50 & f(x_6) &= (2.50)^4 = 39.0625 \\
x_7 &= 2.75 & f(x_7) &= (2.75)^4 \approx 57.1914 \\
x_8 &= 3.00 & f(x_8) &= (3.00)^4 = 81.0000
\end{aligned}$$

Thus,

$$\int_1^3 x^4\,dx \approx \frac{0.25}{2}[1.0000 + 2(2.4414) + \cdots + 2(57.1914) + 81.0000]$$

$$= 48.9414$$

This may be compared with the exact value

$$\int_1^3 x^4\,dx = \left[\frac{x^5}{5}\right]_1^3 = \frac{242}{5} = 48.4000$$

Presumably we could get a better approximation by taking n larger; this would be easy to do using a computer. However, while taking n larger reduces the error of the method, it at least potentially increases the error of calculation. It would be unwise, for example, to take $n = 1{,}000{,}000$, since the potential round-off errors would more than compensate for the fact that the error of the method would be minuscule. We will have more to say about errors shortly. ■

EXAMPLE 2   Use the Trapezoidal Rule with $n = 6, 12, 24, 48, 96$, and 192 to approximate

$$\int_0^1 e^{-x^2} \, dx$$

Solution   For $n = 6$, we have $h = \frac{1}{6}$. Thus,

$$
\begin{aligned}
x_0 &= 0.0000 & f(x_0) &= 1.0000 \\
x_1 &\approx 0.1667 & f(x_1) &\approx 0.9726 \\
x_2 &\approx 0.3333 & f(x_2) &\approx 0.8948 \\
x_3 &= 0.5000 & f(x_3) &\approx 0.7788 \\
x_4 &\approx 0.6667 & f(x_4) &\approx 0.6412 \\
x_5 &\approx 0.8333 & f(x_5) &\approx 0.4994 \\
x_6 &= 1.0000 & f(x_6) &\approx 0.3679
\end{aligned}
$$

$$\int_0^1 e^{-x^2} \, dx \approx \tfrac{1}{12}\left[1.000 + 2(0.9726) + \cdots + 2(0.4994) + 0.3679\right]$$

$$= 0.7451$$

n	Estimate of $\int_0^1 e^{-x^2}\,dx$ using the Trapezoidal Rule
6	0.74512
12	0.74640
24	0.74672
48	0.74680
96	0.74682
192	0.74682

The Trapezoidal Rule is easily implemented on a computer algebra system. We used Mathematica to compute the values in the table. As n gets large, the estimates of $\int_0^1 e^{-x^2} \, dx$ seem to be getting close to 0.74682.

As we mentioned earlier, this integral is important in probability and statistics. Because it cannot be evaluated by a direct application of the Second Fundamental Theorem of Calculus, we must rely on approximations such as those obtained by the Trapezoidal Rule.   ■

The Error in the Trapezoidal Rule
In any practical use of the Trapezoidal Rule, we need to have some idea of the size of the error involved. Fortunately, we can give a formula for the error of the method for functions that are twice differentiable.

Theorem A   Error for the Trapezoidal Rule

Suppose that f'' exists on $[a, b]$. Then

$$\int_a^b f(x)\, dx = \frac{h}{2}\left[f(x_0) + 2f(x_1) + \cdots + 2f(x_{n-1}) + f(x_n)\right] + E_n$$

where the error E_n is given by

$$E_n = -\frac{(b-a)^3}{12n^2} f''(c)$$

and c is some point between a and b.

We omit the proof of this theorem, which may be found in more advanced books. In Example 3 we illustrate its use.

EXAMPLE 3   Give a bound for the possible error in Example 2.

Solution   Since $f(x) = e^{-x^2}, f'(x) = -2xe^{-x^2}$ and $f''(x) = e^{-x^2}(4x^2 - 2)$. We could actually find the maximum value of $f''(x)$ on $[0, 1]$, but it is sufficient to bound it there using properties of absolute value.

$$|f''(x)| = |e^{-x^2}(4x^2 - 2)| = e^{-x^2}|4x^2 - 2|$$

$$\leq e^{-x^2}(4x^2 + 2) \leq 1(4 + 2) = 6$$

Thus,

$$|E_n| = \frac{(b-a)^3}{12n^2}|f''(c)| \le \frac{1^3}{12(6^2)}(6) = \frac{1}{72} \approx 0.0139$$

This is, of course, a bound for the error of the method. However, n is so small that the error of calculation can safely be ignored. We feel confident in reporting that

$$\int_0^1 e^{-x^2}\,dx = 0.7451 \pm 0.0139$$

A slightly more sophisticated analysis, based on the fact that $4x^2 - 2$ is increasing on $[0, 1]$, shows that $|4x^2 - 2| \le 2$ on $[0, 1]$, and hence $|E_n| \le 0.0047$. ∎

EXAMPLE 4 How large must n be to ensure that the error of the method in Example 2 is less than 0.0001?

Solution From Example 3,

$$|E_n| = \frac{(b-a)^3}{12n^2}|f''(c)| \le \frac{1^3 \cdot 6}{12n^2} = \frac{1}{2n^2}$$

We want $1/2n^2 < 0.0001$, which holds if $n \ge 71$. ∎

The Parabolic Rule (Simpson's Rule) In the Trapezoidal Rule, we approximated the curve $y = f(x)$ by line segments. It seems likely that we could do better using parabolic segments. Just as before, partition the interval $[a, b]$ into n subintervals of length $h = (b-a)/n$, but this time with n an *even* number. Then fit parabolic segments to neighboring triples of points, as shown in Figure 3.

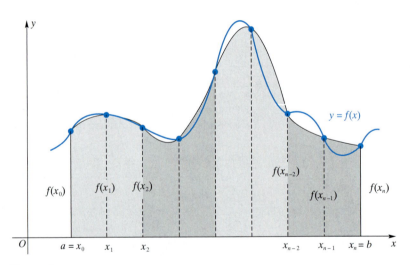

Figure 3

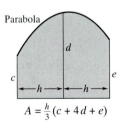

Parabola

$$A = \frac{h}{3}(c + 4d + e)$$

Figure 4

Using the area formula in Figure 4 (see Problem 15 for the derivation) leads to an approximation called the **Parabolic Rule**. It is also called **Simpson's Rule**, after the English mathematician Thomas Simpson (1710–1761).

Parabolic Rule (n even)

$$\int_a^b f(x)\,dx \approx \frac{h}{3}\left[f(x_0) + 4f(x_1) + 2f(x_2) + \cdots + 4f(x_{n-1}) + f(x_n)\right]$$

The pattern of coefficients is $1, 4, 2, 4, 2, 4, 2, \ldots, 2, 4, 1$.

EXAMPLE 5 Use the Parabolic Rule with $n = 8$ to approximate

$$\int_1^3 x^4\,dx$$

Solution   We can make use of the calculations in Example 1. We obtain

$$\int_1^3 x^4\, dx \approx \frac{0.25}{3}\big[1.0000 + 4(2.4414) + 2(5.0625) + 4(9.3789) + 2(16.0000)$$

$$+ 4(25.6289) + 2(39.0625) + 4(57.1914) + 81.0000\big]$$

$$\approx 48.4010 \qquad\blacksquare$$

As expected, the Parabolic Rule gives an answer closer to the true value 48.4000 than did the Trapezoidal Rule, which gave 48.9414. Because it requires so little extra work, we prefer the Parabolic Rule over the Trapezoidal Rule in most problems. That its error term is generally smaller is borne out by the following theorem (note the factor of n^4 in the denominator).

Theorem B   Error for the Parabolic Rule

Suppose that the fourth derivative $f^{(4)}(x)$ exists on $[a, b]$. Then the error E_n in the Parabolic Rule is given by

$$E_n = -\frac{(b-a)^5}{180n^4} f^{(4)}(c)$$

for some c between a and b.

EXAMPLE 6   Use the Parabolic Rule with $n = 8$ to approximate

$$\int_1^2 (1 + x)^{-1}\, dx$$

and find a bound for the error made.

Solution   Since $h = \frac{1}{8}$ and $f(x) = 1/(1 + x)$, we calculate the results shown in the table.

i	x_i	$f(x_i)$	c_i	$c_i f(x_i)$
0	1	0.50000	1	0.50000
1	1.125	0.47059	4	1.88236
2	1.25	0.44444	2	0.88888
3	1.375	0.42105	4	1.68420
4	1.5	0.40000	2	0.80000
5	1.625	0.38095	4	1.52380
6	1.75	0.36364	2	0.72728
7	1.875	0.34783	4	1.39132
8	2	0.33333	1	0.33333
				Sum = 9.73117

Thus,

$$\int_1^2 \frac{dx}{1 + x} \approx \frac{1}{24}(9.73117) \approx 0.4055$$

To find a bound for E_8, we first calculate $f^{(4)}(x)$ to be $24/(1 + x)^5$, and then

$$E_8 = -\frac{(2-1)^5}{180(8^4)} \cdot \frac{24}{(1 + c)^5}$$

Consequently,

$$|E_8| \le \frac{24}{180(8^4)(1 + 1)^5} \approx 0.00000102$$

Assuming that calculation errors are negligible, we feel very confident in reporting that the answer 0.4055 is accurate to four decimal places. $\blacksquare$

Computers and Integration

A computer algebra system like Mathematica or Maple can evaluate a number of definite integrals exactly. It can also use numerical methods to *approximate* definite integrals. It is important to know when your CAS is doing integration exactly and when it is using an approximation. When a software package uses some form of integration, it may use one of the rules in this section, or it may use something more sophisticated. In any case, when it uses approximations, you should not expect the answers that it gives to be exact.

Concepts Review

1. The pattern of coefficients in the Trapezoidal Rule is _____.

2. The pattern of coefficients in the Parabolic Rule is _____.

3. The error in the Trapezoidal Rule has n^2 in the denominator, whereas the error in the Parabolic Rule has _____ in the

denominator, so we expect the latter to give a better approximation to a definite integral.

4. If f is positive and concave up, then the Trapezoidal Rule will always give a value for $\int_a^b f(x)\,dx$ that is too _____.

Problem Set 11.2

In Problems 1–4, calculate the Riemann sum $\sum_{i=1}^{n} f(\bar{x}_i)\,\Delta x_i$ (see Section 5.5) for an equally spaced partition with $n = 8$ subintervals; choose the sample point $\bar{x}_i$ to be the left end point of each subinterval. Then use the Trapezoidal Rule and the Parabolic Rule, both with $n = 8$, to approximate each definite integral. Then use the Second Fundamental Theorem of Calculus to find the exact value of each interval.

1. $\int_1^3 \dfrac{1}{x^2}\,dx$ **2.** $\int_1^3 \dfrac{1}{x}\,dx$

3. $\int_0^2 \sqrt{x}\,dx$ **4.** $\int_1^3 x\sqrt{x^2+1}\,dx$

C **5.** Use the Trapezoidal Rule with $n = 2, 6, 12$ to approximate $\int_0^\pi \sin x\,dx$. Note how these approximations do get closer to the actual value, which is 2.

C **6.** Follow the instructions of Problem 5 using the Parabolic Rule.

C **7.** Approximate π by calculating

$$\int_0^1 \frac{4}{1+x^2}\,dx$$

using the Parabolic Rule with $n = 10$.

C **8.** Use the Trapezoidal Rule with $n = 10$ to approximate $\int_0^1 \cos(\sin x)\,dx$.

C *In Problems 9–12, determine an n so that the Trapezoidal Rule will approximate the integral with an error E_n satisfying $|E_n| \le 0.01$ (see Example 4). Then, using that n, approximate the integral.*

9. $\int_0^1 e^{-x^2}\,dx$ **10.** $\int_0^{0.6} e^{x^2}\,dx$

11. $\int_1^{1.5} \sqrt{\cos x}\,dx$ **12.** $\int_1^2 \cos\sqrt{x}\,dx$

C *In Problems 13 and 14, determine an n so that the Parabolic Rule will approximate the integral with an error E_n satisfying $|E_n| \le 0.005$. Then, using that n, approximate the integral.*

13. $\int_2^6 \dfrac{1+x}{1-x}\,dx$ **14.** $\int_1^3 \ln x\,dx$

15. Let $f(x) = ax^2 + bx + c$. Show that

$$\int_{m-h}^{m+h} f(x)\,dx \text{ and } \frac{h}{3}\left[f(m-h) + 4f(m) + f(m+h)\right]$$

both have the value $(h/3)\left[a(6m^2 + 2h^2) + b(6m) + 6c\right]$. This establishes the area formula on which the Parabolic Rule is based.

16. Show that the Parabolic Rule is exact for any cubic polynomial in two different ways.
(a) By direct calculation.   (b) By showing that $E_n = 0$.

17. We know that $\ln 2 = \int_1^2 (1/x)\,dx$. If we wish to estimate $\ln 2$ using the Trapezoidal Rule with an error less than 10^{-10}, how large should n be?

18. Answer the question of Problem 17 for the Parabolic Rule.

19. Show that the Parabolic Rule gives the exact value of $\int_{-a}^{a} x^k\,dx$ provided that k is odd.

20. It is interesting that a modified version of the Trapezoidal Rule turns out to be in general more accurate than the Parabolic Rule. This version says that

$$\int_a^b f(x)\,dx \approx T - \frac{\left[f'(b) - f'(a)\right]h^2}{12}$$

where T is the standard trapezoidal estimate.

(a) Use this formula with $n = 8$ to estimate $\int_1^3 x^4\,dx$ and note its remarkable accuracy (see Example 1 for T and also the exact value of this integral).

(b) Use this formula with $n = 12$ to estimate $\int_0^\pi \sin x\,dx$ (the true value is 2 and T was calculated in Problem 5).

21. Use the Trapezoidal Rule to approximate the area of the lakeside lot shown in Figure 5. Dimensions are in feet.

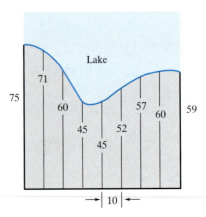

Figure 5

22. Use the Parabolic Rule to approximate the amount of water required to fill a pool shaped like Figure 6 to a depth of 6 feet. All dimensions are in feet.

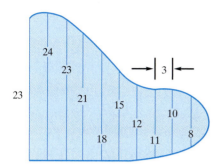

Figure 6

C **23.** Figure 7 shows the depth in feet of the water in a river measured at 20-foot intervals across the width of the river. If the river flows at 4 miles per hour, how much water (in cubic feet) flows past the place where these measurements were taken in one day? Use the Parabolic Rule.

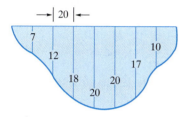

Figure 7

24. Example 9 of Section 5.8 gave the speed of a car every 10 minutes. Use the given table along with the Trapezoidal Rule to estimate how far the car traveled.

25. Rework Problem 69 of Section 5.8 using
(a) the Trapezoidal Rule and   (b) the Parabolic Rule.

26. Rework Problem 70 of Section 5.8 using
(a) the Trapezoidal Rule and   (b) the Parabolic Rule.

27. Rework Problem 34 of Section 6.6 using
(a) the Trapezoidal Rule and   (b) the Parabolic Rule.

28. Rework Problem 35 of Section 6.6 using
(a) the Trapezoidal Rule and   (b) the Parabolic Rule.

29. Rework Problem 36 of Section 6.6 using the Trapezoidal Rule.

C EXPL **30.** Another commonly used numerical integration rule is the **Midpoint Rule**. Let n and h have their usual meanings, but assume that n is even. Then $\int_a^b f(x)\,dx \approx M_n$, where

$$M_n = 2h[f(x_1) + f(x_3) + f(x_5) + \cdots + f(x_{n-1})]$$

(a) Draw a picture to interpret this rule.

(b) Use this rule with $n = 16$ to approximate $\int_1^3 x^4\,dx$ (see Example 1).

*In Problems 31–36, we consider improper integrals that have infinite limits of integration, infinite integrands, or both. We present dif-*ferent situations and develop numerical techniques to approximate the integrals in those situations.

EXPL **31.** We are able to approximate an integral with an infinite limit of integration if we can approximate the integral over a finite interval and get a small bound on the portion of the integral that is neglected. Since

$$\int_0^\infty e^{-x^2}\,dx = \int_0^X e^{-x^2}\,dx + \int_X^\infty e^{-x^2}\,dx$$

we can choose X so that the second integral is small.
(a) Show that

$$\int_X^\infty e^{-x^2}\,dx < \int_X^\infty e^{-Xx}\,dx = \frac{1}{X}e^{-X^2}$$

C (b) Show for $X = 5$ that $\int_X^\infty e^{-x^2}\,dx < 10^{-11}$.

C (c) Use the Parabolic Rule with $n = 10$ to compute $\int_0^5 e^{-x^2}\,dx$, and use this result to approximate $\frac{2}{\sqrt{\pi}}\int_0^\infty e^{-x^2}\,dx$ with an estimate for the error in your result.

EXPL **32.** An integral over an infinite or half-infinite interval can be transformed into one over a finite interval by the technique of substitution. This technique frequently leads to problems where the integrand is infinite at some point in the interval of integration.
(a) Show that the substitution $t = e^{-x}$ transforms the interval $[0,\infty)$ into $(0,1]$, and $t = \dfrac{x}{1+x}$ transforms the interval $[0,\infty)$ into $[0,1)$.

(b) Show that the substitution $t = \dfrac{e^x - 1}{e^x + 1}$ transforms the interval $(-\infty, \infty)$ into $(-1, 1)$.

(c) Convert the integral $\int_0^\infty \dfrac{e^{-x}}{\sqrt{1 + e^{-2x}}}\,dx$ to an integral over a finite interval using the substitution $t = e^{-x}$. Use the Parabolic Rule with $n = 10$ on the resulting integral to get an approximate result with an estimate for the error.

C (d) Use the substitution $t = 1/x$ and the Parabolic Rule with $n = 10$ to evaluate the improper integral $\int_1^\infty \dfrac{x}{1 + x^3}\,dx$.

33. Integrals that are infinite at an end point can sometimes be transformed using integration by parts into proper integrals.
(a) Integrate by parts to show that the improper integral $\int_0^2 \dfrac{dx}{\sqrt{4x + x^2}}$ can be transformed into a proper one.
(b) Integrate by parts to show that the improper integral $\int_1^\infty \dfrac{\sin x}{x}\,dx$ exists.

(c) Integrate by parts to show that $\int_0^1 \dfrac{1}{\sqrt{x}}\dfrac{1}{4 + x}\,dx$ can be turned into a proper integral. *Hint:* Differentiate the factor $1/(4 + x)$.

EXPL **34.** Integration by parts can be used to transform an integral into a form that permits more accurate results using numerical integration routines. As a specific example we consider $\int_0^2 x \cdot x(4 - x^2)^{1/4}\,dx$.

(a) Integrate by parts by differentiating x, to show that the integral is equivalent to $\int_0^2 (2/5)(4 - x^2)^{5/4}\, dx$.

(b) Plot the functions $x^2(4 - x^2)^{1/4}$ and $(2/5)(4 - x^2)^{5/4}$ over the domain $[0, 2]$, and pay particular attention to the slope of each curve. Notice that it is difficult to perform accurate numerical integration when a function has a large and rapidly varying slope. This is reflected in the error estimate that depends on the higher derivatives of the integrand.

C (c) Break the original integral into two parts,

$$\int_0^1 (2/5)(4 - x^2)^{5/4}\, dx + (2/5)\int_1^2 (1/x)\cdot x(4 - x^2)^{5/4}\, dx$$

and integrate the second integral by parts again. Use the Trapezoidal Rule (with $n = 4$) with an estimate for the error to approximate both integrals. Explain why the error estimate would fail to give any reasonable result for the original integral.

EXPL C 35. Approximate the integral $\int_0^{\pi/2} \ln(\sin x)\, dx$ using the Parabolic Rule with $n = 4$. *Hint*: Use $\ln(\sin x) = \ln x + \ln\left(\dfrac{\sin x}{x}\right)$, and then use the Parabolic Rule to integrate $\ln\left(\dfrac{\sin x}{x}\right)$.

C 36. Use the method of truncating the interval of integration to find an X such that $\left|\int_X^\infty \dfrac{\cos x}{1 + x^4}\, dx\right| < 10^{-5}$ by choosing $X = (2n + 1)\pi/2$ and using the fact that the cosine function alternates in sign. Explain how to use this result to approximate the integral $\int_0^\infty \dfrac{\cos x}{1 + x^4}\, dx$. Be sure to indicate how you would estimate the error in your result.

Answers to Concepts Review: **1.** 1, 2, 2, ... , 2, 1
2. 1, 4, 2, 4, 2, ... , 4, 1   **3.** n^4 **4.** large

11.3
Solving Equations Numerically

In mathematics and science, we often need to find the roots (solutions) of an equation $f(x) = 0$. To be sure, if $f(x)$ is a linear or quadratic polynomial, formulas for writing exact solutions exist and are well known. But for other algebraic equations, and certainly for transcendental equations, formulas for exact solutions are rarely available. What can be done in such cases?

There is a general method of solving problems known to all resourceful people. Given a cup of tea, we add sugar a bit at a time until it tastes just right. Given a stopper too large for a hole, we whittle it down until it fits. We change the solution a bit at a time, improving the accuracy, until we are satisfied. Mathematicians call this the *method of successive approximations*, or the *method of iterations*.

In this section, we present two such methods for solving equations: the Bisection Method and Newton's Method. Both are designed to find the real roots of $f(x) = 0$. Both require many computations. You will want to keep your calculator handy.

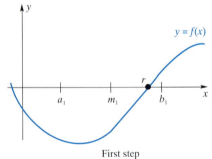

Figure 1

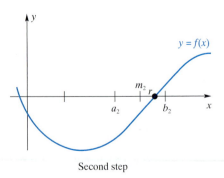

Figure 2

The Bisection Method   In Example 7 of Section 2.9 we saw how to use the Intermediate Value Theorem to approximate a solution of $f(x) = 0$ by successively bisecting an interval known to contain a solution. This Bisection Method has two great virtues—simplicity and reliability. It also has a major vice—the large number of steps needed to achieve desired accuracy (otherwise known as slowness of convergence).

Begin the process by sketching the graph of f, which is assumed to be a continuous function (see Figure 1). A real root r of $f(x) = 0$ is a point (technically, the x-coordinate of a point) where the graph crosses the x-axis. As a first step in pinning down this point, locate two points, $a_1 < b_1$, at which you are sure that f has opposite signs; if f has opposite signs at a_1 and b_1, then the product $f(a_1)\cdot f(b_1)$ will be negative. (Try choosing a_1 and b_1 on opposite sides of your best guess at r.) The Intermediate Value Theorem guarantees the existence of a root between a_1 and b_1. Now evaluate f at the midpoint $m_1 = (a_1 + b_1)/2$ of $[a_1, b_1]$. The number m_1 is our first approximation to r.

Either $f(m_1) = 0$, in which case we are done, or $f(m_1)$ differs in sign from $f(a_1)$ or $f(b_1)$. Denote the one of the subintervals $[a_1, m_1]$, or $[m_1, b_1]$ on which the sign change occurs by the symbol $[a_2, b_2]$, and evaluate f at its midpoint $m_2 = (a_2 + b_2)/2$ (Figure 2). The number m_2 is our second approximation to r.

Repeat the process, thus determining a sequence of approximations $m_1, m_2, m_3, ...$ and subintervals $[a_1, b_1], [a_2, b_2], [a_3, b_3], ...$, each subinterval containing the root r and each half the length of its predecessor. Stop when r is determined to desired accuracy, that is, when $(b_n - a_n)/2$ is less than the allowable error, which we will denote by E.

Algorithm   Bisection Method

Let $f(x)$ be a continuous function, and let a_1 and b_1 be numbers satisfying $a_1 < b_1$ and $f(a_1) \cdot f(b_1) < 0$. Let E denote the desired bound for the error $|r - m_n|$. Repeat steps 1 to 5 for $n = 1, 2, \ldots$ until $h_n < E$:

1. Calculate $m_n = (a_n + b_n)/2$.
2. Calculate $f(m_n)$, and if $f(m_n) = 0$, then STOP.
3. Calculate $h_n = (b_n - a_n)/2$.
4. If $f(a_n) \cdot f(m_n) < 0$, then set $a_{n+1} = a_n$ and $b_{n+1} = m_n$.
5. If $f(a_n) \cdot f(m_n) > 0$, then set $a_{n+1} = m_n$ and $b_{n+1} = b_n$.

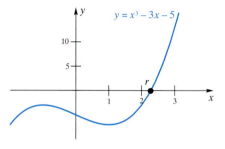

Figure 3

EXAMPLE 1   Determine the real root of $f(x) = x^3 - 3x - 5$ to accuracy within 0.0000001.

Solution   We first sketch the graph of $y = x^3 - 3x - 5$ (Figure 3) and, noting that it crosses the x-axis between 2 and 3, we begin with $a_1 = 2$ and $b_1 = 3$.

Step 1: $m_1 = (a_1 + b_1)/2 = (2 + 3)/2 = 2.5$

Step 2: $f(m_1) = f(2.5) = 2.5^3 - 3 \cdot 2.5 - 5 = 3.125$

Step 3: $h_1 = (b_1 - a_1)/2 = (3 - 2)/2 = 0.5$

Step 4:   Since

$$f(a_1) \cdot f(m_1) = f(2)f(2.5) = (-3)(3.125) = -9.375 < 0$$

we set $a_2 = a_1 = 2$ and $b_2 = m_1 = 2.5$.

Step 5:   The condition $f(a_n) \cdot f(m_n) > 0$ is false.

Next we increment n so that it has the value 2 and repeat these steps. We can continue this process to obtain the entries in the following table:

n	h_n	m_n	$f(m_n)$
1	0.5	2.5	3.125
2	0.25	2.25	−0.359
3	0.125	2.375	1.271
4	0.0625	2.3125	0.429
5	0.03125	2.28125	0.02811
6	0.015625	2.265625	−0.16729
7	0.0078125	2.2734375	−0.07001
8	0.0039063	2.2773438	−0.02106
9	0.0019532	2.2792969	0.00350
10	0.0009766	2.2783203	−0.00878
11	0.0004883	2.2788086	−0.00264
12	0.0002442	2.2790528	0.00043
13	0.0001221	2.2789307	−0.00111
14	0.0000611	2.2789918	−0.00034
15	0.0000306	2.2790224	0.00005
16	0.0000153	2.2790071	−0.00015
17	0.0000077	2.2790148	−0.00005
18	0.0000039	2.2790187	−0.000001
19	0.0000020	2.2790207	0.000024
20	0.0000010	2.2790197	0.000011
21	0.0000005	2.2790192	0.000005
22	0.0000003	2.2790189	0.0000014
23	0.0000002	2.2790187	−0.0000011
24	0.0000001	2.2790188	0.0000001

We conclude that $r = 2.2790188$ with an error of at most 0.0000001.   ∎

Example 1 illustrates the shortcoming of the Bisection Method. The approximations $m_1, m_2, m_3, \ldots$ converge very slowly to the root r. But they do converge; that is, $\lim_{n \to \infty} m_n = r$. The method works, and we have at step n a good bound for the error $E_n = r - m_n$, namely, $|E_n| \le h_n$.

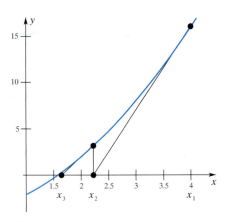

Figure 4

Newton's Method We are still considering the problem of solving the equation $f(x) = 0$ for a root r. Suppose that f is differentiable, so the graph of $y = f(x)$ has a tangent line at each point. If we can find a first approximation x_1 to r by graphing or any other means (see Figure 4), then a better approximation x_2 ought to lie at the intersection of the tangent at $(x_1, f(x_1))$ with the x-axis. Using x_2 as an approximation, we can then find a still better approximation x_3, and so on.

The process can be mechanized so that it is easy to do on a calculator. The equation of the tangent line at $(x_1, f(x_1))$ is

$$y - f(x_1) = f'(x_1)(x - x_1)$$

and its x-intercept x_2 is found by setting $y = 0$ and solving for x. The result is

$$x_2 = x_1 - \frac{f(x_1)}{f'(x_1)}$$

More generally, we have the following algorithm, also called a *recursion formula* or an *iteration scheme*.

Algorithm Newton's Method

Let $f(x)$ be a differentiable function and let x_1 be an initial approximation to the root r of $f(x) = 0$. Let E denote a bound for the error $|r - m_n|$.
Repeat the following step for $n = 1, 2, \ldots$ until $|x_{n+1} - x_n| < E$:

1. $x_{n+1} = x_n - \dfrac{f(x_n)}{f'(x_n)}$

Algorithms

Algorithms have been part of mathematics since people first learned to do long division, but it is computer science that has given algorithmic thinking its present popularity. What is an algorithm? Donald Knuth, dean of computer scientists, responds,

"An algorithm is a precisely defined sequence of rules telling how to produce specified output information from given input information in a finite number of steps."

And what is computer science? According to Knuth,

"It is the study of algorithms."

EXAMPLE 2 Use Newton's Method to find the real root r of $f(x) = x^3 - 3x - 5 = 0$ to seven decimal places.

Solution This is the same equation considered in Example 1. Let's use $x_1 = 2.5$ as our first approximation to r, as we did there. Since $f(x) = x^3 - 3x - 5$ and $f'(x) = 3x^2 - 3$, the algorithm is

$$x_{n+1} = x_n - \frac{x_n^3 - 3x_n - 5}{3x_n^2 - 3} = \frac{2x_n^3 + 5}{3x_n^2 - 3}$$

We obtain the following table.

n	x_n
1	2.5
2	2.30
3	2.2793
4	2.2790188
5	2.2790188

After just four steps, we get a repetition of the first eight digits. We feel confident in reporting that $r \approx 2.2790188$, with perhaps some question about the last digit. ∎

EXAMPLE 3 Use Newton's Method to find the real root r of $f(x) = x - e^{-x} = 0$ to seven decimal places.

Solution The graph of $y = x - e^{-x}$ is sketched in Figure 5. We use $x_1 = 0.5$ and $x_{n+1} = x_n - (x_n - e^{-x_n})/(1 + e^{-x_n}) = (x_n + 1)/(e^{x_n} + 1)$ to obtain the following table:

n	x_n
1	0.5
2	0.566
3	0.56714
4	0.5671433
5	0.5671433

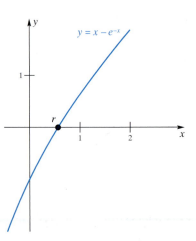

Figure 5

After just four steps, we get a repetition of the seven digits after the decimal point. We conclude that $r \approx 0.5671433$. ◼

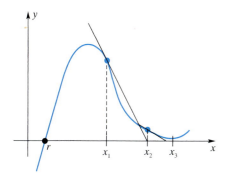

Figure 6

Convergence of Newton's Method It is not always true that Newton's Method yields approximations that converge to the root r, as the diagram in Figure 6 shows (see also Problem 21). In this case, the difficulty is that x_1 is not close enough to r to get a convergent process started. Another obvious difficulty arises if $f'(x)$ is zero at or near r, since $f'(x_n)$ occurs in the denominator of the algorithm. However, we have the following theorem:

> **Theorem A**
>
> Let f be a function that is twice differentiable on an interval I, having as its midpoint a root r of $f(x) = 0$. Suppose that there are positive numbers m and M such that $|f'(x)| \geq m$ and $|f''(x)| \leq M$ on I. If x_1 is in I and sufficiently close to r $(|x_1 - r| < 2m/M$ will do), then
>
> (i) $|x_{n+1} - r| \leq \dfrac{M}{2m}(x_n - r)^2$;
>
> (ii) $\displaystyle\lim_{n \to \infty} x_n = r$

Proof From Taylor's Formula with Remainder (Theorem 10.8B), there is a number c between x_n and r such that

$$f(r) = f(x_n) + f'(x_n)(r - x_n) + \frac{f''(c)}{2}(r - x_n)^2$$

After dividing both sides by $f'(x_n)$ and using the fact that $f(r) = 0$, we obtain

$$0 = \frac{f(x_n)}{f'(x_n)} + r - x_n + \frac{f''(c)}{2f'(x_n)}(r - x_n)^2$$

and then, successively,

$$x_n - \frac{f(x_n)}{f'(x_n)} - r = \frac{f''(c)}{2f'(x_n)}(r - x_n)^2$$

$$|x_{n+1} - r| = \left| \frac{f''(c)}{2f'(x_n)} \right| (x_n - r)^2$$

$$|x_{n+1} - r| \leq \frac{M}{2m}(x_n - r)^2$$

which is (i).

From (i), one may show by induction (Problem 18) that

$$|x_n - r| \leq \frac{2m}{M} \left(\frac{M}{2m}|x_1 - r| \right)^{2^{n-1}}$$

Since $(M/2m)|x_1 - r| < 1$, the right side of the last inequality approaches 0 as $n \to \infty$. This implies that $|x_n - r|$ also tends to 0 as $n \to \infty$, which is equivalent to (ii). ◆

The speed of convergence of Newton's Method is truly remarkable, tending in fact to double the number of decimal places of accuracy at each step. To see why this is so, suppose that $M/2m \leq 2$. Then, if the error $|x_n - r|$ at the nth step is less than 0.005, the error $|x_{n+1} - r|$ at the next step satisfies (by (i))

$$|x_{n+1} - r| \leq \frac{M}{2m}|x_n - r|^2 \leq 2(0.005)^2 = 0.00005$$

Thus, the accuracy of x_n to two decimal places is doubled to an accuracy of x_{n+1} to four decimal places. Of course, we should not expect quite such spectacular results if $M/2m$ is substantially greater than 2.

Concepts Review

1. The virtues of the Bisection Method are its simplicity and reliability; its vice is its _____.

2. If f is continuous on $[a, b]$, and $f(a)$ and $f(b)$ have opposite signs, then there is a _____ of $f(x) = 0$ between a and b. This follows from the _____ Theorem.

3. Both the Bisection Method and Newton's Method are examples of _____; that is, they provide a finite sequence of steps that, if followed, will produce a root of an equation to desired accuracy.

4. Newton's Method can fail to yield a root of $f(x) = 0$. This can happen if _____ is too far from the root r or if _____.

Problem Set 11.3

[C] *In Problems 1–4, use the Bisection Method to approximate the real root of the given equation on the given interval. Each answer should be accurate to two decimal places.*

1. $x^3 + 2x - 6 = 0$; $[1, 2]$ **2.** $x^4 + 5x^3 + 1 = 0$; $[-1, 0]$

3. $2\cos x - e^{-x} = 0$; $[1, 2]$ **4.** $x - 2 + 2\ln x = 0$; $[1, 2]$

[C] *In Problems 5–14, use Newton's Method to approximate the indicated root of the given equation accurate to five decimal places. Begin by sketching a graph.*

5. The largest root of $x^3 + 6x^2 + 9x + 1 = 0$

6. The real root of $7x^3 + x - 5 = 0$

7. The root of $x - 2 + 2\ln x = 0$ (see Problem 4)

8. The smallest positive root of $2\cos x - e^{-x} = 1$ (see Problem 3)

9. The root of $\cos x = 2x$ **10.** The root of $x \ln x = 2$

11. All real roots of $x^4 - 8x^3 + 22x^2 - 24x + 8 = 0$

12. All real roots of $x^4 + 6x^3 + 2x^2 + 24x - 8 = 0$

13. The positive root of $2x^2 - \sin^{-1} x = 0$

14. The positive root of $2\tan^{-1} x = x$

[C] **15.** Use Newton's Method to calculate $\sqrt[3]{6}$ to five decimal places. *Hint:* Solve $x^3 - 6 = 0$.

[C] **16.** Use Newton's Method to calculate $\sqrt[4]{47}$ to five decimal places.

[C] **17.** Where on $(\pi, 2\pi)$ does $(\sin x)/x$ attain a minimum and what is its minimum value?

18. Show by induction that if

$$|x_{n+1} - r| \le \frac{M}{2m}(x_n - r)^2, \qquad n = 1, 2, \ldots$$

then

$$|x_n - r| \le \frac{2m}{M}\left(\frac{M}{2m}|x_1 - r|\right)^{2^{n-1}}, \qquad n = 1, 2, \ldots$$

[C] **19.** Suppose we use Newton's Method to find the positive root of $x^2 - 2 = 0$, that is, to approximate $\sqrt{2}$. Suppose further that we know this root is on the interval $[1, 2]$. Calculate m and M of Theorem A. Use the second inequality of Problem 18 to estimate $|x_6 - \sqrt{2}|$, given that $x_1 = 1.5$.

[C] **20.** How large should we take n in Problem 19 to make sure that $|x_n - \sqrt{2}| \le 5 \times 10^{-41}$?

[C] **21.** Consider finding the real root of $(1 + \ln x)/x = 0$ by Newton's Method. Show that this leads to the algorithm

$$x_{n+1} = 2x_n + \frac{x_n}{\ln x_n}$$

Apply this algorithm with $x_1 = 1.2$. Next try it with $x_1 = 0.5$. Finally, graph $y = (1 + \ln x)/x$ to understand your results.

22. Sketch the graph of $y = x^{1/3}$. Obviously, its only x-intercept is zero. Convince yourself that Newton's Method fails to converge. Explain this failure.

23. In installment buying, one would like to figure out the real interest rate (effective rate), but unfortunately this involves solving a complicated equation. If one buys an item worth $\$P$ today and agrees to pay for it with payments of $\$R$ at the end of each month for k months, then

$$P = \frac{R}{i}\left[1 - \frac{1}{(1+i)^k}\right]$$

where i is the interest rate per month. Tom bought a used car for $\$2000$ and agreed to pay for it with $\$100$ payments at the end of each of the next 24 months.

(a) Show that i satisfies the equation

$$20i(1 + i)^{24} - (1 + i)^{24} + 1 = 0$$

(b) Show that Newton's Method for this equation reduces to

$$i_{n+1} = i_n - \left[\frac{20i_n^2 + 19i_n - 1 + (1 + i_n)^{-23}}{500i_n - 4}\right]$$

[C] (c) Find i accurate to five decimal places starting with $i = 0.012$, and then give the annual rate r as a percent ($r = 1200i$).

24. In applying Newton's Method to solve $f(x) = 0$, one can usually tell by simply looking at the numbers $x_1, x_2, x_3, \ldots$ whether the sequence is converging. But even if it converges, say to $\bar{x}$, can we be sure that $\bar{x}$ is a solution? Show that the answer is yes provided f and f' are continuous at $\bar{x}$ and $f'(\bar{x}) \ne 0$.

25. Experiment with the algorithm

$$x_{n+1} = 2x_n - ax_n^2$$

using several different values of a.
(a) Make a conjecture about what this algorithm computes.
(b) Prove your conjecture.

[CAS] *Some computer packages implement Newton's Method. Experiment with your software on some of the earlier problems in this Problem Set. Then find all real roots of the following equations.*

26. $x^6 - 4 = 0$ **27.** $x^3 - 3x + 1 = 0$

28. $x^2 - 2x = \cos 3x$ **29.** $x^3 - 3x + 1 = 2\sin 4x$

30. $\sqrt{|x - 1|} = 0$

Answers to Concepts Review: **1.** slowness of convergence **2.** root; Intermediate Value **3.** algorithms **4.** x_1; $f'(r) = 0$

11.4
The Fixed-Point Algorithm

We offer next a method of solving equations that is so simple that it has no right to work. Yet it does work in a large number of cases. Moreover, this method has numerous applications in advanced mathematics.

Suppose that an equation that interests us can be written in the form $x = g(x)$. To solve this equation is to find a number r that is unchanged by the function g. We call such a number a **fixed point** of g. To find this number, we propose the following algorithm. Make a first guess x_1. Then let $x_2 = g(x_1)$, $x_3 = g(x_2)$, and so on. If we are lucky, x_n will converge to the root r as $n \to \infty$.

> **Algorithm**   Fixed-point Algorithm
>
> Let $g(x)$ be a continuous function, and let x_1 be an initial approximation to the root r of $x = g(x)$. Let E denote a bound for the error $|r - m_n|$.
>   Repeat the following step for $n = 1, 2, \ldots$ until $|x_{n+1} - x_n| < E$:
>
> 1. $x_{n+1} = g(x_n)$

The Method Illustrated   We begin with an example treated in the previous section (Example 3).

EXAMPLE 1   Solve $x - e^{-x} = 0$ using the Fixed-point Algorithm.

Solution   We write the equation as $x = e^{-x}$ and apply the algorithm $x_{n+1} = e^{-x_n}$ with $x_1 = 0.5$. The results are shown in the accompanying table.

n	x_n	n	x_n	n	x_n
1	0.5	10	0.5675596	19	0.5671408
2	0.6065307	11	0.5669072	20	0.5671447
3	0.5452392	12	0.5672772	21	0.5671425
4	0.5797031	13	0.5670674	22	0.5671438
5	0.5600646	14	0.5671864	23	0.5671430
6	0.5711722	15	0.5671189	24	0.5671435
7	0.5648630	16	0.5671572	25	0.5671432
8	0.5684381	17	0.5671354	26	0.5671433
9	0.5664095	18	0.5671478	27	0.5671433

Although it took 27 steps to get a repetition of the first seven digits, the process did produce a sequence that converges and to the right value. Moreover, the process was very easy to carry out.   ∎

EXAMPLE 2   Solve $x = 2 \cos x$.

Solution   Note first that solving this equation is equivalent to solving the pair of equations $y = x$ and $y = 2 \cos x$. Thus, to get our initial seed value, we graph these two equations (Figure 1) and observe that the two curves cross at approximately $x = 1$. Taking $x_1 = 1$ and applying the algorithm $x_{n+1} = 2 \cos x_n$, we obtain the results in the following table.

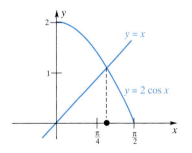

Figure 1

n	x_n	n	x_n
1	1	6	1.4394614
2	1.0806046	7	0.2619155
3	0.9415902	8	1.9317916
4	1.1770062	9	−0.7064109
5	0.7673820	10	1.5213931

Quite clearly the process is unstable, even though our initial guess is very close to the actual root.

Let's take a different tack. Rewrite the equation $x = 2 \cos x$ as $x = (x + 2 \cos x)/2$ and use the algorithm

$$x_{n+1} = \frac{x_n + 2 \cos x_n}{2}$$

This process produces a convergent sequence, shown in the following table. (The oscillation in the last digit is probably due to round-off errors.)

n	x_n	n	x_n	n	x_n
1	1	7	1.0298054	13	1.0298665
2	1.0403023	8	1.0298883	14	1.0298666
3	1.0261107	9	1.0298588	15	1.0298665
4	1.0312046	10	1.0298693	16	1.0298666
5	1.0293881	11	1.0298655		
6	1.0300374	12	1.0298668		

Now we raise an obvious question. Why did the second algorithm yield a convergent sequence, whereas the first one failed to do so? Also, can we be sure that we have obtained a correct result in the second case? To this latter question we can offer an affirmative answer. The nature of the calculation suggests that when we get a repetition of the first seven digits we have a solution to at least six-digit accuracy. We will answer the first question after we have considered another example.

EXAMPLE 3   Solve $x^3 + 6x - 3 = 0$ using the Fixed-point Algorithm.

Solution   The given equation is equivalent to $x = (-x^3 + 3)/6$, so we use the algorithm

$$x_{n+1} = \frac{-x_n^3 + 3}{6}$$

The graph in Figure 2 suggests an initial value of $x_1 = 0.5$, but we also consider what happens with $x_1 = 1.5$, $x_1 = 2.2$, and $x_1 = 2.7$. The results are shown in the table.

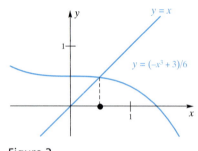

Figure 2

n	x_n	n	x_n	n	x_n	n	x_n
1	0.5	1	1.5	1	2.2	1	2.7
2	0.4791667	2	−0.0625	2	−1.2744667	2	−2.7805
3	0.4816638	3	0.5000407	3	0.8451745	3	4.0827578
4	0.4813757	4	0.4791616	4	0.3993792	4	−10.842521
5	0.4814091	5	0.4816644	5	0.4893829	5	212.9416
6	0.4814052	6	0.4813756	6	0.4804658	6	−16909274.5
7	0.4814057	7	0.4814091	7	0.4815143		
8	0.4814056	8	0.4814052	8	0.4813930		
9	0.4814056	9	0.4814057	9	0.4814071		
		10	0.4814056	10	0.4814054		
		11	0.4814056	11	0.4814056		
				12	0.4814056		

It appears that if our initial guess x_1 is close enough to the fixed point r the sequence will converge, but if we begin too far away from r, it will diverge.

Convergence of the Method
Sometimes the method works; sometimes it fails. We can get a pretty good idea about which is which by executing an appropriate number of steps of the algorithm. But wouldn't it be nice to have a way of telling in advance whether there will be convergence? And to be sure of our conclusion?

To get a feeling for the problem, let's look at it geometrically. Note that we can get x_2 from x_1 by locating $g(x_1)$, sending it horizontally to the line $y = x$, and projecting down to the x-axis (see Figure 3). When we use this process repeatedly, we are faced with one of the situations in Figure 4.

What determines success or failure, convergence or divergence? It appears to depend on the slope of the curve $y = g(x)$, that is, $g'(x)$, near the root r. If $|g'(x)|$ is too large, the method fails; if $|g'(x)|$ is small enough, the method works. Here is a general result.

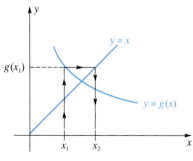

Figure 3

Theorem A   Fixed-point Theorem

Let g be a continuous function taking $[a, b]$ into itself, that is, a function that satisfies $a \le g(x) \le b$ whenever $a \le x \le b$. Then g has at least one fixed point r

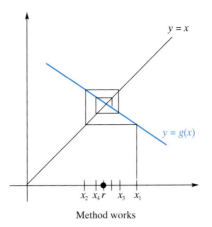

Method works

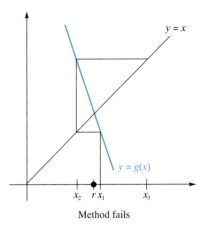

Method fails

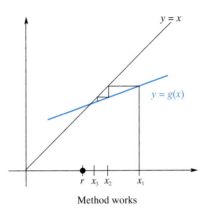

Method works

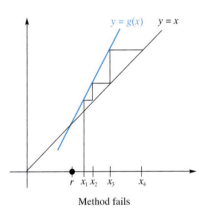

Method fails

Figure 4

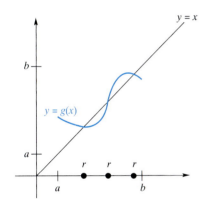

Figure 5

Making It Work

Can we make the Fixed-point algorithm work on every equation? No. First, it may be impossible to write the equation in the form $x = g(x)$. Second, even if the equation can be written in the right form, the Fixed-point algorithm may produce a divergent sequence. However, equations can be written in many equivalent ways, as Example 4 illustrates. Our goal then is to find one for which $|g'(x)| < 1$. Moreover, we would prefer to have this derivative as small as possible, since the smaller it is, the faster is the convergence. In summary, making the Fixed-point algorithm work requires more ingenuity than a first impression suggests.

on $[a, b]$. If, in addition, g is differentiable and satisfies $|g'(x)| \leq M < 1$ for all x in $[a, b]$, M a constant, then the fixed point is unique and the algorithm

$$x_{n+1} = g(x_n), \qquad x_1 \text{ in } [a, b]$$

yields a sequence that converges to r as $n \to \infty$.

Proof A typical graph of a continuous function mapping $[a, b]$ into $[a, b]$ is shown in Figure 5. If either $g(a) = a$ or $g(b) = b$, we have our fixed point, so suppose that neither is true. Let $h(x) = g(x) - x$ and note that $h(a) > 0$ and $h(b) < 0$. By the Intermediate Value Theorem, there is a point r (possibly several points) such that $h(r) = 0$, that is, $r = g(r)$. The first assertion of our theorem is proved.

Next suppose that $|g'(x)| \leq M < 1$ for all x in $[a, b]$ and let r be a fixed point of g. By the Mean Value Theorem for Derivatives, we may write

$$g(x) - g(r) = g'(c)(x - r)$$

with c some point between x and r. Thus,

$$|g(x) - g(r)| = |g'(c)||x - r| \leq M|x - r|$$

Applying this inequality successively to $x_1, x_2, \ldots$ yields

$$|x_2 - r| = |g(x_1) - g(r)| \leq M|x_1 - r|$$
$$|x_3 - r| = |g(x_2) - g(r)| \leq M|x_2 - r| \leq M^2|x_1 - r|$$
$$|x_4 - r| = |g(x_3) - g(r)| \leq M|x_3 - r| \leq M^3|x_1 - r|$$
$$\vdots$$
$$|x_n - r| = |g(x_{n-1}) - g(r)| \leq M|x_{n-1} - r| \leq M^{n-1}|x_1 - r|$$

Since $M^{n-1} \to 0$ as $n \to \infty$, we conclude that $x_n \to r$ as $n \to \infty$.

Finally, if r and s are two fixed points of g, we have just shown that $x_n \to r$ and $x_n \to s$ as $n \to \infty$. This is impossible unless $r = s$. Thus, there is only one fixed point. ◆

Now we can understand the behavior in Example 2. If $g(x) = 2 \cos x$, then $|g'(x)| = |-2 \sin x|$, which is greater than 1 in a neighborhood of the fixed point $x \approx 1.03$. On the other hand, if $g(x) = (x + 2 \cos x)/2$, then $|g'(x)| = |\frac{1}{2} - \sin x| < 1$ near $x = 1.03$. We should not expect convergence in the first case; we can guarantee it in the second.

In Example 3, $g(x) = (-x^3 + 3)/6$ and $g'(x) = -x^2/2$. Clearly, $|g'(x)| \le 1$ near the fixed point $x \approx 0.48$. We can be confident of convergence as long as we choose x_1 on the interval where $|g'(x)| \le 1$. Actually, our experiments in Example 3 show that we can start as far away as $x_1 = 2.2$ (which is more than we should expect), but $x_1 = 2.7$ is too far away.

One final remark: The closer $|g'(x)|$ is to zero near the root, the faster will be the convergence of the fixed-point algorithm.

EXAMPLE 4 The equation $x^3 - 3x + 1 = 0$ has three real roots (see Figure 6). Use the Fixed-point Algorithm to find the root between 1 and 2.

Solution Proceeding as in Example 3, we write

$$x = \frac{x^3 + 1}{3} = g(x)$$

but unfortunately $g'(x) = x^2 \ge 1$ on $[1, 2]$. Another way to write the equation is

$$x = -\frac{1}{x^2 - 3} = g(x)$$

Now $g'(x) = 2x/(x^2 - 3)^2$. This too can be greater than 1 on the interval $[1, 2]$ (e.g., $g'(1.5) \approx 5.33$). But there are still other possibilities. Consider

$$x = \frac{3}{x} - \frac{1}{x^2} = g(x)$$

for which $g'(x) = (-3x + 2)/x^3$. After a bit of work (we study g''), we see that g' is increasing on $[1, 2]$, varying from $g'(1) = -1$ to $g'(2) = -\frac{1}{2}$. Thus, $|g'(x)|$ is strictly less than 1 as long as we stay strictly away from the left end point $x = 1$.

With the algorithm

$$x_{n+1} = \frac{3}{x_n} - \frac{1}{x_n^2}$$

we obtain the data in the table in the margin. ∎

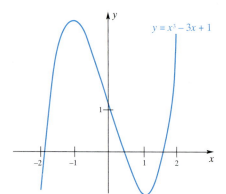

Figure 6

n	x_n
1	1.5
2	1.5555556
3	1.5153061
⋮	⋮
21	1.5320411
22	1.5321234
⋮	⋮
33	1.5320871
34	1.5320902
⋮	⋮
43	1.5320888
44	1.5320889
45	1.5320889

Concepts Review

1. A point x satisfying $g(x) = x$ is called a _____ of g.

2. The Fixed-point Algorithm for g is _____ $= g(x_n)$.

3. The critical condition on g needed to make the Fixed-point Algorithm converge is that _____ in an interval containing the fixed point.

4. The equation $x = g(x) = x^2 - 2$ has 2 as a root. Yet the algorithm $x_{n+1} = x_n^2 - 2$ may not converge to this root because _____ .

Problem Set 11.4

$\boxed{\text{C}}$ *In Problems 1–4, use the Fixed-point Algorithm with x_1 as indicated to solve the equations to five decimal places.*

1. $x = \dfrac{1}{9} e^{-2x}; x_1 = 1$

2. $x = 2 \tan^{-1} x; x_1 = 2$

3. $x = \sqrt{2.7 + x}; x_1 = 1$

4. $x = \sqrt{3.2 + x}; x_1 = 47$

$\boxed{\text{GC}}$ **5.** Consider the equation $x = 2(x - x^2) = g(x)$.
(a) Sketch the graph of $y = x$ and $y = g(x)$ using the same coordinate system, and thereby approximately locate the positive root of $x = g(x)$.
(b) Try solving the equation by the Fixed-point Algorithm starting with $x_1 = 0.7$.
(c) Solve the equation algebraically.
(d) Find $g'(x)$ and evaluate it at the root.

$\boxed{\text{GC}}$ **6.** Follow the directions of Problem 5 for $x = 5(x - x^2) = g(x)$. Explain your results.

$\boxed{\text{GC}}$ **7.** Consider the equation $x = (3/2) \sin \pi x = g(x)$.
(a) Sketch the graphs of $y = x$ and $y = g(x)$.
(b) Try solving the equation by the Fixed-point Algorithm.
(c) Find $g'(x)$ and use it to explain your results.

8. Follow the directions for Problem 7 for $x = \frac{1}{2} \sin \pi x = g(x)$

9. Consider $x = (3/2) \sin \pi x$ of Problem 7 again.
(a) Show that it can be written as $x = \frac{5}{6} x + \frac{1}{4} \sin \pi x = g(x)$.
(b) Now solve the latter equation by the Fixed-point Algorithm.
(c) Why is the convergence so rapid? *Hint*: Evaluate $g'(x)$ at the root.

10. Consider $x = 5(x - x^2) = g(x)$ of Problem 6.
(a) Rewrite this equation so that the Fixed-point Algorithm will converge (see Problem 9).
(b) Use this algorithm to solve the equation.

$\boxed{\text{C}}$ **11.** Find the positive root of $x^3 - x^2 - x - 1 = 0$. *Hint*: See Example 4.

$\boxed{\text{C}}$ **12.** Consider $x = \sqrt{5 + x}$.
(a) Apply the Fixed-point Algorithm starting with $x_1 = 0$ to find x_2, x_3, x_4, and x_5.
(b) Algebraically solve for x in $x = \sqrt{5 + x}$.
(c) Evaluate $\sqrt{5 + \sqrt{5 + \sqrt{5 + \cdots}}}$.

$\boxed{\text{C}}$ **13.** Consider $x = \sqrt{1 + x}$.
(a) Apply the Fixed-point Algorithm starting with $x_1 = 0$ to find x_2, x_3, x_4, and x_5.
(b) Algebraically solve for x in $x = \sqrt{1 + x}$.
(c) Evaluate $\sqrt{1 + \sqrt{1 + \sqrt{1 + \cdots}}}$.

$\boxed{\text{C}}$ **14.** Consider $x = 1 + \dfrac{1}{x}$.
(a) Apply the Fixed-point Algorithm starting with $x_1 = 1$ to find x_2, x_3, x_4, and x_5.

(b) Algebraically solve for x in $x = 1 + \dfrac{1}{x}$.

(c) Evaluate the following expression. (An expression like this is called a **continued fraction**.)

$$1 + \cfrac{1}{1 + \cfrac{1}{1 + \cfrac{1}{1 + \cdots}}}$$

$\boxed{\text{C}}$ **15.** Note that $\sqrt{a}$ is a solution of $x = \frac{1}{2}(x + a/x) = g(x)$. Use the Fixed-point Algorithm to find $\sqrt{\pi}$. Calculate $g'(a)$ to see why the convergence is so rapid.

$\boxed{\text{C}}$ **16.** Kepler's equation $x = m + E \sin x$ is important in astronomy. Use the Fixed-point Algorithm to solve this equation when $m = 0.8$ and $E = 0.2$.

$\boxed{\text{C}}$ **17.** If an item selling today for P dollars is purchased on a credit plan with monthly payments of R dollars at the end of each of the next k months with interest at the rate of i per month, then

$$P = \frac{R}{i}\left[1 - (1 + i)^{-k}\right]$$

(a) A new car costing \$10,000 is purchased on a credit plan with payments of R dollars at the end of each of the next 48 months and stated interest of 18% (which means $i = 0.18/12 = 0.015$). Use algebra to find R.

(b) Suppose that the monthly payment in part (a) is \$300. Then what is i? *Hint*: Use the Fixed-point Algorithm with $i_1 = 0.015$.

$\boxed{\text{C}}$ **18.** A television set costing \$500 is purchased on a credit plan with payments of \$30 at the end of each of the next 24 months. What is i, the monthly interest rate? See Problem 17.

$\boxed{\text{EXPL}}$ **19.** Consider the equation $x = x - f(x)/f'(x)$ and suppose that $f'(x) \neq 0$ in an interval $[a, b]$.
(a) Show that if r is in $[a, b]$ then r is a root if and only if $f(r) = 0$.
(b) Show that Newton's Method is a special case of the Fixed-point Algorithm, in which $g'(r) = 0$.

$\boxed{\text{GC}}$ **20.** Sketch graphs to convince yourself that each of the following equations has a unique solution. Decide whether the Fixed-point Algorithm will work and, if so, use it. Otherwise, solve by Newton's Method.

(a) $\sin^{-1} x = \dfrac{1}{\sin x}$ (b) $\cos^{-1} x = \dfrac{1}{\cos x}$

(c) $\tan^{-1} x = \dfrac{1}{\tan x}$

Answers to Concepts Review: **1.** fixed point **2.** x_{n+1}
3. $|g'(x)| \leq M < 1$ **4.** $|2x| > 1$ near $x = 2$

11.5
Approximations for Differential Equations

A Function of Two Variables

The function f depends on two variables. Since $y'(x) = f(x, y)$, the slope of a solution depends on *both* the x- and y-coordinates. Functions of two or more variables were introduced in Section 2.1. We will study them further in Chapter 15.

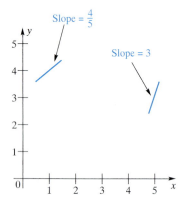

Figure 1

In Chapter 7 we studied a number of differential equations that arise from physical applications. For each equation, we were always able to find an **analytic solution**; that is, we found an explicit function that satisfies the equation. Many differential equations do not have such analytic solutions, so for these equations we must settle for approximations. In this section, we will study two ways to approximate a solution to a differential equation; one method is graphical and the other is numerical.

Slope Fields   Consider a first-order differential equation of the form

$$y' = f(x, y)$$

This equation says that at the point (x, y) the slope of a solution is given by $f(x, y)$. For example, the differential equation $y' = y$ says that the slope of the curve passing through the point (x, y) is equal to y.

For the differential equation $y' = \frac{1}{5}xy$, at the point $(5, 3)$ the slope of the solution is $y' = \frac{1}{5} \cdot 5 \cdot 3 = 3$; at the point $(1, 4)$ the slope is $y' = \frac{1}{5} \cdot 1 \cdot 4 = \frac{4}{5}$. We can indicate graphically this latter result by drawing a small line segment through the point $(1, 4)$ having slope $\frac{4}{5}$ (see Figure 1).

If we repeat this process for a number of ordered pairs (x, y), we obtain a **slope field**. Since plotting a slope field is a tedious job if done by hand, the task is best suited for computers; Mathematica and Maple are capable of plotting slope fields. Figure 2 shows a slope field for the differential equation $y' = \frac{1}{5}xy$. Given an initial condition, we can follow the slopes to get at least a rough approximation to the particular solution. We can often see from the slope field the behavior of all solutions to the differential equation.

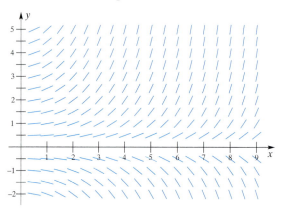

Figure 2

EXAMPLE 1   Suppose that the size y of a population satisfies differential equation $y' = 0.2y(16 - y)$. The slope field for this differential equation is shown in Figure 3.

(a) Sketch the solution that satisfies the initial condition $y(0) = 3$. Describe the behavior of solutions when

(b) $y(0) > 16$, and   (c) $0 < y(0) < 16$.

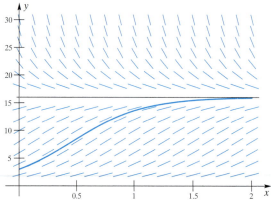

Figure 3

Solution

(a) The solution that satisfies the initial condition $y(0) = 3$ contains the point $(0, 3)$. From that point to the right, the solution follows the slope lines. The curve in Figure 3 shows a graph of the solution.

(b) If $y(0) > 16$, then the solution decreases toward the horizontal asymptote $y = 16$.

(c) If $0 < y(0) < 16$, then the solution increases toward the horizontal asymptote $y = 16$.

Parts (b) and (c) indicate that the size of the population will converge toward the value 16 for any initial population size. ∎

Euler's Method   We again consider differential equations of the form $y' = f(x, y)$ with initial condition $y(x_0) = y_0$. Keep in mind that y is a function of x, whether we write it explicitly or not. The initial condition $y(x_0) = y_0$ tells us that the ordered pair (x_0, y_0) is a point on the graph of the solution. We also know just a bit more about the unknown solution: the slope of the tangent line to the solution at x_0 is $f(x_0, y_0)$. This information is summarized in Figure 4.

If h is positive, but small, we would expect the tangent line (or, equivalently, the Taylor polynomial of order 1 based at x_0), which has equation

$$P_1(x) = y_0 + y'(x_0)(x - x_0) = y_0 + f(x_0, y_0)(x - x_0)$$

to be "close" to the solution $y(x)$ over the interval $[x_0, x_0 + h]$. Let $x_1 = x_0 + h$. Then, at x_1, we have

$$P_1(x_1) = y_0 + hy'(x_0) = y_0 + hf(x_0, y_0)$$

Setting $y_1 = y_0 + hf(x_0, y_0)$, we now have an approximation for the solution at x_1. See Figure 5.

Since $y' = f(x, y)$, we know that the slope of the solution when $x = x_1$ is $f(x_1, y(x_1))$. At this point, we do not know $y(x_1)$, but we do have the approximation y_1 for it. Thus, we repeat the process to obtain the estimate $y_2 = y_1 + hf(x_1, y_1)$ for the solution at the point $x_2 = x_1 + h$. This process, when continued in this fashion, is called **Euler's Method**, named after the Swiss mathematician Leonhard Euler (1707–1783). (Euler is pronounced "oiler.") The parameter h is often called the **step size**.

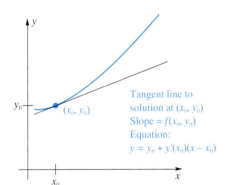

Figure 4

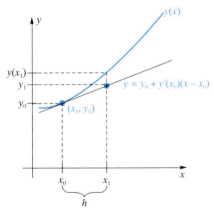

Figure 5

Algorithm   Euler's Method

To approximate the solution of the differential equation $y' = f(x, y)$ with initial condition $y(x_0) = y_0$, choose a step size h and repeat the following steps for $n = 1, 2, \dots$.

1. Set $x_n = x_{n-1} + h$.
2. Set $y_n = y_{n-1} + hf(x_{n-1}, y_{n-1})$.

Remember, the solution to a differential equation is a *function*. Euler's Method does not yield a function; rather, it gives a set of ordered pairs (x_i, y_i) that approximates the solution y. Often, this set of ordered pairs is enough to describe the solution to the differential equation.

Notice the difference between $y(x_n)$ and y_n; $y(x_n)$ (usually unknown) is the value of the exact solution at x_n, and y_n is our approximation to the exact solution at x_n. In other words, y_n is our approximation to $y(x_n)$.

EXAMPLE 2   Use Euler's Method with $h = 0.2$ to approximate the solution to

$$y' = y, \qquad y(0) = 1$$

over the interval $[0, 1]$.

12. With a computer and the Parabolic Rule, one can always approximate $\int_a^b f(x)\,dx$ to any desired degree of accuracy by taking h small enough.

13. The function $f(x) = e^{-x^2} + x^2 + \sin(x + 1)$ satisfies $|f(x)| \le 6$ on $[-1, 2]$.

14. If $f(x) = ax^2 + bx + c$, then
$$\int_{-2}^2 f(x)\,dx = \tfrac{2}{3}\big[f(-2) + 4f(0) + f(2)\big].$$

15. If f is continuous on $[a, b]$ and $f(a)f(b) < 0$, then $f(x) = 0$ has a root between a and b.

16. One of the virtues of the Bisection Method is its rapid convergence.

17. Newton's Method will produce a convergent sequence for the function $f(x) = x^{1/3}$.

18. If $f'(x) > 1$ on an open interval containing a root r of $x = f(x)$, then the Fixed-point Method will fail to produce a sequence converging to r (unless the first guess happens to be r).

19. The Fixed-point Method will work to find the largest root of $x = 5(x - x^2) + 0.01$.

20. The Fixed-point Method will produce a convergent sequence for $x = \dfrac{1}{2}\left(x + \dfrac{a}{x}\right)$ if $a > 0$ and the first guess is greater than $\sqrt{a/3}$.

21. The solution to the differential equation $y' = 2y$ that passes through the point $(2, 1)$ has slope 2 at that point.

22. Euler's Method will always overestimate the solution of the differential equation $y' = 2y$ with initial condition $y(0) = 1$.

Sample Test Problems

[C] **1.** Find the Maclaurin polynomial of order 1 for $f(x) = x \cos x^2$ and use it to approximate $f(0.2)$.

[C] **2.** Find the Maclaurin polynomial of order 4 for $f(x)$, and use it to approximate $f(0.1)$.

(a) $f(x) = xe^x$ (b) $f(x) = \cosh x$

3. Find the Taylor polynomial of order 3 based at 2 for $g(x) = x^3 - 2x^2 + 5x - 7$, and show that it is an exact representation of $g(x)$.

4. Use the result of Problem 3 to calculate $g(2.1)$.

5. Find the Taylor polynomial of order 4 based at 1 for $f(x) = 1/(x + 1)$.

6. Obtain an expression for the error term $R_4(x)$ in Problem 5, and find a bound for it if $x = 1.2$.

7. Find the Maclaurin polynomial of order 4 for $f(x) = \sin^2 x = \tfrac{1}{2}(1 - \cos 2x)$, and find a bound for the error $R_4(x)$ if $|x| \le 0.2$. *Note*: A better bound is obtained if you observe that $R_4(x) = R_5(x)$ and then bound $R_5(x)$.

[C] **8.** If $f(x) = \ln x$, then $f^{(n)}(x) = (-1)^{n-1}(n-1)!/x^n$. Thus, the Taylor polynomial of order n based at 1 for $\ln x$ is

$$\ln x = (x - 1) - \frac{1}{2}(x - 1)^2 + \frac{1}{3}(x - 1)^3 + \cdots$$

$$+ \frac{(-1)^{n-1}}{n}(x - 1)^n + R_n(x)$$

How large would n have to be for us to know that $|R_n(x)| \le 0.00005$ if $0.8 \le x \le 1.2$?

[C] **9.** Refer to Problem 8. Use the Taylor polynomial of order 5 based at 1 to approximate

$$\int_{0.8}^{1.2} \ln x\,dx$$

and give a good bound for the error that is made.

[C] **10.** Use the Trapezoidal Rule with $n = 8$ to approximate

$$\int_{0.8}^{1.2} \ln x\,dx$$

and give an error bound.

[C] **11.** Use the Parabolic Rule with $n = 8$ to approximate

$$\int_{0.8}^{1.2} \ln x\,dx$$

and give a good error bound.

[C] **12.** Compute

$$\int_{0.8}^{1.2} \ln x\,dx$$

using the Fundamental Theorem of Calculus.
Hint: $D_x[x \ln x - x] = \ln x$.

[C] **13.** Use Newton's Method to solve $3x - \cos 2x = 0$ accurate to six decimal places. Use $x_1 = 0.5$.

[C] **14.** Use the Fixed-point Method to solve $3x - \cos 2x = 0$, starting with $x_1 = 0.5$.

[C] **15.** Use Newton's Method to find the solution of $x - \tan x = 0$ in the interval $(\pi, 2\pi)$ accurate to four decimal places. *Hint*: Sketch graphs of $y = x$ and $y = \tan x$ using the same axes to get a good initial guess for x_1.

[C] **16.** Try the Fixed-point Method on the equation of Problem 15. Why doesn't it work?

[C] **17.** Use Newton's Method to find the largest solution of $e^x - \sin x = 0$. *Hint*: Begin by sketching $y = e^x$ and $y = \sin x$ to get an initial guess, x_1.

[C] **18.** Use Euler's Method with $h = 0.2$ to approximate the solution to the differential equation $y' = xy$ with initial condition $y(1) = 2$ over the interval $[1, 2]$.

[C] **19.** Use the Improved Euler Method with $h = 0.2$ to approximate the solution to the differential equation $y' = 3y$ with initial condition $y(0) = 2$ over the interval $[0, 2]$.

Maclaurin Polynomials

I. Preparation

Exercise 1 Find the Maclaurin polynomial of order 4 for each of the following:

(a) $2x^3 - 3x^2 + x$
(b) $-x^4 + 2x^3 - 3x^2 + x$
(c) $x^5 - x^4 + 2x^3 - 3x^2 + x$
(d) $x^6 + x^5 - x^4 + 2x^3 - 3x^2 + x$

(e) $\dfrac{1}{1-x}$

(f) $\tan^{-1} x$

(g) $\dfrac{1}{1-x^2}$

(h) $\tan^{-1} x + \dfrac{1}{1-x}$

Exercise 2 Based on your answers to Exercise 1 (and perhaps to other examples), what can you say about the Maclaurin polynomial of order 4 when f is itself a polynomial? What can you say about the Maclaurin polynomial of order 4 for the function $f + g$?

II. Using Technology

Exercise 3 Suppose that we wish to approximate $f(x) = \sin x$ for x in the interval $[0, 2\pi]$. Use your technology to plot the Maclaurin series of order $n = 3, 5, 7, 9$ on the same graph, along with a graph of $y = \sin x$. Also, plot the errors $\sin x - P_n(x)$ over the interval $[0, 2\pi]$.

Exercise 4 Continuing with Exercise 3, gradually increase n until you have an error that is at most 0.002. Show your graphs and explain your reasoning. Next, use the error formula from Section 11.1 to determine how large n must be to guarantee that the maximum error is at most 0.002.

Exercise 5 Consider the two functions $\sin(\tan x)$ and $\tan(\sin x)$.

(a) Obtain series approximations to these functions about $x = 0$, and determine the first power of x for which the series differ.

(b) Since these two series look very much alike, you may think that they represent essentially the same function. Plot both functions over the intervals $[0, \pi], [0, \pi/2]$, and $[0, \pi/4]$. Explain the behavior near $\pi/2$.

III. Reflection

Exercise 6 To approximate the sine function, as well as other trigonometric, logarithmic, and exponential functions, computers and calculators usually use some kind of polynomial approximation. For periodic functions like the sine or cosine, it is not reasonable to suppose that we could find a polynomial function that could approximate the function for *all* real numbers x. (Remember, the Taylor series are usually good approximations only near the point x_0.) For example, we should not expect to use the same polynomial to approximate both $\sin(0.02)$ and $\sin(14.02)$. In this exercise, you will exploit the symmetry of the sine function to find rules to approximate $\sin x$ for any x.

(a) Because the sine function is periodic with period 2π, an approximation for $\sin x$ over $[0, 2\pi]$ can be used to approximate $\sin x$ (theoretically, at least) for any x. Give a method of approximating $\sin x$ for any value of x using only the Maclaurin series over $[0, 2\pi]$. How large must n be in order to be sure that the maximum error is smaller than 0.000001?

(b) We can actually do a little better. Use the symmetry of the sine function about the point $(\pi, 0)$ to give a method of approximating $\sin x$ for any value of x using only the Maclaurin series over $[0, \pi]$. How large must n be in order to be sure that the maximum error is smaller than 0.000001?

(c) We can do better still! Use a further property about the symmetry of the sine function to give a method of approximating $\sin x$ for any value of x using only the Maclaurin series over $[0, \pi/2]$. How large must n be in order to be sure that the maximum error is smaller than 0.000001?

Numerical Integration

I. Preparation

Exercise 1 Approximate the integral

$$\int_1^2 \frac{\sin x}{x}\, dx$$

using $n = 4$ subintervals with the following methods:

(a) Left Riemann sum   (b) Right Riemann sum
(c) Midpoint Riemann sum   (d) Trapezoidal Rule
(e) Parabolic Rule

The exact value of this integral is $\mathrm{Si}(2) - \mathrm{Si}(1)$, where Si is the sine-integral function, which was described in Technology Project 5.2. $\mathrm{Si}(2) - \mathrm{Si}(1)$ is approximately 0.6593299064355120.

II. Using Technology

Exercise 2 Let L_n, R_n, and M_n denote, respectively, the left, right, and midpoint Riemann sums using n subintervals. Exercise 4 asks you to show that the Trapezoidal Rule and the Parabolic Rule can be obtained from these three Riemann sums as follows:

$$T_n = \frac{1}{2}\left[L_n + R_n\right]$$

$$P_{2n} = \frac{1}{3}\left[T_n + 2M_n\right]$$

For now, assume these to be true. Use these results to approximate

$$\int_1^2 \frac{\sin x}{x}\, dx$$

using all five methods with $n = 4, 8, 16, 32, 64, 128, 256, 512,$ and 1024. Use the approximation 0.6593299064355120 as if it were exact to compute the errors in the five methods. Create and complete the following tables.

	Approximations to $\int_1^2 \frac{\sin x}{x}\, dx$				
n	L_n	R_n	M_n	T_n	P_n
4					
8					
16					
32					
64					
128					
256					
512					
1024					

	Errors in Approximations to $\int_1^2 \frac{\sin x}{x}\, dx$				
n	L_n	R_n	M_n	T_n	P_n
4					
8					
16					
32					
64					
128					
256					
512					
1024					

Exercise 3 Consider now the definite integral

$$(2) \qquad A = \int_1^2 e^x\, dx$$

The exact value of A is $e^2 - e$, which is approximately 4.670774270471605. In each part that follows, fill in the blanks and justify your answers.

(a) When approximating A, we will obtain approximately the same accuracy by using the left Riemann sum with $n = $ ____ as we would by using the Trapezoidal Rule with $n = 20$.

(b) When approximating A, we will obtain approximately the same accuracy by using the Trapezoidal Rule with $n = $ ____ as we would by using the Parabolic Rule with $n = 25$.

(c) When approximating A, we will obtain approximately the same accuracy by using the midpoint Riemann sum with $n = $ ____ as we would by using the Trapezoidal Rule with $n = 1000$.

(d) When approximating A, we will obtain approximately the same accuracy by using the left Riemann sum with $n = $ ____ as we would by using the Parabolic Rule with $n = 40$.

III. Reflection

Exercise 4 Show that the Trapezoidal Rule and Parabolic Rule can be obtained from the three Riemann sums as follows:

$$T_n = \frac{1}{2}\left[L_n + R_n\right] \qquad P_{2n} = \frac{1}{3}\left[T_n + 2M_n\right]$$

Exercise 5 The error in most numerical integration methods is proportional to some power of h. By inspecting the second table above, determine whether each method has an error proportional to h, h^2, h^3, or some other power of h. Explain the connection with the errors stated in Theorems 11.2A and 11.2B.

Bisection, Newton's, and Fixed-Point Methods

I. Preparation

Exercise 1 Apply the Bisection Method, Newton's Method, and the Fixed-point Method to approximate the solution to $x - \cos x = 0$ to two decimal places.

II. Using Technology

Exercise 2 Use a CAS to approximate the solution to $x - \cos x = 0$.

Exercise 3 Implement the Fixed-point Algorithm on your CAS. Use your program to approximate the solution of $x = \cos x$.

Exercise 4 The Fixed-point Algorithm leads to an area of contemporary research and serves as a possible model for turbulence, one of the least understood phenomena in science. The problems in this exercise will introduce you to this exciting area. Each problem deals with the equation

$$(1) \qquad x = \lambda x(1 - x)$$

as we gradually increase λ from 2 to 5.

(a) $\lambda = 2.5$. Sketch the graphs of $y = x$ and $y = 2.5x(1 - x)$ using the same axes, and solve equation (1) by iteration. Next, solve equation (1) by simple algebra, confirming your answer.

(b) $\lambda = 3.1$. Sketch $y = x$ and $y = f(x) = 3.1x(1 - x)$ using the same axes and attempt to solve (1) by fixed-point iteration. (Notice that $|f'(x)| > 1$ at the root.) You will find that x_n bounces back and forth, but gets closer to two values r_1 and r_2, called **attractors**. Find r_1 and r_2

to six decimal places. Let $f(x) = 3.1x(1 - x)$. Superimpose the graph of

$$y = g(x) = f(f(x))$$
$$= 9.61x - 39.401x^2 + 59.582x^3 - 29.791x^4$$

on your earlier graph, and observe that r_1 and r_2 appear to be the two roots of $x = g(x)$, where $|g'(x)| < 1$.

(c) $\lambda = 3.1$, continued. Note that $f(r_1) = r_2$ and $f(r_2) = r_1$. Use this to show that $g'(r_1) = g'(r_2)$.

(d) $\lambda = 3.5$. In this case, use the iteration to find four attractors: $s_1, s_2, s_3,$ and s_4. Guess what equation they are the solution of.

(e) $\lambda = 3.56$. Use the iteration to find eight attractors.

(f) $\lambda = 3.57$. As you keep increasing λ by smaller and smaller amounts, you will double the number of attractors at each stage, until at approximately 3.57 you should get *chaos*. Beyond $\lambda = 3.57$, other strange things happen.

III. Reflection

Exercise 5 Perform experiments similar to Exercise 4 on

$$x = \lambda \sin \pi x$$

Summarize your findings in a report.

For very readable accounts of the strange phenomenon described in Exercises 4 and 5, see:

James Gleick, *Chaos: Making a New Science*. New York: Penguin Books, 1987.

Douglas R. Hofstadter, "Strange attractors: mathematical patterns delicately poised between order and chaos" *Scientific American*, vol. 245 (November 1981), 22–43.

If we take an arbitrary point $P(x, y)$ on the ellipse, then, from the condition $|PF| = e|PL|$ applied first to the left focus and directrix and then to the right ones, we get

$$|PF'| = e\left(x + \frac{a}{e}\right) = ex + a \qquad |PF| = e\left(\frac{a}{e} - x\right) = a - ex$$

and so

$$\boxed{|PF'| + |PF| = 2a}$$

Next consider the hyperbola with $P(x, y)$ on its right branch, as shown in the right part of Figure 3. Then

$$|PF'| = e\left(x + \frac{a}{e}\right) = ex + a \qquad |PF| = e\left(x - \frac{a}{e}\right) = ex - a$$

and so $|PF'| - |PF| = 2a$. If $P(x, y)$ had been on the left branch, we would have gotten $-2a$ in place of $2a$. In either case,

$$\boxed{\big||PF'| - |PF|\big| = 2a}$$

EXAMPLE 1   Find the equation of the set of points the sum of whose distances from $(\pm 3, 0)$ is equal to 10.

Solution   This is a horizontal ellipse with $a = 5$ and $c = 3$. Thus, $b = \sqrt{a^2 - c^2} = 4$, and the equation is

$$\frac{x^2}{25} + \frac{y^2}{16} = 1 \qquad \blacksquare$$

EXAMPLE 2   Find the equation of the set of points the difference of whose distances from $(0, \pm 6)$ is equal to 4.

Solution   This is a vertical hyperbola with $a = 2$ and $c = 6$. Thus, $b = \sqrt{c^2 - a^2} = \sqrt{32} = 4\sqrt{2}$, and the equation is

$$-\frac{x^2}{32} + \frac{y^2}{4} = 1 \qquad \blacksquare$$

Optical Properties   Consider two mirrors, one with the shape of an ellipse and the other with the shape of a hyperbola. If a light ray emanating from one focus strikes the mirror, it will be reflected back to the other focus in the case of the ellipse and directly away from the other focus in the case of the hyperbola. These facts are shown in Figure 4.

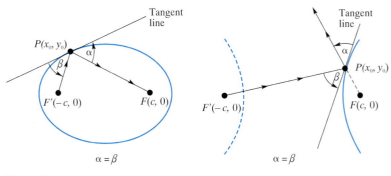

Figure 4

To demonstrate these optical properties (i.e., to show that $\alpha = \beta$ in both parts of Figure 4), we suppose the curves to be in standard position so that their equations are $x^2/a^2 + y^2/b^2 = 1$ and $x^2/a^2 - y^2/b^2 = 1$, respectively. For the ellipse, we differentiate implicitly and then substitute (x_0, y_0) thereby obtaining the slope m of the tangent line.

$$\frac{2x}{a^2} + \frac{2yy'}{b^2} = 0$$

$$y' = -\frac{b^2}{a^2}\frac{x}{y}$$

$$m = -\frac{b^2}{a^2}\frac{x_0}{y_0}$$

The equation of the tangent line may be written successively as

$$y - y_0 = -\frac{b^2 x_0}{a^2 y_0}(x - x_0)$$

$$\frac{x_0}{a^2}(x - x_0) + \frac{y_0}{b^2}(y - y_0) = 0$$

$$\frac{x_0 x}{a^2} + \frac{y_0 y}{b^2} = \frac{x_0^2}{a^2} + \frac{y_0^2}{b^2} = 1$$

A similar derivation for the hyperbola leads to similar results. We summarize in the following table:

	Ellipse	Hyperbola
Equation	$\dfrac{x^2}{a^2} + \dfrac{y^2}{b^2} = 1$	$\dfrac{x^2}{a^2} - \dfrac{y^2}{b^2} = 1$
Slope of tangent at (x_0, y_0)	$m = \dfrac{-b^2 x_0}{a^2 y_0}$	$m = \dfrac{b^2 x_0}{a^2 y_0}$
Equation of tangent at (x_0, y_0)	$\dfrac{x_0 x}{a^2} + \dfrac{y_0 y}{b^2} = 1$	$\dfrac{x_0 x}{a^2} - \dfrac{y_0 y}{b^2} = 1$

To calculate $\tan \alpha$ for the ellipse, we recall (Problem 40 of Section 2.3) a formula for the tangent of the counterclockwise angle from one line ℓ_1 to another ℓ in terms of their respective slopes m_1 and m:

$$\tan \alpha = \frac{m - m_1}{1 + mm_1}$$

Now refer to Figure 4 and let ℓ_1 be the line FP and ℓ be the tangent line at P. Then

$$\tan \alpha = \frac{\dfrac{-b^2 x_0}{a^2 y_0} - \dfrac{y_0 - 0}{x_0 - c}}{1 + \left(\dfrac{-b^2 x_0}{a^2 y_0}\right)\left(\dfrac{y_0 - 0}{x_0 - c}\right)} = \frac{-b^2 x_0(x_0 - c) - a^2 y_0^2}{a^2 y_0(x_0 - c) - b^2 x_0 y_0}$$

$$= \frac{b^2 c x_0 - (b^2 x_0^2 + a^2 y_0^2)}{(a^2 - b^2)x_0 y_0 - a^2 c y_0} = \frac{b^2 c x_0 - a^2 b^2}{c^2 x_0 y_0 - a^2 c y_0}$$

$$= \frac{b^2(c x_0 - a^2)}{c y_0(c x_0 - a^2)} = \frac{b^2}{c y_0}$$

The same calculation with c replaced by $-c$ gives

$$\tan(-\beta) = \frac{b^2}{-c y_0}$$

and so $\tan \beta = b^2/c y_0$. We conclude that $\tan \alpha = \tan \beta$, and consequently $\alpha = \beta$. A similar derivation establishes the corresponding result for the hyperbola.

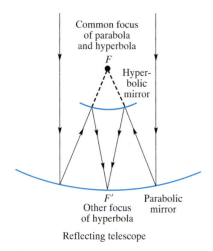

Common focus of parabola and hyperbola

F

Hyperbolic mirror

F' Other focus of hyperbola

Parabolic mirror

Reflecting telescope

Figure 5

Applications The reflecting property of the ellipse is the basis of the *whispering gallery* effect that can be observed, for example, in the U.S. Capitol, the Mormon Tabernacle, and many science museums. A speaker standing at one focus can be heard whispering by a listener at the other focus, even though his or her voice is inaudible in other parts of the room.

The optical properties of the parabola and hyperbola are combined in one design for a reflecting telescope (Figure 5). The parallel rays from a star are finally focused at the eyepiece at F'.

The string property of the hyperbola is used in navigation. A ship at sea can determine the difference $2a$ in its distance from two fixed transmitters by measuring the difference in reception times of synchronized radio signals. This puts its path on a hyperbola, with the two transmitters F and F' as foci. If another pair of transmitters G and G' are used, the ship must lie at the intersection of the two corresponding hyperbolas (see Figure 6). LORAN, a system of long-range navigation, is based on this principle.

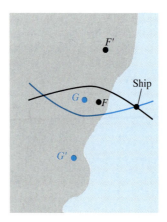

Figure 6

Concepts Review

1. An ellipse is the set of points P satisfying $|PF| + |PF'| = 2a$, where F and F' are fixed points called the _____ of the ellipse.

2. Similarly, a hyperbola is the set of points P satisfying _____.

3. A ray from a light source at one focus of an elliptical mirror will be reflected _____.

4. A ray from a light source at one focus of a hyperbolic mirror will be reflected _____.

Problem Set 12.3

In Problems 1–4, find the equation of the set of points P satisfying the given conditions.

1. The sum of the distances of P from $(0, \pm 9)$ is 26.

2. The sum of the distances of P from $(\pm 4, 0)$ is 14.

3. The difference of the distances of P from $(\pm 7, 0)$ is 12.

4. The difference of the distances of P from $(0, \pm 6)$ is 10.

In Problems 5–12, find the equation of the tangent line to the given curve at the given point.

5. $\dfrac{x^2}{27} + \dfrac{y^2}{9} = 1$ at $(3, \sqrt{6})$

6. $\dfrac{x^2}{24} + \dfrac{y^2}{16} = 1$ at $(3\sqrt{2}, -2)$

7. $\dfrac{x^2}{27} + \dfrac{y^2}{9} = 1$ at $(3, -\sqrt{6})$

8. $\dfrac{x^2}{2} - \dfrac{y^2}{4} = 1$ at $(\sqrt{3}, \sqrt{2})$

9. $x^2 + y^2 = 169$ at $(5, 12)$

10. $x^2 - y^2 = -1$ at $(\sqrt{2}, \sqrt{3})$

11. The curve of Problem 1 at $(0, 13)$

12. The curve of Problem 2 at $(7, 0)$

13. If two tangent lines to the ellipse $9x^2 + 4y^2 = 36$ intersect the y-axis at $(0, 6)$, find the points of tangency.

14. If the tangent lines to the hyperbola $9x^2 - y^2 = 36$ intersect the y-axis at $(0, 6)$, find the points of tangency.

15. The slope of the tangent to the hyperbola

$$2x^2 - 7y^2 - 35 = 0$$

at two points on the hyperbola is $-\frac{2}{3}$. What are the coordinates of the points of tangency?

16. Find the equations of the tangents to the ellipse $x^2 + 2y^2 - 2 = 0$ that are parallel to the line

$$3x - 3\sqrt{2}y - 7 = 0$$

17. Find the area of the ellipse $b^2 x^2 + a^2 y^2 = a^2 b^2$.

18. Find the volume of the solid obtained by revolving the ellipse $b^2 x^2 + a^2 y^2 = a^2 b^2$ about the y-axis.

19. The region bounded by the hyperbola

$$b^2 x^2 - a^2 y^2 = a^2 b^2$$

and a vertical line through a focus is revolved about the x-axis. Find the volume of the resulting solid.

20. If the ellipse of Problem 18 is revolved about the x-axis, find the volume of the resulting solid.

21. Find the dimensions of the rectangle having the greatest possible area that can be inscribed in the ellipse $b^2 x^2 + a^2 y^2 = a^2 b^2$. Assume that the sides of the rectangle are parallel to the axes of the ellipse.

22. Show that the point of contact of any tangent to a hyperbola is midway between the points in which the tangent intersects the asymptotes.

23. Find the point in the first quadrant where the two hyperbolas $25x^2 - 9y^2 = 225$ and $-25x^2 + 18y^2 = 450$ intersect.

24. Find the points of intersection of $x^2 + 4y^2 = 20$ and $x + 2y = 6$.

25. Sketch a design for a reflecting telescope that uses a parabola and an ellipse rather than a parabola and a hyperbola as described in the text.

26. A ball placed at a focus of an elliptical billiard table is shot with tremendous force so that it continues to bounce off the cushions indefinitely. Describe its ultimate path? *Hint*: Draw a picture.

27. If the ball of Problem 26 is initially on the major axis between a focus and the neighboring vertex, what can you say about its path?

28. Show that an ellipse and a hyperbola with the same foci intersect at right angles. *Hint*: Draw a picture and use the optical properties.

29. Describe a string apparatus for constructing a hyperbola. (There are several possibilities.)

30. Sound travels at u feet per second and a rifle bullet at $v > u$ feet per second. The sound of the firing of a rifle and the impact of the bullet hitting the target were heard simultaneously. If the rifle was at $A(-c, 0)$, the target was at $B(c, 0)$, and the listener was at $P(x, y)$, find the equation of the curve on which P lies (in terms of u, v, and c).

31. Listeners $A(-8, 0)$, $B(8, 0)$, and $C(8, 10)$ recorded the exact times at which they heard an explosion. If B and C heard the explosion at the same time and A heard it 12 seconds later, where was the explosion? Assume that distances are in kilometers and that sound travels $\frac{1}{3}$ kilometer per second.

Answers to Concepts Review: **1.** foci **2.** $\|PF\| - \|PF'\| = 2a$ **3.** to the other focus **4.** directly away from the other focus

---

12.4
Translation of Axes

So far we have placed the conics in the coordinate system in very special ways—always with the major axis along one of the coordinate axes and either the vertex (in the case of a parabola) or the center (in the case of an ellipse or hyperbola) at the origin. Now we place our conics in a more general position, though we still require that the major axis be parallel to one of the coordinate axes. Even this restriction will be removed in Section 12.5.

The case of a circle is instructive. The circle of radius 5 centered at $(2, 3)$ has equation

$$(x - 2)^2 + (y - 3)^2 = 25$$

or, in equivalent expanded form,

$$x^2 + y^2 - 4x - 6y = 12$$

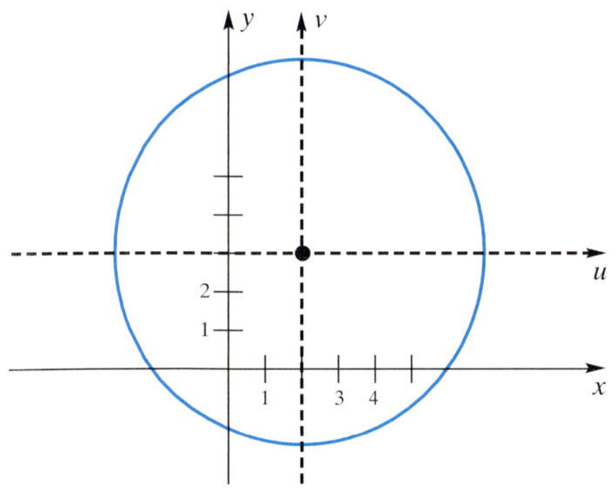

The same circle with its center at the origin of the uv-coordinate system (Figure 1) has the simple equation

$$u^2 + v^2 = 25$$

$(x - 2)^2 + (y - 3)^2 = 25$

or

$u^2 + v^2 = 25$

Figure 1

The introduction of new axes does not change the shape or size of a curve, but it may greatly simplify its equation. It is this *translation* of axes and the corresponding change of variables in an equation that we wish to investigate.

Rotations in the Plane

I. Preparation

This project deals with rotations, which are described in Section 12.5. Review that section, especially Examples 2 and 3.

I. Using Technology

Exercise 1 Consider the function $f(x, y) = x^2 + 24xy + 8y^2 - 136$. Graph the relationship $f(x, y) = 0$.

Exercise 2 Next, define $\theta = \pi/12$, and define the variables u and v as

$$x = u \cos \theta - v \sin \theta$$
$$y = u \sin \theta + v \cos \theta$$

Vary the angle θ and plot $f(u \cos \theta - v \sin \theta, u \sin \theta + v \cos \theta) = 0$. *Hint:* You will have to define θ and then make an implicit plot over some domain $a \leq u \leq b$ and $c \leq v \leq d$.

Exercise 3 Vary the angle θ again. For each value of θ, display the expression for

$$f(u \cos \theta - v \sin \theta, u \sin \theta + v \cos \theta)$$

and concentrate on the uv term. What values of θ make the coefficient of uv nearly equal to zero?

Exercise 4 Animate the rotations described above by choosing $\theta = 0$, $\pi/12, 2\pi/12, \ldots, 24\pi/12$. Describe what happens.

Exercise 5 Repeat Exercises 1–4 for $g(x, y) = x^2 - 2x + 3xy + 8y^2 - 26$.

III. Reflection

Exercise 6 For the function g defined in Exercise 5, the relationship $g(x, y) = 0$ describes an ellipse. Find the lengths of the major and minor axes, and find both foci.

Another Kind of Rose

I. Preparation

This project deals with a special type of rose in polar coordinates. These patterns arise from connecting points with lines. (See Peter M. Maurer. "A Rose Is a Rose," *American Mathematical Monthly*, Volume 94, pp. 631–645, 1987.)

Exercise 1 Play "connect the dots" with the following points. (The ordered pairs are in polar coordinates, except the angles are in *degrees*, not radians. Arguments to the sine function are also in *degrees*.)

$$(k, \sin 2k) \qquad k = 0, 1, 2, 3, 4, 5$$

Describe what happens if you continue this process through $(360, \sin(2 \cdot 360))$.

Exercise 2 Play "connect the dots" again, this time with $(19k, \sin(2 \cdot 19k)), k = 0, 1\ 2, 4, 6, 8, 10, 12$.

II. Using Technology

A **Maurer rose** consists of lines connecting the points

$$(k, \sin(n \cdot k)) \qquad k = 0, d, 2d, 3d, \ldots, 360d$$

A curve is said to be **closed** if the starting point and the ending point coincide. Figure 1 shows a Maurer rose for $n = 7$ and $d = 29$.

Exercise 3 Prove that a Maurer rose is closed.

Exercise 4 Use your CAS to construct the Maurer rose for $n = 2, d = 1$. Explain how this is related to Exercise 1.

Exercise 5 Use your CAS to construct roses for $n = 2$ and $d = 1, 2, 3, 4, 5, 6, 7, 11, 13, 19, 23, 31, 37, 41, 61, 133, 191$.

Exercise 6 Use your CAS to construct roses for $n = 3$ and $d = 1, 5, 7, 11, 13, 19, 23, 31, 37, 41, 61, 133, 191$.

Exercise 7 Use your CAS to construct roses for $n = 4$ and $d = 1, 5, 7, 11, 13, 19, 23, 31, 37, 41, 61, 133, 191$.

Exercise 8 Construct a few more roses with $n = 5$ and $n = 6$. Explain the effect of n on the shape of the rose.

III. Reflection

Exercise 9 You probably noticed that when d is a multiple of 2, 3, or 5 there are very few lines in the rose, but when you use a number like 7 or 19, there are plenty of lines. When d and 360 are **relatively prime**, that is, when d and 360 have no common factors except 1, you get the greatest number of lines. Explain why this is the case.

Exercise 10 Experiment with a few more roses. Print out and turn in the rose that you think is most attractive.

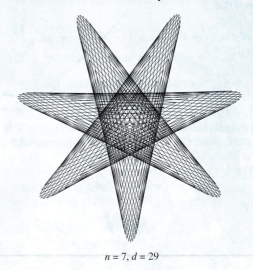

$n = 7, d = 29$

Figure 1

13

Geometry in the Plane, Vectors

13.1
Plane Curves: Parametric Representation

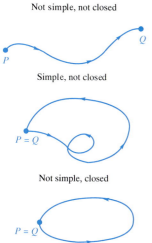

Not simple, not closed

Simple, not closed

Not simple, closed

Simple and closed

Figure 1

We gave the general definition of a plane curve in Section 6.4 in connection with our derivation of the arc length formula. A **plane curve** is determined by a pair of parametric equations

$$x = f(t), \qquad y = g(t), \qquad t \text{ in } I$$

with f and g continuous on the interval I. Usually I is a closed interval $[a, b]$. Think of t, called the **parameter**, as measuring time. As t advances from a to b, the point (x, y) traces out the curve in the xy-plane. When I is the closed interval $[a, b]$, the points $P = (x(a), y(a))$ and $Q = (x(b), y(b))$ are called the **initial** and **final end points**. If the curve has end points that coincide, then we say that the curve is **closed**. If distinct values of t yield distinct points in the plane (except possibly for $t = a$ and $t = b$), we say the curve is a **simple** curve (Figure 1). The pair of relationships $x = f(t)$, $y = g(t)$, together with the interval I is called the **parametrization** of the curve.

Eliminating the Parameter   To recognize a curve given by parametric equations, it may be desirable to eliminate the parameter. Sometimes this can be accomplished by solving one equation for t and substituting in the other (Example 1). Often we can make use of a familiar identity, as in Example 2.

EXAMPLE 1   Eliminate the parameter in

$$x = t^2 + 2t, \qquad y = t - 3, \qquad -2 \le t \le 3$$

Then identify the corresponding curve and sketch its graph.

Solution   From the second equation, $t = y + 3$. Substituting this expression for t in the first equation gives

$$x = (y + 3)^2 + 2(y + 3) = y^2 + 8y + 15$$

or

$$x + 1 = (y + 4)^2$$

This we recognize as a parabola with vertex at $(-1, -4)$ and opening to the right.

In graphing the given equation, we must be careful to display only that part of the parabola corresponding to $-2 \leq t \leq 3$. A table of values and the graph are shown in Figure 2. The arrowhead indicates the curve's *orientation*, that is, the direction of increasing t.   ■

t	x	y
-2	0	-5
-1	-1	-4
0	0	-3
1	3	-2
2	8	-1
3	15	0

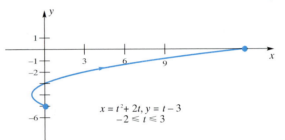

$x = t^2 + 2t, y = t - 3$
$-2 \leq t \leq 3$

Figure 2

EXAMPLE 2   Show that

$$x = a \cos t, \qquad y = b \sin t, \qquad 0 \leq t \leq 2\pi$$

represents the ellipse shown in Figure 3.

Solution   We solve the equations for $\cos t$ and $\sin t$, then square, and add.

$$\left(\frac{x}{a}\right)^2 + \left(\frac{y}{b}\right)^2 = \cos^2 t + \sin^2 t = 1$$

$$\frac{x^2}{a^2} + \frac{y^2}{b^2} = 1$$

A quick check of a few values for t convinces us that we do get the complete ellipse. In particular, $t = 0$ and $t = 2\pi$ give the same point, namely, $(a, 0)$.

If $a = b$, we get the circle $x^2 + y^2 = a^2$.   ■

Different pairs of parametric equations may have the same graph. In other words, a given curve can have more than one parametrization.

EXAMPLE 3   Show that each of the following pairs of parametric equations has the same graph, namely, the semicircle shown in Figure 4.

(a) $x = \sqrt{1 - t^2}, y = t, -1 \leq t \leq 1$

(b) $x = \cos t, y = \sin t, -\dfrac{\pi}{2} \leq t \leq \dfrac{\pi}{2}$

(c) $x = \dfrac{1 - t^2}{1 + t^2}, y = \dfrac{2t}{1 + t^2}, -1 \leq t \leq 1$

Solution   In each case, we discover that

$$x^2 + y^2 = 1$$

It is then just a matter of checking a few values of t to make sure that the given intervals for t yield the same section of the circle.   ■

EXAMPLE 4   Show that each of the following pairs of parametric equations yields one branch of a hyperbola.

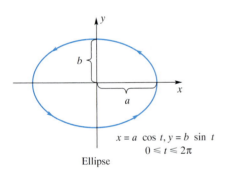

$x = a \cos t, y = b \sin t$
$0 \leq t \leq 2\pi$

Ellipse

Figure 3

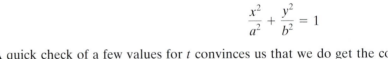

Semicircle

Figure 4

(a) $x = a\sec t, y = b\tan t, -\dfrac{\pi}{2} < t < \dfrac{\pi}{2}$

(b) $x = a\cosh t, y = b\sinh t, -\infty < t < \infty$

Assume in both cases that $a > 0$ and $b > 0$.

Solution

(a) In the first case,

$$\left(\frac{x}{a}\right)^2 - \left(\frac{y}{b}\right)^2 = \sec^2 t - \tan^2 t = 1$$

(b) In the second case,

$$\left(\frac{x}{a}\right)^2 - \left(\frac{y}{b}\right)^2 = \cosh^2 t - \sinh^2 t$$

$$= \left(\frac{e^t + e^{-t}}{2}\right)^2 - \left(\frac{e^t - e^{-t}}{2}\right)^2 = 1$$

Checking a few t-values shows that, in both cases, we obtain the branch of the hyperbola $x^2/a^2 - y^2/b^2 = 1$ shown in Figure 5. ◼

Notice that in Example 4 we have in part (a) a parametric curve defined on the *open* interval $(-\pi/2, \pi/2)$, and in part (b) we have a curve defined on the *infinite* interval $(-\infty, \infty)$. Since the curve does not contain end points, it is not closed.

The Cycloid A **cycloid** is the curve traced by a point P on the rim of a wheel as the wheel rolls along a straight line without slipping (Figure 6). The Cartesian equation of a cycloid is quite complicated, but simple parametric equations are readily found, as shown in the next example.

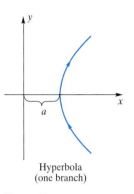

Hyperbola
(one branch)

Figure 5

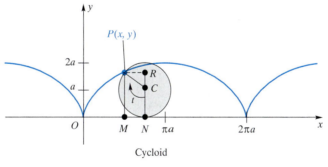

Cycloid

Figure 6

EXAMPLE 5 Find parametric equations for the cycloid.

Solution Let the wheel roll along the x-axis with P initially at the origin. Denote the center of the wheel by C, and let a be its radius. Choose for a parameter the radian measure t of the clockwise angle through which the line segment CP has turned from its vertical position when P was at the origin. All of this is shown in Figure 6.
Since $|ON| = \text{arc } PN = at$,

$$x = |OM| = |ON| - |MN| = at - a\sin t = a(t - \sin t)$$

and

$$y = |MP| = |NR| = |NC| + |CR| = a - a\cos t = a(1 - \cos t)$$

Thus, the parametric equations for the cycloid are

$$x = a(t - \sin t), \qquad y = a(1 - \cos t) \qquad \blacksquare$$

The cycloid has a number of interesting applications, especially in mechanics. It is the "curve of fastest descent." If a particle, acted on only by gravity, is allowed

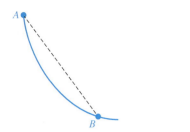

Figure 7

Figure 8

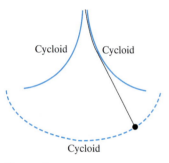

Figure 9

to slide down some curve from a point A to a lower point B not on the same vertical line, it completes its journey in the shortest time when the curve is an inverted cycloid (Figure 7). Of course, the shortest *distance* is along the straight line segment AB, but the *least time* is used when the path is along a cycloid; this is because the acceleration when it is released depends on the steepness of descent, and along a cycloid it builds up velocity much more quickly than it does along a straight line.

Another interesting property is this: If L is the lowest point on an arch of an inverted cycloid, the time that it takes a particle P to slide down the cycloid to L is the same no matter where P starts from on the inverted arch; thus, if several particles, P_1, P_2, and P_3, in different positions on the cycloid (Figure 8), start to slide at the same instant, all will reach the low point L at the same time.

In 1673, the Dutch astronomer Christian Huygens (1629–1695) published a description of an ideal pendulum clock. Because the bob swings between cycloidal "cheeks," the path of the bob is a cycloid (Figure 9). This means that the period of the swing is independent of the amplitude, and so the period does not change as the clock's spring unwinds.

A surprising fact is that the three results just mentioned all date from the seventeenth century. To demonstrate them is a nontrivial task, as you may discover by looking at any book on the history of calculus.

Calculus for Curves Defined Parametrically

Can we find the slope of the tangent line to a curve given parametrically without first eliminating the parameter? The answer is yes, according to the following theorem:

Theorem A

Let f and g be continuously differentiable with $f'(t) \neq 0$ on $\alpha < t < \beta$. Then the parametric equations

$$x = f(t), \qquad y = g(t)$$

define y as a differentiable function of x and

$$\frac{dy}{dx} = \frac{dy/dt}{dx/dt}$$

Proof Since $f'(t) \neq 0$ for $\alpha < t < \beta$, f is strictly monotonic and so has a differentiable inverse f^{-1} (see the Inverse Function Theorem (Theorem 7.2B)). Define F by $F = g \circ f^{-1}$ so that

$$y = g(t) = g(f^{-1}(x)) = F(x) = F(f(t))$$

Then, by the Chain Rule,

$$\frac{dy}{dt} = F'(f(t)) \cdot f'(t) = \frac{dy}{dx} \cdot \frac{dx}{dt}$$

Since $dx/dt \neq 0$, we have

$$\frac{dy}{dx} = \frac{dy/dt}{dx/dt} \qquad \blacklozenge$$

EXAMPLE 6 Find the first two derivatives dy/dx and d^2y/dx^2 for the function determined by

$$x = 5 \cos t, \qquad y = 4 \sin t, \qquad 0 < t < 3$$

and evaluate them at $t = \pi/6$ (see Example 2).

Solution Let y' denote dy/dx. Then

$$\frac{dy}{dx} = \frac{dy/dt}{dx/dt} = \frac{4 \cos t}{-5 \sin t} = -\frac{4}{5} \cot t$$

$$\frac{d^2y}{dx^2} = \frac{dy'}{dx} = \frac{dy'/dt}{dx/dt} = \frac{\frac{4}{5} \csc^2 t}{-5 \sin t} = -\frac{4}{25} \cos^3 t$$

At $t = \pi/6$,

$$\frac{dy}{dx} = \frac{-4\sqrt{3}}{5}, \qquad \frac{d^2y}{dx^2} = \frac{-4}{25}(8) = \frac{-32}{25}$$

The first value is the slope of the tangent line to the ellipse $x^2/25 + y^2/16 = 1$ at the point $(5\sqrt{3}/2, 2)$. You can check that this is so by implicit differentiation.   ■

Sometimes a definite integral involves two variables, such as x and y, in the integrand and differential, and y may be defined as a function of x by equations that give x and y in terms of a parameter such as t. In such cases, it is often convenient to evaluate the definite integral by expressing the integrand and the differential in terms of t and dt and adjusting the limits of integration before integrating with respect to t.

EXAMPLE 7   Evaluate (a) $\displaystyle\int_1^3 y\,dx$ and (b) $\displaystyle\int_1^3 xy^2\,dx$, where $x = 2t - 1$ and $y = t^2 + 2$.

Solution   From $x = 2t - 1$, we have $dx = 2\,dt$ when $x = 1, t = 1$ and when $x = 3$, $t = 2$.

(a) $\displaystyle\int_1^3 y\,dx = \int_1^2 (t^2 + 2)2\,dt = 2\left[\frac{t^3}{3} + 2t\right]_1^2 = \frac{26}{3}$

(b) $\displaystyle\int_1^3 xy^2\,dx = \int_1^2 (2t - 1)(t^2 + 2)^2 2\,dt$

$$= 2\int_1^2 \left(2t^5 - t^4 + 8t^3 - 4t^2 + 8t - 4\right)dt = 86\tfrac{14}{15}$$ ■

EXAMPLE 8   Find the area A under one arch of a cycloid (Figure 10) and the length L of this arch.

Solution   From Example 5, we know that we may represent one arch of the cycloid by

$$x = a(t - \sin t), \qquad y = a(1 - \cos t), \qquad 0 \le t \le 2\pi$$

Thus, $dx = a(1 - \cos t)\,dt$. The area A is therefore

$$A = \int_0^{2\pi a} y\,dx$$

$$= a^2\int_0^{2\pi}(1 - \cos t)(1 - \cos t)\,dt$$

$$= a^2\int_0^{2\pi}(1 - 2\cos t + \cos^2 t)\,dt$$

$$= a^2\int_0^{2\pi}\left(1 - 2\cos t + \tfrac{1}{2} + \tfrac{1}{2}\cos 2t\right)dt$$

$$= a^2\left[\tfrac{3}{2}t - 2\sin t + \tfrac{1}{4}\sin 2t\right]_0^{2\pi} = 3\pi a^2$$

To calculate L, we recall the arc-length formula from Section 6.4:

$$L = \int_\alpha^\beta \sqrt{\left(\frac{dx}{dt}\right)^2 + \left(\frac{dy}{dt}\right)^2}\,dt$$

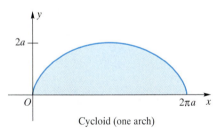

Cycloid (one arch)

Figure 10

Two Fleas on a Trike

Two fleas are arguing about who will get the longest ride when Jenny pedals her tricycle home from the park. A will ride between the treads of the front tire; B will ride between the treads of one of the rear tires. Settle the argument by showing that their paths will have equal lengths. Example 8 should help.

In our case this reduces to

$$L = \int_0^{2\pi} \sqrt{a^2(1 - \cos t)^2 + a^2(\sin^2 t)}\, dt$$

$$= a \int_0^{2\pi} \sqrt{2(1 - \cos t)}\, dt$$

$$= a \int_0^{2\pi} \sqrt{4 \sin^2 \frac{t}{2}}\, dt$$

$$= 2a \int_0^{2\pi} \sin \frac{t}{2}\, dt$$

$$= \left[-4a \cos \frac{t}{2} \right]_0^{2\pi} = 8a \qquad \blacksquare$$

Concepts Review

1. A circle is a premier example of a curve that is both _____ and _____; a figure eight is an example of a closed curve that is not _____.

2. We call two equations $x = f(t)$ and $y = g(t)$ a _____ representation of a curve, and t is called a _____.

3. The path of a point on the rim of a rolling wheel is called a _____.

4. The formula for dy/dx, given the representation $x = f(t)$ and $y = g(t)$, is $dy/dx = $ _____.

Problem Set 13.1

In each of the Problems 1–20, a parametric representation of a curve is given.

(a) *Graph the curve.*

(b) *Is the curve closed? Is it simple?*

(c) *Obtain the Cartesian equation of the curve by eliminating the parameter (see Examples 1–4).*

1. $x = 3t, y = 2t; t$ in $\mathbb{R}$

2. $x = 2t, y = 3t; t$ in $\mathbb{R}$

3. $x = 3t - 1, y = t; 0 \le t \le 4$

4. $x = 4t - 2, y = 2t; 0 \le t \le 3$

5. $x = 4 - t, y = \sqrt{t}; 0 \le t \le 4$

6. $x = t - 3, y = \sqrt{2t}; 0 \le t \le 8$

7. $x = \dfrac{1}{s}, y = s; 1 \le s < 10$

8. $x = s, y = \dfrac{1}{s}; 1 \le s \le 10$

9. $x = t^3 - 4t, y = t^2 - 4; -3 \le t \le 3$

10. $x = t^3 - 2t, y = t^2 - 2t; -3 \le t \le 3$

11. $x = 2\sqrt{t - 2}, y = 3\sqrt{4 - t}; 2 \le t \le 4$

12. $x = 3\sqrt{t - 3}, y = 2\sqrt{4 - t}; 3 \le t \le 4$

13. $x = 2 \sin t, y = 3 \cos t; 0 \le t \le 2\pi$

14. $x = 3 \sin r, y = -2 \cos r; 0 \le r \le 2\pi$

15. $x = -2 \sin r, y = -3 \cos r; 0 \le r \le 4\pi$

16. $x = 2 \cos^2 r, y = 3 \sin^2 r; 0 \le r \le 2\pi$

17. $x = 9 \sin^2 \theta, y = 9 \cos^2 \theta; 0 \le \theta \le \pi$

18. $x = 9 \cos^2 \theta, y = 9 \sin^2 \theta; 0 \le \theta \le \pi$

19. $x = \cos \theta, y = -2 \sin^2 2\theta; \theta$ in $\mathbb{R}$

20. $x = \sin \theta, y = 2 \cos^2 2\theta; \theta$ in $\mathbb{R}$

In Problems 21–30, find dy/dx and d^2y/dx^2 without eliminating the parameter.

21. $x = 3\tau^2, y = 4\tau^3; \tau \neq 0$

22. $x = 6s^2, y = -2s^3; s \neq 0$

23. $x = 2\theta^2, y = \sqrt{5}\theta^3; \theta \neq 0$

24. $x = \sqrt{3}\theta^2, y = -\sqrt{3}\theta^3; \theta \neq 0$

25. $x = 1 - \cos t, y = 1 + \sin t; t \neq n\pi$

26. $x = 3 - 2 \cos t, y = -1 + 5 \sin t; t \neq n\pi$

27. $x = 3 \tan t - 1, y = 5 \sec t + 2; t \neq \dfrac{(2n + 1)\pi}{2}$

28. $x = \cot t - 2, y = -2 \csc t + 5; 0 < t < \pi$

29. $x = \dfrac{1}{1 + t^2}, y = \dfrac{1}{t(1 - t)}; 0 < t < 1$

30. $x = \dfrac{2}{1 + t^2}, y = \dfrac{2}{t(1 + t^2)}; t \neq 0$

In Problems 31–34, find the equation of the tangent to the given curve at the given point without eliminating the parameter. Make a sketch.

31. $x = t^2, y = t^3; t = 2$

32. $x = 3t, y = 8t^3; t = -\frac{1}{2}$

33. $x = 2\sec t, y = 2\tan t; t = -\dfrac{\pi}{6}$

34. $x = 2e^t, y = \frac{1}{3}e^{-t}; t = 0$

In Problems 35–46, find the length of the parametric curve defined over the given interval.

35. $x = 2t - 1, y = 3t - 4; 0 \le t \le 3$

36. $x = 2 - t, y = 2t - 3; -3 \le t \le 3$

37. $x = t, y = t^{3/2}; 0 \le t \le 3$

38. $x = 2\sin t, y = 2\cos t; 0 \le t \le \pi$

39. $x = 3t^2, y = t^3; 0 \le t \le 2$

40. $x = t + \dfrac{1}{t}, y = \ln t^2; 1 \le t \le 4$

41. $x = 2e^t, y = 3e^{3t/2}; \ln 3 \le t \le 2\ln 3$

42. $x = \sqrt{1 - t^2}, y = 1 - t; 0 \le t \le \dfrac{1}{4}$

43. $x = 4\sqrt{t}, y = t^2 + \dfrac{1}{2t}; \dfrac{1}{4} \le t \le 1$

44. $x = \tanh t, y = \ln(\cosh^2 t); -3 \le t \le 3$

45. $x = \cos t, y = \ln(\sec t + \tan t) - \sin t; 0 \le t \le \dfrac{\pi}{4}$

46. $x = \sin t - t\cos t, y = \cos t + t\sin t; \dfrac{\pi}{4} \le t \le \dfrac{\pi}{2}$

47. Find the length of the curve with the given parametric equations
(a) $x = \sin\theta, y = \cos\theta$ for $0 \le \theta \le 2\pi$
(b) $x = \sin 3\theta, y = \cos 3\theta$ for $0 \le \theta \le 2\pi$
(c) Explain why the lengths in parts (a) and (b) are not equal.

You can generate surfaces by revolving smooth curves, given parametrically, about a coordinate axis. As t increases from a to b, a smooth curve $x = F(t)$ and $y = G(t)$ is traced out exactly once. Revolving this curve about the x-axis for $y \ge 0$ gives the surface of revolution with surface area

$$S = \int_a^b 2\pi y \sqrt{\left(\frac{dx}{dt}\right)^2 + \left(\frac{dy}{dt}\right)^2}\, dt$$

See Section 6.4. Problems 48–54 relate to such surfaces.

48. Derive a formula for the surface area generated by the rotation of the curve $x = F(t), y = G(t)$ for $a \le t \le b$ about the y-axis for $x \ge 0$, and show that the result is given by

$$S = \int_a^b 2\pi x \sqrt{\left(\frac{dx}{dt}\right)^2 + \left(\frac{dy}{dt}\right)^2}\, dt$$

49. A parametrization of a circle of radius 1 centered at $(1,0)$ in the xy-plane is given by $x = 1 + \cos t, y = \sin t$, for $0 \le t \le 2\pi$. Find the surface area when this curve is revolved about the y-axis.

50. Find the area of the surface generated by revolving the curve $x = \cos t, y = 3 + \sin t$, for $0 \le t \le 2\pi$ about the x-axis.

51. Find the area of the surface generated by revolving the curve $x = 2 + \cos t, y = 1 + \sin t$, for $0 \le t \le 2\pi$ about the x-axis.

52. Find the area of the surface generated by revolving the curve $x = (2/3)t^{3/2}, y = 2\sqrt{t}$, for $0 \le t \le 2\sqrt{3}$ about the y-axis.

53. Find the area of the surface generated by revolving the curve $x = t + \sqrt{7}, y = t^2/2 + \sqrt{7}t$, for $-\sqrt{7} \le t \le \sqrt{7}$ about the y-axis.

54. Find the area of the surface generated by revolving the curve $x = t^2/2 + at, y = t + a$, for $-\sqrt{a} \le t \le \sqrt{a}$ about the x-axis.

Evaluate the integrals in Problems 55 and 56.

55. $\displaystyle\int_0^1 (x^2 - 4y)\, dx$, where $x = t + 1, y = t^3 + 4$.

56. $\displaystyle\int_1^{\sqrt{3}} xy\, dy$, where $x = \sec t, y = \tan t$.

57. Find the area of the region between the curve $x = e^{2t}$, $y = e^{-t}$ and the x-axis from $t = 0$ to $t = \ln 5$. Make a sketch.

58. The path of a projectile fired from level ground with a speed of v_0 feet per second at an angle α with the ground is given by the parametric equations

$$x = (v_0\cos\alpha)t, \qquad y = -16t^2 + (v_0\sin\alpha)t$$

(a) Show that the path is a parabola.
(b) Find the time of flight.
(c) Show that the range (horizontal distance traveled) is $(v_0^2/32)\sin 2\alpha$.
(d) For a given v_0, what value of α gives the largest possible range?

59. Modify the text discussion of the cycloid (and its accompanying diagram) to handle the case where the point P is $b < a$ units from the center of the wheel. Show that the corresponding parametric equations are

$$x = at - b\sin t, \qquad y = a - b\cos t$$

Sketch the graph of these equations (called a **curtate cycloid**) when $a = 8$ and $b = 4$.

60. Follow the instructions of Problem 59 for the case $b > a$ (a flanged wheel, as on a train), showing that you get the same parametric equations. Sketch the graph of these equations (called a **prolate cycloid**) when $a = 6$ and $b = 8$.

61. Let a circle of radius b roll, without slipping, inside a fixed circle of radius $a, a > b$. A point P on the rolling circle traces out a curve called a **hypocycloid**. Find parametric equations of the hypocycloid. *Hint:* Place the origin O of Cartesian coordinates at the center of the fixed, larger circle, and let the point $A(a, 0)$ be one position of the tracing point P. Denote by B the moving point of tangency of the two circles, and let t, the radian measure of the angle AOB, be the parameter (see Figure 11).

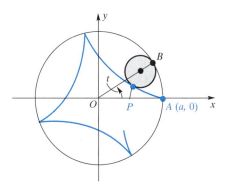

Figure 11

62. Show that if $b = a/4$ in Problem 61 the parametric equations of the hypocycloid may be simplified to

$$x = a\cos^3 t, \qquad y = a\sin^3 t$$

This is called a **hypocycloid of four cusps**. Sketch it carefully and show that its Cartesian equation is $x^{2/3} + y^{2/3} = a^{2/3}$.

63. The curve traced by a point on a circle of radius b as it rolls without slipping on the outside of a fixed circle of radius a is called an **epicycloid**. Show that it has parametric equations

$$x = (a + b)\cos t - b\cos\frac{a + b}{b}t$$

$$y = (a + b)\sin t - b\sin\frac{a + b}{b}t$$

(See the hint in Problem 61.)

64. If $b = a$, the equations in Problem 63 are

$$x = 2a\cos t - a\cos 2t$$

$$y = 2a\sin t - a\sin 2t$$

Show that this special epicycloid is the cardioid $r = 2a(1 - \cos\theta)$, where the pole of the polar coordinate system is the point $(a, 0)$ in the Cartesian system and the polar axis has the direction of the positive x-axis. *Hint*: Find a Cartesian equation of the epicycloid by eliminating the parameter t between the equations. Then show that the equations connecting the Cartesian and polar systems are

$$x = r\cos\theta + a, \qquad y = r\sin\theta$$

and use these equations to transform the Cartesian equation into $r = 2a(1 - \cos\theta)$.

65. If $b = a/3$ in Problem 61, we obtain a hypocycloid of three cusps, called a **deltoid**, with parametric equations

$$x = \left(\frac{a}{3}\right)(2\cos t + \cos 2t), \qquad y = \left(\frac{a}{3}\right)(2\sin t - \sin 2t)$$

Find the length of the deltoid.

66. Consider the ellipse $x^2/a^2 + y^2/b^2 = 1$.
(a) Show that its perimeter is

$$P = 4a\int_0^{\pi/2}\sqrt{1 - e^2\cos^2 t}\,dt,$$

where e is the eccentricity.

[C] (b) The integral in part (a) is called an *elliptic integral*. It has been studied at great length, and it is known that the integrand does not have an elementary antiderivative, so we must turn to approximate methods to evaluate P. Do so when $a = 1$ and $e = \frac{1}{4}$ using the Parabolic Rule with $n = 4$. (Your answer should be near 2π. Why?)

[CAS] (c) Repeat part (b) using $n = 20$.

[CAS] **67.** The parametric curve given by $x = \cos 3t$ and $y = \sin 5t$ is difficult to sketch by hand. The result is known as a *Lissajous* figure. The x-coordinate oscillates three times between 1 and -1 as t goes from 0 to 2π, while the y-coordinate oscillates five times over the same t interval. This behavior is repeated over every interval of length 2π. The entire motion takes place in a unit square. Plot the following *Lissajous* figures for a range of t that ensures that the resulting figure is a closed curve. In each case, count the number of times that the curve touches the horizontal and vertical borders of the unit square.

(a) $x = \sin t, y = \cos t$
(b) $x = \sin 3t, y = \cos 5t$
(c) $x = \cos 5t, y = \sin 15t$
(d) $x = \sin 2t, y = \cos 9t$

[CAS] **68.** Sometimes the *Lissajous* figures do not appear to be closed. This is because the curve retraces itself. For example, the curves $x = \cos 2t, y = \sin 7t$, and $x = \cos(2t + 0.1), y = \sin 7t$ are plotted as Figures 12 and 13.

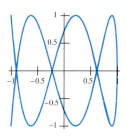

Figure 12

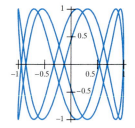

Figure 13

Using this knowledge, and the following *Lissajous* figures associated with $x = \cos at$ and $y = \sin bt$, match the appropriate figure with the correct ratio for a/b.

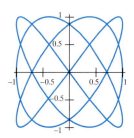

Figure 14

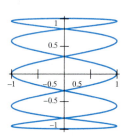

Figure 15

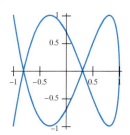

Figure 16

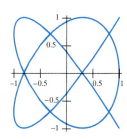

Figure 17

(a) 5/1   (b) 2/5   (c) 4/5   (d) 3/4

Hint: If a curve touches a corner of a square, it counts as one-half a contact.

[CAS] **69.** Plot the following parametric curves. Describe in words how the point moves around the curve in each case.

(a) $x = \cos(t^2 - t), y = \sin(t^2 - t)$
(b) $x = \cos(2t^2 + 3t + 1), y = \sin(2t^2 + 3t + 1)$
(c) $x = \cos(-2\ln t), y = \sin(-2\ln t)$
(d) $x = \cos(\sin t), y = \sin(\sin t)$

[CAS] **70.** Using a computer algebra system, plot the following parametric curves for $0 \le t \le 2$. Describe the shape of the curve in each case and the similarities and differences among all the curves.

(a) $x = t, y = t^2$
(b) $x = t^3, y = t^6$
(c) $x = -t^4, y = -t^8$
(d) $x = t^5, y = t^{10}$

CAS EXPL **71.** Plot the graph of the hypocycloid (see Problem 61)

$$x = (a - b)\cos t + b\cos\frac{a - b}{b}t,$$

$$y = (a - b)\sin t - b\sin\frac{a - b}{b}t$$

for appropriate values of t in each of the following cases:

(a) $a = 4, b = 1$ (b) $a = 3, b = 1$

(c) $a = 5, b = 2$ (d) $a = 7, b = 4$

Experiment with other positive integer values of a and b and then make conjectures about the length of the t-interval required for the curve to return to its starting point and about the number of cusps. What can you say if a/b is irrational?

CAS EXPL **72.** Draw the graph of the epicycloid (see Problem 63)

$$x = (a + b)\cos t - b\cos\frac{a + b}{b}t,$$

$$y = (a + b)\sin t - b\sin\frac{a + b}{b}t$$

for various values of a and b. What conjectures can you make (see Problem 71)?

73. Draw the **Folium of Descartes** $x = 3t/(t^3 + 1)$, $y = 3t^2/(t^3 + 1)$. Then determine the values of t for which this graph is in each of the four quadrants.

Answers to Concepts Review: **1.** simple; closed; simple
2. parametric; parameter **3.** cycloid **4.** $(dy/dt)/(dx/dt)$

13.2
Vectors in the Plane: Geometric Approach

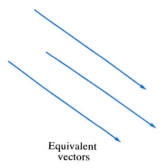

Figure 1

Tail   Head

Figure 2

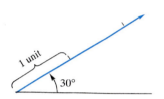

Equivalent
vectors

Figure 3

Many quantities that occur in science (e.g., length, mass, volume, and electric charge) can be specified by giving a single number. These quantities (and the numbers that measure them) are called **scalars**. Other quantities, such as velocity, force, torque, and displacement, require both a magnitude and a direction for complete specification. We call such quantities **vectors** and represent them by arrows (directed line segments). The length of the arrow represents the **magnitude**, or length, of the vector; its direction is the **direction** of the vector. The vector in Figure 1 has length 2.3 units and direction 30° north of east (or 30° from the positive x-axis).

Arrows that we draw, like those shot from a bow, have two ends. There is the feather end (the initial point), called the **tail**, and the pointed end (the terminal point), called the **head**, or tip (Figure 2). Two vectors are considered to be **equivalent** if they have the same magnitude and direction (Figure 3). We shall symbolize vectors by boldface letters, such as **u** and **v**. Since this is hard to accomplish in normal writing, you might use $\vec{u}$ and $\vec{v}$. The magnitude of a vector **u** is symbolized by $|\mathbf{u}|$.

Operations on Vectors   To find the **sum**, or **resultant**, of **u** and **v**, move **v** without changing its magnitude or direction until its tail coincides with the head of **u**. Then **u** + **v** is the vector connecting the tail of **u** to the head of **v**. This method (called the *Triangle Law*) is illustrated in the left half of Figure 4.

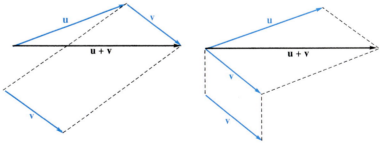

Two equivalent ways of adding vectors

Figure 4

As an alternative way to find **u** + **v**, move **v** so that its tail coincides with that of **u**. Then **u** + **v** is the vector with this common tail and coinciding with the diagonal of the parallelogram that has **u** and **v** as sides. This method (called the *Parallelogram Law*) is illustrated on the right in Figure 4.

These two methods are equivalent ways to define what we mean by the sum of two vectors. You should convince yourself that vector addition is commutative and associative; that is,

$$\mathbf{u} + \mathbf{v} = \mathbf{v} + \mathbf{u}$$

$$(\mathbf{u} + \mathbf{v}) + \mathbf{w} = \mathbf{u} + (\mathbf{v} + \mathbf{w})$$

on the tires must at least balance the centrifugal force pulling out-ward. The force F satisfies $F = \mu mg$, where μ is the *coefficient of friction*, m is the mass of the car, and g is the acceleration of gravity. Thus, $\mu mg \geq mv^2/R$. Show that v_R, the speed beyond which skidding will occur, satisfies

$$v_R = \sqrt{\mu g R}$$

and use this to determine v_R for a curve with $R = 400$ feet and $\mu = 0.4$. Use $g = 32$ feet per second per second.

56. Consider again the car of Problem 55. Suppose that the curve is icy at its worst spot ($\mu = 0$), but is banked at angle θ from the horizontal (Figure 11). Let $\mathbf{F}$ be the force exerted by the road on the car. Then, at the critical speed v_R, $mg = |\mathbf{F}|\cos\theta$ and $mv_R^2/R = |\mathbf{F}|\sin\theta$.
(a) Show that $v_R = \sqrt{Rg\tan\theta}$.
(b) Find v_R for a curve with $R = 400$ feet and $\theta = 10°$.

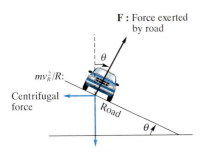

F : Force exerted by road

θ

mv_R^2/R:
Centrifugal force

Road

θ

mg: Weight of car

Figure 11

EXPL **57.** Derive the polar coordinate curvature formula

$$\kappa = \frac{|r^2 + 2(r')^2 - rr''|}{(r^2 + (r')^2)^{3/2}}$$

where the derivatives are with respect to θ.

In Problems 58–63, use the formula in Problem 57 to find the curvature κ of the following:

58. Circle $r = 4\cos\theta$

59. Cardioid $r = 1 + \cos\theta$ at $\theta = 0$

60. $r = \theta$ at $\theta = 1$

61. $r = 4(1 + \cos\theta)$ at $\theta = \pi/2$

62. $r = e^{3\theta}$ at $\theta = 1$

63. $r = 4(1 + \sin\theta)$ at $\theta = \pi/2$

64. Show that the curvature of the polar curve $r = e^{6\theta}$ is proportional to $1/r$.

65. Show that the curvature of the polar curve $r^2 = \cos 2\theta$ is directly proportional to r for $r > 0$.

66. Derive the first curvature formula in Theorem A by working directly with $\kappa = |\mathbf{T}'(t)|/|\mathbf{r}'(t)|$.

GC **67.** Draw the graph of $x = 4\cos t$, $y = 3\sin(t + 0.5)$, $0 \leq t \leq 2\pi$. Estimate its maximum and minimum curvature by looking at the graph (curvature is the reciprocal of the radius of curvature). Then use a graphing calculator or a CAS to approximate these two numbers to four decimal places.

CAS **68.** Draw the graph of $x = 2.5\cos t + 0.5\cos(t - 2\theta)$, $y = 2.5\sin t - 0.5\sin(t - 2\theta)$, $0 \leq t \leq 2\pi$, for $\theta = 0$, $\theta = \pi/6$, $\theta = \pi/3$, $\theta = \pi/2$, and $\theta = 3\pi/4$. Make a conjecture about the shape of these curves. Use this conjecture and Example 4 to determine the maximum and minimum curvature.

CAS **69.** Generalize Problem 68 by drawing graphs for various values of a, b, and θ of

$$x = \frac{a + b}{2}\cos t + \frac{a - b}{2}\cos(t - 2\theta)$$

$$y = \frac{a + b}{2}\sin t - \frac{a - b}{2}\sin(t - 2\theta)$$

$0 \leq t \leq 2\pi$. Make a conjecture about the shape of these graphs. Can you prove your conjecture? (See Problem 40 of Section 12.2 for one idea.)

Answers to Concepts Review: **1.** $|d\mathbf{T}/ds|$ **2.** $1/a; 0$ **3.** $\kappa = 1/R$
4. $(d^2s/dt^2)\mathbf{T} + (ds/dt)^2\kappa\mathbf{N}$

13.6 Chapter Review

Concepts Test

Respond with true or false to each of the following assertions. Be prepared to justify your answer.

1. The parametric representation of a curve is unique.

2. The graph of $x = 2t^3$, $y = t^3$ is a line.

3. If $x = f(t)$ and $y = g(t)$, then we can find a function h such that $y = h(x)$.

4. The curve with parametric representation $x = \ln t$ and $y = t^2 - 1$ passes through the origin.

5. If $x = f(t)$ and $y = g(t)$ and if both f'' and g'' exist, then $d^2y/dx^2 = g''(t)/f''(t)$ wherever $f''(t) \neq 0$.

6. A curve may have more than one tangent line at a point on the curve.

7. The vectors $2\mathbf{i} - 3\mathbf{j}$ and $6\mathbf{i} + 4\mathbf{j}$ are perpendicular.

8. If $\mathbf{u}$ and $\mathbf{v}$ are unit vectors, then the angle θ between them satisfies $\cos\theta = \mathbf{u} \cdot \mathbf{v}$.

9. The dot product for vectors satisfies the associative law.

10. If $\mathbf{u}$ and $\mathbf{v}$ are any two vectors, then $|\mathbf{u} \cdot \mathbf{v}| \leq |\mathbf{u}||\mathbf{v}|$.

11. $|\mathbf{u} \cdot \mathbf{v}| = |\mathbf{u}||\mathbf{v}|$ for nonzero vectors $\mathbf{u}$ and $\mathbf{v}$ if and only if $\mathbf{u}$ is a scalar multiple of $\mathbf{v}$.

12. If $|\mathbf{u}| = |\mathbf{v}| = |\mathbf{u} + \mathbf{v}|$, then $\mathbf{u} = \mathbf{v} = \mathbf{0}$.

13. If $\mathbf{u} + \mathbf{v}$ and $\mathbf{u} - \mathbf{v}$ are perpendicular, then $|\mathbf{u}| = |\mathbf{v}|$.

14. For any two vectors $\mathbf{u}$ and $\mathbf{v}$, $|\mathbf{u} + \mathbf{v}|^2 = |\mathbf{u}|^2 + |\mathbf{v}|^2 + 2\mathbf{u} \cdot \mathbf{v}$.

15. The vector-valued function $\langle f(t), g(t) \rangle$ is continuous at $t = a$ if and only if both f and g are continuous at $t = a$.

16. $D_t[\mathbf{F}(t) \cdot \mathbf{F}(t)] = 2\mathbf{F}(t) \cdot \mathbf{F}'(t)$.

17. The curvature of the curve determined by $x = 3t + 4$ and $y = 2t - 1$ is zero for all t.

18. The curvature of the curve determined by $x = 2\cos t$ and $y = 2\sin t$ is 2 for all t.

19. If $\mathbf{T} = \mathbf{T}(t)$ is a unit vector tangent to a smooth curve, then $\mathbf{T}(t)$ and $\mathbf{T}'(t)$ are perpendicular.

20. If $v = |\mathbf{v}|$ is the speed of a particle moving along a smooth curve, then $|dv/dt|$ is the magnitude of the acceleration.

21. If $y = f(x)$ and $y'' = 0$ everywhere, then the curvature of this curve is zero.

22. If $y = f(x)$ and y'' is a constant, then the curvature of this curve is a constant.

23. If $\mathbf{u} \cdot \mathbf{v} = 0$, then either $\mathbf{u} = \mathbf{0}$ or $\mathbf{v} = \mathbf{0}$, or both $\mathbf{u}$ and $\mathbf{v}$ are $\mathbf{0}$.

24. If $|\mathbf{r}(t)| = 1$ for all t, then $|\mathbf{r}'(t)| = $ constant.

25. If $\mathbf{v} \cdot \mathbf{v} = $ constant, then $\mathbf{v} \cdot \mathbf{v}' = 0$.

Sample Test Problems

In Problems 1–4, a parametric representation of a curve is given. Eliminate the parameter to obtain the corresponding Cartesian equation. Sketch the given curve.

1. $x = 6t + 2$, $y = 2t$; $-\infty < t < \infty$

2. $x = 4t^2$, $y = 4t$; $-1 \le t \le 2$

3. $x = 4\sin t - 2$, $y = 3\cos t + 1$; $0 \le t \le 2\pi$

4. $x = 2\sec t$, $y = \tan t$; $-\dfrac{\pi}{2} < t < \dfrac{\pi}{2}$

In Problems 5 and 6, find the equations of the tangent line and the normal line at $t = 0$.

5. $x = 2t^3 - 4t + 7$, $y = t + \ln(t + 1)$

6. $x = 3e^{-t}$, $y = \frac{1}{2}e^t$

7. Find the length of the curve

$$x = \cos t + t\sin t$$

$$y = \sin t - t\cos t$$

from 0 to 2π. Make a sketch.

8. Find $\mathbf{u}$ and $|\mathbf{u}|$ if $\mathbf{u}$ is the vector from P_1 to P_2.

(a) $P_1 = (2, 4)$, $P_2 = (-1, 5)$ (b) $P_1 = (-3, 0)$, $P_2 = (-4, 5)$

9. Let $\mathbf{a} = \langle 2, -5\rangle$, $\mathbf{b} = \langle 1, 1\rangle$, and $\mathbf{c} = \langle -6, 0\rangle$. Find each of the following:

(a) $3\mathbf{a} - 2\mathbf{b}$ (b) $\mathbf{a} \cdot \mathbf{b}$
(c) $\mathbf{a} \cdot (\mathbf{b} + \mathbf{c})$ (d) $(4\mathbf{a} + 5\mathbf{b}) \cdot 3\mathbf{c}$
(e) $|\mathbf{c}|\mathbf{c} \cdot \mathbf{b}$ (f) $\mathbf{c} \cdot \mathbf{c} - |\mathbf{c}|$

10. Find the cosine of the angle between $\mathbf{a}$ and $\mathbf{b}$ and make a sketch.

(a) $\mathbf{a} = 3\mathbf{i} + 2\mathbf{j}$, $\mathbf{b} = -\mathbf{i} + 4\mathbf{j}$ (b) $\mathbf{a} = -5\mathbf{i} - 3\mathbf{j}$, $\mathbf{b} = 2\mathbf{i} - \mathbf{j}$
(c) $\mathbf{a} = \langle 7, 0\rangle$, $\mathbf{b} = \langle 5, 1\rangle$

11. Given $\mathbf{a} = -2\mathbf{i}$ and $\mathbf{b} = 3\mathbf{i} - 2\mathbf{j}$ and another vector $\mathbf{r} = 5\mathbf{i} - 4\mathbf{j}$, find scalars k and m such that $\mathbf{r} = k\mathbf{a} + m\mathbf{b}$.

12. Find a vector of length 3 that is parallel to the tangent line to $y = x^2$ at $(-1, 1)$.

13. Find a vector of length 10 that makes an angle of $150°$ with the positive x-axis.

14. Two forces $\mathbf{F}_1 = 2\mathbf{i} - 3\mathbf{j}$ and $\mathbf{F}_2 = 3\mathbf{i} + 12\mathbf{j}$ are applied at a point. What force $\mathbf{F}$ must be applied at the point to counteract the resultant of these two forces?

15. What heading and airspeed are required for an airplane to fly 450 miles per hour due north if a wind of 100 miles per hour is blowing in the direction N $60°$ E?

16. If $\mathbf{r}(t) = \langle e^{2t}, e^{-t}\rangle$ find each of the following:

(a) $\lim_{t \to 0} \mathbf{r}(t)$ (b) $\lim_{h \to 0} \dfrac{\mathbf{r}(0 + h) - \mathbf{r}(0)}{h}$

(c) $\displaystyle\int_0^{\ln 2} \mathbf{r}(t)\, dt$ (d) $D_t[t\mathbf{r}(t)]$

(e) $D_t[\mathbf{r}(3t + 10)]$ (f) $D_t[\mathbf{r}(t) \cdot \mathbf{r}'(t)]$

17. Find $\mathbf{r}'(t)$ and $\mathbf{r}''(t)$ for each of the following:

(a) $\mathbf{r}(t) = (\ln t)\mathbf{i} - 3t^2\mathbf{j}$ (b) $\mathbf{r}(t) = \sin t\,\mathbf{i} + \cos 2t\,\mathbf{j}$
(c) $\mathbf{r}(t) = \tan t\,\mathbf{i} - t^4\mathbf{j}$

18. Find the length of the arc $\mathbf{r}(t) = 4t^{3/2}\mathbf{i} + 3t\mathbf{j}$ from $t = 0$ to $t = 2$.

In Problems 19 and 20, the position of a moving particle at time t is given by $\mathbf{r}(t)$. Find the velocity and acceleration vectors, $\mathbf{v}(t)$ and $\mathbf{a}(t)$, and their values at the given time $t = t_1$. Also, find the speed of the particle for $t = t_1$.

19. $\mathbf{r}(t) = 2t^2\mathbf{i} + (4t + 2)\mathbf{j}$; $t_1 = -1$

20. $\mathbf{r}(t) = 4(1 - \sin t)\mathbf{i} + 4(t + \cos t)\mathbf{j}$; $t_1 = \frac{2}{3}\pi$.

21. Find the curvature κ of the given curve at P.

(a) $y = x^2 - x$ at $P(1, 0)$.
(b) $\mathbf{r}(t) = (t + t^3)\mathbf{i} + (t + t^2)\mathbf{j}$ at $P(2, 2)$
(c) $y = a\cosh(x/a)$ at $P(a, a\cosh 1)$

22. Find the unit tangent vector $\mathbf{T}(t)$ for the curve $\mathbf{r}(t) = t\mathbf{i} + \frac{1}{3}t^3\mathbf{j}$. At the point $P(1)$ on the curve where $t = 1$, find $\mathbf{T}(1)$. Find the curvature $\kappa(1)$ of the curve at $P(1)$. Sketch the curve and draw the unit tangent vector $\mathbf{T}(1)$ with its initial point at $P(1)$.

23. If $\mathbf{r}(t) = (1 - t^2)\mathbf{i} + 2t\mathbf{j}$, find the tangential and the normal components, a_T and a_N, of the acceleration $\mathbf{a}$ at $P(0, 2)$.

Hypocycloids

I. Preparation

In this project, you will study properties of parametric curves, specifically, the unit circle and the hypocycloid.

The equations $x = \cos t$, $y = \sin t$ with the parameter t in the range $[0, 2\pi]$ give a familiar parametric representation of the unit circle $x^2 + y^2 = 1$. If the position of an object at time t is given by $\mathbf{r}(t) = \langle \cos t, \sin t \rangle$, then the object's velocity is $\mathbf{v}(t) = \mathbf{r}'(t) = \langle -\sin t, \cos t \rangle$, and its speed is

$$v(t) = |\mathbf{v}(t)| = \sqrt{(x'(t))^2 + (y'(t))^2}$$
$$= \sqrt{(-\sin t)^2 + (\cos t)^2} = 1$$

Thus, this parametrization of the unit circle leads to a transversal of the circle with a constant speed of 1.

Exercise 1 Show that the parametric equations

$$x_1(t) = \frac{1 - t^2}{1 + t^2}, \qquad y_1(t) = \frac{2t}{1 + t^2}, \qquad -\infty < t < \infty$$

give another representation of the unit circle. (Technically, the point $(-1, 0)$ is missing.)

II. Using Technology

Exercise 2
(a) Use a CAS to plot the curve $\mathbf{r}(t) = \langle \cos t, \sin t \rangle$, $0 \le t \le 2\pi$.
(b) Use your CAS to make a parametric plot of

$$x_1(t) = \frac{1 - t^2}{1 + t^2}, \qquad y_1(t) = \frac{2t}{1 + t^2}, \qquad -A \le t \le A$$

for $A = 1, 2, 5, 10, 20, 40, 50,$ and 100. Discuss the "missing point" in terms of limits.

Exercise 3 For the parametrization $(x_1(t), y_1(t))$ make a table showing the values of x, y and the speed v for the points $t = -5, -4, \ldots, 3, 4, 5$. Describe how the speed varies as the point $P(x_1(t), y_1(t))$ moves around the unit circle.

The *hypocycloid* is the curve generated by a circle of radius b rolling on the inside of a larger circle of radius a. See Figure 1. The equations of the hypocycloid are given parametrically as

$$x(t) = (a - b)\cos t + b\cos\left(\frac{a - b}{b}t\right)$$

$$y(t) = (a - b)\sin t - b\sin\left(\frac{a - b}{b}t\right), \qquad a > b$$

We could factor out the scaling factor a in each equation and study the family of hypocycloids in terms of the single parameter b/a. Instead, we just set $a = 1$, but label our plots in terms of the ratio b/a (see Figures 1 and 2). We say that a po-

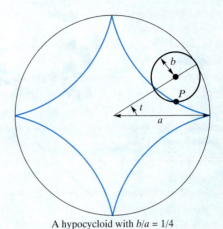

A hypocycloid with $b/a = 1/4$

Figure 1

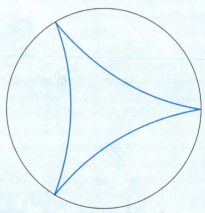

A hypocycloid with $b/a = 1/3$

Figure 2

sition function $\mathbf{r}(t)$ has *period* P if $\mathbf{r}(t + P) = \mathbf{r}(t)$ and P is the smallest positive number with this property. In the next exercise, you will investigate the period of a hypocycloid.

Exercise 4 Experiment by plotting the hypocycloid to determine its period for the case where $a = 1$ and $b = 1/k$, with $k = 3, 4, 5, 6, \ldots$. What are the periods?

Hint: If you suspect, for example, that the period is 4π, then plot the hypocycloid over the intervals $[0, 3.5\pi]$, $[0, 3.8\pi]$, $[0, 3.9\pi]$ and $[0, 4\pi]$ to see if the curve is "closing up."

Exercise 5 Next, plot the hypocycloids for $a = 1$ and $b = 1/k$, with $k = 3, 4, 5, 6, \ldots, 12$; in each case, use a domain that shows one full period. Describe the nature of the graphs. What happens as $k \to \infty$?

Exercise 6 Determine the arc lengths of the hypocycloid for $a = 1, b = 1/k$ with $k = 3, 4, 5, 6, 7,$ and 8.

Hint: It is better to find the arc length of one "lobe" and then multiply by the number of lobes. Have your CAS simplify the integrand before trying to evaluate it.

Make a conjecture about the arc length for a general k. What is the limit of the arc length as $k \to \infty$?

Exercise 7   Perform some graphical experiments to determine the period of the hypocloid when $a = 1$ and $b = j/k$, for $k = 16$ and $j = 1, 2, 3, 4, 5, 6, 7, 9, 10, 11, 12, 13, 14$, and 15. (Notice that we skipped $j = 8$. Do you see why?) Make a table with three columns like the one following:

j	$\dfrac{j}{k}$ in lowest terms	Period P
1	$\frac{1}{16}$	2π
2	$\frac{1}{8}$	$\vdots$
$\vdots$	$\vdots$	

Make a conjecture about a formula for the period of the hypocloid with $a = 1$ and $b = j/16$.

Hint: Look at the numerator in the reduced fraction j/k.

Repeat this exercise for $a = 1$ and $b = j/k$, $k = 20$, $j = 1, 2, 3, 4, 5, 6, 7, 8, 9, 11, 12, 13, 14, 15, 16, 17, 18$, and 19.

Exercise 8   Using your results from Exercise 7, give a formula for the period of the hypocloid with $a = 1$ and $b = j/k$, where j and k are integers with $j < k$.

Exercise 9   Plot the hypocloid for $a = 1$ and $b = 1/\pi$ over various domains of the form $[0, 2L\pi]$. Vary L and turn in your *prettiest* graph.

III. Reflection

Exercise 10   Explain the limiting result of Exercise 6 in light of the fact that the circumference of the unit circle is 2π.

Exercise 11   Hypocloid pairs corresponding to j/k's given by $1/3$ and $2/3$, or $1/4$ and $3/4$, or in general, $1/k$ and $1 - 1/k$, have the same graph, but there are differences. What are they? Be as specific as possible.

Exercise 12   What can you say about the hypocloid if a/b is irrational?

Measuring Home Run Distance

I. Preparation

On May 16, 1998, Mark McGwire of the St. Louis Cardinals hit a 545-foot home run in Busch Stadium (St. Louis) in a game against the Florida Marlins. The figure of 545 feet is actually an estimate of how far the ball *would have traveled* had it returned to field level. Most home runs go into the stands or hit part of the stadium above field level. In the case of McGwire's home run on May 16, the ball hit a center field sign that was hanging from the upper deck. In this project, you will study how home run distances are estimated.

Exercise 1   Review Example 5 in Section 13.4. Show that if the height of the ball when it makes contact with the bat is c then the height y as a function of horizontal position x is

(1)
$$y(x) = -\frac{16}{v_0^2 \cos^2 \theta} x^2 + (\tan \theta)x + c$$

where v_0 is the initial velocity of the ball after contact with the bat and θ is the initial angle of the ball's path after contact with the bat. Let d be the horizontal distance that the ball travels before it strikes some part of the stadium, and let h be the height of the ball when it strikes the stadium. Finally, let ϕ be the angle that the ball's path makes with the horizontal when it hits the stadium. See Figure 3.
Explain why

$$y(d) = h$$

and

(2)
$$y'(d) = -\tan \phi$$

Exercise 2   To measure the home run distance, that is, the horizontal distance that the ball would have traveled had it returned to field level, many ball clubs estimate distances d and h and the angle ϕ. Many clubs have a chart that gives the distances and heights of each seat in the outfield. Given d, h, and ϕ, they estimate the home run distance D. The initial velocity v_0 and the initial angle θ are unknown. We can, however, determine these quantities given d, h, and ϕ.
   Find y' from (1) and substitute d for x. Use (2) and solve for $v_0^2 \cos^2 \theta$ to obtain

$$v_0^2 \cos^2 \theta = \frac{32d}{\tan \phi + \tan \theta}$$

Then substitute this expression into (1) and solve for θ. The derivation is a bit long, but you should end up with

$$\theta = \tan^{-1}\left(\tan \phi + \frac{2(h - c)}{d} \right)$$

(*Note*: In most computer systems, $\tan^{-1}$ is denoted arctan.) Then solve for v_0.

Exercise 3   To find the home run distance, we need to solve $y(x) = 0$. Apply the quadratic formula to find x. Explain the significance of both solutions from the quadratic equation. Which one do you choose? Why?

II. Using Technology

Exercise 4   Estimate the home run distance for a ball that strikes an object 360 feet from home plate at a height of 30 feet and with an angle of 30 degrees. Assume that the player makes contact with the ball when it is 3 feet off the ground.

Exercise 5   If a home run ball strikes a center field sign 430 feet from home plate and 80 feet above the field, what angle would the ball's path have to make with the horizontal at the point of impact in order for the home run distance to be 545 feet?

Hint: Fix $d = 430$ and $h = 80$ and vary ϕ until the home run distance is about 545. What is the initial velocity v_0?

Exercise 6   In Busch Stadium the center field wall is 8 feet high and 402 feet from home plate. Give five combinations of θ and v_0 that will lead to a home run over the center field wall.

Exercise 7   In Boston's Fenway Park, the left field wall, called the Green Monster, is 37 feet high and 310 feet from home plate. Give five combinations of θ and v_0 that will lead to a home run over the Green Monster.

Exercise 8   Give an example of a pair (θ, v_0) that would lead to a center field home run in Busch Stadium, but not a left field home run in Fenway Park, and vice versa.

III. Reflection

Exercise 9   In Exercise 4, vary the angle ϕ and observe the results. What is the effect of a 5° error in estimating ϕ? Next vary the initial height c. What is the effect of an error of 2 feet in estimating c?

Exercise 10   Another interpretation of "home run distance" would be the arc length of the ball's trajectory. According to this definition, find the distance of Mark McGwire's May 16 home run using the information from Exercise 5.

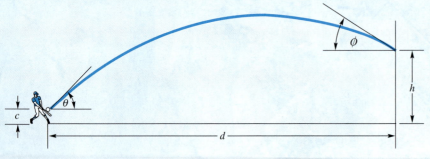

Figure 3

14

Geometry in Space, Vectors

14.1
Cartesian Coordinates in Three-Space

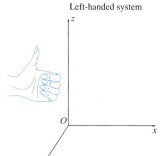

Right-handed system

Left-handed system

Figure 1

We have reached an important transition point in our study of calculus. Until now, we have been traveling across that broad flat expanse known as the Euclidean plane, or two-space. The concepts of calculus have been applied to functions of a single variable, functions whose graphs can be drawn in the plane.

The mountains lie ahead. Our charted course winds through three-space and occasionally into n-space. We are going to study *multiple variable calculus*, the calculus that applies to functions of two or more variables. All the familiar ideas (such as limit, derivative, integral) are to be explored again from a loftier perspective.

To begin, consider three mutually perpendicular coordinate lines (the x-, y-, and z-axes) with their zero points at a common point O, called the *origin*. Although these lines can be oriented in any way one pleases, we follow a custom in thinking of the y- and z-axes as lying in the plane of the paper with their positive directions to the right and upward, respectively. The x-axis is then perpendicular to the paper, and we suppose its positive end to point toward us, thus forming a **right-handed system**. We call it right-handed because, if the fingers of the right hand are curled so that they curve from the positive x-axis toward the positive y-axis, the thumb points in the direction of the positive z-axis (Figure 1).

The three axes determine three planes, the yz-, xz-, and xy-planes, which divide space into eight octants (Figure 2). To each point P in space corresponds an ordered triple of numbers (x, y, z), its **Cartesian coordinates**, which measure its directed distances from the three planes (Figure 3).

Plotting points in the first octant (the octant where all three coordinates are positive) is relatively easy. In Figures 4 and 5, we illustrate something more difficult by plotting two points from other octants, the points $P(2, -3, 4)$ and $Q(-3, 2, -5)$.

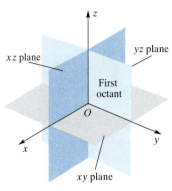

Figure 2

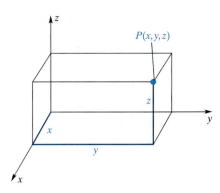

Figure 3

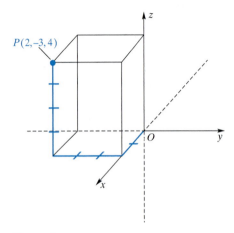

Figure 4

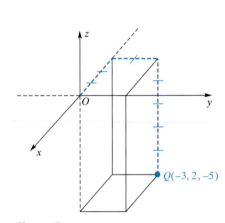

Figure 5

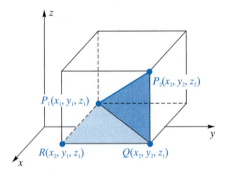

Figure 6

The Distance Formula Consider two points $P_1(x_1, y_1, z_1)$ and $P_2(x_2, y_2, z_2)$ in three-space $(x_1 \neq x_2, y_1 \neq y_2, z_1 \neq z_2)$. They determine a **parallelepiped** (i.e., a rectangular box) with P_1 and P_2 as opposite vertices and with edges parallel to the coordinate axes (Figure 6). The triangles P_1QP_2 and P_1RQ are right triangles and, by the Pythagorean Theorem,

$$|P_1P_2|^2 = |P_1Q|^2 + |QP_2|^2$$

and

$$|P_1Q|^2 = |P_1R|^2 + |RQ|^2$$

Thus,

$$|P_1P_2|^2 = |P_1R|^2 + |RQ|^2 + |QP_2|^2$$
$$= (x_2 - x_1)^2 + (y_2 - y_1)^2 + (z_2 - z_1)^2$$

This gives us the **Distance Formula** in three-space.

$$\boxed{|P_1P_2| = \sqrt{(x_2 - x_1)^2 + (y_2 - y_1)^2 + (z_2 - z_1)^2}}$$

The formula is correct even if some of the coordinates are identical.

EXAMPLE 1 Find the distance between the points $P(2, -3, 4)$ and $Q(-3, 2, -5)$, which were plotted in Figures 4 and 5.

Solution

$$|PQ| = \sqrt{(-3 - 2)^2 + (2 + 3)^2 + (-5 - 4)^2} = \sqrt{131} \approx 11.45 \quad \blacksquare$$

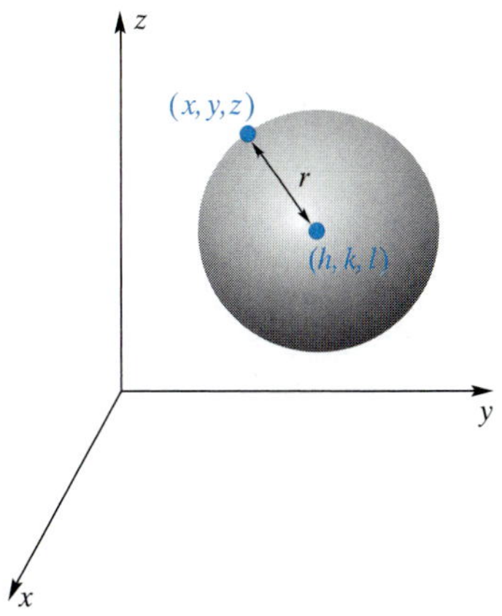

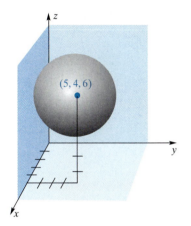

Figure 7

Spheres and Their Equations   It is a small step from the Distance Formula to the equation of a sphere. By a **sphere**, we mean the set of all points in three-dimensional space that are a constant distance (the radius) from a fixed point (the center). (Recall that a circle is defined as the set of points *in a plane* that are a constant distance from a fixed point.) In fact, if (x, y, z) is a point on the sphere of radius r centered at (h, k, l), then (see Figure 7).

$$(x - h)^2 + (y - k)^2 + (z - l)^2 = r^2$$

We call this the **standard equation of a sphere**.

In expanded form, the boxed equation may be written as

$$x^2 + y^2 + z^2 + Gx + Hy + Iz + J = 0$$

Conversely, the graph of any equation of this form is either a sphere, a point (a degenerate sphere), or the empty set. To see why, consider the following example.

EXAMPLE 2   Find the center and radius of the sphere with equation

$$x^2 + y^2 + z^2 - 10x - 8y - 12z + 68 = 0$$

and sketch its graph.

Solution   We use the process of completing the square.

$$\left(x^2 - 10x + \quad\right) + \left(y^2 - 8y + \quad\right) + \left(z^2 - 12z + \quad\right) = -68$$
$$\left(x^2 - 10x + 25\right) + \left(y^2 - 8y + 16\right) + \left(z^2 - 12z + 36\right) = -68 + 25 + 16 + 36$$
$$(x - 5)^2 + (y - 4)^2 + (z - 6)^2 = 9$$

Thus, the equation represents a sphere with center at $(5, 4, 6)$ and radius 3. Its graph is shown in Figure 8. ∎

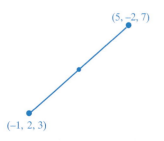

Figure 8

If, after completing the square in Example 2, the equation had been

$$(x - 5)^2 + (y - 4)^2 + (z - 6)^2 = 0$$

then the graph would be the single point $(5, 4, 6)$; if the right side were negative, the graph would be the empty set.

Another simple result that follows from the Distance Formula is the **Midpoint Formula**. If $P_1(x_1, y_1, z_1)$ and $P_2(x_2, y_2, z_2)$ are end points of a line segment, then the midpoint $M(m_1, m_2, m_3)$ has coordinates

$$m_1 = \frac{x_1 + x_2}{2}, \qquad m_2 = \frac{y_1 + y_2}{2}, \qquad m_3 = \frac{z_1 + z_2}{2}$$

In other words, to find the coordinates of the midpoint of a segment, simply take the average of corresponding coordinates of the end points.

EXAMPLE 3   Find the equation of the sphere that has the line segment joining $(-1, 2, 3)$ and $(5, -2, 7)$ as a diameter (Figure 9).

Solution   The center of this sphere is at the midpoint of the segment, that is, at $(2, 0, 5)$; the radius r satisfies

$$r^2 = (5 - 2)^2 + (-2 - 0)^2 + (7 - 5)^2 = 17$$

We conclude that the equation of the sphere is

$$(x - 2)^2 + y^2 + (z - 5)^2 = 17$$ ∎

Figure 9

Graphs in Three-Space   It was natural to consider a quadratic equation first because of its relation to the Distance Formula. But, presumably, a **linear equation** in x, y, and z, that is, an equation of the form

$$Ax + By + Cz = D, \qquad A^2 + B^2 + C^2 \neq 0$$

should be even easier to analyze. As a matter of fact, we will show in the next section that the graph of a linear equation is a plane. Taking this for granted for now, let's consider how we might graph such an equation.

If, as will often be the case, the plane intersects the three axes, we begin by finding these intersection points; that is, we find the x-, y-, and z-intercepts. These three points determine the plane and allow us to draw the (coordinate-plane) **traces**, which are the lines of intersection of that plane with the coordinate planes. Then, with just a bit of artistry, we can shade in the plane.

EXAMPLE 4   Sketch the graph of $3x + 4y + 2z = 12$.

Solution   To find the x-intercept, set y and z equal to zero and solve for x, obtaining $x = 4$. The corresponding point is $(4, 0, 0)$. Similarly, the y- and z-intercepts are $(0, 3, 0)$ and $(0, 0, 6)$. Next, connect these points by line segments to get the traces. Then shade in (the first octant part of) the plane, thereby obtaining the result shown in Figure 10.   ■

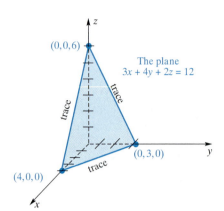

Figure 10

What if the plane does not intersect all three axes? This will happen, for example, if one of the variables in the equation of the plane is missing (i.e., has a zero coefficient).

EXAMPLE 5   Sketch the graph of the linear equation

$$2x + 3y = 6$$

in three-space.

Solution   The x- and y-intercepts are $(3, 0, 0)$ and $(0, 2, 0)$, respectively, and these points determine the trace in the xy-plane. The plane never crosses the z-axis (x and y cannot both be 0), and so the plane is parallel to the z-axis. We have sketched the graph in Figure 11.   ■

Notice that in each of our examples the graph of an equation in three-space was a *surface*. This contrasts with the two-space case, where the graph of an equation was usually a *curve*. We will have a good deal more to say about graphing equations and the corresponding surfaces in Section 14.6.

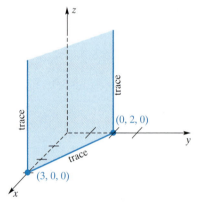

Figure 11

Concepts Review

1. The numbers x, y, and z in (x, y, z) are called the _____ of a point in three-space.

2. The distance between the points $(-1, 3, 5)$ and (x, y, z) is _____.

3. The equation $(x + 1)^2 + (y - 3)^2 + (z - 5)^2 = 16$ determines a sphere with center _____ and radius _____.

4. The graph of $3x - 2y + 4z = 12$ is a _____ with x-intercept _____, y-intercept _____, and z-intercept _____.

Problem Set 14.1

1. Plot the points whose coordinates are $(1, 2, 3)$, $(2, 0, 1)$, $(-2, 4, 5)$, $(0, 3, 0)$, and $(-1, -2, -3)$. If appropriate, show the "box" as in Figures 4 and 5.

2. Follow the directions of Problem 1 for $(\sqrt{3}, -3, 3)$, $(0, \pi, -3)$, $(-2, \frac{1}{3}, 2)$, and $(0, 0, e)$.

3. What is peculiar to the coordinates of all points in the yz-plane? On the z-axis?

4. What is peculiar to the coordinates of all points in the xz-plane? On the y-axis?

5. Find the distance between the following pairs of points.

(a) $(6, -1, 0)$ and $(1, 2, 3)$ (b) $(-2, -2, 0)$ and $(2, -2, -3)$

(c) $(e, \pi, 0)$ and $(-\pi, -4, \sqrt{3})$

6. Show that $(4, 5, 3)$, $(1, 7, 4)$, and $(2, 4, 6)$ are vertices of an equilateral triangle.

7. Show that $(2, 1, 6)$, $(4, 7, 9)$, and $(8, 5, -6)$ are vertices of a right triangle. *Hint*: Only right triangles satisfy the Pythagorean Theorem.

8. Find the distance from $(2, 3, -1)$ to

(a) the xy-plane,   (b) the y-axis, and

(c) the origin.

9. A rectangular box has its faces parallel to the coordinate planes and has $(2, 3, 4)$ and $(6, -1, 0)$ as the end points of a main diagonal. Sketch the box and find the coordinates of all eight vertices.

10. $P(x, 5, z)$ is on a line through $Q(2, -4, 3)$ that is parallel to one of the coordinate axes. Which axis must it be and what are x and z?

11. Write the equation of the sphere with the given center and radius.

(a) $(1, 2, 3); 5$ (b) $(-2, -3, -6); \sqrt{5}$

(c) $(\pi, e, \sqrt{2}); \sqrt{\pi}$

12. Find the equation of the sphere whose center is $(2, 4, 5)$ and that is tangent to the xy-plane.

In Problems 13–16, complete the squares to find the center and radius of the sphere whose equation is given (see Example 2).

13. $x^2 + y^2 + z^2 - 12x + 14y - 8z + 1 = 0$

14. $x^2 + y^2 + z^2 + 2x - 6y - 10z + 34 = 0$

15. $4x^2 + 4y^2 + 4z^2 - 4x + 8y + 16z - 13 = 0$

16. $x^2 + y^2 + z^2 + 8x - 4y - 22z + 77 = 0$

In Problems 17–24, sketch the graphs of the given equations. Begin by sketching the traces in the coordinate planes (see Examples 4 and 5).

17. $2x + 6y + 3z = 12$ **18.** $3x - 4y + 2z = 24$

19. $x + 3y - z = 6$ **20.** $-3x + 2y + z = 6$

21. $x + 3y = 8$ **22.** $3x + 4z = 12$

23. $x^2 + y^2 + z^2 = 9$ **24.** $(x - 2)^2 + y^2 + z^2 = 4$

25. Find the equation of the sphere that has the line segment joining $(-2, 3, 6)$ and $(4, -1, 5)$ as a diameter (see Example 3).

26. Find the equations of the tangent spheres of equal radii whose centers are $(-3, 1, 2)$ and $(5, -3, 6)$.

27. Find the equation of the sphere that is tangent to the three coordinate planes if its radius is 6 and its center is in the first octant.

28. Find the equation of the sphere with center $(1, 1, 4)$ that is tangent to the plane $x + y = 12$.

29. Describe the graph in three-space of each equation.

(a) $z = 2$ (b) $x = y$

(c) $xy = 0$ (d) $xyz = 0$

(e) $x^2 + y^2 = 4$ (f) $z = \sqrt{9 - x^2 - y^2}$

30. The sphere $(x - 1)^2 + (y + 2)^2 + (z + 1)^2 = 10$ intersects the plane $z = 2$ in a circle. Find the circle's center and radius.

31. A point P moves so that its distance from $(1, 2, -3)$ is twice its distance from $(1, 2, 3)$. Show that P is on a sphere and find its center and radius.

32. A point P moves so that its distance from $(1, 2, -3)$ equals its distance from $(2, 3, 2)$. Find the equation of the plane on which P lies.

33. The balls $(x - 1)^2 + (y - 2)^2 + (z - 1)^2 \le 4$ and $(x - 2)^2 + (y - 4)^2 + (z - 3)^2 \le 4$ intersect in a solid. Find its volume.

34. Do Problem 33 assuming that the second ball is $(x - 2)^2 + (y - 4)^2 + (z - 3)^2 \le 9$.

Answers to Concepts Review: **1.** coordinates

2. $\sqrt{(x + 1)^2 + (y - 3)^2 + (z - 5)^2}$ **3.** $(-1, 3, 5); 4$

4. plane; $4; -6; 3$

14.2
Vectors in Three-Space

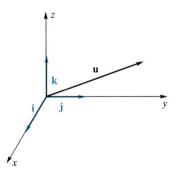

Figure 1

The material from Sections 13.2 and 13.3 on vectors in the plane can be repeated almost word for word for vectors in space. About the only difference is that a vector $\mathbf{u}$ now has three components; that is,

$$\mathbf{u} = \langle u_1, u_2, u_3 \rangle = u_1\mathbf{i} + u_2\mathbf{j} + u_3\mathbf{k}$$

Here, $\mathbf{i} = \langle 1, 0, 0 \rangle$, $\mathbf{j} = \langle 0, 1, 0 \rangle$, and $\mathbf{k} = \langle 0, 0, 1 \rangle$, are the standard unit vectors, called **basis vectors**, in the directions of the three positive coordinate axes (Figure 1). From the Distance Formula, the **length** of $\mathbf{u}$, denoted by $|\mathbf{u}|$, is given by

$$\boxed{|\mathbf{u}| = \sqrt{u_1^2 + u_2^2 + u_3^2}}$$

Vectors in space are added, multiplied by scalars, and subtracted just as in the plane, and the algebraic laws that are satisfied agree with those studied earlier. The **dot product** of $\mathbf{u} = \langle u_1, u_2, u_3 \rangle$ and $\mathbf{v} = \langle v_1, v_2, v_3 \rangle$ is defined by

$$\boxed{\mathbf{u} \cdot \mathbf{v} = u_1 v_1 + u_2 v_2 + u_3 v_3}$$

and it has the geometric interpretation noted in the previous chapter,

$$\boxed{\mathbf{u} \cdot \mathbf{v} = |\mathbf{u}||\mathbf{v}|\cos\theta}$$

where θ is the angle between $\mathbf{u}$ and $\mathbf{v}$. Consequently, it continues to be true that two vectors are perpendicular if and only if their dot product is zero.

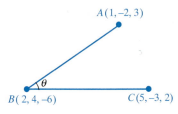

Figure 2

EXAMPLE 1 Find the angle ABC if $A = (1, -2, 3)$, $B = (2, 4, -6)$, and $C = (5, -3, 2)$ (Figure 2).

Solution First we determine vectors $\mathbf{u}$ and $\mathbf{v}$ (emanating from the origin) equivalent to $\overrightarrow{BA}$ and $\overrightarrow{BC}$. This is done by subtracting the coordinates of the initial points from those of the terminal points, that is,

$$\mathbf{u} = \langle 1 - 2, -2 - 4, 3 + 6 \rangle = \langle -1, -6, 9 \rangle$$

$$\mathbf{v} = \langle 5 - 2, -3 - 4, 2 + 6 \rangle = \langle 3, -7, 8 \rangle$$

Thus,

$$\cos\theta = \frac{\mathbf{u} \cdot \mathbf{v}}{|\mathbf{u}||\mathbf{v}|} = \frac{(-1)(3) + (-6)(-7) + (9)(8)}{\sqrt{1 + 36 + 81}\,\sqrt{9 + 49 + 64}} \approx 0.9251$$

$$\theta = 0.3894 \quad (\text{about } 22.31°) \qquad \blacksquare$$

EXAMPLE 2 Express $\mathbf{u} = \langle 2, 4, 5 \rangle$ as the sum of a vector $\mathbf{m}$ parallel to $\mathbf{v} = \langle 2, -1, -2 \rangle$ and a vector $\mathbf{n}$ perpendicular to $\mathbf{v}$.

Solution Figure 3 tells the story. First, we find $\mathbf{m} = \mathrm{pr}_{\mathbf{v}}\mathbf{u}$, the projection of $\mathbf{u}$ on $\mathbf{v}$ (see Section 13.3).

$$\mathbf{m} = \frac{\mathbf{u} \cdot \mathbf{v}}{|\mathbf{v}|^2}\mathbf{v}$$

$$= \frac{\langle 2, 4, 5 \rangle \cdot \langle 2, -1, -2 \rangle}{|\langle 2, -1, -2 \rangle|^2}\langle 2, -1, -2 \rangle$$

$$= \frac{(2)(2) + (4)(-1) + (5)(-2)}{4 + 1 + 4}\langle 2, -1, -2 \rangle$$

$$= \left\langle \frac{-20}{9}, \frac{10}{9}, \frac{20}{9} \right\rangle$$

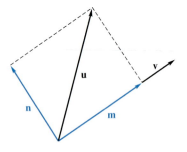

Figure 3

Then

$$\mathbf{n} = \mathbf{u} - \mathbf{m} = \left\langle \frac{38}{9}, \frac{26}{9}, \frac{25}{9} \right\rangle$$

If you doubt that $\mathbf{m}$ and $\mathbf{n}$ are perpendicular, compute their dot product. You will get zero. $\qquad \blacksquare$

Direction Angles and Cosines The (smallest nonnegative) angles between a nonzero vector $\mathbf{a}$ and the basis vectors $\mathbf{i}$, $\mathbf{j}$, and $\mathbf{k}$ are the **direction angles** of $\mathbf{a}$; they are designated by α, β, and γ, respectively (Figure 4). It is generally more convenient to work with the **direction cosines** $\cos\alpha$, $\cos\beta$, and $\cos\gamma$. If $\mathbf{a} = a_1\mathbf{i} + a_2\mathbf{j} + a_3\mathbf{k}$, then

$$\cos\alpha = \frac{\mathbf{a} \cdot \mathbf{i}}{|\mathbf{a}||\mathbf{i}|} = \frac{a_1}{|\mathbf{a}|}$$

and similarly

$$\cos\beta = \frac{a_2}{|\mathbf{a}|}, \qquad \cos\gamma = \frac{a_3}{|\mathbf{a}|}$$

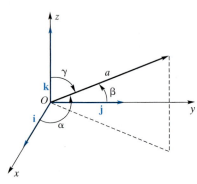

Figure 4

Notice that

$$\cos^2\alpha + \cos^2\beta + \cos^2\gamma = 1$$

In fact, the vector $\langle \cos\alpha, \cos\beta, \cos\gamma \rangle$ is a unit vector with the same direction as the original vector $\mathbf{a}$.

EXAMPLE 3   Find the direction angles for the vector $\mathbf{a} = 4\mathbf{i} - 5\mathbf{j} + 3\mathbf{k}$.

Solution   Since $|\mathbf{a}| = \sqrt{4^2 + (-5)^2 + 3^2} = 5\sqrt{2}$,

$$\cos\alpha = \frac{4}{5\sqrt{2}} = \frac{2\sqrt{2}}{5}, \qquad \cos\beta = \frac{-\sqrt{2}}{2}, \qquad \cos\gamma = \frac{3\sqrt{2}}{10}$$

and

$$\alpha \approx 55.55°, \qquad \beta = 135°, \qquad \gamma \approx 64.90° \qquad\qquad \blacksquare$$

EXAMPLE 4   Find a vector 5 units long that has $\alpha = 32°$ and $\beta = 100°$ as two of its direction angles.

Solution   First, we note that the third direction angle γ must satisfy

$$\cos^2\gamma = 1 - \cos^2 32° - \cos^2 100° \approx 0.25066$$

Thus,

$$\cos\gamma \approx \pm 0.50066$$

Two vectors meet the requirements of the problem. They are

$$5\langle\cos\alpha, \cos\beta, \cos\gamma\rangle \approx 5\langle 0.84805, -0.17365, 0.50066\rangle$$
$$= \langle 4.2403, -0.8683, 2.5033\rangle$$

and $\langle 4.2403, -0.8683, -2.5033\rangle$. $\blacksquare$

Planes   One fruitful way to describe a plane is by using vector language. Let $\mathbf{n} = \langle A, B, C\rangle$ be a fixed nonzero vector and $P_1(x_1, y_1, z_1)$ be a fixed point. The set of points $P(x, y, z)$ satisfying $\overrightarrow{P_1 P} \cdot \mathbf{n} = 0$ is the **plane** through P_1 perpendicular to $\mathbf{n}$. Since every plane contains a point and is perpendicular to some vector, a plane can be characterized in this way.

To get the Cartesian equation of the plane, write the vector $\overrightarrow{P_1 P}$ in component form; that is,

$$\overrightarrow{P_1 P} = \langle x - x_1, y - y_1, z - z_1\rangle$$

Then $\overrightarrow{P_1 P} \cdot \mathbf{n} = 0$ is equivalent to

$$\boxed{A(x - x_1) + B(y - y_1) + C(z - z_1) = 0}$$

This equation (in which at least one of A, B, and C is different from zero) is called the **standard form for the equation of a plane.**

If we remove the parentheses and simplify, the boxed equation takes the form of the general linear equation

$$Ax + By + Cz = D, \qquad A^2 + B^2 + C^2 \neq 0$$

Thus, every plane has a linear equation. Conversely, the graph of a linear equation in three-space is always a plane. To see the latter, let (x_1, y_1, z_1) satisfy the equation; that is,

$$Ax_1 + By_1 + Cz_1 = D$$

When we subtract this equation from the one above, we have the boxed equation, which we know represents a plane.

EXAMPLE 5   Find the equation of the plane through $(5, 1, -2)$ perpendicular to $\mathbf{n} = \langle 2, 4, 3\rangle$. Then find the angle between this plane and the one with equation $3x - 4y + 7z = 5$.

Solution   To perform the first task, simply apply the standard form for the equation of a plane to the problem at hand, which gives

$$2(x - 5) + 4(y - 1) + 3(z + 2) = 0$$

or, equivalently,

$$2x + 4y + 3z = 8$$

A vector **m** perpendicular to the second plane is $\mathbf{m} = \langle 3, -4, 7 \rangle$. The angle θ between two planes is the angle between their normals (Figure 5). Thus,

$$\cos \theta = \frac{\mathbf{m} \cdot \mathbf{n}}{|\mathbf{m}||\mathbf{n}|} = \frac{(3)(2) + (-4)(4) + (7)(3)}{\sqrt{9 + 16 + 49}\sqrt{4 + 16 + 9}} \approx 0.2375$$

$$\theta \approx 76.26°$$

Actually, there are two angles between two planes, but they are *supplementary*. The process just described will lead to one of them. The other, if desired, is obtained by subtracting the first value from 180°. In our case, it would be 103.74°. ■

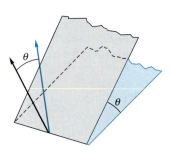

Figure 5

EXAMPLE 6   Show that the distance L from the point (x_0, y_0, z_0) to the plane $Ax + By + Cz = D$ is given by the formula

$$L = \frac{|Ax_0 + By_0 + Cz_0 - D|}{\sqrt{A^2 + B^2 + C^2}}$$

Solution   Let (x_1, y_1, z_1) be a point on the plane, and let $\mathbf{m} = \langle x_0 - x_1, y_0 - y_1, z_0 - z_1 \rangle$ be the vector from (x_1, y_1, z_1) to (x_0, y_0, z_0), as in Figure 6. Now $\mathbf{n} = \langle A, B, C \rangle$ is a vector perpendicular to the given plane, though it might point in the opposite direction of that in our figure. The number L that we seek is the length of the projection of **m** on **n**. Thus,

$$L = \|\mathbf{m}|\cos \theta| = \frac{|\mathbf{m} \cdot \mathbf{n}|}{|\mathbf{n}|}$$

$$= \frac{|A(x_0 - x_1) + B(y_0 - y_1) + C(z_0 - z_1)|}{\sqrt{A^2 + B^2 + C^2}}$$

$$= \frac{|Ax_0 + By_0 + Cz_0 - (Ax_1 + By_1 + Cz_1)|}{\sqrt{A^2 + B^2 + C^2}}$$

But (x_1, y_1, z_1) is on the plane, and so

$$Ax_1 + By_1 + Cz_1 = D$$

Substitution of this result in the expression for L yields the desired formula. ■

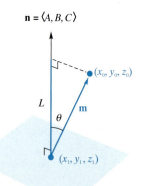

$\mathbf{n} = \langle A, B, C \rangle$

Figure 6

EXAMPLE 7   Find the distance between the parallel planes $3x - 4y + 5z = 9$ and $3x - 4y + 5z = 4$.

Solution   The planes are parallel, since the vector $\langle 3, -4, 5 \rangle$ is perpendicular to both of them (Figure 7). The point $(1, 1, 2)$ is easily seen to be on the first plane. We find the distance L from $(1, 1, 2)$ to the second plane using the formula of Example 6.

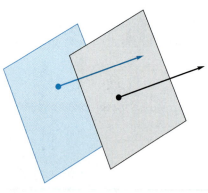

Figure 7

$$L = \frac{|3(1) - 4(1) + 5(2) - 4|}{\sqrt{9 + 16 + 25}} = \frac{5}{5\sqrt{2}} \approx 0.7071$$ ■

Concepts Review

1. Let $\mathbf{u} = \langle 2, -3, \sqrt{3} \rangle$ and $\mathbf{v} = \langle 3, 2, -2\sqrt{3} \rangle$ be two vectors. The length of $\mathbf{u}$ is $|\mathbf{u}| = $ _____ , and the dot product of $\mathbf{u}$ and $\mathbf{v}$ is $\mathbf{u} \cdot \mathbf{v} = $ _____ .

2. Two vectors are perpendicular if and only if their _____ is _____ .

3. $3(x - 2) - 2(y + 1) + 4z = 0$ is the equation of a plane through the point _____ , with _____ being a vector perpendicular to the plane.

4. The (smallest nonnegative) angle θ between the vectors $\mathbf{u}$ and $\mathbf{v}$ can be found from the geometric formula for the dot product, $\mathbf{u} \cdot \mathbf{v} = $ _____ . This gives $\theta = $ _____ .

Problem Set 14.2

1. For each pair of points P_1 and P_2 given below, sketch the directed line segment $\overrightarrow{P_1 P_2}$ and then write the corresponding vector in the form $a\mathbf{i} + b\mathbf{j} + c\mathbf{k}$.
(a) $P_1(1, 2, 4), P_2(4, 5, 6)$
(b) $P_1(-1, -3, 204), P_2(-14, 52, 26)$

2. Follow the directions of Problem 1.
(a) $P_1(-2, -2, -2), P_2(-3, -4, 5)$
(b) $P_1(0, -1, e), P_2(-\sqrt{14}, -5, \pi)$

3. Find the length of and direction cosines for each of the following vectors:
(a) $4\mathbf{i} + \mathbf{j} + 2\mathbf{k}$
(b) $-2\mathbf{i} - 3\mathbf{j} + 7\mathbf{k}$

4. Follow the directions for Problem 3.
(a) $\langle 2, -1, -2 \rangle$
(b) $\langle -1, 2, -2 \rangle$

5. Find the unit vector with the same direction as $\langle 3, -4, 5 \rangle$. Also, find a vector of length 5 oriented in the opposite direction.

6. Find a vector of length 10 with direction opposite to $-4\mathbf{i} + 3\mathbf{j} + -2\mathbf{k}$.

7. Find the angle between $\langle 4, -3, -1 \rangle$ and $\langle -2, -3, 5 \rangle$.

8. Find the angle between $-4\mathbf{i} + 2\mathbf{j} + 3\mathbf{k}$ and $2\mathbf{i} + \mathbf{j} + 5\mathbf{k}$.

9. Find two vectors of length 10, each of which is perpendicular to both $-4\mathbf{i} + 5\mathbf{j} + \mathbf{k}$ and $4\mathbf{i} + \mathbf{j}$.

10. Find all the vectors perpendicular to both $\langle 1, -2, -3 \rangle$ and $\langle -3, 2, 0 \rangle$.

11. Find the angle ABC if $A = (1, 2, 3), B = (-4, 5, 6)$, and $C = (1, 0, 1)$ (see Example 1).

12. Show that the triangle ABC is a right triangle if $A = (6, 3, 3), B = (3, 1, -1)$, and $C = (-1, 10, -2.5)$. *Hint*: Check the angle at B.

13. Find the *scalar projection* of $\mathbf{u} = -\mathbf{i} + 5\mathbf{j} + 3\mathbf{k}$ on $\mathbf{v} = -\mathbf{i} + \mathbf{j} - \mathbf{k}$. The scalar projection is the signed magnitude of the vector projection (Example 2); that is, it is $|\mathbf{u}| \cos \theta = \mathbf{u} \cdot \mathbf{v}/|\mathbf{v}|$.

14. Find the scalar projection of $\mathbf{u} = 5\mathbf{i} + 5\mathbf{j} + 2\mathbf{k}$ on $\mathbf{v} = -\sqrt{5}\mathbf{i} + \sqrt{5}\mathbf{j} + \mathbf{k}$.

15. If $\mathbf{u} = -3\mathbf{i} + 2\mathbf{j} + \mathbf{k}$ and $\mathbf{v} = -3\mathbf{i} + 5\mathbf{j} - 3\mathbf{k}$, express $\mathbf{u}$ as the sum of a vector $\mathbf{m}$ parallel to $\mathbf{v}$ and a vector $\mathbf{n}$ perpendicular to $\mathbf{v}$ (see Example 2).

16. Follow the directions of Problem 15 for $\mathbf{u} = e\mathbf{i} + \pi\mathbf{j} + \mathbf{k}$ and $\mathbf{v} = \mathbf{i} + \mathbf{j}$.

17. Find the direction angles for each vector.
(a) $\mathbf{u} = -3\mathbf{i} + 2\mathbf{j} + \mathbf{k}$
(b) $\mathbf{u} = 3\mathbf{i} + 6\mathbf{j} - \mathbf{k}$

18. If $\alpha = 46°$ and $\beta = 108°$ are direction angles for a vector $\mathbf{u}$, find two possible values for the third angle (see Example 4).

19. A vector $\mathbf{u} = 2\mathbf{i} + 3\mathbf{j} + z\mathbf{k}$ emanating from the origin points into the first octant. If $|\mathbf{u}| = 5$, find z.

20. If $\mathbf{u} = 2\mathbf{i} + 3\mathbf{j} + z\mathbf{k}$ and $\mathbf{v} = 2\mathbf{i} + 6\mathbf{j} - 3\mathbf{k}$ are perpendicular, find z.

21. Find two perpendicular vectors $\mathbf{u}$ and $\mathbf{v}$ such that each is also perpendicular to $\mathbf{w} = \langle -4, 2, 5 \rangle$.

22. Find the vector emanating from the origin whose terminal point is the midpoint of the segment joining $(3, 2, -1)$ and $(5, -7, 2)$.

23. Which of the following do *not* make sense?
(a) $\mathbf{u} \cdot (\mathbf{v} \cdot \mathbf{w})$
(b) $(\mathbf{u} \cdot \mathbf{w}) + \mathbf{w}$
(c) $|\mathbf{u}|(\mathbf{v} \cdot \mathbf{w})$
(d) $(\mathbf{u} \cdot \mathbf{v})\mathbf{w}$
(e) $(|\mathbf{u}|\mathbf{v}) \cdot \mathbf{w}$
(f) $|\mathbf{u}| \cdot \mathbf{v}$

24. Let $\mathbf{a}, \mathbf{b}, \mathbf{c}$, and $\mathbf{d}$ be vectors emanating from the origin and terminating at A, B, C, and D, respectively. Use vector notation to express a necessary and sufficient condition that the figure $ABCD$ be a parallelogram.

25. Find the equation of the plane passing through P and perpendicular to $\mathbf{n}$ (see Example 5).
(a) $P(1, 2, -3), \mathbf{n} = 2\mathbf{i} - 4\mathbf{j} + 3\mathbf{k}$
(b) $P(-2, -3, 4), \mathbf{n} = 3\mathbf{i} - 2\mathbf{j} - \mathbf{k}$

26. Find the smaller of the angles between the planes $3x - 2y + 5z = 7$ and $4x - 2y - 3z = 2$ (see Example 5).

27. Find the smaller of the angles between the two planes of Problem 25.

28. Find the equation of a plane through $(-1, 2, -3)$ and parallel to the plane $2x + 4y - z = 6$.

29. Find the equation of the plane through $(-4, -1, 2)$ and parallel
(a) to the xy-plane,
(b) to the plane $2x - 3y - 4z = 0$

30. Find the distance from $(1, -1, 2)$ to the plane
$$x + 3y + z = 7$$
(see Example 6).

31. Find the distance from $(2, 6, 3)$ to the plane

$$-3x + 2y + z = 9$$

32. Find the distance between the parallel planes

$$-3x + 2y + z = 9 \quad \text{and} \quad 6x - 4y - 2z = 19$$

(see Example 7).

33. Find the distance between the parallel planes

$$5x - 3y - 2z = 5 \quad \text{and} \quad -5x + 3y + 2z = 7$$

34. Find the equation of the plane each of whose points is equidistant from $(-2, 1, 4)$ and $(6, 1, -2)$.

35. Prove that $|\mathbf{u} + \mathbf{v}|^2 + |\mathbf{u} - \mathbf{v}|^2 = 2|\mathbf{u}|^2 + 2|\mathbf{v}|^2$. *Hint:* $|\mathbf{w}|^2 = \mathbf{w} \cdot \mathbf{w}$.

36. Prove that $\mathbf{u} \cdot \mathbf{v} = \frac{1}{4}|\mathbf{u} + \mathbf{v}|^2 - \frac{1}{4}|\mathbf{u} - \mathbf{v}|^2$.

37. Find the angle between a main diagonal of a cube and one of its faces.

38. Find a unit vector whose direction angles are equal.

39. Find the smallest angle between the main diagonals of a rectangular box 4 feet by 6 feet by 10 feet.

40. Find the angles formed by the diagonals of a cube.

41. A constant force of $\mathbf{F} = -4\mathbf{k}$ newtons is applied to an object in moving it from $(0, 0, 8)$ to $(4, 4, 0)$, where coordinates are given in meters. Find the work done. (Recall that $W = \mathbf{F} \cdot \mathbf{D}$ (see Section 13.3).)

42. A constant force of $\mathbf{F} = 3\mathbf{i} - 6\mathbf{j} + 7\mathbf{k}$ pounds is applied to an object in moving it from $(2, 1, 3)$ to $(9, 4, 6)$, coordinates given in feet. Find the work done.

43. How much work is done by a force of 5 newtons acting in the direction $2\mathbf{i} + 2\mathbf{j} - \mathbf{k}$ in moving an object from $(0, 1, 2)$ to $(3, 5, 7)$, distances being measured in meters? (See Problem 41.)

44. A weight of 30 pounds is suspended by three wires with resulting tensions $3\mathbf{i} + 4\mathbf{j} + 15\mathbf{k}$, $-8\mathbf{i} - 2\mathbf{j} + 10\mathbf{k}$, and $a\mathbf{i} + b\mathbf{j} + c\mathbf{k}$. Determine a, b, and c, assuming that $\mathbf{k}$ points straight up.

45. Find the point one-fifth of the way from $(2, 3, -1)$ to $(7, -2, 9)$.

46. Suppose that the three coordinate planes bounding the first octant are mirrors. A light ray with direction $a\mathbf{i} + b\mathbf{j} + c\mathbf{k}$ is reflected successively from the xy-plane, the xz-plane, and the yz-plane. Determine the direction of the ray after each reflection, and state a nice conclusion concerning the final reflected ray.

47. Find the distance from the sphere $x^2 + y^2 + z^2 + 2x + 6y - 8z = 0$ to the plane $3x + 4y + z = 15$.

48. Refine the method of Example 7 by showing that the distance L between the parallel planes $Ax + By + Cz = D$ and $Ax + By + Cz = E$ is

$$L = \frac{|D - E|}{\sqrt{A^2 + B^2 + C^2}}$$

49. Let $\mathbf{a} = \langle a_1, a_2, a_3 \rangle$ and $\mathbf{b} = \langle b_1, b_2, b_3 \rangle$ be fixed vectors. Show that $(\mathbf{x} - \mathbf{a}) \cdot (\mathbf{x} - \mathbf{b}) = 0$ is the equation of a sphere, and find its center and radius.

50. Show that the work done by a constant force $\mathbf{F}$ on an object that moves completely around a closed polygonal path is 0.

51. The medians of a triangle meet in a point P (the centroid by Problem 30 of Section 6.6) that is two-thirds of the way from a vertex to the midpoint of the opposite edge. Show that P is the head of the position vector $(\mathbf{a} + \mathbf{b} + \mathbf{c})/3$, where $\mathbf{a}$, $\mathbf{b}$, and $\mathbf{c}$ are the position vectors of the vertices, and use this to find P if the vertices are $(2, 6, 5)$, $(4, -1, 2)$, and $(6, 1, 2)$.

52. Let $\mathbf{a}$, $\mathbf{b}$, $\mathbf{c}$, and $\mathbf{d}$ be the position vectors of the vertices of a tetrahedron. Show that the lines joining the vertices to the centroids of the opposite faces meet in a point P, and give a nice vector formula for it, thus generalizing Problem 51.

Answers to Concepts Review: **1.** 4; -6 **2.** dot product; 0 **3.** $(2, -1, 0)$; $\langle 3, -2, 4 \rangle$ **4.** $|\mathbf{u}||\mathbf{v}|\cos\theta$; $\cos^{-1}(\mathbf{u} \cdot \mathbf{v}/|\mathbf{u}||\mathbf{v}|)$

14.3
The Cross Product

The dot product of two vectors is a scalar. We have explored some of its uses in earlier sections. Now we introduce the **cross product** (or vector product); it will also have many uses. The cross product $\mathbf{u} \times \mathbf{v}$ of $\mathbf{u} = \langle u_1, u_2, u_3 \rangle$ and $\mathbf{v} = \langle v_1, v_2, v_3 \rangle$ is defined by

$$\mathbf{u} \times \mathbf{v} = \langle u_2 v_3 - u_3 v_2, u_3 v_1 - u_1 v_3, u_1 v_2 - u_2 v_1 \rangle$$

In this form, the formula is hard to remember and its significance is not obvious. Note the one thing that is obvious. The cross product of two vectors is a vector.

To help us remember the formula for the cross product, we recall a subject from an earlier mathematics course, namely, *determinants*. First, the value of a 2×2 determinant is

$$\begin{vmatrix} a & b \\ c & d \end{vmatrix} = ad - bc$$

Then the value of a 3×3 determinant is (expanding along to the top row)

$$\begin{vmatrix} a_1 & a_2 & a_3 \\ b_1 & b_2 & b_3 \\ c_1 & c_2 & c_3 \end{vmatrix} = a_1 \begin{vmatrix} a_1 & a_2 & a_3 \\ b_1 & b_2 & b_3 \\ c_1 & c_2 & c_3 \end{vmatrix} - a_2 \begin{vmatrix} a_1 & a_2 & a_3 \\ b_1 & b_2 & b_3 \\ c_1 & c_2 & c_3 \end{vmatrix} + a_3 \begin{vmatrix} a_1 & a_2 & a_3 \\ b_1 & b_2 & b_3 \\ c_1 & c_2 & c_3 \end{vmatrix}$$

$$= a_1 \begin{vmatrix} b_2 & b_3 \\ c_2 & c_3 \end{vmatrix} - a_2 \begin{vmatrix} b_1 & b_3 \\ c_1 & c_3 \end{vmatrix} + a_3 \begin{vmatrix} b_1 & b_2 \\ c_1 & c_2 \end{vmatrix}$$

Torque

The cross product plays an important role in mechanics. Let O be a *fixed point* in a body, and suppose that a force $\mathbf{F}$ is applied at another point P of the body. Then $\mathbf{F}$ tends to rotate the body about an axis through O and perpendicular to the plane of OP and $\mathbf{F}$. The vector

$$\tau = \overrightarrow{OP} \times \mathbf{F}$$

is called the **torque**. It points in the direction of the axis and has magnitude $|\overrightarrow{OP}||\mathbf{F}|\sin\theta$, which is just the moment of force about the axis due to $\mathbf{F}$.

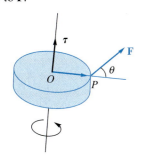

Using determinants, we may write the definition of $\mathbf{u} \times \mathbf{v}$ as

$$\mathbf{u} \times \mathbf{v} = \begin{vmatrix} \mathbf{i} & \mathbf{j} & \mathbf{k} \\ u_1 & u_2 & u_3 \\ v_1 & v_2 & v_3 \end{vmatrix} = \begin{vmatrix} u_2 & u_3 \\ v_2 & v_3 \end{vmatrix}\mathbf{i} - \begin{vmatrix} u_1 & u_3 \\ v_1 & v_3 \end{vmatrix}\mathbf{j} + \begin{vmatrix} u_1 & u_2 \\ v_1 & v_2 \end{vmatrix}\mathbf{k}$$

Note that the components of the left vector $\mathbf{u}$ go in the second row, and those of the right vector $\mathbf{v}$ go in the third row. This is important, because if we interchange the positions of $\mathbf{u}$ and $\mathbf{v}$, we interchange the second and third rows of the determinant, and this changes the sign of the determinant's value, as you may check. Thus,

$$\mathbf{u} \times \mathbf{v} = -(\mathbf{v} \times \mathbf{u})$$

which is sometimes called the *anticommutative law*.

EXAMPLE 1 Let $\mathbf{u} = \langle 1, -2, -1 \rangle$ and $\mathbf{v} = \langle -2, 4, 1 \rangle$. Calculate $\mathbf{u} \times \mathbf{v}$ and $\mathbf{v} \times \mathbf{u}$ using the determinant definition.

Solution

$$\mathbf{u} \times \mathbf{v} = \begin{vmatrix} \mathbf{i} & \mathbf{j} & \mathbf{k} \\ 1 & -2 & -1 \\ -2 & 4 & 1 \end{vmatrix} = \mathbf{i}\begin{vmatrix} -2 & -1 \\ 4 & 1 \end{vmatrix} - \mathbf{j}\begin{vmatrix} 1 & -1 \\ -2 & 1 \end{vmatrix} + \mathbf{k}\begin{vmatrix} 1 & -2 \\ -2 & 4 \end{vmatrix}$$

$$= 2\mathbf{i} + \mathbf{j} + 0\mathbf{k}$$

$$\mathbf{v} \times \mathbf{u} = \begin{vmatrix} \mathbf{i} & \mathbf{j} & \mathbf{k} \\ -2 & 4 & 1 \\ 1 & -2 & -1 \end{vmatrix} = \mathbf{i}\begin{vmatrix} 4 & 1 \\ -2 & -1 \end{vmatrix} - \mathbf{j}\begin{vmatrix} -2 & 1 \\ 1 & -1 \end{vmatrix} + \mathbf{k}\begin{vmatrix} -2 & 4 \\ 1 & -2 \end{vmatrix}$$

$$= -2\mathbf{i} - \mathbf{j} + 0\mathbf{k} \qquad \blacksquare$$

Geometric Interpretation of $\mathbf{u} \times \mathbf{v}$ Like the dot product, the cross product gains significance from its geometric interpretation.

Theorem A

Let $\mathbf{u}$ and $\mathbf{v}$ be vectors in three-space and θ be the angle between them. Then:

1. $\mathbf{u} \cdot (\mathbf{u} \times \mathbf{v}) = 0 = \mathbf{v} \cdot (\mathbf{u} \times \mathbf{v})$, that is, $\mathbf{u} \times \mathbf{v}$ is perpendicular to both $\mathbf{u}$ and $\mathbf{v}$;
2. $\mathbf{u}, \mathbf{v}$, and $\mathbf{u} \times \mathbf{v}$ form a right-handed triple;
3. $|\mathbf{u} \times \mathbf{v}| = |\mathbf{u}||\mathbf{v}|\sin\theta$.

Figure 1

Proof Let $\mathbf{u} = \langle u_1, u_2, u_3 \rangle$ and $\mathbf{v} = \langle v_1, v_2, v_3 \rangle$.

1. $\mathbf{u} \cdot (\mathbf{u} \times \mathbf{v}) = u_1(u_2 v_3 - u_3 v_2) + u_2(u_3 v_1 - u_1 v_3) + u_3(u_1 v_2 - u_2 v_1)$. When we remove parentheses, the six terms cancel in pairs. A similar event occur's when we expand $\mathbf{v} \cdot (\mathbf{u} \times \mathbf{v})$.

2. The meaning of right-handedness for the triple $\mathbf{u}, \mathbf{v}, \mathbf{u} \times \mathbf{v}$ is illustrated in Figure 1. There θ is the angle between $\mathbf{u}$ and $\mathbf{v}$, and the fingers of the right hand are curled in the direction of the rotation through θ that makes $\mathbf{u}$ coincide with $\mathbf{v}$. It is difficult to establish analytically that the indicated triple is right-handed, but you might check it with a few examples. Note in particular that $\mathbf{i} \times \mathbf{j} = \mathbf{k}$, and by definition we know that the triple $\mathbf{i}, \mathbf{j}, \mathbf{k}$ is right-handed.

3. We need Lagrange's Identity,

$$|\mathbf{u} \times \mathbf{v}|^2 = |\mathbf{u}|^2|\mathbf{v}|^2 - (\mathbf{u} \cdot \mathbf{v})^2$$

whose proof is a simple algebraic exercise (Problem 25). Using this identity, we may write

$$|\mathbf{u} \times \mathbf{v}|^2 = |\mathbf{u}|^2|\mathbf{v}|^2 - (|\mathbf{u}||\mathbf{v}|\cos\theta)^2$$
$$= |\mathbf{u}|^2|\mathbf{v}|^2(1 - \cos^2\theta)$$
$$= |\mathbf{u}|^2|\mathbf{v}|^2 \sin^2\theta$$

Since $0 \le \theta \le \pi$, $\sin\theta \ge 0$. Taking principal square roots yields

$$|\mathbf{u} \times \mathbf{v}| = |\mathbf{u}||\mathbf{v}|\sin\theta \quad \blacklozenge$$

It is important that we have geometric interpretations of both $\mathbf{u} \cdot \mathbf{v}$ and $\mathbf{u} \times \mathbf{v}$. While both products were originally defined in terms of components that depend on a choice of coordinate system, they are actually independent of coordinate systems. They are intrinsic geometric quantities, and you will get the same results for $\mathbf{u} \cdot \mathbf{v}$ and $\mathbf{u} \times \mathbf{v}$ no matter how you introduce the coordinates used to compute them.

Here is a simple consequence of Theorem A (part 3) and the fact that vectors are parallel if and only if the angle θ between them is either 0° or 180°.

> **Theorem B**
>
> Two vectors $\mathbf{u}$ and $\mathbf{v}$ in three-space are parallel if and only if $\mathbf{u} \times \mathbf{v} = \mathbf{0}$.

Applications   Our first application is to find the equation of the plane through three noncollinear points.

EXAMPLE 2   Find the equation of the plane (Figure 2) through the three points $P_1(1, -2, 3)$, $P_2(4, 1, -2)$, and $P_3(-2, -3, 0)$.

Solution   Let $\mathbf{u} = \overrightarrow{P_2P_1} = \langle -3, -3, 5\rangle$ and $\mathbf{v} = \overrightarrow{P_2P_3} = \langle -6, -4, 2\rangle$. From the first part of Theorem A we know that

$$\mathbf{u} \times \mathbf{v} = \begin{vmatrix} \mathbf{i} & \mathbf{j} & \mathbf{k} \\ -3 & -3 & 5 \\ -6 & -4 & 2 \end{vmatrix} = 14\mathbf{i} - 24\mathbf{j} - 6\mathbf{k}$$

is perpendicular to both $\mathbf{u}$ and $\mathbf{v}$ and thus to the plane containing them. The plane through $(4, 1, -2)$ with normal $14\mathbf{i} - 24\mathbf{j} - 6\mathbf{k}$ has equation (see Section 14.2)

$$14(x - 4) - 24(y - 1) - 6(z + 2) = 0$$

or

$$14x - 24y - 6z = 44 \qquad \blacksquare$$

EXAMPLE 3   Show that the area of a parallelogram with $\mathbf{a}$ and $\mathbf{b}$ as adjacent sides is $|\mathbf{a} \times \mathbf{b}|$.

Solution   Recall that the area of a parallelogram is the product of the base times the height. Now look at Figure 3 and use the fact that $|\mathbf{a} \times \mathbf{b}| = |\mathbf{a}||\mathbf{b}|\sin\theta$. $\blacksquare$

EXAMPLE 4   Show that the volume of the parallelepiped determined by the vectors $\mathbf{a}$, $\mathbf{b}$, and $\mathbf{c}$ is

$$V = |\mathbf{a} \cdot (\mathbf{b} \times \mathbf{c})| = \left|\begin{vmatrix} a_1 & a_2 & a_3 \\ b_1 & b_2 & b_3 \\ c_1 & c_2 & c_3 \end{vmatrix}\right|$$

Solution   Refer to Figure 4 and regard the parallelogram determined by $\mathbf{b}$ and $\mathbf{c}$ as the base of the parallelepiped. The area of this base is $|\mathbf{b} \times \mathbf{c}|$ by Example 3; the height h of the parallelepiped is the absolute value of the scalar projection of $\mathbf{a}$ on $\mathbf{b} \times \mathbf{c}$. Thus,

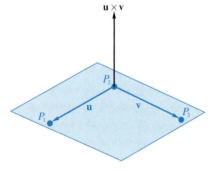

Figure 2

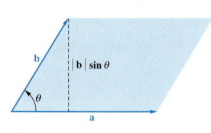

Figure 3

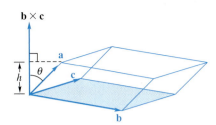

Figure 4

$$h = |\mathbf{a}||\cos\theta| = \frac{|\mathbf{a}||\mathbf{a} \cdot (\mathbf{b} \times \mathbf{c})|}{|\mathbf{a}||\mathbf{b} \times \mathbf{c}|} = \frac{|\mathbf{a} \cdot (\mathbf{b} \times \mathbf{c})|}{|\mathbf{b} \times \mathbf{c}|}$$

and

$$V = h|\mathbf{b} \times \mathbf{c}| = |\mathbf{a} \cdot (\mathbf{b} \times \mathbf{c})|$$

That V can also be expressed as a determinant is established by expanding $|\mathbf{a} \cdot (\mathbf{b} \times \mathbf{c})|$ in terms of components and then comparing it with the value of the indicated determinant.   ■

Suppose that the vectors **a**, **b**, and **c** from the previous example are in the *same* plane. In this case, the parallelepiped has height zero, so the volume should be zero. Does the formula for the volume yield $V = 0$? If **a** is in the plane determined by **b** and **c**, then any vector perpendicular to **b** and **c** will be perpendicular to **a** as well. The vector $\mathbf{b} \times \mathbf{c}$ is perpendicular to both **b** and **c**; hence $\mathbf{b} \times \mathbf{c}$ is perpendicular to **a**. Thus, $\mathbf{a} \cdot (\mathbf{b} \times \mathbf{c}) = 0$.

Algebraic Properties   The rules for calculating with cross products are summarized in the following theorem. Proving this theorem is a matter of writing everything out in terms of components and will be left as an exercise.

> ### Theorem C
>
> If **u**, **v**, and **w** are vectors in three-space and k is a scalar, then:
>
> 1. $\mathbf{u} \times \mathbf{v} = -(\mathbf{v} \times \mathbf{u})$ (anticommutative law);
> 2. $\mathbf{u} \times (\mathbf{v} + \mathbf{w}) = (\mathbf{u} \times \mathbf{v}) + (\mathbf{u} \times \mathbf{w})$ (left distributive law);
> 3. $k(\mathbf{u} \times \mathbf{v}) = (k\mathbf{u}) \times \mathbf{v} = \mathbf{u} \times (k\mathbf{v})$;
> 4. $\mathbf{u} \times \mathbf{0} = \mathbf{0} \times \mathbf{u} = \mathbf{0}, \mathbf{u} \times \mathbf{u} = \mathbf{0}$;
> 5. $(\mathbf{u} \times \mathbf{v}) \cdot \mathbf{w} = \mathbf{u} \cdot (\mathbf{v} \times \mathbf{w})$;
> 6. $\mathbf{u} \times (\mathbf{v} \times \mathbf{w}) = (\mathbf{u} \cdot \mathbf{w})\mathbf{v} - (\mathbf{u} \cdot \mathbf{v})\mathbf{w}$.

Once the rules in Theorem C are mastered, complicated calculations with vectors can be done with ease. We illustrate by calculating a cross product in a new way. We will need the following simple but important products.

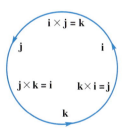

Figure 5

$$\boxed{\mathbf{i} \times \mathbf{j} = \mathbf{k}, \qquad \mathbf{j} \times \mathbf{k} = \mathbf{i}, \qquad \mathbf{k} \times \mathbf{i} = \mathbf{j}}$$

These results have a cyclic order, which can be remembered by appealing to Figure 5.

EXAMPLE 5   Calculate $\mathbf{u} \times \mathbf{v}$ if $\mathbf{u} = 3\mathbf{i} - 2\mathbf{j} + \mathbf{k}$ and $\mathbf{v} = 4\mathbf{i} + 2\mathbf{j} - 3\mathbf{k}$.

Solution   We appeal to Theorem C, especially the distributive law and the anticommutative law.

$$\begin{aligned}
\mathbf{u} \times \mathbf{v} = {} & (3\mathbf{i} - 2\mathbf{j} + \mathbf{k}) \times (4\mathbf{i} + 2\mathbf{j} - 3\mathbf{k}) \\
= {} & 12(\mathbf{i} \times \mathbf{i}) + 6(\mathbf{i} \times \mathbf{j}) - 9(\mathbf{i} \times \mathbf{k}) - 8(\mathbf{j} \times \mathbf{i}) - 4(\mathbf{j} \times \mathbf{j}) \\
& + 6(\mathbf{j} \times \mathbf{k}) + 4(\mathbf{k} \times \mathbf{i}) + 2(\mathbf{k} \times \mathbf{j}) - 3(\mathbf{k} \times \mathbf{k}) \\
= {} & 12(\mathbf{0}) + 6(\mathbf{k}) - 9(-\mathbf{j}) - 8(-\mathbf{k}) - 4(\mathbf{0}) \\
& + 6(\mathbf{i}) + 4(\mathbf{j}) + 2(-\mathbf{i}) - 3(\mathbf{0}) \\
= {} & 4\mathbf{i} + 13\mathbf{j} + 14\mathbf{k}
\end{aligned}$$

Experts would do most of this in their heads; novices might find the determinant method easier.   ■

Concepts Review

1. The cross product of $\mathbf{u} = \langle -1, 2, 1 \rangle$ and $\mathbf{v} = \langle 3, 1, -1 \rangle$ is given by a specific determinant; evaluation of this determinant gives $\mathbf{u} \times \mathbf{v} = \underline{\hspace{1cm}}$.

2. Geometrically, $\mathbf{u} \times \mathbf{v}$ is a vector perpendicular to the plane of $\mathbf{u}$ and $\mathbf{v}$ and has length $|\mathbf{u} \times \mathbf{v}| = \underline{\hspace{1cm}}$.

3. The cross product is anticommutative; that is, $\mathbf{u} \times \mathbf{v} = \underline{\hspace{1cm}}$.

4. Two vectors are $\underline{\hspace{1cm}}$ if and only if their cross product is $\mathbf{0}$.

Problem Set 14.3

1. Let $\mathbf{a} = -3\mathbf{i} + 2\mathbf{j} - 2\mathbf{k}$, $\mathbf{b} = -\mathbf{i} + 2\mathbf{j} - 4\mathbf{k}$, and $\mathbf{c} = 7\mathbf{i} + 3\mathbf{j} - 4\mathbf{k}$. Find each of the following:
(a) $\mathbf{a} \times \mathbf{b}$ (b) $\mathbf{a} \times (\mathbf{b} + \mathbf{c})$
(c) $\mathbf{a} \cdot (\mathbf{b} + \mathbf{c})$ (d) $\mathbf{a} \times (\mathbf{b} \times \mathbf{c})$

2. If $\mathbf{a} = \langle 3, 3, 1 \rangle$, $\mathbf{b} = \langle -2, -1, 0 \rangle$, and $\mathbf{c} = \langle -2, -3, -1 \rangle$, find each of the following:
(a) $\mathbf{a} \times \mathbf{b}$ (b) $\mathbf{a} \times (\mathbf{b} + \mathbf{c})$
(c) $\mathbf{a} \cdot (\mathbf{b} \times \mathbf{c})$ (d) $\mathbf{a} \times (\mathbf{b} \times \mathbf{c})$

3. Find all vectors perpendicular to both of the vectors $\mathbf{a} = \mathbf{i} + 2\mathbf{j} + 3\mathbf{k}$ and $\mathbf{b} = -2\mathbf{i} + 2\mathbf{j} - 4\mathbf{k}$.

4. Find all vectors perpendicular to both of the vectors $\mathbf{a} = -2\mathbf{i} + 5\mathbf{j} - 2\mathbf{k}$ and $\mathbf{b} = 3\mathbf{i} - 2\mathbf{j} + 4\mathbf{k}$.

5. Find the unit vectors perpendicular to the plane determined by the three points $(1, 3, 5)$, $(3, -1, 2)$, and $(4, 0, 1)$.

6. Find the unit vectors perpendicular to the plane determined by the three points $(-1, 3, 0)$, $(5, 1, 2)$, and $(4, -3, -1)$.

7. Find the area of the parallelogram with $\mathbf{a} = -\mathbf{i} + \mathbf{j} - 3\mathbf{k}$ and $\mathbf{b} = 4\mathbf{i} + 2\mathbf{j} - 4\mathbf{k}$ as the adjacent sides.

8. Find the area of the parallelogram with $\mathbf{a} = 2\mathbf{i} + 2\mathbf{j} - \mathbf{k}$ and $\mathbf{b} = -\mathbf{i} + \mathbf{j} - 4\mathbf{k}$ as the adjacent sides.

9. Find the area of the triangle with $(3, 2, 1)$, $(2, 4, 6)$, and $(-1, 2, 5)$ as vertices.

10. Find the area of the triangle with $(1, 2, 3)$, $(3, 1, 5)$, and $(4, 5, 6)$ as vertices.

11. Find the equation of the plane through $(1, 3, 2)$, $(0, 3, 0)$, and $(2, 4, 3)$ (see Example 2).

12. Find the equation of the plane through $(1, 1, 2)$, $(0, 0, 1)$, and $(-2, -3, 0)$.

13. Find the equation of the plane through $(-1, -2, 3)$ and perpendicular to both the planes $x - 3y + 2z = 7$ and $2x - 2y - z = -3$.

14. Find the equation of the plane through $(2, -3, 2)$ and parallel to the plane of the vectors $4\mathbf{i} + 3\mathbf{j} - \mathbf{k}$ and $2\mathbf{i} - 5\mathbf{j} + 6\mathbf{k}$.

15. Find the equation of the plane through $(6, 2, -1)$ and perpendicular to the line of intersection of the planes $4x - 3y + 2z + 5 = 0$ and $3x + 2y - z + 11 = 0$.

16. Let $\mathbf{a}$ and $\mathbf{b}$ be nonparallel vectors, and let $\mathbf{c}$ be any nonzero vector. Show that $(\mathbf{a} \times \mathbf{b}) \times \mathbf{c}$ is a vector in the plane of $\mathbf{a}$ and $\mathbf{b}$.

17. Find the volume of the parallelepiped with edges $\langle 2, 3, 4 \rangle$, $\langle 0, 4, -1 \rangle$, and $\langle 5, 1, 3 \rangle$ (see Example 4).

18. Find the volume of the parallelepiped with edges $3\mathbf{i} - 4\mathbf{j} + 2\mathbf{k}$, $-\mathbf{i} + 2\mathbf{j} + \mathbf{k}$, and $3\mathbf{i} - 2\mathbf{j} + 5\mathbf{k}$.

19. Let K be the parallelepiped determined by $\mathbf{u} = \langle 3, 2, 1 \rangle$, $\mathbf{v} = \langle 1, 1, 2 \rangle$, and $\mathbf{w} = \langle 1, 3, 3 \rangle$.
(a) Find the volume of K.
(b) Find the area of the face determined by $\mathbf{u}$ and $\mathbf{v}$.
(c) Find the angle between $\mathbf{u}$ and the plane containing the face determined by $\mathbf{v}$ and $\mathbf{w}$.

20. The formula for the volume of a parallelepiped derived in Example 4 should not depend on the choice of which one of the three vectors we call $\mathbf{a}$, which one we call $\mathbf{b}$, and which one we call $\mathbf{c}$. Use this result to explain why $|\mathbf{a} \cdot (\mathbf{b} \times \mathbf{c})| = |\mathbf{b} \cdot (\mathbf{a} \times \mathbf{c})| = |\mathbf{c} \cdot (\mathbf{a} \times \mathbf{b})|$.

21. Which of the following do *not* make sense?
(a) $\mathbf{u} \cdot (\mathbf{v} \times \mathbf{w})$ (b) $\mathbf{u} + (\mathbf{v} \times \mathbf{w})$
(c) $(\mathbf{a} \cdot \mathbf{b}) \times \mathbf{c}$ (d) $(\mathbf{a} \times \mathbf{b}) + k$
(e) $(\mathbf{a} \cdot \mathbf{b}) + k$ (f) $(\mathbf{a} + \mathbf{b}) \times (\mathbf{c} + \mathbf{d})$
(g) $(\mathbf{u} \times \mathbf{v}) \times \mathbf{w}$ (h) $(k\mathbf{u}) \times \mathbf{v}$

22. Show that if $\mathbf{a}, \mathbf{b}, \mathbf{c}$, and $\mathbf{d}$ all lie in the same plane then
$$(\mathbf{a} \times \mathbf{b}) \times (\mathbf{c} \times \mathbf{d}) = \mathbf{0}$$

23. The volume of a tetrahedron is known to be $\frac{1}{3}$(area of base)(height). From this, show that the volume of the tetrahedron with edges $\mathbf{a}, \mathbf{b}$, and $\mathbf{c}$ is $\frac{1}{6}|\mathbf{a} \cdot (\mathbf{b} \times \mathbf{c})|$.

24. Find the volume of the tetrahedron with vertices $(-1, 2, 3)$, $(4, -1, 2)$, $(5, 6, 3)$, and $(1, 1, -2)$ (see Problem 23).

25. Prove **Lagrange's Identity**,
$$|\mathbf{u} \times \mathbf{v}|^2 = |\mathbf{u}|^2|\mathbf{v}|^2 - (\mathbf{u} \cdot \mathbf{v})^2$$
without using Theorem A.

26. Prove the left distributive law,
$$\mathbf{u} \times (\mathbf{v} + \mathbf{w}) = (\mathbf{u} \times \mathbf{v}) + (\mathbf{u} \times \mathbf{w})$$

27. Use Problem 26 and the anticommutative law to prove the right distributive law.

28. If both $\mathbf{u} \times \mathbf{v} = \mathbf{0}$ and $\mathbf{u} \cdot \mathbf{v} = 0$, what can you conclude about $\mathbf{u}$ or $\mathbf{v}$?

29. Use Example 3 to develop a formula for the area of the triangle with vertices $P(a, 0, 0), Q(0, b, 0)$, and $R(0, 0, c)$ shown in the top half of Figure 6.

30. Show that the triangle in the plane with vertices (x_1, y_1), (x_2, y_2), and (x_3, y_3) has area equal to one-half the absolute value of the determinant
$$\begin{vmatrix} x_1 & y_1 & 1 \\ x_2 & y_2 & 1 \\ x_3 & y_3 & 1 \end{vmatrix}$$

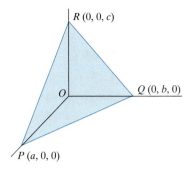

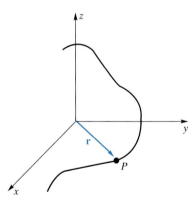

Figure 6

31. A Pythagorean Theorem in Three-Space   As in Figure 6, let P, Q, R, and O be the vertices of a (right-angled) tetrahedron, and let A, B, C, and D be the areas of the opposite faces, respectively. Show that $A^2 + B^2 + C^2 = D^2$.

32. Let vectors $\mathbf{a}, \mathbf{b}$, and $\mathbf{c}$ with common initial point determine a tetrahedron, and let $\mathbf{m}, \mathbf{n}, \mathbf{p}$, and $\mathbf{q}$ be vectors perpendicular to the four faces, pointing outward, and having length equal to the area of the corresponding face. Show that $\mathbf{m} + \mathbf{n} + \mathbf{p} + \mathbf{q} = \mathbf{0}$.

33. Let $\mathbf{a}, \mathbf{b}$, and $\mathbf{a} - \mathbf{b}$ denote the three edges of a triangle with lengths a, b, and c, respectively. Use Lagrange's Identity together with $2\mathbf{a} \cdot \mathbf{b} = |\mathbf{a}|^2 + |\mathbf{b}|^2 - |\mathbf{a} - \mathbf{b}|^2$ to prove **Heron's Formula** for the area A of a triangle,

$$A = \sqrt{s(s - a)(s - b)(s - c)}$$

where s is the semiperimeter $(a + b + c)/2$.

34. Use the method of Example 5 to show directly that, if $\mathbf{u} = u_1\mathbf{i} + u_2\mathbf{j} + u_3\mathbf{k}$ and $\mathbf{v} = v_1\mathbf{i} + v_2\mathbf{j} + v_3\mathbf{k}$, then

$$\mathbf{u} \times \mathbf{v} = (u_2 v_3 - u_3 v_2)\mathbf{i} + (u_3 v_1 - u_1 v_3)\mathbf{j} + (u_1 v_2 - u_2 v_1)\mathbf{k}$$

Answers to Concepts Review: **1.** $\langle -3, 2, -7 \rangle$ or $-3\mathbf{i} + 2\mathbf{j} - 7\mathbf{k}$
2. $|\mathbf{u}||\mathbf{v}|\sin\theta$ **3.** $-(\mathbf{v} \times \mathbf{u})$ **4.** parallel

---

14.4
Lines and Curves in Three-Space

Our study of lines and curves in the plane extends easily to three-space. A **space curve** is determined by a triple of parametric equations

$$x = f(t), \qquad y = g(t), \qquad z = h(t), \qquad t \in I$$

with f, g, and h continuous on the interval I. In vector language, a curve is specified by giving the position vector $\mathbf{r} = \mathbf{r}(t)$ of a point $P = P(t)$; that is,

$$\mathbf{r} = \mathbf{r}(t) = \langle f(t), g(t), h(t) \rangle = f(t)\mathbf{i} + g(t)\mathbf{j} + h(t)\mathbf{k}$$

The tip of $\mathbf{r}$ traces out the curve as t ranges over the interval I, as we see in Figure 1.

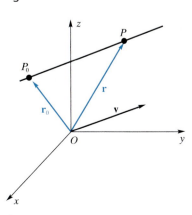

Figure 1

Lines   The simplest of all curves is a line. A line is determined by a fixed point P_0 and a fixed vector $\mathbf{v} = a\mathbf{i} + b\mathbf{j} + c\mathbf{k}$. It is the set of all points P such that $\overrightarrow{P_0 P}$ is parallel to $\mathbf{v}$, that is, that satisfy

$$\overrightarrow{P_0 P} = t\mathbf{v}$$

for some real number t (Figure 2). If $\mathbf{r} = \overrightarrow{OP}$ and $\mathbf{r}_0 = \overrightarrow{OP_0}$ are the position vectors of P and P_0, respectively, then $\overrightarrow{P_0 P} = \mathbf{r} - \mathbf{r}_0$, and the equation of the line can thus be written

$$\mathbf{r} = \mathbf{r}_0 + t\mathbf{v}$$

If we write $\mathbf{r} = \langle x, y, z \rangle$ and $\mathbf{r}_0 = \langle x_0, y_0, z_0 \rangle$ and equate components in the last equation above, we obtain

$$x = x_0 + at, \qquad y = y_0 + bt, \qquad z = z_0 + ct$$

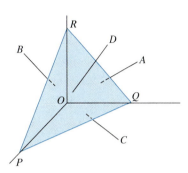

Figure 2

These are **parametric equations** of the line through (x_0, y_0, z_0) and parallel to $\mathbf{v} = \langle a, b, c \rangle$. The numbers a, b, and c are called **direction numbers** for the line. They are not unique; any nonzero constant multiples ka, kb, and kc are also direction numbers.

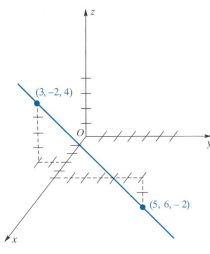

Figure 3

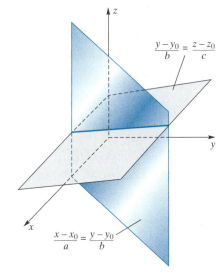

Figure 4

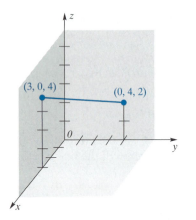

Figure 5

EXAMPLE 1 Find parametric equations for the line through $(3, -2, 4)$ and $(5, 6, -2)$ (see Figure 3).

Solution A vector parallel to the given line is

$$\mathbf{v} = \langle 5 - 3, 6 + 2, -2 - 4 \rangle = \langle 2, 8, -6 \rangle$$

If we choose (x_0, y_0, z_0) as $(3, -2, 4)$, we obtain the parametric equations

$$x = 3 + 2t, \qquad y = -2 + 8t, \qquad z = 4 - 6t$$

Note that $t = 0$ determines the point $(3, -2, 4)$, whereas $t = 1$ gives $(5, 6, -2)$. In fact, $0 \le t \le 1$ corresponds to the segment joining these two points. ∎

If we solve each of the parametric equations for t (assuming that a, b, and c are all different from zero) and equate the results, we obtain the **symmetric equations** for the line through (x_0, y_0, z_0) with direction numbers a, b, c; that is,

$$\frac{x - x_0}{a} = \frac{y - y_0}{b} = \frac{z - z_0}{c}$$

This is the conjunction of the two equations

$$\frac{x - x_0}{a} = \frac{y - y_0}{b} \quad \text{and} \quad \frac{y - y_0}{b} = \frac{z - z_0}{c}$$

both of which are the equations of planes (Figure 4); and, of course, the intersection of two planes is a line.

EXAMPLE 2 Find the symmetric equations of the line that is parallel to the vector $\langle 4, -3, 2 \rangle$ and goes through $(2, 5, -1)$.

Solution

$$\frac{x - 2}{4} = \frac{y - 5}{-3} = \frac{z + 1}{2}$$ ∎

EXAMPLE 3 Find the symmetric equations of the line of intersection of the planes

$$2x - y - 5z = -14 \quad \text{and} \quad 4x + 5y + 4z = 28$$

Solution We begin by finding two points on the line. Any two points would do, but we choose to find the points where the line pierces the yz-plane and the xz-plane (Figure 5). The former is obtained by setting $x = 0$ and solving the resulting equations $-y - 5z = -14$ and $5y + 4z = 28$ simultaneously. This yields the point $(0, 4, 2)$. A similar procedure with $y = 0$ gives the point $(3, 0, 4)$. Consequently, a vector parallel to the required line is

$$\langle 3 - 0, 0 - 4, 4 - 2 \rangle = \langle 3, -4, 2 \rangle$$

Using $(3, 0, 4)$ for (x_0, y_0, z_0), we get

$$\frac{x - 3}{3} = \frac{y - 0}{-4} = \frac{z - 4}{2}$$

An alternative solution is based on the fact that the line of intersection of two planes is perpendicular to both of their normals. The vector $\mathbf{u} = \langle 2, -1, -5 \rangle$ is normal to the first plane; $\mathbf{v} = \langle 4, 5, 4 \rangle$ is normal to the second. Since

$$\mathbf{u} \times \mathbf{v} = \begin{vmatrix} \mathbf{i} & \mathbf{j} & \mathbf{k} \\ 2 & -1 & -5 \\ 4 & 5 & 4 \end{vmatrix} = 21\mathbf{i} = 28\mathbf{j} + 14\mathbf{k}$$

the vector $\mathbf{w} = \langle 21, -28, 14 \rangle$ is parallel to the required line. This implies that $\frac{1}{7}\mathbf{w} = \langle 3, -4, 2 \rangle$ also has this property. Next, find any point on the line of intersection, for example, $(3, 0, 4)$, and proceed as in the earlier solution. ∎

EXAMPLE 4   Find parametric equations of the line through $(1, -2, 3)$ that is perpendicular to both the x-axis and the line

$$\frac{x-4}{2} = \frac{y-3}{-1} = \frac{z}{5}$$

Solution   The x-axis and the given line have directions $\mathbf{u} = \langle 1, 0, 0 \rangle$ and $\mathbf{v} = \langle 2, -1, 5 \rangle$, respectively. A vector perpendicular to both $\mathbf{u}$ and $\mathbf{v}$ is

$$\mathbf{u} \times \mathbf{v} = \begin{vmatrix} \mathbf{i} & \mathbf{j} & \mathbf{k} \\ 1 & 0 & 0 \\ 2 & -1 & 5 \end{vmatrix} = 0\mathbf{i} - 5\mathbf{j} - \mathbf{k}$$

The required line is parallel to $\langle 0, -5, -1 \rangle$ and so also to $\langle 0, 5, 1 \rangle$. Since the first direction number is zero, the line does not have symmetric equations. Its parametric equations are

$$x = 1, \qquad y = -2 + 5t, \qquad z = 3 + t \qquad \blacksquare$$

Tangent Line to a Curve   Let

$$\mathbf{r} = \mathbf{r}(t) = f(t)\mathbf{i} + g(t)\mathbf{j} + h(t)\mathbf{k}$$

be the position vector determining a curve in three-space (Figure 6). In complete analogy with what we did in the plane (Section 13.4), we define $\mathbf{r}'(t)$ by

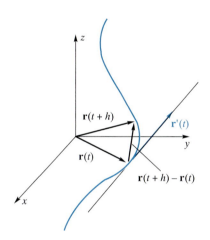

$$\mathbf{r}'(t) = \lim_{h \to 0} \frac{\mathbf{r}(t+h) - \mathbf{r}(t)}{h}$$

It follows that $\mathbf{r}'(t)$, if it exists, has the direction of the tangent line to the curve at the point $P(t)$ corresponding to t. Moreover, $\mathbf{r}'(t)$ exists if and only if $f'(t)$, $g'(t)$, and $h'(t)$ exist and, in this case,

Figure 6

$$\mathbf{r}'(t) = f'(t)\mathbf{i} + g'(t)\mathbf{j} + h'(t)\mathbf{k}$$

Thus, $f'(t)$, $g'(t)$, and $h'(t)$ are direction numbers for the tangent line at P.

EXAMPLE 5   Find the symmetric equations for the tangent line to the curve determined by

$$\mathbf{r}(t) = t\mathbf{i} + \tfrac{1}{2}t^2\mathbf{j} + \tfrac{1}{3}t^3\mathbf{k}$$

at $P(2) = \left(2, 2, \tfrac{8}{3}\right)$.

Solution

$$\mathbf{r}'(t) = \mathbf{i} + t\mathbf{j} + t^2\mathbf{k}$$

and

$$\mathbf{r}'(2) = \mathbf{i} + 2\mathbf{j} + 4\mathbf{k}$$

so the tangent line has direction $\langle 1, 2, 4 \rangle$. Its symmetric equations are

$$\frac{x-2}{1} = \frac{y-2}{2} = \frac{z - \tfrac{8}{3}}{4} \qquad \blacksquare$$

Concepts Review

1. The parametric equations for a line through $(1, -3, 2)$ parallel to the vector $\langle 4, -2, -1 \rangle$ are $x = \underline{\hspace{1cm}}$, $y = \underline{\hspace{1cm}}$, $z = \underline{\hspace{1cm}}$.

2. The symmetric equations for the line of Question 1 are $\underline{\hspace{1cm}}$.

3. If $\mathbf{r}(t) = t^2 \mathbf{i} - 3t \mathbf{j} + t^3 \mathbf{k}$, then $\mathbf{r}'(t) = \underline{\hspace{1cm}}$.

4. A vector parallel to the tangent line at $t = 1$ of the curve determined by the position vector $\mathbf{r}(t)$ of Question 3 is $\underline{\hspace{1cm}}$. This tangent line has symmetric equations $\underline{\hspace{1cm}}$.

Problem Set 14.4

In Problems 1–4, find the parametric equations of the line through the given pair of points.

1. $(1, -2, 3), (4, 5, 6)$ **2.** $(2, -1, -5), (7, -2, 3)$

3. $(4, 2, 3), (6, 2, -1)$ **4.** $(5, -3, -3), (5, 4, 2)$

In Problems 5–8, write both the parametric equations and the symmetric equations for the line through the given point parallel to the given vector.

5. $(4, 5, 6), \langle 3, 2, 1 \rangle$ **6.** $(-1, 3, -6), \langle -2, 0, 5 \rangle$

7. $(1, 1, 1), \langle -10, -100, -1000 \rangle$ **8.** $(-2, 2, -2), \langle 7, -6, 3 \rangle$

In Problems 9–12, find the symmetric equations of the line of intersection of the given pair of planes.

9. $4x + 3y - 7z = 1, 10x + 6y - 5z = 10$

10. $x + y - z = 2, 3x - 2y + z = 3$

11. $x + 4y - 2z = 13, 2x - y - 2z = 5$

12. $x - 3y + z = -1, 6x - 5y + 4z = 9$

13. Find the symmetric equations of the line through $(4, 0, 6)$ and perpendicular to the plane $x - 5y + 2z = 10$.

14. Find the symmetric equations of the line through $(-5, 7, -2)$ and perpendicular to both $\langle 2, 1, -3 \rangle$ and $\langle 5, 4, -1 \rangle$.

15. Find the parametric equations of the line through $(5, -3, 4)$ that intersects the z-axis at right angles.

16. Find the symmetric equations of the line through $(2, -4, 5)$ that is parallel to the plane $3x + y - 2z = 5$ and perpendicular to the line

$$\frac{x + 8}{2} = \frac{y - 5}{3} = \frac{z - 1}{-1}$$

17. Find the equation of the plane that contains the parallel lines

$$\begin{cases} x = -2 + 2t \\ y = 1 + 4t \\ z = 2 - t \end{cases} \quad \text{and} \quad \begin{cases} x = 2 - 2t \\ y = 3 - 4t \\ z = 1 + t \end{cases}$$

18. Show that the lines

$$\frac{x - 1}{-4} = \frac{y - 2}{3} = \frac{z - 4}{-2}$$

and

$$\frac{x - 2}{-1} = \frac{y - 1}{1} = \frac{z + 2}{6}$$

intersect, and find the equation of the plane that they determine.

19. Find the equation of the plane containing the line $x = 1 + 2t, y = -1 + 3t, z = 4 + t$ and the point $(1, -1, 5)$.

20. Find the equation of the plane containing the line $x = 3t$, $y = 1 + t$, $z = 2t$ and parallel to the intersection of the planes $2x - y + z = 0$ and $y + z + 1 = 0$.

21. Find the distance between the skew (nonintersecting and nonparallel) lines $x = 2 - t, y = 3 + 4t, z = 2t$ and $x = -1 + t$, $y = 2, z = -1 + 2t$ by using the following steps.
(a) Note by putting $t = 0$ that $(2, 3, 0)$ is on the first line.
(b) Find the equation of the plane π through $(2, 3, 0)$ parallel to both given lines (i.e., with normal perpendicular to both).
(c) Find a point Q on the second line.
(d) Find the distance from Q to the plane π. (See Example 6 of Section 14.2.)
See Problem 30 for another way to do this problem.

22. Find the distance between the skew lines $x = 1 + 2t$, $y = -3 + 4t, z = -1 - t$ and $x = 4 - 2t, y = 1 + 3t, z = 2t$ (see Problem 21).

23. Find the symmetric equations of the tangent line to the curve with equation

$$\mathbf{r}(t) = 2 \cos t \mathbf{i} + 6 \sin t \mathbf{j} + t \mathbf{k}$$

at $t = \pi/3$.

24. Find the parametric equations of the tangent line to the curve $x = 2t^2, y = 4t, z = t^3$ at $t = 1$.

25. Find the equation of the plane perpendicular to the curve $x = 3t, y = 2t^2, z = t^5$ at $t = -1$.

26. Find the equation of the plane perpendicular to the curve

$$\mathbf{r}(t) = t \sin t \mathbf{i} + 3t \mathbf{j} + 2t \cos t \mathbf{k}$$

at $t = \pi/2$.

27. Consider the curve $\mathbf{r}(t) = \sin t \cos t \mathbf{i} + \sin^2 t \mathbf{j} + \cos t \mathbf{k}$, $0 \le t \le 2\pi$.
(a) Show that the curve lies on a sphere centered at the origin.
(b) Where does the tangent line at $t = \pi/6$ intersect the xy-plane?

28. Point to Plane Let P be a point on a plane with normal $\mathbf{n}$ and Q be a point off the plane (Figure 7). Show that the distance d from Q to the plane is given by

$$d = \frac{|\overrightarrow{PQ} \cdot \mathbf{n}|}{|\mathbf{n}|}$$

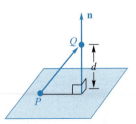

Figure 7

and use this result to find the distance from $(4, -2, 3)$ to the plane $4x - 4y + 2z = 2$. Compare your result with Example 6 in Section 14.2.

29. Point to Line Let P be a point on a line with direction $\mathbf{n}$ and Q a point off the line (Figure 8). Show that the distance d from Q to the line is given by

$$d = \frac{|\overrightarrow{PQ} \times \mathbf{n}|}{|\mathbf{n}|}$$

and use this result to find each distance in parts (a) and (b).

(a) From $Q(1, 0, -4)$ to the line $\dfrac{x - 3}{2} = \dfrac{y + 2}{-2} = \dfrac{z - 1}{1}$

(b) From $Q(2, -1, 3)$ to the line $x = 1 + 2t, y = -1 + 3t, z = -6t$

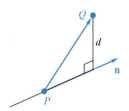

Figure 8

30. Line to Line Let P and Q be points on nonintersecting skew lines with directions $\mathbf{n}_1$ and $\mathbf{n}_2$, and let $\mathbf{n} = \mathbf{n}_1 \times \mathbf{n}_2$ (Figure 9). Show that the distance d between these lines is given by

$$d = \frac{|\overrightarrow{PQ} \cdot \mathbf{n}|}{|\mathbf{n}|}$$

and use this result to find the distance between each pair of lines in parts (a) and (b).

(a) $\dfrac{x - 3}{1} = \dfrac{y + 2}{1} = \dfrac{z - 1}{2}$ and $\dfrac{x + 4}{3} = \dfrac{y + 5}{4} = \dfrac{z}{5}$

(b) $x = 1 + 2t, y = -2 + 3t, z = -4t$ and $x = 3t, y = 1 + t, z = -5t$

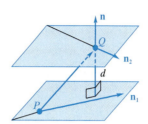

Figure 9

Answers to Concepts Review: **1.** $1 + 4t; -3 - 2t; 2 - t$

2. $\dfrac{x - 1}{4} = \dfrac{y + 3}{-2} = \dfrac{z - 2}{-1}$ **3.** $2t\mathbf{i} - 3\mathbf{j} + 3t^2\mathbf{k}$

4. $\langle 2, -3, 3 \rangle; \dfrac{x - 1}{2} = \dfrac{y + 3}{-3} = \dfrac{z - 1}{3}$

14.5
Velocity, Acceleration, and Curvature

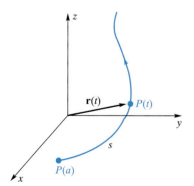

Figure 1

All that we did with curvilinear motion in the plane (Sections 13.4 and 13.5) generalizes in a natural way to three-space. Let

$$\mathbf{r}(t) = f(t)\mathbf{i} + g(t)\mathbf{j} + h(t)\mathbf{k}, \qquad a \le t \le b$$

be the position vector for a point $P = P(t)$ that is tracing out a curve as t increases (Figure 1). We suppose that $\mathbf{r}'(t)$ exists and is continuous and $\mathbf{r}'(t) \ne \mathbf{0}$, in which case the curve is said to be **smooth**. The length s of the arc from $P(a)$ to $P(t)$ is given by

$$s = \int_a^t |\mathbf{r}'(u)|\, du = \int_a^t \sqrt{[f'(u)]^2 + [g'(u)]^2 + [h'(u)]^2}\, du$$

If t measures time, we may define velocity, speed, and acceleration of the moving point P by

Velocity: $\mathbf{v}(t) = \mathbf{r}'(t)$

Speed: $\dfrac{ds}{dt} = |\mathbf{r}'(t)| = |\mathbf{v}(t)|$

Acceleration: $\mathbf{a}(t) = \mathbf{r}''(t)$

An Example: The Circular Helix
Suppose that a point P moves so that its position vector at time t is

$$\mathbf{r}(t) = a\cos t\,\mathbf{i} + a\sin t\,\mathbf{j} + ct\,\mathbf{k}$$

where a and c are positive constants. Then P traces a curve that winds around the right circular cylinder with parametric equations $x = a\cos t, y = a\sin t$, but spirals higher and higher because $z = ct$ increases with t. The curve is called a **circular helix**; part of it is shown in Figure 2.

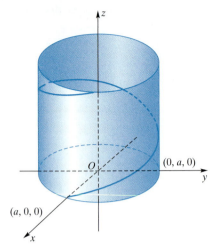

Figure 2

EXAMPLE 1   Find the arc length of the circular helix for $0 \leq t \leq 2\pi$.

Solution

$$s = \int_0^{2\pi} \sqrt{(-a\sin t)^2 + (a\cos t)^2 + c^2}\, dt$$

$$= \int_0^{2\pi} \sqrt{a^2 + c^2}\, dt = 2\pi\sqrt{a^2 + c^2}$$

Note that when $c = 0$ this reduces to $2\pi a$ (the circumference of a circle of radius a), as it should.   ∎

EXAMPLE 2   For the motion $\mathbf{r}(t) = a\cos t\,\mathbf{i} + a\sin t\,\mathbf{j} + ct\,\mathbf{k}$ described above, calculate the acceleration $\mathbf{a}$ at $t = 2\pi$.

Solution

$$\mathbf{v}(t) = \mathbf{r}'(t) = -a\sin t\,\mathbf{i} + a\cos t\,\mathbf{j} + c\,\mathbf{k}$$

$$\mathbf{a}(t) = \mathbf{r}''(t) = -a\cos t\,\mathbf{i} - a\sin t\,\mathbf{j}$$

$$\mathbf{a}(2\pi) = -a\,\mathbf{i}$$ ∎

EXAMPLE 3   Beginning at $t = 0$, a bee flew so that its position vector was $\mathbf{r} = t\cos t\,\mathbf{i} + t\sin t\,\mathbf{j} + t\,\mathbf{k}$ until $t = 4\pi$, at which time it flew off on a tangent at the speed then attained. What was the total length of its flight path on the interval $0 \leq t \leq 4\pi + 3$?

Solution   The path consists of a spiral part and a straight-line part with lengths L_1 and L_2, respectively. On the spiral part,

$$\mathbf{r}'(t) = (\cos t - t\sin t)\mathbf{i} + (\sin t + t\cos t)\mathbf{j} + \mathbf{k}$$

and

$$\left|\mathbf{r}'(t)\right| = \left[(\cos t - t\sin t)^2 + (\sin t + t\cos t)^2 + 1\right]^{1/2} = \sqrt{2 + t^2}$$

Using Formula 44 at the end of the book, we find that

$$L_1 = \int_0^{4\pi} \sqrt{2 + t^2}\, dt$$

$$= \left[\frac{t}{2}\sqrt{2 + t^2} + \ln\left|t + \sqrt{2 + t^2}\right|\right]_0^{4\pi} \approx 82.336$$

Also,

$$L_2 = 3\left|\mathbf{r}'(4\pi)\right| = 3\sqrt{2 + 16\pi^2} \approx 37.937$$

We conclude that $L_1 + L_2 \approx 120.273$.   ∎

Curvature   As we noted in Section 14.4, $\mathbf{v}(t) = \mathbf{r}'(t)$ is a vector with the same direction as the tangent to the curve at $P(t)$. Thus,

$$\mathbf{T} = \mathbf{T}(t) = \frac{\mathbf{v}(t)}{\left|\mathbf{v}(t)\right|} = \frac{\mathbf{r}'(t)}{\left|\mathbf{r}'(t)\right|}$$

is a **unit tangent vector** at $P(t)$. Since s denotes the arc length measured from some fixed point in the direction of increasing t, $d\mathbf{T}/ds$ measures the rate of change of direction of the tangent with respect to the distance along the curve. By the Chain Rule,

$$\frac{d\mathbf{T}}{ds} = \frac{d\mathbf{T}}{dt}\frac{dt}{ds} = \frac{\mathbf{T}'(t)}{\left|\mathbf{v}(t)\right|}$$

Thus, just as for plane curves, we may define the **curvature** κ of a space curve by

$$\kappa = \kappa(t) = \left|\frac{d\mathbf{T}}{ds}\right| = \frac{\left|\mathbf{T}'(t)\right|}{\left|\mathbf{v}(t)\right|}$$

EXAMPLE 4   Find the curvature of the circular helix

$$\mathbf{r}(t) = a\cos t\,\mathbf{i} + a\sin t\,\mathbf{j} + ct\,\mathbf{k}, \qquad a > 0$$

Solution

$$\mathbf{v}(t) = -a\sin t\,\mathbf{i} + a\cos t\,\mathbf{j} + c\,\mathbf{k}$$

$$\mathbf{T}(t) = \frac{\mathbf{v}(t)}{|\mathbf{v}(t)|} = \frac{1}{\sqrt{a^2 + c^2}}(-a\sin t\,\mathbf{i} + a\cos t\,\mathbf{j} + c\,\mathbf{k})$$

$$\kappa(t) = \frac{|\mathbf{T}'(t)|}{|\mathbf{v}(t)|} = \frac{1}{a^2 + c^2}\left|-a\cos t\,\mathbf{i} - a\sin t\,\mathbf{j}\right|$$

$$= \frac{a}{a^2 + c^2}$$

Thus, κ is a constant for the circular helix.   ■

The radius of curvature R is the reciprocal of κ. In the example above, $R = (a^2 + c^2)/a$. This reduces to $R = a$ when $c = 0$, which corresponds to the fact that the motion is then along the circle $\mathbf{r} = a\cos t\,\mathbf{i} + a\sin t\,\mathbf{j}$ in the xy-plane. When c is large, R is large as we should expect.

Components of Acceleration   Just as in the plane case, we define the **principal unit normal vector N** at P by

$$\mathbf{N} = \frac{d\mathbf{T}/ds}{|d\mathbf{T}/ds|} = \frac{1}{\kappa}\frac{d\mathbf{T}}{ds}$$

It is obvious that $\mathbf{N}$ is a unit vector. That $\mathbf{N}$ is normal (perpendicular) to the curve follows by differentiating $\mathbf{T}\cdot\mathbf{T} = 1$ with respect to s. This gives

$$2\mathbf{T}\cdot\frac{d\mathbf{T}}{ds} = 0$$

which implies that $d\mathbf{T}/ds$ is perpendicular to $\mathbf{T}$. With these facts in hand, we can mimic a derivation given in the plane case (Section 13.5) to obtain (Figure 3)

$$\boxed{\mathbf{a} = \frac{d^2s}{dt^2}\mathbf{T} + \left(\frac{ds}{dt}\right)^2\kappa\mathbf{N} = a_T\mathbf{T} + a_N\mathbf{N}}$$

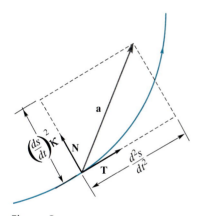

Figure 3

If we take the dot product of this last result with $\mathbf{T}$, we get

$$\mathbf{T}\cdot\mathbf{a} = a_T\mathbf{T}\cdot\mathbf{T} + a_N\mathbf{T}\cdot\mathbf{N} = a_T$$

or

$$\boxed{a_T = \mathbf{T}\cdot\mathbf{a} = \frac{\mathbf{r}'\cdot\mathbf{r}''}{|\mathbf{r}'|}}$$

If we take the cross product of $\mathbf{T}$ with $\mathbf{a}$, we get

$$\mathbf{T}\times\mathbf{a} = a_T(\mathbf{T}\times\mathbf{T}) + a_N(\mathbf{T}\times\mathbf{N}) = a_N(\mathbf{T}\times\mathbf{N})$$

and so

$$|\mathbf{T}\times\mathbf{a}| = a_N|\mathbf{T}\times\mathbf{N}| = a_N|\mathbf{T}||\mathbf{N}|\sin\theta = a_N$$

or

$$\boxed{a_N = |\mathbf{T}\times\mathbf{a}| = \frac{|\mathbf{r}'\times\mathbf{r}''|}{|\mathbf{r}'|}}$$

Cartesian Coordinates   Cylindrical Coordinates   Spherical Coordinates

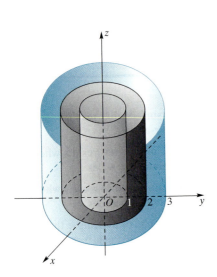

Figure 1

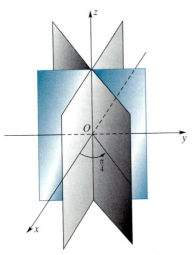

The cylinders $r = 1, r = 2, r = 3$

Figure 2

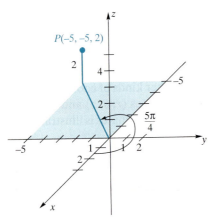

The planes $\theta = 0$, $\theta = \frac{\pi}{4}$, $\theta = \frac{\pi}{2}$

Figure 3

xy-plane, and ϕ is the angle between the positive z-axis and the line segment OP. We require that

$$\rho \geq 0, \quad 0 \leq \theta < 2\pi, \quad 0 \leq \phi \leq \pi$$

Cylindrical Coordinates   If a solid or a surface has an axis of symmetry, it is often wise to orient it so this axis is the z-axis and then use cylindrical coordinates. Note in particular the simplicity of the equation of a circular cylinder with z-axis symmetry (Figure 2) and also of a plane containing the z-axis (Figure 3). In Figure 3, we have allowed $r < 0$.

Cylindrical and Cartesian coordinates are related by the following equations:

Cylindrical to Cartesian	**Cartesian to Cylindrical**
$x = r\cos\theta$	$r = \sqrt{x^2 + y^2}$
$y = r\sin\theta$	$\tan\theta = y/x$
$z = z$	$z = z$

With these relationships, we can go back and forth between the two coordinate systems.

EXAMPLE 1   Find

(a) the Cartesian coordinates of the point with cylindrical coordinates $(4, 2\pi/3, 5)$ and
(b) the cylindrical coordinates of the point with Cartesian coordinates $(-5, -5, 2)$.

Solution

(a) $x = 4\cos\dfrac{2\pi}{3} = 4 \cdot \left(-\dfrac{1}{2}\right) = -2$

$y = 4\sin\dfrac{2\pi}{3} = 4 \cdot \left(\dfrac{\sqrt{3}}{2}\right) = 2\sqrt{3}$

$z = 5$

Thus, the Cartesian coordinates of $(4, 2\pi/3, 5)$ are $(-2, 2\sqrt{3}, 5)$.

(b) $r = \sqrt{(-5)^2 + (-5)^2} = 5\sqrt{2}$

$\tan\theta = \dfrac{-5}{-5} = 1$

$z = 2$

Figure 4 indicates that θ is between $\pi/2$ and π. Since $\tan\theta = 1$, we must have $\theta = 5\pi/4$. The cylindrical coordinates of $(-5, -5, 2)$ are therefore $\left(5\sqrt{2}, 5\pi/4, 2\right)$. ∎

EXAMPLE 2   Find the equations in cylindrical coordinates of the paraboloid and cylinder whose Cartesian equations are $x^2 + y^2 = 4 - z$ and $x^2 + y^2 = 2x$.

$P(-5, -5, 2)$

Figure 4

Solution

Paraboloid: $r^2 = 4 - z$

Cylinder: $r^2 = 2r\cos\theta$ or (equivalently) $r = 2\cos\theta$

Division of an equation by a variable creates the potential for losing a solution. For example, dividing $x^2 = x$ by x gives $x = 1$ and loses the solution $x = 0$. Similarly, dividing $r^2 = 2r\cos\theta$ by r gives $r = 2\cos\theta$ and appears to lose the solution $r = 0$ (the origin). However, the origin satisfies the equation $r = 2\cos\theta$ with coordinates $(0, \pi/2)$. Thus, $r^2 = 2r\cos\theta$ and $r = 2\cos\theta$ have identical polar graphs (see CAUTION in the margin of Section 12.6). ∎

EXAMPLE 3 Find the Cartesian equations of the surfaces whose equations in cylindrical coordinates are $r^2 + 4z^2 = 16$ and $r^2\cos 2\theta = z$.

Solution Since $r^2 = x^2 + y^2$, the surface $r^2 + 4z^2 = 16$ has the Cartesian equation $x^2 + y^2 + 4z^2 = 16$ or $x^2/16 + y^2/16 + z^2/4 = 1$. Its graph is an ellipsoid.

Since $\cos 2\theta = \cos^2\theta - \sin^2\theta$, the second equation can be written $r^2\cos^2\theta - r^2\sin^2\theta = z$. In Cartesian coordinates it becomes $x^2 - y^2 = z$, the graph of which is a hyperbolic paraboloid. ∎

Spherical Coordinates

When a solid or a surface is symmetric with respect to a point, spherical coordinates are likely to play a simplifying role. In particular, a sphere centered at the origin (Figure 5) has the simple equation $\rho = \rho_0$. Also note that the equation of a cone with axis along the z-axis and vertex at the origin (Figure 6) is $\phi = \phi_0$.

It is easy to determine the relationships between spherical and cylindrical coordinates and between spherical and Cartesian coordinates. The following table shows these relationships.

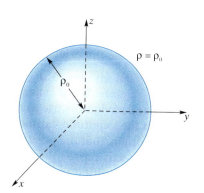

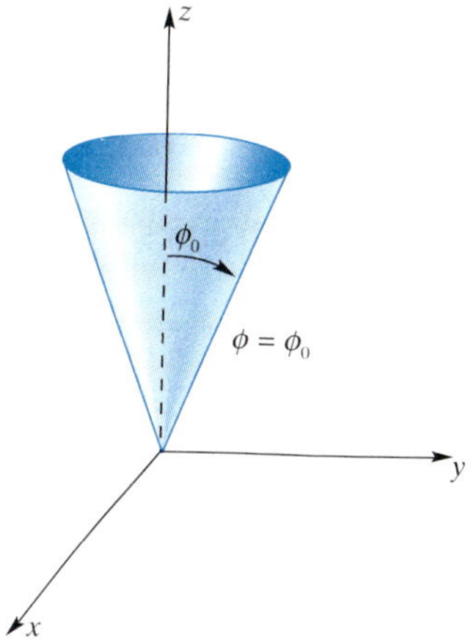

Figure 5

Spherical to Cartesian	**Cartesian to Spherical**
$x = \rho\sin\phi\cos\theta$	$\rho = \sqrt{x^2 + y^2 + z^2}$
$y = \rho\sin\phi\sin\theta$	$\tan\theta = y/x$
$z = \rho\cos\phi$	$\cos\phi = \dfrac{z}{\sqrt{x^2 + y^2 + z^2}}$

EXAMPLE 4 Find the Cartesian coordinates of the point P with spherical coordinates $(8, \pi/3, 2\pi/3)$.

Solution We have plotted the point P in Figure 7.

$$x = 8\sin\frac{2\pi}{3}\cos\frac{\pi}{3} = 8\frac{\sqrt{3}}{2}\frac{1}{2} = 2\sqrt{3}$$

$$y = 8\sin\frac{2\pi}{3}\sin\frac{\pi}{3} = 8\frac{\sqrt{3}}{2}\frac{\sqrt{3}}{2} = 6$$

$$z = 8\cos\frac{2\pi}{3} = 8\left(-\frac{1}{2}\right) = -4$$

Thus, P has Cartesian coordinates $\left(2\sqrt{3}, 6, -4\right)$. ∎

EXAMPLE 5 Describe the graph of $\rho = 2\cos\phi$.

Solution We change to Cartesian coordinates. Multiply both sides by ρ to obtain

$$\rho^2 = 2\rho\cos\phi$$

$$x^2 + y^2 + z^2 = 2z$$

$$x^2 + y^2 + (z - 1)^2 = 1$$

Figure 6

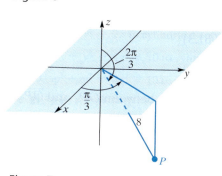

Figure 7

Solution We note first that

$$\frac{\partial f}{\partial x} = e^y + 2xy, \qquad \frac{\partial f}{\partial y} = xe^y + x^2$$

Both of these functions are continuous everywhere and so, by Theorem A, f is differentiable everywhere. The gradient is

$$\nabla f(x, y) = (e^y + 2xy)\mathbf{i} + (xe^y + x^2)\mathbf{j} = \langle e^y + 2xy, xe^y + x^2 \rangle$$

Thus,

$$\nabla f(2, 0) = \mathbf{i} + 6\mathbf{j} = \langle 1, 6 \rangle$$

and the equation of the tangent plane is

$$\begin{aligned} z &= f(2, 0) + \nabla f(2, 0) \cdot \langle x - 2, y \rangle \\ &= 2 + \langle 1, 6 \rangle \cdot \langle x - 2, y \rangle \\ &= 2 + x - 2 + 6y = x + 6y \end{aligned}$$

∎

EXAMPLE 2 For $f(x, y, z) = x \sin z + x^2 y$, find $\nabla f(1, 2, 0)$.

Solution The partial derivatives are

$$\frac{\partial f}{\partial x} = \sin z + 2xy, \qquad \frac{\partial f}{\partial y} = x^2, \qquad \frac{\partial f}{\partial z} = x \cos z$$

At $(1, 2, 0)$, these partials have the values 4, 1, and 1, respectively. Thus,

$$\nabla f(1, 2, 0) = 4\mathbf{i} + \mathbf{j} + \mathbf{k}$$

∎

Rules for Gradients In many respects, gradients behave like derivatives. Recall that D considered as an operator is linear. The operator ∇ is also linear.

Theorem B Properties of ∇

∇ is a linear operator; that is,

(i) $\nabla[f(\mathbf{p}) + g(\mathbf{p})] = \nabla f(\mathbf{p}) + \nabla g(\mathbf{p})$
(ii) $\nabla[\alpha f(\mathbf{p})] = \alpha \nabla f(\mathbf{p})$

Also, we have the product rule.

(iii) $\nabla[f(\mathbf{p})g(\mathbf{p})] = f(\mathbf{p})\nabla g(\mathbf{p}) + g(\mathbf{p})\nabla f(\mathbf{p})$

Proof All three results follow from the corresponding facts for partial derivatives. We prove (iii) in the two-variable case, suppressing the point $\mathbf{p}$ for brevity.

$$\begin{aligned} \nabla fg &= \frac{\partial(fg)}{\partial x}\mathbf{i} + \frac{\partial(fg)}{\partial y}\mathbf{j} \\ &= \left(f\frac{\partial g}{\partial x} + g\frac{\partial f}{\partial x} \right)\mathbf{i} + \left(f\frac{\partial g}{\partial y} + g\frac{\partial f}{\partial y} \right)\mathbf{j} \\ &= f\left(\frac{\partial g}{\partial x}\mathbf{i} + \frac{\partial g}{\partial y}\mathbf{j} \right) + g\left(\frac{\partial f}{\partial x}\mathbf{i} + \frac{\partial f}{\partial y}\mathbf{j} \right) \\ &= f\nabla g + g\nabla f \quad \blacklozenge \end{aligned}$$

Continuity versus Differentiability Recall that for functions of one variable, differentiability implies continuity, but not vice versa. The same is true here.

Theorem C

If f is differentiable at $\mathbf{p}$, then f is continuous at $\mathbf{p}$.

Proof Since f is differentiable at $\mathbf{p}$,

$$f(\mathbf{p} + \mathbf{h}) - f(\mathbf{p}) = \nabla f(\mathbf{p}) \cdot \mathbf{h} + \varepsilon(\mathbf{h}) \cdot \mathbf{h}$$

Thus,

$$|f(\mathbf{p} + \mathbf{h}) - f(\mathbf{p})| \le |\nabla f(\mathbf{p}) \cdot \mathbf{h}| + |\boldsymbol{\varepsilon}(\mathbf{h}) \cdot \mathbf{h}|$$
$$= |\nabla f(\mathbf{p})||\mathbf{h}||\cos\theta| + |\boldsymbol{\varepsilon}(\mathbf{h}) \cdot \mathbf{h}|$$

Both of the latter terms approach 0 as $\mathbf{h} \to \mathbf{0}$, and so

$$\lim_{\mathbf{h} \to \mathbf{0}} f(\mathbf{p} + \mathbf{h}) = f(\mathbf{p})$$

This last equality is one way of formulating the continuity of f at $\mathbf{p}$. ◆

The Gradient Field   The gradient ∇f associates with each point $\mathbf{p}$ in the domain of f a vector $\nabla f(\mathbf{p})$. The set of all these vectors is called the **gradient field** for f. In Figures 5 and 6, we show graphs of the surface $z = x^2 - y^2$ and the corresponding gradient field. Do these figures suggest something about the direction in which the gradient vectors point? We explore this subject in the next section.

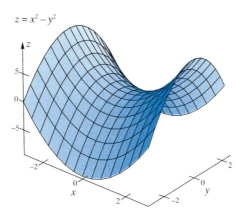

Figure 5   Figure 6

Concepts Review

1. The analog of the derivative $f'(x)$ for a function of more than one variable is the _____ denoted by $\nabla f(\mathbf{p})$.

2. The function $f(x, y)$ is differentiable at (a, b) if and only if f is _____ at (a, b).

3. For a function f of two variables, the gradient is $\nabla f(\mathbf{p}) = $ _____. Thus, if $f(x, y) = xy^2, \nabla f(x, y) = $ _____.

4. $f(x, y)$ being differentiable at (x_0, y_0) is equivalent to the existence of a _____ to the graph at this point.

Problem Set 15.4

In Problems 1–10, find the gradient ∇f.

1. $f(x, y) = x^2 y + 3xy$

2. $f(x, y) = x^3 y - y^3$

3. $f(x, y) = xe^{xy}$

4. $f(x, y) = x^2 y \cos y$

5. $f(x, y) = x^2 y / (x + y)$

6. $f(x, y) = \sin^3(x^2 y)$

7. $f(x, y, z) = \sqrt{x^2 + y^2 + z^2}$

8. $f(x, y, z) = x^2 y + y^2 z + z^2 x$

9. $f(x, y, z) = x^2 y e^{x-z}$

10. $f(x, y, z) = xz \ln(x + y + z)$

In Problems 11–14, find the gradient vector of the given function at the given point $\mathbf{p}$. Then find the equation of the tangent plane at $\mathbf{p}$ (see Example 1).

11. $f(x, y) = x^2 y - xy^2, \mathbf{p} = (-2, 3)$

12. $f(x, y) = x^3 y + 3xy^2, \mathbf{p} = (2, -2)$

13. $f(x, y) = \cos\pi x \sin\pi y + \sin 2\pi y, \mathbf{p} = (-1, \frac{1}{2})$

14. $f(x, y) = \dfrac{x^2}{y}, \mathbf{p} = (2, -1)$

In Problems 15 and 16, find the equation $w = T(x, y, z)$ of the tangent "hyperplane" at $\mathbf{p}$.

15. $f(x, y, z) = 3x^2 - 2y^2 + xz^2, \mathbf{p} = (1, 2, -1)$

16. $f(x, y, z) = xyz + x^2, \mathbf{p} = (2, 0, -3)$

17. Show that

$$\nabla\left(\frac{f}{g}\right) = \frac{g\nabla f - f\nabla g}{g^2}$$

18. Show that

$$\nabla(f^r) = rf^{r-1}\nabla f$$

19. Find all points (x, y) at which the tangent plane to the graph of $z = x^2 - 6x + 2y^2 - 10y + 2xy$ is horizontal.

20. Find all points (x, y) at which the tangent plane to the graph of $z = x^3$ is horizontal.

21. Find parametric equations of the line tangent to the surface $z = y^2 + x^3y$ at the point $(2, 1, 9)$ whose projection on the xy-plane is

(a) parallel to the x-axis;   (b) parallel to the y-axis;
(c) parallel to the line $x = y$.

22. Find parametric equations of the line tangent to the surface $z = x^2y^3$ at the point $(3, 2, 72)$ whose projection on the xy-plane is

(a) parallel to the x-axis;   (b) parallel to the y-axis;
(c) parallel to the line $x = -y$.

23. Refer to Figure 1. Find the equation of the tangent plane to $z = -10\sqrt{|xy|}$ at $(1, -1)$. *Recall*: $d|x|/dx = |x|/x$ for $x \neq 0$.

24. Mean Value Theorem for Several Variables If f is differentiable at each point of the line segment from **a** to **b**, then there exists on that line segment a point **c** between **a** and **b** such that

$$f(\mathbf{b}) - f(\mathbf{a}) = \nabla f(\mathbf{c}) \cdot (\mathbf{b} - \mathbf{a})$$

Assuming that this result is true, show that, if f is differentiable on a convex set S and if $\nabla f(\mathbf{p}) = \mathbf{0}$ on S, then f is constant on S. *Note*: A set S is convex if each pair of points in S can be connected by a line segment in S.

25. Use the result of Problem 24 to show that if $\nabla f(\mathbf{p}) = \nabla g(\mathbf{p})$ for all **p** in a convex set S then f and g differ by a constant on S.

26. Find the most general function $f(\mathbf{p})$ satisfying $\nabla f(\mathbf{p}) = \mathbf{p}$.

$\boxed{\text{CAS}}$ **27.** Plot the graph of $f(x, y) = -|xy|$ together with its gradient field.

(a) Based on this and Figures 5 and 6, make a conjecture about the direction in which a gradient vector points.

(b) Is f differentiable at the origin? Justify your answer.

$\boxed{\text{CAS}}$ **28.** Plot the graph of $f(x, y) = \sin x + \sin y - \sin(x + y)$ on $0 \leq x \leq 2\pi, 0 \leq y \leq 2\pi$. Also draw the gradient field to see if your conjecture in Problem 27 (a) holds up.

29. Prove Theorem B for

(a) the three-variable case and

(b) the n-variable case. *Hint*: Denote the standard unit vectors by $\mathbf{i}_1, \mathbf{i}_2, \dots, \mathbf{i}_n$.

Answers to Concepts Review: **1.** gradient **2.** locally linear **3.** $\dfrac{\partial f(\mathbf{p})}{\partial x}\mathbf{i} + \dfrac{\partial f(\mathbf{p})}{\partial y}\mathbf{j}$; $y^2\mathbf{i} + 2xy\mathbf{j}$ **4.** tangent plane

---

15.5
Directional Derivatives and Gradients

Consider again a function $f(x, y)$ of two variables. The partial derivatives $f_x(x, y)$ and $f_y(x, y)$ measure the rate of change (and the slope of the tangent line) in directions parallel to the x- and y-axes. Our goal now is to study the rate of change of f in an arbitrary direction. This leads to the concept of the directional derivative, which in turn is related to the gradient.

It will be convenient to use vector notation. Let $\mathbf{p} = (x, y)$, and let **i** and **j** be the unit vectors in the positive x- and y-directions. Then the two partial derivatives at **p** may be written as follows:

$$f_x(\mathbf{p}) = \lim_{h \to 0} \frac{f(\mathbf{p} + h\mathbf{i}) - f(\mathbf{p})}{h}$$

$$f_y(\mathbf{p}) = \lim_{h \to 0} \frac{f(\mathbf{p} + h\mathbf{j}) - f(\mathbf{p})}{h}$$

To get the concept we are after, all we have to do is replace **i** or **j** by an arbitrary unit vector **u**.

Definition

For any unit vector **u**, let

$$D_{\mathbf{u}}f(\mathbf{p}) = \lim_{h \to 0} \frac{f(\mathbf{p} + h\mathbf{u}) - f(\mathbf{p})}{h}$$

This limit, if it exists, is called the **directional derivative** of f at **p** in the direction **u**.

Thus, $D_{\mathbf{i}}f(\mathbf{p}) = f_x(\mathbf{p})$ and $D_{\mathbf{j}}f(\mathbf{p}) = f_y(\mathbf{p})$. Since $\mathbf{p} = (x, y)$, we also use the notation $D_{\mathbf{u}}f(x, y)$. Figure 1 gives the geometric interpretation of $D_{\mathbf{u}}f(x_0, y_0)$. The vector **u** determines a line L in the xy-plane through (x_0, y_0). The plane through L perpendicular to the xy-plane intersects the surface $z = f(x, y)$ in a curve C. Its tangent at the point $(x_0, y_0, f(x_0, y_0))$ has slope $D_{\mathbf{u}}f(x_0, y_0)$. Another useful interpretation is that $D_{\mathbf{u}}f(x_0, y_0)$ measures the rate of change of f with respect to distance in the direction **u**.

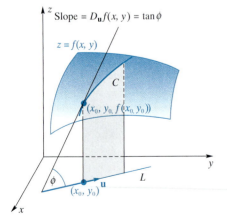

Slope $= D_{\mathbf{u}}f(x, y) = \tan\phi$

$z = f(x, y)$

C

$(x_0, y_0, f(x_0, y_0))$

L

(x_0, y_0)

Figure 1

Connection with the Gradient

Recall from Section 15.4 that $\nabla f(\mathbf{p})$ is given by

$$\nabla f(\mathbf{p}) = f_x(\mathbf{p})\mathbf{i} + f_y(\mathbf{p})\mathbf{j}$$

> **Theorem A**
>
> Let f be differentiable at $\mathbf{p}$. Then f has a directional derivative at $\mathbf{p}$ in the direction of the unit vector $\mathbf{u} = u_1\mathbf{i} + u_2\mathbf{j}$ and
>
> $$D_{\mathbf{u}}f(\mathbf{p}) = \mathbf{u} \cdot \nabla f(\mathbf{p})$$
>
> That is,
>
> $$D_{\mathbf{u}}f(x, y) = u_1 f_x(x, y) + u_2 f_y(x, y)$$

Proof   Since f is differentiable at $\mathbf{p}$,

$$f(\mathbf{p} + h\mathbf{u}) - f(\mathbf{p}) = \nabla f(\mathbf{p}) \cdot (h\mathbf{u}) + \boldsymbol{\varepsilon}(h\mathbf{u}) \cdot (h\mathbf{u})$$

where $\boldsymbol{\varepsilon}(h\mathbf{u}) \to \mathbf{0}$ as $h \to 0$. Thus,

$$\frac{f(\mathbf{p} + h\mathbf{u}) - f(\mathbf{p})}{h} = \nabla f(\mathbf{p}) \cdot \mathbf{u} + \boldsymbol{\varepsilon}(h\mathbf{u}) \cdot \mathbf{u}$$

The conclusion follows by taking limits as $h \to 0$. ◆

EXAMPLE 1   If $f(x, y) = 4x^2 - xy + 3y^2$, find the directional derivative of f at $(2, -1)$ in the direction of the vector $\mathbf{a} = 4\mathbf{i} + 3\mathbf{j}$.

Solution   The unit vector $\mathbf{u}$ in the direction of $\mathbf{a}$ is $\left(\frac{4}{5}\right)\mathbf{i} + \left(\frac{3}{5}\right)\mathbf{j}$. Also, $f_x(x, y) = 8x - y$ and $f_y(x, y) = -x + 6y$; thus, $f_x(2, -1) = 17$ and $f_y(2, -1) = -8$. Consequently, by Theorem A,

$$D_{\mathbf{u}}f(2, -1) = \left\langle \tfrac{4}{5}, \tfrac{3}{5} \right\rangle \cdot \left\langle 17, -8 \right\rangle = \tfrac{4}{5}(17) + \tfrac{3}{5}(-8) = \tfrac{44}{5}$$ ∎

Although we will not go through the details, we assert that what we have done is valid for functions of three or more variables, with obvious modifications. We illustrate.

EXAMPLE 2   Find the directional derivative of the function $f(x, y, z) = xy \sin z$ at the point $(1, 2, \pi/2)$ in the direction of the vector $\mathbf{a} = \mathbf{i} + 2\mathbf{j} + 2\mathbf{k}$.

Solution   The unit vector $\mathbf{u}$ in the direction of $\mathbf{a}$ is $\frac{1}{3}\mathbf{i} + \frac{2}{3}\mathbf{j} + \frac{2}{3}\mathbf{k}$. Also, $f_x(x, y, z) = y \sin z$, $f_y(x, y, z) = x \sin z$, and $f_z(x, y, z) = xy \cos z$, and so $f_x(1, 2, \pi/2) = 2, f_y(1, 2, \pi/2) = 1$, and $f_z(1, 2, \pi/2) = 0$. We conclude that

$$D_{\mathbf{u}}f\left(1, 2, \frac{\pi}{2}\right) = \frac{1}{3}(2) + \frac{2}{3}(1) + \frac{2}{3}(0) = \frac{4}{3}$$ ∎

Maximum Rate of Change

For a given function f at a given point $\mathbf{p}$, it is natural to ask in what direction the function is changing most rapidly, that is, in what direction is $D_{\mathbf{u}}f(\mathbf{p})$ the largest? From the geometric formula for the dot product (Section 14.2), we may write

$$D_{\mathbf{u}}f(\mathbf{p}) = \mathbf{u} \cdot \nabla f(\mathbf{p}) = |\mathbf{u}||\nabla f(\mathbf{p})|\cos \theta = |\nabla f(\mathbf{p})|\cos \theta$$

where θ is the angle between $\mathbf{u}$ and $\nabla f(\mathbf{p})$. Thus, $D_{\mathbf{u}}f(\mathbf{p})$ is maximized when $\theta = 0$ and minimized when $\theta = \pi$. We summarize as follows.

> **Theorem B**
>
> A function increases most rapidly at $\mathbf{p}$ in the direction of the gradient (with rate $|\nabla f(\mathbf{p})|$) and decreases most rapidly in the opposite direction (with rate $-|\nabla f(\mathbf{p})|$).

EXAMPLE 3   Suppose that a bug is located on the hyperbolic paraboloid $z = y^2 - x^2$ at the point $(1, 1, 0)$, as in Figure 2. In what direction should it move for the steepest climb and what is the slope as it starts out?

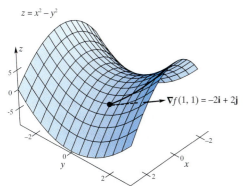

$z = x^2 - y^2$

$\nabla f(1, 1) = -2\mathbf{i} + 2\mathbf{j}$

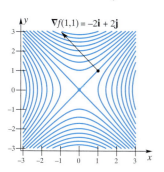

$\nabla f(1,1) = -2\mathbf{i} + 2\mathbf{j}$

Figure 2

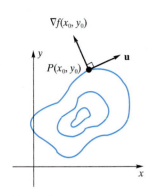

Figure 3

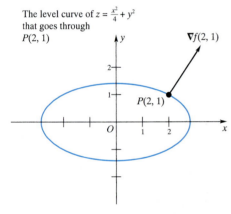

The level curve of $z = \frac{x^2}{4} + y^2$
that goes through
$P(2, 1)$

Figure 4

Solution   Let $f(x, y) = y^2 - x^2$. Since $f_x(x, y) = -2x$ and $f_y(x, y) = 2y$,
$$\nabla f(1, 1) = f_x(1, 1)\mathbf{i} + f_y(1, 1)\mathbf{j} = -2\mathbf{i} + 2\mathbf{j}$$
Thus, the bug should move from $(1, 1, 0)$ in the direction $-2\mathbf{i} + 2\mathbf{j}$, where the slope will be $|-2\mathbf{i} + 2\mathbf{j}| = \sqrt{8}$.   ∎

Level Curves and Gradients   Recall from Section 15.1 that the *level curves* of a surface $z = f(x, y)$ are the projections onto the xy-plane of the curves of intersection of the surface with planes $z = k$ that are parallel to the xy-plane. The value of the function at all points on the same level curve is constant (Figure 3).

Denote by L the level curve of $f(x, y)$ that passes through an arbitrarily chosen point $P(x_0, y_0)$ in the domain of f, and let the unit vector $\mathbf{u}$ be tangent to L at P. Since the value of f is the same at all points on the level curve L, its directional derivative $D_\mathbf{u}f(x_0, y_0)$, which is the rate of change of $f(x, y)$ in the direction $\mathbf{u}$, is zero when $\mathbf{u}$ is tangent to L. (This statement, which seems very clear intuitively, requires justification, which we omit since the result we want also follows from an argument to be given in Section 15.7.) Since
$$0 = D_\mathbf{u}f(x_0, y_0) = \nabla f(x_0, y_0) \cdot \mathbf{u}$$
we conclude that ∇f and $\mathbf{u}$ are perpendicular, a result worthy of theorem status.

Theorem C

The gradient of f at a point P is perpendicular to the level curve of f that goes through P.

EXAMPLE 4   For the paraboloid $z = x^2/4 + y^2$, find the equation of its level curve that passes through the point $P(2, 1)$ and sketch it. Find the gradient vector of the paraboloid at P, and draw the gradient with its initial point at P.

Solution   The level curve of the paraboloid that corresponds to the plane $z = k$ has the equation $x^2/4 + y^2 = k$. To find the value of k belonging to the level curve through P, we substitute $(2, 1)$ for (x, y) and obtain $k = 2$. Thus, the equation of the level curve that goes through P is the ellipse
$$\frac{x^2}{8} + \frac{y^2}{2} = 1$$
Next let $f(x, y) = x^2/4 + y^2$. Since $f_x(x, y) = x/2$ and $f_y(x, y) = 2y$, the gradient of the paraboloid at $P(2, 1)$ is
$$\nabla f(2, 1) = f_x(2, 1)\mathbf{i} + f_y(2, 1)\mathbf{j} = \mathbf{i} + 2\mathbf{j}$$
The level curve and the gradient at P are shown in Figure 4.   ∎

To provide additional illustration of Theorems B and C, we asked our computer to draw the surface $z = |xy|$, together with its contour map and gradient field. The results are shown in Figure 5. Note that the gradient vectors are perpendicular to the level curves and that they do point in the direction of greatest increase of z.

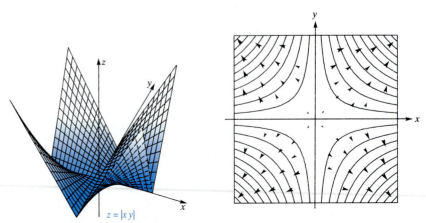

Figure 5

Higher Dimensions   The concept of level curves for functions of two variables generalizes to level surfaces for functions of three variables. If f is a function of three variables, the surface $f(x, y, z) = k$, where k is a constant, is called a **level surface** for f. At all points on a level surface, the value of the function is the same, and the gradient vector of $f(x, y, z)$ at a point $P(x, y, z)$ in its domain is normal to the level surface of f that goes through P.

In problems of heat conduction in a homogeneous body, where $w = f(x, y, z)$ gives the temperature at the point (x, y, z), the level surface $f(x, y, z) = k$ is called an *isothermal surface* because all points on it have the same temperature k. At any given point of the body, heat flows in the direction opposite to the gradient (i.e., in the direction of the greatest *decrease* in temperature) and therefore perpendicular to the isothermal surface through the point. If $w = f(x, y, z)$ gives the electrostatic potential (voltage) at any point in an electric potential field, the level surfaces of the function are called *equipotential surfaces*. All points on an equipotential surface have the same electrostatic potential, and the direction of flow of electricity is along the negative gradient, that is, in the direction of greatest drop in potential.

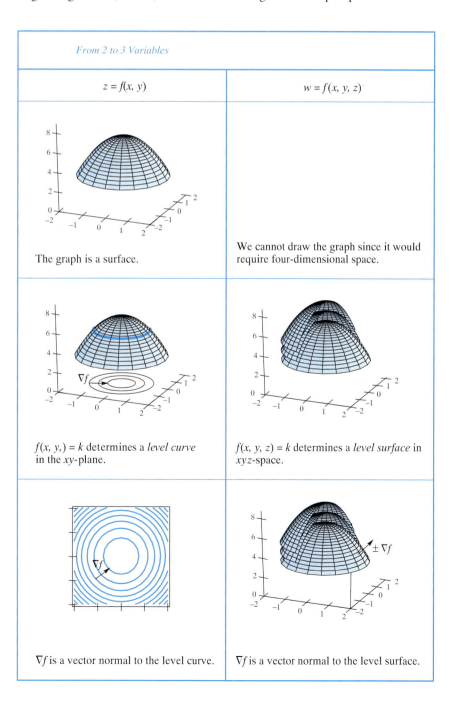

From 2 to 3 Variables

$z = f(x, y)$	$w = f(x, y, z)$
The graph is a surface.	We cannot draw the graph since it would require four-dimensional space.
$f(x, y,) = k$ determines a *level curve* in the xy-plane.	$f(x, y, z) = k$ determines a *level surface* in xyz-space.
∇f is a vector normal to the level curve.	∇f is a vector normal to the level surface.

Theorem A Chain Rule

Let $x = x(t)$ and $y = y(t)$ be differentiable at t, and let $z = f(x, y)$ be differentiable at $(x(t), y(t))$. Then $z = f(x(t), y(t))$ is differentiable at t and

$$\frac{dz}{dt} = \frac{\partial z}{\partial x}\frac{dx}{dt} + \frac{\partial z}{\partial y}\frac{dy}{dt}$$

Proof We mimic the one-variable proof of Appendix A.2, Theorem B. To simplify notation, let $\mathbf{p} = (x, y)$, $\Delta\mathbf{p} = (\Delta x, \Delta y)$, and $\Delta z = f(\mathbf{p} + \Delta\mathbf{p}) - f(\mathbf{p})$. Then, since f is differentiable,

$$\Delta z = f(\mathbf{p} + \Delta\mathbf{p}) - f(\mathbf{p}) = \nabla f(\mathbf{p}) \cdot \Delta\mathbf{p} + \boldsymbol{\varepsilon}(\Delta\mathbf{p}) \cdot \Delta\mathbf{p}$$
$$= f_x(\mathbf{p})\,\Delta x + f_y(\mathbf{p})\,\Delta y + \boldsymbol{\varepsilon}(\Delta\mathbf{p}) \cdot \Delta\mathbf{p}$$

with $\boldsymbol{\varepsilon}(\Delta\mathbf{p}) \to \mathbf{0}$ as $\Delta\mathbf{p} \to \mathbf{0}$.

When we divide both sides by Δt, we obtain

(1) $$\frac{\Delta z}{\Delta t} = f_x(\mathbf{p})\frac{\Delta x}{\Delta t} + f_y(\mathbf{p})\frac{\Delta y}{\Delta t} + \boldsymbol{\varepsilon}(\Delta\mathbf{p}) \cdot \left\langle \frac{\Delta x}{\Delta t}, \frac{\Delta y}{\Delta t} \right\rangle$$

Now, $\left\langle \dfrac{\Delta x}{\Delta t}, \dfrac{\Delta y}{\Delta t} \right\rangle$ approaches $\left\langle \dfrac{dx}{dt}, \dfrac{dy}{dt} \right\rangle$ as $\Delta t \to 0$. Also, when $\Delta t \to 0$, both Δx and Δy approach 0 (remember that $x(t)$ and $y(t)$ are continuous, being differentiable). It follows that $\Delta\mathbf{p} \to \mathbf{0}$, and hence $\boldsymbol{\varepsilon}(\Delta\mathbf{p}) \to \mathbf{0}$ as $\Delta t \to 0$. Consequently, when we let $\Delta t \to 0$ in (1), we get

$$\frac{dz}{dt} = f_x(\mathbf{p})\frac{dx}{dt} + f_y(\mathbf{p})\frac{dy}{dt}$$

a result equivalent to the claimed assertion. ◆

Beauty and Generality

Does the general analog of the one variable Chain Rule (Theorem A, Section 3.5) hold? Yes, and here is a particularly elegant statement of it. Let $\mathbb{R}^n$ denote Euclidean *n*-space, let g be a function from $\mathbb{R}$ to $\mathbb{R}^n$, and let f be a function from $\mathbb{R}^n$ to $\mathbb{R}$. If g is differentiable at t and if f is differentiable at $g(t)$, then the composite function $f \circ g$ is differentiable at $g(t)$ and

$$(f \circ g)'(t) = \nabla f(g(t)) \cdot g'(t)$$

All the machinery needed to demonstrate this is available; see if you can give the proof.

EXAMPLE 1 Suppose that $z = x^3 y$, where $x = 2t$ and $y = t^2$. Find dz/dt.

Solution

$$\frac{dz}{dt} = \frac{\partial z}{\partial x}\frac{dx}{dt} + \frac{\partial z}{\partial y}\frac{dy}{dt}$$
$$= (3x^2 y)(2) + (x^3)(2t)$$
$$= 6(2t)^2(t^2) + 2(2t)^3(t)$$
$$= 40t^4$$

We could have done Example 1 without use of the Chain Rule. By direct substitution,

$$z = x^3 y = (2t)^3 t^2 = 8t^5$$

and so $dz/dt = 40t^4$. However, the direct substitution method is often not available or not convenient—witness the next example.

The Chain Rule: Two-Variable Case

Here is a device that may help you to remember the Chain Rule.

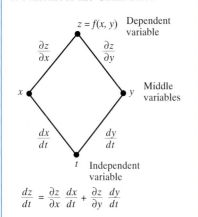

$$\frac{dz}{dt} = \frac{\partial z}{\partial x}\frac{dx}{dt} + \frac{\partial z}{\partial y}\frac{dy}{dt}$$

EXAMPLE 2 As a solid right circular cylinder is heated, its radius r and height h increase; hence, so does its surface area S. Suppose that at the instant when $r = 10$ centimeters and $h = 100$ centimeters r is increasing at 0.2 centimeter per hour and h is increasing at 0.5 centimeter per hour. How fast is S increasing at this instant?

Solution The formula for the total surface area of a cylinder (Figure 1) is

$$S = 2\pi rh + 2\pi r^2$$

Thus,

$$\frac{dS}{dt} = \frac{\partial S}{\partial r}\frac{dr}{dt} + \frac{\partial S}{\partial h}\frac{dh}{dt}$$
$$= (2\pi h + 4\pi r)(0.2) + (2\pi r)(0.5)$$

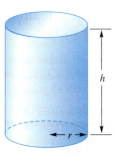

Figure 1

At $r = 10$ and $h = 100$,

$$\frac{dS}{dt} = (2\pi \cdot 100 + 4\pi \cdot 10)(0.2) + (2\pi \cdot 10)(0.5)$$

$$= 58\pi \text{ square centimeters per hour} \qquad \blacksquare$$

The result in Theorem A extends readily to a function of three variables, as we now illustrate.

EXAMPLE 3   Suppose that $w = x^2 y + y + xz$, where $x = \cos\theta$, $y = \sin\theta$, and $z = \theta^2$. Find $dw/d\theta$ and evaluate it at $\theta = \pi/3$.

Solution

$$\frac{dw}{d\theta} = \frac{\partial w}{\partial x}\frac{dx}{d\theta} + \frac{\partial w}{\partial y}\frac{dy}{d\theta} + \frac{\partial w}{\partial z}\frac{dz}{d\theta}$$

$$= (2xy + z)(-\sin\theta) + (x^2 + 1)(\cos\theta) + (x)(2\theta)$$

$$= -2\cos\theta\sin^2\theta - \theta^2\sin\theta + \cos^3\theta + \cos\theta + 2\theta\cos\theta$$

At $\theta = \pi/3$,

$$\frac{dw}{d\theta} = -2 \cdot \frac{1}{2} \cdot \frac{3}{4} - \frac{\pi^2}{9} \cdot \frac{\sqrt{3}}{2} + \left(\frac{1}{4} + 1\right)\frac{1}{2} + \frac{2\pi}{3} \cdot \frac{1}{2}$$

$$= -\frac{1}{8} - \frac{\pi^2\sqrt{3}}{18} + \frac{\pi}{3} \qquad \blacksquare$$

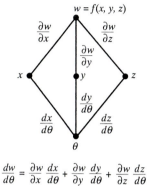

The Chain Rule: Three-Variable Case

$w = f(x, y, z)$

$$\frac{dw}{d\theta} = \frac{\partial w}{\partial x}\frac{dx}{d\theta} + \frac{\partial w}{\partial y}\frac{dy}{d\theta} + \frac{\partial w}{\partial z}\frac{dz}{d\theta}$$

Second Version   Suppose next that $z = f(x, y)$, where $x = x(s, t)$ and $y = y(s, t)$. Then it makes sense to ask for $\partial z/\partial s$ and $\partial z/\partial t$.

Theorem B   Chain Rule

Let $x = x(s, t)$ and $y = y(s, t)$ have first partial derivatives at (s, t), and let $z = f(x, y)$ be differentiable at $(x(s, t), y(s, t))$. Then $z = f(x(s, t), y(s, t))$ has first partial derivatives given by

(i) $\dfrac{\partial z}{\partial s} = \dfrac{\partial z}{\partial x}\dfrac{\partial x}{\partial s} + \dfrac{\partial z}{\partial y}\dfrac{\partial y}{\partial s};$ (ii) $\dfrac{\partial z}{\partial t} = \dfrac{\partial z}{\partial x}\dfrac{\partial x}{\partial t} + \dfrac{\partial z}{\partial y}\dfrac{\partial y}{\partial t}.$

Proof   If s is held fixed, then $x(s, t)$ and $y(s, t)$ become functions of t alone, which means that Theorem A applies. When we use this theorem with ∂ replacing d to indicate that s is fixed, we obtain the formula in (ii) for $\partial z/\partial t$. The formula for $\partial z/\partial s$ is obtained in a similar way by holding t fixed. ◆

EXAMPLE 4   If $z = 3x^2 - y^2$, where $x = 2s + 7t$ and $y = 5st$, find $\partial z/\partial t$ and express it in terms of s and t.

Solution

$$\frac{\partial z}{\partial t} = \frac{\partial z}{\partial x}\frac{\partial x}{\partial t} + \frac{\partial z}{\partial y}\frac{\partial y}{\partial t}$$

$$= (6x)(7) + (-2y)(5s)$$

$$= 42(2s + 7t) - 10st(5s)$$

$$= 84s + 294t - 50s^2 t \qquad \blacksquare$$

Here is the corresponding result for three intermediate variables illustrated in an example.

EXAMPLE 5   If $w = x^2 + y^2 + z^2 + xy$, where $x = st$, $y = s - t$, and $z = s + 2t$, find $\partial w/\partial t$.

and use this result to find $F'(\sqrt{2})$, where

$$F(t) = \int_{\sin\sqrt{2}\,\pi t}^{t^2} \sqrt{9 + u^4}\, du$$

32. Call a function $f(x, y)$ *homogeneous of degree 1* if $f(tx, ty) = tf(x, y)$ for all $t > 0$. For example, $f(x, y) = x + ye^{y/x}$ satisfies this criterion. Prove **Euler's Theorem** that such a function satisfies

$$f(x, y) = x\frac{\partial f}{\partial x} + y\frac{\partial f}{\partial y}$$

Note: Let $f(x, y)$ denote the value of production from x units of capital and y units of labor. Then f is a homogeneous function (e.g., doubling capital and labor doubles production). Euler's Theorem then asserts an important law of economics that may be phrased as follows: The value of production $f(x, y)$ equals the cost of capital plus the cost of labor provided that they are paid for at their respective marginal rates $\partial f/\partial x$ and $\partial f/\partial y$.

C **33.** Leaving from the same point P, airplane A flies due east while airplane B flies N 50° E. At a certain instant, A is 200 miles from P flying at 450 miles per hour, and B is 150 miles from P flying at 400 miles per hour. How fast are they separating at that instant?

34. Recall Newton's Law of Gravitation, which asserts that the magnitude F of the force of attraction between objects of masses M and m is $F = GMm/r^2$, where r is the distance between them and G is a universal constant. Let an object of mass M be located at the origin, and suppose that a second object of changing mass m (say from fuel consumption) is moving away from the origin so that its position vector is $\mathbf{r} = x\mathbf{i} + y\mathbf{j} + z\mathbf{k}$. Obtain a formula for dF/dt in terms of the time derivatives of m, x, y, and z.

Answers to Concepts Review: **1.** $\dfrac{\partial z}{\partial x}\dfrac{dx}{dt} + \dfrac{\partial z}{\partial y}\dfrac{dy}{dt}$

2. $y^2\cos t + 2xy(-\sin t) = \cos^3 t - 2\sin^2 t \cos t$

3. $\dfrac{\partial z}{\partial x}\dfrac{\partial x}{\partial t} + \dfrac{\partial z}{\partial y}\dfrac{\partial y}{\partial t}$ **4.** 12

15.7
Tangent Planes, Approximations

We introduced the notion of a tangent plane to a surface in Section 15.4, but dealt only with surfaces determined by equations of the form $z = f(x, y)$ (Figure 1). Now we want to consider the more general situation of a surface determined by $F(x, y, z) = k$. (Note that $z = f(x, y)$ can be written as $F(x, y, z) = f(x, y) - z = 0$.) Consider a curve on this surface passing through the point (x_0, y_0, z_0). If $x = x(t)$, $y = y(t)$, and $z = z(t)$ are parametric equations for this curve, then, for all t,

$$F\big(x(t), y(t), z(t)\big) = k$$

By the Chain Rule,

$$\frac{dF}{dt} = \frac{\partial F}{\partial x}\frac{dx}{dt} + \frac{\partial F}{\partial y}\frac{dy}{dt} + \frac{\partial F}{\partial z}\frac{dz}{dt} = \frac{dk}{dt} = 0$$

We can express this in terms of the gradient of F and the derivative of the vector expression for the curve $\mathbf{r}(t) = x(t)\mathbf{i} + y(t)\mathbf{j} + z(t)\mathbf{k}$ as

$$\nabla F \cdot \frac{d\mathbf{r}}{dt} = 0$$

As we learned earlier (Section 14.4), $d\mathbf{r}/dt$ is tangent to the curve. In summary, the gradient at (x_0, y_0, z_0) is perpendicular to the tangent line at this point.

The argument just given is valid for any curve through (x_0, y_0, z_0) that lies in the surface $F(x, y, z) = k$ (Figure 2). This suggests the following general definition.

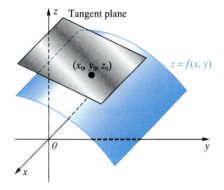

Figure 1

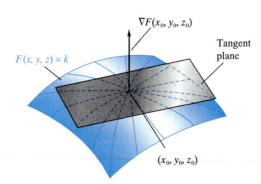

Figure 2

Definition

Let $F(x, y, z) = k$ determine a surface, and suppose that F is differentiable at a point $P(x_0, y_0, z_0)$ of this surface, with $\nabla F(x_0, y_0, z_0) \neq \mathbf{0}$. Then the plane through P perpendicular to $\nabla F(x_0, y_0, z_0)$ is called the **tangent plane** to the surface at P.

As a consequence of this definition and Section 14.2, we can write the equation of the tangent plane.

> **Theorem A** Tangent Planes
>
> For the surface $F(x, y, z) = k$, the equation of the tangent plane at (x_0, y_0, z_0) is $\nabla F(x_0, y_0, z_0) \cdot \langle x - x_0, y - y_0, z - z_0 \rangle = 0$; that is,
>
> $$F_x(x_0, y_0, z_0)(x - x_0) + F_y(x_0, y_0, z_0)(y - y_0) + F_z(x_0, y_0, z_0)(z - z_0) = 0$$
>
> In particular, for the surface $z = f(x, y)$, the equation of the tangent plane at $(x_0, y_0, f(x_0, y_0))$ is
>
> $$z - z_0 = f_x(x_0, y_0)(x - x_0) + f_y(x_0, y_0)(y - y_0)$$

Proof The first statement is immediate, and the second follows from it by considering $F(x, y, z) = f(x, y) - z$. ♦

If z is a function of x and y, say $z = f(x, y)$, then, from the second part of Theorem A, we can write the equation of the tangent plane as

$$z - f(x_0, y_0) = f_x(x_0, y_0)(x - x_0) + f_y(x_0, y_0)(y - y_0)$$

Letting $\mathbf{p} = (x, y)$ and $\mathbf{p}_0 = (x_0, y_0)$, we see that the equation of the tangent plane is

$$z = f(x_0, y_0) + (f_x(x_0, y_0), f_y(x_0, y_0)) \cdot (x - x_0, y - y_0)$$
$$= f(\mathbf{p}_0) + \nabla f(\mathbf{p}_0) \cdot (\mathbf{p} - \mathbf{p}_0)$$

Thus, our definition in this section agrees with the definition of a tangent plane given in Section 15.4.

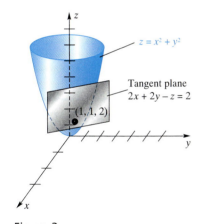

$z = x^2 + y^2$

Tangent plane
$2x + 2y - z = 2$

$(1, 1, 2)$

Figure 3

EXAMPLE 1 Find the equation of the tangent plane (Figure 3) to $z = x^2 + y^2$ at the point $(1, 1, 2)$.

Solution Let $f(x, y) = x^2 + y^2$, and note that $\nabla f(x, y) = 2x\mathbf{i} + 2y\mathbf{j}$. Thus, $\nabla f(1, 1) = 2\mathbf{i} + 2\mathbf{j}$ and, from Theorem A, the required equation is

$$z - 2 = 2(x - 1) + 2(y - 1)$$

or

$$2x + 2y - z = 2 \qquad \blacksquare$$

EXAMPLE 2 Find the equation of the tangent plane and the normal line to the surface $x^2 + y^2 + 2z^2 = 23$ at $(1, 2, 3)$.

Solution Let $F(x, y, z) = x^2 + y^2 + 2z^2$ so that $\nabla F(x, y, z) = 2x\mathbf{i} + 2y\mathbf{j} + 4z\mathbf{k}$ and $\nabla F(1, 2, 3) = 2\mathbf{i} + 4\mathbf{j} + 12\mathbf{k}$. According to Theorem A, the equation of the tangent plane at $(1, 2, 3)$ is

$$2(x - 1) + 4(y - 2) + 12(z - 3) = 0$$

Similarly, the symmetric equations of the normal line through $(1, 2, 3)$ are

$$\frac{x - 1}{2} = \frac{y - 2}{4} = \frac{z - 3}{12} \qquad \blacksquare$$

Differentials and Approximations We suggest that you review Section 3.10, where the topics of differentials and approximations are treated for functions of one variable.

15. Show that the surfaces $x^2 + 4y + z^2 = 0$ and $x^2 + y^2 + z^2 - 6z + 7 = 0$ are tangent to each other at $(0, -1, 2)$; that is, show that they have the same tangent plane at $(0, -1, 2)$.

16. Show that the surfaces $z = x^2 y$ and $y = \frac{1}{4}x^2 + \frac{3}{4}$ intersect at $(1, 1, 1)$ and have perpendicular tangent planes there.

17. Find a point on the surface $x^2 + 2y^2 + 3z^2 = 12$ where the tangent plane is perpendicular to the line with parametric equations: $x = 1 + 2t$, $y = 3 + 8t$, $z = 2 - 6t$.

18. Show that the equation of the tangent plane to the ellipsoid

$$\frac{x^2}{a^2} + \frac{y^2}{b^2} + \frac{z^2}{c^2} = 1$$

at (x_0, y_0, z_0) can be written in the form

$$\frac{x_0 x}{a^2} + \frac{y_0 y}{b^2} + \frac{z_0 z}{c^2} = 1$$

19. Find the parametric equations of the line that is tangent to the curve of intersection of the surfaces

$$f(x, y, z) = 9x^2 + 4y^2 + 4z^2 - 41 = 0$$

and

$$g(x, y, z) = 2x^2 - y^2 + 3z^2 - 10 = 0$$

at the point $(1, 2, 2)$. *Hint*: This line is perpendicular to $\nabla f(1, 2, 2)$ and $\nabla g(1, 2, 2)$.

20. Find the parametric equations of the line that is tangent to the curve of intersection of the surfaces $x = z^2$ and $y = z^3$ at $(1, 1, 1)$ (see Problem 19).

21. In determining the specific gravity of an object, its weight in air is found to be $A = 36$ pounds and its weight in water is $W = 20$ pounds, with a possible error in each measurement of 0.02 pound. Find, approximately, the maximum possible error in calculating its specific gravity S, where $S = A/(A - W)$.

22. Use differentials to find the approximate amount of copper in the four sides and bottom of a rectangular copper tank that is 6 feet long, 4 feet wide, and 3 feet deep *inside*, if the sheet copper is $\frac{1}{4}$ inch thick. *Hint*: Make a sketch.

23. The radius and height of a right circular cone are measured with errors of at most 2% and 3%, respectively. Use differ-

entials to estimate the maximum percentage error in the calculated volume (see Example 4).

24. The period T of a pendulum of length L is given by $T = 2\pi\sqrt{L/g}$, where g is the acceleration of gravity. Show that $dT/T = \frac{1}{2}[dL/L - dg/g]$, and use this result to estimate the maximum percentage error in T due to an error of 0.5% in measuring L and 0.3% in measuring g.

25. The formula $1/R = 1/R_1 + 1/R_2$ determines the combined resistance R when resistors of resistance R_1 and R_2 are connected in parallel. Suppose that R_1 and R_2 were measured at 25 and 100 ohms, respectively, with possible errors in each measurement of 0.5 ohm. Calculate R and give an estimate for the maximum error in this value.

26. A bee sat at the point $(1, 2, 1)$ on the ellipsoid $x^2 + y^2 + 2z^2 = 6$ (distances in feet). At $t = 0$, it took off along the normal line at a speed of 4 feet per second. Where and when did it hit the plane $2x + 3y + z = 49$?

27. Show that a plane tangent at any point of the surface $xyz = k$ forms with the coordinate planes a tetrahedron of fixed volume and find this volume.

28. Find and simplify the equation of the tangent plane at (x_0, y_0, z_0) to the surface $\sqrt{x} + \sqrt{y} + \sqrt{z} = a$. Then show that the sum of the intercepts of this plane with the coordinate axes is a^2.

[C] **29.** For the function $f(x, y) = \sqrt{x^2 + y^2}$, find the second-order Taylor approximation based at $(x_0, y_0) = (3, 4)$. Then estimate $f(3.1, 3.9)$ using

(a) the first-order approximation,

(b) the second-order approximation, and

(c) your calculator directly.

Answers to Concepts Review: **1.** perpendicular **2.** $\langle 3, 1, -1 \rangle$

3. $x - 2 + 4(y - 1) + 6(z - 1) = 0$ **4.** $\dfrac{\partial f}{\partial x}\,dx + \dfrac{\partial f}{\partial y}\,dy$

15.8
Maxima and Minima

Our goal is to extend the notions of Chapter 4 to functions of several variables; a quick review of that chapter, especially Sections 4.1 and 4.3, will be helpful. The definitions given there extend almost without change, but for clarity we repeat them. In what follows, let $\mathbf{p} = (x, y)$ and $\mathbf{p}_0 = (x_0, y_0)$ be a variable point and a fixed point, respectively, in two-space (they could just as well be points in n-space).

Definition

Let f be a function with domain S, and let $\mathbf{p}_0$ be a point in S.

1. $f(\mathbf{p}_0)$ is a **global maximum value** of f on S if $f(\mathbf{p}_0) \geq f(\mathbf{p})$ for all $\mathbf{p}$ in S.
2. $f(\mathbf{p}_0)$ is a **global minimum value** of f on S if $f(\mathbf{p}_0) \leq f(\mathbf{p})$ for all $\mathbf{p}$ in S.
3. $f(\mathbf{p}_0)$ is a **global extreme value** of f on S if $f(\mathbf{p}_0)$ is either a global maximum value or a global minimum value.

We obtain definitions for **local maximum value** and **local minimum value** if in (1) and (2) we require only that the inequalities hold on $N \cap S$, where N is some neighborhood of $\mathbf{p}_0$. $f(\mathbf{p}_0)$ is a **local extreme value** of f on S if $f(\mathbf{p}_0)$ is either a local maximum value or a local minimum value.

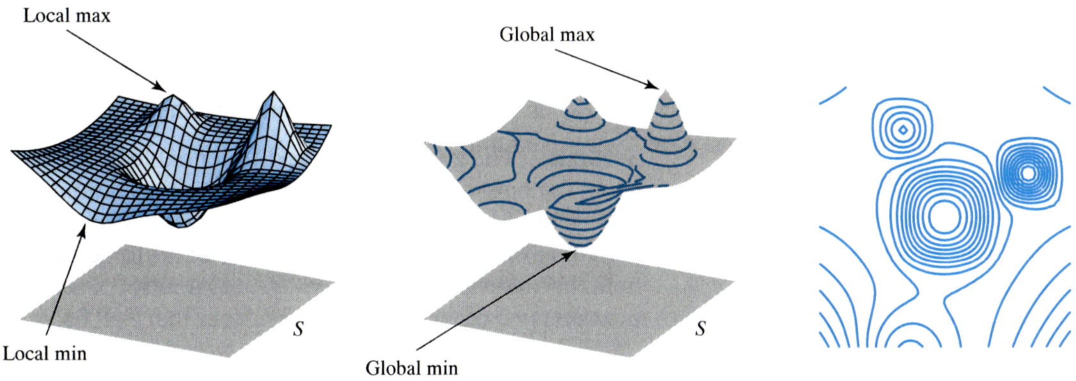

Figure 1

Figure 1 gives a geometric interpretation of the concepts we have defined. Note that a global maximum (or minimum) is automatically a local maximum (or minimum).

Our first theorem is a big one—difficult to prove, but intuitively clear.

Theorem A   Max–Min Existence Theorem

If f is continuous on a closed bounded set S, then f attains both a (global) maximum value and a (global) minimum value there.

The proof may be found in most books on advanced calculus.

Where Do Extreme Values Occur?   The situation is analogous to the one-variable case. The **critical points** of f on S are of three types.

1. **Boundary points**. See Section 15.3.
2. **Stationary points**. We call $\mathbf{p}_0$ a stationary point if $\mathbf{p}_0$ is an interior point of S where f is differentiable and $\nabla f(\mathbf{p}_0) = \mathbf{0}$. At such a point, the tangent plane is horizontal.
3. **Singular points**. We call $\mathbf{p}_0$ a singular point if $\mathbf{p}_0$ is an interior point of S where f is not differentiable, for example, a point where the graph of f has a sharp corner.

Now we can state another big theorem; we can actually prove this one.

Theorem B   Critical Point Theorem

Let f be defined on a set S containing $\mathbf{p}_0$. If $f(\mathbf{p}_0)$ is an extreme value, then $\mathbf{p}_0$ must be a critical point; that is, either $\mathbf{p}_0$ is

(i)   a boundary point of S; or
(ii) a stationary point of f; or
(iii) a singular point of f.

Proof   Suppose that $\mathbf{p}_0$ is neither a boundary point nor a singular point (so that $\mathbf{p}_0$ is an interior point where ∇f exists). We will be done if we can show that $\nabla f(\mathbf{p}_0) = \mathbf{0}$. For simplicity, set $\mathbf{p}_0 = (x_0, y_0)$; the higher-dimensional cases will follow in a similar fashion.

Since f has an extreme value at (x_0, y_0), the function $g(x) = f(x, y_0)$ has an extreme value at x_0. Moreover, g is differentiable at x_0 since f is differentiable at (x_0, y_0) and therefore, by the Critical Point Theorem for functions of one variable (Theorem 4.1 B),

$$g'(x_0) = f_x(x_0, y_0) = 0$$

Similarly, the function $h(y) = f(x_0, y)$ has an extreme value at y_0 and satisfies

$$h'(y_0) = f_y(x_0, y_0) = 0$$

The gradient is $\mathbf{0}$ since both partials are 0. ◆

Visualizing the Directional Derivative

I. Preparation

In this project, we will explore the concept of the directional derivative.

Exercise 1 By hand, find the directional derivative of $f(x, y) = 3x + 4y + 4$ at the point $(1, -1)$ in each of the directions

(a) $\mathbf{i}$ (b) $\mathbf{i} + \mathbf{j}$ (c) $\mathbf{j}$
(d) $-\mathbf{i} + \mathbf{j}$ (e) $-\mathbf{i}$ (f) $-\mathbf{i} - \mathbf{j}$
(g) $-\mathbf{j}$ (h) $\mathbf{i} - \mathbf{j}$ (i) $\mathbf{i} + t\mathbf{j}$
(j) $\cos t\, \mathbf{i} + \sin t\, \mathbf{j}$

Exercise 2 Repeat Exercise 1 for the function $f(x, y) = x^3 + 4y^3$.

II. Using Technology

In the first part of the Technology section, we will visualize the results of Exercises 1 and 2.

Exercise 3

(a) Create a three-dimensional plot and a contour plot of $f(x, y) = 3x + 4y + 4$ for the domain $S = \{(x, y): -2 \le x \le 2, -2 \le y \le 2\}$. Now imagine that you are a bug located *on* the surface at the point $(1, -1, f(1, -1))$. If you take one step (a bug's step is 0.2 unit) in the direction $\mathbf{i}$, have you gone uphill or downhill? Before you took the step, your altitude (i.e., your z-coordinate) was $f(1, -1) = 3$; after your step, what is your altitude? What is the slope in taking this step?

(b) Repeat part (a) for the direction $\mathbf{j}$.

(c) Repeat part (a) for the direction $-\mathbf{i} - \mathbf{j}$.

(d) Do the slopes you found in parts (a), (b), and (c) agree with the results that you obtained by hand in Exercise 1? Explain why your results agree or disagree.

Exercise 4 Let $f(x, y) = 3x + 4y + 4$, $x_0 = 1$, $y_0 = -1$, and $e = 0.2$. For t in the interval $[0, 2\pi]$, define

$$h(t) = f(x_0 + e \cos t, y_0 + e \sin t)$$

$$a(t) = \frac{f(x_0 + e \cos t, y_0 + e \sin t) - f(x_0, y_0)}{e}$$

(a) Plot the graph of $h(t)$. Explain what this graph represents.

(b) Plot the graph of $a(t)$. Explain what each of the following represents and how it relates to a part of Exercise 3: $a(0)$, $a(\pi/2)$, $a(5\pi/4)$, and $a(t)$.

Exercise 5

(a) Does the graph of $a(t)$ look like it is related to the trigonometric functions? Without actually calculating it, can you conjecture what the value of $\int_0^{2\pi} a(t)\, dt$ is?

(b) Zoom in on the graph of $a(t)$ to approximate, to two decimal places, the value of t that maximizes $a(t)$? Again, suppose that you are a bug on the surface of $f(x, y) = 3x + 4y + 4$ at the point $(1, -1, 3)$. What direction should you go if you want to go as high as possible in one step?

(c) Derive a formula for $a(t)$ and use your technology to evaluate $\int_0^{2\pi} a(t)\, dt$.

Exercise 6 Now let $f(x, y) = x^3 + 4y^3$, $x_0 = 1$, $y_0 = -1$, and $e = 0.2$. Repeat Exercise 3 for this function, and compare with your results from Exercise 2.

Exercise 7 Repeat Exercise 4 for $f(x, y) = x^3 + 4y^3$, $x_0 = 1$, $y_0 = -1$, and $e = 0.2$. Then change e to $e = 0.002$ and repeat Exercise 4.

Exercise 8 The function $a(t)$ depends on the choice of e; hence, the definite integral $\int_0^{2\pi} a(t)\, dt$ depends on e. Use your technology to approximate $\int_0^{2\pi} a(t)\, dt$ for $e = 0.2$, $0.02, 0.002, 0.0002$, and 0.00002. Make a conjecture regarding $\lim\limits_{e \to 0} \int_0^{2\pi} a(t)\, dt$.

III. Reflection

Exercise 9 Let $f(x, y) = x^3 + 4y^3$, $x_0 = 2$, and $y_0 = 2$. Define $h(t) = f(x_0 + \cos t, y_0 + \sin t)$. (Note that this is the same h defined previously, but with $e = 1$.)

(a) Make a three-dimensional surface plot of $y = f(x, y)$ and a three-dimensional plot of the curve defined parametrically by $(x_0 + \cos t, y_0 + \sin t, h(t))$ in the same graph window.

(b) Let $0 = t_0 < t_1 < \cdots < t_{n-1} < t_n = 2\pi$ be a *regular* (i.e., equally spaced) partition of the interval $[0, 2\pi]$. Let Δs_i be the length of the line segment connecting $(x_0 + \cos t_{i-1}, y_0 + \sin t_{i-1})$ and $(x_0 + \cos t_i, y_0 + \sin t_i)$. Consider the sum

$$\sum_{i=1}^{n} h(t_i)\, \Delta s_i$$

This is a Riemann sum for what definite integral? (We will study integrals such as this in Chapter 17.)

(c) Use your technology to evaluate the definite integral from part (b).

(d) Suppose that you wanted to hang a curtain that stretched from the xy-plane to the surface. The bottom of the curtain should lie along the circle $(x, y) = (x_0 + \cos t, y_0 + \sin t)$, $0 \le t \le 2\pi$, in the xy-plane. How many square units of material would be needed?

16

The Integral in *n*-Space

16.1
Double Integrals Over Rectangles

Differentiation and integration are the major processes of calculus. We have studied differentiation in *n*-space (Chapter 15); it is time to consider integration in *n*-space. The theory and the applications of single (Riemann) integrals are to be generalized to multiple integrals. In Chapter 6 we used single integrals to calculate the area of curved planar regions, to find the length of planar curves, and to determine the center of mass of straight wires of variable density. In this chapter we use multiple integrals to find the volume of general solids, the area of general surfaces, and the center of mass of laminas and solids of variable density.

The intimate connection between integration and differentiation was enunciated in the Fundamental Theorems of Calculus; these theorems provided the principal theoretical tools for evaluating single integrals. Here we reduce multiple integration to a succession of single integrations where again the Second Fundamental Theorem will play a central role. The integration skills that you learned in Chapters 5 through 8 will be tested.

The Riemann integral for a function of one variable was defined in Section 5.5, a section worth reviewing. Recall that we formed a partition P of the interval $[a, b]$ into subintervals of length Δx_k, $k = 1, 2, \ldots, n$, picked a sample point $\bar{x}_k$ from the kth subinterval, and then wrote

$$\int_a^b f(x)\, dx = \lim_{|P| \to 0} \sum_{k=1}^n f(\bar{x}_k)\, \Delta x_k$$

We proceed in a very similar fashion to define the integral for a function of two variables.

1. The double integral is linear; that is,

(a) $\iint\limits_{R} kf(x, y)\,dA = k \iint\limits_{R} f(x, y)\,dA;$

(b) $\iint\limits_{R} [f(x, y) + g(x, y)]\,dA = \iint\limits_{R} f(x, y)\,dA + \iint\limits_{R} g(x, y)\,dA.$

2. The double integral is additive on rectangles (Figure 6) that overlap only on a line segment.

$$\iint\limits_{R} f(x, y)\,dA = \iint\limits_{R_1} f(x, y)\,dA + \iint\limits_{R_2} f(x, y)\,dA$$

3. The comparison property holds. If $f(x, y) \le g(x, y)$ for all (x, y) in R, then

$$\iint\limits_{R} f(x, y)\,dA \le \iint\limits_{R} g(x, y)\,dA$$

All of these properties hold on more general sets than rectangles, but that is a matter we take up in Section 16.3.

Evaluation of Double Integrals   This topic will receive major attention in the next section, where we will develop a powerful tool for evaluating double integrals. However, we can already evaluate a few integrals, and we can approximate others.

Note first that if $f(x, y) = 1$ on R then the double integral is the area of R, and from this it follows that

$$\iint\limits_{R} k\,dA = k \iint\limits_{R} 1\,dA = kA(R)$$

EXAMPLE 1   Let f be the staircase function of Figure 5; that is, let

$$f(x, y) = \begin{cases} 1 & 0 \le x \le 3, 0 \le y \le 1 \\ 2 & 0 \le x \le 3, 1 \le y \le 2 \\ 3 & 0 \le x \le 3, 2 \le y \le 3 \end{cases}$$

Calculate $\iint\limits_{R} f(x, y)\,dA$, where $R = \{(x, y): 0 \le x \le 3, 0 \le y \le 3\}$.

Solution   Introduce rectangles R_1, R_2, and R_3 as follows:

$$R_1 = \{(x, y): 0 \le x \le 3, 0 \le y \le 1\}$$
$$R_2 = \{(x, y): 0 \le x \le 3, 1 \le y \le 2\}$$
$$R_3 = \{(x, y): 0 \le x \le 3, 2 \le y \le 3\}$$

Then, using the additivity property of the double integral, we obtain

$$\iint\limits_{R} f(x, y)\,dA = \iint\limits_{R_1} f(x, y)\,dA + \iint\limits_{R_2} f(x, y)\,dA + \iint\limits_{R_3} f(x, y)\,dA$$

$$= 1A(R_1) + 2A(R_2) + 3A(R_3)$$

$$= 1 \cdot 3 + 2 \cdot 3 + 3 \cdot 3 = 18$$

In this derivation, we also used the fact that the value of f on the boundary of a rectangle does not affect the value of the integral.   ■

Example 1 was a minor accomplishment, and to be honest we cannot do much more without more tools. However, we can always approximate a double integral

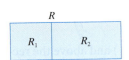

R

Figure 6

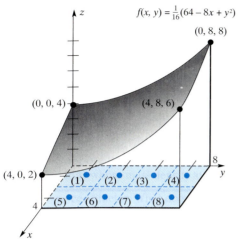

$f(x, y) = \frac{1}{16}(64 - 8x + y^2)$

$(0, 8, 8)$

$(0, 0, 4)$ $(4, 8, 6)$

$(4, 0, 2)$

Figure 7

by calculating a Riemann sum. In general, we can expect the approximation to be better the finer the partition we use.

EXAMPLE 2 Approximate $\iint\limits_{R} f(x, y)\, dA$, where

$$f(x, y) = \frac{64 - 8x + y^2}{16}$$

and

$$R = \{(x, y): 0 \le x \le 4, 0 \le y \le 8\}$$

Do this by calculating the Riemann sum obtained by dividing R into eight equal squares and using the center of each square as the sample point (Figure 7).

Solution The required sample points and the corresponding values of the function are as follows:

$$(\bar{x}_1, \bar{y}_1) = (1, 1), \qquad f(\bar{x}_1, \bar{y}_1) = \frac{57}{16}$$

$$(\bar{x}_2, \bar{y}_2) = (1, 3), \qquad f(\bar{x}_2, \bar{y}_2) = \frac{65}{16}$$

$$(\bar{x}_3, \bar{y}_3) = (1, 5), \qquad f(\bar{x}_3, \bar{y}_3) = \frac{81}{16}$$

$$(\bar{x}_4, \bar{y}_4) = (1, 7), \qquad f(\bar{x}_4, \bar{y}_4) = \frac{105}{16}$$

$$(\bar{x}_5, \bar{y}_5) = (3, 1), \qquad f(\bar{x}_5, \bar{y}_5) = \frac{41}{16}$$

$$(\bar{x}_6, \bar{y}_6) = (3, 3), \qquad f(\bar{x}_6, \bar{y}_6) = \frac{49}{16}$$

$$(\bar{x}_7, \bar{y}_7) = (3, 5), \qquad f(\bar{x}_7, \bar{y}_7) = \frac{65}{16}$$

$$(\bar{x}_8, \bar{y}_8) = (3, 7), \qquad f(\bar{x}_8, \bar{y}_8) = \frac{89}{16}$$

Thus, since $\Delta A_k = 4$,

$$\iint\limits_{R} f(x, y)\, dA \approx \sum_{k=1}^{8} f(\bar{x}_k, \bar{y}_k)\, \Delta A_k$$

$$= 4 \sum_{k=1}^{8} f(\bar{x}_k, \bar{y}_k)$$

$$= \frac{4(57 + 65 + 81 + 105 + 41 + 49 + 65 + 89)}{16} = 138$$

In Section 16.2, we shall learn how to find the exact value of this integral. It is $138\frac{2}{3}$. ■

14. Find the moment of inertia of the lamina of Problem 13 with respect to the *x*-axis.

15. Find the area of the surface of the cylinder $z^2 + y^2 = 9$ lying in the first octant between the planes $y = x$ and $y = 3x$.

16. Evaluate by changing to cylindrical or spherical coordinates.

(a) $\displaystyle\int_0^3 \int_0^{\sqrt{9-x^2}} \int_0^2 \sqrt{x^2 + y^2}\, dz\, dy\, dx$

(b) $\displaystyle\int_0^2 \int_0^{\sqrt{4-x^2}} \int_0^{\sqrt{4-x^2-y^2}} z\sqrt{4 - x^2 - y^2}\, dz\, dy\, dx$

17. Find the mass of the solid between the spheres $x^2 + y^2 + z^2 = 1$ and $x^2 + y^2 + z^2 = 9$ if the density is proportional to the distance from the origin.

18. Find the center of mass of the homogeneous lamina bounded by the cardioid $r = 4(1 + \sin\theta)$.

19. Find the mass of the solid in the first octant under the plane $x/a + y/b + z/c = 1$ (a, b, c positive) if the density is $\delta(x, y, z) = kx$.

20. Compute the volume of the solid bounded by $z = x^2 + y^2$, $z = 0$, and $x^2 + (y - 1)^2 = 1$.

TECHNOLOGY PROJECT 16.1

Newton's Law of Gravitation

I. Preparation

According to **Newton's Law of Gravitation**, the force between two mass particles with masses M and m, respectively, and separated by a distance d has magnitude

(1) $$|F_{\text{point}}| = \frac{GMm}{d^2}$$

where G is the universal gravitational constant. We call this F_{point} because it is the force between the two *particles*, which are assumed to be concentrated at *points* in space. The force F_{point} acts along the line connecting the two particles. If the objects are not particles, then the natural question is this: Is the force between two objects the same as the force between two point masses separated by a distance equal to the distance of the mass centers of the original objects? (*Note*: Isaac Newton showed that this *is* true when one object is a spherical shell and the other is a particle.)

In this project, you will investigate the force between a particle of mass m located at the point $Q(0, 0, q)$ (i.e., on the z-axis) and a thin square plate of thickness h and of constant density ρ filling the region

$$R = \left\{(x, y, z): -\frac{a}{2} \le x \le \frac{a}{2}, -\frac{a}{2} \le y \le \frac{a}{2}, -\frac{h}{2} \le z \le \frac{h}{2}\right\}$$

By symmetry, the resultant force is wholly in the z-direction.

Now partition the region $-\dfrac{a}{2} \le x \le \dfrac{a}{2}, -\dfrac{a}{2} \le y \le \dfrac{a}{2}$ with a regular partition. Consider a typical *volume element* with length Δx, width Δy, center $(\bar{x}_k, \bar{y}_k, 0)$, and height h. The vertical force acting on the particle due to this volume element is

$$\Delta F = -\frac{Gm\,\Delta M \cos\alpha}{d^2}$$

where $d = \sqrt{\bar{x}_k^2 + \bar{y}_k^2 + q^2}$, $\Delta M = \rho h\, \Delta x\, \Delta y$, and $\cos\alpha = q/d$. Note that if an object having the same mass as the plate were concentrated at the origin (which, by symmetry, is the center of mass of the square plate) then the force on the object at Q would be given by

(2) $$F_{\text{point}} = -\frac{GMm}{q^2}$$

Exercise 1 Draw a sketch and justify the formulas given above for d, ΔM, and $\cos\alpha$.

Exercise 2 Assuming that the vertical variation in force is negligible over the small thickness h, show that the force acting on the particle is

$$F = -Gmh\rho q \int_{-a/2}^{a/2} \int_{-a/2}^{a/2} \frac{1}{d^3}\, dx\, dy$$

Exercise 3 To make our formula for the force look more like the force between two point masses given in (2), eliminate the density ρ and show that the resulting expression is

(3) $$F = -\frac{GMmq}{a^2} \int_{-a/2}^{a/2} \int_{-a/2}^{a/2} \frac{1}{d^3}\, dx\, dy$$

II. Using Technology

Exercise 4 Use your CAS to evaluate the integral in Exercise 3. (Set $a = 1$ if necessary, and if your CAS still will not evaluate the integral, then set $q = 1$.) Is $F = F_{\text{point}}$?

Exercise 5 Repeat Exercises 1 through 4 assuming that the plate is a circular disk with radius $a/2$ and thickness h.

III. Reflection

Exercise 6 In Exercise 4, take $a = 1$ and compute the ratio F/F_{point} for $q = 1, 2, 4, 8, 16, 32, 64, 128$, and 256. Create and fill in a table like the one below. Do you think that $F \to F_{\text{point}}$ as $q \to \infty$? (Large q results are referred to as *far field* results in the scientific literature.)

q	F using (3)	F_{point}	$\dfrac{F \text{ using (3)}}{F_{\text{point}}}$
1			
2			
4			
⋮			

Monte Carlo Integration

I. Preparation

Many of the numerical methods for integration discussed in Chapter 11 can be generalized to double, triple, and even n-dimensional integrals. For high dimensions, however, generalizations of the methods of Section 11.2 can be inefficient. In this project, we introduce **Monte Carlo integration**, which has been used successfully to approximate high-dimensional integrals.

We begin by looking at the Monte Carlo method for single integrals.

Exercise 1 Sketch the graph $f(x) = x^2$ over the interval $[0, 2]$. The region S below this curve, above the x-axis, and between the lines $x = 0$ and $x = 2$ is completely contained in the rectangle $R = \{(x, y): 0 \le x \le 2, 0 \le y \le 4\}$. What is the ratio of the area of S to the area of R?

Exercise 2 If we choose a point at random from the rectangle R, what is the probability that it will fall in the region S? If we were to choose 5000 points (x_i, y_i), $i = 1, 2, \ldots,$ 5000, at random from the rectangle, how many would we expect to fall in S?

II. Using Technology

The name, Monte Carlo integration comes from the use of random (really, pseudorandom) numbers. The idea is to approximate the area of a region by observing how many random ordered pairs fall inside the region and how many fall outside.

Monte Carlo integration is not needed to approximate an integral like the one in Exercise 1, which can easily be evaluated using the Second Fundamental Theorem of Calculus. The method is most useful when the integrand has many oscillations, because such an integral is difficult to approximate with the methods in Section 11.2. For Exercise 3, we will consider the integral

$$\int_0^1 x(1 - x)\sin^2(100x(1 - x))\,dx$$

The curve in Figure 1 (ignore for now the dots) is a graph of the integrand. We can see that the region S under the curve and above the x-axis lies completely in the rectangle $R = \{(x, y): 0 \le x \le 1, 0 \le y \le 0.25\}$. The Monte Carlo method then involves the following steps.

1. Choose random ordered pairs (x_i, y_i), $i = 1, 2, \ldots, n$, in the rectangle R.
2. Determine whether each ordered pair is *in* the region S.
3. Count how many of the n ordered pairs (x_i, y_i) lie in the region S.

4. The estimate for the integral is then

estimate of $A(S)$

$$= \text{area of rectangle} \cdot \frac{\text{number of points in } S}{\text{number of ordered pairs}}$$

We used Mathematica to approximate the area of S. We used $n = 1000$ and found that 273 of the 1000 points were in S. The dots in Figure 1 show our random ordered pairs. Thus, our estimate for $A(S) = \int_0^1 x(1 - x)\sin^2(100x(1 - x))\,dx$ is

$$\text{estimate of } A(S) = (1 \cdot 0.25)\frac{273}{1000} = 0.06825$$

Exercise 3 Run a Monte Carlo program ten times, each with 1000 random points for the integral given above. Because of the *random* generation of the ordered pairs, your answers will vary.

The first step in applying Monte Carlo integration to n-dimensional, or n-fold, integrals is to enclose the region of integration in an n-dimensional box whose volume can be easily computed. We then take points at random in the box and count the number that fall inside the region. The ratio of the number of points that fall inside the region to the total number of points is thus an estimate of the ratio of the volume of the region to the volume of the box. In the next exercise you will approximate the volume of a sphere of radius 1 in three-space.

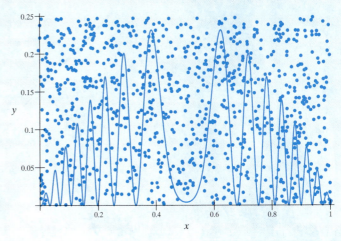

Figure 1

Exercise 4 Generate random points in the three-dimensional cube

$$C = \{(x, y, z): -1 \le x \le 1, -1 \le y \le 1, -1 \le z \le 1\}$$

For each random point (x_i, y_i, z_i), compute $x_i^2 + y_i^2 + z_i^2$. If $x_i^2 + y_i^2 + z_i^2 < 1$, then (x_i, y_i, z_i) is inside the sphere; otherwise, it is outside (or on) the sphere. Repeat this process ten times, each with at least 1000 random points. Record your estimates for the volume of the sphere.

Exercise 5 The volume of the sphere of radius $r = 1$ is $\frac{4}{3}\pi$. Divide your approximations from Exercise 4 by π and you should get roughly $\frac{4}{3}$. How do your numbers compare?

Exercise 6 Use Monte Carlo integration to estimate the volume of a four-dimensional sphere of radius 1. The actual volume is a multiple of π^2. What multiple do you think it is?

Exercise 7 Set up and evaluate (using your CAS, if necessary) a single integral to find the area of a circle. (You may find it helpful to find the area of a semicircle and double it.) Next, set up and evaluate a double integral to find the volume of a sphere. Finally, set up and evaluate a triple integral to find the volume of a four-dimensional sphere. *Bonus*: Set up and evaluate a four-dimensional integral to find the volume of a five-dimensional sphere.

Exercise 8 Here is another version of Monte Carlo integration for approximating $\int_a^b f(x)\,dx$.

1. Take n random points $x_1, x_2, \ldots, x_n$ on the interval $[a, b]$.
2. Compute $y_1 = f(x_1), y_2 = f(x_2), \ldots, y_n = f(x_n)$.
3. Find the average $\dfrac{1}{n} \sum_{i=1}^{n} y_i$.
4. The estimate for $\int_a^b f(x)\,dx$ is then $(b - a)\dfrac{1}{n}\sum_{i=1}^{n} y_i$.

Explain why this method should give a reasonable estimate for the integral.

Hint: Consider the *average value* of the function on the interval $[a, b]$.

Exercise 9 Use the version of Monte Carlo integration described in Exercise 8 to approximate

$$\int_0^1 x(1 - x)\sin^2(100x(1 - x))\,dx$$

Repeat this exercise ten times, each with $n = 1000$.

III. Reflection

Exercise 10 Explain how the version of Monte Carlo integration described in Exercise 8 can be generalized to
(a) double integrals and
(b) triple integrals.

Exercise 11 Compare your ten runs of Exercise 3 with your ten runs of Exercise 9. In which set of ten estimates is there less variation? Which method seems to be more efficient?

17

Vector Calculus

17.1
Vector Fields

All that we have done relates in some way to the concept of a function. This concept and the associated calculus have been steadily generalized as we have moved along.

1. Real-valued functions of one real variable (Chapters 1–12).
2. Vector-valued functions of one real variable (Chapters 13 and 14).
3. Real-valued functions of several real variables (Chapters 15 and 16).

You will correctly guess that the next step is to study vector-valued functions of several real variables, which is in fact the principal task of the present chapter. This is the final step in a first calculus course; more advanced courses will carry the process of generalization further—to complex-valued functions of complex variables, to real-valued functions of infinitely many variables, and so on.

Consider then a function **F** that associates with each point **p** in n-space a vector **F(p)**. A typical example in two-space is

$$\mathbf{F}(\mathbf{p}) = \mathbf{F}(x, y) = -\tfrac{1}{2}y\mathbf{i} + \tfrac{1}{2}x\mathbf{j}$$

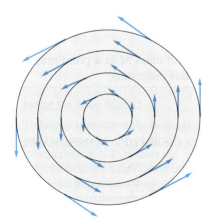

Figure 1

For historical reasons, we refer to such a function as a **vector field**, a name arising from a visual image that we now describe. Imagine that to each point **p** in a region of space is attached a vector **F(p)** emanating from **p**. We cannot draw all these vectors, but a representative sample can give us a good intuitive picture of a field. Figure 1 is just such a picture for the vector field $\mathbf{F}(x, y) = -\tfrac{1}{2}y\mathbf{i} + \tfrac{1}{2}x\mathbf{j}$ mentioned earlier. It is the velocity field of a wheel spinning at a constant rate of $\tfrac{1}{2}$ radian per unit of time (see Example 2). Figure 2 might represent the velocity field for water flowing in a curved pipe.

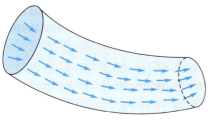

Figure 2

Theorem A   Fundamental Theorem for Line Integrals

Let C be a piecewise smooth curve given parametrically by $\mathbf{r} = \mathbf{r}(t)$, $a \le t \le b$, which begins at $\mathbf{a} = \mathbf{r}(a)$ and ends at $\mathbf{b} = \mathbf{r}(b)$. If f is continuously differentiable on an open set containing C, then

$$\int_C \nabla f(\mathbf{r}) \cdot d\mathbf{r} = f(\mathbf{b}) - f(\mathbf{a})$$

Proof   We suppose first that C is smooth. Then

$$\int_C \nabla f(\mathbf{r}) \cdot d\mathbf{r} = \int_a^b \left[\nabla f(\mathbf{r}(t)) \cdot \mathbf{r}'(t) \right] dt$$

$$= \int_a^b \frac{d}{dt} f(\mathbf{r}(t)) \, dt = f(\mathbf{r}(b)) - f(\mathbf{r}(a))$$

$$= f(\mathbf{b}) - f(\mathbf{a})$$

Note how we first wrote the line integral as an ordinary definite integral, then applied the Chain Rule, and finally used the Second Fundamental Theorem of Calculus.

If C is not smooth but only piecewise smooth, we simply apply the above result to the individual pieces. We leave the details to you.   ◆

EXAMPLE 1   Recall from Example 4 of Section 17.1 that

$$f(x, y, z) = f(\mathbf{r}) = \frac{c}{|\mathbf{r}|} = \frac{c}{\sqrt{x^2 + y^2 + z^2}}$$

is a potential function for the inverse square law field $\mathbf{F}(\mathbf{r}) = -c\mathbf{r}/|\mathbf{r}|^3$. Calculate $\int_C \mathbf{F}(\mathbf{r}) \cdot d\mathbf{r}$, where C is any simple piecewise smooth curve from $(0, 3, 0)$ to $(4, 3, 0)$ that misses the origin.

Solution   Since $\mathbf{F}(\mathbf{r}) = \nabla f(\mathbf{r})$,

$$\int_C \mathbf{F}(\mathbf{r}) \cdot d\mathbf{r} = \int_C \nabla f(\mathbf{r}) \cdot d\mathbf{r} = f(4, 3, 0) - f(0, 3, 0)$$

$$= \frac{c}{\sqrt{16 + 9}} - \frac{c}{\sqrt{9}} = \frac{-2c}{15}$$ ∎

Now compare Example 1 with Example 4 of the previous section. There we calculated the same integral, but for a specific curve C, the line segment from $(0, 3, 0)$ to $(4, 3, 0)$. Surprisingly, we will get the same answer no matter what curve we take from $(0, 3, 0)$ to $(4, 3, 0)$. We say that the given line integral is independent of path.

Criteria for Independence of Path   Call a set D **connected** if any two points in D can be joined by a piecewise smooth curve lying entirely in D (Figure 1). Then call $\int_C \mathbf{F}(\mathbf{r}) \cdot d\mathbf{r}$ **independent of path in** D if for any two points A and B in D the line integral has the same value for every path C in D that is positively oriented from A to B.

One consequence of Theorem A is that if $\mathbf{F}$ is the gradient of another function f then $\int_C \mathbf{F}(\mathbf{r}) \cdot d\mathbf{r}$ is independent of path. The converse is also true.

Theorem B   Independence of Path Theorem

Let $\mathbf{F}(\mathbf{r})$ be continuous on an open connected set D. Then the line integral $\int_C \mathbf{F}(\mathbf{r}) \cdot d\mathbf{r}$ is independent of path if and only if $\mathbf{F}(\mathbf{r}) = \nabla f(\mathbf{r})$ for some scalar function f; that is, if and only if $\mathbf{F}$ is a conservative vector field on D.

A Connected Set

A Disconnected Set

Figure 1

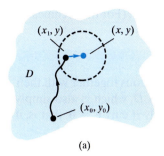

(a)

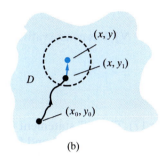

(b)

Figure 2

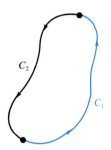

Figure 3

Proof   Theorem A takes care of the "if" statement. Suppose then that $\int_C \mathbf{F}(\mathbf{r}) \cdot d\mathbf{r}$ is independent of path in D. Our task is to construct a function f satisfying $\nabla f = \mathbf{F}$; that is, we must find a potential for the vector field $\mathbf{F}$. For simplicity, we restrict ourselves to the two-dimensional case, where D is a plane set and $\mathbf{F}(\mathbf{r}) = M(x, y)\mathbf{i} + N(x, y)\mathbf{j}$.

Let (x_0, y_0) be a fixed point of D, and let (x, y) be any other point of D. Choose a third point (x_1, y) in D and slightly to the left of (x, y), and join it to (x, y) by a horizontal segment in D. Then join (x_0, y_0) to (x_1, y) by a curve in D. (All this is possible because D is both open and connected; see Figure 2a). Finally, let C denote the path from (x_0, y_0) to (x, y) composed of these two pieces, and define f by

$$f(x, y) = \int_C \mathbf{F}(\mathbf{r}) \cdot d\mathbf{r} = \int_{(x_0, y_0)}^{(x_1, y)} \mathbf{F}(\mathbf{r}) \cdot d\mathbf{r} + \int_{(x_1, y)}^{(x, y)} \mathbf{F}(\mathbf{r}) \cdot d\mathbf{r}$$

That we get a unique value is clear from the assumed independence of path.

The first integral on the right above does not depend on x; the second, which has y fixed, can be written as an ordinary definite integral using, for example, t as a parameter. It follows that

$$\frac{\partial f}{\partial x} = 0 + \frac{\partial}{\partial x} \int_{x_1}^{x} M(t, y)\, dt = M(x, y)$$

The last equality is a consequence of the First Fundamental Theorem of Calculus (Theorem 5.6A).

A similar argument using Figure 2b shows that $\partial f/\partial y = N(x, y)$. We conclude that $\nabla f = M(x, y)\mathbf{i} + N(x, y)\mathbf{j} = \mathbf{F}$, as desired.   ◆

The condition that $\int_C \mathbf{F}(\mathbf{r}) \cdot d\mathbf{r}$ be independent of path in D implies that if C is any closed, oriented curve in D then $\int_C \mathbf{F}(\mathbf{r}) \cdot d\mathbf{r} = 0$. To see this, consider C to be composed of two oriented curves C_1 and C_2, as shown in Figure 3. Let $-C_2$ denote the curve C_2 with opposite orientation. Since C_1 and $-C_2$ have the same initial and terminal points, the independence of path guarantees that

$$\int_C \mathbf{F}(\mathbf{r}) \cdot d\mathbf{r} = \int_{C_1} \mathbf{F}(\mathbf{r}) \cdot d\mathbf{r} + \int_{C_2} \mathbf{F}(\mathbf{r}) \cdot d\mathbf{r}$$

$$= \int_{C_1} \mathbf{F}(\mathbf{r}) \cdot d\mathbf{r} - \int_{-C_2} \mathbf{F}(\mathbf{r}) \cdot d\mathbf{r}$$

$$= \int_{C_1} \mathbf{F}(\mathbf{r}) \cdot d\mathbf{r} - \int_{C_1} \mathbf{F}(\mathbf{r}) \cdot d\mathbf{r} = 0$$

The argument just given is reversible. Hence, we have three equivalent conditions.

1. $\mathbf{F} = \nabla f$ for some function f ($\mathbf{F}$ is conservative).

2. $\int_C \mathbf{F}(\mathbf{r}) \cdot d\mathbf{r}$ is independent of path.

3. $\int_C \mathbf{F}(\mathbf{r}) \cdot d\mathbf{r} = 0$ for every closed path.

There is an interesting physical interpretation of Condition 3. The work done by a conservative force field as it moves a particle around a closed path is zero. In particular, this is true of both gravitational fields and electric fields, since they are conservative.

While Conditions 2 and 3 each imply that $\mathbf{F}$ is the gradient of a scalar function f, they are not particularly useful in this connection. A more useful criterion is given in the following theorem. We need, however, to impose the additional condition on D, that it is **simply connected**. In two-space, this means that D has no "holes" and

On the other hand,

$$\text{curl } \mathbf{F} = \nabla \times \mathbf{F} = \begin{vmatrix} \mathbf{i} & \mathbf{j} & \mathbf{k} \\ \dfrac{\partial}{\partial x} & \dfrac{\partial}{\partial y} & \dfrac{\partial}{\partial z} \\ M & N & 0 \end{vmatrix} = \left(\dfrac{\partial N}{\partial x} - \dfrac{\partial M}{\partial y} \right) \mathbf{k}$$

so that

$$(\text{curl } \mathbf{F}) \cdot \mathbf{k} = \left(\dfrac{\partial N}{\partial x} - \dfrac{\partial M}{\partial y} \right)$$

Green's Theorem thus takes the form

$$\oint_C \mathbf{F} \cdot \mathbf{T} \, ds = \iint_S (\text{curl } \mathbf{F}) \cdot \mathbf{k} \, dA$$

which is sometimes called Stokes's Theorem in the plane.

If we apply this result to a small circle C_r centered at (x_0, y_0), we obtain

$$\oint_{C_r} \mathbf{F} \cdot \mathbf{T} \, ds \approx \left(\text{curl } \mathbf{F}(x_0, y_0) \right) \cdot \mathbf{k} \left(\pi r^2 \right)$$

This says that the flow in the direction of the tangent to C_r (the *circulation* of $\mathbf{F}$ around C_r) is measured by the curl of $\mathbf{F}$. In other words, curl $\mathbf{F}$ measures the tendency of the fluid to rotate about (x_0, y_0). If curl $\mathbf{F} = \mathbf{0}$ in a region S, the corresponding fluid flow is said to be *irrotational*.

EXAMPLE 5   The vector field $\mathbf{F}(x, y) = -\frac{1}{2} y \mathbf{i} + \frac{1}{2} x \mathbf{j} = M \mathbf{i} + N \mathbf{j}$ is the velocity field of a steady counterclockwise rotation of a wheel about the z-axis (see Example 2 of Section 17.1). Calculate $\oint_C \mathbf{F} \cdot \mathbf{n} \, ds$ and $\oint_C \mathbf{F} \cdot \mathbf{T} \, ds$ for any closed curve C in the xy-plane.

Solution   If S is the region enclosed by C,

$$\oint_C \mathbf{F} \cdot \mathbf{n} \, ds = \iint_S \text{div } \mathbf{F} \, dA = \iint_S \left(\dfrac{\partial M}{\partial x} + \dfrac{\partial N}{\partial y} \right) dA = 0$$

$$\oint_C \mathbf{F} \cdot \mathbf{T} \, ds = \iint_S (\text{curl } \mathbf{F}) \cdot \mathbf{k} \, dA = \iint_S \left(\dfrac{\partial N}{\partial x} - \dfrac{\partial M}{\partial y} \right) dA$$

$$= \iint_S \left(\dfrac{1}{2} + \dfrac{1}{2} \right) dA = A(S) \qquad \blacksquare$$

Concepts Review

1. Let C be a simple closed curve bounding a region S in the xy-plane. Then, by Green's Theorem, $\oint_C M \, dx + N \, dy = \iint_S$ _____ dA.

2. Thus, if C is the boundary of the square $S = \{(x, y): 0 \le x \le 1, 0 \le y \le 1\}$, $\oint_C y \, dx - x \, dy = \iint_S$ _____ $dA =$ _____.

3. The div $\mathbf{F}(x, y)$ measures the rate at which a homogeneous fluid flow with velocity field $\mathbf{F}$ diverges away from (x, y). If div $\mathbf{F}(x, y) > 0$, there is a(n) _____ of fluid at (x, y): if div $\mathbf{F}(x, y) < 0$, there is a(n) _____ at (x, y).

4. On the other hand, curl $\mathbf{F}(x, y)$ measures the tendency of the fluid to _____ about (x, y). If curl $\mathbf{F}(x, y) = \mathbf{0}$ in a region, the flow is _____.

Problem Set 17.4

In Problems 1–6, use Green's Theorem to evaluate the given line integral. Begin by sketching the region S.

1. $\oint_C 2xy\, dx + y^2\, dy$, where C is the closed curve formed by $y = x/2$ and $y = \sqrt{x}$ between $(0, 0)$ and $(4, 2)$

2. $\oint_C \sqrt{y}\, dx + \sqrt{x}\, dy$, where C is the closed curve formed by $y = 0$, $x = 2$, and $y = x^2/2$

3. $\oint_C (2x + y^2)\, dx + (x^2 + 2y)\, dy$, where C is the closed curve formed by $y = 0$, $x = 2$, and $y = x^3/4$

4. $\oint_C xy\, dx + (x + y)\, dy$, where C is the triangle with vertices $(0, 0)$, $(2, 0)$, and $(0, 1)$

5. $\oint_C (x^2 + 4xy)\, dx + (2x^2 + 3y)\, dy$, where C is the ellipse $9x^2 + 16y^2 = 144$

6. $\oint_C (e^{3x} + 2y)\, dx + (x^2 + \sin y)\, dy$, where C is the rectangle with vertices $(2, 1)$, $(6, 1)$, $(6, 4)$, and $(2, 4)$.

In Problems 7 and 8, use the result of Example 2 to find the area of the indicated region S. Make a sketch.

7. S is bounded by the curves $y = 4x$ and $y = 2x^2$.

8. S is bounded by the curves $y = \frac{1}{2}x^3$ and $y = x^2$.

In Problems 9–12, use the vector forms of Green's Theorem to calculate (a) $\oint_C \mathbf{F} \cdot \mathbf{n}\, ds$ and (b) $\oint_C \mathbf{F} \cdot \mathbf{T}\, ds$.

9. $\mathbf{F} = y^2\mathbf{i} + x^2\mathbf{j}$; C is the boundary of unit square with vertices $(0, 0)$, $(1, 0)$, $(1, 1)$, and $(0, 1)$.

10 $\mathbf{F} = ay\mathbf{i} + bx\mathbf{j}$; C as in Problem 9.

11. $\mathbf{F} = y^3\mathbf{i} + x^3\mathbf{j}$; C is the unit circle.

12. $\mathbf{F} = x\mathbf{i} + y\mathbf{j}$; C is the unit circle.

13. Suppose that the integrals $\oint_C \mathbf{F} \cdot \mathbf{T}\, ds$ taken counterclockwise around the circles $x^2 + y^2 = 36$ and $x^2 + y^2 = 1$ are 30 and -20, respectively. Calculate $\iint_S (\text{curl } \mathbf{F}) \cdot \mathbf{k}\, dA$, where S is the region between the circles.

14. If $\mathbf{F} = (x^2 + y^2)\mathbf{i} + 2xy\mathbf{j}$, find the flux of $\mathbf{F}$ across the boundary C of the unit square with vertices $(0, 0)$, $(1, 0)$, $(1, 1)$, and $(0, 1)$; that is, calculate $\oint_C \mathbf{F} \cdot \mathbf{n}\, ds$.

15. Find the work done by $\mathbf{F} = (x^2 + y^2)\mathbf{i} - 2xy\mathbf{j}$ in moving a body counterclockwise around the curve C of Problem 14.

16. If $\mathbf{F} = (x^2 + y^2)\mathbf{i} + 2xy\mathbf{j}$, calculate the circulation of $\mathbf{F}$ around C of Problem 14; that is, calculate $\oint_C \mathbf{F} \cdot \mathbf{T}\, ds$.

17. Show that the work done by a constant force $\mathbf{F}$ in moving a body around a simple closed curve is 0.

18. Use Green's Theorem to prove the plane case of Theorem 17.3C; that is, show that $\partial N/\partial x = \partial M/\partial y$ implies that $\oint_C M\, dx + N\, dy = 0$, which implies that $\mathbf{F} = M\mathbf{i} + N\mathbf{j}$ is conservative.

19. Let

$$\mathbf{F} = \frac{y}{x^2 + y^2}\mathbf{i} - \frac{x}{x^2 + y^2}\mathbf{j} = M\mathbf{i} + N\mathbf{j}$$

(a) Show that $\partial N/\partial x = \partial M/\partial y$.

(b) Show, by using the parametrization $x = \cos t$, $y = \sin t$, that $\oint_C M\, dx + N\, dy = -2\pi$, where C is the unit circle.

(c) Why doesn't this contradict Green's Theorem?

20. Let $\mathbf{F}$ be as in Problem 19. Calculate $\oint_C M\, dx + N\, dy$, where

(a) C is the ellipse $x^2/9 + y^2/4 = 1$

(b) C is the square with vertices $(1, -1)$, $(1, 1)$, $(-1, 1)$, and $(-1, -1)$

(c) C is the triangle with vertices $(1, 0)$, $(2, 0)$, and $(1, 1)$.

21. Let the piecewise smooth, simple closed curve C be the boundary of a region S in the xy-plane. Modify the argument in Example 2 to show that

$$A(S) = \oint_C (-y)\, dx = \oint_C x\, dy$$

22. Let S and C be as in Problem 21. Show that the moments M_x and M_y about the x- and y-axes are given by

$$M_x = -\frac{1}{2}\oint_C y^2\, dx, \qquad M_y = \frac{1}{2}\oint_C x^2\, dy$$

23. Calculate the area of the asteroid $x^{2/3} + y^{2/3} = a^{2/3}$. *Hint:* Parametrize by $x = a\cos^3 t$, $y = a\sin^3 t$, $0 \le t \le 2\pi$.

24. Calculate the work done by $\mathbf{F} = 2y\mathbf{i} - 3x\mathbf{j}$ in moving an object around the asteroid of Problem 23.

It follows from Theorem A that

$$\iint_G \mathbf{F} \cdot \mathbf{n} \, dS = \iint_R (M\mathbf{i} + N\mathbf{j} + P\mathbf{k}) \cdot \frac{-f_x\mathbf{i} - f_y\mathbf{j} + \mathbf{k}}{\sqrt{f_x^2 + f_y^2 + 1}} \sqrt{f_x^2 + f_y^2 + 1} \, dx \, dy$$

$$= \iint_R (-Mf_x - Nf_y + P) \, dx \, dy \quad \blacklozenge$$

You might try reworking Example 5 using Theorem B. We offer a different example.

EXAMPLE 6   Evaluate the flux for the vector field $\mathbf{F} = x\mathbf{i} + y\mathbf{j} + z\mathbf{k}$ across the part G of the paraboloid $z = 1 - x^2 - y^2$ that lies above the xy-plane, taking $\mathbf{n}$ to be the upward normal.

Solution

$$f(x, y) = 1 - x^2 - y^2, \qquad f_x = -2x, \qquad f_y = -2y$$

$$-Mf_x - Nf_y + P = 2x^2 + 2y^2 + z$$

$$= 2x^2 + 2y^2 + 1 - x^2 - y^2 = 1 + x^2 + y^2$$

$$\iint_G \mathbf{F} \cdot \mathbf{n} \, dS = \iint_R (1 + x^2 + y^2) \, dx \, dy$$

$$= \int_0^{2\pi} \int_0^1 (1 + r^2) r \, dr \, d\theta = \frac{3}{2}\pi \quad \blacksquare$$

Concepts Review

1. A _____ generalizes the ordinary double integral similar to the way a line integral generalizes the definite integral.

2. If G is a surface, $\displaystyle\iint_G g(x, y, z) \, dS = \lim_{|P| \to 0} \underline{\hspace{1cm}}$.

3. Let G be a surface given by $z = f(x, y)$, where (x, y) is in R. Then $\displaystyle\iint_G g(x, y, z) \, dS = \iint_R g(x, y, f(x, y)) \underline{\hspace{1cm}} \, dy \, dx$.

4. Consider the cone with axis along the z-axis, with vertex at the origin and making an angle of $30°$ with the z-axis. If G is the portion of this cone above the set $R = \{(x, y): x^2 + y^2 \le 9\}$, then

$$\iint_G dS = \iint_R \underline{\hspace{1cm}} \, dy \, dx = \underline{\hspace{1cm}}.$$

Problem Set 17.5

In Problems 1–8, evaluate $\displaystyle\iint_G g(x, y, z) \, dS$.

1. $g(x, y, z) = x^2 + y^2 + z$; G:$z = x + y + 1$, $0 \le x \le 1, 0 \le y \le 1$

2. $g(x, y, z) = x$; G:$x + y + 2z = 4, 0 \le x \le 1, 0 \le y \le 1$

3. $g(x, y, z) = x + y$; G: $z = \sqrt{4 - x^2}, 0 \le x \le \sqrt{3}$, $0 \le y \le 1$

4. $g(x, y, z) = 2y^2 + z$; G: $z = x^2 - y^2, 0 \le x^2 + y^2 \le 1$

5. $g(x, y, z) = \sqrt{4x^2 + 4y^2 + 1}$; G is the part of $z = x^2 + y^2$ below $y = z$

6. $g(x, y, z) = y$; G:$z = 4 - y^2, 0 \le x \le 3, 0 \le y \le 2$

7. $g(x, y, z) = x + y$; G is the surface of the cube $0 \le x \le 1, 0 \le y \le 1, 0 \le z \le 1$.

8. $g(x, y, z) = z$; G is the tetrahedron bounded by the coordinate planes and the plane $4x + 8y + 2z = 16$.

In Problems 9–12, use Theorem B to calculate the flux of $\mathbf{F}$ across G.

9. $\mathbf{F}(x, y, z) = -y\mathbf{i} + x\mathbf{j}$; G is the part of the plane $z = 8x - 4y - 5$ above the triangle with vertices $(0, 0, 0), (0, 1, 0)$, and $(1, 0, 0)$.

10. $\mathbf{F}(x, y, z) = (9 - x^2)\mathbf{j}$; G is the part of the plane $2x + 3y + 6z = 6$ in the first octant.

11. $\mathbf{F}(x, y, z) = y\mathbf{i} - x\mathbf{j} + 2\mathbf{k}$; G is the surface determined by $z = \sqrt{1 - y^2}, 0 \le x \le 5$.

12. $\mathbf{F}(x, y, z) = 2\mathbf{i} + 5\mathbf{j} + 3\mathbf{k}$; G is the part of the cone $z = (x^2 + y^2)^{1/2}$ that is inside the cylinder $x^2 + y^2 = 1$.

13. Find the mass of the triangle with vertices $(a, 0, 0)$, $(0, a, 0)$, and $(0, 0, a)$ if its density δ satisfies $\delta(x, y, z) = kx^2$.

14. Find the mass of the surface $z = 1 - \frac{1}{2}(x^2 + y^2)$ over $0 \le x \le 1, 0 \le y \le 1, z = 0$, if $\delta(x, y, z) = kxy$.

15. Find the center of mass of the homogeneous triangle with vertices $(a, 0, 0)$, $(0, a, 0)$, and $(0, 0, a)$.

16. Refer to Example 3. The hemispherical surface $z = f(x, y) = \sqrt{9 - x^2 - y^2}$ has a thin metal covering with density $\delta(x, y, z) = z$. Find the mass of this covering. Note that Theorem A does not apply directly, since f_x and f_y are undefined on the boundary $x^2 + y^2 = 9$ of R. Therefore, proceed by letting R_ε be the region $0 \le x^2 + y^2 \le (3 - \varepsilon)^2$, make the calculation, and then let $\varepsilon \to 0$. Discover that you get the same answer as you would if you ignored this subtle point.

17. Let G be the sphere $x^2 + y^2 + z^2 = a^2$. Evaluate each of the following:

(a) $\displaystyle\iint\limits_{G} z \, dS$ (b) $\displaystyle\iint\limits_{G} \frac{x + y^3 + \sin z}{1 + z^4} \, dS$

(c) $\displaystyle\iint\limits_{G} (x^2 + y^2 + z^2) \, dS$ (d) $\displaystyle\iint\limits_{G} x^2 \, dS$

(e) $\displaystyle\iint\limits_{G} (x^2 + y^2) \, dS$

Hint: Use symmetry properties to make this a trivial problem.

18. The sphere $x^2 + y^2 + z^2 = a^2$ has constant area density k. Find each moment of inertia.

(a) About a diameter

(b) About a tangent line (assume the Parallel Axis Theorem from Problem 20 of Section 16.5).

19. Find the total force against the surface of a tank full of a liquid of weight density k for each tank shape.

(a) Sphere of radius a

(b) Hemisphere of radius a with a flat base

(c) Vertical cylinder of radius a and height h

Hint: The force against a small patch of area ΔG is approximately $kd \, \Delta G$, where d is the depth of the water at the patch.

20. Find the center of mass of the part of the sphere $x^2 + y^2 + z^2 = a^2$ between the planes $z = h_1$ and $z = h_2$, where $0 \le h_1 \le h_2 \le a$. Do this by the methods of this section and then compare with Problem 15 of Section 16.6.

Answers to Concepts Review: **1.** surface integral

2. $\displaystyle\sum_{i=1}^{n} g(\bar{x}_i, \bar{y}_i, \bar{z}_i)\Delta S_i$ **3.** $\sqrt{f_x^2 + f_y^2 + 1}$ **4.** $2; 18\pi$

17.6
Gauss's Divergence Theorem

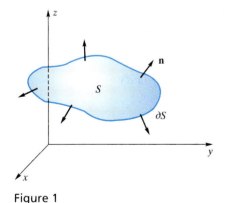

Figure 1

The theorems of Green, Gauss, and Stokes all relate an integral over a set S to another integral over the boundary of S. To emphasize the similarity in these theorems, we introduce the notation ∂S to stand for the boundary of S. Thus, one form of Green's Theorem (Section 17.4) can be written as

$$\oint_{\partial S} \mathbf{F} \cdot \mathbf{n} \, ds = \iint\limits_{S} \text{div } \mathbf{F} \, dA$$

It says that the flux of $\mathbf{F}$ across the boundary ∂S of a closed bounded plane region S is equal to the double integral of div $\mathbf{F}$ over that region. Gauss's Theorem (also called the *Divergence Theorem*) lifts this result up a dimension.

Gauss's Theorem Let S be a closed bounded solid in three-space that is completely enclosed by a piecewise smooth surface ∂S (Figure 1).

Theorem A Gauss's Theorem

Let $\mathbf{F} = M\mathbf{i} + N\mathbf{j} + P\mathbf{k}$ be a vector field such that M, N, and P have continuous first-order partial derivatives on a solid S with boundary ∂S. If $\mathbf{n}$ denotes the outer unit normal to ∂S, then

$$\iint\limits_{\partial S} \mathbf{F} \cdot \mathbf{n} \, dS = \iiint\limits_{S} \text{div } \mathbf{F} \, dV$$

In other words, the flux of $\mathbf{F}$ across the boundary of a closed region in three-space is the triple integral of its divergence over the region.

It is useful both for some applications and for the proof to state the conclusion to Gauss's Theorem in its Cartesian (nonvector) form. We may write

$$\mathbf{n} = \cos\alpha \, \mathbf{i} + \cos\beta \, \mathbf{j} + \cos\gamma \, \mathbf{k}$$

where α, β, and γ are the direction angles for $\mathbf{n}$. Thus

$$\mathbf{F} \cdot \mathbf{n} = M\cos\alpha + N\cos\beta + P\cos\gamma$$

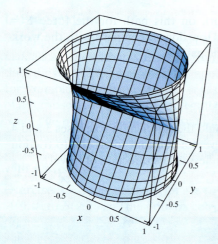

Figure 1

Exercise 5 The surface area of a parametrically defined surface,

$$x = x(s, t), \qquad y = y(s, t), \qquad z = z(s, t), \qquad (s, t) \in R$$

can be computed from the formula

(2)
$$\iint_R \sqrt{\left(\frac{\partial y}{\partial s}\frac{\partial z}{\partial t} - \frac{\partial z}{\partial s}\frac{\partial y}{\partial t}\right)^2 + \left(\frac{\partial z}{\partial s}\frac{\partial x}{\partial t} - \frac{\partial x}{\partial s}\frac{\partial z}{\partial t}\right)^2 + \left(\frac{\partial x}{\partial s}\frac{\partial y}{\partial t} - \frac{\partial y}{\partial s}\frac{\partial x}{\partial t}\right)^2} \, dA$$

Use this formula to find the surface area for the pinched cylinder in Exercise 4.

Exercise 6 Use the formula in (2) to find the surface area of a sphere of radius r.

Exercise 7 Suppose that the pinched cylinder is covered at the top and bottom by flat lids. Find the total volume enclosed by this surface. *Hint*: You can do this using a simple single integral. Consider just the upper half of the region. For a fixed height z, we have $x = \sin s$, $y = z \cos s$. What is the shape of the cross section at a height of z? What is its cross-sectional area?

Exercise 8 The pinched cylinder is an example of a surface that intersects itself, although this intersection has a rather simple form. Plot the surface given parametrically as

$$x = \sin s \cos t, \qquad y = \sin s \sin t, \qquad z = \cos t$$

Notice that the parametric equations are nearly the same as for the sphere, but the surface is completely different. This surface has a complicated pattern of self-intersection. Examine this surface from a number of viewpoints and explain how it intersects itself.

III. Reflection

Exercise 9 Explain why the graph of the parametric equations

$$x = \sin s \cos t, \qquad y = \sin s \sin t, \qquad z = \cos s, \qquad \text{where } 0 \le s \le \pi \quad 0 \le t \le 2\pi$$

is a sphere. *Hint*: Refer to spherical coordinates.

Exercise 10 Derive equation (2). *Hint*: Partition the region R, find the **u** and **v** vectors that form the sides of a parallelogram over rectangle R_k, take the cross product, and integrate.

18

Differential Equations

18.1
Linear Homogeneous Equations

We call an equation involving one or more derivatives of an unknown function a **differential equation**. In particular, an equation of the form

$$F\left(x, y, y^{(1)}, y^{(2)}, \ldots, y^{(n)}\right) = 0$$

in which $y^{(k)}$ denotes the k th derivative of y with respect to x, is called an **ordinary differential equation of order n**. Examples of orders 1, 2, and 3 are

$$y' + 2 \sin x = 0$$

$$\frac{d^2 y}{dx^2} + 3x \frac{dy}{dx} - 2y = 0$$

$$\frac{d^3 y}{dx^3} + \left(\frac{dy}{dx}\right)^2 - e^x = 0$$

If, when $f(x)$ is substituted for y in the differential equation, the resulting equation is an identity for all x in some interval, then $f(x)$ is called a **solution** of the differential equation. Thus, $f(x) = 2 \cos x + 10$ is a solution to $y' + 2 \sin x = 0$ since

$$f'(x) + 2 \sin x = -2 \sin x + 2 \sin x = 0$$

for all x. We call $2 \cos x + C$ the **general solution** of the given equation, since it can be shown that every solution can be written in this form. In contrast, $2 \cos x + 10$ is called a **particular solution** of the equation.

Differential equations appeared earlier in this book, principally in three sections. In Section 5.2, we introduced the technique called *separation of variables* and used it to solve a wide variety of first-order equations. In Section 7.5, we solved the

21. $\dfrac{2x^5}{1 + x^4} + \dfrac{x^2}{1 + x^2}$

25. 10; **27.** 4;

29. (a) 0 **(b)** $\frac{1}{5}x^5 + x + C$ **(c)** $\frac{1}{5}x^5 + x$

33. Lower bound 20; upper bound 276

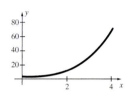

35. Lower bound $\frac{68}{5}$; upper bound 20

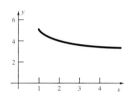

37. Lower bound 20π; upper bound $\frac{101}{5}\pi$

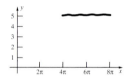

39. $\frac{1}{2}$ **41.** 2 **43.** $\sqrt{x}/2$

45. True **47.** False **49.** True

Problem Set 5.7

1. 4 **3.** 15 **5.** $\frac{3}{4}$ **7.** $\frac{16}{3}$ **9.** $\frac{1783}{96}$ **11.** 1

13. $\frac{22}{5}$ **15.** $\frac{2047}{11}$ **17.** $\frac{4}{5}$ **19.** $\frac{122}{9}$ **21.** 0 **23.** $\frac{1}{3}$

25. $\dfrac{\pi^2}{4} + 1$ **27.** 14 **29.** $\frac{38}{15}$ **31.** $\frac{1}{4}x^4 - \frac{1}{4}$ **33.** 40

35. $\frac{17}{6}$ **37.** 0

39. $\approx 1250\pi$ **41.** ≈ 3.2

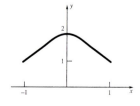

43. $\frac{115}{81}$ **45.** 1 **47.** 9 **49.** 2

51. $\frac{77}{200} = 0.385; \frac{1}{3} \approx 0.333$

59. (a) positive, **(b)** negative, **(c)** negative, **(d)** positive

63. ≈ 25

Problem Set 5.8

1. $\frac{2}{9}(3x + 2)^{3/2} + C$ **3.** $\frac{4}{27}(6x - 7)^{9/8} + C$

5. $\frac{1}{3}\sin(3x + 2) + C$ **7.** $-\frac{1}{6}\cos(6x - 7) + C$

9. $\frac{1}{3}(x^2 + 4)^{3/2} + C$ **11.** $-\frac{7}{10}(x^2 + 3)^{-5/7} + C$

13. $-\frac{1}{2}\cos(x^2 + 4) + C$ **15.** $-\frac{1}{18}\cos(6x^3 - 7) + C$

17. $-\cos\sqrt{x^2 + 4} + C$ **19.** $\frac{1}{27}\sin[(x^3 + 5)^9] + C$

21. $\frac{1}{3}[\sin(x^2 + 4)]^{3/2} + C$ **23.** $-\frac{1}{30}\cos^{10}(x^3 + 5) + C$

25. $\sec^{1/2}(x^2 + 2x) + C$ **27.** $\frac{85}{4}$ **29.** $\frac{2}{117}$

31. $\frac{5}{12}$ **33.** $\frac{1}{64}$ **35.** $\frac{\sin 3}{3}$ **37.** $\frac{1}{\pi}$

39. 1 **41.** $1 - \cos 1$ **43.** $\dfrac{1 - \cos^4 1}{8}$ **45.** $\frac{5}{36}$

47. 0 **49.** 0 **51.** 2π **53.** $\frac{8}{3}$ **55.** $\frac{1}{2}$

57. Even: $\displaystyle\int_{-b}^{-a} f(x)\,dx = \int_{a}^{b} f(x)\,dx$;

Odd: $\displaystyle\int_{-b}^{-a} f(x)\,dx = -\int_{a}^{b} f(x)\,dx$

59. 8 **61.** $\displaystyle\int_{1}^{1+\pi} |\sin x|\,dx = 2$ **63.** 883.2 **65.** $\dfrac{\pi^2}{4} - 2$

67. (a) Even; **(b)** 2π

(c)

Interval	Value of Integral
$\left[0, \frac{\pi}{2}\right]$	0.46
$\left[-\frac{\pi}{2}, \frac{\pi}{2}\right]$	0.92
$\left[0, \frac{3\pi}{2}\right]$	−0.46
$\left[-\frac{3\pi}{2}, \frac{3\pi}{2}\right]$	−0.92
$\left[0, 2\pi\right]$	0
$\left[\frac{\pi}{6}, \frac{13\pi}{6}\right]$	0
$\left[\frac{\pi}{6}, \frac{4\pi}{3}\right]$	−0.44
$\left[\frac{13\pi}{6}, \frac{10\pi}{3}\right]$	−0.44

69. With a left Rieman sum, $d \approx 14.2$ miles

Chapter Review 5.9

Concepts Test

1. True **3.** True **5.** True **7.** False **9.** True
11. True **13.** True **15.** True **17.** True **19.** True
21. False **23.** True **25.** True **27.** False **29.** False
31. False **33.** True **35.** False **37.** True **39.** False
41. True **43.** True

Sample Test Problems

1. $\frac{5}{4}$ **3.** $\frac{1}{3}y^3 + 9\cos y - \frac{26}{y} + C$

5. $\frac{3}{16}(2z^2 - 3)^{4/3} + C$ **7.** $\frac{1}{18}\tan^3(3\pi^2)$ **9.** 46.9

11. $\frac{5}{24}(2y^3 + 3y^2 + 6y)^{4/5} + C$ **13.** $y = 2\sqrt{x + 1} + 14$

15. $y = \frac{1}{3}(2t - 1)^{3/2} - 1$ **17.** $y = \sqrt{3x^2 - \frac{1}{4}x^4 + 9}$

19. $y = \dfrac{x^3}{6} + 1$ **21.** 7 s; −176 ft/s

23. $\frac{7}{4}$

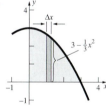

25. $\frac{5}{6}$ **27.** $\frac{39}{4}$ **29.** 1870   **31. (a)** $\sum_{n=2}^{78} \frac{1}{n}$;   **(b)** $\sum_{n=1}^{50} n x^{2n}$

33. (a) -2;   **(b)** -4;   **(c)** 6;   **(d)** -12;   **(e)** 2

35. (a) -8;   **(b)** 8;   **(c)** 0;   **(d)** -16;   **(e)** -2 **(f)** -5

37. $c = -\sqrt{7}$ **39. (a)** $\sin^2 x$;   **(b)** $f(x + 1) - f(x)$

(c) $-\dfrac{1}{x^2} \displaystyle\int_0^x f(z)\,dz + \dfrac{1}{x} f(x)$;   **(d)** $\displaystyle\int_0^x f(t)\,dt$;

(e) $g'(g(x)) g'(x)$;   **(f)** $-f(x)$

Additional Problems 5.10

1. Figure 1

3. (a) 11,010 ft;   **(b)** 8370 ft;   **(c)** 9690 ft;   **(d)** (a) uses the largest and (b) the smallest value of each pair. (c) averages each pair so the total value is the average of (a) and (b).

7. (a) Local minima at 0, ≈3.8, ≈5.8, ≈7.9, ≈9.9

(b) local maxima at ≈3.1, ≈5, ≈7.1, ≈9

9. (a) positive;   **(b)** negative;   **(c)** negative;   **(d)** positive

Problem Set 6.1

1. 6   **3.** $\frac{40}{3}$ **5.** $\frac{9}{2}$ **7.** $\frac{253}{12}$ **9.** $\frac{9}{2}$

11.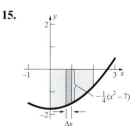

6

13.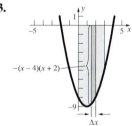

24

15.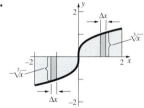

$\frac{17}{6}$

17.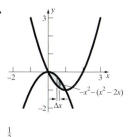

$3\sqrt[3]{2}$

19.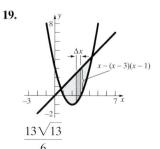

$\dfrac{13\sqrt{13}}{6}$

21.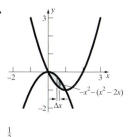

$\frac{1}{3}$

23.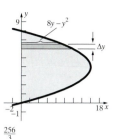

$\frac{256}{3}$

25.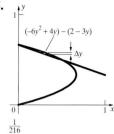

$\frac{1}{216}$

27.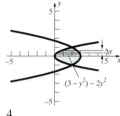

4

29.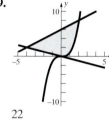

22

31. 130 ft; 194 ft   **33.** 6 s; $2 + 2\sqrt{2}$ s

35. $A(A) = 9$; $A(B) = \frac{37}{6}$; $A(C) = \frac{37}{6}$; $A(D) = \frac{44}{3}$; $A(A + B + C + D) = 36$

Problem Set 6.2

1. $\frac{206\pi}{15}$ **3. (a)** $\frac{256\pi}{15}$;   **(b)** 8π

5.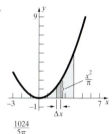

$\frac{1024}{5\pi}$

7.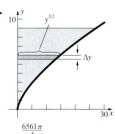

$\frac{\pi}{4}$

9.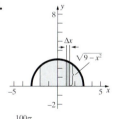

$\frac{100\pi}{3}$

11.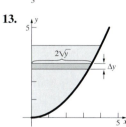

$\frac{243\pi}{5}$

13.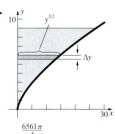

32π

15.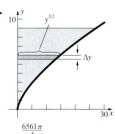

$\frac{6561\pi}{4}$

17. $\frac{4}{3} ab^2 \pi$ **19.** $\frac{512\pi}{3}$ **21.** $\frac{2\pi}{3}$ **23.** $\frac{128}{3}$

25. 2   **27.** $\frac{2}{3}$ **29.** $2\pi r^2 L - \frac{16}{3} r^3$

31. $\pi r^2 (L_1 + L_2) - \frac{8}{3} r^3$

33. (a) $\frac{1024\pi}{35}$;   **(b)** $\frac{704\pi}{5}$ **35.** $2\pi + \frac{16}{3}$

37. $\frac{2}{3} r^3 \tan\theta$ **39. (a)** $\frac{1}{3}\pi r^2 h$;   **(b)** $\dfrac{\sqrt{2}}{12} r^3$ **41.** $\frac{2}{3}\pi r^3$

Problem Set 6.3

1. (a), (b)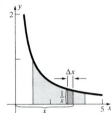

(c) $\Delta V \approx 2\pi \Delta x$;

(d) $2\pi \int_1^4 \frac{1}{x} dx$;

(e) 6π

3. (a), (b)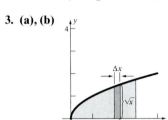

(c) $\Delta V \approx 2\pi x^{3/2} \Delta x$;

(d) $2\pi \int_0^3 x^{3/2} dx$;

(e) $\frac{36\sqrt{3}}{5} \pi$

5. (a), (b)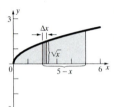

(c) $\Delta V \approx 2\pi(5x^{1/2} - x^{3/2})\Delta x$

(d) $2\pi \int_0^5 (5x^{1/2} - x^{3/2})dx$

(e) $\frac{40\sqrt{5}}{3} \pi$

7. (a), (b)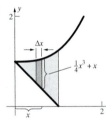

(c) $\Delta V \approx 2\pi(\frac{1}{4}x^4 + x^2)\Delta x$;

(d) $2\pi \int_0^1 (\frac{1}{4}x^4 + x^2)dx$

(e) $\frac{23\pi}{30}$

9. (a), (b)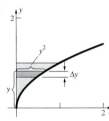

(c) $\Delta V \approx 2\pi y^3 \Delta y$;

(d) $2\pi \int_0^1 y^3 dy$

(e) $\frac{\pi}{2}$

11. (a), (b)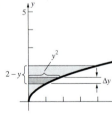

(c) $\Delta V \approx 2\pi(2y^2 - y^3)\Delta y$

(d) $2\pi \int_0^2 (2y^2 - y^3)dy$;

(e) $\frac{8\pi}{3}$

13. (a) $\pi \int_a^b [f(x)^2 - g(x)^2] dx$;

(b) $2\pi \int_a^b x[f(x) - g(x)] dx$;

(c) $2\pi \int_a^b (x - a)[f(x) - g(x)] dx$;

(d) $2\pi \int_a^b (b - x)[f(x) - g(x)] dx$

15.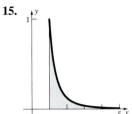

(a) $\int_1^3 \frac{1}{x^3} dx$;

(b) $2\pi \int_1^3 \frac{1}{x^2} dx$;

(c) $\pi \int_1^3 \left(\frac{1}{x^6} + \frac{2}{x^3}\right) dx$;

(d) $2\pi \int_1^3 \left(\frac{4}{x^3} - \frac{1}{x^2}\right) dx$

17. $\frac{64\pi}{5}$ **19.** $\frac{4\pi}{3}(b^2 - a^2)^{3/2}$ **21.** $\pi(\sqrt{2} - 1)$

23. (a) $\frac{2\pi}{15}$;   **(b)** $\frac{\pi}{6}$;   **(c)** $\frac{\pi}{60}$ **25.** $\frac{1}{3}rS$

Problem Set 6.4

1. (a) 10.2462   **(b)** 11.0897

3. (a) 6.2832   **(b)** 7.4484

5. $2\sqrt{5}$ **7.** $\frac{1}{54}(181\sqrt{181} - 13\sqrt{13})$ **9.** 9   **11.** $\frac{595}{144}$

13.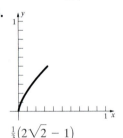

15.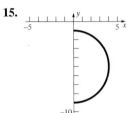

$\frac{1}{3}(2\sqrt{2} - 1)$

4π

17.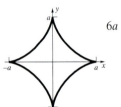

$6a$

19. $8a$ **21. (a)** $\frac{2}{5}(4\sqrt{2} - 1)$;   **(b)** 16   **23.** $6\sqrt{37}\pi$

25. $28\sqrt{2}\pi$ **27.** $\frac{\pi}{27}(10\sqrt{10} - 1)$ **29.** $4\pi r^2$

33. (b) $\frac{64}{3}\pi a^2$

35. (a)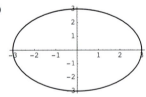

(b)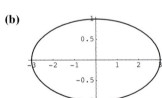

(c)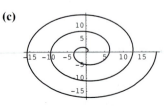

(d)

(e)

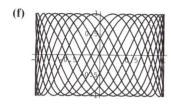

(f)

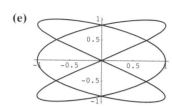

37.

$n = 1; L \approx 1.41; n = 2: L \approx 1.48; n = 4: L \approx 1.60$
$n = 10: L \approx 1.75; n = 100: L \approx 1.95$

Problem Set 6.5

1. 1.5 ft-lb   **3.** 0.012 Joules   **7.** 18 ft-lb   **9.** 52,000 ft-lb

11. 76,128 ft-lb   **13.** 125,664 ft-lb   **17.** 2075.83 in.-lb

19. 350,000 ft-lb   **21.** 952,381 mi-lb   **23.** 43,200 ft-lb

25. $\frac{3mh}{4} + 15m$ **27.** 86,934 ft-lb

Problem Set 6.6

1. $\frac{5}{21}$ **3.** $\frac{21}{5}$ **5.** $M_y = 17, M_x = -3; \bar{x} = 1, \bar{y} = -\frac{3}{17}$

9.

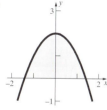

$\bar{x} = 0, \bar{y} = \frac{4}{5}$

11.

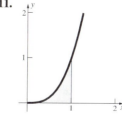

$\bar{x} = \frac{4}{5}, \bar{y} = \frac{2}{7}$

13.

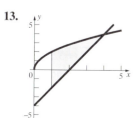

$\bar{x} = \frac{192}{95}, \bar{y} = \frac{27}{19}$

15.

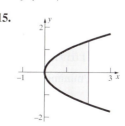

$\bar{x} = \frac{6}{5}, \bar{y} = 0$

17. $m(R_1) = \frac{1}{2}\delta, \bar{x}_1 = \frac{2}{3}, \bar{y} = \frac{1}{3}, M_y(R_1) = \frac{1}{3}\delta, M_x(R_1) = \frac{1}{6}\delta;$
$m(R_2) = 2\delta, \bar{x}_2 = 2, \bar{y}_2 = \frac{1}{2}, M_y(R_2) = 4\delta, M_x(R_2) = \delta.$

21. $\bar{x} = -\frac{3}{14}, \bar{y} = \frac{1}{14}$ **23.** $\bar{x} = \frac{9}{16}, \bar{y} = \frac{31}{16}$ **25.** $\frac{2\pi}{5}$

27. The centroid is $\frac{4a}{3\pi}$ units perpendicular from the center of the diameter. $\left(\bar{y} = \frac{4a}{3\pi}, \bar{x} = 0\right)$

29. (a) $V = 2\pi \int_c^d (e - y)w(y)\,dy$

31. (a) $4\pi r^3 n \sin\frac{\pi}{2n} \cos^2\frac{\pi}{2n}$ **35.** $\bar{x} = -7, \bar{y} \approx 0.669$

Chapter Review 6.7
Concepts Test

1. False   **3.** False   **5.** True   **7.** False   **9.** False
11. False   **13.** True   **15.** True   **17.** True

Sample Test

1. $\frac{1}{6}$ **3.** $\frac{\pi}{6}$ **5.** $\frac{5\pi}{6}$

7. $V(S_1) = \frac{\pi}{30}; V(S_2) = \frac{\pi}{6}; V(S_3) = \frac{7\pi}{10}; V(S_4) = \frac{5\pi}{6}$

9. 205,837 ft-lb   **11. (a), (b)** $\frac{32}{3}$ **13.** $\frac{2048\pi}{15}$

15. $\frac{53}{6}$ **17.** 36   **19.** $\pi \int_a^b \left[f^2(x) - g^2(x)\right]dx$

21. $M_y = \delta \int_a^b x[f(x) - g(x)]\,dx$

$M_x = \frac{\delta}{2} \int_a^b \left[f^2(x) - g^2(x)\right]dx$

23. $2\pi \int_a^b f(x)\sqrt{1 + \left[f'(x)\right]^2}\,dx$

$+2\pi \int_a^b g(x)\sqrt{1 + \left[g'(x)\right]^2}\,dx$

$+\pi\left[f^2(a) - g^2(a)\right] + \pi\left[f^2(b) - g^2(b)\right]$

Additional Problems 6.8

1. (a) ft/sec; $F(t)$ is the change in velocity from 6 seconds to T seconds;   **(b)** ft-lb; $F(s)$ is the change in work from 3 feet to s feet.;   **(c)** in.; $F(r)$ is the center of mass of an object r inches long whose mass at x is $f(x)$.

3. (a) $\frac{1}{3}$;   **(b)** $\frac{6}{343}$;   **(c)** $\frac{1}{27}$ **5. (b)** 2

7. $F = 0.2496$ lb $P = 624$ lb/ft^2

9. (c) $F \approx 12,009$ lb   **11.** 4602.96 lb   **13.** 4.54 ft
15. (a) 19 hp;   **(b)** 88 hp;   **(c)** 1,898,424 ft-lb

Problem Set 7.1

1. (a) 1.792;   **(b)** 0.406;   **(c)** 4.396;   **(d)** 0.3465;
(e) −3.584;   **(f)** 3.871

3. $\dfrac{2x + 3}{x^2 + 3x + \pi}$ **5.** $\dfrac{3}{x - 4}$ **7.** $\dfrac{3}{x}$

9. $2x + 4x \ln x + \frac{3}{x}(\ln x)^2$ **11.** $\dfrac{1}{\sqrt{x^2 + 1}}$ **13.** $\frac{1}{243}$

15. $\frac{1}{2}\ln|2x + 1| + C$ **17.** $\ln|3v^2 + 9v| + C$

19. $(\ln x)^2 + C$ **21.** $\frac{1}{10}\left[\ln(486 + \pi) - \ln\pi\right]$

23. $\ln\dfrac{(x + 1)^2}{x}$ **25.** $\ln\dfrac{x^2(x - 2)}{x + 2}$

55. (a)

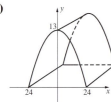

(b) 42,200 ft³;
(c) 5640 ft²

61.

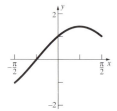

$y = \sinh x$ and
$y = \ln\left(x + \sqrt{x^2 + 1}\right)$
are inverse functions.

Chapter Review 7.9

Concepts Test

1. False	**3.** True	**5.** True	**7.** False	**9.** True
11. True	**13.** True	**15.** True	**17.** False	**19.** True
21. False	**23.** True	**25.** False	**27.** False	**29.** True
31. False	**33.** True	**35.** True	**37.** True	**39.** True
41. True	**43.** False	**45.** True		

Sample Test Problems

1. $\dfrac{4}{x}$ **3.** $(2x - 4)e^{x^2 - 4x}$ **5.** $\sec^2 x$ **7.** $\dfrac{\operatorname{sech}^2 \sqrt{x}}{\sqrt{x}}$

9. $|\sec x|$ **11.** $\dfrac{1}{\sqrt{e^{2x} - 1}}$ **13.** $\dfrac{15e^{5x}}{e^{5x} + 1}$

15. $-\dfrac{e^{\sqrt{x}} \sin e^{\sqrt{x}}}{2\sqrt{x}}$ **17.** $-\dfrac{1}{\sqrt{x - x^2}}$

19. $-\dfrac{\csc \sqrt{x} \cot \sqrt{x}}{\sqrt{x}}$ **21.** $20 \sec 5x(2 \sec^2 5x - 1)$

23. $x^{1+x}\left(\ln x + 1 + \frac{1}{x}\right)$ **25.** $\frac{1}{3}e^{3x-1} + C$

27. $-\cos e^x + C$ **29.** $\dfrac{\ln(e^{x+3} + 1)}{e} + C$

31. $2 \sin^{-1} 2x + C$ **33.** $-\tan^{-1}(\ln x) + C$

35. Increasing: $\left[-\frac{\pi}{2}, -\frac{\pi}{4}\right]$; decreasing: $\left[\frac{\pi}{4}, \frac{\pi}{2}\right]$; concave up: $\left(-\frac{\pi}{2}, -\frac{\pi}{4}\right)$; concave down: $\left(-\frac{\pi}{4}, \frac{\pi}{2}\right)$; inflection point: $\left(-\frac{\pi}{4}, 0\right)$; global minimum: $\left(-\frac{\pi}{2}, -1\right)$ global maximum: $\left(\frac{\pi}{4}, \sqrt{2}\right)$

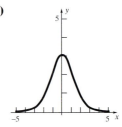

37. (b) 1   **(c)** $\frac{1}{15}$

39. (a) \$112;   **(b)** \$112.68;   **(c)** \$112.75;   **(d)** \$112.75

41. $y = 1$ **43.** $y = Cx^{-1}$

45. $y = 1 + 2e^{-x^2}$ **47.** $y = -e^x + Ce^{2x}$

Additional Problems 7.10

1. \$1051.27

Problem Set 8.1

1. $\frac{1}{6}(x - 2)^6 + C$ **3.** 1302   **5.** $\frac{1}{12}\tan^{-1}\left(\frac{x}{2}\right) + C$

7. $\frac{1}{2}\ln(x^2 + 4) + C$ **9.** $2(4 + z^2)^{3/2} + C$ **11.** $\frac{1}{2}\tan^2 z + C$

13. $-2 \cos \sqrt{t} + C$ **15.** $\tan^{-1}\dfrac{\sqrt{2}}{2}$

17. $\frac{3}{2}x^2 - x + \ln|x + 1| + C$ **19.** $-\frac{1}{2}\cos(\ln 4x^2) + C$

21. $6 \sin^{-1}(e^x) + C$ **23.** $-3\sqrt{1 - e^{2x}} + C$ **25.** $1/\ln 3$

27. $x - \ln|\sin x| + C$ **29.** $\ln|\sec e^x + \tan e^x| + C$

31. $\tan x + e^{\sin x} + C$ **33.** $-\dfrac{1}{3 \sin(t^3 - 2)} + C$

35. $-\frac{1}{3}\left[\cot(t^3 - 2) + t^3\right] + C$ **37.** $\frac{1}{2}e^{\tan^{-1} 2t} + C$

39. $\frac{1}{6}\sin^{-1}\left(\dfrac{3y^2}{4}\right) + C$ **41.** $\frac{1}{3}\cosh x^3 + C$

43. $\frac{1}{3}\sin^{-1}\left(\dfrac{e^{3t}}{2}\right) + C$ **45.** $\frac{1}{4}\tan^{-1}\left(\frac{1}{4}\right)$

47. $\frac{1}{2}\tan^{-1}\left(\dfrac{x + 1}{2}\right) + C$ **49.** $\frac{1}{3}\tan^{-1}(3x + 3) + C$

51. $\frac{1}{18}\ln|9x^2 + 18x + 10| + C$ **53.** $\frac{1}{3}\sec^{-1}\left(\dfrac{|\sqrt{2}t|}{3}\right) + C$

55. $\frac{2}{135}(9x - 4)(3x + 2)^{3/2} + C$ **57.** $\frac{1}{24}\ln\left|\dfrac{4x + 3}{4x - 3}\right| + C$

59. $\dfrac{x}{16}(4x^2 - 9)\sqrt{9 - 2x^2} + \dfrac{81\sqrt{2}}{23}\sin^{-1}\left(\dfrac{\sqrt{2}x}{3}\right) + C$

61. $\dfrac{1}{\sqrt{3}}\ln\left|\sqrt{3}x + \sqrt{3x^2 + 5}\right| + C$

63. $\ln|t + 1 + \sqrt{t^2 + 2t - 3}| + C$

65. $\frac{2}{27}(3 \sin t - 10)\sqrt{3 \sin t + 5} + C$ **67.** $\ln|\sqrt{2} + 1|$

69. π^2

Problem Set 8.2

1. $\frac{1}{2}x - \frac{1}{4}\sin 2x + C$ **3.** $-\cos x + \frac{1}{3}\cos^3 x + C$

5. $\frac{8}{15}$ **7.** $-\frac{1}{12}\cos^3 4x + \frac{1}{10}\cos^5 4x - \frac{1}{28}\cos^7 4x + C$

9. $-\frac{1}{3}\csc 3\theta - \frac{1}{3}\sin 3\theta + C$

11. $\frac{3}{128}t - \frac{1}{384}\sin 12t + \frac{1}{3072}\sin 24t + C$

13. $\frac{1}{2}\cos y - \frac{1}{18}\cos 9y + C$

15. $\frac{1}{16}w - \frac{1}{32}\sin 2w - \frac{1}{24}\sin^3 w + C$

Problem Set 8.1 (top right)

3. (a) \$1232.61;   **(b)** $\dfrac{100e^{0.05 \cdot (30/365)} - 100e^{0.05 \cdot (390/365)}}{1 - e^{0.05 \cdot (30/365)}}$

5. (c) 10 years

11. (a) In order of increasing slope: $y = 2^x$; $y = 3^x$, $y = 4^x$;

(b) $\ln y$ is linear with respect to x;   **(c)** $b = 2^{5/2}$; $c = 2^{3/2}$

13. (b)

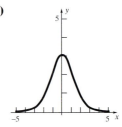

17. $\frac{1}{3}\tan^3 x - \tan x + x + C$ **19.** $\frac{1}{2}\tan^2 x + \ln|\cos x| + C$

21. $\frac{1}{2}\tan^4(\frac{\theta}{2}) - \tan^2(\frac{\theta}{2}) - 2\ln|\cos\frac{\theta}{2}| + C$

23. $-\frac{1}{2}\tan^{-2} x + \ln|\tan x| + C$ **25.** $\frac{1}{4}\sec^4 x - \frac{1}{2}\sec^2 x + C$

27. 0 for $m \neq n$, since $\sin k\pi = 0$ for all integers k.

29. $\dfrac{\pi^4}{3} + \dfrac{5\pi^2}{2}$

Problem Set 8.3

1. $\frac{2}{5}(x+1)^{5/2} - \frac{2}{3}(x+1)^{3/2} + C$

3. $\frac{2}{27}(3t+4)^{3/2} - \frac{8}{9}(3t+4)^{1/2} + C$

5. $2\sqrt{2} - 2 - 2e\ln\left(\dfrac{\sqrt{2}+e}{1+e}\right)$

7. $\frac{2}{63}(3t+2)^{7/2} - \frac{4}{45}(3t+2)^{5/2} + C$

9. $2\ln\left|\dfrac{2 - \sqrt{4-x^2}}{x}\right| + \sqrt{4-x^2} + C$

11. $\dfrac{x}{4\sqrt{x^2+4}} + C$ **13.** $-\dfrac{\sqrt{2}}{9} - \dfrac{1}{2}\sec^{-1}(-3) + \dfrac{\sqrt{3}}{8} + \dfrac{\pi}{3}$

15. $-2\sqrt{1-z^2} - 3\sin^{-1} z + C$

17. $\ln|\sqrt{x^2+2x+5} + x + 1| + C$

19. $3\sqrt{x^2+2x+5} - 3\ln|\sqrt{x^2+2x+5} + x + 1| + C$

21. $\frac{9}{2}\sin^{-1}\left(\dfrac{x+2}{3}\right) + \dfrac{x+2}{2}\sqrt{5-4x-x^2} + C$

23. $\sin^{-1}\left(\dfrac{x-2}{2}\right) + C$

25. $\ln|x^2+2x+2| - \tan^{-1}(x+1) + C$

27. $\frac{\pi}{16}\left(\frac{1}{10} + \frac{\pi}{4} - \tan^{-1}\frac{1}{2}\right)$ **29.** $\frac{1}{2}\ln|x^2+9| + C$

31. $2\ln\left|\dfrac{2-\sqrt{4-x^2}}{x}\right| + \sqrt{4-x^2} + C$

35. $y = -\sqrt{a^2-x^2} - a\ln\left|\dfrac{a-\sqrt{a^2-x^2}}{x}\right|$

Problem Set 8.4

1. $xe^x - e^x + C$ **3.** $\frac{1}{5}te^{5t+\pi} - \frac{1}{25}e^{5t+\pi} + C$

5. $x\sin x + \cos x + C$

7. $(t-3)\sin(t-3) + \cos(t-3) + C$

9. $\frac{2}{3}t(t+1)^{3/2} - \frac{4}{15}(t+1)^{5/2} + C$ **11.** $x\ln 3x - x + C$

13. $x\arctan x - \frac{1}{2}\ln(1+x^2) + C$ **15.** $-\frac{\ln x}{x} - \frac{1}{x} + C$

17. $\frac{2}{9}(e^{3/2} + 2)$ **19.** $\frac{1}{4}z^4\ln z - \frac{1}{16}z^4 + C$

21. $t\arctan(\frac{1}{t}) + \frac{1}{2}\ln(1+t^2) + C$

23. $-\frac{x}{3}\cos^3 x + \frac{1}{3}\sin x - \frac{1}{9}\sin^3 x + C$ **25.** $\dfrac{\pi}{2\sqrt{3}} + \ln 2$

27. $\dfrac{\pi}{4} - \dfrac{\pi}{6\sqrt{3}} + \dfrac{1}{2}\ln\dfrac{2}{3}$

29. $\frac{2}{9}x^3(x^3+4)^{3/2} - \frac{4}{45}(x^3+4)^{5/2} + C$

31. $\dfrac{t^4}{6(7-3t^4)^{1/2}} + \frac{1}{9}(7-3t^4)^{1/2} + C$

33. $\dfrac{z^4}{4(4-z^4)} + \frac{1}{4}\ln|4-z^4| + C$

35. $x\cosh x - \sinh x + C$

37. $2\sqrt{x}\ln x - 4\sqrt{x} + C$ **39.** $\dfrac{x}{\ln 2}2^x - \dfrac{1}{(\ln 2)^2}2^x + C$

41. $x^2e^x - 2xe^x + 2e^x + C$ **43.** $z\ln^2 z - 2z\ln z + 2z + C$

45. $\frac{1}{2}e^t(\sin t + \cos t) + C$

47. $x^2\sin x + 2x\cos x - 2\sin x + C$

49. $\frac{x}{2}\left[\sin(\ln x) - \cos(\ln x)\right] + C$

51. $x\ln^3 x - 3x\ln^2 x + 6x\ln x - 6x + C$ **69.** 1

71. $9 - \dfrac{9}{e^3}$

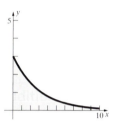

73. $\dfrac{\sqrt{2}\pi}{4} - 1$ **75.** $\bar{x} = \dfrac{e^2+1}{4}, \bar{y} = \dfrac{e-2}{4}$

77. (a) $(x^3 - 2x)e^x - (3x^2 - 2)e^x + 6xe^x - 6e^x + C$

(b) $(x^2 - 3x + 1)(-\cos x) - (2x - 3)(-\sin x) + 2\cos x + C$

79. $\begin{cases} 0 & \text{if } n \text{ is odd} \\ \dfrac{1\cdot 3\cdot 5\cdots(n-1)}{2\cdot 4\cdot 6\cdots n}2\pi & \text{if } n \text{ is even} \end{cases}$

93. $e^x(3x^4 - 12x^3 + 38x^2 - 76x + 76)$

Problem Set 8.5

1. $\ln|x| - \ln|x+1| + C$

3. $-\frac{3}{2}\ln|x+1| + \frac{3}{2}\ln|x-1| + C$

5. $3\ln|x+4| - 2\ln|x-1| + C$

7. $4\ln|x+5| - \ln|x-2| + C$

9. $2\ln|2x-1| - \ln|x+5| + C$

11. $\frac{5}{3}\ln|3x-2| + 4\ln|x+1| + C$

13. $2\ln|x| - \ln|x+1| + \ln|x-2| + C$

15. $\ln|2x-1| - \ln|x+3| + 3\ln|x-2| + C$

17. $\frac{1}{2}x^2 - x + \frac{8}{3}\ln|x+2| + \frac{1}{3}\ln|x-1| + C$

19. $\frac{1}{2}x^2 - 2\ln|x| + 7\ln|x+2| + 7\ln|x-2| + C$

21. $\ln|x-3| - \dfrac{4}{x-3} + C$ **23.** $-\dfrac{3}{x+1} + \dfrac{1}{2(x+1)^2} + C$

25. $2\ln|x| + \ln|x-4| + \dfrac{1}{x-4} + C$

27. $-2\ln|x| + \frac{1}{2}\tan^{-1}(\frac{x}{2}) + 2\ln|x^2+4| + C$

29. $-2\ln|2x-1| + \frac{3}{2}\ln|x^2+9| + C$

31. $-\dfrac{2}{125}\ln|x-1| - \dfrac{1}{25(x-1)}$

$\qquad + \dfrac{2}{125}\ln|x+4| - \dfrac{1}{25(x+4)} + C$

33. $\sin t - \frac{50}{13}\ln|\sin t + 3| - \frac{68}{13}\tan^{-1}(\sin t - 2)$

$\qquad\qquad - \frac{41}{26}\ln|\sin^2 t - 4\sin t + 5| + C$

Problem Set 14.4

1. $x = 1 + 3t, y = -2 + 7t, z = 3 + 3t$

3. $x = 4 + t, y = 2, z = 3 - 2t$

5. $x = 4 + 3t, y = 5 + 2t, z = 6 + t; \dfrac{x-4}{3} = \dfrac{y-5}{2} = \dfrac{z-6}{1}$

7. $x = 1 + t, y = 1 + 10t, z = 1 + 100t;$

$$\dfrac{x-1}{1} = \dfrac{y-1}{10} = \dfrac{z-1}{100}$$

9. $\dfrac{x-4}{27} = \dfrac{y+5}{-50} = \dfrac{z}{-6}$ **11.** $\dfrac{x+8}{10} = \dfrac{y}{2} = \dfrac{z+\frac{21}{2}}{9}$

13. $\dfrac{x-4}{1} = \dfrac{y}{-5} = \dfrac{z-6}{2}$ **15.** $x = 5t, y = -3t, z = 4$

17. $x + y + 6z = 11$ **19.** $3x - 2y = 5$

21. (b) $2x + y - z = 7$;   **(c)** $(-1, 2, -1)$;   **(d)** $\sqrt{6}$

23. $\dfrac{x-1}{-\sqrt{3}} = \dfrac{y-3\sqrt{3}}{3} = \dfrac{z-\frac{\pi}{3}}{1}$ **25.** $3x - 4y + 5z = -22$

27. (b) $\left(\dfrac{3\sqrt{3}}{4}, \dfrac{7}{4}, 0\right)$ **29. (a)** $\dfrac{8\sqrt{2}}{3}$;   **(b)** $\dfrac{3\sqrt{26}}{7}$

Section 14.5

1. $\mathbf{v}(1) = 4\mathbf{i} + 10\mathbf{j} + 2\mathbf{k}; \mathbf{a}(1) = 10\mathbf{j}; s(1) = 2\sqrt{30}$

3. $\mathbf{v}(2) = -\frac{1}{4}\mathbf{i} - \frac{4}{9}\mathbf{j} + 80\mathbf{k}; \mathbf{a}(2) = \frac{1}{4}\mathbf{i} + \frac{26}{27}\mathbf{j} + 160\mathbf{k};$

$$s(2) = \dfrac{\sqrt{8,294,737}}{36}$$

5. $\mathbf{v}(2) = 4\mathbf{j} + \dfrac{2^{2/3}}{3}\mathbf{k}; \mathbf{a}(2) = 4\mathbf{j} - \dfrac{1}{9\sqrt[3]{2}}\mathbf{k};$

$$s(2) = \sqrt{16 + \dfrac{2^{4/3}}{9}}$$

7. $\mathbf{v}(\pi) = -\mathbf{j} + \mathbf{k}; \mathbf{a}(\pi) = \mathbf{i}, s(\pi) = \sqrt{2}$

9. $\mathbf{v}(\frac{\pi}{4}) = 2\mathbf{i} + 3e^{\pi/4}\mathbf{j}; \mathbf{a}(\frac{\pi}{4}) = 4\mathbf{i} + 3e^{\pi/4}\mathbf{j} + 16\mathbf{k};$

$$s(\tfrac{\pi}{4}) = \sqrt{4 + 9e^{\pi/2}}$$

11. $\mathbf{v}(2) = 2\pi\mathbf{i} + \mathbf{j} - e^{-2}\mathbf{k}; \mathbf{a}(2) = 2\pi\mathbf{i} - 2\pi^2\mathbf{j} + e^{-2}\mathbf{k};$

$$s(2) = \sqrt{4\pi^2 + 1 + e^{-4}}$$

15. $2\sqrt{2}$ **17.** 144 **19.** $\sqrt{41}$ **21.** $\frac{3}{2}$

23. $\sqrt{2} \sinh \pi$

25. $\kappa = \dfrac{\sqrt{6}}{10\sqrt{5}}; \mathbf{T} = \dfrac{2}{\sqrt{5}}\mathbf{i} + \dfrac{1}{\sqrt{5}}\mathbf{j};$

$$\mathbf{N} = \dfrac{1}{\sqrt{30}}\mathbf{i} - \dfrac{2}{\sqrt{30}}\mathbf{j} + \dfrac{5}{\sqrt{30}}\mathbf{k};$$

$$\mathbf{B} = \dfrac{1}{\sqrt{6}}\mathbf{i} - \dfrac{2}{\sqrt{6}}\mathbf{j} - \dfrac{1}{\sqrt{6}}\mathbf{k}$$

27. $\dfrac{\sqrt{11}}{21\sqrt{7}}; \mathbf{T} = \dfrac{2}{\sqrt{21}}\mathbf{i} + \dfrac{1}{\sqrt{21}}\mathbf{j} + \dfrac{4}{\sqrt{21}}\mathbf{k};$

$$\mathbf{N} = -\dfrac{5}{\sqrt{77}}\mathbf{i} - \dfrac{6}{\sqrt{77}}\mathbf{j} + \dfrac{4}{\sqrt{77}}\mathbf{k};$$

$$\mathbf{B} = \dfrac{4}{\sqrt{33}}\mathbf{i} - \dfrac{4}{\sqrt{33}}\mathbf{j} - \dfrac{1}{\sqrt{33}}\mathbf{k}$$

29. $\kappa = \frac{9}{91}; \mathbf{T} = \left\langle -\dfrac{3}{\sqrt{13}}, 0, \dfrac{2}{\sqrt{13}} \right\rangle; \mathbf{N} = \langle 0, 1, 0 \rangle;$

$$\mathbf{B} = \left\langle -\dfrac{2}{\sqrt{13}}, 0, -\dfrac{3}{\sqrt{13}} \right\rangle$$

31. $\kappa = \frac{1}{3}\operatorname{sech}^2\frac{1}{3}; \mathbf{T} = \tanh\frac{1}{3}\mathbf{i} + \operatorname{sech}\frac{1}{3}\mathbf{j};$

$$\mathbf{N} = \operatorname{sech}\frac{1}{3}\mathbf{i} - \tanh\frac{1}{3}\mathbf{j}; \mathbf{B} = -\mathbf{k}$$

33. $\kappa = \dfrac{1}{2\sqrt{2}}; \mathbf{T} = -\dfrac{1}{2}\mathbf{i} + \dfrac{1}{2}\mathbf{j} + \dfrac{1}{\sqrt{2}}\mathbf{k};$

$$\mathbf{N} = \dfrac{1}{\sqrt{2}}\mathbf{i} + \dfrac{1}{\sqrt{2}}\mathbf{j}; \mathbf{B} = -\dfrac{1}{2}\mathbf{i} + \dfrac{1}{2}\mathbf{j} - \dfrac{1}{\sqrt{2}}\mathbf{k}$$

35. $a_T(t) = \dfrac{4t}{\sqrt{10 + 4t^2}}; a_N(t) = 2\sqrt{\dfrac{5}{5 + 2t^2}}$

37. $a_T(t) = \dfrac{e^{2t} - e^{-2t}}{\sqrt{e^{2t} + 4 + e^{-2t}}}; a_N(t) = 2\sqrt{\dfrac{e^{2t} + 1 + e^{-2t}}{e^{2t} + 4 + e^{-2t}}}$

39. $a_T(t) = \dfrac{4t^3}{\sqrt{2t^4 + 3}}; a_N(t) = \dfrac{2\sqrt{6}t}{\sqrt{2t^4 + 3}}$

41. $a_T(t) = \dfrac{\tan t \sec^2 t - \cot t \csc^2 t}{\sqrt{1 + \cot^2 t + \tan^2 t}};$

$$a_N(t) = \dfrac{\sqrt{\csc^4 t + 4\csc^2 t \sec^2 t + \sec^4 t}}{\sqrt{1 + \cot^2 t + \tan^2 t}}$$

43. $\mathbf{T} = \langle \frac{1}{3}, \frac{2}{3}, \frac{2}{3} \rangle; \mathbf{N} = \langle -\frac{2}{3}, -\frac{1}{3}, \frac{2}{3} \rangle; \mathbf{B} = \langle \frac{2}{3}, -\frac{2}{3}, \frac{1}{3} \rangle$

45. $\mathbf{T} = \dfrac{1}{\sqrt{\cosh^2\frac{\pi}{6c} + 1}}\left(\cosh\dfrac{\pi}{6c}\mathbf{i} + \mathbf{k} \right);$

$$\mathbf{N} = \dfrac{1}{\sqrt{\cosh^2\frac{\pi}{6c} + 1}}\left(\mathbf{i} - \cosh\dfrac{\pi}{6c}\mathbf{k} \right);$$

$$\mathbf{B} = \mathbf{j}$$

47. $\mathbf{T} = \dfrac{1}{\sqrt{1 + 5\pi^2}}(-\mathbf{i} - \pi\mathbf{j} + 2\pi\mathbf{k});$

$$\mathbf{N} = \dfrac{[(5\pi^3 + 6\pi)\mathbf{i} + (-2 - 5\pi^2)\mathbf{j} + 2\mathbf{k}]}{\sqrt{(8 + 16\pi^2 + 5\pi^4)(1 + 5\pi^2)}};$$

$$\mathbf{B} = \dfrac{1}{\sqrt{8 + 16\pi^2 + 5\pi^4}}[2\pi\mathbf{i} + (2 + 2\pi^2)\mathbf{j} + (2 + \pi^2)\mathbf{k}]$$

49. $\mathbf{T} = \dfrac{1}{\sqrt{3}}\mathbf{i} + \dfrac{1}{2}\left(\dfrac{1}{\sqrt{3}} - 1 \right)\mathbf{j} + \dfrac{1}{2}\left(\dfrac{1}{\sqrt{3}} + 1 \right)\mathbf{k};$

$$\mathbf{N} = -\dfrac{1 + \sqrt{3}}{2\sqrt{2}}\mathbf{j} + \dfrac{1 - \sqrt{3}}{2\sqrt{2}}\mathbf{k};$$

$$\mathbf{B} = \sqrt{\tfrac{2}{3}}\mathbf{i} + \tfrac{1}{12}(3\sqrt{2} - \sqrt{6})\mathbf{j} - \tfrac{1}{12}(3\sqrt{2} + \sqrt{6})\mathbf{k}$$

51. $\mathbf{T} = \dfrac{1}{\sqrt{2}}\mathbf{i} + \dfrac{1}{\sqrt{2}}\operatorname{sech}\dfrac{\pi}{3}\mathbf{j} + \dfrac{1}{\sqrt{2}}\tanh\dfrac{\pi}{3}\mathbf{k};$

$$\mathbf{N} = -\tanh\tfrac{\pi}{3}\mathbf{j} + \operatorname{sech}\tfrac{\pi}{3}\mathbf{k};$$

$$\mathbf{B} = \dfrac{1}{\sqrt{2}}\mathbf{i} - \dfrac{1}{\sqrt{2}}\operatorname{sech}\dfrac{\pi}{3}\mathbf{j} - \dfrac{1}{\sqrt{2}}\tanh\dfrac{\pi}{3}\mathbf{k}$$

53. (b) $R_p = 10 R_m; t = \frac{\pi}{9}$

55. (a) Winding upward around the right circular cylinder
$x = \sin t, y = \cos t$, as t increases.

(b) Same as part (a), but winding much faster by a factor of $3t^2$.

(c) With standard orientation of the axes, the motion is winding to the right around the right circular cylinder $x = \sin t$, $z = \cos t$.

(d) Spiraling upward, with increasing radius, along the spiral $x = t \sin t$, $y = t \cos t$.

(e) Spiraling upward, with decreasing radius, along the spiral $x = \dfrac{1}{t^2} \sin t$, $y = \dfrac{1}{t^2} \cos t$.

(f) Spiraling to the right, with increasing radius along the spiral $x = t^2 \sin(\ln t)$, $z = t^2 \cos(\ln t)$.

57. $P_5(x) = 10x^3 - 15x^4 + 6x^5$ **63.** $(6, 0, 8); 8\sqrt{9\pi^2 + 1}$

65. (a) $\mathbf{r}(t) = \mathbf{r}_0$;   **(b)** $\mathbf{r}(t) = \mathbf{c}t + \mathbf{r}_0$;

(c) $\mathbf{r}(t) = \frac{1}{2}\mathbf{c}t^2 + \mathbf{v}_0 t + \mathbf{r}_0$;   **(d)** $\mathbf{r}(t) = \mathbf{r}_0 e^{ct}$

67. $\left(\dfrac{16\pi^2 + 30}{4\pi + 1}, \dfrac{120\pi - 16\pi^2}{4\pi + 1}, \dfrac{16\pi^2 + 30}{4\pi + 1} \right)$

Problem Set 14.6

1. Elliptic cylinder

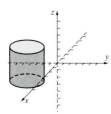

3. Plane

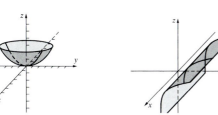

5. Circular cylinder

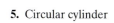

7. Ellipsoid

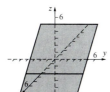

9. Elliptic paraboloid

11. Cylinder

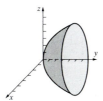

13. Hyperbolic paraboloid   **15.** Elliptic paraboloid

17. Plane

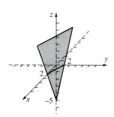

19. Hemisphere

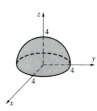

21. (a) Replacing x by $-x$ results in an equivalent equation.
(b) Replacing x by $-x$ and y by $-y$ results in an equivalent equation.   **(c)** Replacing y by $-y$ and z by $-z$ results in an equivalent equation.   **(d)** Replacing x by $-x$, y by $-y$, and z by $-z$ results in an equivalent equation.

23. $y = 2x^2 + 2z^2$ **25.** $4x^2 + 3y^2 + 4z^2 = 12$

27. $(0, \pm 2\sqrt{5}, 4)$ **29.** $\dfrac{\pi ab(c^2 - h^2)}{c^2}$

31. Major diameter 4; minor diameter $2\sqrt{2}$

33. $x^2 + 9y^2 - 9z^2 = 0$

Problem Set 14.7

1. Cylindrical to Spherical: $\rho = \sqrt{r^2 + z^2}$, $\cos \phi = \dfrac{z}{\sqrt{r^2 + z^2}}$, $\theta = \theta$

Spherical to Cylindrical: $r = \rho \sin \phi$, $z = \rho \cos \phi$, $\theta = \theta$

3. (a) $(3\sqrt{3}, 3, -2)$;   **(b)** $(-2, -2\sqrt{3}, -8)$

5. (a) $\left(4\sqrt{2}, \frac{5\pi}{3}, \frac{\pi}{4}\right)$;   **(b)** $\left(4, \frac{3\pi}{4}, \frac{\pi}{6}\right)$

7.

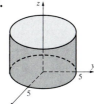

9.

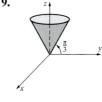

11.

13.

15.

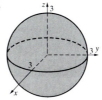

17. $r = 3$ **19.** $r^2 + 4z^2 = 10$ **21.** $\cos^2 \phi = \frac{1}{3}$

23. $\rho^2 = \dfrac{4}{1 + \cos^2 \phi}$ **25.** $r = \dfrac{4}{\sin \theta + \cos \theta}$

27. $\rho^2 \sin^2 \phi = 9$ **29.** $x^2 - y^2 = z$ **31.** $z = 2r^2$

33. 4029 mi   **35.** 4552 mi   **37.** 2252 mi

41. (a) 3485 mi; **(b)** 4552 mi; **(c)** 9331 mi;
(d) 7798 mi; **(e)** 12,441 mi

Chapter Review 14.8

Concepts Test

1. True	**3.** True	**5.** False	**7.** True	**9.** True
11. True	**13.** False	**15.** True	**17.** True	**19.** True
21. True	**23.** False	**25.** True	**27.** False	**29.** True
31. True	**33.** False	**35.** False	**37.** False	

Sample Test Problems

1. $(x - 1)^2 + (y - 2)^2 + (z - 4)^2 = 11$

3.

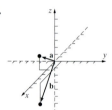

(a) $3; \sqrt{35};$ **(b)** $\dfrac{2}{3}, -\dfrac{1}{3}, \dfrac{2}{3}; \dfrac{5}{\sqrt{35}}, \dfrac{1}{\sqrt{35}}, -\dfrac{3}{\sqrt{35}};$

(c) $\frac{2}{3}\mathbf{i} - \frac{1}{3}\mathbf{j} + \frac{2}{3}\mathbf{k};$ **(d)** $\cos^{-1} \dfrac{1}{\sqrt{35}}$

5. $c\langle 10, -11, -3 \rangle, c$ in $\mathbb{R}$

7. (a) $y = 7;$ **(b)** $x = -5;$ **(c)** $z = -2;$
(d) $3x - 4y + z = -45$

9. 1   **11.** $x = -2 + 8t, y = 1 + t, z = 5 - 8t$

13. $x = 2t, y = 25 + t, z = 16$ **15.** $\langle 2, -2, 1 \rangle + t\langle 5, -4, -3 \rangle$

17. Tangent line: $\dfrac{x - 2}{1} = \dfrac{y - 2}{2} = \dfrac{z - \frac{8}{3}}{4}$

Normal Plane: $3x + 6y + 12z = 50$

19. $\sqrt{3}(e^5 - e)$ **21.** $a_T = \dfrac{22}{\sqrt{14}}; a_N = \dfrac{2\sqrt{19}}{\sqrt{14}}$

23. Sphere   **25.** Circular paraboloid

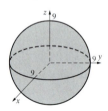   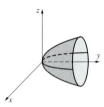

27. Plane   **29.** Ellipsoid

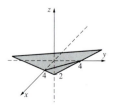   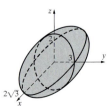

31. (a) $r = 3;$ **(b)** $r^2 = \dfrac{16}{1 + 3\sin^2\theta};$ **(c)** $r^2 = 9z;$

(d) $r^2 + 4z^2 = 10$

33. (a) $\rho = 2;$ **(b)** $\cos^2\phi = \frac{1}{2}$ (Other forms are possible.);

(c) $\rho^2 = \dfrac{1}{2\sin^2\phi\cos^2\theta - 1};$ **(d)** $r = \cot\phi\csc\phi$

35. 1.25

Problem Set 15.1

1. (a) 5; **(b)** 0; **(c)** 6; **(d)** $a^6 + a^2;$ **(e)** $2x^2;$
(f) $(2, -4)$ is not in the domain of f. Domain is set of all (x, y)
such that $y > 0$.

3. (a) 0; **(b)** 2; **(c)** 16; **(d)** $-4.2469;$ **(e)** 0.6311

5. t^2

7.   **9.**

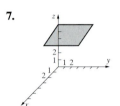   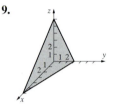

11.   **13.**

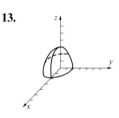   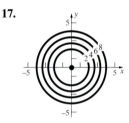

15.   **17.**

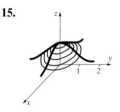   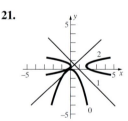

19.   **21.**

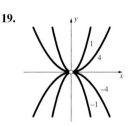   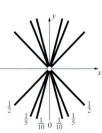

23.

25. (a) San Francisco **(b)** northwest; southeast
(c) southwest or northeast
27. The set of all points on and outside the sphere
$x^2 + y^2 + z^2 = 16$.
29. The set of all points on and inside the ellipsoid
$x^2/9 + y^2/16 + z^2/1 = 1$.
31. All points in $\mathbb{R}^3$ except the origin $(0, 0, 0)$.
33. The set of all spheres with centers at the origin.

35. A set of hyperboloids of revolution about the z-axis.

37. A set of hyperbolic cylinders parallel to the z-axis.

39. All points in $\mathbb{R}^4$ except the origin $(0,0,0,0)$.

41. (a) gentle climb, steep climb;   **(b)** 6700 ft, 3040 ft

43.

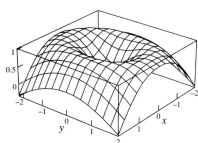

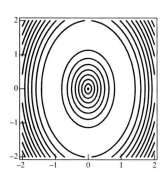

45.

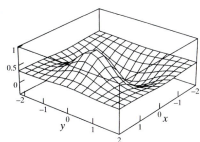

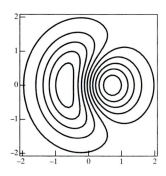

Problem Set 15.2

1. $f_x(x, y) = 8(2x - y)^3; f_y(x, y) = -4(2x - y)^3$

3. $f_x(x, y) = (x^2 + y^2)/(x^2y); f_y(x, y) = -(x^2 + y^2)/(xy^2)$

5. $f_x(x, y) = e^y \cos x; f_y(x, y) = e^y \sin x$

7. $f_x(x, y) = x(x^2 - y^2)^{-1/2}; f_y(x, y) = -y(x^2 - y^2)^{-1/2}$

9. $g_x(x, y) = -ye^{-xy}; g_y(x, y) = -xe^{-xy}$

11. $f_x(x, y) = 4/[1 + (4x - 7y)^2];$
$f_y(x, y) = -7/[1 + (4x - 7y)^2]$

13. $f_x(x, y) = -2xy \sin(x^2 + y^2);$
$f_y(x, y) = -2y^2 \sin(x^2 + y^2) + \cos(x^2 + y^2)$

15. $F_x(x, y) = 2 \cos x \cos y; F_y(x, y) = -2 \sin x \sin y$

17. $f_{xy}(x, y) = 12xy^2 - 15x^2y^4 = f_{yx}(x, y)$

19. $f_{xy}(x, y) = -6e^{2x} \sin y = f_{yx}(x, y)$

21. $F_x(3, -2) = \frac{1}{9}; F_y(3, -2) = -\frac{1}{2}$

23. $f_x(\sqrt{5}, -2) = -\frac{4}{21}; f_y(\sqrt{5}, -2) = -4\sqrt{5}/21$

25. 1   **27.** 3   **29.** 120π **31.** $k/100$

33. $\partial^2 f/\partial x^2 = 6xy; \partial^2 f/\partial y^2 = -6xy$ **35.** $180x^4y^2 - 12x^2$

37. (a) $\partial^3 f/\partial y^3$;   **(b)** $\partial^3 f/\partial y\,\partial x^2$;   **(c)** $\partial^4 f/\partial y^3\,\partial x$

39. (a) $6xy - yz$;   **(b)** 8;   **(c)** $6x - z$

41. $-yze^{-xyz} - y(xy - z^2)^{-1}$ **43.** $(1, 0, 29)$

45. $\{(x, y): x < \frac{1}{2}, y > \frac{1}{2}, y < x + \frac{1}{2}\}$
$\cup \{(x, y): x > \frac{1}{2}, y < \frac{1}{2}, x < y + \frac{1}{2}\}, \{z: 0 < z \le \sqrt{3}/36\}$

47. (a) -4;   **(b)** $\frac{2}{3}$;   **(c)** $\frac{2}{5}$;   **(d)** $\frac{8}{3}$

49. (a) $f_y(x, y, z) = \lim\limits_{\Delta y \to 0} \dfrac{f(x, y + \Delta y, z) - f(x, y, z)}{\Delta y}$

(b) $f_z(x, y, z) = \lim\limits_{\Delta z \to 0} \dfrac{f(x, y, z + \Delta z) - f(x, y, z)}{\Delta z}$

(c) $G_x(w, x, y, z) = \lim\limits_{\Delta x \to 0} \dfrac{G(w, x + \Delta x, y, z) - G(w, x, y, z)}{\Delta x}$

(d) $\dfrac{\partial}{\partial z} \lambda(x, y, z, t) = \lim\limits_{\Delta z \to 0} \dfrac{\lambda(x, y, z + \Delta z, t) - \lambda(x, y, z, t)}{\Delta z}$

(e) $\dfrac{\partial}{\partial b_2} S(b_0, b_1, b_2, \ldots, b_n) =$

$= \lim\limits_{\Delta b_2 \to 0} \left(\dfrac{S(b_0, b_1, b_2 + \Delta b_2, \ldots, b_n) - S(b_0, b_1, b_2, \ldots, b_n)}{\Delta b_2} \right)$

Problem Set 15.3

1. -18 **3.** $2 - \frac{1}{2}\sqrt{3}$ **5.** $\frac{1}{3}$ **7.** Does not exist.

9. Entire plane.   **11.** $\{(x, y): y \ne x^2\}$

13. $\{(x, y): y \le x + 1\}$

15. $\lim\limits_{x \to 0} f(x, 0) = \lim\limits_{x \to 0} [0/(x^2 + 0)] = 0;$

$\lim\limits_{x \to 0} f(x, x) = \lim\limits_{x \to 0} [x^2/(x^2 + x^2)] = \frac{1}{2}$

17. (a) $\lim\limits_{x \to 0} f(x, mx) = \lim\limits_{x \to 0} mx^3/(x^4 + m^2x^2)$

$= \lim\limits_{x \to 0} mx/(x^2 + m^2) = 0;$

(b) $\lim\limits_{x \to 0} f(x, x^2) = \lim\limits_{x \to 0} x^4/(x^4 + x^4) = \frac{1}{2};$

(c) $\lim\limits_{(x, y) \to (0, 0)} f(x, y)$ does not exist.

19. The boundary consists of the line segments that form the outer edges of the given rectangle; the set is closed.

21. Boundary: $\{(x, y): x^2 + y^2 = 4\} \cup \{(0,0)\}$; the set is neither open nor closed.

23. Boundary: $\{(x, y): y = \sin(1/x), x > 0\} \cup \{(x, y): x = 0, y \le 1\}$; the set is open.

25. $g(x) = 2x$

27. (a) Continuous;   **(b)** discontinuous;   **(c)** continuous;   **(d)** continuous;   **(e)** continuous;   **(f)** discontinuous

29. (a) $\{(x, y, z): x^2 + y^2 = 1, 1 \le z \le 2\};$

(b) $\{(x, y, z): x^2 + y^2 = 1, z = 1\};$ **(c)** $\{(x, y, z): z = 1\};$

(d) empty set.

31. (a) $\{(x, y): x > 0, y = 0\}$;

(b) $\{(u, v, x, y): \langle x, y \rangle = k \langle u, v \rangle, k > 0, \langle u, v \rangle \neq \langle 0, 0 \rangle\}$

33.

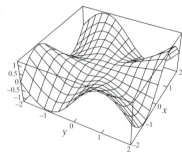

35.

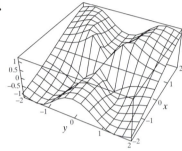

Problem Set 15.4

1. $(2xy + 3y)\mathbf{i} + (x^2 + 3x)\mathbf{j}$ **3.** $e^{xy}(1 + xy)\mathbf{i} + x^2 e^{xy}\mathbf{j}$

5. $(x + y)^{-2}[(x^2y + 2xy^2)\mathbf{i} + x^3\mathbf{j}]$

7. $(x^2 + y^2 + z^2)^{-1/2}(x\mathbf{i} + y\mathbf{j} + z\mathbf{k})$

9. $xe^{x-z}[(yx + 2y)\mathbf{i} + x\mathbf{j} - xy\mathbf{k}]$

11. $\langle -21, 16 \rangle, z = -21x + 16y - 60$

13. $\langle 0, -2\pi \rangle, z = -2\pi y + \pi - 1$

15. $w = 7x - 8y - 2z + 3$

21. (a) $x = 2 + t, y = 1, z = 9 + 12t$

(b) $x = 2, y = 1 + 10t, z = 9 + 10t$

(c) $x = 2 - t, y = 1 - t, z = 9 - 22t$

23. $z = -5x + 5y$

27.

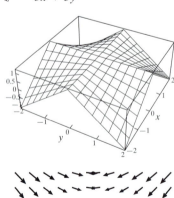

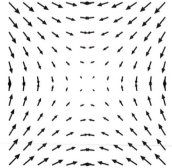

(a) The gradient points in the direction of greatest increase of the function. **(b)** No.

Problem Set 15.5

1. $\frac{8}{5}$ **3.** $3\sqrt{2}/2$ **5.** $(\sqrt{2} + \sqrt{6})/4$ **7.** $\frac{52}{3}$

9. 13   **11.** $\sqrt{21}$ **13.** $(1/\sqrt{5})(-\mathbf{i} + 2\mathbf{j})$

15. $\nabla f(\mathbf{p}) = -4\mathbf{i} + \mathbf{j}$ is perpendicular to the tangent line at $\mathbf{p}$.

17. $\frac{2}{3}$ **19. (a)** $(0, 0, 0)$;   **(b)** $-\mathbf{i} + \mathbf{j} - \mathbf{k}$;   **(c)** yes.

21. $(x^2 + y^2 + z^2)^{-1/2} \cos \sqrt{x^2 + y^2 + z^2} \, \langle x, y, z \rangle$

23. N 63.43°E   **25.** Descend; $-300\sqrt{2}\,e^{-3}$ **27.** $x = -2y^2$

29. (a) $-10/\sqrt{2 + \pi^2}$ deg/m;   **(b)** -10 deg/s

31. (a) $(100, 120)$;   **(b)** $(190, 25)$;   **(c)** $-\frac{1}{3}, 0, \frac{2}{5}$

33. Leave at about $(-0.1, -5)$.   **35.** Leave at about $(3, 5)$.

Problem Set 15.6

1. $12t^{11}$ **3.** $e^{3t}(3 \sin 2t + 2 \cos 2t) + e^{2t}(3 \cos 3t + 2 \sin 3t)$

5. $7t^6 \cos(t^7)$ **7.** $2s^3t - 3s^2t^2$

9. $(2s^2 \sin t \cos t + 2t \sin^2 s) \exp(s^2 \sin^2 t + t^2 \sin^2 s)$

11. $s^4t(1 + s^4t^2)^{-1/2}$ **13.** 72   **15.** $-\frac{1}{2}(\pi + 1)$

17. 244.35 board ft per year   **19.** $\sqrt{20}$ ft/s

21. $(3x^2 + 4xy)/(3y^2 - 2x^2)$

23. $(y \sin x - \sin y)/(x \cos y + \cos x)$

25. $(yz^3 - 6xz)/(3x^3 - 3xyz^2)$

27. $\partial T/\partial s = (\partial T/\partial x)(\partial x/\partial s) + (\partial T/\partial y)(\partial y/\partial s)$
$$+ (\partial T/\partial z)(\partial z/\partial s) + (\partial T/\partial w)(\partial w/\partial s)$$

31. $10\sqrt{2} - 3\pi\sqrt{2}$ **33.** 288 mi/h

Problem Set 15.7

1. $2(x - 2) + 3(y - 3) + \sqrt{3}(z - \sqrt{3}) = 0$

3. $(x - 1) - 3(y - 3) + \sqrt{7}(z - \sqrt{7}) = 0$

5. $x + y - z = 2$ **7.** $z + 1 = -2\sqrt{3}(x - \frac{1}{3}\pi) - 3y$

9. $0.08; 0.08017992$ **11.** $-0.03; -0.03015101$

13. $(3, -1, -14)$

15. $\langle 0, 1, 1 \rangle$ is normal to both surfaces at $(0, -1, 2)$

17. $(1, 2, -1)$ and $(-1, -2, 1)$

19. $x = 1 + 32t; y = 2 - 19t; z = 2 - 17t$ **21.** 0.004375 lb

23. 7%   **25.** 20 ± 0.34 **27.** $V = 9|k|/2$

29. (a) 4.98;   **(b)** 4.98196;   **(c)** 4.9819675

Problem Set 15.8

1. $(2, 0)$: local minimum point.

3. $(0, 0)$: saddle point; $(\pm\frac{1}{2}, 0)$: local minimum points.

5. $(0, 0)$: saddle point (special argument needed).

7. $(1, 2)$: local minimum point.   **9.** No critical points.

11. Global maximum of 7 at $(1, 1)$; global minimum of -4 at $(0, -1)$.

13. Global maximum of 2 at $(\pm 1, 0)$; global minimum of 0 at $(0, \pm 1)$.

15. Each of the three numbers is $N/3$.   **17.** A cube.

19. Base 8 ft by 8 ft; depth 4 ft.   **21.** $3\sqrt{3}(\mathbf{i} + \mathbf{j} + \mathbf{k})$

23. $2\pi/3$; 4 in.   **25. (a)** 8;   **(b)** -11

27. Maximum of 3 at $(1, 2)$; minimum of $-\frac{12}{5}$ at $\left(\frac{8}{5}, -\frac{2}{5}\right)$.

29. $y = \frac{7}{10}x + \frac{1}{10}$

31. $\left(\pm\sqrt{3}/2, -\frac{1}{2}\right)$ where $T = 9/4$; $(0, 1/2)$, where $T = -1/4$

33. Equilateral triangle.

35. Local maximum: $f(1.75, 0) = 1.15$;
global maximum: $f(-3.8, 0) = 2.30$

37. Global minimum: $f(0, 1) = f(0, -1) = -0.12$.

39. Global maximum $f(1.13, 0.79) = f(1.13, -0.79) = 0.53$
global minimum $f(-1.13, 0.79) = f(-1.13, -0.79) = -0.53$.

41. Global maximum $f(3, 3) = f(-3, 3) \approx 74.9225$
global minimum $f(1.5708, 0) = f(-1.5708, 0) = -8$.

43. Global maximum: $f(0.67, 0) = 5.06$;
global minimum: $f(-0.75, 0) = -3.54$.

45. Global maximum: $f(2.1, 2.1) = 3.5$;
global minimum: $f(4.2, 4.2) = -3.5$.

Problem Set 15.9

1. $f(\sqrt{3}, \sqrt{3}) = f(-\sqrt{3}, -\sqrt{3}) = 6$

3. $f(2/\sqrt{5}, -1/\sqrt{5}) = f(-2/\sqrt{5}, 1/\sqrt{5}) = 5$

5. $f\left(\frac{6}{7}, \frac{18}{7}, -\frac{12}{7}\right) = \frac{72}{7}$ **7.** Base is 4 by 4; depth is 2.

9. $10\sqrt{5}$ ft^3 **11.** $8abc/3\sqrt{3}$

13. $x/a + y/b + z/c = 3, V = \frac{9}{2}abc$

15. $x = \alpha d/a, y = \beta d/b, z = \gamma d/c$

17. $f(-1, 1, 0) = 3, f(1, -1, 1) = -1$

19. $x_i = a_i/\sqrt{a_1^2 + \cdots + a_n^2}, w = \sqrt{a_1^2 + \cdots + a_n^2}$

21. $f(4, 0) = -4$ **23.** $f(0, 3) = f(0, -3) = -0.99$

Chapter Review 15.10

Concepts Test

1. True **3.** True **5.** True **7.** False **9.** True
11. True **13.** True **15.** True **17.** True **19.** False

Sample Test Problems

1. (a) $\left\{(x, y): x^2 + 4y^2 \geq 100\right\}$

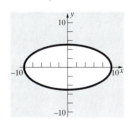

(b) $\left\{(x, y): 2x - y \geq 1\right\}$

3. $12x^3y^2 + 14xy^7; 36x^2y^2 + 14y^7; 24x^3y + 98xy^6$

5. $e^{-y}\sec^2 x; 2e^{-y}\sec^2 x \tan x; -e^{-y}\sec^2 x$

7. $450x^2y^4 - 42y^5$ **9.** 1 **11.** Does not exist.

13. (a) $-4\mathbf{i} - \mathbf{j} + 6\mathbf{k}; -4(\cos 1\mathbf{i} + \sin 1\mathbf{j} - \cos 1\mathbf{k})$

15. $\sqrt{3} + 2$ **17 (a)** $x^2 + 2y^2 = 18$; **(b)** $4\mathbf{i} + 2\mathbf{j}$

19. $(x^2 + 3y - 4z)/x^2yz; (-x^2 - 4x)/xy^2z; (3y - x^2)/xyz^2$

21. $15xy\sqrt{t}/z^3 + 5x^2/tz^3 - 45x^2ye^{3t}/z^4$

23. $18\mathbf{i} + 16\mathbf{j} - 18\mathbf{k}; 9x + 8y - 9z = 34$

25. 0.7728 **27.** $16\sqrt{3}/3$ **29.** Radius 2; height 4.

Problem Set 16.1

1. 14 **3.** 12 **5.** 4 **7.** 3 **9.** 168 **11.** 520

13. 52.57 **15.** 5.5 **19.** $c = 15.30, C = 30.97$

21. (a) -6; **(b)** 6

23. Number of cubic inches of rain that fell on all Colorado in 1999; average rainfall in Colorado during 1999.

25. Approximately 458.

Problem Set 16.2

1. $\frac{32}{3}$ **3.** $\frac{55}{4}$ **5.** 1 **7.** $\pi/2 - 1$

9. $\frac{4}{15}\left[31 - 9\sqrt{3}\right] \approx 4.110$ **11.** $1 - \frac{1}{2}\ln 3 \approx 0.4507$

13. 0 **15.** 2

17.

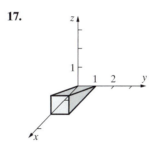

19.

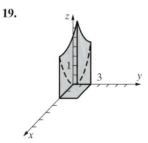

21. 7 **23.** $\frac{10}{3}$ **27.** $\frac{1}{4}(e - 1)^2$

29. (a) $\frac{8}{3}$ **31.** $5 - \sqrt{3} - \sqrt{2}$

Problem Set 16.3

1. $\frac{3}{4}$ **3.** 240 **5.** $\frac{1}{2}\left(e^{27} - e\right)$ **7.** $-\sqrt{2}/(2\pi)$

9. $(3\ln 2 - \pi)/9$ **11.** $\frac{16}{3}$ **13.** 0 **15.** $\frac{27}{70}$

17. $4\tan^{-1}2 - \ln 5$ **19.** 6 **21.** 20 **23.** 10

25. $\frac{4}{15}$ **27.** $-\frac{1}{2}\ln(\cos 1)$ **29.** 3π

31. $\displaystyle\int_0^1\int_y^1 f(x, y)\,dx\,dy$ **33.** $\displaystyle\int_0^1\int_{y^4}^{\sqrt{y}} f(x, y)\,dx\,dy$

35. $\displaystyle\int_{-1}^0\int_{-x}^1 f(x, y)\,dy\,dx + \int_0^1\int_x^1 f(x, y)\,dy\,dx$ **37.** $\frac{256}{15}$

39. $\frac{1}{3}(1 - \cos 8)$ **41.** $15\pi/4$

43. approximately 4,133,000 ft^3

Problem Set 16.4

1. $\frac{1}{12}$ **3.** $\frac{4}{9}$ **5.** $2\sqrt{3} + \frac{4}{3}\pi \approx 7.653$ **7.** $\pi a^2/8$

9. $8\pi + 6\sqrt{3} \approx 35.525$ **11.** $\pi(e^4 - 1) \approx 168.384$

13. $(\pi \ln 2)/8 \approx 0.272$ **15.** $\pi(2 - \sqrt{3})/2 \approx 0.421$

17. $\frac{1}{12}$ **19.** $81\pi/8 \approx 31.809$

21. $625(3\sqrt{3} + 1)/12 \approx 322.716$ **23.** $\frac{2}{3}\pi d^2(3a - d)$

25. $\frac{2}{9}a^3(3\pi - 4)$

Problem Set 16.5

1. $m = 30; \bar{x} = 2; \bar{y} = 1.8$

3. $m = \pi/4; \bar{x} = \pi/2; \bar{y} = 16/(9\pi)$

5. $m \approx 0.1056; \bar{x} \approx 0.281; \bar{y} \approx 0.581$

7. $m = 32/9; \bar{x} = 0; \bar{y} = 6/5$

9. $I_x \approx 269; I_y \approx 5194; I_z \approx 5463$

11. $I_x = I_y = 5a^5/12; I_z \approx 5a^5/6$

13. $\bar{r} = \sqrt{5/12}\,a \approx 0.6455a$ **15.** $I_x = \pi\delta a^4/4; \bar{r} = a/2$

17. $5\pi\delta a^4/4$ **19.** $\bar{x} = 0, \bar{y} = (15\pi + 32)a/(6\pi + 48)$

21. **(a)** a^3;   **(b)** $7a/12$;   **(c)** $11a^5/144$

23. $I_x = \pi k a^4/2, I_y = 17k\pi a^4/2, I_z = 9\pi k a^4$

Problem Set 16.6

1. $\sqrt{61}/3$ **3.** $\pi/3$ **5.** $9\sin^{-1}\left(\frac{2}{3}\right)$ **7.** $8\sqrt{2}$

9. $4\pi a(a - \sqrt{a^2 - b^2})$ **11.** $2a^2(\pi - 2)$

13. $\frac{1}{6}\pi a^2(5\sqrt{5} - 1)$ **15.** $\bar{x} = \bar{y} = 0, \bar{z} = \frac{1}{2}(h_1 + h_2)$

17. $A = \pi b^2, B = 2\pi a^2[1 - \cos(b/a)], C = \pi b^2,$

$\quad D = \pi b^2[2a/(a + \sqrt{a^2 - b^2})], B < A = C < D$

Problem Set 16.7

1. -40 **3.** $\frac{189}{2}$ **5.** $\frac{2}{3}$ **7.** 156

9. $\displaystyle\int_0^1 \int_0^3 \int_0^{(1/6)(12-3x-2y)} f(x, y, z)\, dz\, dy\, dx$

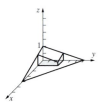

11. $\displaystyle\int_0^2 \int_0^4 \int_0^{y/2} f(x, y, z)\, dx\, dy\, dz$

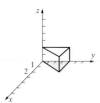

13. $\displaystyle\int_0^{12/5} \int_{x/3}^{(4-x)/2} \int_0^{4-x-2z} f(x, y, z)\, dy\, dz\, dx$

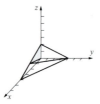

15. $\displaystyle\int_0^3 \int_{2x/3}^{(9-x)/3} \int_0^{(18-2x-6y)/9} f(x, y, z)\, dz\, dy\, dx$

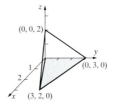

17. $\displaystyle\int_1^4 \int_0^1 \int_0^{\sqrt{1-z^2}} f(x, y, z)\, dy\, dz\, dx$

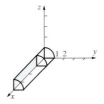

19. $\frac{128}{15}$ **21.** $4\displaystyle\int_0^1 \int_{x^2}^1 \int_0^{\sqrt{y}} dz\, dy\, dx = 2$

23. $\bar{x} = \bar{y} = \bar{z} = \frac{4}{15}$ **25.** $\bar{x} = \bar{y} = \bar{z} = 3a/8$

27. $\displaystyle\int_0^1 \int_0^{\sqrt{1-x^2}} \int_0^{\sqrt{1-x^2-y^2}} f(x, y, z)\, dz\, dy\, dx$

29. $\displaystyle\int_0^2 \int_0^{2-z} \int_0^{9-x^2} f(x, y, z)\, dy\, dx\, dz$ **31.** 4

33. Ave $T = 29.54, \bar{z} = \frac{11}{24}$ **35.** $\frac{8}{15}abc(ab + ac + bc)$

Problem Set 16.8

1. 8π **3.** $2\pi(5\sqrt{5} - 4)/3 \approx 15.038$

5. $\bar{x} = \bar{y} = 0; \bar{z} = \frac{16}{3}$ **7.** $k\pi(b^4 - a^4)$

9. $\bar{x} = \bar{y} = 0; \bar{z} = 2a/5$ **11.** $k\pi^2 a^6/16$ **13.** $\pi/9$

15. $\pi/32$ **17.** **(a)** $3a/4$;   **(b)** $3\pi a/16$;   **(c)** $6a/5$

19. **(a)** $3\pi a \sin\alpha/16\alpha$;   **(b)** $3\pi a/16$

21. $(a + b)(c - 1)/(c + 1)$

Chapter Review 16.9

Concepts Test

1. True   **3.** True   **5.** True   **7.** False
9. True   **11.** True   **13.** False   **15.** True

Sample Test Problems

1. $\frac{1}{24}$ **3.** $\frac{2}{3}$ **5.** $\displaystyle\int_0^1 \int_0^y f(x, y)\, dx\, dy$

7. $\int_0^{1/2} \int_0^{1-2y} \int_0^{1-2y-z} f(x, y, z)\, dx\, dz\, dy$

9. (a) $8\int_0^a \int_0^{\sqrt{a^2-x^2}} \int_0^{\sqrt{a^2-x^2-y^2}} dz\, dy\, dx$;

(b) $8\int_0^{\pi/2} \int_0^a \int_0^{\sqrt{a^2-r^2}} r\, dz\, dr\, d\theta$;

(c) $8\int_0^{\pi/2} \int_0^{\pi/2} \int_0^a \rho^2 \sin\phi\, d\rho\, d\phi\, d\theta$

11. 0.8857 **13.** $\bar{x} = \frac{13}{6}; \bar{y} = \frac{3}{2}$ **15.** 6

17. $80\pi k$ **19.** $ka^2bc/24$

Problem Set 17.1

1.   **3.**

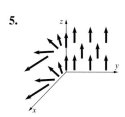

5.

7. $(2x - 3y)\mathbf{i} - 3x\mathbf{j} + 2\mathbf{k}$ **9.** $x^{-1}\mathbf{i} + y^{-1}\mathbf{j} + z^{-1}\mathbf{k}$

11. $e^y \cos z\, \mathbf{i} + xe^y \cos z\, \mathbf{j} - xe^y \sin z\, \mathbf{k}$ **13.** $2yz; z^2\mathbf{i} - 2y\mathbf{k}$

15. $0; 0$ **17.** $2e^x \cos y + 1; 2e^x \sin y\, \mathbf{k}$

19. (a) Meaningless;   **(b)** vector field;   **(c)** vector field;
(d) scalar field;   **(e)** vector field;   **(f)** vector field;
(g) vector field;   **(h)** meaningless;   **(i)** meaningless;
(j) scalar field;   **(k)** meaningless.

25. (a) div $\mathbf{F} = 0$, div $\mathbf{G} < 0$, div $\mathbf{H} = 0$, div $\mathbf{L} > 0$;

(b) clockwise for $\mathbf{H}$, not at all for others.

(c) div $\mathbf{F} = 0$, curl $\mathbf{F} = \mathbf{0}$, div $\mathbf{G} = -2ye^{-y^2}$, curl $\mathbf{G} = \mathbf{0}$, div $\mathbf{H} = 0$, curl $\mathbf{H} = -2xe^{-x^2}\mathbf{k}$, div $\mathbf{L} = 1/\sqrt{x^2 + y^2}$, curl $\mathbf{L} = \mathbf{0}$

Problem Set 17.2

1. $14(2\sqrt{2} - 1)$ **3.** $2\sqrt{5}$ **5.** $\frac{1}{6}(14\sqrt{14} - 1)$

7. $\frac{100}{3}$ **9.** 144 **11.** 0 **13.** $\frac{17}{6}$ **15.** 19

17. $k(17\sqrt{17} - 1)/6$ **19.** $-\frac{7}{44}$ **21.** $-\frac{1}{2}(a^2 + b^2)$

23. $2 - 2/\pi$ **25.** 2.25 gal   **27.** $2\pi a^2$ **29.** $4a^2$

31. (a) 27;   **(b)** $-297/2$

Problem Set 17.3

1. $f(x, y) = 5x^2 - 7xy + y^2 + C$ **3.** Not conservative.

5. $f(x, y) = \frac{2}{5}x^3 y^{-2} + C$ **7.** $f(x, y) = 2xe^y - ye^x + C$

9. $f(x, y, z) = x^3 + 2y^3 + 3z^3 + C$ **11.** 14 **13.** 6

15. $-\pi$ **19.** $f(x, y, z) = \frac{1}{2}k(x^2 + y^2 + z^2)$

21. $\int_C \mathbf{F} \cdot d\mathbf{r} = \int_a^b m\mathbf{r}''(t) \cdot \mathbf{r}'(t)\, dt$

$= \frac{1}{2}m\int_a^b (d/dt)[\mathbf{r}'(t) \cdot \mathbf{r}'(t)]\, dt = \frac{1}{2}m\int_a^b (d/dt)|\mathbf{r}'(t)|^2\, dt$

$= \left[\frac{1}{2}m|\mathbf{r}'(t)|^2\right]_a^b = \frac{1}{2}m[|\mathbf{r}'(b)|^2 - |\mathbf{r}'(a)|^2]$

23. $f(x, y, z) = -gmz$

Problem Set 17.4

1. $-\frac{64}{15}$ **3.** $\frac{72}{35}$ **5.** 0 **7.** $\frac{8}{3}$

9. (a) 0;   **(b)** 0 **11. (a)** 0;   **(b)** 0 **13.** 50

15. -2 **19. (c)** M and N have a discontinuity at $(0, 0)$.

23. $3\pi a^2/8$ **27. (a)** div $\mathbf{F} = 4$;   **(b)** 144

29. (a) div $\mathbf{F} < 0$ in quadrants I and III;
div $\mathbf{F} > 0$ in quadrants II and IV;

(b) $0; -2(1 - \cos 3)^2$

Problem Set 17.5

1. $8\sqrt{3}/3$ **3.** $2 + \pi/3$ **5.** $5\pi/8$ **7.** 6 **9.** 2

11. 20 **13.** $\sqrt{3}ka^4/12$ **15.** $\bar{x} = \bar{y} = \bar{z} = a/3$

17. (a) 0;   **(b)** 0;   **(c)** $4\pi a^4$;   **(d)** $4\pi a^4/3$;   **(e)** $8\pi a^4/3$

19. (a) $4k\pi a^3$;   **(b)** $2k\pi a^3$;   **(c)** $hk\pi a(a + h)$

Problem Set 17.6

1. 0 **3.** $3a^2b^2c^2/4$ **5.** $64\pi/3$ **7.** 4π **9.** 1176π

11. $\nabla \cdot \mathbf{F} = 3$ and so $\iint\limits_{\partial S} \mathbf{F} \cdot \mathbf{n}\, dS = \iiint\limits_S 3\, dV = 3V(S)$.

15. (a) $20\pi/3$;   **(b)** 4π;   **(c)** $16\pi/3$;   **(d)** 1;
(e) 36;   **(f)** $12\pi/5$;   **(g)** $32\pi \ln 2$

Problem Set 17.7

1. 0 **3.** -2 **5.** -48π **7.** 8π **9.** 2 **11.** $\pi/4$

15. $1/3$ **17.** $\frac{4}{3}a^2$ joules

Chapter Review 17.8

Concepts Test

1. True   **3.** False   **5.** True   **7.** False
9. True   **11.** True.

Sample Test Problems

3. curl $(f\nabla f) = f\,\text{curl}(\nabla f) + \nabla f \times \nabla f = \mathbf{0} + \mathbf{0} = \mathbf{0}$

5. (a) $\pi/4$;   **(b)** $(3\pi - 5)/6$ **7.** 47

9. (a) $\frac{1}{2}$;   **(b)** $\frac{4}{3}$;   **(c)** 0 **11.** 6π **13.** 0

15. $9\pi(3a - 2)/\sqrt{a^2 + b^2 + 1}$

Problem Set 18.1

1. $y = C_1e^{2x} + C_2e^{3x}$ **3.** $y = \frac{1}{2}e^x - \frac{1}{2}e^{-7x}$

5. $y = (C_1 + C_2x)e^{2x}$ **7.** $y = e^{2x}(C_1e^{\sqrt{3}x} + C_2e^{-\sqrt{3}x})$

9. $y = 3\sin 2x + 2\cos 2x$

11. $y = e^{-x}(C_1 \cos x + C_2 \sin x)$

13. $y = C_1 + C_2x + C_3e^{-4x} + C_4e^x$

15. $y = C_1e^x + C_2e^{-x} + C_3 \cos 2x + C_4 \sin 2x$

17. $y = D_1 \cosh 2x + D_2 \sinh 2x$

19. $y = e^{-x/2}[(C_1 + C_2 x)\cos(\sqrt{3}/2)x$
$\qquad + (C_3 + C_4 x)\sin(\sqrt{3}/2)x]$

21. $y = (C_1 + C_2 \ln x)x^{-2}$

27. $y = 0.5e^{5.16228x} + 0.5e^{-1.162278x}$

29. $y = 1.29099e^{-0.25x}\sin(0.968246x)$

Problem Set 18.2

1. $y = C_1 e^{3x} + C_2 e^{-3x} - \frac{1}{9}x$

3. $y = (C_1 + C_2 x)e^x + x^2 + 5x + 8$

5. $y = C_1 e^{2x} + C_2 e^{3x} + \frac{1}{2}e^x$

7. $y = C_1 e^{-3x} + C_2 e^{-x} - \frac{1}{2}xe^{-3x}$

9. $y = C_1 e^{2x} + C_2 e^{-x} - \frac{3}{5}\sin x + \frac{1}{5}\cos x$

11. $y = C_1 \cos 2x + C_2 \sin 2x + \frac{1}{2}x\sin 2x$

13. $y = C_1 \cos 3x + C_2 \sin 3x + \frac{1}{8}\sin x + \frac{1}{13}e^{2x}$

15. $y = e^{2x} - e^{3x} + e^x$ **17.** $y = C_1 e^{2x} + C_2 e^x + \frac{5}{2}x + \frac{19}{4}$

19. $y = C_2 \sin x + C_3 \cos x - x\sin x - \cos x \ln|\sin x|$

21. $y = D_1 e^x + D_2 e^{2x} + (e^x + e^{2x})\ln(1 + e^{-x})$

Problem Set 18.3

1. $y = 0.1\cos 5t; 2\pi/5$ **3.** 0.5 m/s

5. $y \approx e^{-0.16t}(\cos 8t + 0.02\sin 8t)$ **7.** 14.4 s

9. $Q = 10^{-6}(1 - e^{-t})$

11. (a) $Q = 2.4 \times 10^{-4}\sin 377t;$
(b) $I = 9.05 \times 10^{-2}\cos 377t$

13. $I \approx 12 \times 10^{-2}\sin 377t$ **17.** $d^2\theta/dt^2 = -(g/L)\sin\theta$

Chapter Review 18.5

Concepts Test

1. False   **3.** True   **5.** True   **7.** True   **9.** False

Sample Test Problems

1. $y = \frac{1}{4}e^x + C_1 e^{-3x} + C_2$ **3.** $y = 3e^{2x} - 3e^x$

5. $y = C_1 e^x + C_2 e^{-x} - 1$ **7.** $y = (C_1 + C_2 x + \frac{1}{2}x^2)e^{-2x}$

9. $y = e^{-3x}(C_1 \cos 4x + C_2 \sin 4x)$

11. $y = C_1 + C_2 e^{-4x} + C_3 e^{2x}$

13. $y = (C_1 + C_2 x)e^{\sqrt{2}x} + (C_3 + C_4 x)e^{-\sqrt{2}x}$

15. $y = -\cos 4t; 1; \pi/2$ **17.** $I = e^{-t}\sin t$

Problem Set A.1

9. $N = 4$ **11.** $N = 5$ **13.** $P_5, P_7, P_9,\ldots$ are true

15. $P_{30}, P_{29}, P_{28},\ldots$ are true   **17.** P_i is true for all $i \geq 1$

19. P_i is true for all $i \geq 1$

21. True for $n = 1, 3, 5,\ldots$. Proof is by induction.

23. True for all $n \geq 3$. Proof is by induction.

25. True for all $n \geq 2$. Proof is by induction.

27. True for all $n \geq 0$. Proof is by induction.

Photo Credits

Front Endpapers

Descartes	Frans Hals/Louvre
Newton	Library of Congress
Leibniz	The Granger Collection, New York
Euler	Library of Congress
Kepler	Library of Congress
Pascal	Courtesy of International Business Machines Corporation. Unauthorized use not permitted.
L'Hôpital	The Granger Collection, New York
Bernoulli	The Granger Collection, New York
Lagrange	New York Public Library
Gauss	The Granger Collection, New York
Cauchy	Corbis
Riemann	The Granger Collection, New York
Lebesgue	The Granger Collection, New York
Agnesi	Library of Congress
Weierstrass	Corbis
Kovalevsky	Library of Congress
Gibbs	Corbis

Text

p. 219	Kennedy Space Center/NASA
p. 224	Susan Van Etten/PhotoEdit
p. 362	David Frazier/Photo Researchers, Inc.
p. 418	Scala/Art Resource, NY

TEACHING OUTLINES

The material in this section summarizes the content of each section (except the Chapter Review and Additional Problems sections), gives the theorems, and points out what should be remembered for future use. Specifically, we give for each section:

1. *Topics to Cover*   The main topics are listed, along with suggestions on what to emphasize.

2. *Theorems*   The titles (if any) of the theorems are given, along with a statement regarding our proof. Because many of the seemingly obvious theorems have difficult proofs, we do not prove all of the theorems in the book. In most cases, when we do not give a proof we give an intuitive explanation of the theorem's result and we *say that we are not going to prove the theorem*. In many cases, we prove only one or two parts of a multipart theorem. There are several instances where students are asked to prove the theorem in the exercises.

3. *Homework*   We suggest a homework assignment of moderate length. Often we give problems in pairs, an odd and an even problem. This way, students can attempt the odd problem and check their answer in the back of the book before they attempt an even problem for which they will not have the answer.

4. *Things to Remember*   It isn't always apparent to a student, or to a beginning teacher or teaching assistant, what material will be needed later in the book. We try to point out what topics will be needed and why they will be needed.

For each Technology Project, we state the objectives of the project and we give the technology required to complete the project.

Experienced teachers may ignore much of the notes in this section. We hope that some of this material, such as the listing of theorems and their disposition, will be useful even to the most experienced teachers. To beginning teachers and teaching assistants the topics to cover, and suggested homework may be helpful.

Students learn by doing. In calculus, this means working homework problems and writing their solutions. We encourage you to assign and collect homework assignments. Quick return of carefully graded homework can help students assess their progress in your course. You should demand quality writing from your students. Expect them to define their variables, honor the equal sign (i.e. say things are equal when they are, and don't say things are equal when they aren't), align the equal signs, explain their reasoning (in English!), and use parentheses when appropriate. Your work on the chalkboard can be a model for what you can expect of students.

Emphasize continually the connection between geometry and algebra. Try to see every result from both perspectives. For example the graph of $y = x^2$ increases without bound as x becomes large (geometric interpretation) and the quantity x^2 can be made large by taking the value of x to be large (algebraic interpretation).

CHAPTER 1   PRELIMINARIES

1.1   The Real Number System

Topics to Cover   Number systems, including integers, rational numbers, and real numbers. Logic: "*P* implies *Q*," the converse and contrapositive. Order on the real line. Quantifiers "for all ..." and "there exists ..." Students should know most of the material in this section, although for many, the material on logic, or at least the terms converse and contrapositive, will be new.

Theorems   None

Homework   1, 2, 7, 8, 21, 22, 33, 34, 41 , 42, 55

Things to Remember   Properties of real numbers will be used throughout the course. Understanding the logic ("if then," contrapositive, converse) is key to understanding the theorems in the book.

1.2   Decimals, Calculators, Estimation

Topics to Cover   Repeating and nonrepeating decimals. Calculators. Estimating answers.

Theorems   None

Homework   1, 2, 9, 10, 13, 16, 21, 22, 30, 31

Things to Remember   The ability to make quick mental estimates of an answer will come in handy throughout the course. Emphasize the "reasonableness check" to avoid obviously wrong answers.

1.3   Inequalities

Topics to Cover   Interval notation for closed, open, and half-open intervals. Solving inequalities.

Theorems   None

Homework   1–4, 13, 14, 27, 28, 33, 35

Things to Remember   Interval notation will be used throughout the course. The ability to work with inequalities like $2.9 < 1/x < 3.1$ will come in handy in Chapters 2 and 3.

1.4   Absolute Values, Square Roots, Squares

Topics to Cover   Absolute value function (emphasize that $|b - a|$ represents the distance between a and b on the number line). Roots and the Quadratic Formula. Example 3 prepares students for the definition of the limit.

Theorems   None

Homework   1–4, 15–18, 21, 22, 25, 26, 31

Things to Remember   Emphasize that $|b - a|$ represents the distance between a and b on the number line

1.5   The Rectangular Coordinate System

Topics to Cover   Cartesian coordinates, the four quadrants, the Distance Formula, the equation of a circle, midpoint formula. Students will have seen many of the topics in this section. This section may be covered quickly or you may assign some problems and have the students read the section on their own.

Theorems   None

Homework   1, 2, 7, 8, 11–16, 21, 22, 31, 35, 36 (challenging)

Things to Remember   Cartesian coordinates are used throughout the book

1.6   The Straight Line

Topics to Cover   Line, slope of a line, point-slope and slope-intercept form for the equation of a line, parallel and perpendicular lines. Again, many students will have seen this material so this section may be covered quickly or you may assign some problems and have the students read the section on their own.

Theorems   None

Homework   1, 2, 9–12, 21, 27, 28, 31, 32, 39, 40, 47

Things to Remember   Lines are one of the simplest geometrical objects that one can study. In Section 3.10 students will encounter the linear approximation to a function. In that section we emphasize that many functions will look linear when you zoom in on them.

1.7   Graphs of Equations

Topics to Cover   Graphing points, symmetry. Using technology (e.g., a graphing calculator or a computer algebra system) to plot equations. Shapes of quadratic and cubic graphs (Figure 6). Intersections of graphs. Students may have seen much of the material in this section. It may be helpful to point out that a graphing calculator or a CAS will graph a function in much the same way that we suggest in the text: make a table and connect the points. Point out the inherent limitations in this method, e.g., with the function $f(x) = 1/x$ on the interval $[-2, 2]$.

Theorems   None

Homework   1–6, 18–21, 29–32, 39

Things to Remember Graphing equations will be helpful throughout the course. Even with a graphing calculator, it is worthwhile to make some plots by hand; doing this often gives you some insight into the nature of the graph.

Technology Project 1.1 Graphing

Objectives Probably the most important task that a graphing calculator or a CAS can do is to graph equations. This project asks students to become familiar with their technology and to graph a number of equations. Students are asked to examine the effect of a parameter on the graph of an equation.

Technology Required Graphing calculator or CAS.

Technology Project 1.2 Solving Equations by Zooming

Objectives Students often view an exercise like "solve $2x^2 + x - 2 = 0$" as a purely algebraic problem. This project emphasizes the connection between algebra and geometry. The solution to this equation is seen to be the point where the graph of $y = 2x^2 + x - 2$ crosses the x-axis. By zooming in on the root, students should see this connection and they can find the root with reasonable accuracy.

Technology Required Graphing calculator or CAS.

CHAPTER 2   FUNCTIONS AND LIMITS

2.1   Functions and Their Graphs

Topics to Cover Definition of a function. Domain and range. Graphs of functions. Symmetry, even and odd functions.

Theorems None

Homework 3, 4, 7–14, 15–20, 31, 38, 39 (Problems 38 and 39 will get students thinking in terms of accumulation functions, i.e., functions defined as area under a curve. We emphasize accumulation functions in Chapter 5 on the definite integral.)

Things to Remember Functions play a key role in calculus. For the first 12 chapters, the reader sees almost exclusively real-valued functions of one variable. Point out that the idea of a function can be generalized to include cases where the input is an ordered pair (or ordered n-tuple) and to cases where the output is an ordered pair (or ordered n-tuple).

2.2   Operations on Functions

Topics to Cover Adding, subtracting, multiplying, dividing functions. Composition of functions. Translations. Types of functions: constant, identity, polynomial, rational.

Theorems None

Homework 1, 2, 9, 10, 11–14, 17, 18, 23, 29, 34

Things to Remember In Section 3.5 when we cover the Chain Rule, students will need to understand function composition. Problems like 13 and 14 will help students in Section 3.5. Notation like $f^2(x)$ will be used throughout the book.

2.3   The Trigonometric Functions

Topics to Cover Definition of sine and cosine function. Properties and graphs of sine and cosine. Period and amplitude of sine and cosine functions. Modeling with the sine and cosine function. Other trigonometric functions: tangent, cotangent, secant, and cosecant functions. Most students will have seen the trigonometric functions, at least in their angle form. For some students, radians will be new. Emphasize that in calculus we will almost *always* deal with radian measure for angles.

Theorems None. The table on page 56 lists a number of trigonometric identities. These identities can also be found on the tear-out sheet from the back of the book.

Homework 1, 2, 8–11, 14, 15, 16, 19, 24, 25, 26, 35, 36, 40–43.

Things to Remember Students should understand the six trigonometric functions with radian measure. Problems 40 (tangent of the angle between two lines) and 42 (area of a circular sector) are referred to later in the text.

2.4   Introduction to Limits

Topics to Cover An intuitive notion of limit. Graphical and tabular approaches to determining the limit. Left and right hand limits

Theorems

Theorem A (A limit exists if and only if both left and right hand limits exist)

The proof is not given.

Homework 1–4, 9–14, 19, 20, 29, 30, 33, 34

Things to Remember   There are no specific results that need to be remembered. This section is a warm-up for the next section. After the two sections, the students should begin to understand the concept of limit, although reinforcement throughout the term (e.g., the derivative is the *limit* of the difference quotient) is helpful.

2.5   Rigorous Study of Limits

Topics to Cover   Definition of the limit (Figure 3). Limit proofs (Optional).

Theorems   None

Homework   1, 2, 7–10, 20 (a squeeze theorem), 23, 27

Things to Remember   Calculus is the study of limits, so an understanding of the concept of limit will be beneficial throughout the calculus sequence.

2.6   Limit Theorems

Topics to Cover   Theorem A gives 9 statements regarding the limit of a sum, difference, product, etc. This theorem is referred to a number of times throughout the book. The proof of Theorem A is optional. Evaluating limits using the Main Limit Theorem. The Squeeze Theorem.

Theorems

Theorem A   Main Limit Theorem

Parts 1–5 are proved in the text. Problems 31 and 32 ask the student to prove parts 6 and 7. The others are proved in the Appendix.

Theorem B   Substitution Theorem

The proof of Theorem B follows from repeated applications of Theorem A.

Theorem C   Squeeze Theorem

An ε–δ proof is given, but it is optional.
All theorems in this section are also valid for right- and left-hand limits.

Homework   1–10, 13–16, 23, 24, 27–30 (these prepare students for the idea of the derivative), 37, 38

Things to Remember   Use the Main Limit Theorem to evaluate limits.

2.7   Limits Involving Trigonometric Functions

Topics to Cover   Limits of trigonometric functions. Combining the results of this section with the Main Limit Theorem from Section 2.6.

Theorem

Theorem A   Limits of Trigonometric Functions

Proofs are given for limits of the sine and cosine functions. The proofs of the others are left as exercises (Problems 15 and 16).

Theorem B   Special Trigonometric Limits $(\lim_{t \to 0} \dfrac{\sin t}{t} = 1$ and $\lim_{t \to 0} \dfrac{1 - \cos t}{t} = 0)$

Both statements are proved using a geometric argument.

Homework   1–6, 15, 17

Things to Remember   Theorem A says, in essence, that the trigonometric functions are continuous on their domains, a result that will covered in Section 2.9. The special trigonometric limits will be needed in the next chapter when we discuss the derivatives of sine and cosine.

2.8   Limits at Infinity, Infinite Limits

Topics to Cover   Definition of limits as $x \to \infty$ or $x \to -\infty$ and examples. Definitions of $\lim_{x \to c+} f(x) = \infty$ and $\lim_{x \to c+} f(x) = -\infty$. Asymptotes.

Theorems   None

Homework   1–8, 15–18, 20, 37–40, 43, 44, 49

Things to Remember   Chapter 9 on l'Hôpital's rule and improper integrals uses the ideas of limits at infinity. Asymptotes are helpful in graphing a function.

2.9   Continuity of Functions

Topics to Cover   Definition of continuity at a point, continuity of polynomial and rational functions, continuity under function operations. Definition of continuity on an open interval and on a closed interval. Intermediate Value Theorem.

Theorems

Theorem A   Continuity of Polynomial and Rational Functions

The proofs follow directly from Theorem 2.6B.

Theorem B   Continuity of Absolute Value and nth Root Functions

The proofs follow directly from Theorem 2.6A.

Theorem C   (f and g continuous at c implies kf, $f + g$, $f - g$, $f \cdot g$, and f/g as long as $g(x) \neq 0$ are continuous at c)

A proof is given for the product. The proofs of the others follow from Theorem 2.6A.

Theorem D (Continuity of Trigonometric Functions)

The proof is given for the sine, cosine, and tangent. A similar argument applies to the other trigonometric functions.

Homework   1–8, 16, 17, 18–21, 24–27, 38, 39

Things to Remember   The concept of continuity.

Technology Project   2.1   Shifting and Scaling the Graph of a Function

Objectives   Students will do enough graphical experiments to understand the relationships among the graphs of $y = f(x)$ and $f(x) + a$, $f(x + a)$, $bf(x)$, and $f(bx)$.

Technology Required   Graphing calculator or CAS.

Technology Project 2.2   Limits

Objectives   Students will use graphical, tabular, and algebraic approaches to finding limits. This project can be assigned after Section 2.4, but before Section 2.5.

Technology Required   Graphing calculator CAS. The software must be capable of plotting and factoring polynomials.

CHAPTER 3   THE DERIVATIVE

3.1   Two Problems with One Theme

Topics to Cover   Slope of the tangent line. Average and instantaneous velocity. Rate of change

Theorems   None

Homework   1–4, 7, 8, 13, 14, 17–21, 24, 25

Things to Remember   Instantaneous rate of change is the limit of a quotient.

3.2   The Derivative

Topics to Cover   Definition of the derivative, examples of finding derivatives from the definition. Equivalent forms of the derivative. A differentiable function must be continuous, but not conversely.

Theorems

Theorem A   Differentiability Implies Continuity

The proof is given.

Homework   1, 2, 5–10, 27–30, 37–40, 44, 45

Things to Remember   The derivative is an instantaneous rate of change, that is, a limit of a difference quotient.

3.3   Rules for Finding Derivatives

Topics to Cover   Rules for finding derivatives: power, sum, difference, product, and quotient rules. Examples.

Theorems

Theorem A   Constant Function Rule

Theorem B   Identity Function Rule

Theorem C   Power Rule (Positive integral exponents)

Theorem D   Constant Multiple Rule

Theorem E   Sum Rule (Derivative of a sum is the sum of derivatives.)

Theorem F   Difference Rule (Derivative of a difference is the difference of derivatives.)

Theorem G   Product Rule

Theorem H   Quotient Rule

Theorems A, B, C, D, E, G, and H are proved. The proof of theorem F is left as an exercise (Problem 54).

Homework   1–10, 21–28, 37–40, 45–50

Things to Remember   All of the rules of this section are needed in the rest of the book. Despite all of these rules, don't let students forget that the derivative is still the limit of a quotient, and therefore is an instantaneous rate of change.

3.4   Derivatives of Trigonometric Functions

Topics to Cover   Derivative formulas for sine and cosine functions (derivations require Theorem 2.7B Special trigonometric limits). Derivative formulas for tangent, cotangent, secant cosecant (derivations require rules for derivatives from Section 3.3)

Theorems

Theorem A   (Derivatives of sine and cosine)

The derivations of these two formulas are given before the theorem is stated.

Theorem B (Derivatives of tangent, cotangent, secant and cosecant)

We say that the formulas for these functions can be obtained by applying the quotient rule, but the details are left as exercises (Problems 5–8).

Homework 1–8, 11–13, 15–18, 21, 22, 24, 28

Things to Remember Although the rules for derivatives of sine and cosine are easy to remember, and are certainly the most important, the derivatives of the other four trigonometric functions should also be committed to memory.

3.5 The Chain Rule

Topics to Cover Operators and the D_x notation. Chain rule using prime notation and operator notation. Examples. Applying the chain rule more than once.

Theorems

Theorem A Chain Rule

A partial proof of the chain rule is given in Section 3.6. A complete proof is given in the appendix.

Homework 1–6, 9, 10, 23, 24, 27–30, 33–36, 41, 42

Things to Remember The Chain Rule will be needed for almost every derivative from here on out. Students should know the Chain Rule so well, that they can use it in reverse when you get to substitution for indefinite integrals (Chapter 5).

3.6 Leibniz Notation

Topics to Cover Increments, Δx and Δy, the difference quotient as $\Delta y/\Delta x$. Leibniz's dy/dx notation for the derivative. The Chain Rule in Leibniz notation. A partial proof of the Chain Rule.

Theorems None

Homework 1–12, 17, 18, 21, 22, 29–32

Things to Remember Leibniz notation is nothing new, just a new way of indicating the derivative.

3.7 Higher-Order Derivatives

Topics to Cover Definitions of second, third, and higher-order derivatives. Prime, operator, Leibniz notation for higher-order derivatives. Velocity and acceleration as first and second derivatives, respectively, of position. Mathematical modeling with derivatives.

Theorems None

Homework 1–6, 9, 10, 17–20, 23, 24, 33, 34, 39

Things to Remember Definitions of higher-order derivatives and their representations in prime, operator and Leibniz notation. Acceleration is the rate of change of velocity, which is the rate of change of position. Therefore, acceleration is the second derivative of position.

3.8 Implicit Differentiation

Topics to Cover Functions defined implicitly. Differentiation of functions defined implicitly. Examples. The Power Rule for rational exponents.

Theorems

Theorem A Power Rule (Rational exponents)

The proof, using implicit differentiation, is given.

Homework 1–6, 11–14, 19, 20, 35, 36, 43, 44

Things to Remember Implicit differentiation is needed in the next section on related rates, in Chapter 7 where we derive formulas for the derivatives of the inverse trigonometric functions, and in a few other places.

3.9 Related Rates

Topics to Cover Rates of change. Related rates. Examples. The systematic procedure.

Theorems None

Homework 1–4, 7, 8, 11, 12, 15, 16, 19, 20

Things to Remember Related rates is not used much in the remainder of the book, but the method is a good application of differentiation and it reinforces the idea that the derivative is a rate of change.

3.10 Differentials and Approximations

Topics to Cover Differentials. Relation to differentiation. Approximations. Absolute and relative errors.

Theorems None

Homework 1, 2, 5, 6, 10, 11, 17, 21, 22

Things to Remember Students should remember to distinguish between derivatives and differentials. They will often write dy when they mean dy/dx.

Technology Project 3.1 Secant and Tangent Lines

Objectives   Students will use technology to compute the difference quotients related to derivatives, thereby gaining some understanding of the derivative as a rate of change.

Technology Required   Graphing calculator or CAS.

Technology Project 3.2   Linear Approximation to a Function

Objectives   Students will investigate how closely a linear approximation to a function approximates the function.

Technology Required   Graphing calculator or CAS.

CHAPTER 4   APPLICATIONS OF THE DERIVATIVE

4.1   Maxima and Minima

Topics to Cover   Definitions of maximum, minimum, extremum, objective function, critical point, stationary point, singular point. Examples.

Theorems

Theorem A   Max-Min Existence Theorem

The proof is difficult, so it is omitted.

Theorem B   Critical Point Theorem

The proof is given.

Homework   1–4, 13–16, 19, 20, 23, 24, 29, 30

Things to Remember   Definitions

4.2   Monotonicity and Concavity

Topics to Cover   Definitions of increasing, decreasing, monotonic, concavity. Examples. Theorems on monotonicity and concavity.

Theorems

Theorem A   Monotonicity Theorem

The proof is given in Section 4.7.

Theorem B   Concavity Theorem

The proof follows directly from Theorem A

Homework   1–4, 11–14, 19, 20, 29, 30, 33, 34, 43, 49, 50

Things to Remember   Definitions. The maximum of a function is a number in its range; an inflection point is an ordered pair.

4.3   Local Maxima and Minima

Topics to Cover   Definitions of local maximum, local minimum, local extremum. Where do local extrema occur? First derivative test, second derivative test. Examples.

Theorems

Theorem A   First Derivative Test

The proof of part (i) is given. Parts (ii) and (iii) are similar.

Theorem B   Second Derivative Test

The proof is given.

Homework   1–4, 7–10, 15–18, 20, 21, 25, 26, 29

Things to Remember   First and second derivative tests for local extrema.

4.4   More Max-Min Problems

Topics to Cover   Extrema on open intervals. Examples. More examples. Least squares (optional).

Theorems   None

Homework   1, 2, 7, 8, 13, 14, 19, 20, 29, 30

Things to Remember   The important skill for this section is translating a problem to mathematical terms so we can use the methods of calculus.

4.5   Economic Applications

Topics to Cover   This section is optional. Definitions of economic terms. Mathematical models of economic phenomena. Marginal analysis. Examples.

Theorems   None

Homework   1–4, 13, 14, 16, 17

Things to Remember   Marginal means derivative.

4.6   Sophisticated Graphing

Topics to Cover Using information regarding increasing/decreasing, concave up/down, intercepts, max-min, inflection points, etc. to sketch the graph of a function. Horizontal, vertical, and oblique asymptotes.

Theorems None

Homework 1–6, 11, 12, 17, 18, 25, 29, 30, 34, 35, 41, 42

Things to Remember How the properties of a function (e.g., increasing, concave down, etc.) affect its graph.

4.7 The Mean Value Theorem

Topics to Cover The Mean Value Theorem for Derivatives. Proof and illustration of the theorem.

Theorems

> **Theorem A** Mean Value Theorem for Derivatives

The proof is given. The proof of the Monotonicity Theorem (Theorem 4.2A) is given in this section.

> **Theorem B** $(F'(x) = G'(x) \Rightarrow F(x) = G(x) + C)$

The proof is given.

Homework 1–6, 12–15, 20–23, 29, 30, 52, 53

Things to Remember The geometric interpretation of the Mean Value Theorem for Derivatives. Functions with the same derivative differ by a constant.

Technology Project 4.1 Reflection and Refraction of Light

Objectives Use the symbolic, numerical, and graphical capabilities of a CAS to solve some challenging max-min problems.

Technology Required CAS. The symbolic capabilities of a CAS will be needed in the derivation of (2). In Exercise 3, students will need some way to program Newton's method.

Technology Project 4.2 An Optimization Problem

Objectives This project leads students through the solution of some challenging max-min problems. The ladder-around-the-corner is counterintuitive in the sense that it seems like we get a maximum, when we set out to get a minimum.

Technology Required Graphing calculator.

CHAPTER 5 THE INTEGRAL

5.1 Antiderivatives (Indefinite Integrals)

Topics to Cover Definition of antiderivative. Power Rule.

Theorems

> **Theorem A** Power Rule
>
> **Theorem B** (Integrals of sine and cosine)
>
> **Theorem C** Indefinite Integral is a Linear Operator
>
> **Theorem D** Generalized Power Rule

All four theorems are proved. The proofs are simple: just take the derivative of the right-hand side and show that it is equal to the integrand.

Homework 1–8, 11, 12, 17–22, 27–30, 35, 36, 39, 40

Things to Remember Antidifferentiation is the inverse of differentiation. Always check your answer by differentiation.

5.2 Introduction to Differential Equations

Topics to Cover Definitions and examples. Separable equations. Problems involving velocity and acceleration.

Theorems None

Homework 1, 2, 5–10, 15–18, 21, 22, 31, 32

Things to Remember The solution of a differential equation is a *function*. Differential equations involve an unknown function and some of its derivatives.

5.3 Sums and Sigma Notation

Topics to Cover Sigma notation, linearity of $\sum$, collapsing sums, special sum formulas, proofs of special sum formulas (optional).

Theorems

> **Theorem A** Linearity of $\sum$

The proofs of the four parts are easy. A proof is given for part (i) only.

Homework   1–4, 7–18, 21, 22, 25, 26, 31, 32, 35, 41–44, 55

Things to Remember   Sigma notation will be used in Chapters 5 and 10.

5.4   Introduction to Area

Topics to Cover   Areas of rectangles, triangles, parallelograms, circles. Limits of area of an inscribed or circumscribed polygon.

Theorems   None

Homework   3–14, 17–19

Things to Remember   Area of a region with a curved boundary is the limit of inscribed or circumscribed rectangles.

5.5   The Definite Integral

Topics to Cover   Riemann sums. Definition of definite integral. $\int_a^b f(x)\,dx = A_{up} - A_{down}$. Integrability. Definite integral is additive over disjoint intervals.

Theorems

   Theorem A   Integrability Theorem

   Theorem B   Interval Additive Property

The proofs are difficult and are left to more advanced textbooks.

Homework   1, 2, 5–10, 13–16, 14–21, 24, 25

Things to Remember   The definite integral is the limit of a sum. Thinking of the integral this way is the key to understanding the applications that come in the next section.

5.6   The First Fundamental Theorem of Calculus

Topics to Cover   Accumulation functions, examples where the derivative of an accumulation function is the integrand. First Fundamental Theorem of Calculus. Comparison Property. Boundedness Property. Linearity of the Definite Integral.

Theorems

   Theorem A   First Fundamental Theorem of Calculus $\left(\dfrac{d}{dx}\int_a^x f(t)\,dt = f(x)\right)$

We first give a sketch of the proof. Once Theorems B, C, and D are established, a proof of the First Fundamental Theorem of Calculus is given. After the examples, the result of the theorem should seem reasonable.

   Theorem B   Comparison Property
   Theorem C   Boundedness Property
   Theorem D   Linearity of the Definite Integral

Proofs of Theorems B–D are given.

Homework   1–6, 11–20, 25, 26, 29, 31, 33–36

Things to Remember   The First Fundamental Theorem of Calculus relates the two major kinds of limits (the derivative and the integral) that students have studied so far.

5.7   The Second Fundamental Theorem of Calculus and the Mean Value Theorem for Integrals

Topics to Cover   Second Fundamental Theorem of Calculus. Examples. More examples. Mean Value Theorem for Integrals.

Theorems

   Theorem A   Second Fundamental Theorem of Calculus

The proof, which is not difficult, is given.

   Theorem B   Mean Value Theorem for Integrals

The proof is given.

Homework   1–6, 11–16, 21–24, 31–34, 43–46, 59, 60

Things to Remember   The definite integral is defined as the limit of a Riemann sum, but we evaluate most definite integrals using the Second Fundamental Theorem of Calculus.

5.8   Evaluating Definite Integrals

Topics to Cover   Substitution for indefinite and definite integrals. Examples. Using symmetry. Functions defined by a table.

Theorems

   Theorem A   Substitution Rule for Indefinite Integrals

   Theorem B   Substitution Rule for Definite Integrals

   Theorem C   Symmetry Theorem

   Theorem D   (For a function f with period p, $\int_{a+p}^{b+p} f(x)\,dx = \int_a^b f(x)\,dx$)

The proofs for Theorems A, B, and D are given. The proof of Theorem C for even functions is given; the proof for odd functions is an exercise.

Homework   1–10, 15, 16, 21, 22, 27–32, 39–42, 47–50, 67–70

Things to Remember   Substitution is the most useful technique of integration.

Technology Project 5.1   Riemann Sums

Objectives   To reinforce the idea that a definite integral is the limit of a sum, and to understand the "order" of the error in terms of powers of n.

Technology Required   Graphing calculator or CAS. There must be a way to evaluate left-, right-, and midpoint-Riemann sums.

Technology Project 5.2   Accumulation Functions

Objectives   To familiarize students with a number of functions defined as accumulation functions, and to reinforce the First Fundamental Theorem of Calculus.

Technology Required   CAS.

CHAPTER 6   APPLICATIONS OF THE INTEGRAL

6.1   The Area of a Plane Region

Topics to Cover   Area under a curve and above the x-axis. Regions below the x-axis. Regions between curves. *Slice, approximate, integrate*. Distance and displacement.

Theorems   None

Homework   1–4, 7–10, 13–16, 25–28, 31, 32, 36, 37

Things to Remember   *Slice, approximate, integrate*.

6.2   Volumes of Solids: Slabs, Disks, Washers

Topics to Cover   Volumes of cylinders. Solids of revolution: disks, washers. Other solids with known cross-sectional areas.

Theorems   None

Homework   1–6, 9–11, 15–18, 21–22, 27–28

Things to Remember   The idea of *slice, approximate, integrate* can be used to find a number of quantities.

6.3   Volumes of Solids of Revolution: Shells

Topics to Cover   Volume of a shell. Method of shells. Examples.

Theorems   None

Homework   1, 2, 5, 6, 9–14, 17, 18, 24

Things to Remember   *Slice, approximate, integrate.*

6.4   Length of a Plane Curve

Topics to Cover   Parametric equation of a curve in the plane. Smooth curves. Arc length. Examples. Differential of arc length. Area of a surface of revolution (optional).

Theorems   None

Homework   1, 2, 5–8, 13, 14, 17–20, 23, 24, 31

Things to Remember $ds = \sqrt{1 + (y')^2}$ and the other similar results are needed later in the book.

6.5   Work

Topics to Cover   Definition of work as $F \cdot D$ for a constant force, and the integral of force times distance for a variable force. Application to springs and to pumping a liquid.

Theorems   None

Homework   3, 4, 7, 8, 9–14, 19, 20

Things to Remember   Work, as $F \cdot D$, is needed in Chapters 13, 14, and 17.

6.6   Moments, Center of Mass

Topics to Cover   Moment of a particle with respect to a point. Continuous mass distribution. Mass distributions in the plane. Centroid. Pappus's Theorem (optional).

Theorems

Theorem A   Pappus's Theorem

The proof is left as an exercise (Problem 28).

Homework   1–9, 14, 15, 17–20, 23, 24, 28, 29, 33, 34

Things to Remember   Moments and center of mass. These ideas arise again in Chapter 16.

Technology Project 6.1   Volume in an Elliptical Cylinder

Objectives   Use the capabilities of a CAS to solve some challenging volume problems.

Technology Required   CAS. You must be able to evaluate definite integrals with a parameter and with a variable upper limit.

Technology Project 6.2   Arc Length

Objectives   Use software to approximate arc length by summing the lengths of line segments and to see another limiting approach to the integral for arc length.

Technology Required   Graphing calculator or CAS. You must be able to numerically evaluate sums.

CHAPTER 7   TRANSCENDENTAL FUNCTIONS

7.1   The Natural Logarithm Function

Topics to Cover   Is there a function whose derivative is $1/x$? Definition of the natural logarithm as an accumulation function. Derivative of natural logarithm function. Examples. Properties of the natural logarithm function. Logarithmic differentiation.

Theorems

> **Theorem A**   (Properties of the Natural Logarithm: $\ln 1 = 0$, $\ln ab = \ln a + \ln b$, etc.)

All four parts are proved.

Homework   1–6, 9–12, 15–20, 23, 24, 27, 28, 35, 36

Things to Remember   The natural logarithm function, defined as an accumulation function, is a function whose derivative is $1/x$.

7.2   Inverse Functions and Their Derivatives

Topics to Cover   Definition and existence of the inverse of a function. Derivatives of inverse functions.

Theorems

> **Theorem A**   (A monotonic function has an inverse.)
>
> **Theorem B**   Inverse Function Theorem

The proofs of these two theorems are difficult and are omitted.

Homework   1–4, 7, 8, 11–13, 15–18, 27–30, 31–34

Things to Remember   Inverses exist for monotonic functions.

7.3   The Natural Exponential Function

Topics to Cover   Definition of natural exponential function as inverse of natural logarithm function. The number e. Properties of the exponential function. $D_x e^x = e^x$.

Theorems

> **Theorem A** $\left(e^a e^b = e^{a+b} \text{ and } e^a/e^b = e^{a-b}\right)$

The first statement is proved; the proof of the second is similar.

Homework   1–6, 11–14, 19–21, 23–32, 48, 53

Things to Remember   The natural exponential function is the inverse of the natural logarithm function, and is e^x where e is the real number satisfying $\ln e = 1$.

7.4   General Exponential and Logarithmic Functions

Topics to Cover   Definition of a^x. Properties of a^x. Definition and properties of $\log_a x$. Comparison of a^x, x^a, and x^x.

Theorems

> **Theorem A**   Properties of Exponents ($a^x a^y = a^{x+y}$, etc.)

Two parts are proved. The proofs of the rest are left as exercises.

> **Theorem B**   Exponential Function Rules (Derivative and integral of exponential functions)

The derivative formula is proved; the integral formula follows immediately from the derivative formula.

Homework   1–10, 13–18, 21–30, 35–37, 43

Things to Remember   We have come full circle (Figure 1 in the text). Our definition of $\ln x$ as an accumulation function has led us to $\log_e x = \ln x$.

7.5   Exponential Growth and Decay

Topics to Cover   Differential equations. Solving differential equations by separation of variables. Exponential growth and decay, compound interest. Examples.

Theorems

> **Theorem A** $\left(\lim_{h \to 0}(1+h)^{1/h} = e\right)$

The proof is given.

Homework   1, 2, 5–8, 11–15, 24, 25, 27, 32, 33

Things to Remember   Simple differential equations like $y' = ky$ lead to exponential growth or decay.

7.6   First-Order Linear Differential Equations

Topics to Cover   Classification of differential equations. Solving first-order linear equations using integrating factors. Examples and applications

Theorems   None

Homework   1, 2, 4, 5, 11, 12, 15, 16, 19, 20

Things to Remember   Multiplying both sides by the integrating factor is just what is needed to make the left-hand side the derivative of a product. This trick allows us to integrate both sides and thereby solve the equation.

7.7   The Inverse Trigonometric Functions and Their Derivatives

Topics to Cover   To obtain an inverse for the trigonometric functions, we must restrict the domain. Inverses of the six trigonometric functions. Derivatives of trigonometric and inverse trigonometric functions.

Theorems

Theorem A $\left(\sin\left(\cos^{-1} x\right) = \sqrt{1 - x^2},\ \text{etc.}\right)$

The proof of part (i) is given; the others are similar.

Theorem B   Derivatives of Four Inverse Trigonometric Functions ($\sin^{-1}$, $\cos^{-1}$, $\tan^{-1}$, and $\sec^{-1}$)

A proof is given for the inverse sine function. Proofs for inverse cosine and inverse tangent are similar. The proof for the inverse secant involves a twist, and it is given.

Homework   1–4, 11–16, 19, 20, 23–27, 33–35, 38, 39, 42, 43, 47, 59, 61

Things to Remember   Restricting the domain of the trigonometric functions allows us to obtain inverse functions. Remember the formulas for the derivatives of the trigonometric and inverse trigonometric functions.

7.8   The Hyperbolic Functions and Their Inverses

Topics to Cover   Definitions and derivatives of the six hyperbolic functions. Inverse hyperbolic functions. Applications.

Theorems

Theorem A   Derivatives of Hyperbolic Functions

The derivatives of $\sinh x$ and $\cosh x$ are derived. The derivatives of the others are just straightforward applications of the Quotient Rule.

Homework   1, 2, 5, 6, 9, 11, 13–20, 23, 24, 27, 28, 38–43, 47, 48, 53

Things to Remember   Definitions and derivatives of hyperbolic functions.

Technology Project 7.1   Special Functions

Objectives   To give students some experience working with special functions such as the error function and the standard normal cumulative distribution function.

Technology Required   CAS.

Technology Project 7.2   Population Growth and Least Squares

Objectives   Postulate a model, estimate parameters, assess the fit, and make predictions for population growth.

Technology Required   Graphing calculator or CAS.

CHAPTER 8   TECHNIQUES OF INTEGRATION

8.1   Integration by Substitution

Topics to Cover   Standard integral forms (students should these commit these to memory). Substitutions in indefinite integrals, examples. Substitutions in definite integrals, examples.

Theorems

Theorem A   Substitutions in Indefinite Integrals

This is just a restatement of Theorem 5.8A

Homework   1–10, 25–30, 33–36, 49–52, 55–58

Things to Remember   Substitution, combined with the standard integral forms, is a powerful tool for evaluating integrals. A substitution in a definite integral requires three things:
   1. substitution in the integrand
   2. substitution for the differential
   3. changing the limits of integration

8.2 Some Trigonometric Integrals

Topics to Cover   Review some trigonometric identities. Integrals of the type $\int \sin^n x \, dx$, $\int \cos^n x \, dx$, $\int \sin^m x \cos^n x \, dx$, $\int \sin mx \cos nx \, dx$, $\int \sin mx \sin nx \, dx$, $\int \cos mx \cos nx \, dx$. Examples.

Theorems   None

Homework   1–8, 11–15, 17, 18, 23, 24

Things to Remember   Remember the trigonometric identities and how they lead to evaluation of an integral.

8.3 Rationalizing Substitutions

Topics to Cover   Integrands involving $\sqrt[n]{ax + b}$. Integrands involving $\sqrt{a^2 - x^2}$, $\sqrt{x^2 + a^2}$, and $\sqrt{x^2 - a^2}$. Completing the square. Examples.

Theorems   None

Homework   1–6, 9–13, 17–21, 27, 29, 32, 33

Things to Remember   Remember what substitutions work in each case.

8.4 Integration by Parts

Topics to Cover   Integration by parts for indefinite and definite integrals. Repeated integration by parts. Reduction Formulas. Examples.

Theorems   None.

Homework   1–6, 9–11, 16–19, 34–47, 41–44, 53, 54, 69, 73

Things to Remember   Next to substitution, integration by parts is the most frequently used technique of integration.

8.5 Integration of Rational Functions

Topics to Cover   Long division of polynomials. Partial fraction decomposition. Examples.

Theorems   None

Homework   1–6, 9, 10, 13, 14, 17, 18, 23, 24, 29, 30, 33, 34, 41–43

Things to Remember   Partial fraction decomposition is a useful technique for solving differential equations.

Technology Project 8.1   Integration Using a Computer Algebra System

Objectives   See the strengths and weaknesses of, and the pitfalls involved with, using a CAS to evaluate indefinite or definite integrals.

Technology Required   CAS.

Technology Project 8.2   The Logistic Differential Equation

Objectives   Investigate the effect of each parameter on the behavior of the solution of the logistic differential equation.

Technology Required   Graphing calculator or CAS.

CHAPTER 9   INDETERMINATE FORMS AND IMPROPER INTEGRALS

9.1 Indeterminate Forms of Type 0/0

Topics to Cover   Definitions of indeterminate forms. L'Hôpital's Rule. Examples. Cauchy's Mean Value Theorem.

Theorems

    Theorem A   L'Hôpital's Rule

The proof is given after Cauchy's Mean Value Theorem

    Theorem B   Cauchy's Mean Value Theorem

The proof is given.

Homework   1–6, 9–12, 19, 20, 23, 27

Things to Remember   When and how to apply l'Hôpital's Rule.

9.2 Other Indeterminate Forms

Topics to Cover   L'Hôpital's Rule for forms of type ∞/∞. The indeterminate forms 0^0, ∞^0, and 1^∞. Examples

Theorems

    Theorem A   L'Hôpital's Rule for Forms of Type ∞/∞

The proof is omitted, but a plausibility argument is given.

Homework   1–6, 11, 12, 15, 16, 19, 20, 27, 39, 40

Things to Remember   When and how to apply l'Hôpital's Rule.

9.3 Improper Integrals: Infinite Limits of Integration

Topics to Cover A single infinite limit of integration. Definitions. Two infinite limits, examples. Probability density functions.

Theorems None

Homework 1–10, 13, 15, 17, 20, 21, 24, 24, 31, 32.

Things to Remember An integrand with an infinite limit of integration is the limit of a proper integral.

9.4 Improper Integrals: Infinite Integrands

Topics to Cover Definition of integral over interval where the integrand is infinite at one end point. Integrands that are infinite at an interior point. Examples.

Theorems None

Homework 1–4, 7, 8, 11, 17, 18, 23, 24, 29, 33, 34, 53, 54

Things to Remember Apply the correct limit to evaluate an improper integral.

Technology Project 9.1 Probability Density Functions

Objectives Study properties of the normal and Cauchy distributions.

Technology Required Graphing calculator or CAS.

Technology Project 9.2 The Normal Distribution

Objectives Study the normal distribution with arbitrary mean μ and variance σ^2, and its relationship to the normal distribution with mean 0 and variance 1.

Technology Required Graphing calculator or CAS.

CHAPTER 10   INFINITE SERIES

10.1   Infinite Sequences

Topics to Cover Notation and terminology for infinite sequences. Definitions of convergence and divergence.

Theorems

Theorem A Properties of Limits of Sequences

Theorem B Squeeze Theorem

The proofs of Theorems A and B are similar to the analogous theorems for limits of functions of a single variable.

Theorem C ($\lim |a_n| = 0 \Rightarrow \lim a_n = 0$)

The proof follows from the Squeeze Theorem

Theorem D Monotonic Sequence Theorem

The proof is difficult, and so it is omitted.

Homework 1–4, 11–14, 21–26, 31–35, 45

Things to Remember Definition of convergence of a sequence.

10.2   Infinite Series

Topics to Cover Definitions of infinite series, partial sums, convergence, and divergence. Examples. The harmonic series. Collapsing series. Linearity of $\sum$ for convergent series.

Theorems

Theorem A nth Term Test for Divergence

The proof is given.

Theorem B Linearity of Convergent Series

This theorem is proved

Theorem C ($\sum a_k$ diverges and $c \neq 0 \Rightarrow \sum c\, a_k$ diverges)

The proof is left as an exercise (Problem 35).

Theorem D Grouping Terms in an Infinite Series

A sketch of the theorem is given. It is assumed that if $\{T_m\}$ is a subsequence of $\{S_n\}$, and if $S_n \to S$, then $T_m \to S$ as well.

Homework 1–8, 13–16, 19–23, 30, 31, 34

Things to Remember nth term test for divergence can be used to establish divergence only. (Students will often try to apply Theorem A to establish the convergence of a theorem.)

10.3   Positive Series: The Integral Test

Topics to Cover Bounded partial sums and the bounded sum test. Integral test. p-series.

Theorems

Theorem A Bounded Sum Test

The proof, which relies on Theorem 10.1D, is proved

Theorem B Integral Test

The proof is given.

Homework   1–4, 7, 8, 11–16, 21–25

Things to Remember   The p-series converges for $p > 1$ and diverges for $p \leq 1$.

10.4   Positive Series: Other Tests

Topics to Cover   Comparison test. Limit comparison test. Ratio test. Examples. Summary of tests for convergence.

Theorems

Theorem A   Ordinary Comparison Test

The proof is given.

Theorem B   Limit Comparison Test

Most parts of the theorem are proved. The proof of the last part is left as an exercise (Problem 37).

Theorem C   Ratio Test

This theorem is proved.

Homework   1–16, 21, 22, 25, 26, 33

Things to Remember   At this point, students should review all tests for convergence or divergence.

10.5   Alternating Series, Absolute Convergence, and Conditional Convergence

Topics to Cover   Definitions. Alternating series test. Absolute convergence, absolute convergence test, absolute ratio test. Conditional convergence. Rearrangement of terms.

Theorems

Theorem A   Alternating Series Test

The proof is given before the statement of the theorem.

Theorem B   Absolute Convergence Test

This theorem is proved.

Theorem C   Absolute Ratio Test

The proof is given.

Theorem D   Rearrangement Theorem

The proof is omitted.

Homework   1–4, 7–10, 13–17, 21, 23, 29, 35

Things to Remember   Definitions of absolute and conditional convergence, and how to test for these.

10.6   Power Series

Topics to Cover   Definitions of power series, convergence sets. Possible types of convergence sets. Radius of convergence. Power series in $x - a$.

Theorems

Theorem A   (Convergence set of a power series is a point, an interval, or the whole real line.)

The proof is given.

Theorem B   (A power series converges absolutely on the interior of its interval of convergence.)

The proof is contained within the proof of Theorem A.

Homework   1–6, 10–14, 17–19, 21, 25.

Things to Remember   A power series is a series of numbers. Rather than asking *whether* a series converges, we ask *for what values of* x it converges.

10.7   Operations on Power Series

Topics to Cover   Term by term differentiation and integration. Examples based on the geometric series. Multiplying and dividing power series.

Theorems

Theorem A   (On the interior of the convergence set, power series can be differentiated and integrated term by term.)

Theorem B   (Convergence of sums, differences, products, and quotients of convergent power series.)

The proofs of Theorems A and B are difficult and are omitted.

Homework   1–7, 11, 13–18, 25, 31, 35

Things to Remember   Term by term differentiation and integration is allowed under the right conditions.

10.8   Taylor and Maclaurin Series

Topics to Cover   Uniqueness of power series. Definitions of Taylor and Maclaurin series. Taylor's Formula, Taylor's Theorem. Examples. Binomial series. Maclaurin series to remember.

Theorems

Theorem A   Uniqueness Theorem

The proof is given before the statement of the theorem.

Theorem B   Taylor's Formula with Remainder

The proof is given for the case of $n = 4$. The proof for an arbitrary n is left as an exercise (Problem 37).

Theorem C   Taylor's Theorem

The proof follows directly from Theorem B.

Theorem D   Binomial Series

A partial proof is given. Problem 38 gives an entirely different way to prove the theorem.

Homework   1–4, 7–9, 14, 17, 18, 21–23, 27, 32

Things to Remember   Memorize the important Maclaurin series given at the end of the section.

Technology Project 10.1   Using Infinite Series to Approximate π

Objectives   Use infinite series for the inverse tangent function to obtain approximations for π. Students should also investigate how some series converge "faster" than others.

Technology Required   Graphing calculator or CAS. You must be able to numerically evaluate sums.

Technology Project 10.2   Euler's Derivation of $\frac{1}{1^2} + \frac{1}{2^2} + \frac{1}{3^2} + \cdots = \frac{\pi^2}{6}$

Objectives   Students will step through Euler's derivation of this result. After realizing that Euler did all of this by hand, students should appreciate Euler's algebraic ability.

Technology Required   CAS.

CHAPTER 11   NUMERICAL METHODS, APPROXIMATIONS

11.1   The Taylor Approximation to a Function

Topics to Cover   Taylor polynomial of order 1 and of order n. Maclaurin polynomials. Error in the method. Bounding the absolute value of the remainder. The error of calculation.

Theorems   None

Homework   1–4, 9–12, 17, 19, 20, 27, 28, 33–35, 39, 41

Things to Remember   Taylor polynomials are good approximations near a.

11.2   Numerical Integration

Topics to Cover   Examples where the Second Fundamental Theorem of Calculus can't be applied. Trapezoidal rule and its error formula. Parabolic rule and its error formula.

Theorems

Theorem A   Error for the Trapezoidal Rule
Theorem B   Error for the Parabolic Rule

The proofs of Theorems A and B are difficult and are omitted.

Homework   1, 2, 5, 6, 9, 10, 13, 14, 17, 18, 21, 25

Things to Remember   The Parabolic Rule can be used to approximate the value of a definite integral for which you might not be able to apply the Second Fundamental Theorem of Calculus.

11.3   Solving Equations Numerically

Topics to Cover   Bisection Method, Newton's Method, Convergence of Newton's Method. Examples. Advantages and disadvantages of each.

Theorems

Theorem A   (Conditions that guarantee convergence of Newton's Method)

The proof is given.

Homework   1, 2, 5, 7, 11–13, 15, 17, 21

Things to Remember   The advantages and disadvantages of the two methods.

11.4   The Fixed Point Algorithm

Topics to Cover The fixed-point algorithm. Fixed point theorem. Examples

Theorems

Theorem A Fixed-Point Theorem

The proof is given.

Homework 1, 2, 5, 7, 8, 11–14, 19

Things to Remember Some formulations of $x = g(x)$ will lead to convergence to a fixed point, and others will not.

11.5 Approximations for Differential Equations

Topics to Cover Slope fields. Euler's method, the improved Euler method. The effect of the step size h.

Theorems None

Homework 1–12, 15–18, 21, 22, 23, 27

Things to Remember Slope fields can give us a general idea about what solutions of a differential equation will look like. Euler's method, and the improved Euler method, can give an approximation to the solution of a first-order differential equation given an initial condition.

Technology Project 11.1 Maclaurin Polynomials

Objectives To investigate how closely Maclaurin polynomials approximate a function, and to exploit symmetry in the sine function to obtain an efficient method of approximating $\sin x$.

Technology Required CAS.

Technology Project 11.2 Numerical Integration

Objectives To investigate the order (i.e., the rate at which the errors decrease as n increases) of techniques of numerical integration.

Technology Required Graphing calculator or CAS. You must be able to numerically evaluate sums.

Technology Project 11.3 Bisection, Newton's, and Fixed-Point Methods

Objectives To apply and compare the three methods of approximating the solution of an equation, and to investigate the chaotic behavior of a simple iterative method.

Technology Required Graphing calculator or CAS. You will need a fast way to iterate.

CHAPTER 12 CONICS AND POLAR COORDINATES

12.1 The Parabola

Topics to Cover Conics as intersection of a plane and a cone. Conics as points satisfying $|PF| = e\,|PL|$. $e = 1$ leads to the parabola. Optical properties of the parabola.

Theorems None

Homework 1–4, 7, 8, 11–15, 17–22, 30, 31, 41

Things to Remember $|PF| = |PL|$ for the parabola.

12.2 Ellipses and Hyperbolas

Topics to Cover Central conics. Standard equations of ellipse and hyperbola. Eccentricity. Examples.

Theorems None

Homework 1–4, 9–12, 17–19, 22–25, 31, 38

Things to Remember $|PF| = e\,|PL|$ where $e < 1$ for ellipse and $e > 1$ for hyperbola.

12.3 More on Ellipses and Hyperbolas

Topics to Cover String properties of the ellipse and hyperbola. Optical properties of the ellipse and hyperbola. Applications.

Theorems None

Homework 1–10, 13, 17–19, 23, 31

Things to Remember String and optical properties.

12.4 Translation of Axes

Topics to Cover Translations. Completing the square. Examples.

Theorems None

Homework 1–10, 17–23, 31, 32, 35–39

Things to Remember Complete the square to determine the type of conic.

12.5 Rotation of Axes

Topics to Cover Rotation through an angle of θ. Determining the appropriate θ. Rotation through a nonspecial angle.

Theorems None

Homework 1–4, 11, 12, 14, 15, 23

Things to Remember You can always rotate the graph of $Ax^2 + Bx + Cy^2 + Dy + Exy + F = 0$ to eliminate the xy term.

12.6 The Polar Coordinate System

Topics to Cover Identifying points in the plane by r and θ. Polar equations. Graphs of polar equations. Relationship between Cartesian and polar coordinates. Polar equations for lines, circles, and other conics.

Theorems None

Homework 1, 5, 7, 9, 11–14, 17–21, 23–29, 32, 33, 37

Things to Remember Remember the relationships between polar and Cartesian coordinates. These will be needed again in Chapters 16 and 17.

12.7 Graphs of Polar Equations

Topics to Cover Symmetry tests. Cardiods, limaçons, lemniscates, roses, spirals. Intersections of curves in polar coordinates.

Theorems None

Homework 1–4, 9, 10, 13, 14, 17, 18, 21–25, 27–30, 33–35

Things to Remember This section should develop some general skills working with polar equations. This overall familiarity is more important than any specific skill.

12.8 Calculus in Polar Coordinates

Topics to Cover Area of a sector. Area in polar coordinates. Tangents in polar coordinates.

Theorems None

Homework 1–6, 11–13, 15–17, 21–25

Things to Remember Slice, approximate, integrate worked again!

Technology Project 12.1 Rotations in the Plane

Objectives To investigate how θ affects the rotation of a graph.

Technology Required CAS or some software capable of rendering high resolution implicit plots.

Technology Project 12.2 Another Kind of Rose

Objectives To investigate the effect of parameters on the shape of this kind of rose.

Technology Required CAS or some software capable of rendering high resolution polar plots.

CHAPTER 13 GEOMETRY IN THE PLANE, VECTORS

13.1 Plane Curves: Parametric Representation

Topics to Cover Parametric representation of a plane curve. Parametrizations. Eliminating the parameter. The cycloid. Calculus for curves defined parametrically. Arc length of a parametric curve.

Theorems

Theorem A $\left(\dfrac{dy}{dx} = \dfrac{dy/dt}{dx/dt}\right)$

The proof is given.

Homework 1–6, 9–13, 17, 21, 22, 31, 32, 35–37, 61

Things to Remember A general familiarity with curves defined parametrically is needed in this and the next chapters.

13.2 Vectors in the Plane: Geometric Approach

Topics to Cover Direction and magnitude of vectors. Sums, differences, and scalar multiples of vectors. Applications.

Theorems None

Homework 1–4, 7–10, 13

Things to Remember Students should understand the geometric interpretation of vectors.

13.3 Vectors in the Plane: Algebraic Approach

Topics to Cover Operations on vectors. Length and dot product. Angle between two vectors. Basis vectors. Vector and scalar projections.

Theorems

Theorem A (Properties of addition, subtraction, scalar multiplication, etc.)

The proofs of these properties are easy; only the proof for Rule 6 is given.

Theorem B (Properties of dot product)

These properties are easy to establish.

Theorem C   Perpendicularity criterion

The proof is given.

Homework   1, 3, 5–10, 13–15, 19, 22, 23

Things to Remember   Students must see vectors from both a geometric and algebraic point of view. The formula for $\cos\theta$ is needed in a number of places in the rest of the book.

13.4   Vector-Valued Functions and Curvilinear Motion

Topics to Cover   Definitions, limits, continuity, differentiation. Position, speed, velocity. Uniform circular motion.

Theorems

Theorem A $(\lim \mathbf{F}(t) = \lim f(t)\,\mathbf{i} + \lim g(t)\,\mathbf{j})$

The Appendix contains the proof.

Theorem B   Differentiation Formulas

Part 4 is proved. The others are left as exercises.

Homework   1–3, 9–14, 16, 17, 19, 23–25, 31, 32, 35, 41

Things to Remember   The velocity vector is the derivative of the position vector, and the acceleration vector is the derivative of the velocity vector.

13.5   Curvature and Acceleration

Topics to Cover   Definitions, tangent vector, curvature, radius of curvature. Examples. Alternative formulas for the curvature. Normal and tangential components of acceleration.

Theorems

Theorem A   (A formula for the curvature)

The proof is given.

Homework   1–4, 12, 13, 15–17, 29, 30, 32, 35, 36, 47, 51

Things to Remember   Much of this material is repeated in the next chapter when we study vectors and motion in three-space.

Technology Project 13.1   Hypocycloids

Objectives   Practice manipulating parametric curves, and to make, test, and prove conjectures about arc length of the hypocycloid.

Technology Required   CAS. You will need to have high resolution graphs of curves defined parametrically.

Technology Project 13.2   Measuring Home Run Distance

Objectives   Apply the results on projectile motion to estimate how far a baseball would travel if it were to return to field level.

Technology Required   CAS.

CHAPTER 14   GEOMETRY IN SPACE, VECTORS

14.1   Cartesian Coordinates in Three-Space

Topics to Cover   Right-handed coordinate systems. Distance Formula. Graphs in three-space, spheres, planes.

Theorems   None

Homework   1, 3–8, 12–15, 17–21, 27

Things to Remember   An overall familiarity with three-space is the goal of this section. Much of the rest of the book deals with calculus in three-space.

14.2   Vectors in Three-Space

Topics to Cover   Basis vectors $\mathbf{i}$, $\mathbf{j}$, and $\mathbf{k}$. Length, dot product. Examples.

Theorems   None

Homework   1, 3, 4, 6–9, 11, 12, 16, 17, 21, 23, 25, 27, 28, 33, 41

Things to Remember   Students should be comfortable manipulating vectors in angle bracket notation $\langle\ \rangle$ and in $\mathbf{i}$, $\mathbf{j}$, and $\mathbf{k}$ notation.

14.3   The Cross Product

Topics to Cover   Definition of cross product. Geometric interpretation of the cross product. Properties. Applications.

Theorems

Theorem A   (Properties of the cross product)

Properties 1 and 3 are proved. A plausibility argument is given for property 2, which is difficult to prove rigorously.

Theorem B   (**u** and **v** are parallel $\Leftrightarrow$ **u** $\times$ **v** $= 0$)

The proof is given.

Theorem C   (Further properties of the cross product)

The proofs are easy and are left as exercises.

Homework   1, 3–8, 11, 13, 15–18, 20–22

Things to Remember   The cross product **u** $\times$ **v** is perpendicular to both **u** and **v**. The area of a parallelogram whose sides are **u** and **v** is $|\mathbf{u} \times \mathbf{v}|$, a result that is needed in Chapters 16 and 17.

14.4   Lines and Curves in Three-Space

Topics to Cover   Parametric and symmetric equations of a line. Parametric curves in three-space. Tangent line to a curve.

Theorems   None

Homework   1–10, 13–16, 18, 23, 24, 27

Things to Remember   A line in space can be represented parametrically or through its symmetric equations.

14.5   Velocity, Acceleration and Curvature

Topics to Cover   This section generalizes the results of Sections 13.4 and 13.5 to three-space. The circular helix. Curvature. Normal and tangential components of acceleration. The binormal vector.

Theorems   None

Homework   1–5, 11, 14, 15, 17, 21, 25, 27, 35, 43, 53–55

Things to Remember   After reading this section, students should feel comfortable with motion in three-space.

14.6   Surfaces in Three-Space

Topics to Cover   Cross-sections, traces. Cylinders. Quadric surfaces. Examples.

Theorems   None

Homework   1–13, 21, 23, 32

Things to Remember   The ability to visualize and draw graphs in three-space is essential to mastering Chapters 15–17.

14.7   Cylindrical and Spherical Coordinates

Topics to Cover   How to represent points in space by cylindrical or spherical coordinates. Conversion formulas. Graphs in cylindrical and spherical coordinates. Great circles.

Theorems   None

Homework   1–12, 15, 17–21, 35

Things to Remember   Students should be comfortable working in these coordinate systems, which are used to evaluate double and triple integrals in Chapters 16 and 17.

Technology Project 14.1   Curves in Three-Space

Objectives   Visualize curves in three-space.

Technology Required   CAS. You need to plot curves in three-space.

Technology Project 14.2   The Ferris Wheel and Corkscrew Roller Coaster

Objectives   Relate the material on motion, curvature, acceleration, etc., to actual motions that many students have experienced.

Technology Required   Graphing calculator.

CHAPTER 15   THE DERIVATIVE IN n-SPACE

15.1   Functions of Two or More Variables

Topics to Cover   Definitions. 3-D plots. Contour plots, level curves. Applications of contour plots. Functions of three or more variables.

Theorems   None

Homework   1, 5, 7–13, 17–19, 23, 25–27, 33, 39

Things to Remember   The ability to visualize and manipulate functions of two (or more) variables is essential for the remainder of the book.

15.2   Partial Derivatives

Topics to Cover   Partial derivatives as a rate of change. Examples. Geometric and physical interpretations. Higher-order partials.

Theorems   None

Homework   1–4, 8, 9, 13, 17, 19, 24, 29, 33, 34, 47, 49

Things to Remember   Partial derivatives are used throughout the rest of the book.

15.3 Limits and Continuity

Topics to Cover Definition of limit of a function of two variables. Examples where limit does not exist. Continuity at a point and on a set.

Theorems

Theorem A Composition of Functions

The proof is similar to the proof of Theorem 2.9E.

Theorem B Equality of Mixed Partials

The proof is omitted. A counterexample for which the continuity of f_{xy} is lacking is given in Problem 32.

Homework 1–5, 8–12, 15, 16, 19–22, 37

Things to Remember Limits and continuity are trickier than for functions of one variable.

15.4 Differentiability

Topics to Cover Local linearity of a function of one variable. Local linearity of a function of two variables. Definition of differentiable as synonymous with locally linear. Tangent plane. Gradient. Properties of the gradient

Theorems

Theorem A (Continuous partials in a neighborhood $\Rightarrow$ differentiable)

The proof is given.

Theorem B Properties of ∇

The proofs follow from the corresponding rules for partial derivatives. Part (iii) is proved for the two-variable case.

Theorem C (Differentiability $\Rightarrow$ Continuity)

The proof is given.

Homework 1–5, 9, 11, 13, 15, 17, 19, 20, 21

Things to Remember The gradient is the analog of the derivative for functions of more than one variable.

15.5 Directional Derivatives and Gradients

Topics to Cover Concept of rate of change in any direction. Definition of directional derivative. Computing the directional derivative. The gradient always points in the direction of most rapid increase in f. Higher dimensions.

Theorems

Theorem A $(D_{\mathbf{u}}f(\mathbf{p}) = \mathbf{u} \cdot \nabla f(\mathbf{p}))$

The proof is given.

Theorem B (The gradient always points in the direction of most rapid increase in f.)

The proof follows directly from Theorem A and from the rule $\mathbf{u} \cdot \mathbf{v} = |\mathbf{u}|\,|\mathbf{v}|\cos\theta$.

Theorem C (The gradient is perpendicular to the level curve.)

The proof is omitted.

Homework 1–6, 9–13, 15, 17, 19, 20

Things to Remember The gradient always points in the direction of most rapid increase in f.

15.6 The Chain Rule

Topics to Cover Two versions of the Chain Rule. Examples and applications.

Theorems

Theorem A (Chain Rule for $\dfrac{d}{dt}f\left(x(t), y(t)\right)$)

The proof is given.

Theorem B (Chain Rule for $\dfrac{\partial}{\partial s}f\left(x(s,t), y(s,t)\right)$, etc.)

The proof follows from Theorem A by holding one variable fixed.

Homework 1–4, 7–10, 13, 14, 20–22

Things to Remember Remember the patterns in both versions of the Chain Rule.

15.7 Tangent Planes, Approximations

Topics to Cover Tangent plane to a surface defined by $F(x, y, z) = k$. Differentials and approximations. Second-order Taylor approximations.

Theorems

Theorem A Tangent Plane

The proof is given.

Homework 1–5, 8, 9, 11, 13, 15, 18, 22, 29

Things to Remember The formula for the equation of the tangent line to a surface.

15.8 Maxima and Minima

Topics to Cover Definitions. Existence of extrema. Examples. Second Partials Test.

Theorems

Theorem A   Max-Min Existence Theorem

The proof is omitted.

Theorem B   Critical Point Theorem

The proof is given.

Theorem C   Second Partials Test

The proof is omitted.

Homework   1–7, 11, 12, 15–17, 21–23, 28, 29, 34

Things to Remember   The important skill in this section is translating a problem into mathematical terms so we can use the method of calculus to solve it.

15.9   Lagrange's Method

Topics to Cover   Geometric interpretation of the Lagrange multiplier. Examples. Two or more constraints.

Theorems

Theorem A   Lagrange's Method

We forego a rigorous proof for the intuitive explanation given before the statement of the theorem.

Homework   1, 2, 7–9, 11, 13, 15

Things to Remember   Remember the geometric interpretation of the Lagrange multiplier.

Technology Project 15.1   Newton's Method for Two Equations in Two Unknowns

Objectives   Generalize Newton's Method to the case of two equations in two unknowns and apply this method to an optimization problem.

Technology Required   CAS.

Technology Project 15.2   Visualizing the Directional Derivative

Objectives   To visualize the directional derivative. The last few exercises should get students to begin thinking about the line integral.

Technology Required   CAS.

CHAPTER 16   THE INTEGRAL IN n-SPACE

16.1   Double Integrals Over Rectangles

Topics to Cover   Review the single Riemann integral. Definition of double integral. Integrability. Properties of the double integral, linearity, additivity on rectangles, comparison property. Evaluation of double integrals.

Theorems

Theorem A   Integrability Theorem

The proof is difficult, so it is omitted.

Homework   1–6, 9–11, 15, 19, 23

Things to Remember   Just like the single integral, the double integral is a limit of a sum.

16.2   Iterated Integrals

Topics to Cover   Slice, approximate, integrate to obtain the volume of a solid by an iterated integral. Evaluating iterated integrals using the Second Fundamental Theorem of Calculus. Examples of calculating volumes

Theorems   None

Homework   1–9, 13–18, 21–23, 25, 26, 29,

Things to Remember   A double integral can be evaluated by writing it as an iterated integral. Students will need to evaluate iterated integrals in the rest of Chapter 16.

16.3   Double Integrals Over Nonrectangular Regions

Topics to Cover x-simple and y-simple sets. Definition of double integral over nonrectangular region. Evaluating iterated integrals over nonrectangular regions.

Theorems   None

Homework   1–8, 13–16, 19–22, 27, 28, 31, 32, 35, 39

Things to Remember   Choosing the limits is the hardest part of evaluating iterated integrals. Fix one variable and integrate over the other.

16.4   Double Integrals in Polar Coordinates

Topics to Cover   Polar rectangles, $dx\,dy = r\,dr\,d\theta$. r-simple and θ-simple regions. Volume of a solid.

Theorems   None

Homework   1–8, 11–15, 19, 29, 30

Things to Remember $dx\,dy = r\,dr\,d\theta$. Converting to polar coordinates often simplifies the integrand or the limits of integration. An understanding of polar coordinates is needed to understand cylindrical and spherical coordinates in three dimensions.

16.5 Applications of Double Integrals

Topics to Cover   Mass, center of mass, centroid. Moment of inertia, radius of gyration (optional). Examples.

Theorems   None

Homework   1–5, 9–13, 16, 18

Things to Remember   The formulas for the center of mass are generalizations of those given in Chapter 6.

16.6 Surface Area

Topics to Cover   Review area of a parallelogram as $|\mathbf{u} \times \mathbf{v}|$. Slice (the region S), approximate (the surface area above each small rectangle), and integrate (to form a double integral). $A(G) = \iint_S \sqrt{f_x^2 + f_y^2 + 1} \, dA$. Examples.

Theorems   None

Homework   1–6, 9, 13, 16, 21, 22

Things to Remember   Formula for surface area. A similar development will occur in Chapter 17 when we study surface integrals.

16.7 Triple Integrals (Cartesian Coordinates)

Topics to Cover   Riemann sum for function of three variables. Definition of triple integral over rectangular box, and over a general region. x-simple, y-simple, and z-simple regions. Examples. Center of mass.

Theorems   None

Homework   1–5, 9–15, 17, 19, 21, 24, 25, 27, 28, 31

Things to Remember   Emphasize that setting up the limits is usually the hardest part. It is often beneficial to set up a number of iterated integrals and leave it to the students to work them out.

16.8 Triple Integrals (Cylindrical and Spherical Coordinates)

Topics to Cover   Review cylindrical and spherical coordinates. Setting up integrals in these coordinate systems. $dV = r \, dz \, dr \, d\theta = \rho^2 \sin \phi \, d\rho \, d\theta \, d\phi$.

Theorems   None

Homework   1–9, 13, 15, 16, 22–24

Things to Remember   Formulas for dV. Emphasize that dV is not simply $d\rho \, d\theta \, d\phi$.

Technology Project 16.1   Newton's Law of Gravitation

Objectives   Investigate Newton's Law of Gravitation when one object is a point mass and the other is a slab.

Technology Required   CAS. You must be able to evaluate multiple integrals with parameters.

Technology Project 16.2   Monte Carlo Integration

Objectives   Use random numbers to approximate area, volume, etc. Emphasize that Monte Carlo is most useful for higher-dimensional integrals.

Technology Required   CAS. You must be able to evaluate triple integrals, and even a four-dimensional integral for the bonus problem.

CHAPTER 17   VECTOR CALCULUS

17.1 Vector Fields

Topics to Cover   Vector fields, scalar fields. Conservative vector fields, potential function. Divergence and curl of a vector field. Physical interpretation of div and curl.

Theorems   None

Homework   1, 2, 7, 8, 13–16, 19–22, 25

Things to Remember   A working knowledge of vector fields is needed throughout the rest of Chapter 17.

17.2 Line Integrals

Topics to Cover   Parametric curves, parametrizations, orientation. Line integrals of the form $\int_C f(x, y) \, ds$, $\int_C M \, dx + N \, dy$, $\int_C f(x, y, z) \, ds$, $\int_C M \, dx + N \, dy + P \, dz$. Examples and applications.

Theorems   None

Homework   1–3, 7, 9, 11–14, 17, 19–21, 25, 30

Things to Remember   Notation is the key. Line integrals can be written in a number of different ways, so be sure students understand notation.

17.3 Independence of Path

Topics to Cover   Fundamental Theorem of Line Integrals. (Emphasize that to evaluate $\int \mathbf{F} \cdot d\mathbf{r}$, you must be able to find a potential function for **F**.) Independence of path theorem. Recovering a function from its gradient. Conservation of energy (optional).

Theorems

Theorem A   Fundamental Theorem of Line Integrals

The proof is given.

Theorem B   Independence of Path Theorem

The proof is given.

Theorem C   (**F** is conservative $\Leftrightarrow$ curl **F** = 0)

The "if" follows from Green's Theorem in the two-variable case and Stokes's Theorem in the three-variable case. The "only if" part is easy.

Homework   1–5, 9–12, 15, 16, 18, 21, 23

Things to Remember   **F** must be conservative to apply the Fundamental Theorem of Line Integrals to evaluate $\int \mathbf{F} \cdot d\mathbf{r}$. If you know that **F** is conservative, you can find f by the method described in the text.

17.4   Green's Theorem in the Plane

Topics to Cover   Statement and proof of Green's Theorem (the proof is not difficult). Examples and applications. Vector forms of Green's Theorem $\oint \mathbf{F} \cdot \mathbf{n}\, ds = \iint_S \operatorname{div} \mathbf{F}\, dA$, and $\oint \mathbf{F} \cdot T\, ds = \iint_S (\operatorname{curl} \mathbf{F}) \cdot \mathbf{k}\, dA$. Flux in the plane.

Theorems

Theorem A   Green's Theorem

$$\left(\oint M\, dx + N\, dy = \iint_S \left(\frac{\partial N}{\partial x} - \frac{\partial M}{\partial y} \right) dA \right)$$

The proof is given.

Homework   1–3, 5, 7–11, 15–18

Things to Remember   Green's theorem relates a double integral over a region and a line integral along its boundary. If you emphasize the vector forms of Green's Theorem, it will make Gauss's Theorem (Section 17.6) and Stokes's Theorem (Section 17.7) easier to understand.

17.5   Surface Integrals

Topics to Cover   Emphasize that $f(x, y, z)$ is defined only on the surface $z = f(x, y)$. Slice, approximate, integrate. Formula for surface integral. Examples. Flux across a surface. Formula for flux.

Theorems

Theorem A   (Formula for surface integral:

$$\iint_G g(x, y, z)\, dS = \iint_S g(x, y, f(x, y))\sqrt{f_x^2 + f_y^2 + 1}\, dy\, dx)$$

The derivation (slice, approximate, integrate) is given before the theorem..

Theorem B

$$\left(\text{flux } \mathbf{F} = \iint_R [-M f_x - N f_y + P]\, dy\, dx\right)$$

The proof is given.

Homework   1–4, 7, 9, 11, 13–16

Things to Remember   Remember the concept of the surface integral and the formula for evaluating a surface integral, and how to obtain flux with a surface integral.

17.6   Gauss's Divergence Theorem

Topics to Cover   Review vector form of Green's Theorem. Gauss's Divergence Theorem. Flux. Examples.

Theorems

Theorem A   Gauss's Theorem

$$\left(\iint_{\partial S} \mathbf{F} \cdot \mathbf{n}\, dS = \iiint_S \operatorname{div} \mathbf{F}\, dV \right)$$

The proof is long, but not difficult.

Homework   1–5, 9, 11, 12, 15

Things to Remember   Gauss's theorem relates a triple integral over a solid and a surface integral over its boundary. Flux through a solid can be obtained by evaluating a surface integral over the solid's boundary or by evaluating a triple integral over the solid.

17.7   Stokes's Theorem

Topics to Cover   Review vector forms of Green's Theorem. Statement of Stokes's Theorem. Verify Stokes's Theorem for some specific cases. Physical interpretation of curl.

Theorems

Theorem A Stokes's Theorem

$$\left(\oint_{\partial S} \mathbf{F} \cdot \mathbf{T}\,ds = \iint_{S} (\operatorname{curl}\mathbf{F}) \cdot \mathbf{n}\,dS\right)$$

The proof is omitted.

Homework 1–5, 7–10, 19

Things to Remember Stokes's Theorem relates a surface integral and a line integral along the edge of the surface.

Technology Project 17.1 Line Integrals and Work

Objectives Use line integrals to compute work. Experiment with different paths.

Technology Required CAS.

Technology Project 17.2 Parametrized Surfaces

Objectives Understand and work with surfaces defined parametrically (using parameters s and t).

Technology Required CAS. You must be able to plot surfaces defined parametrically.

CHAPTER 18   DIFFERENTIAL EQUATIONS

18.1   Linear Homogeneous Equations

Topics to Cover Definitions. Review material on differential equations from Chapter 7. The auxiliary equation for second and higher-order equations. Examples.

Theorems

Theorem A   Distinct Real Roots

Theorem B   A Single Repeated Root

Theorem C   Complex Conjugate Roots

The proof of Theorem A is easy. Proofs for Theorems B and C are exercises (Problems 23, 24).

Homework 1–5, 9, 10, 13, 14, 19, 23

Things to Remember Remember the method of solving nth-order linear homogeneous equations.

18.2   Nonhomogeneous Equations

Topics to Cover Solutions of a nonhomogeneous equation can be written as $y = y_p + y_h$. Method of undetermined coefficients. Method of variation of parameters.

Theorems

Theorem A (Solutions of a nonhomogeneous equation can be written as $y = y_p + y_h$.)

The proof is given.

Homework 1, 2, 5, 6, 9, 12, 15, 17–19

Things to Remember Remember how to carry out each method.

18.3   Applications of Second Order Equations

Topics to Cover Vibrating spring, simple harmonic motion. Damped vibrations. Electric circuits. Examples.

Theorems None

Homework 1–6, 9, 10, 14

Things to Remember Students should understand the effects of parameters such as the spring constant, damping coefficient, etc., on the resulting motion.

Technology Project 18.1 Vibrating Spring

Objectives Experiment with various parameters, e.g., spring constant, damping coefficient, mass of object, etc., to understand the effect of each on the resulting motion.

Technology Required Graphing calculator or CAS.

Technology Project 18.2 Phase Portraits

Objectives Use phase portraits to understand the behavior of solutions of a second-order equation.

Technology Required CAS. You need to plot phase portraits.

Appendix

On the last day of the last term of calculus, it is worthwhile and satisfying to discuss the full-page table in Section A.3. This will give students a sense of accomplishment and will put things in perspective.